Modern Coconut Management

Palm cultivation and products

Edited by
JOHAN G. OHLER

INTERMEDIATE TECHNOLOGY PUBLICATIONS 1998

Practical Action Publishing Ltd
25 Albert Street, Rugby, CV21 2SD, Warwickshire, UK
www.practicalactionpublishing.com

First published in 1998
Transferred to digital printing in 2008
ISBN 978 1 85339 467 6
ISBN Library Ebook: 9781780445502
Book DOI: http://dx.doi.org/10.3362/9781780445502

A catalogue record for this book is available from the British Library.

Since 1974, Practical Action Publishing has published and disseminated books and information in support of international development work throughout the world. Practical Action Publishing is a trading name of Practical Action Publishing Ltd (Company Reg. No. 1159018), the wholly owned publishing company of Practical Action. Practical Action Publishing trades only in support of its parent charity objectives and any profits are covenanted back to Practical Action (Charity Reg. No. 247257, Group VAT Registration No. 880 9924 76).

Published for the Food and Agricultural Organization (FAO) in Rome in collaboration with Leiden
University in The Netherlands.
Typeset by the LEAD Programme of Leiden University in The Netherlands

Contents

Acknowledgments

The authors are very pleased to have been entrusted by FAO to write this new book on coconuts. They are also grateful for the support received from the Director of AGP, Dr. M.S. Zehni, the former Senior Officer and the Agricultural Officer of the Industrial Crops Group, Ir. G. Blaak, and Mr. P.J. Griffee respectively.

The editor is very grateful for the co-operation and confidence received from the co-authors, each one of them representing the top in their field of specialty. Their contributions are a great added value to this book. The editor is also very greatful for the co-operation received from CIRAD-CP. Not only for their authorization of the participation of two of their prominent scientists in the writing of two chapters, but also for their great hospitality in Montpelier, particularly for the great help received from the library; for the authorization to reproduce a number of their coloured photographs and last but not least for the reading and correction of the manuscript. The authors are very much obliged to Dr. Rognon, head of the Coconut Programme of CIRAD-CP, who has given valuable suggestions about the writing of this book, and to Dr. Wuidart for the reading and improving of the text, which is almost as much work as the writing itself.

Many thanks are also due to the library of the Royal Tropical Institute in Amsterdam for the great help received in obtaining literature, and for the kindness of the staff and the facilities made available.

Valuable help has been received from Ir. P. Poetiray, former Senior Officer of the FAO Industrial Crops Group, in the collection of literature, not available at the library of the Royal Tropical Institute in Amsterdam.

The Editor is greatly indebted to Prof. L. Jan Slikkerveer and Leiden University for the great technical and moral support to realize the publication of this manuscript. Many thanks are also due to Ir. W.L.C.H.M. van den Berg of Leiden University for the financial support and for the provision of the technical facilities to prepare this manuscript to the final stage of "camera ready copy" and to Mrs.drs. Ria Hogervorst for preparing the lay-out.

Special thanks are due to Mrs. Ilonka Barsony for reading the manuscript and improving the text, a difficult and laborious task.

J.G. Ohler,
Editor

List of Contributors

Dr. R. Bourdeix
CIRAD-CP, Selection and Breeding Section,
Marc Delorme Research Station,
Ivory Coast.

Dr. P.K. Das
Principal Scientist (AG. Economics) & Head (Retd),
Central Plantation Crops Research Institute (ICAR),
Kasarcode, Kerala,
India.

D. Mariau
Head Plant Protection Research Unit,
CIRAD - C,P
Montpelier,
France.

Ir. J.G. Ohler
Agriculturist, specialized in coconuts and cashew nuts,
Former Senior Agricultural Officer,
Royal Tropical Institute,
Amsterdam,
The Netherlands

Prof. Dr. F. Opio
Head, Department of Agricultural Economics,
University of the South Pacific,
School of Agriculture, Alafua Campus, Apia,
Western Samoa.

Dr. S.G. Reynolds
Senior Country Project Officer,
Agricultural Operation Division,
FAO, Rome,
Italy.

Eng. T.K.G. Rhanasinghe
Coconut and Agro Processing Consultant,
Colombo,
Sri Lanka.

List of Abbreviations

(Excluding variety codes and standard abbreviations of measures, chemical formulae, currencies, etc.)

AARD	Agency for Agricultural Research and Development (Indonesia)
ACIAR	Australian Centre for International Agricultural Research
AG	Aktiengesellschaft (limited liability company)
APCC	Asian and Pacific Coconut Community
ARC	Albay Research Centre (Philippines)
AU	Animal Unit
BUROTROP	Bureau for the Development of Research in Tropical Perennial Oil Crops (EC)
CBFS	Coconut-based Farming System
CCCV	Coconut Cadang Cadang Viroid
CEC	Cation Exchange Capacity
CDA	Coconut Development Authority (Sri Lanka)
CFTRI	Central Food Technological Research Institute (India)
CGIAR	Consultative Group on International Agricultural Research
CPATC	Centro de Pesquisa Agropecuária dos Tabuleiros Costeiros
CGR	Crop Growth Rate
CIC	Coconut Information Centre (Sri Lanka)
CICY	Centro Investigacion Cientifica de Yucatan A,C (Mexico)
CIP	Cleaning In Place
CIRAD-CP	Centre de Coopération et Recherche Agronomique pour le Développement Département des Cultures Pérennes
COGENT	Coconut Genetic Resources Network
CP	Coloured Plate
CPATC	Centro de Pequisa Agropecuária dos Tabuleiros Costeiros (Brazil)
CPCRI	Central Plantation Crops Research Institute (India)
CPL	Coconut Plus Livestock
CRI	Coconut Research Institute (Sri Lanka)
CSIRO	Commonwealth Scientific and Industrial Research Organization
CTiV	Coconut Tinangaja Viroid
DBT	Dry Bulb Temperature
DC	Desiccated Coconut
DM	Dry Matter Content
DT	Desolventizer-Toaster
EWC	Epicuticular Wax Content
EMBRAPA	Empresa Brasileira de Pesquisa Agricola
FAO	Food and Agriculture Organization of the United Nations
FDD	Foliar Decay Disease
FEDEPALMA	Federation of Colombian Oil Palm Growers
FFA	Free Fatty Acid
FONCOPAL	Fondo Para el Desarollo de Coco y de la Palma Aceitera (Venezuela)
GOS	Genetic Orange Spotting
GTZ	Deutsche Gesellschaft für Technische Zusammenarbeit
HDL	High-Density Lipoprotein cholesterol
IBPGR	International Board for Plant Genetic Resources
IBRD	International Bank for Reconstruction and Development
ICA	Instituto Colombiano Agropecuario
ICAR	Indian Council of Agricultural Research

ICCRA	International Coconut Cultivar Registration Authority
ICRAF	International Council for Research on Agroforestry
IDRC	International Development Research Centre
I/E	Irrigation/cumulative pan Evaporation
IHRO	Institut de Recherches pour les Huiles et Oléagineux
IICA	Inter-American Institute for Cooperation on Agriculture
IICF	Estación Nacional de Frutales (Cuba)
IIG/OOF	Intergovernmental Group on Oilseeds, Oils and Fats
INIAP	Instituto Nacional de Investigaciones Agropecuaria
INIFAP	Instituto Nacional de Investigación Forestal, Agricola e Pecuaria
IPGRI	International Plant Genetic Resources Institute
ITC	International Trade Centre
KARI	Kenya Agricultural Research Institute
LA	Leaf Area
LAI	Leaf Area Index
LAR	Leaf Area Ratio
LBR	Lethal Bole Rot
LD	Lethal Disease
LDL	Low-Density Lipoprotein cholesterol
LEI	Light Efficiency Index
LER	Land Efficiency Ratio
LUE	Light Use Efficiency
LY	Lethal Yellowing
MAC	Multi Agro Corporation
MLO	Mycoplasma Like Organism
NAR	Net Assimilation Rate
MARDI	Malaysian Agricultural Research and Development Institute
NCDP	National Coconut Development Programme (Tanzania)
NFDM	Non-Fat Dry Milk
NIFOR	Nigerian Institute For Oil palm Research
NIST	National Institute of Science and Technology (Philippines)
NRI	Natural Resources Institute (UK)
OP	Open Pollination
ORS	Oral Rehydration Solution
ORSTOM	Institut Français de Recherche Scientifique pour le Developpement et Coopération
PAR	Photosynthetically Active Radiation
PCA	Philippine Coconut Authority
PCR	Polymerase Chain Reaction
PDICC	Production and Dissemination of Improved Coconut Cultivars
PPO	Poly-Phenol Oxidase
RAPD	Randomly Amplified Polymorphic DNA
RARC	Regional Agricultural Research Station (Sri Lanka)
RCRS	Regional Coconut Research Station (India)
RDCIC	Research and Development Centre for Industrial Crops (Indonesia)
RFLP	Restriction Fragment Length Polymorphism
RGR	Relative Growth Rate
RLO	Rickettsia-Like Organism
R&M	Repair and Maintenance
RRD	Red Ring Disease
RRS	Reciprocal Recurrent Selection
RSUP	Riau Satki United Plantations (Indonesia)

RWD	Root Wilt Disease
S1	Selfed generation
SDW	Shoot Dry Weight
SLA	Specific Leaf Area
SM	Skim Milk
SPAAR	Special Programme for African Agricultural Research (World Bank)
SPC	South Pacific Commission
SR	Stocking Rate
STD	Science and Technology for Development Programme
SW	Socorro Wilt
SWD	Soil Water Deficiency
TAC	Technical Advisory Committee
TCDC	Technical Cooperation among Developing Countries
TPI	Tropical Products Institute (UK)
TSFS	Three Strata Forage System
UHT	Ultra High Temperature
UNCTAD	United Nations Conference on Trade and Development
UNDP	United Nations Development Programme
UNIDO	United Nations Industrial Development Organization
UK	United Kingdom
UPB	United Plantation Berhad (Indonesia)
VAM	Vesicular-Arbuscular Mycorrhiza
VLDL	Very-Low-Density Lipoprotein cholesterol
VPD	Vapour Pressure Deficit
WBT	Wet Bulb Temperature
WHO	World Health Organization

Preface

Since the publication of the book 'Coconut, Tree of Life' - FAO, Plant Production and Protection Paper no. 57 (Ohler, 1984), many new developments have taken place in coconut research, out-dating that book in several aspects. FAO saw the necessity to publish a new handbook on coconuts as the livelihood of about 50 million small-scale farmers and processors depend on coconuts.

FAO requested Mr. Ohler to take care of the compilation of a new handbook in 1993. Mr. Ohler is eminently qualified, as he was raised in Indonesia, studied tropical agriculture in The Netherlands and has been actively involved in coconut research and development since 1955. Mr. Ohler was requested to take the resposibility as one of the authors as well as overall editor. Several subject matter specialists provided contributions to achieve in-depth coverage. The subject matter specialists are of different nations and this handbook is a truly international effort in the spirit of the United Nations.

As it is generally accepted that coconut monocropping in many places is no longer economically feasible, a whole part of this book deals with Coconut-Based Farming Systems in which many aspects of the combination of coconuts with other crops and/or animals are discussed. Special attention has been given to the great value of coconut oil as food for human beings, as some very incorrect ideas exist on this matter. Much new information has become available on the drought resistance mechanism of the palm. Research on diseases has revealed much new information and several diseases that formerly had been classified under 'diseases of unknown etiology', could now be grouped according to their causal agent. The chapter on insect pests has been kept as concise as possible. Although a great number of different insects have been observed feeding on the coconut palm, by far the greatest part are only sporadic guests.

The chapter on breeding is of particular interest, as breeding is one of the basic tools to improve the crop's performance and so to improve its ability to compete with other oil crops. It clearly shows the limits of the still much used mass-selection method, using open pollination. It also shows the specific implications caused by the partial autogamy of most tall cultivars. For a crop such as coconuts, the development of successful tissue techniques would be of enormous importance. A breakthrough in this matter is expected within the near future.

Also in the processing of coconut products, many new developments have taken place. The optimal utilization of all coconut products is of great value to the economy of the crop. The village industrial possibilities as well as the manufacturing of new products are of great importance in this respect. 'Nata de coco' is one of the by-products of coconut that recently has shown an extraordinary development. The developments in the processing and the use of the coconut wood, have taken a great flight over the last decade. As adequate processing techniques are as vital for the survival of the coconut industry as adequate farming techniques, this subject has also been given ample space in this book.

This book is not an easy reading textbook but a much broader compilation of knowledge, *i.e.* a reference book. The great effort of the editor and the subject authors of various chapters are much appreciated by FAO in view of its efforts to disseminate know-how on the coconut and its products.

G. Blaak,

Retired Senior Officer Industrial Crops,
FAO, Rome.

Foreword

Around the globe, there is hardly any tree that since the earliest times has been providing humankind with more uses and products than the coconut palm, and it is rather fascinating to learn about the knowledge, experience and practices that has been brought together by a panel of international scientists and experts, focused on modern management of the coconut palm in a wide variety of geographical regions and cultural areas.

Following the Editor's previous book on the coconut palm – *Tree of Life* (1984) – that largely covered many agronomic, species-oriented topics including types of coconut palm in various ecosystems, their diseases and their products, this new handbook not only up-dates scientific knowledge on modern coconut management, but also seeks to assess the great potentiality of the coconut palm tree in combination with the cultivation of other forms of agriculture that links up with indigenous knowledge, experience and practices including multi- and intercropping exemplified in various agroforestry systems in the tropics. Such advanced orientation not only recognises the significance of the newly-developing field of Indigenous Knowledge Systems and Development – particularly in agronomy, agri- and horticulture, botany and agroforestry where local knowledge and practice are no longer ignored – but it also contributes to the evolution of a comprehensive system of truly modern coconut management.

Moreover, such attention for the indigenous context of the breeding, management and use of the coconut palm and its products also links up with the current global interest in the conservation and management of biodiversity, where traditional and indigenous peoples in the tropics have often shown a rather sustainable way of use and conservation of their natural and agricultural resources that generally puts the maintenance of resources before the ever-increasing production schemes characteristic of so many western-based agricultural practices. In this respect, this book provides us with new important information from within a particular sector on relevant starting points for action as identified in the *Global Biodiversity Strategy* (WRI/IUCN/UNEP/ FAO/UNESCO 1992). As many recent studies are indicating, this book also points to the various aspects of coconut palm development that are both conservation-based and people-centred, a prerequisite to remain economically successful while sustaining the resources for the many generations to come.

It was a great pleasure when Mr. Peter Griffee of the Crop and Grassland Service, Plant Production and Protection Division of the Food and Agricultural Organization (FAO) – who from the beginning supported the idea to publish this textbook in a much wider context beyond the 'international food and agricultural arena' and as such recognised the broader interdisciplinary significance of the many related topics and issues in modern coconut management – approached our Leiden Ethnosystems And Development Programme (LEAD) of Leiden University to enter into a joint FAO/LEAD publication in conjunction with Intermediate Technology Publications (ITP) in London. Thanks to the generous financial support of Leiden University's Secretary, Ir. W.L.C.H.M van den Berg, and the great input and kind advice of ITP's Managing Editor, Mr. N. Burton, this aim has been reached in the form of the present publication.

I hope that the great achievement of Mr. Ohler both as General Editor as well as Author to compile, analyse and disseminate this rich and important knowledge base of modern coconut management encapsulated in this handbook will not only find its way to scholars, experts and students in the agricultural sciences in the West, but also to the policy makers and planners and the associated indigenous organisations, institutions and populations in the Developing World.

Dr. L. Jan Slikkerveer,
Associate Professor of Anthropology,
Director of the LEAD Programme,
Institute of Cultural and Social Studies,
Leiden University, The Netherlands.

Introduction

As an oil crop, coconut is facing increasing competition from other oils on the world market, especially from palm oil, palm kernel oil (also a lauric oil), and soy bean oil. Decreasing prices of coconut products seriously affect smallholder incomes. The existence of several levels of copra marketing intermediaries further depresses the prices paid to the producers.

Copra is often prepared in an inadequate way, resulting in a poor quality product offered to the industry or for export. Low quality products fetch low prices, another cause for low prices paid to farmers. Improvement of the quality of coconut products and their marketing position is imperative for the survival of the coconut industry. However, in many cases projects for the improvement of rural copra quality have not led to the expected results. A solution might be the elimination of copra production, working up the fresh kernel directly. For a long time this process has not been sufficiently effective in oil extraction to become accepted. But lately, new very promising so-called 'wet and semi-wet processes' have been developed. These processes can be adapted to a wide range of processing capacity levels. The large quantities of shells and husks that become available can supply enough energy for the processing plants.

The industrial use of more products of the coconut tree, such as coconut shell for meal and activated carbon, etc. could provide added value to the coconut industry. Domestic valorization of production by stimulation of local processing using appropriate technologies could improve the economic situation of coconut smallholders and processors. However, this would require new industries that can survive only when there is a guaranteed supply of adequate volumes of raw materials. In large regions with underdeveloped infrastructure and marketing systems this will be difficult. In small regions, such as some of the Pacific Islands, the volume may be insufficient and transport to a processing centre may be too expensive. It may be expected that unless efficient small-scale processing techniques are developed and subsidized, that can operate close to the production units, coconut smallholdings may shrink to home gardens producing for the family needs only. On the other hand, due to the need of large volumes and a guaranteed supply to new factories, larger coconut plantations may develop in the neighbourhood of such factories, either by the buying up of smallholdings, or by the planting of new, large plantations such as is the case of oil palm. Thus, it may well be possible that in the future the commercial part of coconut production may change considerably. Such developments already take place in some countries.

The only way to improve the difficult position of the coconut smallholder is by improving his cultural techniques, replanting his plantations with high-yielding planting material, making better use of land through intercropping or mixed husbandry, and producing and selling other products to the industry besides copra and oil. However, farmers with a very low income often do not have the required knowledge and capital to improve their production techniques.

Educating coconut smallholders to become good managers, using appropriate techniques and materials is the most difficult part of the problem. Extension and training are as important as new techniques. Coconut improvement projects should be long-term projects, guiding the farmers through the whole period between land preparation and the first years of production. Credit facilities should be made available on conditions suitable to smallholders.

If the coconut sector is not modernized in all its aspects, in many regions commercial coconut production may gradually become a medium and large-scale business, and small farmers may meet difficulties selling their produce even in the local markets.

J.G. Ohler,
Editor

PART I. THE COCONUT PALM AND ITS ENVIRONMENT

1. Historical Background

J.G. Ohler

Origin and Distribution

The coconut palm, *Cocos nucifera*, var. *typica* can be found along the coast and in the interior of almost all tropical countries between the tropics of Cancer and Capricorn. Its wide distribution has been favoured by its usefulness as well as by its adaptability to different ecological conditions and its ability to float in seawater and germinate on the beach when washed ashore. In many tropical countries, coconut is an important part of the daily diet. Its main product is the oil extracted from the kernel. The residue is an important animal feed. The coconut water from the young nuts is a popular beverage. The jelly-like kernel of the young coconut is considered a delicacy. The shredded kernel is sold as desiccated coconut used in food and confectionary. The husk of the nut provides an important fibre, coconut coir, that can be used for ropes, carpets, brushes, etc. The shell of the nut is used for household utensils and the charcoal made from it is an excellent basic material for activated coal. Instead of being used for nut production, the infloresences can be tapped, yielding sap with high sugar content, from which sugar, alcoholic beverages and vinegar can be made. The leaves are used for roof thatching. The mid-ribs of the leaves are used for brooms. Coconut wood is more and more being used for house building and other uses such as furniture or tool handles. It is no wonder that this tree has also been called 'Tree of life'.

In addition to its natural distribution, sailors also have taken the nuts with them for food and drink during their voyages and have planted the remaining nuts when arriving at their destination. There is no part of the palm, except the roots, that has no use in a peasant's household. According to Harries (1990; 1992) research has assumed the commercial plantation coconut to be representative of all coconuts. Its classification is misleading because the commercial coconut is not typical. It is nothing more than a random sample, taken into cultivation, of material that has achieved pan-tropical distribution for totally non-agricultural reasons.

According to Foale (1991b), glaciation has led to great fluctuation in global sea level and as recently as 15,000 years ago it was 140 m below the current level. During one millennium, around 9,000 years ago, the sea level on the coast of New Guinea rose faster than 1 m/100 yr. The consequent rapid drowning of coastal groves of coconut could have allowed a rapid change in the genetic composition of coconut populations as man 'rescued' seeds for planting further inland. There is no lack of theories, but real history may never be revealed.

The country of origin of the coconut is unknown and various regions have been indicated as such by various scientists, from South America to Melanesia, Asia and Madagascar (Ohler 1984). Child (1964) referred to a close association between coconut and the robber crab (*Birgus latro),* which at that time was thought to live specifically on coconuts. Later studies by Davis and Altevogt (1988) showed that the crab is omnivorous and even lives on islands where no coconut palms are growing. Ethnological and entomological indications place the centre of diversity in the area of south east Asia and Melanesia. According to Harries (1990), the coral atoll is considered the world's most stable ecosystem and the coconut palm is its most successful plant. Because of competition from other plants or predation by animals, coconut could not reach or survive inland on larger islands or on continents before it was domesticated. The continental coast and larger islands of Malaysia would be the obvious site for such domestication. This must have happened long before the wild and the domesticated types were both taken independently into agricultural cultivation. Wild forms of coconut have been found in various countries, such as on the island of Java, Indonesia, Queensland, Australia, and in The

Philippines. Wild palms are not necessarily primitive or small fruited. According to Harries (1978), the wild form, or Niu kafa type and the domestic form, or Niu vai type have the following general characteristics:

The Niu kafa has a slender, curved stem, with irregular leaf scars; long leaves that may hang down when still green; long angular fruits with a thick husk with ovoid or spindle shaped, thick shelled nuts with thick endosperm and high oil content. Cross-pollination is not absolute, germination and growth rate are slow, and it is susceptible to windstorms and to MLO diseases;

The Niu vai palm is robust, erect, with a base that can be very large; shorter leaves, rarely hanging down when green; spherical, thin husked fruits with spherical to obovate nuts with a thinner shell, much water, thinner endosperm and lower oil content. It is cross-pollinated but often selfed, it germinates earlier and its growth rate is quicker. It is tolerant of storms and of mycoplasma-like organism (MLO) diseases. The two contrasting types have undergone introgressive hybridization whenever they have been brought into proximity because of their cross-pollinating habit. According to Harries (1992) so many intermediate types developed that calling any of them *typica* is meaningless. Once industrial demand for vegetable oil production accelerated, all forms have been taken into cultivation.

According to Bourdeix (1995), after the work of Vavilov (1951) the narrowly defined meaning of 'centre of origin' has been widened. Harlan (1970) qualifies as 'non-centres' the very large regions (over thousands of kilometres) where domestication of a species has taken place. These regions are characterized by a confrontation of spontaneous and cultivated forms over wide areas. The diversification model proposed by Harries (1978) is based on multiple local introgressions between 'wild' and 'selected' types. The meaning of 'non-centre' could be integrated into this model. Jay *et al.* (1989) studied the genetic diversity of coconut by analysis of leaf polyphenols. They observed a higher within-ecotype variability in the Far East and the Pacific, providing a further argument to support the suggestion that the coconut palm originates from one of these two regions. They concluded that coconut palm dissemination methods have modelled the genetic structure of populations: natural dissemination - through multiple fundamental effects inducing genetic drift - should lead to a mosaic of well differentiated ecotypes, with variable but high degrees of inbreeding, distributed in accordance with geographical gradients. Human intervention combines with natural dissemination; ecotypes located at antipodes from each other may prove to be related, through the effects of human migration. Virtually all the world's coconut groves come from plantings: omnipresent man harvests, selects and plants. The selected forms retain their ability to spread spontaneously. A dynamic balance is established, including successive fundamental effects, multiple introductions, introgression between natural and selected forms, primary and secondary selections, etc. This leads to a structure within the apparently complex diversity.

World Production

Coconut plays an important role in households all over the tropical world. Wherever possible, almost all gardens have a few coconut palms. It is estimated that about 70 per cent of the coconuts are used for domestic consumption in the producing countries, of which just over half the volume is consumed fresh. The balance of the volume is consumed in the form of oil, edible or industrial (Punchihewa 1991). A study in East Kalimantan, Indonesia, showed that per capita consumption per annum of coconut oil by the year 2001 would average about 3.5 kg, (or 4.8 and 2.2 kg per annum for urban and rural consumers respectively) (Darwis 1991). According to Chhabra (1991) per capita nut availability in the four most important producing countries was: India-10, Indonesia-53, Sri Lanka-156, Philippines-282. These figures show that in densely populated countries like India and Indonesia there will be no or almost no copra available for export, whereas in a country like The Philippines the margin for export is much wider. The residue of the oil extraction, the coconut cake, is a valuable feed for domestic animals. Instead of harvesting the nuts, the sap of the infloresences can be tapped for the production of sugar or alcohol. When old, senile palms are cut down, the wood can be used for furniture and building construction and many other purposes.

Plantations predominantly are smallholdings, often not larger than 0.5 ha. In remote areas, these smallholdings mostly cater for local markets, where nuts are sold for consumption. Coconut oil is often prepared in the household kitchen, the shredded and pressed kernel is used as chicken or pig food. CP 74 shows an instrument by which the kernel can be rasped out of the coconut halves. Near towns and tourist resorts many nuts are sold for drinking the water, the so-called 'waternuts'. These waternuts are picked a few months before they are fully ripe, when the water has a sweeter taste. Fresh coconut meat is also sold as a delicacy in the markets. Wherever there is a local oil industry, using copra as a raw material, the surplus coconuts are sold to the factory, either whole or dehusked, or after their processing into copra. In some countries there are important cottage industries, producing coconut oil and selling it in bottles in local markets. Instead of nut production, sugar can be produced by tapping the spadices and collecting the sweet sap or toddy. This sap can be used for the manufacturing of sugar, alcoholic beverages, syrups, etc.

The first incentive to larger-scale production was the use of coconut for soap manufacturing. It was first exported from Sri Lanka to England between 1820 and 1830. Another incentive was the use of coconut oil for margarine, as a result of which vast areas of new coconut plantations were planted by the end of the 19th century. This expansion was more commercially oriented and in general planting was done on a larger scale than before. From a total area planted to coconut of about 3.2 million ha in 1938, this increased to about 8.3 million in 1980 (Das 1985). Nowadays, it may be around 10 million ha. World coconut production increased from about 28 million mt in 1970 to about 42 million mt in 1991. Expansion of the coconut area has continued for many years. According to Ranasinghe (1995), citing Andrew (1972), the export of fresh coconuts from Sri Lanka to England started in the 1880s. The fresh coconut was used as an ingredient for candy. A certain P.V. Appleby experimented with heating grated coconut on steam tables and observed that when treated in this manner, the coconut did not become rancid. The desiccated coconut industry had come into being in 1988.

At the beginning of World War I, coconut oil had become the most important edible oil. However, this position gradually changed with increasing world production of other oils such as palm oil, palm kernel oil, and soybean oil. As these other oils are mostly produced in large-scale operations, they can easily compete with the comparatively labour-intensive coconut product. However, with the development of new processing methods, omitting the manufacturing of copra by the direct processing of the fresh coconut meat, the coconut oil quality will be much improved and production costs will be reduced.

Coconut is widely used as an edible oil, either for cooking or for the manufacturing of margarine and other products, such as non-dairy creamers. It is also used for industrial purposes, such as for the manufacturing of detergents. In both types of final goods, it faces severe competition from vegetable oils less expensively produced. While coconut oil still commands a premium over soybean, rapeseed and palm oils, the trend of prices within the oilseed complex as a whole has been downwards over the past 40 years, and coconut oil prices have followed. A study commissioned by the World Bank estimated the cost of producing a ton of coconut oil in The Philippines at US$ 320-400, whereas the cost of jointly producing palm oil and palm kernel oil in Indonesia was estimated at US$ 200-220 per ton. Additionally, where total vegetal oil consumption has grown rapidly, coconut oil consumption has stagnated. While coconut plantings have been limited, the acreage devoted to oil palms, which produce both palm oil and palm kernel oil, has grown rapidly over the last decade. Production of palm kernel oil was expected to grow at 5-6% annually through the remainder of the century, whereas annual growth of coconut oil production was less than 1%. (Green 1991).

Coconut oil and palm kernel oil have been the most important lauric oils for a long time. But this position may be threatened by the development of other lauric oils. Coconut now provides about 75% of the lauric oils, but over the next 12 years, that share is expected to drop to two-thirds (Green 1991). In 1992, in the US canola [Canada Oil-Low Acid, referring to cultivars of rapeseed (*Brassica napus* and *B. campestris*) which is low in saturated fats and contains a high percentage of mono-unsaturated rather than poly-unsaturated fats] oil has been bio-engineered into a lauric oil. If grown on a large scale, this cultivar could become the first commercial oilseed grown in a temperate climate to produce lauric oil (Anonymous 1993c). Moreover, a new development in reduced-calorie-fat cocoa butter substitute could ultimately shrink the market for coconut-oil-based fats dramatically. It is therefore imperative for the coconut industry to attempt to improve its productivity and reduce its costs at all levels of the production line and to develop new endosperm-based commercial products.

Compared to oil palm, coconut yields much less oil per ha. Under average growing conditions and under good management on estate basis, oil palm may yield about 5-8 tons of oil per ha, against 2-4 tons for coconut. At such yields coconut ranks second only to the oil palm in terms of oil production per hectare, and much exceeds the present potential of any of the annual crops. However, the production level of small-holdings often does not exceed 1 ton of copra, or about 0.65 tons of oil per hectare. Coconut is very well adapted to smallholders' management conditions and vice versa. Usually, some weeding and the picking of the nuts every two or three months is all the labour that the farmer applies to his mature coconuts. Capital investments are very low. Large coconut plantations are scarce. Less than 2 per cent of the holdings are over 20 ha (Friend 1991). However, some very large plantations recently came into production in Indonesia. This may be a new trend in the coconut industry.

Table 1 shows the production of coconut in various countries between 1981-1993. Due to all the different ways of consumption and processing, and the non-availability of written records in remote areas, it is very difficult to estimate the amounts of consumption and production in the coconut growing countries. Most of the figures presented on this crop are estimates, and in some cases only guestimates. From the figures it can be seen that over the past decade some small changes have taken place in the continental participation in coconut production. Percentage-wise, Central and South America increased by more than one per cent of world production, Asia by almost one per cent whereas Oceania decreased over one and a half per cent and Africa remained on the same level. Asia still is by far the largest producer, with almost 85% of total world production. Country-wise some more important changes have occurred.

Indonesia

Indonesia has surpassed The Philippines as the largest producer, in spite of the National Coconut Replanting Scheme launched in The Philippines at the end of the 1970s. Contrary to The Philippines, production in Indonesia increased considerably. From 1980-1986, the 12 members of the Asian and Pacific Coconity (APCC) reported only 776,000 hectares of new area, 65% of which was in Indonesia (Green 1991). About 10 years ago, The Philippines produced 30.1% of world's total production and Indonesia occupied the second position with 29.5%. Indonesia currently is the largest producer, with almost 33% of total world production, whereas The Philippines occupy the second position with a little over 21% only. Indonesian production increased sharply by 16,500 mt in 1987/88 (Anonymous 1989a) and the area under mature coconut increased sharply as a result of a comprehensive planting programme in the late 1970s and early 1980s. Between 1969 and 1989 the coconut area in Indonesia increased by 3.65% per year. The area increases in 1986, 1987 and 1988 were 63,000 ha, 130,000 ha and 150,000 ha respectively, a total increase of more than 10%.

Also qualitywise there has been improvement, as new plantings were done mainly with high-yielding hybrids. (Anonymous 1988). In 1989, the area was estimated at 3,262,000 ha (Androecia and Damanik 1989). In 1922, Mahmud *et al.* (1992) estimated the area in 1992 at 3.3 million ha, of which 3.2 million ha, or 97.12% were smallholders, contrary to the oil palm estates that belong mostly to the government. Of the total area, about 25% are young palms, about 15% are hybrids, about 68% of the palms are in the productive stage and about 7% are senile. On average, smallholder plantations yielded 1.04 tons of copra per ha per year while that of private and government estates was 1.84 and 2.19 tons of copra per ha per year respectively. A national survey in 1987 indicated that with rising farmers' incomes, the consumption of vegetable oil also rose. The growth rate of consumption between 1980 and 1990 was 4.2%. Domestic consumption in Indonesia is almost as large as total production and with a rapidly increasing population the country may become an importing country if the actual production trend cannot be maintained. But the target of the government is to increase coconut production to a level that will meet consumption needs and allow for export of coconut products as well.

Table 1: *World coconut production, 1981 and 1993 (1,000 mt)*

Country	1981	%	1993	%
Indonesia	10,800	29.5	14,219	32.8
Philippines	11,050	30.1	9,300	21.4
India	4,500	12.3	7,700	17.7
Sri Lanka	1,716	4.7	1,597	3.7
Thailand	900	2.4	1,379	3.2
Viet Nam	290	0.8	1,207	2.8
Malaysia	1,204	3.3	1,030	2.4
Others	343	0.9	420	0.9
Asia	30,803	84.0	36,852	84.9
Mexico	827	2.3	990	2.3
Brazil	254	0.7	826	1.9
Venezuela	150	0.4	220	0.5
Dominican Republic	69	0.2	160	0.4
Jamaica	193	0.5	115	0.2
Colombia	64	0.2	89	0.2
Others	517	1.4	470	1.1
C/S America	2,074	5.7	2,870	6.6
Papua New Guinea	800	2.2	790	1.8
Vanuatu	240	0.6	259	0.6
Fiji	225	0.6	200	0.5
Solomon Islands	193	0.6	170	0.4
Pacific Islands	203	0.6	140	0.3
Western Samoa	200	0.6	130	0.3
French Polynesia	129	0.3	86	0.2
Kiribati	72	0.2	65	0.2
Others	213	0.5	103	0.2
Oceania	2,275	6.2	1.943	4.5
Mozambique	420	1.2	425	1.0
Tanzania	320	0.9	360	0.9
Ivory Coast	159	0.4	225	0.5
Ghana	160	0.4	220	0.5
Nigeria	90	0.2	140	0.2
Madagascar	67	0.2	86	0.2
Comoros	53	0.1	50	0.2
Others	246	0.7	215	0.5
Africa	1,515	4.1	1.721	4.0
World	36,667	100.0	43.386	100.0

Source: FAO Production Yearbook 1981 and 1993.

Philippines

The total area planted to coconut in 1991 was 3.112 million ha, employing 1.6 million farmers and 1.9 million farm workers. About one-third of the country's population depends directly or indirectly on the coconut industry as a source of income and a means of employment and livelihood (Magat 1993a). Old age and poor management of plantations may be the most important causes of yield decline. Results of a survey conducted in 1991 showed that 25% of the trees were older than 60 and 19% were between 40 and 60 years old. About 35% of the planted area is unsuitable for coconut. The nuts are often harvested when still immature. Drying of nuts and copra often is inadequate (Aldaba 1995). Domestic consumption in The Philippines in 1986 was around 15% of its production, the remainder was exported. In spite of the large area planted to coconut, the income from this crop in 1986 was only 1.95% of the total national income, the largest part coming from fisheries, forestry and industry. According to Rouziere (1995), coconut represents about 12-15% of total agricultural production and 6-7% of total exports. Although The Philippines is the second largest producer of coconuts, with 52% it is the largest exporter of coconut oil and copra.

The National Coconut Replanting Scheme was launched in The Philippines at the end of the 1970, aimed at replanting about 60,000 ha per year with improved planting material, increasing the production from 2.8 million tons (copra basis) to about 10 million tons in a period of 40 years (Ohler 1984). Apparently, this scheme has not produced the expected results. From the initial planting of hybrids (MYDxWAT and also some local hybrids) from 1976 to date, surviving populations of these high yielding hybrids could be only around 45,000 ha. The country's production has declined substantially, from 2.69 million tons of copra in 1986 to 2.06 million tons in 1990. Rouziere (1995) reported that during the last ten years the area planted with coconut has decreased by 200,000 ha. It is estimated that Philippine coconut production will decline by 2% per annum, mainly due to depressive biotic, environmental and farm management factors (Magat 1993b). This is a dangerous development in a country where coconut is so important, earning a large amount of foreign exchange as an export crop, and providing a living for many smallholders and farm labourers.

Through a new programme, the Small Coconut Farms Development Project, attempts will be made to improve farm management and to rejuvenate areas containing old palms. Participating farmers will receive planting material and fertilizers as grants-in-kind, but will have to provide all labour requirements. The objective is a 25% production increase in 5 years' time.

India

India is the third largest producer in the world and its share in the world production moved up from 12.3% to 15.4%. The increase is a result of a larger planted area as well as improved yields per ha. In 1989, the coconut area in India was estimated at 1.51 million ha (Rethinam 1991). The State of Kerala, which accounts for about 1% of the total land area of India, contributes to about 57% of the coconut area and 47% of the coconut production. The other southern states, Tamil Nadu, Karnataka and Andhra Pradesh together account for 34% of the area under coconut but contribute 44% of the production (George *et al.*, 1991). Coconut in India is essentially a smallholders' crop. There are about 5 million coconut holdings of which 98% are below 2 hectares (Shenoi 1991). Average yields in the three major coconut states, Kerala, Karnataka and Tamil Nadu are about 33, 44 and 54 nuts per palm per year respectively. The comparatively low yield in Kerala is due to the high Root(wilt) disease incidence (19.4%) and the increasing number of senile and unproductive palms in most of the holdings.

Aravindakshan (1991) mentions that the small size of the holdings prevents economic investment and is one of the reasons for the inadequate response of the growers to the transfer of technology programmes. These facts and also a low genetic potential of native palms are important reasons for low average yields. Probably, the latter may turn out to be of less importance when management improves. Domestic demand is higher than production and copra and coconut oil need to be imported. It is estimated that, at the present level of consumption, the demand for coconut oil towards the end of this century would be 10,400 million nuts equivalent. This would mean that efforts should be made to increase the production by at least 50 per cent by the turn of the century. George *et al.*, (1991) reported that the combined efforts of the Central Plantation Crops Research Institute, the Coconut Development Board and the Department of Agriculture and favourable rainfall since 1984 resulted in a steady production increase in the 1980s. There was also considerable expansion in the area under coconut, and India's share in world coconut production rose from 12.3% to 17.7%.

Thailand

In Thailand, coconuts are traditionally planted in small garden plots, often in mixed cultivation with fruit trees. A large proportion of coconuts is thus grown for domestic consumption and few farm families are dependent on the crop. The total acreage is estimated at 417,600 ha and the average farm size is 2.4 ha. Yields are low, being estimated at about 22.1 nuts per palm per year. Specific attention is paid to the combined coconut/cocoa planting system, as a tool for stimulating increased coconut production in Thailand. National production was often insufficient to meet the demand (Dootson *et al.*, 1987).

Sri Lanka
Sri Lanka's production seems to be on the increase after a period of declining acreage between 1962 and 1981 from 466,000 ha to 419,000 ha. This loss of area has been compensated for by an increase in yield per ha from about 4,800 nuts per ha per year in 1978 to about 5,900 nuts per ha per year in 1986 (Liyanage, D.V. 1987). Also in this country it is predominantly a smallholder's crop, 98% of the holdings being smaller than 2 ha. (Chhabra 1991). Consumption exceeds production. About 22% of the calorific intake of Sri Lankans is derived from coconut. The per capita consumption of coconut is estimated at about 120 nuts per year (Nimal 1989).

Vietnam
Vietnam is a new important coconut producing country. Between 1981 and 1993, its share of world production increased sharply from 0.8 to 2.8%.

Central and South America
The second highest coconut producing continent is Central/South America, with a share of 6.6%. Relatively low population density and considerable land availability provide a possibility of a further increase. Development in this direction is seriously hampered by serious disease and pest attack and little motivation among smallholders to improve planting and management practices. As in Indonesia, further growth of production will probably come from new, large estates. Mexico and Brazil are the most important producers of this continent, with 2.3% and 1.9% of world production respectively. Mexico is the eighth largest coconut producer in the world. But the Lethal Yellowing disease is rapidly spreading in this country, posing a serious threat to the crop. The total coconut area in Brazil is estimated at about 300,000 ha, about 82% of which is located in the northeastern states (Ferreira *et al.* 1994).

Oceania
Notwithstanding that coconut by almost any criteria is by far the most important crop in the South Pacific, Oceania, once the second most important producing area with 6.2%, now occupies the third position with 4.5%, with Papua New Guinea as its largest producer, *i.e.* 1.8% of world production. The other countries are generally too small to play an important role, although the coconut itself plays a very important role in the islands' economy. The total area planted to coconut in 1987 was 506,980 ha. The area planted to coconut in Papua New Guinea was about 250,000 ha, more than 60% being owned by smallholders. The second largest coconut area can be found in Vanuatu, *i.e.* 91,300 ha. The area under coconut in Fiji was estimated at about 70,000 ha (Brookfield *et al.* 1985). Many coconut plantings are senile, or nearly so, with many plantings dating back to early 1900s. Most replantings have been from seed from existing old mother palms, sometimes with careful selection, more often not. Some companies are replanting coconuts on a small scale, and almost invariably as a subsidiary crop, providing shade for cocoa. (Cundall 1987).

In the South Pacific, coconut occupies the largest single crop area, and provides the principal source of foreign exchange earnings. It provides a source of livelihood for a dominant part of the population, and provides an important source of food and materials for the population (Ooi 1987). In Western Samoa, Solomon Islands, Vanuatu and Tonga, coconut occupies 83, 75, 63 and 60 per cent of the suitable arable land respectively. Coconut provides 74, 55, and 44 per cent of the export earnings of Vanuatu, Tonga and Western Samoa respectively. Domestic consumption depends very much on copra prices. When copra prices are high, domestic consumption is low and vice versa. Per capita consumption is estimated at ½-1 nut per day. The crop is predominantly grown in small holdings of 2-4 ha, mostly in mixed cropping systems. Inputs in general are low. Yields are estimated at 400-720 kg copra per ha. A significant part of the coconut smallholdings tends to be over-aged and senile, being a particular problem for Papua New Guinea (70%), Western Samoa (48%) and Fiji (42%).

Africa

Africa's share in production has changed very little over this decade; it declined slightly from 4.1% to 4.0% of world production. Within this continent, the increase in Ivory Coast, from 159 mt to 225 mt has been considerable. The situation of the IRHO research station in Port Bouet and the availability of excellent plant material undoubtedly has played an important role in this development. The area under coconut in Ivory Coast in 1989 was 51,000 ha (Taffin and Sangare 1989), about 37% of which were large industrial plantations of >500 ha, 26% semi-industrial plantations of 10-500 ha, and 37% village and family farms of <10 ha. The industrial plantations produce substantial yields of up to an average of about 17,000 nuts per ha, against 5,000 nuts pe ha for the semi-industrial plantations and 2,600-4,000 nuts per ha for the village and family farms (Anonymous 1991b).

In Mozambique, due to internal war strife, little has changed during the past decade. It may be assumed that little planting has been done and that a large part of the planting has suffered from lack of management. Copra production in 1984 was estimated at 215,000 tons as against only 35,000 tons in 1990). In view of the great food shortage it may be assumed that by far the largest part of the production has been used for domestic consumption.

Coconut is an important crop in the entire coastal belt of Tanzania, including the islands of Zanzibar, Pemba and Mafia. The total area of land under coconut is not easy to assess as the crop is scattered and often interplanted with other crops. In 1979 the area throughout the country's coastal belt was estimated at 90,000 ha at varying densities. (Anonymous 1979). According to air photo interpretation data, the National Coconut Development Programme estimated the actual coconut area at 240,000 ha, with a palm population of about 22.4 million palms growing at varying densities (Anonymous 1991b). The potential area is estimated to be between 500,000 and 550,000 ha. The mean annual production is estimated at 22-23 nuts per palm. Oswald and Rashid (1992) estimated the areas under coconut on the islands of Zanzibar and Pemba at 60,000 ha and 18,000 ha respectively. It is estimated that 95% of the crop is cultivated by smallholders. Coconut in Tanzania is mainly intercropped with annual and perennial crops. The crop suffers from Lethal Disease, a mycoplasma disease that appears all over the country, but particularly in the northern half. Moreover, the bug *Pseudotheraptus wayii* seriously damages the nuts and considerably reduces production. Severe droughts are frequent, causing sharp yield declines in certain parts of the country. The National Coconut Development Project has done much to promote coconut growing since 1980. An important research programme was established. New varieties were introduced for comparison with the local East African Tall, and also hybrids were produced for distribution among farmers. Although the full impact of the programme is still to come, more progress would have been made in the production if the Lethal Disease would not have taken such a heavy toll.

2. The Coconut Palm

J.G. Ohler

Botany

Coconut, (*Cocos nucifera* L.), is a monocotyle belonging to the order of the *Palmae*. It is the sole species of the genus *Cocos* within the tribe of *Cocoidae*, to which also the oil palm belongs. Within this genus, two main groups can be recognized, the tall palms, *C. nucifera typica*, and the dwarf palms, *C. nucifera nana*. In strict botanical terms it is not a tree. It has no bark, no branches either and no cambium or secondary growth. It is a woody, perennial monocotyle. Its trunk is called a stem. The chromosome number in coconut is 2n = 32. Pollen fertility is very high in talls and hybrids as compared to that in dwarfs and in semi-talls. Inbreeding depression has been noted in most of the selfed progenies of tall cultivars. Heterosis is reported in almost all cross-breeding trials.

Morphology

The Stem

The coconut palm grows up to a height of about 25 m and in exceptional cases to about 30 m, depending on ecological conditions and age. Its single, light grey, smooth and erect or slightly curved stem rises from a swollen base, the bole. This bole is more developed in some varieties than in others. Dwarf palms have almost no boles, the semi-tall types have a somewhat larger bole development, particularly when growing under very favourable conditions (Figure 1).

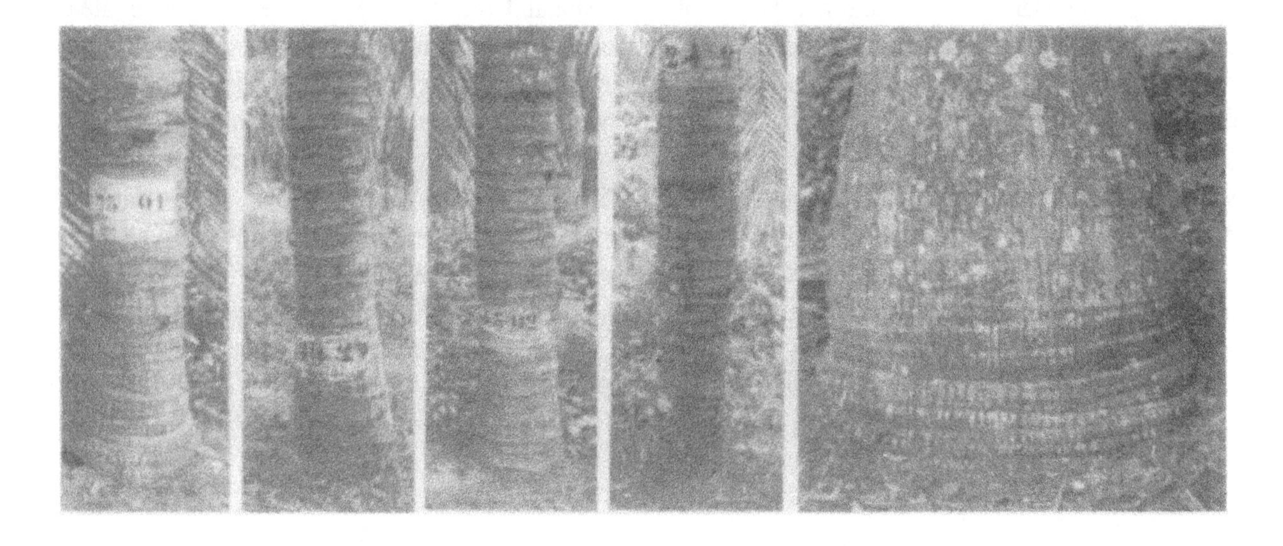

Figure 1: *Stem bases of dwarf and tall coconuts*

Stem growth originates from one terminal bud in the centre at the top of the stem. The first years after germination, only very short internodes are developed from which sprout many adventitious roots. Only when the full width of the stem has been reached (after four years for talls and two to three years for dwarfs) elongation of the stem begins, and the stem emerges from the ground. As the stem has no outer cambium it does not grow laterally and wounds in the stem will always remain visible as they will not be covered by new growth. Steps cut into the stem for climbing will remain for as long as the palm lives. Under normal conditions, the stem diameter remains the same. Under unfavourable conditions such as droughts, bad drainage, mineral deficiencies or diseases, the diameter of the stem will decrease and the stem top will form a so-called 'pencil point.' When conditions improve again, the diameter of the newly formed stem may return to normal and thus a kind of 'bottle-neck' remains visible.

Niu vai type coconuts have an erect stem. Niu kafa types are generally curved. Stems may also curve under the influence of strong winds, lack of light, or after having fallen over due to poor anchoring on stoney or peat soils. Rarely, branched stems can be seen (Figure 2 and C.P. 1). The branching results from slight damage of the growing point, or from severe stress (Foale 1986). Heavy damage of the growing point will kill the tree. Growing point damage can also cause spiralling of further stem growth (C.P. 2). Twin and triple palms result from bicarpelate or tricarpelate nuts, sprouting two or three seedlings (Figure 3). Suckering of coconut palms is very rare and suckers are usually unproductive.

Figure 2: *Coconut with triple crown*

The stem consists of a central cylinder surrounded by a narrow band of tissue, the cortex, also called 'bark', about 1 cm thick, somewhat thicker at the base of the stem. The central cylinder consists of a parenchymatic tissue enclosing vascular and fibrous bundles. The vascular bundles at the centre are more widely spaced than at the periphery. Between the central part of the stem and the periphery, the bundles are closely packed, with just a few layers of thin-walled cells intervening. Since each vascular bundle has a massive, radially extended fibrous sheath external to the phloem, and the ground parenchyma becomes sclerotic, this zone forms the main mechanical support of the stem. The closely packed bundles and the sclerotic parenchyma of the periphery make the outer ring of the central cylinder very hard. Stomata, defined as small, almost isodiametric silica-cells are abundant in the stem. These specialized cells contain as inclusion silica bodies which are more or less spherical, enveloped by the thickened basal wall of the silica cell.

The surface of the cortex shows a pattern of triangular-shaped leaf scars, marking the stem where former leaves have been attached. Between the leaf scars there are unscarred areas, the internodes. The distance between the leaf scars shows the growth rate of the palm. For instance, if a coconut palm produces about 14 leaves per year, the total distance between 14 leaf scars represents 1 year of stem growth. Substantial varietal difference may be observed in yearly leaf production.

Figure 3: *Triple coconut*

The Root System

Compared to the studies on other parts of the palm, root studies were seldom conducted up to about ten years ago. Root studies are difficult to conduct, and are often destructive. They require large numbers of palms and costs involved are high. But it may be stressed that invisible parts of the palm need as much attention as the visible parts, especially when breeding drought-tolerant palm varieties for climates with pronounced dry seasons. A common descriptive method is to count the number of roots per vertical surface unit within a section of soil and to observe their shape, colour and degree of branching. Other methods quantify roots by expressing root density at a given point in terms of dry weight per unit of soil volume.

Being a monocotyledon, the coconut has no taproot. Adventitious roots develop from the bole. The first root is developed at germination and may reach a length of about 5 cm in the first week of growth, during which time also one or more rootlets may have developed. Rootlets are produced in quick succession, radiating in all directions, and breathing roots or pneumatophores are also found on the first root. The second root is formed about 10 days after the first. Six weeks after germination, when the shoot becomes visible above the husk, the seedling may have three roots, the longest measuring about 20 cm. Under conditions of high humidity and wounding of the stem, roots may also develop from the stem above the bole, even at great height. When covered with soil, these roots grow out normally and develop secondary roots and rootlets. Old roots die and new roots develop continuously.

The number of primary roots increases with the age of the palm. Omoti *et al.* (1986b) in Nigeria counted 548 roots on 5-year-old NIFOR Tall coconut palms, increasing to 5,200 roots at the age of 13. The total number of primary roots depends on the variety and health of the palm, as well as on soil conditions such as fertility, humidity and cultural practices. Counts of total primary roots as observed by various authors vary from about 2,000 to 16,500 per palm (Ohler 1984; Avilán *et al.* 1984). Omoti *et al.* (1986b) also found that 5-year-old palms produced significantly more roots under *Pueraria phaseoloides* and *Centrosema pubescens* covers than under bare soil surface conditions. Rootlet development is favoured by soil humidity and fertility. Louis (1989) found that the total bole surface and rooting zone increased with fertilizer application and irrigation of the young palm, indicating that good management right from the early stages of growth may have long-lasting effects. He observed that cultivation of the land increased the number of primary roots by 93 per cent, and cultivation plus fertilizing by 479 per cent. With irrigation, the root number increased even further to 592 per cent. Manuring tended to reduce the number of secondary roots. He suggested that the secondary and tertiary roots are produced only to increase the area of absorption in nutrient-deficient soil. Hoffman, cited by Reyne (1948) observed a strong development of roots under village houses and attributed this to conditions of high humidity and fertility. It appears more likely that the production of secondary and tertiary roots are promoted by favourable coil conditions rather than by unfavourable conditions.

Primary roots are about 1 cm in diameter and of uniform thickness. Mechanical damage does not cause the death of the entire root, although this may occur occasionally. There is a continuous production of secondary roots. The root colour varies from almost white at the growing point through cream colour to scarlet, brown and dark brown on the oldest part. This process takes about 15 years. The exodermis consists of sclerenchyma, causing the brownish colour of these roots. Due to this sclerenchyma and the vascular bundle in the central cylinder, roots can resist strong tractive power. The roots have no root hairs. Their cortex is covered by a thick, hard impervious layer, the hypodermis, which does not allow penetration of water or air. Water and minerals are absorbed only in the white zone about 2-5 cm behind the tip of the root. The tip is covered by a root-cap.

For breathing, the roots have short, whitish outgrowths, so-called breathing roots or pneumatophores. Their size depends on the size and order of the root on which they are produced and may reach a maximum length of about 8 mm and a diameter of about 4 mm. Omoti *et al.* (1986b) observed uniformly distributed pneumatophores among primary, secondary and tertiary roots but there were few or none in the fourth and fifth order roots. Manuring tends to increase the number of breathing roots on primary and secondary roots.

The length a primary root can reach depends on soil conditions and palm variety, explaining the widely differing root lengths mentioned by various authors. The root system development decreases with increasing soil density. According to Pomier and Bonneau (1987), the constraint is not the soil texture itself but resistance of the soil to root penetration. In chemically poor soils, a greater soil volume has to be explored by the roots to find nutritive elements. Thus the palms have to invest more in dry matter in their root system than in richer soils. The author observed a root with a length of 31.75 m on a coconut palm growing is a very loose sandy soil in Mozambique. Thampan (1981) reports a normal root length of 5 m in clay soil and 7 m in sandy soil, but most authors observed greater lengths. It is often assumed that by far the greatest part of the roots can be found within a circle of about 2-3 m from the stem. It is normal that a mass spreading out from a central point decreases in density with distance from that point. But the decrease may be much less in light soils where fertilizers have been applied regularly by broadcasting, than in heavy soils where fertilizers have been applied close to the stem.

Pomier and Bonneau (1987) compared the results of water profile measurements to the useful depth of the root system in different soils in the major coconut growing areas of Ivory Coast. It was found that under non-limiting conditions, the roots reached a depth of 4 m and the proportion of the roots in the first 50 cm of soil was about 50 per cent. The roots did not grow in stagnant water. It was observed that where ground water is moving, roots were also found in the water-bearing layer of the soil. If water fluctuations stay within stable limits, only a few centimetres of soil remaining above the water table year-round are needed for developing an efficient root system. Hence, a shallow water-table is not a limiting factor on production, although the palm will have a lack of anchoring possibilities. In soils with a highly fluctuating water-table, such as coastal soils where the water-table is influenced by the tide, the root system is pushed upwards and a thick mat of roots may be found close to the soil surface. Low soil volume will have a low total nutrient availability unless the groundwater provides a regular fresh supply of nutrients. Compact horizons prevent roots from growing into deeper layers, as a result of which roots do not reach the water table during the dry season, leading to water-stress. On shallow soils with flowing groundwater, the lateral growth of coconut roots was beyond 4 m, but in stagnant water lateral extension was insignificant and the palms were growing poorly.

Soil fertility and soil texture influence the development of the root system. Correction of nutrient deficiencies particularly influences growth of primary roots growing near the soil surface (Anonymous 1989b). Cultural practices such as fertilizing and irrigation within a certain circle around the stem may confine the greatest part of the root system to this area. Cintra *et al.* (1992) observed that the greatest part of the root system concentration of dwarf palms growing on poor sandy soil was found at a depth of 0.2-0.6 m. Despite the small quantity of roots observed in the 0-0.20 m layer, their development needs to be emphasized, since they extended more than 4 m from the stem. About 70% of total roots and 65% of fine roots were found within a radius of 1 m from the stem, and 90% of total roots and 85% of fine roots within a radius of 1.5 m from the stem. The greatest concentration of roots was found at 0.6 m from the stem where the nutrient levels were highest, especially those of potassium and phosphate. At 1.8 m from the stem the nutrient levels were very low. The results of the study confirmed the findings of Ouvrier and Brunin (1974) that the root distribution zone increases with the enlargement of the fertilizer area around the stem.

In a subsequent trial (Cintra *et al.* 1993) they studied the root systems of some tall varieties, growing on a sandy soil under conditions of periodical water-stress. The ground water level during the rainy season was at about one metre depth, whereas at the end of the dry season it was almost three times as deep. They observed that during periods of water-stress new roots grew downwards to greater depths. It was concluded that the root system could act as one of the mechanisms that regulates the water supply to the palm during dry periods. Lateral root development of all varieties studied was limited to about 1.8 m from the stem. It was suggested that this was mainly caused by the nutrient availability within this circle. It was also found that, in addition to other possible influences, the more homogeneous root system throughout the soil profile and the tendency of roots to grow downwards during periods of water-stress of the Polynesia Tall and Brazil Tall (from Forto Beach) varieties suggest a better adaptation of these varieties to conditions of long dry periods than the other varieties studied: Tonga Tall, Rotuma Tall, Rennell Tall and West African Tall. Voleti *et al.* (1993a) observed that during dry periods coconut palms suffered more on red sandy loam soils rich in coarse sand, with low cation exchange capacity and higher hydraulic conductivity, than on laterite soils with low hydraulic conductivity and higher cation exchange capacity.

Tall palms have the longest roots, followed by hybrids and dwarfs. Differences in root distribution between tall and hybrid varieties were studied in The Philippines (Anonymous 1985a). Findings indicated that in all locations the talls showed more roots than any of the local or foreign hybrids. In terms of root distribution around the base of the stem, it was only with the hybrid Malayan Yellow Dwarf (MYD) x West African Tall (WAT) where no significant difference with the talls was noted. The study of Pomier and Bonneau (1987) in Ivory Coast showed that West African Tall palms produced 50 per cent more roots than the PB-121 (MYD x WAT) hybrid, the difference mainly resulting from increased development of main and secondary roots rather than that of root classes 3 and 4. The dwarf hybrid MYD x EGD (Equatorial Guinea Green Dwarf) had 25 per cent more roots than MYD. Also in this case, the difference was mainly due to a greater development of the large roots. The yield of the hybrid was also higher, 41 per cent more than that of MYD. Studies by Avilan *et al.* (1984) also showed that dwarf palms had fewer roots than talls. Cintra *et al.* (1992) studied the difference in root system development between some local and imported dwarf palms in Brazil. They found that the MYD and the Gramane Yellow Dwarf had the most widely developed root system, producing the greatest mass of roots, whereas the Jiqui Green Dwarf and the Cameroon Red Dwarf had the least developed root systems. The Gramane Red Dwarf and the MRD held the middle position. They concluded that in areas subjected to drought periods, the yellow dwarf palms, due to their deeper and wider rooting habits should be preferred to the other varieties. It may be added that the similarity in performance and development of similarly coloured dwarfs indicates that the so-called local dwarfs may be similar to the imported ones, only being imported much earlier. The importance of root studies in relation to variety choice for regions with dry periods is clear.

Pomier and Bonneau (1987) found that to express the coconut root system's actual efficiency, it is necessary to examine not only the quantity of roots per unit of soil, but also the capacity of these roots to take up water and mineral elements. Rajagopal *et al.* (1991) found that in addition to the pattern of transpirational loss of water through the leaf surface, also the pattern of soil moisture depletion by the roots was important. Drought-tolerant genotypes in general extracted more soil moisture from the entire soil profile compared to the susceptible types, despite minor variations within the groups studied.

Leaves

All leaves originate from the growing point at the apex of the stem. The youngest leaves are folded and wrapped together. They form the central spear, or heart of the palm. The central cylinder of the spear, also called the cabbage, is white and soft. It may weigh several kgs and is considered a delicacy. As it can only be obtained by killing the palm, it is also called 'millionaire's salad'. From the spear, the leaves gradually unfold and expand. They are arranged in spirals, running clockwise in some palms, anti-clockwise in others. The two types are distributed more or less equally. The direction of the spirals does not seem to be genetically determined (Darwis 1991). The phyllotaxy of the leaves is about two-fifths, meaning that each consecutive leaf is about 140° around from the former. Starting from one leaf, the sixth leaf reaches a position almost above the first leaf after two complete rounds of the stem. The difference between the position of the first and that of the sixth leaf may be about 25 - 30°. The crown of a mature coconut palm growing under favourable conditions forms an almost full sphere, the leaves standing out at different angles to all sides. The oldest leaves are found in the lowest whorls, and a few withering leaves may hang alongside the stem before dropping. When growing under unfavourable conditions, such as drought or nutrient deficiency, the older leaves may drop off prematurely, the shape of the crown looking like an umbrella, and in severe cases like a broom. Stem growth may be affected by unfavourable conditions, shortening the internodes, as a result of which the leaves are growing from the stem at little vertical distances, flattening the crown.

Leaf primordia are differentiated about two years before appearing as a central unopened spear. The phase of elongation within the spear takes about 6 months and mature leaves may remain on the palm for about three years, sometimes even longer. The total number of unfolded leaves usually varies between 30 and 40, depending on growing conditions and variety. Mean annual leaf production is about 14-16 for mature tall palms and 21 for dwarf palms. However, dwarf palms have only 25-28 unfolded leaves in the crown. High-yielding palms generally have a higher annual leaf production. Leaf production of hybrid palms is more or less the mean of that of the parent palms. Under stable growing conditions, the leaf number per palm will remain constant for many years, until senility sets in.

The length of a leaf depends on variety, growing conditions and age of the palm, reaching up to 7 metres in tall palms, with a leaf area of up to about 10 m^2, and weighing up to 20 kg when green. Dwarf palms have smaller leaves, with a length of up to 4 m. In the spear stage, leaves of seedlings may grow as much as 1-2 cm per day, those of healthy mature palms up to 5-6.5 cm. Once the leaf unfolds fully, there is little increase in area and dry weight (Ramadasan and Mathew 1987).

The length of the petiole is about one-quarter of the total leaf length, but the ratio differs with variety. The petiole continues as the mid-rib of the leaf. It is slightly concave at the upperside, and round at the underside. A strong rounding also indicates a strong leaf that can stand more vertical pressure than a flattish one. Weak leaves may collapse by the weight of heavy fruit bunches resting on the leaf stalk.

The petiole is attached to the stem by means of a sheath in the form of a bracket firmly clasping the stem with its wings almost around it. According to Menon and Pandalai (1958), palms with short leaf stalks have bunches with short stalks as well, keeping the bunch closer to the stem and putting less strain on the leaf bracket.

The leaflets growing out from both sides of the mid-rib have different lengths. The first leaflets at the base are short, the following leaflets gradually increasing in length, reaching a maximum of about 130 cm at about one-third of the mid-rib, gradually becoming smaller again towards the tip of the leaf. The smallest leaflets at the tip may be only 25 cm long. The longest leaflets of dwarf palms may not be longer than 80-110 cm. Length and width of leaflets vary with variety. The total number of leaflets of a mature palm leaf varies from about 200 to 500.

Leaflet halves are divided by a strong mid-rib. In each half, parallel to the mid-rib, 20-25 vascular bundles can be found, surrounded by sclerenchyma. On the underside of each blade, along the mid-rib, a row of cells can be found that lose water easily, resulting in the closing (folding) of the leaflet under conditions of waterstress. Also at night, or in cloudy weather, leaflets are closed tighter than in full sunshine. The lower surface of the leaflet has a number of shield-like outgrowths, or trichomes, their density varying with variety. These trichomes offer a certain resistance to leaf-eating caterpillars, particularly *Setoria nitens* (Taulu *et al.* 1980). The leaf blades are covered by thick cuticles. The lower epidermis is thinner than the upper epidermis. Stomata are located in the lower epidermis and number about 170-220 per mm^2. The stomatal density may be a varietal characteristic. Dwarf varieties have more stomata per mm^2 than tall varieties. The stomatal density of their hybrids is about the mean of that of their parents.

Friend and Corley (1994), when comparing Rennell Tall (RIT), Malayan Red Dwarf (MRD) and the hybrid between them on Russell Islands, found that the dwarf variety has a smaller leaf area than the tall, but both varieties reach their maximum leaf area about five years after planting. The maximum leaf area for the hybrid appears to be comparable to that of the tall parent. The trend with age is not so clear as for the two parental varieties, but its maximum leaf area is probably reached at about the same age.

Flowers

The inflorescence of the coconut palm is a spadix that develops in the axil of each leaf. Thus, the number of leaves produced annually also determines the number of spadices developed. Under unfavourable conditions or in old age, spadices may be aborted at an early stage of development. Initiation of the spadix begins about thirty-three months before its opening. Differentiation of the sheath takes place about two years before its opening and differentiation of the spikes about six months later. Flower primordia are initiated about one year before opening.

Before opening, the spadix has a spear-like shape, with a length of about one to one-and-a-half metres. The inflorescence is enclosed within the sheath, branches or spikes of the inflorescence lying close to the main axis. When the inflorescence develops within the sheath, the latter finally splits open over its entire length, about three months after its appearance, the inflorescence emerging and unfolding. The spadix consists of a main axis with 20-65 branches, or spikes, bearing flowers. Female flowers are located at the base of the spikes, the rest of the spike being fully covered by male flowers. These flowers are sessile. Each spike may have one or more female flowers and generally about two to three hundred male flowers. The total number of male flowers varies with the total number of spikes and their length (CP 5 - 8). A short length of the inflorescence can be a yield-limiting factor, the inflorescence having insufficient room for many nuts. When the main axis, or rachis is very long (CP 21), this may lead to immature drop of bunches that become too heavy due to a high number of nuts developing on it.

The total number of female flowers per inflorescence usually varies between twenty and forty. Dwarf palms generally carry a higher number of female flowers in the spadix. There is a great variability in this characteristic which, apart from being determined genetically, is also strongly influenced by growing conditions. First spadices usually have fewer female flowers per inflorescence than spadices in mature palms. Palms producing large nuts usually have fewer female flowers than palms producing smaller nuts. Abnormalities, such as inflorescences with only male flowers also occur. In all coconut growing regions, palms with inflorescences with numerous female flowers (CP 14) or fully covered with female flowers appear, but very rarely. Sometimes such inflorescences have one or two spikes, and also some male flowers may be found. This phenomenon also occurs in dwarf palms. This form of coconut is called *Spicata*.

In some palms the spadix is transformed into a compressed vegetative shoot with pinnate leaves resembling a seedling. Such bulbil shoots remain attached to the crown much longer than the normal fruit-bearing spadix, sometimes even for several years. The lateral shoots of a spadix can also transform into vegetative shoots. Female flowers developing into bulbils have also been observed. These bulbils were short-lived and dropped when the supporting spadix withered and shed.

Löhr (1993), studied the flowering behaviour of a number of coconut varieties at three greatly differing sites along the coast of Tanzania. All varieties, dwarfs, hybrids and talls, showed substantial fluctuations in the rate of inflorescence emission and in the number of female flowers on the inflorescences. Shorter intervals and higher numbers of female flowers coincided with the onset of the hot season and the first rains in November-December, after the long dry season, which is also the cold season. Average numbers of female flowers during the long dry season were about 20, whereas the highest averages during and immediately after the short rains, the hot season, reached 80-100 per inflorescence. Inflorescence and flower production started to decline in March-April with lowering of temperature. As a result, between 50 and 65 per cent of all female flowers were produced in only four months from December to March, irrespective of of the site. Significant relationships were found at all sites and years between both average daily maximum and minimum temperatures and the number of flowers per inflorescence and the number of flowers per palm per month. This confirms that in addition to drought, temperature also influences seasonal flower production. In this case the negative effect of low temperature was enhanced by the effect of the long dry season occurring during the same period.

High nutlet abortion due to coreid bug damage led to increased female flower production at two sites, low damage at the third site resulted in good nut set and a substantial decrease in female flower production. Varietal differences in inflorescence production were negligible. Löhr concluded that palms can react with increasing flower production and continued high seasonal fluctuation under situations of high pest damage in an attempt to compensate for crop loss.

The female flower is much larger than the male flower and its shape resembles a very small coconut. Its size continues to enlarge between the opening of the sheath and receptivity. Receptivity in tall palms begins about 22-24 days after the opening of the sheath, in most dwarf palms this period is only 2-7 days and in D x T hybrids and some green dwarfs it is about 16-21 days. At that stage the diameter of the female flower is about 3 cm. The ovary is tricarpous and each carpel has a single ovule, of which normally only one develops while the other two degenerate. But bicarpelate and tricarpelate nuts can sometimes be found. When the female flower becomes receptive, it opens at the apex and the three sessile stigmas protrude from it like a three-pointed star. Nectar is produced at their bases and at the three pores on the pericarp towards the top of the ovary. Receptivity of a female flower of a tall palm lasts about three days and of a dwarf palm about two days, after which the stigmas turn from white to pinkish-brown. The female phase, or the period between the beginning of receptivity of the first female flower and the end of receptivity of the last female flower lasts about 5-7 days in tall palms and in some green dwarfs, and twice as long in dwarfs and in the King coconut variety.

Generally, the first male flower opens immediately after the splitting of the sheath, starting with the male flowers located at the sides of the female flowers and those at the tips of the upper spikes. Opening of the flowers continues from the tips of the spikes towards the base. After shedding their pollen, male flowers wither and fall off, usually within two days following their opening. Duration of the male phase of a coconut palm, that is the period between the opening of the first male flower and the shedding of the last, is about 18-22 days, depending on variety and growing conditions. The pollen is roundish when fresh, turning ellipsoidal with a longitudinal grove within a few seconds after exposure to the open air, a shape that is considered favourable for transport by insects. The amount of pollen produced per inflorescence varies with variety. Nasayao and Malasaga (1989) studied flower and pollen characteristics of seven dwarf palms growing in one field in The Philippines. They found a highest pollen yield of 7.19 g per inflorescence for the Camotes variety and a lowest pollen yield of 2.42 g for the Coconiño variety, compared to 5 g for the tall Laguna and Bago-Oshiro varieties and 1-2 g per inflorescence for the Tacunan and Catigan dwarf varieties found in earlier studies. However, according to observations by CIRAD-CP, an inflorescence of the West African Tall variety can yield 8-12 g of pollen, depending also on drying and sieving techniques. Hybrid varieties usually produce more pollen than tall varieties.

In palms with a long female phase, overlapping of the male and female phase is common. In palms with a short female phase, male flowers are usually all withered before the female phase begins. Sangare *et al.* (1978), taking both intra-spadix overlapping and inter-spadix overlapping into account, classified the various coconut types into four categories:

- strict allogamy. Palms with short female phases without overlapping the male phase of the same inflorescence or that of the following. Certain WAT populations show this behaviour (Bourdeix 1994);

- indirect autogamy. Palms with short female phases without overlapping the male phase of the same inflorescence, but widely or totally overlapping the male phase of the following inflorescence. Bourdeix (1994) names this behaviour 'preferential allogamy'. It comprises most of the talls: Polynesia, Vanuatu, Cambodia, Malaysia, Comores, and certain WAT. The Niu Leka dwarf is the only known dwarf of this group;

- direct autogamy. Palms with long female phases completely covered by the male phases of the same inflorescence, with or without overlapping the male phases of the following inflorescence. The MYD and MRD, Cameroon Red Dwarf (CRD), and Polynesia dwarfs belong to this group (Bourdeix 1994);

- semi-direct autogamy. Palms with short female phases overlapping the male phases of the same inflorescence as well as those of the following one. Green dwarfs from Brazil, Sri Lanka, and Thailand, and the brown dwarf from New Guinea belong to this group, which is named 'preferentially autogamous' by Bourdeix (1994).

In general, the dwarfs are preferentially autogamous, whereas talls are preferentially allogamous. There are, however, some exceptions: The Niu Leka dwarf from the Fiji Islands is allogamous, The Rath Tembili Tall (King Coconut) from Sri Lanka is autogamous (Bourdeix 1994). CP 26 shows two overlapping inflorescences.

Overlapping of the male phase of the following inflorescence depends entirely on the number of inflorescences produced annually, which in turn is also influenced by environmental conditions. Autogamy may occur in tall coconuts and it is likely to be higher when ecological conditions are favourable. Löhr (1993) concluded that large variations in intervals between the emission of subsequent inflorescences, due to temperature differences between seasons may have some significance for breeding purposes. According to his observations, the East African Tall may fluctuate between strict allogamy during most of the year to indirect autogamy during the time of high inflorescence production.

Pollination in coconut is effected mainly by insects. Honey bees seem to be the main pollinating insects. However, other insects such as wasps, flies and ants also play a role. Even mites (*Neocypholaelaps ampulula*), transported by bees are considered to be potential pollinators. The flight range of a local honey bee (*Apis indica)* in Indonesia is about 1.6 km, but when in search of pollen or nectar it makes short landings on palms one after another in close proximity. When a bee visits a coconut inflorescence at the end of its male phase, hundreds of mites crawl up its legs and rest on the dorsal part of the thorax. At its next visit to another inflorescence, the bee makes specific vibrating movements and during that process the mites, loaded with pollen grains, crawl down to the flowers. The bee also carries pollen sticking to the legs. When it next visits a female flower, the mites and extra pollen sticking to its legs are shed on the flower during the ritual rotations around the stigma. Thus, a honey bee cannot carry pollen grains of one tree beyond the two or three subsequent trees which it visits. (Liyanaga, D.V. and Azis 1983). Trials in Sri Lanka showed that coconut pollen was not wind-borne through a jungle belt of 315 m. A fair percentage of pollen was found at up to about 180 m.

Fruits

The fruit of the coconut palm is not a real nut but a fibrous drupe. It is developed from a tri-carpellate ovary. Normally, two carpels are aborted. Of the three eyes, the one belonging to the developed carpel covers the embryo. After pollination, one of the ovules of the female flower develops into a fruit, the other two degenerate. In exceptional cases two or three carpels develop (Figure 3).

A normal healthy tall coconut palm produces one mature bunch of coconuts per month on average. The development cycle of a bunch is about 3.5 years from the primordial stage of the inflorescence to fruit maturity. The first stage between primordium of inflorescence and primordium of female flowers lasts about 1.5 years; from primordium of female flowers to spathe opening is one year and from spathe opening to fruit maturity covers another year.

The female flowers as well as the very young nutlets are called buttons. Sometimes only the female flowers are called 'buttons', but this may lead to confusion because of their similarity to very young nuts. Unfertilized female flowers and very young nuts are difficult to distinguish and therefore in this book they are both called buttons, referring more to their size than to their stage of development. There is considerable varietal difference in fruit size, shape and colour and even in the proportion of the components, husk, shell, endosperm and water (CP 9-12). Also, growing conditions have an influence on fruit development. For instance, under conditions of drought or K-deficiency, the fruits, when ripe, are smaller than usual.

Fruits mature in about 12 months, in some varieties somewhat earlier, in others somewhat later. When ripe, in most varieties the fruits drop off, but in some varieties they remain attached to the tree and may even start germinating while still hanging in the air. The volume of a mature fruit may vary between 1 and 4 litres. Fruits of the dwarf palm are generally much smaller than those of the tall palms, although some dwarf varieties have fruits almost the size of those of tall palms. The fruit has a smooth surface, its colour either being green, yellow or reddish brown or one of the shades in between. With time, the ripe fruit turns greyish-brown, starting with some patches and a few weeks later its original colour can no longer be recognized. The shape of the fruit varies from round to ovoid, with three sides separated by three ridges almost always recognizable. Usually, one of the sides is wider and flatter than the other two. Some fruits, particularly the long-shaped ones, when drying out, show a shrinkage ring at the stalk end. Large nuts do not always have large kernels, and vice versa.

According to Harries (1978), the original coconut types, or 'Niu kafa' were triangular and germinated slowly. This type floats well in the sea, and when washed ashore the nuts did not roll down again into the water, so they could sprout and grow on the beach. The roundish type, or 'Niu vai', developed by man, carries more fruits and germinates earlier.

Immediately after pollination, fruit growth begins with a rapid development of the pericarp at the basal region, which remains soft and white until the fruit is almost mature. The endocarp is already differentiated as a soft, creamy white structure long before the time of pollination (Shivashankar 1991). The coloured, smooth, hard outer layer of the husk, the epicarp, is about 0.10 mm thick.
The fibrous mesocarp can be 1 to 5 cm thick, varying with variety. At the stalk end it may even be as thick as 10 cm. The fibres, or coir, are imbedded within a soft mass, the coir pith. For the use of its fibres, nuts are often harvested one month before ripening, as the fibres at this age are softer and

more flexible. Some varieties have very soft mesocarp that can be eaten when the fruit is still young. In the immature stage, the mesocarp is whitish, turning brown at ripening. In some varieties the mesocarp is pinkish or partially pinkish. The husk of mature nuts gradually turn from whitish to brown when drying out. Coconut husk is constituted of about 30% fibre and 70% coir pith. Fibre consists of about 40% bristle fibre and 60% matras fibre (Savithri and Kahn 1994). The fibre covering the soft eye is softer and less compact than elsewhere. Fernandez (1988) isolated various fungi from the mesocarp. However, these fungi did not affect the soft eye when the embryo was sound. Penetration through the soft eye is possible when the embryo has germinated. Coir pith is acidic and contains lignin and cellulose in almost equal amounts, *i.e.* 40-45% and 8-12% soluble tannin-related phenolic compounds. The bulk density is very low due to high porosity coupled with low partical density. The water-holding capacity is very high, ranging from 400 to 600% (Savithri and Kahn 1994).

Like the fruit, the shell, or endocarp varies in shape. It may be long, ovoid to almost round. The size of the kernel is not always related to the fruit size. Large fruits with thick husks may have smaller kernels than smaller fruits with thin husks. Usually three ridges can be found on the shell, corresponding to the three ridges on the fruit. Shell thickness varies between 3 and 6 mm. The shell is very hard. At the stalk end, three so-called 'eyes' can be found, one on each carpel, under one of which the embryo is located, embedded in the endosperm. This eye is soft-textured, the other two are lignified. Often, it is assumed that the germinating eye is located opposite the widest segment. Wickramaratne and Padmasiri (1986) found that in a sample of 736 nuts, 36 per cent had the soft eye just beneath the fruit ridge opposite the widest surface of the nut, not significantly differing from the 1:2 ratio that might be expected when any one of the three eyes may develop into the soft eye irrespective of its position in the nut. The shell begins to form during the fourth month after pollination (Jayasuriya and Perera 1985; Shivashankar 1991). About eight months after pollination, the soft, white shell begins to harden and turns brown. This process starts at the apex, reaching the stalk end of the nut when it is almost ripe. At this stage the fruit's volume is largest and its weight heaviest. The cavity of the embryo sac enlarges considerably during fruit growth.

Coconut water is formed in the third month and its volume reaches a maximum in the eighth month, declining thereafter as the fruit ripens. Jayalekshmy *et al.* (1988) studied the chemical composition of coconut during maturation. They found that a large coconut may contain as much as 0.6 l of water with 30 g of sugar and 2 g of potassium when still immature. The water of the young fruits is under hydrostatic pressure, which might facilitate the dissolution of CO_2 in the water. The appearance of effervescence in the water on the opening of a tender nut is common. Depletion of water on maturation causes an empty space into which the gases escape. Thus, the cavity of a mature fruit is no longer completely filled and ripe fruits splash when shaken.

Major chemical constituents of coconut water include sugars and minerals; minor ones, fat and nitrogenous substances. Enonuya (1988) observed vast differences between the syrups derived from nutwater of mature tall and dwarf coconuts. Fructose was the major sugar in the nutwater syrup of tall coconut whereas sucrose was most abundant in that of the dwarf coconuts. The water of dwarf nuts had a higher sugar content than that of tall nuts. Total fat and total protein show a sharp increase towards maturation. Potassium decreases with maturation, sodium increases. There are no major changes in the other mineral elements. The pleasant taste of tender coconut water can be mainly attributed to the sugar and mineral matter. Fat, free amino-acids, nucleic acids, organic acids and dissolved gases may also contribute to the overall flavour and oral feeling. Changes in their concentration on maturity render the water bland and less relishing. The two major changes in the profile of sugars are the steep fall of total sugars per nut by more than 90 per cent and the disappearance of reducing sugars. Chikkasubbanna *et al.* (1990) concluded that for the use of coconut water as a drink, the fruits should be harvested between the seventh and eighth month of maturity when the amount of sugars and nutrients in the water is highest. As long as the plumule has not protruded through the covering tissue of the soft eye, the nut contents, such as nut water, haustorium and endosperm, are sterile. Coconut water can be used for intravenous injections without producing side effects. (Fernandez 1988).

About 6-7 months after fruit set, the endosperm begins to form against the inner wall of the cavity, first at the apex, and gradually extending towards the stalk end. The endosperm commences to develop about five months from fruit set and at six months appreciable amounts of semi-solid endosperm has been formed. The very young endosperm is a thin and gelatinous layer. When the nut is opened at this stage, the endosperm can be scooped out with a spoon or with a sliver of the coconut husk. It is considered a delicacy. At this stage, the coconut water is also very much appreciated, It is sweet and tasty. The layer of endosperm gradually becomes thicker and harder. Jayasuriya and Perera (1985) observed a rapid increase in endosperm weight up to the tenth month. Accumulation of dry matter in the endosperm ceased after eleven months. With that, also the increase in dry weight of the whole nut ceased, indicating that no assimilates are transported into the nut after full development of the endosperm. Finally, the endosperm becomes hard and white, covered on the outside by a brown testa firmly adhering to it. During the last month of ripening, the fruit loses some weight due to drying out of the husk and a decrease in water content.

The endosperm of tall fruits, when dried to copra, is hard and rigid. The highest oil content can be found near the testa, the lowest near the water. The water content is highest near the water and lowest near the testa. When the endosperm curls during the drying process, it is a sign of bad quality and reduced oil content. It may be caused by harvesting immature nuts, or by spoilage of the endosperm after harvest. However, the endosperm of dwarf fruits always curls in the process of drying. The dried endosperm, or copra, of an average tall fruit may weigh between 200 and 250 g, but the range of variation is as wide as from about 100 g to about 350 g. Nambiar and Rao (1991) observed that the copra content of the fruits was maximal during summer, followed by winter and minimal in the post-monsoon period of October-November. Table 2 shows the dry copra per fruit of some varieties and hybrids (Anonymous 1987b)

Table 2: *Dry copra weight per fruit of some varieties*

Variety	Dry copra g per nut	Variety	Dry copra g per nut
MRD x Rennel Tall	280	Rotuman Tall	337
MRD x Malayan Tall	240	Rennel Tall	330
MRD x Solomon Tall	200	Rangiroa Tall	311
Hybrid dwarf (not specified)	233	Christmas Island Tall	307
MRD	165	Markham Valley Tall;	290
MYD	165	Malayan Tall	250
Niu Leka	260	Solomon Tall	190

The composition of the fresh endosperm shows considerable varietal difference. Table 3 shows the approximate range of the main components in per centage on a wet basis.

Some palms have an abnormal type of endosperm that almost fills the total shell cavity. Instead of hard, crispy endosperm and water, there is an outer portion which is a white and soft substance and a viscous, somewhat transparent liquid. Adriano and Manahan (1931) distinguished three different types of such nuts:

- nuts with a hard outer layer (hard as boiled rice), a soft viscous middle layer and a semi-liquid inner layer of endosperm;
- nuts with a hard outer layer and a soft, viscous inner layer, and
- nuts with one layer of endosperm, hard as boiled rice, filling the whole cavity.

Table 3: *Approximate composition of fresh (wet) endosperm (%)*

Component	%
water	44 - 52
oil	35 - 38
carbohydrates	9 - 11
protein	3 - 4
crude fibre	2 - 4
ash	1

Such nuts are called Makapuno in The Philippines ('filled coconut'), Kelapa Kopjor in Indonesia, or Thairu Thengai ('curd coconut') in India and Dikiri-pol in Sri Lanka. Makapuno is the most frequently used name. The Makapuno endosperm has a peculiar taste. It is considered a delicacy in all countries where it occurs. Various sweets and ice-cream are prepared from it, but it is also consumed fresh, or mixed with some syrup as a drink. Such nuts fetch higher prices than ordinary nuts. The palms cannot be recognized from other palms by eye. Not all nuts of a Makapuno palm are of the Makapuno type, usually only one or two in each bunch. These nuts can be recognized by shaking them, as they do not make the splashing sound that normal nuts do.

Also, when tapping on such nuts they produce a different sound. Makapuno nuts have different cell wall materials and the component sugars of each fraction of the hydrolysis product varies between Makapuno and normal nuts. According to Kovoor (1981), the Makapuno endosperm may well be a case of cell dedifferentiation occurring on the plant itself, and this may have to do with the fact that it carries a triple set of responsible genes. Cytochemical studies by Sebastian *et al.* (1987) confirmed the cellular cytochemical differences between normal coconut and Makapuno endosperm and support the abnormal and tumour-like character of the Makapuno cells. An increasing concentration gradient of oil globules and protein bodies towards the testa suggests the early synthesis of these organelles during endosperm development. According to Rillo (1991), the Makapuno tree is believed to be heterozygous in character. The phenomenon is governed by a single recessive gene, which explains its relative rarity. The tree is propagated by planting the normal fruit from the Makapuno-bearing tree which will bear only 2-25% Makapuno nuts. The appearance of this characteristic in dwarf palms has not been reported yet. Apparently, it is a genetic characteristic which may appear in any tall variety.

The Makapuno nut contains a normal embryo, which under natural conditions does not germinate. When excised and grown *in vitro*, it will develop into a normal tree which will bear up to 80% Makapuno nuts. Assisted pollination (selfing) will result in 100% Makapuno yield. Theoretically, the centre palms in a plantation of *in vitro* grown Makapuno embryos will easily give 100% Makapuno without assistance. With new findings on embryo *in vitro* culture in The Philippines it is now commercially viable and economically feasible to establish Makapuno groves. When done on a large scale, the economic advantage of higher prices may disappear. But a few investors were considering catering to the frozen dessert market in the United States and Europe. Ebert *et al.* (1991) envisage that Makapuno farming could be developed in The Philippines as a non-traditional coconut industry. This would be more imperative, especially now that coconut production areas are getting smaller due to fast urban development and increased coconut timber utilization. Composition of the fruit is different for different varieties. Some have a high husk percentage, others have less husk and a larger kernel with more water and endosperm, as can be seen from Table 4.

Table 4: *Composition of fresh coconut fruit (g)*

Population	fruit	husk	%	shell	%	water	%	kernel	%
	1.109	408	37	158	14	185	17	358	32
MYD x WAT	1.160	300	26	190	16	290	25	380	33
MYD x Rennel Tall	2.209	673	31	311	14	563	25	662	30
Rennel Tall	1.196	543	45	175	15	178	15	300	25
Solomon I. Tall	1.460	670	46	200	14	190	13	400	27
Sri Lanka Tall									

(Based on Foale 1991b)

Any abnormal weather factor coinciding with the critical phase of nut development, 4-7 months after pollination adversely affects the rate of growth and final size of the nut, copra and oil content. It was observed that this second phase of nut development was influenced by surface air temperature (Prasada Rao 1991). The other weather elements had no significant influence on this phase of nut development. See also Bourdeix, Part II, par. 7.2.4.3.

A study by Akpan and Obisesan (1984) in Nigeria indicated that mean temperature, relative humidity and number of dry days were the most important climatic factors affecting nut length. Only the number of dry days and sunshine hours accounted for the variation in the observed percentage of copra to nut and oil to copra. The number of dry days and mean temperature were largely responsible for annual variations in percentage of shell to nut, whereas weight of nut was influenced by annual rainfall and sunshine hours.

The Oil

The oil is the most important product of the coconut palm. The oil content of the dried endosperm varies according to variety, from 65 per cent for the Malayan Yellow Dwarf to 74 per cent for the West African Tall. Nambiar and Rao (1991) studied the oil content of five varieties and their hybrids in India. They found the highest oil percentage, 73%, in Laccadive Ordinary and the lowest, 66%, in Chowgat Orange Dwarf. The percentage of oil content of the hybrid nuts was intermediate between that of the respective parents. They observed that oil content varied with the seasons. The minimum oil content was observed during the summer months and the maximum during winter. Within the nut the highest oil content can be found near the testa.

The fatty acid composition of the coconut varies with variety and growing conditions. Therefore, figures on analyses from different sources may show slight differences. The fatty acid composition also varies with progressing maturity. At 8 months from fruit set, the oil has a high percentage of oleic and linoleic acids, 30 and 20 per cent respectively. With increasing age of the nuts these percentages decrease rapidly, and also the percentage of palmitic acid declines, whereas the percentages of other fatty acids such as capric acid and caprylic acid rise. During the last three months, the oil composition of the kernel does not change any more, justifying the practice of harvesting the last three bunches in one harvesting round (Padua-Resurreccion and Banzon 1979). The next younger bunch, however, has a distinctly different fatty acid pattern, lower in lauric acid and higher in unsaturated acids.

Among the vegetable oils, coconut oil stands out as having one of the highest saturated fatty acid contents. But of these fatty acids, a very high percentage is of short and medium chain length (Table 5). Due to its high level of saturated fatty acids, coconut oil is very resistant to oxidative rancidity. Coconut oil is considered a lauric oil because of the high content of lauric acid, belonging to the same class as palm kernel oil, babassu oil and cahune palm oil. Coconut lacks adequate quantities of essential fatty acids such as linoleic oil.

In 1986, the American Soybean Association, afraid of competition from coconut and other palm oils, launched a campaign to boost soybean sales, by warning the health-conscious American consumers that coconut oil and palm oil were health hazards. According to them, the high percentage of saturated fatty acids in the oil tends to increase the levels of serum cholesterol, one of the causes of coronary heart disease (Intengan 1987). An editorial in the *New York Times* referred to coconut oil, palm oil, and palm kernel oil as ' the cheaper, artery-clogging oils from Malaysia and Indonesia'. (Blackburn *et al.* 1992a).

However, cholesterol is not a vegetative product. The human body obtains part of its cholesterol from the consumption of animal fatty products. But the greatest part of the cholesterol is produced within the human body itself, by the liver, mainly from the consumption of saturated fatty acids. Coconut oil does not contain cholesterol, it is rich in saturated fatty acids. Due to the specific composition of these fatty acids, normal coconut consumption does not cause any elevation of plasma cholesterol. Coconut and coconut oil have been consumed for many centuries by millions of people in India, Sri Lanka, Indonesia, The Philippines and many islands in the Pacific, for whom it was their main or only source of fat. However, the incidence of heart attacks among the villagers in these countries has always been relatively low.

Recent investigations have shown that not all fats rich in saturated fatty acids raise blood cholesterol levels and that this common manner of classifying dietary fats is overly simple, no longer appropriate, and misleading (Chong 1989). Contrary to poly-unsaturated fatty acids that are harder to digest, short chain and medium chain fats such as in coconut oil, are well absorbed.

There are two kinds of cholesterol complexes in the blood, the high-density lipoprotein (HDL) cholesterol, and the low-density lipoprotein (LDL) and very-low-density lipoprotein (VLDL) cholesterol. The latter is the bad cholesterol that should remain at low levels to reduce the risk of heart attacks by coronary heart disease. The HDL is the good cholesterol that keeps the other cholesterols from invading the inner lining of the arteries. Recent studies have showed that replacing saturated fat in the diet by poly-unsaturated fatty acid usually lowers both LDL and HDL cholesterols. The substitution of saturated fatty acid with mono-unsaturated fatty acid in the form of either safflower oil or olive oil reduced LDL cholesterol but maintained the level of HDL cholesterol (Intengan 1987). Coconut oil is also a mono-unsaturated oil.

Table 5: *Fatty acid composition of some oils**

Fatty acid	Carbon chain	Coconut	Soybean	Maize	Saf-flower	Ground-nut	Butter	Herring
	C4						3.6	
Caproic	C6	0.5					2.2	
Caprylic	C8	8.0					1.0	
Capric	C10	6.4					2.5	
Lauric	C12	47.3					2.9	
Myristic	C14	17.6	0.1		0.1	0.1	10.8	7.4
Palmitic	C16	8.4	11.1	12.3	6.5	11.8	26.9	20.4
Stearic, Oleic, Linoleic	C18	10.5	88.4	87.6	93.2	81.0	43.8	22.4
	C20		0.3	0.1	0.2			21.2
	C22		0.1					24.6
	C24					1.0		

Source: Kintanar (1987)

* This table is indicative, not absolute, as it is only one analysis. Contents may vary with varieties, ecological conditions, etc.

Poly-unsaturated fatty acids have been observed to decrease serum cholesterol levels but it has also been shown that excessive intake of poly-unsaturates may be detrimental to health. Evidence has been discovered that these same beneficial oils may well be a primary source of radicals within the cell that causes it to age. Superiority of coconut oil over soybean oil as a dietary oil has been repeatedly proved (Dayrit *et al.* 1992; Kaunitz and Dayrit 1992; Intengan *et al.*, 1992). In addition, in laboratory trials with mice treated with chemicals with chromosome-breaking effects, it was shown that the anti-gentoxic effect of coconut oil was far superior to that of soybean oil (Lim-Sylianco *et al.*, 1992a and 1992b). Laboratory tests also showed that coconut oil, even in low percentage in the diet, showed considerable inhibition of chemical carcinogenesis (Lim-Sylianco 1987).

Trans-fatty acids are found in major amounts in partly hydrogenated oils, a process by which fatty acids in the oils are artificially saturated. These hydrogenated fatty acids become solid fats (Enig 1991). The hydrogenation process, for instance, is used to produce margarine from liquid vegetative oils. The hydrogenation process introduces saturate equivalents which almost never occur naturally in the food supply. Trans-fatty acids are usually 18 carbons long. Typically, soybean oil is hydrogenated when used as a substitute for coconut oil for manufacturing margarine. Recent studies (Anonymous 1993b) have showed that the trans-fatty acids found in margarine, vegetable shortening and a host of products ranging from doughnuts and pies to crackers may also cause heart diseases. It was shown that trans-fatty acids raise the harmful elements in cholesterol while lowering the protective elements.

In a testimony before the US Senate Committee on Labor and Human Resources on S. 1109, a bill to amend the Federal Food, Drug and Cosmetic Act to require new labels on foods containing coconut-, palm-, and palm-kernel oil in 1987, by the Harvard Medical School Nutrition Coordinating Center in Boston, many of the above mentioned characteristics of the various oils and fats were mentioned (Blackburn, *et al.* 1987). In this statement, coconut oil is described as a unique and desirable fat for human use. It concluded that 'it would be particularly unfortunate if consumers were deterred from buying products containing coconut oil on health grounds, when the most recent medical evidence suggests that coconut oil is more beneficial to consumers than the hydrogenated fats that would be exempt from the proposed legislation'. The most recent recommendations on fat consumption from the National Research Council in the US suggests that 30% of the total energy intake should come from fats, of which 33% should come from saturated fatty acids, 40-43% from mono-unsaturated fatty acids, and 23-27% from poly-unsaturated fatty acids.

Growth, Development and Yielding Capacity

In general, a ripe coconut kept in a humid atmosphere starts germinating soon after harvesting. The speed of germination is determined by genetic factors as well as by environmental conditions, such as humidity and temperature. The germination speed is characteristic for coconut varieties. According to Harries (1981), the time taken from reaping to sprouting distinguishes later germinating Niu kafa types from early germinating Niu vai types. According to Shivashankar (1991), as the fruits mature, growth inhibitory factors develop in the water and kernel, functioning in initiating and maintaining the dormancy of the embryo. The fruits contain a carefully poised complement of stimulators and inhibitors. Any shift in their relative proportions could be crucial in the overall regulation of germination and the timing of various metabolic events which occur during germination. Very dry conditions may retard germination. However, it is difficult to determine how much the germination is influenced by genetic factors and how much by environmental conditions.

The cylindrical embryo embedded in the endosperm under the soft eye is divided into two parts by a small circular contraction. As the first morphological sign of germination, the embryo enlarges and the apical part protrudes out of the shell. Simultaneously with the development of the apical part of the embryo, the basal part, or cotyledon, starts growing out and develops into the so-called 'apple' or haustorium. The haustorium is of a pale yellow colour on the outside, the inside is cream-white. It is a spongy tissue and grows steadily until in close contact with the endosperm. Through the haustorium, the young palm absorbs nutrients from the water in the cavity and from the endosperm. The haustorium has a sweet taste. The soluble nutrients in it include a considerable proportion of sugars. The plant is almost entirely supported by the endosperm during the first four months after germination. Thereafter, there is a gradual turnover from internal to external assimilation by photosynthesis. After about 18 months, the endosperm is completely exhausted and decomposed.

Soon after germination, the primary root emerges from the apical mass, followed by the plumule which appears like a conic hump in the opposite direction. As the growth advances, the part of the apical mass encircling their plumule differentiates into a tubular structure, the coleoptile. From the coleoptile, a pointed shoot of leathery leaves develops and grows through the husk. Usually, a coconut is considered to have germinated when the shoot appears above the surface of the husk. In fact, the germinating process may have been underway for two months. Simultaneously with the development of the apical shoot, the primary roots start growing through the husk, soon followed by some adventitious roots developing from the base of the shoot. They grow downwards and sideways through the husk and into the soil.

After emergence of the sprout, a period of about five weeks elapses before the first leaf lamina unfolds. The leaves progressively increase in size. For about one year after sprouting, the developing leaves are undivided. Thereafter, when the seedling has about 7-10 leaves, it will produce leaves gradually splitting more and more into leaflets. This pinnating of young leaves may be retarded under unfavourable conditions such as nitrogen deficiency. There is also varietal difference in the rank of the first leaf in which some leaflets are separate from the 'entire' leaf lamina. Also the rank varies in the first leaf on which all leaflets are separate (Foale 1991d). Ouvrier (1984a), observing the growth and development of PB-121 hybrid seedlings, found that the roots of the seeds grew regularly each year, with a period of stagnation between 52 and 65 months of age. Between 16 months and 65 months the root system grew much more slowly than the rest of the coconut, so there was a temporary imbalance between roots and leaves, expressed by greater sensitivity of the palms to gusts of wind during this period. The number of leaves annually produced increased from about 7-10 in the first year to 13-19 in the course of about four years (Table 6)

Table 6: *Vegetative characteristics of PB-121 (Ouvrier 1984)*

Age (months)	6	16	28	40	52	65	77	89
No. of living leaves	9	11	15	25	37	35	40	40
No. of leaves produced	9	13	16	19	16	18	19	
Girth at collar (cm)	20	36	115	141	150	151	153	160

The number of leaflets in the hybrid seedlings increased regularly up to leaf 14 of the 52 months sample and then stabilized at about 100-120 leaflets on each side. Ouvrier considered this as a valid criterion for characterizing the development of the coconut at an early stage. The length of the leaf (petiole + rachis) increased rapidly up to leaf 19 of the 52 months sample and thereafter very slowly, stabilizing at 77 months. In the planting system used, 8.5 m triangular (160 palms per ha), at 52 months competition for light was considered to have reached its maximum.

The shape of the crown during the early years is broom-like as the leaves continue to grow straight upwards and have not yet reached the stage of hanging down. It takes several years for the number of leaves to reach its maximum. The shape and size of the canopy depends on the length and the number of leaves. All coconut palms show a change in crown shape with advancing age. The result is that light interception falls gradually from the age of about 15 years to a value of about 50 per cent at 25 years, and so on with increasing age. The progressive decline is caused by the inability of the older fronds to retain a horizontal position. This is due to a weakening of the fibres at the base of the frond caused by the expanding thick base portion of the newly emerged frond, forcing the base of adjacent older fronds outwards. This mechanical stress is a consequence of the close spacing of leaf scars of the trunk (Foale 1992). As a consequence of the mechanical stress on the attachment of the frond to the stem it now lacks the strength to retain the frond in a horizontal position. After it is displayed upwards for a few months when young and supported by the fronds immediately below, the frond descends over a few weeks to a downward position, as the support of the older fronds declines. This gives the canopy an x-shaped profile.

In addition to the number of leaves produced per annum, girth size is also used as a measure of vegetative development. It offers the advantage of being accessible at any given moment without prior marking such as is necessary when counting leaves produced within a certain period. Both methods are valuable only within the different treatments in a trial or when comparing vegetative growth of different varieties within one trial with equal treatment of all palms, as these characteristics are very much influenced by environment. Girth is measured, using a tape between 0-10 cm above the ground. On young coconuts it includes the petiole bases which are still green and attached to the stem. It is therefore recommended to calculate the mean of at least 25 palms per 25-50 ha, depending on the homogeneity of the zone considered. Girth size measurement is also worth while to evaluate plantations, taking local conditions into account. In Table 7 (Rognon and Boutin 1988), a range of references is proposed, depending on the water deficit and soil and upkeep conditions. Comparing the results obtained makes it possible to assess the development of the plantations observed.

From bottlenecks in the stem, something of the palm's history may be read. When unfavourable conditions remain for a long time, the stem may become narrower and narrower, forming the so-called 'pencil-point'. Old, senile palms will also develop pencil points.

The base of the stem swells and forms the bole, its size being a varietal characteristic. Most dwarf coconuts do not have boles. However, under very favourable environmental conditions it can happen that even dwarf palms develop slight swellings at the foot of the stem. This was the case with Ghana Yellow Dwarf and Malaysian Yellow Dwarf at Port Bouet, in Ivory Coast, whereas the Equatorial Guinea Green Dwarf and the Cameroon Red Dwarf growing in the same conditions did not have these characteristics. (Nucé de Lamothe and Rognon 1977). Most palms grow erect, but some varieties have a slightly curved stem by nature. Prevailing strong winds from one direction cause the palms to lean towards leeward. Palms on the border of a plantation will lean towards the light. Palms growing in a loose soil, having been blown over by strong winds with still some roots in the ground, will grow upward again, curving the stem.

When palms are young and grow well, leaf scars will be widely separated, about 10-15 per metre; when palms are old or grow under unfavourable conditions, their internodes are much shorter and as many as 30-35 leaf scars may be found per metre. Fruit production also results in a shortening of internodes. The length of the internodes is also a varietal character, dwarf palms having much shorter ones than talls. The Niu leka variety has very short internodes.

From the number of leaf scars the age of the palm may be estimated, the stemless years to be added. Some palms show abnormal elongations, often normalizing after a number of years. Such palms have erect leaves, giving these palms a broom-like appearance. Such developments occur under conditions of too much shade, when young palms are interplanted between old palms at too high density. Usually, such palms bear only few or no nuts. Very old palms have smaller crowns than younger palms and may produce only a few nuts. Generally, coconut palms are considered to have an economic life span of about 60 years. However, this depends very much on growing conditions.

Table 7: *Range of references to evaluate management operations on a PB-121 plantation using girth size (cm); (Rognon and Boutin 1988)*

Pedo-climatic conditions
Reference values for the girth depending on the trees' age (in months of planting). Case of the PB-121 managed satisfactorily (mainly upkeep and fertilizer application)

Water deficit since planting (annual mean)	Type of soil	12	18	24	30	36	42
0 - 200 mm	good	55	83	100	138	150	155
	average	50	75	100	125	140	150
	mediocre	45	68	90	113	130	140
200 - 400 mm	good	50	75	100	125	140	150
	average	45	68	90	113	130	140
	mediocre	40	60	80	100	120	130
> 400 mm	good	45	68	90	113	130	140
	average	40	60	80	100	120	130
	mediocre	35	53	70	88	110	120

Sometimes, very old palms still produce surprisingly high yields. However, in commercial coconut production it is doubtful whether the maintenance of old coconut plantations is economically preferable to the substitution of the plantation by new, improved planting material at an earlier age. Gradual replanting of parts of the plantation over a number of years may be economically preferable and will facilitate future replantings as the plantation consists of different age groups.

The earliest age at which a coconut palm may start flowering is genetically determined, but the actual age of first flowering is also influenced by growing conditions. Under favourable conditions, tall coconut varieties may start flowering at the age of 5 - 7 years, hybrids during the fourth year, and dwarfs during the third year of age. Ouvrier (1984a) obtained the first harvest from the PB-121 hybrids when the palms were 52 months old. Recently, a new variety, Salak, was discovered in Indonesia, that flowers during the first year of age, when the palm still has only a few leaves (CP 23), the inflorescence emerging from below the soil level. When growing conditions are unfavourable, flowering may be delayed by several years.

Usually, when flowering begins, stem elongation has already started. The first bunch of coconuts hangs closely to the ground or, in the case of dwarfs, may even be resting on the ground. Patel (1938) observed that palms with a greater number of leaves flowered earlier and yielded more than palms with fewer leaves. A rapid rate of leaf production and, consequently, the presence of a larger number of leaves in the crown results in a larger total leaf area, which may possibly increase the building-up of an adequate carbohydrate reserve in the stem. There is also a correlation between chlorophyll content in the leaves, the rate of apparent photosynthesis and annual yield of nuts (Narayanan Kutty and Gopalakrishnan 1991).

Growth is a function of dry matter production. Friend and Corley (1994) developed a non-destructive method for estimating above-ground dry matter production by the coconut palm. Due to their shorter and narrower trunks and slightly smaller leaf areas, dwarf palms convert less biomass into vegetative matter than tall palms. They found that Rennel Island Tall (RIT), Malayan Red Dwarf (MRD) and their hybrids had higher proportions of dry matter in nuts than Solomon Islands Tall (SIT). RIT nuts had more copra and less husk and shell than SIT, the dwarf had poorer nut composition than either tall. A combination of a high proportion of dry matter in nuts and good nut composition gave RIT and RIT x MRD the best harvest indices (the relation between copra production and total dry matter production); the higher planting density used gave the hybrid the greatest yield per ha. MRD had a smaller leaf area than the talls, but due the higher number of trees per ha, its leaf area index per ha (L) was comparable to that of the talls. Leaf area of the hybrids was comparable to that of the talls, but L was higher because of higher planting density. Differences in net assimilation rate between varieties were small. Total dry matter production of SIT x MRD was 23 t per ha, which is not high compared to other tropical tree crops. In the Ivory Coast, the dry matter production of MYD x WAT was estimated at 30 t per ha.

Understanding the factors contributing to dry matter production is very important. Yield potential is a function of total light interception. In selection and breeding the greatest yield potential might be obtained through increased light interception by the leaves (Kasturi Bai and Ramadasan 1990). According to Eschbach *et al.*, 1982), coconut palm, like oil palm, has all the characteristics of a plant in C3, with a low net photosynthesis rate, a high compensation point for CO_2, a marked sensitivity of photo-respiration to O_2 concentration, and photo-respiration 3 or 4 times greater than night respiration. Nocturnal dissimilation is about one-third of the net photosynthesis, and according to Foale (1991b), a leaf will become light-saturated at a value less than half the intensity of direct solar radiation. Interception by leaf surfaces oblique to direct-beam radiation therefore favours an increase in light use efficiency (LUE) and permits a large leaf area to be illuminated. Adequate mineral and water availability ensure that the LUE of individual leaves is maintained close to the maximal potential value. Nutrient deficiency is frequently brought about by competition between fruit growth and leaf maintenance. Yield potential is a function of total light interception.

Total leaf area of a palm and the pattern of display of the leaves determine its potential capacity to intercept light. The rate of leaf production, and the amount of leaf area per palm are not much affected by plant density, as palms are highly determinate with respect to leaf production. Moss (1992) observed that palms in a high density treatment had an average of 1.1 more fronds per palm over the year than palms in a low density treatment. Increasing competition for light has been shown to increase partition of the assimilates to the production of leaf area in other crops and this might also be the case with increased frond production. The rate of frond production also sets the potential rate of bunch production. In selection and breeding, the greatest yield potential increase might be obtained through increased light interception by leaves. According to Foale (1991b) lack of knowledge of photosynthetic rates and light distribution within the canopy precludes any immediate progress using these traits. Likewise, there is no information about the relative size and level of activity of the root system of different cultivars. Better partitioning of assimilates therefore still provides the best opportunity for yield improvement, especially where introduced hybrids are not suitable due to lack of appropriate adaptation. Also Ramadasan and Mathew (1987) were of the opinion that leaf area and dry matter production are two plant characters that determine the total biological activity, but that the partitioning of the total biological yield is the most important inherent characteristic that determines the economic yield.

According to Corley (1983) total radiation in the humid tropics is usually between 60 and 75 TJ/ha/year (1 TJ = 10^{12} J) and photosynthetically active radiation (PAR) is approximately 50 per cent of this total. He estimated the potential partitioning of dry matter and energy in harvested product of coconuts as 62 and 69% respectively. The percentage in economic product as 26 for dry matter and 38 for energy. From these figures he calculated the potential dry matter production as 51 t per ha per year and the potential yield of economic product as 13 t per ha per year. Eschbach *et al.* (1982) found that photosynthesis is greater at 20 than at 30°C. At higher temperatures, it drops off sharply, probably due to increased respiration. The rate of apparent photosynthesis is more directly related to chlorophyll-a than to the total chlorophyll content. In India, TxD hybrids showed a 15% increased rate of apparent photosynthesis over the tall and dwarf parents (Anonymous 1983). Increase in plant density increases leaf area progressively to a point where partitioning begins to decrease. Density giving maximum yield in crops like coconut and oil palm is usually below that giving maximum light interception. According to Foale (1991b) the general disposition of the palm crown is familiar enough, but the effect of variation in the disposition of fronds within the crown on light interception is not well understood. The extreme crown forms are hemispherical, with most fronds erectophile, and spherical, with half or more of these fronds in a lax down-pointing position. There is a gradient of light intensity from the upper to the lower leaf within the assembly of coconut crowns that forms the plantation canopy. There appears to be a critical fall-off in light interception beginning at about 20 years, when the crown begins to show an 'x' shape. This shape appears when the crown is viewed from any direction in the same plane, when there are few fronds in the near-horizontal position. The young fronds retain a high angle for about one year and then fall rapidly to a lower position, angled downwards around the trunk, leaving the intermediate zone, around the horizontal position with few fronds present. The change in the intensity of PAR, known as the rate of light extinction 'k', is a useful measure to define the effect of different canopy structures. Physiologists would need to provide data to show the effect of different crown types on 'k', however, before the breeder could make use of this parameter. Moss (1992) found that fractional interception of light varied considerably between months and was associated with frond shedding caused by dry season waterstress. Palms receiving fertilizer intercepted more light and replaced leaf area lost during the dry season more rapidly than non-fertilized palms. Palms planted at a higher density intercepted more light and carried more fronds per palm, but intercepted less light per frond than those planted at a lower density.

Table 8: *Effect of leaf pruning on one year of nut and copra yield*

No. of leaves	nuts per tree	copra per nut (g)	copra per tree (kg)
31 (control)	94.1	166.6	15.6
23	88.9	158.4	13.9
18	91.9	162.0	14.9
13	98.4	157.2	15.4

Apparently, also other factors than position influence the effectiveness of the leaf in its photosynthetic contribution to growth and yield of the palm. Magat and Habana (1991) found that the age of the leaf also may be a very important factor. They conducted a leaf-pruning trial in 35-year-old Laguna Tall palms, growing on good soil under conditions of adequate rainfall. Palms usually carrying an average of 31 leaves were pruned to 23, 18 and 13 leaves by removing 8, 13 or 18 of the oldest leaves respectively. At each harvest (every 45 days) the number of leaves was adjusted to the treatment figure. About 0.75 cm of each leaf was left on the palm, to support the bunches. Table 8 shows the result of pruning on nut and copra yields.

Differences in nut and copra were statistically non-significant. It was suggested that even with the remaining 13 on 18 leaves, the physiological activities of these coconut palms were not impaired, particularly the photosynthesis, respiration, translocation (carbohydrates and fats from functional leaves to developing bunches), and uptake of mineral nutrients. The remaining leaves were apparently adequate to support normal growth and production, providing adequate photosynthesis. The leaves might begin to be less active even before clear signs of senescence, and might therefore be dispensable for the normal crop production. When removing these leaves, the photosynthetic activity of the remaining leaves apparently improved. It was suggested that this might be attributed to mobile elements, such as N, S, K, and Cl, and water becoming readily available to the remaining leaves. However, the results published were obtained after a period of 12 months only. The influence of pruning on production in the subsequent years will probably be much higher. According to Blaak (personal communication), pruning of oil palms in Malaysia from 40 leaves to 24 and 16 leaves reduced bunch yields from 25.2 t per ha to 20.5 and 13.3 t per ha respectively.

Corley (1983) reported that after partial defoliation of oil palms, total dry matter production per tree was reduced, almost entirely at the expense of bunch production. In two progenies, however, vegetative growth was reduced significantly and one progeny was capable of maintaining a larger-than-average harvest index when defoliated. He suggested that such a progeny would also maintain a large harvest index under intense inter-palm competition. Any stress might have a greater than average effect on the vegetative vigour of such a progeny, so it might be suitable for planting only under optimum conditions. However, under these conditions it should be capable of producing exceptionally high yields in dense stands. These findings, especially if they also hold for coconut, may greatly influence some cultural practices such as intercropping, reduction of water uptake during dry periods, spacing and pest and disease control. It is therefore recommended that such trials be repeated with different coconut progenies under different growing conditions.

In a study of one hundred seedlings of open pollinated West Coast Tall (WCT) palms growing under identical conditions, Kasturi Bai and Ramadasan (1990) at the twelve month stage observed a high positive correlation between Relative Growth Rate (RGR), Leaf Area Ratio (LAR) and Net Assimilation Rate (NAR). A similar correlation was observed between Crop Growth Rate (CGR) and Leaf Area Index (LAI) and NAR. Between LAI and NAR a negative correlation was obtained. However, NAR estimates have little value in selection as it changes according to growing conditions. The total LA, Shoot Dry Weight (SDW), and LAI showed significant positive correlation with the LA, SDW, LAI and LAR in the 9th month, as well as the LA, SDW and LAI in the 12th month, indicating that the seedling vigour is expressed at the 6th month growth itself. Ramadasan *et al.* (1980; 1985) were cited, who reported a highly significant correlation in one-year-old seedlings between LA and SDW and a high heritability in the efficiency of dry matter production.

As seedling vigour is decided by the LA and SDW, the possible factors that influence LA development in the early growth stages merit serious consideration. It was observed that in seedlings raised in open conditions (LA) varied from 0.1-0.7 m and Shoot Dry Weight (SDW) from 15.0-175.5 g, whereas under conditions of shade it was 0.06-0.2 m and 8.0-66.0 g respectively, indicating that seedling vigour is highly influenced by nursery management.

Based on the correlation between dry matter production at the 6th month growth stage with that of the 12th month, the possibility was suggested of selection at the 6th month growth stage of seedlings having higher efficiency for dry matter production, confirming the findings of Satyabalan and Mathew (1983).

Raveendran *et al.* (1989) studied the parameters related to the productivity components in East Coast Tall (ECT) and three dwarf varieties and the various T x D and D x T hybrids. They found that the ECT was the most efficient parent. The hybrids were physiologically more efficient than the parents and used the energy to produce a higher number of leaves, spadices and button nuts which are the most important yield components in coconut. However, in ECT the energy translocation was towards the structural part, viz. the stem. A higher yield in hybrids could be associated with a higher photosynthesis rate, chlorophyll content and a lower respiratory rate. Shivashankar and Kasturi Bai (1988) studied the growth and nitrogen accumulation capacity in seedlings of three hybrids of West Coast Tall with three different dwarfs. They found that the cross with Malayan Yellow Dwarf exhibited a higher net assimilation rate, relative growth index, crop growth rate, leaf area index and leaf area ratio compared with the WCT crosses with Malayan Orange Dwarf and Chowghat Orange Dwarf. MYD x WCT showed higher levels of reducing sugars and very low starch in leaves, whereas COD x WCT exhibited an opposite trend. High starch content in leaves has been reported as an end-product inhibitor of photosynthesis. Also, in COD x WCT, very high respiratory rates lead to low rates of dry matter accumulation. Since the nitrate-reducing capability is dependent on photosynthesis, nitrogen reductase was also affected, resulting in a reduced rate of leaf area development and dry matter accumulation in COD x WCT compared to the other two hybrids. The total quantity of nitrate reduced through nitrate reductase activity by the leaves of the MYD x WCT hybrid was 2.6 and 8.2 times higher than in the other two hybrids. The contribution of reduced N by way of shoot nitrate reduction was 6.7%, 10%, and 19.9% respectively. The rate of N accumulation in all hybrids was significantly correlated to dry matter production. The results seem to confirm the conclusion by Shivashankar and Ramadasan (1985) that nitrate reductase activity can be used as an index of the palm's ability to accumulate dry matter and, consequently, high yielding capacity. Further confirmation should be obtained from comparative trials.

Amalu *et al.* (1987) compared growth characteristics of two Nigerian talls, the Nifor Tall and the Badagry Tall. Generally, the Badagry Talls had larger leaf surfaces per palm. The annual dry matter production of 13-year-old palms was 92.1 kg for Badagry Talls against 92 kg for the NIFOR Talls. Net assimilation rates ranged from a minimum of 0.028 g dm^2-1 per wk in the 5-year-old Nifor palms to a peak value of 0.133 g dm^2-1 per wk in the 32-year-old NIFOR palms. In the mature (13-32 years) palms, the net assimilation rate attained a mean of 0.15 g dm^2-1 per wk. The total amount of dry matter produced was 84.8 and 80.4 kg for 13-year-old NIFOR Tall and Badagry palms respectively. The average maximum crop growth rate (whole palms) occurred in the 13-year-old palms and was 18,447 kg per ha per annum, which is low compared to a value of 24 tons per ha per annum for the aerial parts only, reported from Malaysia (Chew and Ooi 1982). Ramadasan and Mathew (1987) measured annual dry matter productions in 30-40-year-old West Coast Talls of 65-85 kg per palm, which at a planting rate of 156 palms per ha would correspond with 10-13 tons per ha only. This is less than half of the dry matter production of MYD x WAT in Ivory Coast mentioned above. In addition to varietal differences, differences in ecological conditions may may greatly influence the observed performances.

Economic evaluation of coconut varieties can be made according to their copra production per hectare, and the oil percentage of the copra. Where water nuts are concerned, the number of nuts, water content and sweetness of the water will be the most important criteria. Total copra production is the result of total nut production and copra yield per nut. Nevertheless, coconut yields are often expressed in number of nuts produced per palm per hectare. For an economic comparison of varieties this is erroneous. For instance, the MAREN hybrid (Malaysian Red Dwarf x Rennell Tall) and the Solomon population have about the same fruit weight, but the MAREN has 32% more kernel, which corresponds with the estimated difference in yield capacity of the two populations (Foale 1991b). Compared to an estimated world average number of 4,500-5,000 nuts required for one ton of copra, this figure for India is about 6,800 (Khan *et al.* 1990). This shows that nut yield is not a good parameter for yielding performance, unless when comparing individual palms of the same variety yielding nuts of equal quality. The performance of varieties may differ under different ecological conditions. Santos *et al.* (1986) observed in a variety trial conducted in three different locations with different climates and soils in The Philippines, that the ranking according to copra production per palm was different in each location for most of the varieties.

There are tall varieties with nuts yielding less copra than some hybrid nuts. However, in general, nuts of tall palms are superior to nuts of hybrids and these, in their turn are better than dwarf nuts.

The comparatively lower copra content of tall x dwarf hybrid nuts is generally compensated for by the much larger number of nuts produced by hybrids. In yield per hectare, this factor is enhanced by the greater number of hybrids planted per hectare compared to tall palms. The earlier yield in hybrids contributes to their superiority to tall palms. Not only economically, but also financially, as hybrids need shorter financing periods when loans are used for plantation establishment. Moreover, the potential response of hybrids to fertilizer application is generally greater than that of talls, especially when the tall has a lower number of female flowers per inflorescence, limiting its potential response expressed in number of nuts produced. On the other hand, the lower copra yield per nut is a disadvantage for hybrids where processing is concerned. Most handling of the coconut is paid for per nut and total cost per ton involves more hybrid nuts than tall nuts, increasing the cost per ton for hybrids compared to talls. The latter also holds for dwarf nuts, which have an additional disadvantage of producing leathery copra with a low oil content.

In general, hybrids yield more than talls, even under conditions of low soil fertility. However, hybrids are more sensitive to drought, probably due to their less extensive and probably shallower root system, combined with a greater number of leaves and bunches. Higher yields result in higher export of minerals from the field and it may be expected that hybrids exhaust the soil sooner than less productive talls, especially when the husks are not returned to the land. When well managed, hybrids can maintain their high level of productivity for many years. The first hybrids planted out in India, in 1940, were still yielding at the same level in 1990 as in 1944 (Pillai 1991). Santos *et al.* (1986) noted vast differences in the performance of any of the four hybrids tested in the different locations, clearly showing the occurrence of genotype x environment interactions. It was concluded that, apparently, hybrid performance can be predicted to be similarly good under favourable conditions and, on the contrary, no amount of management level may change their performance when climatic conditions are not favourable. In Ivory Coast, the potential yield of PB-121 hybrids ranged from 3.8 t of copra per ha per year in areas with a mean water deficit of 500 mm per year to 2.6 t in areas with a water deficit of 600 mm per year (Van *et al.* 1984). The potential yield where there was no deficit at all was estimated at 5.5 t. The highest yield obtained in Ivory Coast under experimental station conditions was 6.3 t. However, the highest yield obtained from the same planting material in The Philippines was 4.3 t copra per ha per year.

In Indonesia, three local hybrids were developed using the Nias Yellow Dwarf as mother tree and the Tenga, Palu and Bali Tall as male parents. These KHINA (Indonesian Hybrid Coconut) hybrids yielded 4.77, 4.46 and 4.70 t copra per ha per yr (Tarigans 1989).

Table 9: *Classification of tall coconut yield levels under normal ecological conditions*

Classification	nuts per palm	copra per palm	copra per ha
Very bad	0 - 10	0 - 2	0 - 300
Bad	11 - 20	2 - 4	300 - 600
Fair	21 - 30	4 - 6	600 - 900
Moderate	31 - 50	6 - 10	900 - 1,500
Good	51 - 70	10 - 14	1,500 - 2,100
Very good	71 - 90	14 - 18	2,100 - 2,700
Excellent	> 90	> 18	> 2,700

In addition to the creation of hybrids, selection and breeding in tall coconut is also beginning to show very promising results. The latest research results in The Philippines showed that full bearing 16-year old outstanding talls such as Baybay, Tagnanan and Laguna under rainfed conditions and moderate levels of fertilization produced 4.0-4.5 t copra per ha per year (Magat 1993b). In North Sulawesi (Indonesia), a selection programme started in 1930 had produced tall palms that under low input conditions produced an average of 31.6 kg copra per palm, with a density of 120 palms per hectare or 3.8 tons of copra per hectare. The highest yielders produced 45 kg of copra per year. A breeding programme with selected selfed progenies resulted in the production of four new cultivars, called Kelapa Baru 1-4 (Kelapa Baru (KB) = New Coconut). These new cultivars under local conditions produce 16 bunches per year. Yields expressed in nuts per palm and in copra per ha per year for KB 1, 2, 3, and 4 are 96 and 3.88 t; 121 and 4.49 t; 124 and 4.66 t; and 118 and 4.07 t, respectively (Tarigans 1989).

The average yield in India is estimated at 28-30 nuts per palm per year, corresponding with about 0.72-0.77 t copra per ha per year, whereas the highest number of nuts per palm in India has been 417. The size of these nuts and their copra content was not mentioned (Khan *et al.* 1990). In most coconut-producing countries yields under conditions of smallholder management are often not higher than 0.5-1.0 t copra per ha per year. In a coconut rehabilitation programme in The Philippines, coconut yield increased from 0.9 t to 1.97 t copra per ha in the third year of fertilizer application. Even the highest yield obtained from hybrids in Ivory Coast is 6.5 t per ha, no more than about only one-half of the estimated 13 tons of estimated maximum yield capacity. The above figures clearly show the great scope for improvement that can be obtained through improved management of existing stands. It indicates that in many cases multiplication of the present yields would be possible through management improvement, planting of improved planting material, and fertilizing. This is also shown by the very high yields of local palms growing near copra kilns and in farmhouse backyards, indicating that often local material has a considerable genetic yield potential and that major yield increases are to be expected from improved management. Table 9 presents a classification of tall coconut palm performance under conditions of adequate rainfall and soil conditions. The relationship between the figures in different colums may differ according to nut size and copra content, the latter also being influenced by soil management.

3. Climate and Soils

J.G. Ohler

Climate

Latitude and Day Length

Coconut is a tropical crop, its growing area is confined between the tropics of Cancer and Capricorn. Where the local climate is influenced by warm sea currents, such as in Florida and Southern Mozambique, coconut can grow and produce at greater distance from the equator. But in such regions, development is less rapid and yields are relatively low. The nuts are mainly sold as waternuts, a luxury product.

Latitude differences bring about day-length differences. The influence of day-length on coconut is unknown and is difficult to find out, as with the change of day-length there are also changes in the angles of the sunrays as well as in temperatures. Long days may have a favourable influence on coconut growth, due to longer periods of sunshine. At these latitudes temperatures during daytime in summer are high and favourably influence fruit set and yield during this season. Lower sunshine intensity may not be a limiting factor as (according to Foale 1991b) a leaf becomes light-saturated at values of less than 50% of direct solar radiation. Short days and lower temperatures in winter months may have an opposite effect. According to Blaak (personnal communication), in all palms the annual yield peak increases with distance from the equator. In Hainan, at latitudes 18-20°N, almost the total oil palm yield is obtained in four months and in the Dominican Republic the yield peak in oil palms is 20-26% of annual yield in one month. This subject certainly deserves more research.

Altitude

The limiting factor determining the maximum altitude at which coconuts can be grown is temperature. Thus, it may be expected that at greater latitudes, maximum altitude at which coconut can be grown will be less. Because of lower temperatures at greater altitudes, coconut development is slower than that at ground level. In northern Sulawesi, Indonesia, near the equator, coconuts planted at an altitude of 600 m come into production at an age of 10-12 years. Yields at higher altitudes are lower and so are the copra content of the nuts and their oil content. In general, coconut is not commercially grown at altitudes above 500 m. However, in home-gardens it can be found at greater altitudes, sometimes as high as above 1,000 m, such as in Indonesia and India (Wuidart, personal communication).

Temperature

It is generally assumed that the optimum temperature for coconut is about 27°C, and the mean diurnal variation between 5-7°C. Coomans (1975) found that fruit set is directly influenced by the monthly minimum temperatures below 23°C, over a period of four months, 18 months before harvest. He also observed a significant positive correlation between the annual yield and the mean annual minimum temperatures over a period of 18 months before harvesting, and a positive correlation between the number of female flowers per inflorescence and insolation and temperature of the 29th and 30th month before harvest. He suggested that the latter correlation was apparently influenced by the growing conditions during the stage of spadix differentiation.

Low temperatures are more limiting than high temperatures. On Hainan Island, China, coconut is grown between 17 and 20 degrees northern latitude. In Haikou City in the north, coconut produces less than on the East Coast, mainly due to cold spells from North China during winter. Normally, an absolute low temperature of 0°C for a short spell will not kill a palm, but cold spells of below 10°C can seriously affect coconut trees. Frost may be lethal to most coconut varieties, especially seedlings. The extent of damage is also determined by the length of the cold period. Blaak (1983) reported heavy losses of MAWA seedlings in a nursery in Hainan as a result of low temperatures. When during one day temperatures dropped below 7°C, leaf damage was caused to about 50 per cent of 2.5-years-old MAWA palms, whereas the adjacent Hainan Tall seedlings were not affected. One cold spell with temperatures below 15°C lasting 57-75 days in the main areas of coconut cultivation seriously damaged coconuts. In the north of the island, losses of up to 90 per cent of the nuts occurred during two consecutive years (Zushun 1986).

Due to low temperatures, coconuts may not only fail to produce nuts, but nut quality also may become seriously affected. Usually, damage caused to the nut by slight cold is not outwardly perceptible but can be seen in the abnormal or incomplete development of its meat, which is unevenly formed with many wrinkles. It was observed that nuts at their fastest development stage, *i.e.* 5-6 months after flowering, are most vulnerable to cold. Cold weather in such a period would cause the young nut to split or fall, after which it turns brown or black and dries up. Zushun (1986) observed that the lower limit for the palm and the leaf to survive the winter is 8°C, whereas that for the nut is 13°C. Besides withering of mature leaves, spear leaves may be killed and non-uniform development of the leaves in the crown may occur. It was stated that, even in the absence of any damage to the palms or nuts, the physiology of the palm is modified by low temperatures and production is non-existent during the winter period.

A new hybrid, WY78F1, has been developed in Hainan, which is characterized by fast growth, good precocity, high yields and good resistance to cold and strong winds. Its yields are similar to those of MAWA hybrids (Zushun 1994).

Water Requirement, Rainfall and Water Regulation

The water requirement of coconut is strongly influenced by such factors as temperature, relative humidity of the air, and frequency and force of winds. Because of these varying conditions in different growing regions, rainfall requirements also differ considerably. Therefore, coconut can be found growing in regions with annual precipitation varying from 1,000 to 4,000 mm. But even in very dry climates, coconut can grow when soil water supply is adequate, such as in valley bottoms, or at the foothills of mountains where water infiltrates into the soil from higher areas.

Different cultivars have different water requirements. Adaptation to water deficit has been recognized when populations from particularly favoured environments were found to perform poorly in areas where a well-defined and extended dry season occurred. It is likely that highly adapted populations are present on some of the drier atolls of the Pacific or in East Africa.

Results of field studies in Sri Lanka showed that an adult coconut palm with 35 leaves (150 m^2 leaf area) transpired 30 to 120 l water per day, depending on the atmospheric evaporative demand and the soil water level (Jayasekara and Jayasekara 1993). Mannil (1989), reported volumes of 28 to 74 l under West Coast conditions in India. Climatic and soil conditions are different everywhere. In addition, different cultivars have different levels of drought tolerance. Volumes of water required by coconut palms differ from place to place, even for the same cultivar.

In addition to total rainfall, the pattern of rainfall is very important. According to Khalfaoui (1985), a certain period of low rainfall may be distributed as either low but throughout the year, or as heavy rainfall separated by dry periods. In the first case, the plant is almost continuously suffering from a certain degree of water-stress, making it impossible for the plant to build up starch reserves. In the second case, the build-up of starch reserves during periods of heavy rainfall may enable the plant to survive during periods of water-stress. Other factors, such as ground water availability play an important role in the need for rainwater supply.

For coconut, adequate water supply throughout the year is one of the major requirements for good coconut production, particularly on soils with a water table that cannot be reached by the roots. The upper limit of rainfall depends on soil drainage and on insufficient insolation due to cloudiness. A monthly rainfall of about 150 mm with a dry period of not more than three months is considered optimal for coconut in India (Mahindapala 1987; Rajagopal *et al.* 1990). Cosico and Fernandez (1983) considered a rainfall of 125-195 mm per month, corresponding to 1,500-2,300 mm per year, ideal for coconut, distribution being more important than annual total. It may be concluded that the suitability of the climate for coconut cultivation as far as water supply is concerned cannot be measured by rainfall alone. Soil water availability, relative humidity of the air, winds and temperatures all play a role in water supply and evapo-transpiration of the palm.

Periods of very high rainfall may be harmful to coconut yield. Peiris (1993) observed in Sri Lanka, that rainfall exceeding 460 mm in a bimonthly period depressed yields. A disadvantage of concentrated rainfall, combined with high relative humidity, relatively low temperature and low insolation is the negative effect on transpiration intensity and, consequently, on water uptake by the palm. Reduced water uptake may lead to insufficient nutrient uptake and malnutrition. On the other hand, periods of drought may cause the palm to shed some leaves and to close its stomata, thus decreasing photosynthesis. In regions with high rainfall in the wet season and very low rainfall in the dry season, the palm is caught between dying of starvation or dying of thirst, as expressed by Wickramaratne (1987b).

Water consumption of coconut varies with ecological conditions and differs between genotypes. Water consumption depends also on the total leaf area of the palm's canopy. Palms of the same variety will have different water consumptions in different environments. Therefore, the figures presented in this paragraph can give only a rough idea about the coconut's water consumption. According to Reyne (1948), citing Tulner (1933) a mature coconut palm under favourable conditions transpires about 200 litres per day. Studies carried out by Jayasekara and Jayasekara (1993) in Sri Lanka showed that an adult coconut palm with 35 leaves (150 m^2 leaf area) transpired 30 to 120 l water per day depending on the atmospheric evaporative demand and the soil water level. Evaporation starts immediately after sunrise, reaching a maximum at about 10 o'clock in the morning, maintaining this level until about 3 o'clock in the afternoon, after which it declines again. Leaves in full sunshine evaporate 2-3 times as much as leaves in the shade. Thus, planting density will also influence the total evaporation by individual palms. When leaves are cut off, stomata close within half an hour. Passos (1994) observed that the highest evapo-transpiration takes place at noon, when temperatures are at their highest and the relative humidity of the air is at its lowest.

According to IRHO (CIRAD-CP) (Anonymous 1989b), there still is no complete picture of the relationship between climate and production in coconut. The water deficit plays an important role but the exact mechanism of its action is not known. First it is necessary to evaluate the water requirement of the coconut palm, particularly the maximum evapo-transpiration under conditions of good nutrition and water availability. Measurements made in representative regions of Ivory Coast during some years have permitted a preliminary estimation. It should be equal to about 110% of the evaporation of the free water surface in a container of class A. The potential evapo-transpiration calculated on the basis of the Penman and Turc formula is too high. A first interpretation, based on results of the dry season of 1986-87, shows a mean daily evapo-transpiration of 2 mm.

As already mentioned above, water supply does not depend on rainfall alone, but also on soil water availability. Jayasekara and Jayasekara (1993) in Sri Lanka observed that with the onset of a dry period, root water uptake was initially confined to the top 50 cm soil layer. After two weeks, roots extracted water mainly from layers below 1 m depth in deep sandy soils. Soil water availability is linked with soil texture, soil water-holding capacity, soil depth, and depth of the soil water-table. The influence of soil texture and depth and, consequently, depth of the root system on production stability is illustrated by observations made in the research station in Port Bouet, Ivory Coast (Anonymous 1989b). Hybrid PB-121 coconuts planted in 1963 yielded an average of 5.2 tons of copra per year during 1978-1982, but only 3.7 tons per year during the much drier period of 1982-1986, when the soil water-table dropped out of reach of the roots. Disappearance of the soil water-table could not be compensated for by the available water content in the soil. In a nearby area, where coconuts had been planted on very deep sandy soil without a water-table within reach of the roots, the yields during the same periods were 3.9 t copra per year and 3.7 t copra per year respectively. The slight drop in yield was due to the very deep rooting of the palms in this soil. In another experimental plantation the tolerance to drought of coconut planted on deep, friable soils was demonstrated by a yield of 3 t copra per year in 1988 under conditions of an average annual water deficit of about 800 mm. In a plantation in Indonesia, in a region with a slight annual water deficit normally, coconuts growing on shallow soil normally yielding 4 tons of copra per year produced less than 1 t copra per year in an abnormally dry year with a water deficit of 500 mm. Voleti *et al.* (1993a) studied the response of three coconut genotypes, *i.e.* West Coast Tall (WCT) and the hybrids WCT x Chowghat Orange Dwarf (COD) and COD x WCT to moisture stress on red sandy loam and laterite soils. The water-holding capacity of the red sandy loam soil was low compared to the laterite soil. The response to moisture stress varied according to soil type and genotype. Variations in water availability in the two soil types greatly influenced the genotypes, as revealed by leaf temperature, stomatal resistance, transpiration rate, water potential and epicuticular wax content. These interrelated stress-sensitive responses ultimately reflected on the performance of genotypes under stress conditions. There was a significant rise in leaf temperature between pre-stress and stress in both soils; the hybrid COD x WCT registering significantly higher leaf temperatures than the other two genotypes. These changes could be related to the transpiration rate, which in turn was controlled by stomatal resistance. Although the general tendency of palms was to show increased stomatal resistance with the onset of stress, there was significant genotype variation, which again differed depending on soil types. Maintenance of higher leaf turgor potential by all genotypes in laterite rather than in sandy soil showed different responses of the same genotype to moisture stress occurring in different soil types. Compared to WCT, the hybrid COD x WCT had a significantly high soil water deficit and stomatal resistance with low transpiration rate and high leaf turgor potential in laterite soil, while WCT x COD was intermediate. The epicuticular wax content showed marginal differences among the genotypes in the sandy loam soil, whereas in laterite the content was higher in COD x WCT than in the two other types, which may have contributed to enable this hybrid to withstand drought in laterite. Conversely, relatively low stomatal resistance with high transpiration, resulting in low leaf turgor potential appeared to make COD x WCT susceptible to stress in red sandy loam soil, indicating that both high rates of transpiration rate and soil water deficit in sandy loams cause an imbalance in the water economy of palms, as revealed by the turgor. It was concluded that under such conditions, a genotype with good stomatal regulation and high turgor, such as WCT, withstands stress better than others. Apart from differences in water-holding capacity of the soil, differences in root development in different soils may also play a role. Variation in the root system of the genotypes in different soils was only indirectly indicated by the soil water deficit pattern at different depths, representing root growth, but it was concluded that the picture on the root system was not conclusive on the presented data. Further conclusions of the study were that it became evident that the capacity of the same genotype to withstand moisture stress depends on the type of soil and the operation of stomatal mechanism. Secondly, drought screening should be in relation to the type of soil on which palms are grown.

Several other studies on water consumption of coconut palms at different ages and under different conditions have been conducted, such as the studies of Liyanage, L.V.K. (1987b), Saseendran and Jayakumar (1988), Jayasekara and Mahindapala (1988) and Rao (1989). However, the different findings of water consumption and regulation of different coconut genotypes under different conditions, and studied with different methods, lead to confusion. It is highly recommended that standard trials be developed, using standard planting material in various climatological conditions in different countries, to obtain a better basic knowledge of the palm's water requirement and the consequences of water deficit, and also on the influences of different morphological characteristics of the different genotypes, including the root system development in different types of soil. Varying climatic conditions such as wind velocity may still influence the results. There are indications that genotypes that perform very well in one region may do worse elsewhere. With standard, multi-local trials it may be possible to indicate the most suitable ecological conditions for each cultivar, which might prevent errors with long-lasting effects in planting schemes.

Palms suffering from water-stress first show drooping of the leaves (CP 28). The leaflets may be folded slightly, permitting more light to penetrate through the foliage. Then, the oldest leaves may yellow, hang down and finally die prematurely. The first symptom of a loss of turgor noted by Pomier and Taffin (1982a) was a constriction which formed on the underside of the leaf stalk. The leaf folds at this point and dries up in two to three weeks. Generally, it remains attached to the stem until the beginning of the rainy season. Nuts ripening during a period of water-stress remain smaller than normal. Female flowers and buttons are shed, sometimes even unopened inflorescences are aborted. In a later stage, nuts in various stages of immaturity drop as well. (Bourdeix 1994) mentions an example of a cultivar hyper-susceptible to drought, surviving a drought because it had dropped all its fruits during an early stage, whereas high-yielding, less drought-susceptible cultivars that had retained their fruits, finally collapsed. The rachis of the older bunches, losing their support from the leaf petioles, may break or may be torn and hang down along the stem with the older leaves. Nuts on such bunches are dropped after the first rains. According to Schuiling and Mpunami (1992), the symptoms of the Tanga disease in Tanzania (Ohler 1984) may have been caused by a prolonged drought.

Although coconut can tolerate six months or more of drought, yields are seriously affected when the water deficit amounts to more than 300 mm per year. After a severe drought, yields may be affected during two following yields (Ohler 1984). Prasada Rao (1986) reported that after severe droughts in Sri Lanka, in 1931 and in 1983, nut yields were affected for a period of two years, with the maximum effect at about thirteen months after the conclusion of the drought. In the second period, no yield reduction was observed during the year of the drought. The first drought effect was noted only eight months after the conclusion of the drought, lasting for 12 months. The time between the drought and the visibility of its effects probably depends on soil water-holding capacity, soil depth and plantation management.

Pomier and Taffin (1982a) found that the local West African Tall was very sensitive to drought, whereas the hybrid PB-121 that has a WAT male parent showed remarkable tolerance. In some hybrids drought had little influence on leaves but more on nut loss, whereas the opposite effect occurred in other hybrids. Shivashankar *et al.* (1993) observed that from three hybrids with West Coast Tall as their female parent, the cross with Malaysian Yellow Dwarf had the highest drought tolerance. Apparently the MYD has a favourable influence in this aspect, confirming observations of Cintra *et al.* (1992). From the above, it is clear that there are substantial differences in drought tolerance between varieties, the differences not always caused by the same factor(s) that contribute to a palm's tolerance to drought conditions. Daniel (1991) concluded that drought tolerance, that is the maintenance of high yield levels even under drought conditions, is obtained by the conjunction of several apparently contradictory characteristics. This would confirm that behaviour in drought conditions depends on several factors, which must all be taken into account.

The integrated anatomical and physiological mechanisms through which plants defend themselves against water-stress have been classified into two groups: avoidance and tolerance mechanisms (Repellin *et al.* 1994). In the case of avoidance mechanisms, the plant reduces its development cycle and the size of the transpiration surfaces, increases its root volume and controls water loss through stomatal regulation. In case of tolerance mechanisms, plants maintain normal physiological activity despite a reduction in tissue hydration. Drought tolerance thus occurs in the cells. It may result in an increase in abscisic acid, the hormone that induces stomatal closure but also seems to play a role in numerous processes in cell adaptation to water deficit. Another drought tolerance mechanism is the accumulation of solubles such as proline and sugars, contributing to the osmotic adjustment of cell content and enabling turgidity maintenance in leaf tissue. Water-stress can also cause a drop in cell protein and lipid contents, which are the main constituents of membranes. The coconut palm has several mechanisms against water-stress, such as:

- effective soil water depletion by the roots;
- stomatal closure and osmotic adjustment;
- increased cuticular resistance;
- change of leaf angle, leaf senescence and shedding of leaves to reduce evaporation;
- apparently, early dropping of the fruits is also a mechanism of drought tolerance;
- sugar and proline accumulation;
- enzyme activity;
- cell lipid content reduction.

Effective soil water depletion

Studies by Rajagopal *et al.* (1991) indicated that drought-tolerant coconut genotypes, in general, extract more soil moisture from the entire soil profile compared to drought-susceptible types. The efficiency of the root system was attributed to its total mass as well as its activity. In fertilizer trials conducted in Mozambique, on a poor sandy soil, in a climate with a dry season of about six months, the author observed that adequately fertilized palms were much more tolerant to drought than inadequately fertilized palms. They remained green and retained their older leaves, whereas the other palms yellowed and shed their oldest leaves. This could be explained by a much better development of the root system as a result of fertilizing (applied by broadcasting), as well as by improved stomatal function (see next paragraph), reducing water loss.

There is varietal difference in root-system development and water uptake between genotypes. A comparative study conducted in Ivory Coast (Anonymous 1992a) showed that the hybrid PB-121 (West African Tall x Malayan Yellow Dwarf) had a larger leaf crown than its WAT parent. The hybrid had higher stomatal conductance and the water potential of the hybrids was often higher than for WAT, although leaflet relative water contents were comparable. During the dry season, despite higher transpiration and a lower suction capacity, PB-121 had a similar water content to WAT. It therefore had a better water supply, enabling it to withstand drought and show less damage, whilst producing more than WAT. It was suggested that this advantage may be linked to a more efficient root system in the hybrid. A previous study had shown that tertiary and quarternary roots made up a higher percentage of the root mass in the hybrid than in the WAT. On the contrary, at the seedling stage talls managed water losses more effectively than hybrids or dwarfs when watering was stopped. This could be explained by the fact that in the field water-stress is very different as relative water contents generally remained above 90%, but in the seedlings, the stress was intense and led to a drop in relative water content to values below 78%. However, the possibility that for certain parameters, coconut seedling performance is not comparable to that of the adult tree, as is the case with oil palm, was not excluded. It was also observed that dwarfs consume more water than other varieties. During water-stress, they will be the first to reach the point of irreversible wilting.

Cintra *et al.* (1992) when comparing the root system distribution of several dwarf coconuts concluded that in view of the deep rooting habit of the yellow dwarfs (MYD and Gramame Yellow Dwarf), hence their greater water uptake ability, these varieties can be considered the most promising of the varieties studied for regions subject to a water deficit.

Stomatal closure and osmotic adjustment

Water potential in a plant, which is the energy-level of water, is controlled by the availability of water from the soil, the demand of water imposed by the atmosphere and the resistance to water movement within the plant. Changes in water status of plants depend on evaporative demand in the atmosphere (Voletti *et al.* 1993a). Maintenance of a higher leaf water potential under stress conditions is a desirable trait, as that would enable tissues to maintain favourable metabolic activities to withstand desiccation. Leaf water potential was differently affected between coconut genotypes, which appeared to have been related to a great extent to stomatal activity. Thus, low stomatal resistance resulted in high transpiration and lowering of leaf water potential. Some of the talls and hybrids exhibited a protective mechanism through effective stomatal regulation, whereas dwarfs seemed to lack this mechanism. Some of the talls with low stomatal frequency per mm^2 reacted to drought differently from dwarfs through the adaptive mechanism provided by high leaf diffusive resistance, further aided by high epicuticular wax content resulting in effective water conservation in the tissues. The most drought-tolerant genotypes showed not only effective stomatal regulation but also efficient soil water extraction and favourable leaf characteristics, resulting in a well balanced water economy (Rajagopal *et al.* 1990).

Under conditions of atmospheric or osmotic water-stress, coconut palms regulate their water balance by stomatal closures. Even if the soil-water supply is adequate, coconut palms close their stomata when the relative humidity of the air drops below 60 per cent. Ollagnier (1985) underlined the important role of K and Cl ions in drought tolerance. These elements are never included in the living matter and they are the only ions that form no part at all of the organic compounds of the tissues. A major characteristic common to them is their salinity and the role they play in turgor. A movement of K in and out of the guard cells coincides exactly with the opening and closing of the stomata. These cell movements, although much slower, also occur in the absence of K. K improves the water and carbohydrate balances by shortening the period during which the state of the stomata lags behind the development of the surrounding conditions. Apparently, the K in the stomata guard cells needs to be counterbalanced by a negatively charged ion, which can be either Cl or a malate, synthetisized within the guard cells.

Absence of starch from guard cells deprives them of the ability to produce malate in amounts of osmotic consequence. In such cases, the presence of absorbable Cl^- is necessary for stomatal opening. Analysis of leaves of Coconut, Nipa and Areca palms showed that they contained no chloroplasts in their guard cells. The same had been found in the leaves of oil palms. This suggests that for stomatal movement the entire family of palmae depends on the presence of adequate Cl^-. (Uexkull 1985).

Studies by Braconnier and d'Auzac (1985; 1990) confirmed that K ion influx is associated with stomatal opening and the accumulation in guard cells of open stomata. Chlorine plays a double role in stomatal osmotic movement. When stomata are closed, it is found in the subsidiary lateral cells, helping to maintain cell turgidity which prohibits stomatal opening. When stomata are open, Cl ions migrate towards the guard cells, simultaneously reducing subsidiary lateral cell turgidity and increasing that of the guard cells, which encourages stomatal opening. Reduction of Cl ions in the treatment medium reduces stomatal opening. In Cl-deficient palms, the absence of Cl influx to guard cells is balanced by another mechanism not yet determined. The latter, however, acts more slowly. Braconnier and d'Auzac (1990) also found that when leaves contain less than 0.25% (dry weight) of Cl, coconut responds to fertilizers such as KCl. Trials with young coconuts showed that Cl deficiency led to delayed stomatal opening. The trials also showed that, with time, the coconut plantlets were adapting to stress.

Under osmotic stress, Cl deficiency led to stomatal conductance reduction throughout the day, which probably led to a marked reduction in photosynthetic gas exchanges. This reduction was due to a fall in osmotic potential values. Osmotic stress resulted in lowering of leaf water potential, more in non-deficient than in Cl-deficient plantlets. Osmotic potential of deficient plantlets rapidly reached a minimum level, whereas that of non-deficient plantlets could continue to decrease, suggesting that the latter were somewhat slow to re-establish their water balance. Therefore, the Cl-deficient plantlets were much less capable of accumulating osmotica than non-deficient plantlets. Stress led to more significant and more rapid reduction of turgidity in Cl-deficient plantlets than in non-deficient plantlets. Thus, Cl deficiency led to a significant reduction in osmo-regulation capacity of coconut, whereas the maintenance of cellular turgidity is regarded as the major factor for continuation of growth and production in water-stress periods. How Cl affects osmo-regulation is still not known. Studies in Ivory Coast (Ollagnier *et al.* 1983) clearly indicated that Cl application considerably increased yields of nuts and copra per nut as well as improving the palm's tolerance to drought conditions. Also in Indonesia, results of trials showed that chlorine is an essential nutrient for adult and young coconuts, right from the nursery stage. Chlorine had a predominant effect on resistance to high water-stress in young and adult coconut palms (Bonneau, *et al.* 1993a).

Trials conducted by Rajagopal *et al.* (1989; 1990) indicated that in dwarf coconut palms stomatal frequency (number of stomata per square cm) was higher than in talls and hybrids. All the dwarfs and one of the hybrids had significantly lower stomatal resistance, which reflected a higher loss of water through transpiration than talls and other hybrids. This was further confirmed by leaf temperature, which is correlated with diffusive resistance, showing less difference between the pre-stress and stress periods than among the talls and hybrids. Thus, with the onset of drought caused by atmospheric and soil conditions, the genotypes showed differences by closing their stomata to different degrees as a mechanism of adaptability to low water availability. Kasturibai *et al.* (1988) observed increases in radiation, temperature and vapour pressure deficit (VPD) between 10 and 12 hours. During this period stomatal resistance reached a maximum and the leaf water potential was reduced. There was a sixfold increase in stomatal resistance between the wet and the dry seasons. Under conditions of their experiment, stomatal closure set in when coconut palms were exposed to an environmental situation with irradiation around 265 W per m^2, temperature of 33°C and VPD of 26 mbar. Field values of the variables exceeding those mentioned above definitely lead to severe stress in coconut palms.

Voletti *et al.* (1993b) observed that the water potential of the spear leaf was significantly higher throughout the day. During the early hours, the difference between the spear leaf and the twentieth leaf was substantial under rain-fed as well as irrigated conditions, while within the spear leaf the difference was negligible. Soon after sunrise, the water potential began to decrease in all leaf positions, particularly in the spear leaf. Leaf-water potential declined rapidly up to noon and rose again thereafter. The results indicated significant differences between hours of sampling, leaf positions, rainfed versus irrigated conditions and all their interactions, except hours of sampling versus different leaf positions. Although the overall differences in leaf-water potential were the same between the rainfed and irrigated conditions by 10.00 hours and 13.00 hours, the gradient was much steeper in irrigated conditions, presumably because of the very high leaf-water potential in the irrigated palms during the early hours. Characteristic mid-day depression in leaf-water potential was evident in both the spear leaf and the first leaf, and beyond this leaf-water potential remained constant up to 16.00 hours as compared to the first leaf. The study indicated the existence of a vertical profile in leaf-water potential over a period of time, and also confirmed the importance of fixing the spear leaf as the index leaf, as well as the optimum time for leaf-water potential measurements (between 10.00 hours and 12.00 hours) in coconut palms.

Repellin *et al.* (1994) studied drought effects in young palms of five different coconut varieties. Results showed that cellular tolerance to dehydration is an essential factor in adaptation to drought. In the case of young coconuts subjected to edaphic water-stress, stomatal closure seemed to be the main carbon assimilation limiting factor. Therefore, young coconuts adopt a dehydration, avoidance strategy. After dehydration, stomatal conductance was no longer the main factor limiting photosynthesis. This may be explained by chloroplasma malfunctioning after the series of disruptions occurring when stomata begin to close. It seems that prolonged stress brings non-stomatal factors into play in photosynthesis regulation. They also observed that the first moves to close the stomata depended neither on the relative water content, nor on leaf-water potential. Therefore, the water status of the leaf tissues is not the signal triggering stomatal closure. They concluded that transpiration and stomatal conductance diminish identically under the effects of water-stress, but these parameters do not provide any clear distinction between varieties, and thus are not suitable for selecting young coconuts, unlike oil palm, although stomatal conductance can be a discriminative parameter for adult coconuts. Nevertheless, the existence was indicated of a root signal responsible for reduced transpiration and stomatal conductance, especially in Cameroon Red Dwarf. This signal could be abscisic acid. Reduction of carbon dioxide assimilation occurred in the same way for all varieties and thus is not a discriminative parameter. But it indicated factors that determine changes in carbon assimilation under water-stress which seem to be, in order of importance: stomatal conductance, water potential of the leaves and, to a lesser degree, their relative water content. The parameters relative to dehydration rate are discriminative when water-stress is severe. Differences between dehydration rates do not appear unless stomata conductance and transpiration are nil. They may be due to variability in cuticular transpiration.

Epicuticular wax content

Formation of a wax layer on the leaf surface is an adaptive mechanism to withstand water deficit situations. Kurup *et al.* (1993) observed a clear indication of a negative relationship between the epicuticular wax content (EWC) of leaves and transpiration rate. In all types studied, the wax content of the sixth leaf exhibited a sharp increase during periods of water-stress and a decline in post-stress periods. The peak in EWC content coincided with high light intensity, high temperature and low relative humidity. The epicuticular wax content was higher among drought-tolerant than among susceptible genotypes. The precise mechanisms of these changes in wax content are not clear.

Change of leaf angle and senescence
Drooping of the leaves and folding of leaflets change the position of the leaf blades in relation to incoming sunrays, reducing their effect on these leaves. Pomier and Taffin (1982a) when comparing the drought tolerance of some hybrids, found that the number of dry leaves (n) in relation to the number of green leaves on the crown (N) can provide a very useful index of drought tolerance for coconuts. This drought index was expressed in the formula

$$\frac{n \times 100}{N}.$$

Daniel (1991) reported that in the Ivory Coast, at the end of the dry season, the hybrids PB-121 and Sri Lanka Tall x Green Sri Lanka Dwarf and others showed proportions of dry leaves and hanging green leaves of up to more than 30%, whereas others, such as Malayan Tall x Green Sri Lanka Dwarf had only 12%.

Early fruit drop
As mentioned above, early fruit drop of drought-susceptible varieties may reduce the burden on trees, whereas trees that retain their fruits may finally collapse.

Sugar and proline accumulation
Shivashankar *et al.* (1993) observed that changes in leaf-water status of some varieties was reflected in changes in the levels of chlorophylls, free amino nitrogen, epicuticular wax and total soluble sugars. Irrigated palms in general contained higher chlorophyll a and b than the rain-fed palms, whereas the levels of free amino acids, epicuticular wax and total soluble sugars were higher in rainfed palms compared to irrigated palms.

Drought-tolerant genotypes were found to possess high photosynthetic rates and instantaneous water use efficiency compared to drought-susceptible genotypes (Rajagopal *et al.* 1993). Jayasekara *et al.* (1993) when analysing leaf sugars and starch content in leaves during wet and dry months found no variation in palms that had been selected for their yield stability over the years. Voleti *et al.* (1990) induced artificial water-stress to coconut leaflets of tall, hybrid and dwarf coconuts, using an osmoticum and air desiccation. In general, the relative water content of leaves of all genotypes was reduced more due to air desiccation rather than to osmotic stress. Accumulation of proline did not differ much between the two types of stress simulation, except in dwarfs, which exhibited a higher proline content in osmotic stress leaves than in air-desiccated ones. There was an inverse relationship between relative water content and proline accumulation. It was concluded that proline content was not associated with drought tolerance in coconut. However, Jayasekara *et al.* (1993) observed a slightly higher concentration of proline during dry months in palms selected as drought tolerant. Some hybrids and one tall showed higher relative water content than the other genotypes, indicating the degree of stress tolerance, while the low relative water content in dwarfs indicated their susceptibility. Thus, relative water content determination can be reliably used for screening large numbers of samples for drought tolerance.

Enzyme activity
Shivashankar (1988), observed that the leaf polyphenol oxidase (PPO) activity in rain-fed palms increased with the development of water-stress, whereas there was little change in the activity of irrigated palms. The increase in activity was gradual from October to April, followed by a sudden jump between April and May, at which time soil moisture was significantly lower and temperature markedly higher. The occurrence of both soil and atmospheric droughts may thus be responsible for the increase in PPO activity. Coconut cultivars showed differential responses to a given level of stress condition. The relatively more drought-tolerant variety WCT showed much less enhancement of PPO

activity (58%) as opposed to CDO x WCT and WCT x CDO, which showed nearly 80% increase in activity.

The activity of acid phosphatase did not differ much between a tolerant and a susceptible genotype under pre-stress conditions, whereas with the onset of stress, activities increased by 43% in the tolerant against 79% in the susceptible genotypes. Moisture-stress enhanced the activity of glutamate oxalacetic transaminase two to threefold among the studied genotypes, but the difference in the increase between the tolerants and the susceptibles was less than that observed in the case of acid phosphatase. It was concluded that the measurement of enzyme activities during stress development could serve to monitor the degree of tolerance to water-stress.

Through an applied molecular biology study of cell responses to dehydration, Repellin *et al.* (1994) indicated the importance of this type of response, which ensures the continuity of cell physiological activity during drought and guarantees rapid resumption after dehydration. In this respect, two hydrolytic enzymes were discovered whose activities are stimulated in response to drought and the greater the susceptibility of the variety to drought, the greater the stimulation. Further studies of these enzymes may provide powerful molecular tools for selecting drought-tolerant coconut parents, and for studying the regulation of the expression of genes coding for these enzymes, with the ultimate aim of transforming coconuts and obtaining clones with low hydrolytic activity and great tolerance to drought. They concluded that this strategy, considered here for a lipase and a protease, which have been shown to be molecular markers for drought tolerance, can be adapted to any other enzyme playing a role in cellular responses to water-stress.

The above-described mechanisms of the coconut palm to obtain sufficient water for its requirements and to reduce water loss during periods of water-stress, may be useful tools for breeders to develop drought-resistant coconut genotypes. It appears that breeders will co-operate more and more with physiologists in order to select the proper criteria for their breeding programmes.

Cell lipid content reduction

In India it was found that drought-tolerant coconut cultivars showed significantly low lipid peroxidation levels compared to drought-susceptible palms (Anonymous, 1993a). Repellin *et al.* (1994) observed that water-stress induced a reduction in leaf lipid contents, especially that of chloroplast lipids. It was shown that membrane fatty acid reduction was greater the less the variety was tolerant to dehydration. Plant rehydration led to a rise in the fatty acid contents of lipids. This phenomenon reflects resumed synthesis and/or a slowing down of deteriorating phenomena. After four days of rehydration, the overall fatty acid content was 80% of the initial content in the five varieties studied (2-year-old plants). Differences in membrane lipid composition highlighted the high tolerance of PB-121 compared to that of its parents, and compared to CRD, which has the lowest membrane tolerance of the five varieties. The different tendencies between varieties listed by Repellin *et al.* (1994) were:

- West African Tall easily conserves its stock of water during water-stress, but it has a relatively high membrane fatty acid reduction;
- MYD, and especially CRD, stand out through their high water losses during water-stress. In addition, they have high membrane fatty acid reduction under conditions of water-stress. Both dwarf varieties therefore have low drought resistance, as has been observed before;
- PB-121 has a dehydration tendency generally between that of its two parents, WAT and MYD, but it stands out from them through its high membrane tolerance of water-stress.

Relative Humidity

Relative humidity is one of the factors determining the transpiration rate and, consequently, the water and nutrient uptake by the palm. Even in the case of sufficient water supply from the soil, low relative humidity of the air may induce stomatal closure in the palm, reducing its photosynthetic capacity. In Madagascar it was observed that coconut reduces its stomata opening when relative humidity drops below 60%, even if soil humidity is close to field capacity.

Insolation

It is generally accepted that the coconut palm requires at least 2,000 hours of sunshine per year to exploit its production potential fully (Ochs 1977). Murry (1977), cited by Rajagopal *et al.* (1990) estimated that 120 hours of sunshine per month would be favourable for coconut. However, climates with favourable total numbers of sunshine hours are not necessarily favourable for coconut when there are considerable seasonal differences. This could mean an excess during the dry season and a shortage during the rainy season. Stomata may close when leaf temperatures become too high due to intensive sunshine. According to Passos (1994), stomatal conductance in coconut is governed principally by solar irradiation, followed by a decline of leaf-water potential. It was observed that the stomata opened at an irradiation intensity between 200 and 300 W/m^2, reaching their maximum opening at 500-900 W per m^2. Rapid closure occurs at 16.00 hours in the afternoon and is completed by 18.00 hours. Ziller (1960) observed that the copra content of the nut is positively correlated to the rate of insolation during the last four months before the ripening of the nut. Mean copra content of the nuts showed a minimum at the end of the rainy period and a progressive increase towards the end of the dry season. Coomans (1975) found a positive correlation between the rate of insolation, 29 and 30 months before harvest and the number of female flowers produced per inflorescence. Löhr (1993) attributed this effect to higher temperatures during the dry season. Possibly both factors may have a similar influence on female flower production, enhancing the effect.

Wind

Some wind is favourable for coconuts when sufficient water is available, as it increases transpiration and, consequently, water and nutrient uptake. The drying effect may also help to prevent the development of certain fungus diseases. Strong winds, and especially cyclones may do considerable damage, distorting leaves, inflorescences and bunches, even tearing them off. Marty *et al.* (1986) studied the effect of cyclones on coconut palms in the Vanuatu archipelago, and observed the following effects: On light soils, quickly drenched by accompanying heavy rainfall, many palms were blown over and partly uprooted. These palms could often continue growing, producing vertical stems perpendicular to the downed stem. Trees that were not too old could be put in a right position again. On heavier soils, providing good anchorage, coconut stems were twisted to such an extent that they broke, either at the bole or at the upper third part of the stem. Heavy damage was done to leaves and bunches of the palms that remained standing. Post-cyclonic trauma observed eight months later included emission of abnormally shaped leaves and premature fall of bunches that had lost their leaf support. It was observed that a critical age exists, concerning all varieties observed. Although varietal difference in this aspect exists, this critical age is generally between 5 and 7 years. Damage to the palm increased from this age onwards due to increased height, large crown size and heavier nutload. Among the remaining palms, the damage was light. Leaf damage tended to disappear quickly. The palm's adaptability is linked with the stem's anatomy. Stem rigidity is not ensured by wood and fibre cellular tissues, like that of dicotyledonous plants, but by an assemblage of sclerenchymatic tissue, which gives coconut its elasticity. This enables the palm to orient its crown towards the point of least wind resistance. Dwarf palms are more fragile and sensitive to cyclone damage. This is explained by the absence of a prominent bole and its slender stem in relation to crown size. Talls and hybrids with large boles and root systems and thicker stems are much more difficult to uproot. Hybrids suffer more than talls, due to their heavier nutload and greater number of leaves, characters that increase wind resistance.

These findings were more or less confirmed by a study on the effects of a hurricane that struck Jamaica in 1988 (Johnston *et al.* 1994). Eight-year-old Malayan Yellow Dwarf palms and 11 hybrids of MYD and different male parents showed great differences in tolerance. The highest proportion of palms killed outright by the hurricane was in MYD, 37%. The lowest was in the hybrid MYD x Panama Tall, 7%. In general, the percentage of palms killed was correlated to the bole girth, with the exception of the hybrid MYD x Cambodia Tall that had the second largest bole girth but was fourth in tolerance. This was due to its larger stem height. The general conclusion was that the larger the bole girth the lower the mortality, whereas the taller the tree, the higher the mortality. Irrespective of variety, those which survived the hurricane were trees with the larger bole girth. The results suggest that the canopy may be of less importance in determining wind damage than the trunk height and the bole diameter.

Magat *et al.*(1988b) observed over a period of 7 years that the mean yield of PB-121 hybrids in the various regions of The Philippines was negatively correlated to cyclone frequency. Alforja *et al.* (1985), reported that frequent typhoons have a greater negative effect on the performance of hybrids than on local talls. It may be concluded that with greater cyclone frequency, the choice of coconut varieties to be planted should focus on palms with heavy stems and large boles, such as the Malayan Tall and the Panama Tall. Also, for hybrid breeding programmes, male (tall) parents should be selected from such varieties. However, trees with large bole withstand strong winds only when they are planted in soils that provide ample anchorage for the roots (not on shallow soil or peat soil).

The effect of sea winds on the first rows of coconuts growing along the shore is well known to anyone who has walked along the beach. These first rows are subject to sea sprays, especially during periods of strong wind and large waves breaking against the shore. Very fine seawater droplets are carried by the wind and deposited on the coconut leaves. As a result, the first rows of palms may have scorched leaves, caused by the salt in the seawater droplets. Remison (1988) observed that in addition to scorching of the leaves, sea sprays reduced the number of bunches and fruits per palm, especially in the first three rows of palms. Most inflorescences of palms subjected to such sea sprays were without button-nuts. There were indications of premature button fall and damage to secondary axes of the inflorescences in rows close to the sea. Weight of mature nuts was, however, not statistically different from nuts produced further inland. It was concluded that leaf desiccation and scorching may result from increased osmotic pressure caused by accumulation of salt water and subsequent adsorption of water from the cell-sap of the leaves.

Soils

The soil provides physical support to the palm, and supplies water, minerals and oxygen that are taken up by the roots. The uptake of water and minerals may be facilitated by micro-organism activity where effective micro-organisms are available. Coconut adapts well to a wide range of soil types, from pure sand to clay soils, provided sufficient water, nutrients and oxygen can be extracted. Coconut is also tolerant to a wide range of soil pH. It can be grown on a swampy soil with pH of 3.5 to alkaline soils with pH of 8.5. However, the most favourable conditions are found within the pH range of 5.5 - 7.

The best physical support the coconut can find is in a soil where the root system can grow out to its full potential and where the soil has a certain degree of firmness, providing strong anchorage for the palm. For instance, on a deep, pure organic soil, roots may grow well, but the soil during rains may provide insufficient anchorage for the palm during heavy rainstorms. The best soil type for coconut might be a sandy loam soil, with a good cation-exchange-capacity (CEC) and a soil water level at about 4 m depth. Such soils are good for many other tree crops as well.

Soil texture determines the soil water-holding capacity. Heavy soils have a higher water-holding capacity than light soils. Comparison of the performance of PB-121 hybrid in different regions of The Philippines showed that highest nut and copra yields were obtained on soils with the highest water-holding capacity (Magat *et al.* 1988b), underlining the importance of an adequate soil moisture supply for coconuts. Coconut grows well on soils with a very high water table, provided the shallow soil is rich enough for an adequate mineral supply. In swampy lands, drainage canals can be made, using the dug-out soil to raise the land between the drainage canals. Where swamp soils have a very acid sub-soil, care should be taken not to throw too much of this acid material on the land. According to the height of the water table, drainage canals can be dug between every two, three or more rows of palms.

Usually, not all good characteristics are combined in a coconut soil. Sandy soils may have a very good aeration but an excessive drainage, a low water-holding capacity and a low cation exchange capacity (CEC). But due to extensive root system development in such soils, the large volume of soil exploited by the root system may compensate for its low quality as far as water and mineral supply are concerned. In Ivory Coast, commercial plantations on very deep, sandy soils (4m and deeper) gave the best yields. These deep sandy soils enable the coconuts to gradually produce a very deep root system, increasing the amount of available water. This means that the trees are capable of withstanding dry periods and make better use of very sunny periods. The existence of a water table in certain places seems to play a role, although an ideal water table (height, amplitude of level variations, etc.) could not yet be defined (Taffin *et al.* 1991). Stones and laterite concretions may considerably reduce root development, with negative effects on mineral and water uptake. In a climate with low rainfall periods, water-stress may be greatly enhanced in such soils. Impenetrable soil layers have the same effect. Heavy clays may crack during the dry season, rupturing the roots. On hilly lands, heavy soils are more subject to erosion than light, friable soils, due to their impermeability, causing run-off of rainwater. A good internal and external drainage of the soil is very important for coconut. External drainage depends on topography, and on flat lands might require special drainage works. Drainage canals near the coast may have sluices with a hanging door, opening outwards only. Thus, at high tide, the sluice is pushed tight from the outside, and at low tide pushed open by accumulated water from the inside. Such sluices should be regularly inspected for possible accumulated debris impeding their functioning. Occasional floods will not harm coconut roots, provided the water is drained off within 48 hours. Moving soil-water with continuous oxygen supply will not affect roots, but coconut roots will die in stagnant water. When part of the roots has died off due to water-logging, this may cause physiological drought even during the rainy season, which is not uncommon in coconut gardens in India (Prasada Rao 1991). Coarse sands are excessively drained, also causing serious leaching of minerals. Such soils have a low water-holding capacity. They require special fertilizing techniques, such as incorporation of organic material and split applications of fertilizers.

Where the soil water-table is several metres deep, young palms may suffer during the dry season, and the development of these palms may be slower than usual. However, once the roots of the palms reach the water-table or the moist zone close to the water-table, growth and production will normalize. In a relatively deep soil, palms will have more soil volume from which water and minerals can be obtained, but when soils are so deep that the roots cannot reach the humid soil near the water-table, palms may suffer considerably from water-stress during the dry season. In general, tall palms may root deeper than hybrids, which in turn have a larger root system than dwarfs. On deep soils, dwarfs suffer more from water-stress than hybrids and hybrids may suffer more than talls. Among the dwarfs, the yellow and the red dwarfs suffer more from waterstress than the green and the brown dwarfs. It is likely that coconuts growing in a climate with a severe dry season may have adapted to these conditions by developing extensive root systems and leaves that have a low evaporation rate. Strong fluctuation of the water table is unfavourable, as it may cause the dying off of roots grown into deep soil layers during the dry season. According to Darbin *et al.* (1983), when selecting land for coconut growing, soils should be chosen without signs of hydromorphism in the top 50 cm, and with a water table situated at a depth of more than 3 m during the dry season.

Organic Soils

Organic soils cover large stretches in the humid tropics. Their friability and looseness pose a problem for coconuts due to lack of good anchorage. These peat soils are chemically poor. Peat is almost pure organic matter (less than 5% ash after calcination), has a very high waterholding capacity (3-11 times its dry weight) and has a very low density. The bulk density of peat, even when compacted, is 10 to 15 times lower than that of mineral soils. It is therefore a specific material requiring specific techniques (Ochs *et al.* 1992).

A deep peat soil planted with coconuts in Sumatra (Indonesia) showed severe deficiency of K and P. N levels were normal but combined with the very high C:N ratio might be deficient after drainage and the development of aerobic bacterial activity. Trace elements were also low, only the Mg level was satisfactory. This was probably due to the high Mg content of the original substrate. Mineral poverty of the peat can be partially explained by gradual dilution; the vegetation is no longer able to draw nutrients from the mineral substrate, which has become inaccessible, and draws on reserves built up originally, in a closed cycle (Bonneau *et al.* 1993b). Usually, the water-table in peat soils is close to the surface. When drainage is practised and the water-table is kept at a lower depth, serious subsidence of the topsoil and lowering of the soil surface level occurs. The water-table should be kept sufficiently low to prevent asphyxiation of the coconut root system. A mean depth of 70 cm below the surface of compacted peat appears about right, provided the imperfections of the system mean that the actual depth is between 40 and 100 cm. Experience has shown that coconut can tolerate a water-table at a depth of 40 cm, provided it is kept stable most, if not all, of the time. In Ivory Coast, coconuts suffer more from seasonal fluctuations of the water table (alternate waterlogging/drying out) than from a permanently high water-table (Ochs *et al.* 1992).

Atoll Soils

According to Trewren (1991) 'The structure of all atoll soils is based on aragonite coral with secondary calcite overlying a core of volcanic origin. The soil cover is formed largely from calcareous sand, mostly calcium and magnesium carbonates derived from the shells of marine algae and foraminifera. On the lagoon side of atolls the parent material may be almost exclusively composed of such sand, *i.e.* it is virtually non-coraline.. In marked contrast, the outer windward ramparts will comprise boulder-sized coral and coraline algal rubble thrown up from the reef by periodic storms. Between the two, the soils contain varying quantities of loose angular stones of coraline origin. Some of the poorer 'soils' consist almost entirely of this material.'

Thickness of the humus determines the suitability for coconut cultivation. The surface layer of the sands contains a very variable quantity of humus, but is usually about 25 cm deep, has a pH of 8.5 and is dark greyish-brown to black in colour. This rapidly gives way to a very coarse white and pink gravelly sand which consists almost exclusively of Ca and Mg carbonates. This subsoil layer contains no feeder roots of coconuts, although thick structural roots may be found at depths of more than one metre. Feeder roots of coconuts are generally concentrated in the top 7 cm of the soil. These soils are very highly permeable and have a low moisture-retaining capacity. Due to their highly permeable nature, potassium reserves are extremely low, whereas nitrogen reserves are proportional to the organic matter content. Phosphorus is rapidly fixed as tricalcium phosphate and insoluble compounds due to high calcium carbonate content. The levels of sulphur may be critical in some areas. Trace elements, *i.e.* iron, manganese, zinc and copper are also fixed in insoluble compounds (Mataora, 1987). By rectifying iron and other trace element deficiencies in coconuts, the need for applied nitrogen plus phosphorus is greatly reduced on normal atoll soils (Ubaitoi 1987). Humus provides colloidal binding for the formation of aggregates necessary for the fixation of mineral elements. It also has a high water-holding capacity and organic acids reduce the soil pH level. The adsorption complex is almost entirely saturated by Ca. Logically, on such soils it is very important to protect the humus layer. All plant material, such as leaves and husks should be left or returned to the plantation. On the atolls in the Pacific, yields increased two to threefold by slashing the undergrowth and using the organic material as a mulch instead of the traditional practice of burning it down periodically. Additional fertilizing will be required to maintain soil fertility.

Beach Soils

Beaches are a favoured area for coconut growing. Often, it can be observed that coconut roots are washed by seawater at high tide. But this is only temporarily. On the beach, fresh water from inland, being lighter than salt water, floats on top of the salt ground water table. It is mainly from this fresh or somewhat brackish water that the coconut gets its water supply. It may be assumed that different varieties have different salt tolerance levels. Formerly, it was generally believed that coconut required salt, as it was mostly growing along beaches and in coastal areas. The use of sea salt as a fertilizer has been practised in various countries. When more and more coconut was grown inland, the thought of salt requirement was abandoned, until Ollagnier and Ochs (1971) found that for coconut development Na is not important, but Cl is essential. A trial conducted by Remison *et al.* (1988) in Nigeria revealed that salination influenced the nutrient contents of the leaves. In this trial, coconut seedlings growing in polybags containing 45 kg of soil (polybags usually contain 15-20 kg soil), were treated with 2, 4, 6, 8, 10 and 12 g of common salt every fortnight, starting at the two-leaf stage of growth. After 12 months of growth, the Na and Cl contents of the leaves had increased appreciably while N, K, Ca and, to a lesser extent, P decreased with salinity. S decreased with salinity and Mg was not affected by salt application. N and K contents were higher in leaves of coconut palms grown in forest soil, whereas Na content was higher in those grown in maritime soil. Antagonistic effects also occurred between Na and K, and Ca and P in both soils. Salinity did not affect the ratio of monovalent to divalent cations. The ratio was generally greater in palms grown in forest soil. In a trial conducted in Indonesia, seedlings treated with doses of 30, 60, 90 and 120 g of salt, the high salinity in soil with doses over 90 g per seedling caused plasmolysis and leaves faded and dried (Darwis 1991).

Microbiological Activity

Microbiological activity in the root zone of the coconut and intercrops can be directly or indirectly beneficial to these crops. Such a symbiosis may considerably influence plant nutrition and production. In the process of root formation, plants suffer a continual loss of organic compounds via root exudates or tissue degeneration. This loss is broadly the controlling factor in micro-organism development and activities in the roots. Estimation of this loss is difficult, but is reckoned to be in the range of 2-20% of total plant dry weight. The amounts lost differ between plant species and are influenced by environmental conditions (Ikram 1990). The micro-flora influence on plant growth may be:

- the fixing of atmospheric nitrogen;
- the breakdown of minerals and organic matter;
- suppression of plant pathogens;
- influencing absorption of inorganic elements;
- production of growth regulators, toxins and enzymes;
- possible detoxification of allelopathic substances.

The plant-microbial relationship can play a very important role in crop performance, particularly where management is aimed at low input. Often the relationship is symbiotic. According to Bopaiah *et al.* (1987), root exudates may play a role in neutralizing the soil pH and altering the microclimate of the rhizosphere through liberation of water and carbon dioxide. The roots exudate large quantities of sugars and other compounds into the rhizosphere, which may directly or indirectly influence the number and quality of micro-organisms in the rhizosphere. Another important aspect of plant-microbial relationship is frequent formation of a fungal mantle around the root system of many plant species. Some of these fungi play an important role in extracting nutrients in available form from different substrata, while others confer protection to the plant from invading pathogens. Using an *in vitro* study of the antagonistic effect of *Trichoderma lignorum*, the most common fungus they isolated from the coconut root itself, it was found that this fungus partially inhibited the growth of many fungi occurring in the coconut rhizosphere and rhizoplane. The growth of *Aspergillus flavus* and *A. fumigatus* was not inhibited, indicating a certain degree of specificity in the activity of the fungus. This is of special interest, as these fungi are beneficial to coconut palm by solubilizing inorganic phosphate and producing gibberellin-like substances in the coconut rhizosphere.

Among the active micro-organisms in the soil there are five main groups that are of more or less direct importance to the palm's nutrient supply:

- nitrogen-fixing bacteria in the rhizosphere;
- nitrogen-fixing *Rhizobium* bacteria living in symbiosis with legumes;
- phosphate solubilizing bacteria;
- phosphate solubilizing fungi;
- vesicular-arbuscular mycorrhizal (VAM) association fungi that help in the absorption of phosphorus and other immobile elements.

Bacterial nitrogen fixation

Nitrogen fixation by micro-organisms is particularly important for crops growing in conditions where the greater part of the applied nitrogenous fertilizers are lost due to leaching or denitrification.

Populations of the non-symbiotic nitrogen-fixing bacteria are favoured by intercropping coconut with green manures, grasses or other crops and also by mixed cropping of coconut with other tree crops and by mixed farming. In addition to nitrogen fixation, the bacteria produce a substantial amount of polysaccharides, which may be important in soil aggregation under field conditions. There are effective and less effective strains of bacteria in the rhizosphere and roots of the coconut. Introduction of efficient cultures can therefore be important to obtaining the full benefit from the fixing potential of bacteria (Thomas *et al.* 1991).

Bacteria which reduce gaseous atmospheric nitrogen to a biologically usable form can be a source of nitrogen to the palm. These include the non-symbiotic bacteria *Beijerinckia* and *Azotobacter* and the associated bacterium *Azospirillum*. *Beijerinckia* is a non-symbiotic nitrogen-fixing bacterium capable of growing and fixing nitrogen in acidic soil conditions. Merilyn and Thomas (1992) found that the rhizosphere soil samples harboured significantly higher populations than non-rhizosphere samples. The results also showed significantly higher populations of *Beijerinckia* in sandy soils compared to laterite soils. It was also found that rhizosphere soils had a higher pH than non-rhizosphere soils. Among the six locations studied, rhizosphere samples from four locations yielded greater proportion of *Beijerinckia* in the total bacterial population than in the corresponding non-rhizosphere samples. It was suggested that root exudates of coconut may provide a better micro-environment for proliferation of *Beijerinckia* in coconut rhizosphere, or that low nitrogen environment availability in the root zone enables nitrogen-fixing bacteria to compete with the great bulk of soil micro-organisms. Another reason for higher *Beijerinckia* populations could be the favourable pH of 5.0 - 6.0 found in most of these coconut soils.

Higher populations of *Beijerinckia* were observed in coconut under basin management with green manure legumes and multi-storeyed cropping when compared to monocropping of coconut. Similar findings were mentioned by Bopaiah (1988) in coconuts grown in a mixed farming system. Multi-storeyed cropping and mixed farming systems were found to be superior to neglected conditions, farmers' gardens and plots receiving fertilizer plus manure and tillage. The cultures of *Beijerinckia* showed growth and nitrogen fixation at a wide range of pH from 3 - 10. The best range was between a pH levels of 4 - 7. *Beijerinckia* isolates from coconut-based cropping systems fixed 7.0 - 15.6 mg N per g carbon source when tested by the Kjeldahl method. The studies revealed the occurrence of a mixed population of effective and less effective strains of bacteria in the rhizosphere and roots of coconut. Isolates obtained from rhizosphere soil had better nitrogen-fixing efficiency than those isolated from non-rhizosphere soil, indicating that exudates of the coconut root stimulated the proliferation of efficient strains of *Beijerinckia* in its rhizosphere. This indicates the possibility of increasing the population of *Beijerinckia* in coconut nurseries by inoculation with efficient strains.

Azospirillum spp. have been found in association with roots of various crops, particularly grasses. They have been found to be more efficient than the non-symbiotic nitrogen-fixers. *Azospirillum* has also been found in roots of coconuts and in roots of intercrops. The principal mechanisms by which it increases crop growth are nitrogen-fixation, hormonal effects and bacterial nitrate reductase activity in roots. Its effect on coconut is not yet known. From the results of a study in various coconut-based farming systems, Ghai and Thomas (1989) concluded that the large variation in the extent of association and nitrogenase activity of isolates from the different crops indicate the need for inoculation with efficient cultures in a number of crops in a coconut-based cropping system.

Nitrogen fixation by rhizobia in symbiosis with legumes

The nitrogen-fixing potential of *Rhizobium* bacteria can be exploited by planting legumes as green manures or as cover crops. These bacteria can be of different effectiveness and some strains are more effective in one crop than in others. Inoculation of legume seed by pelleting the seed with a substance inoculated with an effective strain can substantially increase the beneficial effects of the legume cover crop and green manures. Introduced strains of rhizobia can be protected against acidic soil conditions by pelleting with such alkaline substances as lime and rock phosphate.

Phosphate-solubilizing bacteria

Bacteria are the predominant micro-organisms that contribute to the dissolution of phosphate from insoluble P sources in the soil, followed by fungi and actinomycetes. Bacteria found by Thomas and Shantaram (1986), solubilizing phosphate in coconut soils in India were *Pseudomonas* sp., *Micrococcus* sp., *Micrococcus roseus, Bacillus subtilis, Corynebacterium* sp. and *Alcaligenes* sp. The population of bacteria did not show much variation in laterite, sandy and alluvial soil types, whereas clayey soils harboured lower populations as compared to the three other soil types. The maximum P-solubility observed was due to inoculation with *B. subtilis* in a sandy soil whereas inoculation with *M. roseus* resulted in maximum increase in laterite soil. There was an increase in the available efficient P-solubilizing bacteria after the addition of rock phosphate and farmyard manure. *In vitro* estimation of the P-solubilizing ability of the isolates revealed solubilization of 19.5 - 54 per cent of the insoluble phosphates supplied in the culture broth. *M. roseus* and *B. subtilis* possessed better capacity to survive in unamended soils and have a greater potential as inoculants in crops for better utilization of insoluble P.

Phosphate-solubilizing fungi

Thomas *et al.* (1985) found that the P-solubilizing fungi in India belonged to three genera viz: *Aspergillus, Penicillium* and *Phialotubus*, the dominant ones being the former two genera. The fungi belonging to the *Aspergillus* group were more widely distributed in the coconut plantations studied and exhibited a better activity in P-solubilizing (62%) on average compared to the mean activity of isolates of *Penicillium* (49%). The laterite, alluvial and clayey soils harboured more P-solubilizing fungi than the sandy soils.

VAM association fungi

Coconut also forms associations with Vesicular-arbuscular mycorrhiza (VAM) fungi. This symbiotic association, in addition to its role in suppressing root pathogens, helps plants in the absorption of phosphorus and other mobile elements, such as sulphur, calcium, zinc and copper, particularly from low fertility soils. The VAM mycorrhiza hyphae penetrate only the root cortex cells and branch by dichotomic branching inside the cell to a structure called arbuscule. These arbuscules disappear after some time and the highly activated cortex cells return to their original metabolic level. The mycorrhizal mycelium can grow as far as 8 cm from the invaded roots, as a result substantially increasing the exploited soil volume, and increasing P uptake. The phosphate is transferred through the mycelium and via the arbuscules to the host (Blaak 1986). Mycorrhizal association would be one of the factors contributing to the establishment and survival of coconut in nutrient-poor soils and in drought conditions.

In India, a number of fungi belonging to four genera viz., *Glomus, Gigaspora, Sclerocystis*, and *Acaulospora* have been found to form mycorrhizal associations with coconut (Thomas *et al.* 1991). Also the VAM fungus identified as *Endogone fasciculata* has been found in coconut roots in various parts of Kerala, India. Tall coconuts are superior to dwarfs and hybrids in harbouring VAM in their roots. Mycorrhizal activity increases up to an optimum level of fertilizer application and gradually decreases thereafter. Concentration of nutrients in the host tissue also decides the VAM species colonizing the plant and its extent. Thus, the dose of fertilizer applied has a direct bearing on the mycorrhizal activity, and this may vary with host-fungus combinations. A good number of VAM fungi can withstand only low concentrations of soil nutrients. VAM infection is negatively related to the available soil phosphorus. Results of a trial conducted by Harikumar and Thomas (1991) showed that the colonization percentage was significantly lower in palms which received the recommended level of fertilizers compared to that of non-fertilized palms. The colonization observed under conditions of a double dose of fertilizers was significantly lower than in the single dose treatment. The spore counts in the two treatments did not differ significantly.

Iyer *et al.* (1993) assessed the changes in VAM status (percentage of root infection incidence) and the extent of root colonization in a multi-species coconut intercropping trial with reference to season and soil fertility. Crops were fertilized with one-third, two-thirds and full doses of recommended fertilizers. It was found that the peaks of infection in certain crops differed in time of the year for different fertilizer treatments. When infection-grading as an index of mycorrhizal activity was considered, pepper and coffee grown with one-third doses of fertilizer application supported the maximum activity, whereas for banana, clove, and pineapple a two-third dose was found to influence maximum activity. Coconut harboured maximum activity in both one and two-third doses of fertilizer.

Harikumar and Thomas (1991) observed that mycorrhizal colonization was significantly higher in West Coast Tall (WCT) than in two WCT x dwarf hybrids (WCT x Chowghat Orange Dwarf and COD x WCT). It was suggested that the variation in VAM colonization could arise from differences in the physiological and biochemical characteristics of the root system, which are controlled by the genetic constitution of a particular cultivar. The study also revealed a favourable influence of irrigation during summer months on VAM symbiosis in coconut. It was suggested that this may be due to the movement of the fertilizers from higher to lower soil layers. Thomas (1991) observed a higher level of mycorrhiza colonization in drought-tolerant genotypes than in drought-sensitive ones. Percentage incidence and infection grading increased with the commencement of the monsoon, declined during the heavy rain, and again picked up when the heavy rains were over. This trend was similar in both groups of palms. This observation confirms the genotypic variation, but not the role of fertilizer movement in the soil.

In a study of the effects of various weed control methods on the VAM colonization in weed plants and coconuts, Sasidharan *et al.* (1991) found that weed control in coconut plantations by cultural means generally reduced the VAM colonization. However, the mycorrhizal colonization was not reduced by chemical weed control. Incorporation of 2,4-D sodium salt stimulated VAM intensity in the root and rhizosphere soil of weeds and coconuts. Bopaiah (1990) found an activity gradient in the root zone and in inter-space soils of coconuts and areca palms. He attributed the reduction in micro-flora, enzyme activities and C and N mineralization at lower depths to acidity and lower organic C content of the soil.

Although VAM fungi do not favour high nutrient contents in the soil, there are always some remaining fungi in soils in which these higher concentrations occur. These fungi merit some further observation and their isolates may be used for inoculation in soils with higher nutrient levels.

Infected roots can be used for mycorrhiza inoculation. Small pieces of these roots can be placed in the seedbed of leguminous crops. Infected soil can be used as well. It has the advantage of a high spore concentration, resulting in good competition with existing mycorrhizas (Blaak 1986). This can be produced on about 25 m^2 of land in 4-6 months, using a commercially available inoculum as a starter. A disadvantage is the volume of soil to be displaced, 2-4 t per ha. By inoculating both, legumes and coconut with an effective mycorrhiza and by using an effective *Rhizobium* strain for the legume as well, soil minerals could be much better exploited and fertilizer costs reduced. In a long-term trial conducted by Harikumar and Thomas (1991), plots with no fertilizer yielded 35 nuts per palm only, compared to 79.4 and 86.5 nuts respectively for two fertilizer treatments. Non-fertilizer plots had a higher mycorrhizal colonization than the fertilized plots, indicating a higher level of dependence of palms on VAM in the absence of fertilizer additions. Mycorrhizal colonizations in infection grading were significantly different in the three genotypes of coconuts planted, *i.e.* West Coast Tall, Chowgat Orange Dwarf x West Coast Tall and West Coast Tall x Chowgat Orange Dwarf. WCT was superior to the two hybrids in harbouring VAM in the roots. It was mentioned that the results were in agreement with an earlier report by Thomas and Ghai (1987) on higher proportion of roots with VAM in seedlings of tall cultivars compared to the dwarfs and hybrids. Compared to fertilizer use, the effect of VAM fungi in that trial was only small.

A combination of VAM with phosphate-solubilizing bacteria, light doses of fertilizers low in P, wider spacing and the establishment of a leguminous covercrop would probably provide better results than VAM alone. More research is needed to perfect the methods that make use of soil microbiological activity to improve soil fertility and yield.

For the time being, it appears that farmers especially, those who cannot afford fertilizers, could benefit from fungi colonization in their plantations. However, such landowners usually have neither the knowledge nor the managerial skill to apply these techniques, and depend on a well functioning extension service.

4. Disorders

J.G. Ohler

Mineral Nutrient Deficiencies

Mineral nutrient deficiency symptoms are generally fairly typical for each mineral element. Recognition of the symptoms can be a handy tool in field work, especially where laboratory analyses are not readily available. However, once the symptoms appear, the deficiency is already at a rather advanced stage. Mineral deficiency symptoms tend to become enhanced during dry periods, when less water and, consequently, fewer mineral nutrients, are available to the palm. This is especially the case for potassium, but not for N and Mg (Wuidart, personal communication). Such symptoms sometimes disappear completely during the rainy season. Leaf analyses are recommended in such cases. In palms, the visual symptoms are mainly in the leaves. When the growing point of the stem is affected, this becomes visible only after the leaves have fallen off from the affected part of the stem. Sometimes deficiencies are inherent in the soil type, but more often deficiencies result from continuous cropping and poor management. Especially in the case of micro-elements, which are not available in common commercial fertilizers, recycling of these elements, through incorporation of all organic matter that is not being used for other purposes, is very important.

Diagnosis of deficiencies is not always easy, notwithstanding their typical character. Often, other factors play a role in discolouring leaves or deforming the growing point. Drought, but also inadequate drainage, may be such factors. Sometimes more than one mineral element is deficient, resulting in a general yellowing of the leaves without a typical pattern. Also, differences in original leaf colour may result in different hues of discolouration caused by the deficiency of the same element.

Diseases or pest attacks may cause leaf discolouration that might at first sight look like a mineral deficiency. Not all mineral deficiencies cause a typical discolouration or other symptom. Mineral deficiencies may also enhance disease or pest attacks, which may blur the deficiency symptoms. On the other hand, diseases may also cause nutritional disturbances, showing unbalanced mineral relations. Such cases where diseases are involved with nutritional disturbances are dealt with under the discussion of these diseases. In all cases, leaf analysis is the best method to determine the palm's nutritional status.

When fertilizer applications of so-called macro-elements are changed or increased, a check by leaf analysis also to verify the nutrient balance of the other elements is very useful. Fertilizer applications not only bring about changes in the plant's mineral balance, they may also affect the soil and cause important changes in the soil pH and/or mineral balance which, in its turn, may affect the palm's growing conditions. In the following description of the symptoms, the elements are grouped, as far as possible, according to some general characteristics, facilitating the diagnosis of symptoms observed in the field.

Symptoms Visible on all Leaves

Nitrogen deficiency

Nitrogen deficiency (CP 39) may occur in all soils. However, some soils are more prone to the deficiency than others. Light sandy soils, poor in organic matter, are most likely to be poor in nitrogen. Such soils are often planted with coconut. Also, corraline limestone and hydromorphic soils are often poor in nitrogen. The nitrogen supply of the soil can be greatly affected by the plantation management. Continuous grazing under coconuts without additional nitrogen supply may cause serious nitrogen deficiency. Grasses are heavy feeders on nitrogen. It is often thought that the grazing animals by their droppings enrich the soil with nitrogen, not realizing that all nitrogen excreted comes from the same soil to where it is returned. However, an important part of the consumed nitrogen is taken up in the animal's body and exported when the animal is sold or slaughtered. Another part of the nitrogen is lost from the droppings by evaporation. Clean weeding stimulates nitrification processes and may destroy organic matter, the formed nitrates being leached out by rainwater. Bad drainage and low soil pH may also lead to nitrogen deficiency. Poorly distributed rainfall, slowing down bacterial activity may be another cause of nitrogen deficiency.

Nitrogen is indispensable as a constituent of amino-acids, proteins and nucleic acids. Shortage of nitrogen makes itself felt throughout the coconut's physiology and provokes a substantial yield decline (Manciot *et al.* 1980). Nitrogen deficiency reduces the chlorophyll content of leaves. In the early stage of the deficiency, the crown of the palm loses its glossy appearance and turns pale green, followed by yellowing of the leaves. Certain compounds of nitrogen are very mobile in plants, enabling the plant to mobilize the element to vital growing points. Such a transference usually occurs under conditions of deficiency, and moves from older leaves to young growing points. Thus, in the more advanced stages, the older leaves may turn from golden yellow near the petiole to orange-yellow, and light brown towards the end of the leaves. Shortly afterwards, the leaf dries out completely. Pomier and Benard (1988) consider the deficiency severe when 10 per cent of the palms have developed such symptoms. Yellowing of the younger leaves intensifies. Yellowing of the leaf follows a typical pattern. It starts at the tip of the leaf and the leaflets, progressing inwards along the mid-ribs, the margins remaining pale green. There is no sharp division between the yellow and the pale green parts of the leaflets. New leaves will not grow out to their full length and older leaves will drop early, reducing the number of leaves in the crown and giving the crown an umbrella-like shape. Young nuts may drop prematurely and the number of female flowers per inflorescence is reduced. Yields will gradually reduce. In severe cases, whole bunches may abort. According to Magat (1993a), shortage of proteins as a result of N deficiency increases the C/N ratio, resulting in excessive carbohydrates and increased cellulose and lignin content. Such conditions thicken the cell membrane and increase lignified tissues. Thus, coconuts suffering from N deficiency look abnormal, dry and non-succulent. Root growth is also affected by nitrogen deficiency, seriously increasing the effects of water-stress in such affected palms during dry periods. In young seedlings, growth may be seriously affected, showing significant retardation in height and girth development. Pinnating of the leaves may also be retarded. The stem may develop a so-called 'pencil-point' but this is not characteristic of nitrogen deficiency only. The deficiency can be corrected either by fertilizers or a the planting of a legume green manure, the latter method to be preferred when there is a lack of funds to buy fertilizers, or where the soil does not retain the fertilizers well. The changes in leaf colour become visible within one year if ample fertilizer has been applied. Pomier and Benard (1988) recommends drainage of waterlogged soils and reduction of grass growth before applying nitrogen fertilizers. Recycling of organic waste into the plantation is also important. When using chemical fertilizers, ammonium sulphate is to be preferred in areas where sulphur deficiency may occur. Ammonium sulphate should be dug in slightly, to reduce nitrogen evaporation. When high doses of nitrogen are used, the coconut leaves may turn dark green within a period of a few months.

Sulphur deficiency

Sulphur deficiency in coconuts has been observed in the Pacific islands, Papua New Guinea, the Comoro Islands, Madagascar, on savannah soils in Ivory Coast, and on the coastal plains of Mozambique. Sulphur in plants occurs as a constituent of proteins and is involved in chlorophyll formation. It is essential in the amino acids system and methianine that are part of many enzymes (Blaak personal communication). The greatest proportion of assimilated sulphur is exported by the endosperm. It is immobile in the plant. According to Taysum (1981), it has often been stated that symptoms of S deficiency are also observed on coconuts growing in swampy conditions or where there is a high water-table for long periods of the year, the effect being particularly strong where the water is slightly brackish. Under such conditions, sulphur and sulphate ions are often reduced to sulphite, thiosulphate or sulphide, which cannot be assimilated by the plant or are toxic to it. Even if sulphur is present in the soil in organic form, it would seem to be easily leached during mineralization into sulphate in humid tropical soils, and natural restitution by hidden supplies is not clearly understood (Wuidart 1994a).

In sulphur-deficient palms, all leaves tend to become chlorotic, even the young ones (CP 29 and 34). Velasco (1960) observed that mature leaves progressively lost their green colour from the leaflet margins towards the mid-rib, so that the green seems to form a haze on a yellow background. The rachis became weak and pliant. Manciot *et al.* (1980) observed that leaves turned yellow and orange on young sulphur-deficient palms, becoming necrotic and grey at the tips. On a new plantation in Ivory Coast, Ollagnier and Ochs (1972) observed that due to sulphur deficiency, the growth of young palms stopped almost entirely; the length of the leaves emitted since field planting was half of that of the leaves already present in the nursery, with abnormally early separation of the leaflets. The colour of these leaves ranged from pale yellow to bright red-bronze, and the tips became grey, necrosed and curled. This necrosis spread rapidly over the entire leaf, which dried out completely.

In adult palms, deficiency symptoms first occur on the oldest leaves, rapidly affecting the entire foliage (Wuidart 1994a). The total number of living leaves is strongly reduced and leaves are yellowish green. According to Southern (1967a), sulphur-deficient palms may be distinguished from nitrogen-deficient palms in that the young leaves as well as old leaves are discoloured. Sulphur-deficient palms also have a marked tendency to retain their dead leaves. There is a premature bending of the leaves above the normal abscission layer and, consequently, a considerable number of dead leaves are hanging along the stem.

Nut production is severely reduced, the nuts being small, producing a poor copra quality. The endosperm is of normal thickness, but on drying it collapses to a thin rubbery copra. This copra is difficult to break and may clog machines. Oil contents as low as 36 per cent on a dry basis have been found for extremely rubbery copra. The copra has a capacity for absorbing moisture more readily than normal copra, leading to rapid deterioration. Higher sugar contents are connected with lower oil contents, pointing to a breakdown in fat synthesis in maturing nuts.

The deficiency can be corrected by using fertilizers in their sulphate form, such as ammonium sulphate, and potassium sulphate. The most frequently recommended fertilizer will be ammonium sulphate, which raises N and S levels at the same time. (Wuidart 1994a). Within six months after sulphur application the foliage has turned green and the copra quality has improved.

Phosphorus deficiency

Phosphorus deficiency in coconut palms is rare and very difficult to recognize as it shows hardly any visible symptoms. Lateritic soils are prone to phosphorus deficiency, and so are poor sandy soils. Phosphorus is an important constituent of nucleic acids and it is abundant in young tissues of the plant. It has a great influence on initial flowering of coconuts. It plays an important role in the efficient functioning of nitrogen. Strong interactions between these two elements often occur in crops. The ratio between N, P, and K is very important. Application of potassium or magnesium fertilizer may cause a drop in leaf phosphorus. Nitrogen and phosphorus have a strong interaction, especially in young palms, and a shortage of one element may also cause a drop in the level of the other element. Application of K may cause a drop in P levels. At maturity, nitrogen fertilizing may raise the leaf the phosphorus levels, but during immaturity nitrogen applications have practically no effect. Khan *et al.* (1985a) found that phosphorus contents of younger leaves were always higher when compared to older leaves. P contents were lower in summer, irrespective whether the palms had been fertilized or not, indicating a minimal flux during the summer period.

According to Manciot *et al.* (1980) there are no particularly characteristic visual symptoms apart from slowing down of growth and shortening of the leaves. Only in severe cases may leaves turn yellow before dying prematurely. In young seedlings, leaves with shorter petioles and rosetting of the leaves may occur and growth may be restricted, although this symptom may become evident much later than in nitrogen-deficient palms. Severe drying of old leaves in seedlings may also occur. Phosphorus deficiency retards root growth and delays flowering and also the ripening of nuts. In deficiency situations, P fertilizers have been found to increase the girth at the collar, the number of leaves and rate of leaf production in seedlings. (Thampan 1981)

According to Wahid (1984) in many experiments P levels of none of the leaf ranks correlated significantly with yield in coconut. Failure to obtain yield responses to phosphatic fertilization, and the absence of adverse effects on yield upon discontinuation of P application for 5-10 years, casts doubt on the direct role of this major nutrient in coconut production, notwithstanding the two so-called exceptional cases, one in Ivory Coast and one in Sri Lanka, where positive correlations for leaf P with yield were obtained. The suggestion that P does not play a direct role in coconut yield may be questioned. Phosphorus is an important element for almost all crops, and for coconuts, although in lower quantities than other major elements. The lack of effect in many fertilizer trials may have had several causes that have nothing to do with its function in the coconut palm itself. For instance, the soil may have become saturated with phosphoric acid after many years of application, or soils may neutralize its effect by fixation. In addition, many more trials than only the two trials mentioned have shown its effect on coconut yields. (De Geus 1973). According to Wuidart (personal communication) it was observed in a trial conducted in the State of Ceara, in Brazil, that the ratio between N and P was important, but the ratio between K and P was of very little importance. Palms that did not receive P showed weak development: they had a poorly developed bole, a thin stem and they showed retarded growth, leaves were short and they yielded very little or nothing at all. Apart from their slow development, no visual symptom was observed on the foliage of these palms that could be referred to P deficiency.

Symptoms Usually Occurring on Older, Lower Leaves

Potassium deficiency

Potassium deficiency may occur on light sandy and lateritic soils with low cation exchange capacity (C.E.C). It also occurs on soils with a high Ca/K ratio. The deficiency is often a result of intercropping without fertilizer use, with crops needing high level of potassium, such as as cassava. This element is very much subject to leaching. An antagonism exists between potassium, calcium and magnesium. Potassium in fairly large quantities is present in all parts of the plant, particularly in leaves and growing points. In mature coconut palms the largest quantity may be found in the stem. According to Ollivier (1993b), K plays an important role in coconut physiology, as it is involved in the metabolism, the acceleration of stomatal movement, enzyme activation, metabolite transfer and cellular division. In coconuts, the husk of the nuts contains a considerable quantity of potassium. Luxury absorption of potassium often takes place. The role of potassium in stomata functioning was discussed in paragraph 3.1.4.2. Potassium also influences the number of inflorescences produced, the number of female flowers, fruit set and the number of nuts produced. All these factors do not need to be directly influenced but may be just the result of the improvement of the palm's condition when supplied with potassium in case of a deficiency. Potassium is very mobile within plant tissues. Potassium and magnesium in older leaves may be translocated to younger leaves in case of deficiency. As a result, the older leaves wither and turn brown earlier than usual. Trees with green leaves hanging down do not suffer from K or Mg deficiency. Potassium may increase the plant's resistance to fungus diseases and pest attack. Potassium deficiency symptoms rarely appear during the first year of growth. It was first assumed that this was due to the high potassium content of the coconut. However, Nathanael (1961) observed that even seedlings amputated from the nut growing in pot culture with K-deficient solution did not show deficiency symptoms. Deficiency symptoms are not clearly visible until leaf potassium contents drop below 0.5%, *i.e.* once trees are already suffering from severe deficiency (Ollivier 1993b).

The first symptoms of potassium deficiency (CP 35 and 36) begin in the older functional leaves. They are characterized by yellowing of the leaflets, followed by necrosis. Yellowing starts at the tip of the leaflets, progressing along the margins towards the base. This characteristic distinguishes it from nitrogen deficiency, where the yellowing is more pronounced along the mid-rib. In the case of potassium deficiency, the yellow colour usually has an orange tinge, in this respect also differing from nitrogen deficiency. The tip of the leaflet withers and becomes necrotic, necrotic spots also appearing in the yellow part of the leaflet, mainly along the margins. In severe cases, the necrotic spots coalesce, giving leaves a scorched appearance. In advanced stages only a strip along the mid-rib remains green, widening towards the base of the leaflet. The same pattern can be observed in the entire leaf. The apex of the leaf yellowing first and the yellowing progressing along the margins of the leaf, because the lower leaflets start yellowing later than the leaflets at the tip of the leaf. When holding such a leaf against the light, a green triangle in the leaf with its base in the lowest leaflets, narrowing towards the tip of the leaf may be observed. The demarcation between the yellow and green areas are more sharply defined than in nitrogen-deficient palms. The oldest leaves gradually wither and drop earlier than normal, the middle leaves start yellowing, the younger leaves remaining green. The numbers of infloresences and nuts produced gradually decrease, while the copra content of the nuts is also affected. The growth of the palm slows down, the trunk narrows and the internodes become shorter. In addition to a deficiency correction by fertilizers, recycling of husks into the plantation can greatly contribute to improving the soil potassium content.

Magnesium deficiency

The occurrence of magnesium in soils is fairly similar to that of calcium. It occurs as a carbonate and in a variety of minerals. Like calcium, it is readily brought into the soil solution from the carbonate, and it is held in soils as an exchangeable base. It is very easily leached (Wallace *et al.* 1961). Magnesium deficiency occurs on light sandy and acid, leached laterite soils, with exchangeable Mg in meq/100 g below 0.40. The deficiency may be accentuated by excess acidity and continuous use of ammonium fertilizers. It may also result from heavy potassium fertilizing as is often practised in hybrid coconut plantations, especially seed gardens, requiring such treatments to obtain high yields. Dwarf coconuts are more sensitive to magnesium deficiency than talls and hybrids.

Magnesium plays a vital role as a constituent of the chlorophyll, and deficiency leads to a loss of chlorophylls. It is an activator of many enzyme reactions, acting as a carrier for phosphorus. It is of special importance to the formation of seeds with a high oil content. It is very mobile in the plant, being readily transferred from older to younger leaves. According to Ollivier (1993a), there is a correlation between magnesium levels and the number of green leaves in the crown, once any potassium deficiency has been corrected. As far as production is concerned, magnesium affects only the number of nuts per tree.

The first deficiency symptom is the intervascular yellowing of older leaves. The petiole of the leaf and the mid-ribs of the leaflets remain green. In the early stages a small margin on either side of the mid-rib of the leaflets also remains green. Yellowing starts at the tip and spreads to the base. Gradually, the yellowing proceeds to younger leaves and older leaves turn to a yellow orange and even to an orange colour. The youngest leaves remain green. Mg-deficient leaves are more sensitive to sunlight. The yellowing occurs principally in those parts of the leaves exposed to sunlight, the shaded parts rarely showing chlorosis. This shade effect is more accentuated in red and yellow Dwarfs than in talls or hybrids. Localization and the typical colour clearly distinguishes magnesium deficiency from nitrogen and potassium deficiency. Old leaves may assume a bronzed and dry appearance. In severe cases, the extremities of leaflets show necrosis and turn to a characteristic reddish-brown. Translucent spots may appear on the yellow parts of the leaf. Young palms may have a higher density of tender primary roots, a possible compensatory mechanism for the absorption of more magnesium. Silva *et al.* (1973) observed a magnesium and phosphorus deficiency in the leaves of these palms. Viewed in relation to the observed accumulation of phosphorus in the roots, this seemed to indicate that translocation of phosphorus from root to shoot can be adversely affected by a low magnesium supply. Delayed flowering was thus attributed to induced lowering of phosphorus in the leaves. Whereas potassium influences the number of nuts per tree and copra per nut, magnesium affects only the number of nuts per tree. The deficiency can be corrected by occasional use of magnesium limestone or by dressings of magnesium sulphate. It is very important to keep the right balance between Mg and K supply. Magnesium deficiency can be corrected by application of kieserite.

Chlorine deficiency
Chlorine ions are not held by soil colloids and therefore easily leache out from soils. In high rainfall areas the chlorine content of all well drained soils is therefore very low. Chlorine is also one of the first elements to be removed in the process of weathering and this is the reason why most of the world's chlorine is found either in oceans or in salt deposits (Uexkull 1985). Cl deficiency is rather common in The Philippines. Uexkull (1985) was of the opinion that for palms grown in inland areas Cl- is probably the most common yield-limiting element.

The importance of chlorine in the nutrition of coconuts was first reported by Ollagnier and Ochs in 1972 and confirmed by Uexkull in The Philippines in 1972 and by Daniel and Manciot in the New Hebrides in 1973. Magat and Oguis (1979) observed that Cl application to young palms significantly increased the number of leaves in these palms. Cl can hardly be called a micro-element, as the quantities taken up by some plants may be of the same order as those for nitrogen, phosphorus and potassium. Chlorine plays an important role in the stomatal conduction of coconut. Studies by Ollagnier *et al.* (1983) and Braconnier and d'Auzac (1989) in Ivory Coast, 200 km from the sea, showed that chlorine and potassium had a very significant effect on the number of green leaves per palm. Cl-deficient palms had many dry and broken leaves. Chlorine-deficient palms were also much more susceptible to fungus diseases. There was no significant correlation between K content and production, but production was closely correlated with Cl content. Cl deficiency leads to early stomatal closing during the day, and in the dry season to reduce water losses by transpiration. It also leads to the disadvantage of reduced photosynthesis, hence lower dry matter production. This balance between transpiration and photosynthesis is the very factor a plant has to optimize to be drought-tolerant. In this case, apparently early stomatal closing was a handicap. Leaf Cl also enabled the coconut to achieve more negative water potentials, thereby ensuring a sufficient water-drawing capacity to maintain tissue turgidity. Together with the opening of the stomata for a longer time, these actions provide for optimization of the balance between transpiration and assimilation. It was concluded that in coconut Cl appears to be one of the major drought tolerance factors.

Chlorine deficiency is characterized by yellowing and/or orange mottling of the older leaves, with a drying-up of the outer edges and the tips of the leaflets - a symptom very similar to that of potassium deficiency. The leaves are droopier and the leaflets may be folded slightly, similar to palms suffering from drought. The palms' leaf production is smaller and has a lower number of leaves growing slower than normal. Chlorine deficiency seriously reduces the number of nuts produced as well as the size of the nuts and the copra per nut. This finding is in accordance with the observation that in general nut sizes are larger along the seashore, tending to decrease with increasing distance from the sea. (Uexkull 1972).

According to Uexkull (1985), coconut trees low in Cl^- (below 0.25% Cl in dry matter) have been associated with:

- a reduced growth rate;
- a reduced number of nuts set;
- a greatly reduced nut size;
- a reduced nitrogen content in the foliage;
- droopy leaves and signs of moisture stress during mid-days;
- severe moisture stress resulting in frond breakage if plants go from a wet phase with fast growth into a pronounced dry season;
- stem cracking and frequent occurrence of stem bleeding when palms go from a pronounced wet to a pronounced dry season;
- a high incidence of leaf diseases, especially Grey Leaf Blight *(Pestalozzia palmarum)*.

However, the visual symptoms as presented by Uexkull have never been confirmed.

Symptoms Usually Occurring on Young Leaves

Calcium deficiency

Calcium deficiency, in addition to alkali and saline soils with high sodium content, is confined to acid soils. These soils are often derived from basic material low in calcium, such as silicious sandstone and acid igneous rock. Light sandy soils are frequently acid and deficient in calcium due to an initial low calcium content and low cation exchange capacity. On the other hand, coconut can adapt to a wide range of leaf calcium content of 0.20-0.50 and over, without adverse consequences.

Calcium is a constituent of the cell walls. The element is involved in enzymatic actions in growing points, being of special importance to root development. It does not seem to move freely in the plant, this accounting for younger tissues containing lower proportions of calcium than older ones.

Velasco (1960) observed that calcium-deficient seedlings were tall and slender with sparse leaves taking on a 'tucked in' appearance. The petioles of these leaves were of a deep yellow or orange colour, orange blotches along the mid-rib occurring frequently. The leaflets showed angular patches of orange-yellow, before withering and dying. Calcium deficiency seems to appear especially when palms are very young. The symptoms in the seed garden gradually disappeared after yearly applications of tricalcium phosphate. At 5 years of age, the symptoms had disappeared altogether but the young palms were heterogenous and poorly developed, with slender stems. In a trial in Ivory Coast, Dufour *et al.* (1984) observed calcium deficiency in coconut seedlings growing on soils with low calcium content. The first symptoms were visible on leaves no 1, 2, 3, and sometimes 4, and even on the spears. They consisted of rounded yellow spots 3-8 mm in diameter, becoming brown in the centre. These spots were isolated at first, then widened and became coalescent, and finally dried out. On young leaflets these symptoms were fairly evenly distributed over the leaflets, whereas on leaves above no 4 the symptoms were localized at the base of the leaf. The first 2 or 3 leaves of a young coconut displaying these symptoms were generally yellow, followed by 2 or 3 dried-up leaves, and 5 or 6 leaves still green, but spotted with characteristic yellow patches. Finally, the outer side of the petiole of leaves of order 4 or 5 displayed coalescent brown patches, more or less in alignment. In addition to the easily visible symptoms on the leaves abnormalities occurred in the development of the root system for all coconut varieties. Root systems had a low number of primary roots, which did not exceed 25 cm in length two years after planting and most of them were rotten at the end. The same was true for secondary and tertiary roots. On certain primary roots a succession of swellings was observed, each corresponding to the destruction of tissues, followed by the emission of one or several roots on the side. Magnesium and potassium fertilizers significantly increased these symptoms, whereas phosphocalcium in particular, and also nitrogen, significantly decreased them, but only the effect of calcium phosphate remained significant.

When the symptoms appeared, calcium contents in the leaves were extremely low, potassium levels were high and magnesium was low. When drying-out symptoms appeared, calcium contents in leaf 4 at 20 months, were less than 0.150 per cent. Symptoms may become very serious at levels lower than 0.085 per cent. The symptoms appeared when the mean Ca^{2+} content of the surface layer of the soil (0-30 cm) was below 0.15 meq/100 g.

Iron deficiency

Iron deficiency is found mainly in soils with a high lime content, such as atoll soils. Excessive levels of calcium block iron, which becomes difficult for coconut roots to assimilate (Wuidart 1994b). It may even occur in soils containing a substantial amount of iron. It is not common on acid soils, but it also occurs in peat soils. When phosphate fertilizers are applied, insoluble iron-phosphate is formed. Iron deficiency in coconuts was well known to the local population of the Polynesian atolls, who, before starting a new plantation, used old tins in the plant hole (Eschbach and Manicot 1981). High concentrations of manganese, zinc, copper cobalt, nickel, chromium and phosphorus, as well as deficiencies of calcium, magnesium and potassium in the nutrient solution may also cause iron deficiency. Availability of iron is gradually reduced by higher pH values and under neutral or alkaline conditions phosphorus may play an important role in preventing the absorption of iron (Wallace *et al.* 1961).

Iron is closely connected with chlorophyll formation but is not a constituent of chlorophyll. Its role seems to be that of a catalyst. It is a constituent of enzymes connected with respiration and oxidation systems. It is relatively immobile in plant tissues. Potassium may contribute to the mobility of iron, thereby alleviating iron deficiency and vice versa. Southern and Dick (1967) observed that the iron content of the leaflets was highest in the tips and lowest in the base portions. Where iron was available, the iron content of the leaves increased with age but where iron was deficient, the iron content hardly varied with leaf position.

Manciot *et al.* (1980) described the symptoms as a general chlorosis, with all leaves discolouring to pale green or dark yellow. Iron sulphate at rates of 5 to 10 g per plant applied in the husk had a striking effect on re-greening of seedlings and increased their weight considerably. Ochs and Bonneau (1988) observed iron deficiency on a peat soil in Indonesia. The deficiency appeared very early, at the nursery stage, and occurred during the first year after planting, then diminished naturally. Leaf analyses showed that without iron fertilization iron contents of leaf 4 varied between 30 and 50 ppm in the first planting year. They describe the symptoms as follows: 'Gradual yellowing of the leaflet, in longitudinal strips parallel to the veins. In its advanced stage the leaf becomes completely yellow. The rachis and leaflets become shorter, giving the coconut a runty appearance. There is no necrosis on any part of the leaflet. At first sight, iron deficiency symptoms are similar to those of nitrogen deficiency. However, in the case of nitrogen deficiency, yellowing is uniform, whereas with iron deficiency there is a discolouration in strips: yellowing becomes total only in a very advanced stage.' The deficiency can be corrected with iron sulphate. Each year, up to 2 years in young coconut palms 10 g. can be injected into the husk of the seed nut. In adult palms, 200 to 400 g can be injected into the stem each year. Within 2 to 3 months after this treatment, the spear turns green again, followed by a general improvement in the foliage as a whole as the old leaves, which do not react to iron applications, die off (Wuidart 1994b).

Manganese deficiency

Like iron, manganese is a rather immobile element. Deficiency may occur on alkaline soils. There are not many references to Mn nutrition or to malnutrition of coconut. According to Manciot *et al.* (1980) manganese, like the other metallic elements, iron, zinc and copper, is blocked by and rendered non-assimilable by coconut roots in soils very rich in calcium carbonate. The action of manganese is very subtle and much less than that of iron. Manganese sulphate had no action in the absence of Fe fertilization. For manganese, the lowest contents are on soils of coral origin and the highest are on the quaternary sands in Africa.

Copper deficiency

Copper deficiency occurs on minerally poor acid sand and gravels and on peat soils, but also on highly alkaline calcareous sands. Some of the soils with high organic matter content appear to fix large quantities of added copper salts in forms unavailable to the plants. Liming also reduces the availability of copper in deficient soils (Wallace *et al.* 1961). Southern and Dick (1967) observed Cu deficiency in palms growing on alkaline soils. They found that copper contents of the first leaves were the highest and the values decreased with age, so that the oldest leaves contained one-third of the young leaf copper content. Copper injections were not successful, which could have been caused by incompatibility with other ions.

Copper deficiency (CP 37 and 38) was observed by Ochs and Bonneau (1988) on a peat soil in East Sumatra, Indonesia. The deficiency occurred 8-10 months after planting. The copper content of leaf 4 was very low: around 2 ppm. The symptoms were very characteristic and have been called 'peripheral leaf desiccation'. They involve severe bending of the rachis of the youngest leaves, accompanied by yellowing and desiccation of the leaf tip, which appears to be rimmed with brown and yellow, whilst the central part remains green. The graduated colouring from green through yellow to brown is very characteristic. As the symptoms develop, the dried-out part spreads and new, much shorter and deformed leaves emerge, giving the coconut a runty and sagging appearance. In the terminal phase, hardly any green tissue remains. Ochs *et al.* (1993a) reported that on peat soil, the application of copper sulphate granules to the soil surface in a circle around the root bulb was effective, whereas applications in liquid form into leaf axils and through root uptake did not give significant results. The first leaf produced by affected trees treated with copper sulphate granules is virtually normal. The coconuts either recovered completely or not at all. The palms that did not recover had apparently reached an irreversible deficiency stage. At the start of the trial, although all coconuts were highly deficient, they had relatively high Cu contents; 3-3.2 ppm. Six months after copper sulphate application, the Cu content fell from 3 to 1 ppm, with a more substantial drop for the control coconuts. This was probably due to a dilution effect on the amount of copper taken up in the mass of newly formed plant matter. It was assumed that the growth abnormalities start out in the terminal bud in the leaf primordia, which lose their normal development capability for good. The malformations become visible when these leaf primordia become young leaves some six months later. In the meantime, they have received a certain amount of copper, although less than normal, sufficient to bring their content up to 3 ppm given the limited development of the plant mass in which it is diluted.

Boron deficiency

Boron deficiency, also called Crown Rot disease or Crown Choke disease, is not uncommon in coconut palms. It occurs more frequently on coarse textured leached sands than on heavy soils. Soils with a high Ca content may also be boron-deficient, and liming of acid soils may decrease boron availability. Adding potassium may increase boron availability due to the formation of the highly soluble potassium tetraborate, but this does not happen on soils with a high pH value. Soils derived from marine sediments usually contain good supplies of available boron. Borates move freely in soils, and dressings are thus quickly effective but not lasting.

Boron deficiency is generally observed on small groups or on individual palms only. These palms stand side by side with healthy palms. This may be due to local differences in the soil, individual differences in sensitivity to boron deficiency, or individual differences in root system development, the palms with a more extensive root system being able to obtain a sufficient supply whilst those with a smaller root system suffer from deficiency.

In the plants, boron has several functions such as enhancing tissue respiration, and translocating sugars. It is also involved in the reproduction of plants, the germination of pollen, and the water relation in cells, etc. Due to its important role in plant life, deficiency may cause a sudden collapse of the growth processes and drastic de-arrangements of metabolism. In particular, the meristematic tissue is seriously affected by boron deficiency. As the element is only slightly mobile in the plant, deficiency symptoms first appear in the growing points of the palms.

Studies in Assam, India (Baranwal *et al.* 1989), indicated that boron content in diseased palms was 5.4 ppm, and 7.4 ppm in healthy palms. The deficient palms had higher contents of nitrogen, phosphorus, calcium, magnesium, zinc and manganese, and lower contents of boron, copper and potassium. The manganese content was higher in deficient palms, although the difference was not significant. Analyses of the soils from healthy and deficient palms showed a higher content of all nutritional elements in the soils in which the healthy palms were growing, significantly in the cases of P, K, Cu, Zn, Mn and Fe, but not for B. Baranwal *et al.* (1989) emphasize the relation between Ca and B. The Ca/B ratio in the leaves was significantly lower for healthy plants compared to deficient plants. It seems as if certain balances between nutrients in the palms are of particular importance in the case of boron. Southern and Dick (1967) observed a sharp decrease in boron content from the tip of the leaflets to their base. Injected boron was translocated only to certain leaf positions, with peaks in leaves 3-4 and 8-10.

The first symptoms of B deficiency are the reduction of the elongation of young leaves (CP 32). The leaflets, when unfolding, are crinkled and shorter than normal, sometimes showing yellow or yellowish-orange discoloration. In more advanced stages, round white transparent spots, a few mm in diameter, may appear on these leaflets. Terminal leaflets remain fused. The tips of these leaflets may be 'bayonet-shaped'. This symptom is also called 'hook-leaf', and it is caused by difficulties in unfolding the spear while the growth of the leaf pushes it upwards. The basal part of the petiole may be without leaflets. Gradually, the spear shortens and the leaves are reduced to an embryonic mid-rib without leaflets. This is the so-called 'little-leaf' stage. In the ultimate stage, the palm has only a central bulbous bud that does not develop any more, surrounded by some petioles without leaflets, and finally, the palm dies. However, palms seldom die from B deficiency. Drought may aggravate boron deficiency and in some cases one can observe seasonal boron deficiency, the symptoms appearing in the dry season and disappearing in the wet season, giving the crown an appearance of tufts of leaves separated by parts of stem with very short or no leaves (CP 31).

Observations indicate that the critical stage at which the palms will not respond to treatment is when the leaves are withered and have a severely stunted apical leaf, crinkled leaves, and when some leaves lack leaflets. Rethinam *et al.* (1990) observed that the stem does not taper below the crown. The death of the affected palm is not sudden, but it slowly loses vitality and finally succumbs within 3-4 years. Roots of affected palms remain healthy and normal. They also observed that mealy bugs are often associated with the suffering palms. They mention that Chakra Borthy *et al.* (1970) reported that surgery of the sheath is necessary to release the pressure on crowded leaves within the spear. Depending on the severity of the disorder, one or two leaf sheaths are to be incised longitudinally.

Zinc deficiency

No descriptions are available on zinc deficiency. Bonneau *et al.* (1993b) applied zinc sulphate (22%) to coconut growing on peat soil in Indonesia. Although significant quantities of zinc were absorbed by the coconuts, it had no visible effect. They concluded that the critical level of zinc on peat soil can be considered less than 10 ppm, which is much lower than the optimum level for mineral soils, estimated at 15 ppm.

Mineral Toxicities

The occurrence of mineral toxicities in coconut is very rare. When they occur, they are usually the result of human errors in the doses of mineral salts applied to correct certain deficiencies.

Aluminium Toxicity

Aluminium toxicity in coconuts has not yet been observed in natural conditions. It has been provoked in coconut by applying doses of alum to the trees. Mercado and Velasco (1961) did so with coconut seedlings. For six months they applied doses of 40 and 80 grams of alum at weekly intervals after which these treatments were continued for six weeks at two-weekly intervals. From then on 100 and 200 ppm Al as aluminium sulphate were used instead of alum. After three months the seedlings at the highest concentration level had a higher ratio of rootlets to primary root than plants in the other treatments. After eight months plants in both low and high concentration treatments had rootlets which were stiff, seemingly dry and with numerous dried branches, whereas rootlets of control plants were whitish, fresh and sparingly branched. They suggested that the high concentrations of aluminium kills the cells of the absorption region, and as this region dies, a branch root is produced just behind the killed root. Production of branch roots to replace the dead roots, however, cannot keep pace with the rate at which the other roots die, explaining the rather large number of dead rootlets. Eschbach and Manciot (1981) added aluminium at various concentrations of up to 200 per cent of the total amount of K, Ca, Na, and Mg together to nutrient solutions in which coconut seedlings were grown. At the end of the two-month trial, no sign of toxicity had been observed yet, although analyses of the plant material showed a marked accumulation of aluminium in these roots. In the other organs of the plant, the levels of aluminium were somewhat modified, indicating that the element had not migrated in these plants.

Boron Toxicity

In Davao, Philippines, MYD x WAT hybrid seedlings were treated with increasing doses of borax of 1, 2, 4 and 8 grams of borax (36.5% B_2CO_3) per seedling. First application of 1/6th of the doses was applied one month after polybagging, 2/6ths at three months and 3/6ths at seven months. These applications did not influence seedling development, showing there was no deficiency. However, two weeks after the first application tip burning of the leaflets started in the 2, 4 and 6 g treatments. No symptoms appeared on the seedlings in the lowest treatment, not even after the second application. Two weeks after the last application, burning of leaflet tips increased significantly, affecting 88, 94 and 95 per cent of the leaves in the highest treatments respectively. Phosphorus uptake was significantly enhanced by the application of 2 g of borax. Beyond this rate, a slight reduction in P concentration in the leaves was noted. This indicates some sort of positive interaction between B and P up to a certain concentration. Except for phosphorus, no other elements were affected by these borax treatments. The critical level of boron in leaves (leaf rank 3) appeared to be somewhere between 13 and 14 ppm. (Margate *et al.* 1979b).

In the Ivory Coast, where boron deficiency had been observed in coconuts, preventive application of borax to the plant hole and surface dressings, six months after planting, sometimes provoked boron toxicity symptoms in seedlings. The first symptoms appeared about ten days after treatment. Leaf extremities turned to a reddish colour, becoming necrotic within a few days. This effect was called 'leaf-tip burn'. The necrotic area rarely surpassed a width of more than a few cm. New leaves usually looked normal. Red Malayan Dwarf coconuts were more sensitive than talls, hybrids showed an intermediate reaction (Brunin and Coomans 1973). According to Wuidart (personal communication) the application of boron fertilizers should be done when no rain or irrigation is expected for two days. Boron toxicity affects old leaves in the first place and is not lethal. After some time, young trees recover their vigour. B toxicity has not been observed on coconuts over 4 years old. Due to the small volume of soil available to each young seedling, the application of B in the nursery cannot be recommended.

Chlorine Toxicity

Although for proper growth and development the coconut palm requires a certain amount of chlorine, toxicity becomes apparent when leaf sap chlorine content increases to above 0.5% (Cassidy 1968). The level at which a palm may die from chlorine toxicity has been estimated at about 0.75%. It is very difficult to determine Cl toxicity, as high Cl content is often linked with an excess of seasalt in soils along the coast. In a plantation affected by sea salt, surviving palms may carry a few short leaves standing almost vertically, with others hanging dead around the trunk. Less affected palms may have small leaves with whitish margins along the leaflets.

Copper Toxicity

Schut (1975) reported on a possible phytotoxic effect of copper on dwarf coconut palms, following spraying with copper fungicides as prevention against bud-rot in Surinam. The number of leaves per palm and the number of female flowers per bunch in particular seemed to be affected. Copper toxicity is observed on acid soils in particular, as was also the case with the coconuts in Surinam, growing in soils with pH values of 3.5-4.8. Ochs *et al.* (1993a) observed toxicity symptoms after applying 5 g or more of copper sulphate to the planting hole in copper-deficient peat soils. The leaves showed typical longitudinal beige stripes along the lamina. These stripes coalesce and an overdose can kill young plants.

Fluorine Toxicity

Fluorine toxicity in coconuts has been observed in a coconut nursery fertilized by single super phosphate containing fluorine from natural Togolese phosphates (Eschbach and Manciot 1981; Brunin and Ouvrier 1973). The symptoms became visible along the veins and consisted of round, oily looking brown spots. These coalesced and formed brown patches which dried out. These symptoms only appeared in plants raised in white, leached soil. This type of leaf necrosis was hardly ever found in plants growing on sandy clay soil. The phytotoxicity is a function of pH. Application of lime lowered the number of affected leaves. The symptoms occurred in the nurseries only, not in plantations where the rate of phosphate applied was too low compared to total soil volume available to the plant.

Manganese Toxicity

Manganese toxicity may occur in soils with a low pH value, or when an overdose of manganese sulphate has been applied. The toxicity symptoms are characterized by the development of brown spots on the oldest leaves, and an uneven chlorophyll distribution. This toxicity is often accompanied by iron deficiency (Eschbach and Manciot 1981).

Rare Earth Toxicity

Rare earths are similar to micro-nutrients in being beneficial at very low concentrations and detrimental to plants at slightly higher concentrations. However, the group differs from micro-nutrients in that it is not essential. Moreover, the same low concentration which is beneficial in the very early stage of plant life may kill the plant later on. Either the group accumulates in the plants to toxic concentration, or its detrimental effect builds up with age. Velasco *et al.* (1993) conducted an experiment with young coconut palms growing in a mineral solution to which lanthanum, cerium or chromium (not a rare earth) had been added. Chromium was included as an additional treatment because it was noted to be present in several localities which are seriously affected by the disease.

They found that the lanthanum treatment had more profuse and persistent stipules, and the homologous leaf scars on the trunk were further apart. The profuse and persistent stipules were also found in the cerium treatment. Although it showed an initial spurt of growth, it soon showed weakness and became debilitated. It simulated the changes occurring in palms affected by Cadang-cadang. The chromium treatment was unique in having straight, upright leaves. The initial stimulating effect of chromium tended to be followed by a debilitating effect. The combined effects of lanthanum and chromium set back the plant conspicuously. Cerium seems to be a more potent agent than lanthanum in stimulating growth in the early stage and in inhibiting growth in the later stage. Although late, lanthanum seems to start showing inhibitory effects in the seventh year.

Due to technical circumstances, the experiment was continued after one year with one plant per treatment only and final conclusions cannot be drawn. Moreover, similar symptoms in plants are not always evidence of similar causes.

Drought Effects

With increasing severity of a drought period, the following symptoms develop in the coconut palm. First, the oldest leaves wither and drop off or remain hanging along the stem (CP 28). Middle and young leaves start drooping, no longer giving support to the bunches, which hang down, resulting in the loss of nuts. In severe cases, the mid-ribs may break. The leaflets may be folded slightly, permitting more light to penetrate through the foliage. Buttons and young nuts are dropped and in some cases even unopened inflorescences are aborted. At a later stage, nuts in various stages of immaturity may fall as well. The rachis of the bunches that have lost their support from the petioles may break or may be torn and hang along the stem with the older leaves. Nuts on such bunches may drop after the first rains. The canopy is reduced and in severe cases may collapse entirely. Water deficit retards leaf emergence and hastens senescence. Stem growth is reduced and the apex of the stem may narrow. Petioles may break, the leaves hanging down along the stem, leaving only a small upright tuft of the central leaves until finally the palm succumbs.

Under conditions of prolonged drought, the outer cells in the absorbing region of roots develop thickened walls through which no water will enter. Such roots remain more or less in resting condition. When palms have suffered from severe drought, it may take one or two years for complete recuperation. Some palms remain runts. If the stem has developed a bottleneck, the palm's production potential may be affected and cannot be restored to its original level. As it takes the nuts about one year after pollination to ripen and be harvested, in climates with seasonal rainfall alternating with a pronounced dry season, yields during the dry season are considerably lower than during the rainy season, due to less fruit set and some shedding of buttons the year before. But severe droughts may also affect the formation of the flower primordia, about one year before spathe opening, and thus may influence yields in the second year after drought as well. Lower leaf production results in a lower number of inflorescences produced during the dry period. Drought occurring 15-16 months before opening of the spadices might lead to abortion of these spadices, explaining why this occurs mainly in the rainy season. Chlorine deficiency considerably enhances the palm's susceptibility to drought stress. According to Ollagnier (1985), stem bleeding may sometimes be a physiological disorder resulting from drought. However, stem bleeding is usually caused by fungi.

Stem Tapering

Stem tapering, or narrowing of the stem, also called 'pencil point disease' can have different causes, all of them unfavourable to the growth and development of the palm, such as drought, disease, heavy pest attack, repeated fire damage, mineral deficiencies, inadequate drainage or general negligence and high weed growth. All these factors diminish the palm's vigour and affect its development. Recovery is possible when the causal factors are removed. Then, the stem regains its old diameter. After the leaves have fallen, the stem shows a bottleneck.

Lightning Damage

Coconut palms are frequently struck by lightning, causing the death of one or more palms in one patch and affecting the health of a number of others. The total damage depends on the intensity of the electrical discharge. Plantations in areas where rainstorms are frequent, often show various gaps where palms are missing as a result of lightning damage.

The visible symptoms of lightning strike on coconuts depend on the intensity of the discharge and on the distance between the affected palms and the one directly hit. Directly-hit palms may be killed instantly, their stems being charred. Fire setting to a palm is very rare. It may occur on palms with dry leaves hanging along the stem and the lightning hitting the palm before the rain has started to fall. Splitting of the stem and/or decapitation of the palm is also very rare. Sometimes the central spear of the palm may collapse after the flash, showing no signs of charring, the central leaves being withered. Where the strike has been less intensive, the leaves start drooping after a few days and all leaves gradually dry up with the rachis turning brown, leaving a burnt appearance.

Usually, the visible symptoms are much less spectacular, but even so, directly hit palms and neighbouring palms may die within a week. The leaves of affected palms may be partially scorched, the outer leaves hanging down. In some leaves, the mid-rib may be broken at a short distance from the tip, the leaflets beyond this point turning brown and dying off. Mature leaves may die off completely, the process starting at the tip of the leaf, progressing towards the base. Sometimes all leaves die in this way and finally the palm dies.

Newly transplanted seedlings may show breaking of leaves at the middle region or distal end. When closely observed, large reddish oily patches can be seen on the rachis (Ramaiah 1990). In palms that are not killed instantly, a striking phenomenon may develop after a few days, the exudation of sap from the stem. The brownish liquid oozes from innumerable cracks in the bark, forming froth masses that run down in streaks. This symptom is somewhat similar to that caused by fire heat, but the bleeding is much stronger, giving the impression that the whole of the internal tissues are undergoing rapid fermentation. The affected tissues are pale brown, uniformly coloured and full of sap. Sometimes a crack running along the entire length of the stem may be observed. Sometimes the stem collapses after the strike. Neighbouring palms may have several leaves snapped near the tip, with the broken part hanging vertically down, opposite the palm which received the discharge. The latter is one of the most indicative symptoms of lightning strike. Depending on the intensity of the discharge, one, a few, or even a large number of palms may succumb to the effects, some even after 10 months. The affected palms at greater distance from the strike first may appear to continue their growth normally but develop visible symptoms after a few months only. Other palms, having suffered lightly, may recover.

Premature Button and Nut Shedding

Premature nutfall, or abortion, includes the fall of non-pollinated or pollinated female flowers or nuts during any stage prior to maturity. Button shedding is often caused by defective pollination of the female flowers. Various factors may cause premature nutfall, such as drought, pests and diseases. The phenomenon is discussed in the chapters dealing with these factors.

According to Mathes (1988), experiments in India have indicated that only about one-third or less of the potential nuts remain up to maturity; the rest are lost at various stages of development. About 65% of the nuts are lost during the first four months after opening of the spathe. Physiological nutfall occurs when palms cannot cope with the demand for assimilates by a high number of nuts. The shedding of exceptionally high numbers of nuts may be caused by a strong disturbance in the palm's productivity, such as tapping. When tapping ceases, the common response is to produce many female flowers (Foale 1991b). Ecological conditions may also lead to a high percentage of fruit set. Not always will the palm be capable of carrying these fruits to maturity, especially when growing conditions are unfavourable and, consequently, assimilation is insufficient. Such growing conditions may include lack of water, nutrients, or sunshine, or any other deficiency limiting the palm's production capacity. It was observed in India that immature nutfall is heaviest during periods of moisture stress (Mathes 1988). Irrigation of coconut on the west coast of India increased the number of female flowers and arrested button shedding (Rao 1989). Severe droughts may have a negative effect on female flower primordia initiation, which takes place about one year before the opening of the spathe, affecting fruit set one year later. Intensive rainfall can also lead to button shedding, probably due to the lack of pollen and pollinating agents during heavy rains, as well as waterlogging in coconut gardens, leading to physiological drought due to hypodermal thickening of roots up to the rootcap, reducing the area of absorption. High levels of nitrogen fertilization may increase the female flower production and reduce the setting percentage. In two localities in India, trials were conducted with the objective of making more plant food available for the development of all female flowers in one bunch by removing the other bunches. In one place a significant increase in the setting percentage was obtained, in the other place the results did not show a significant difference (Sudhakara 1990). In the second place, another factor other than nutrition alone probably played a role in the shedding of the nuts. Premature nutfall is usually heaviest during the first two months after spathe opening and, in addition to premature nutfall caused by diseases or pests, drought may be the most common cause.

Vijayaraghavan *et al.* (1989) studied the effect of growth regulators in preventing nut shedding in irrigated and manured tall coconut palms. They found that NAA at 20 ppm was the most effective, followed by 2,4-D 10 ppm. The chemicals were applied by hand-spraying of the inflorescences, 30 days after spathe opening when fertilizing was complete in all buttons. In the NAA treatment, 60.7% of the buttons formed were retained, in the 2,4-D treatment this percentage was 45.7. In the control plot only 26% of the buttons matured into fruits. The nut quality was not affected by these treatments. In Indonesia, premature nutfall was avoided by the use of 2,4-D. Application of 2,4-D at a concentration of more than 20 ppm improved fruit-set by about fifty per cent per bunch but the size of these fruits was abnormal and they were of low quality (Darwis 1991). Abnormally high fruit numbers usually result in smaller nuts with less copra per nut. Applications of 2,4-D are risky, as too high doses or careless applications may cause leaf and fruit deformation (Wuidart, personal communication).

5. Diseases

J.G. Ohler

As time progressed, more and more has become known of coconut diseases and also 'new' diseases have been recognized of which it is not known if they have caused damage to coconut in earlier years. The causal agents of various diseases formerly regarded as diseases of unknown etiology have been discovered. However, this is not always the case with the vectors that helps the spreading of the diseases. In general, curative treatments against diseases are unknown or if they are known they are very costly and uneconomic. Susceptibility to diseases may differ widely between different varieties of coconuts. In general, selection and breeding of tolerant species remains the best solution to counter the disease effects, but sometimes this may be a long and difficult path. Plantation sanitation and management can be important instruments in avoiding or reducing disease attack. According to Renard and Dollet (1991) the as-yet uncontrollable diseases believed to be caused by MLOs pose the greatest threat. Not only as a threat to existing plantations but also as a reason for farmers to plant or replant coconuts.

Soil and nutritional factors as well as climatic conditions may have influenced on the spread, development and intensity of the diseases. Lecoustre and de Reffye (1986) drew attention to statistical methods which enable the description and structural analysis of spatial variables more effectively than methods involving only means and variances (soil type, spread of disease, extent of pest attacks, interaction between soil factors and production), for possible applications to agronomic research, in particular to oil palm and coconut, with respect to epidemiology.

Disease symptoms are sometimes misleading. Different pathogens may provoke similar symptoms as they attack the same tissues in the palm. Good examples are Lethal Yellowing, Hartrot, and Red Ring, caused by a mycoplasma, a trypanosomatid flagellate and a nematode respectively, all of them blocking the sieve system of the palm. Stem bleeding can also have different causal agents, such as lightning or a fungus.

Ollagnier *et al.* (1983) observed in Ivory Coast and Indonesia that drought resistance seems to go hand in hand with resistance to leaf diseases. They noted that *Pestalozzia palmarum* attacks were worst in the K-0 treatments of a fertilizer trial. In another trial, the same was observed with *Helminthosporium* attacks, which increased in proportion to urea applications but dropped with increasing levels of KCL applications. Also from The Philippines, it was reported that the application of seawater at the base of 6-month-old MRD palms, at a rate of 200-1000 ml or 20 g of sea-salt reduced the incidence of *Pestalotiopsis palmarum* on leaves.

Diseases caused by Nematodes

Red Ring Disease

Red Ring Disease, formerly called 'Root Disease' in Trinidad, is widespread in Central America, the Caribbean area and South America, from Mexico to the State of Bahia, in Brazil. It does not occur in the northern Caribbean islands, Cuba or Florida. The disease also affects oil palm and various other palm species. Coconut palms are most susceptible to Red Ring Disease (RRD) between the ages of 3 and 10, but also older palms may be affected. The disease is always fatal and in neglected plantations it can cause very serious damage and cause substantial financial losses.

External and internal symptoms of the disease may vary according to palm species, variety and environmental conditions (Giblin-Davis 1991). Usually, the first symptom of the disease is yellowing of the leaves. Commonly, the lowest living leaves are first affected. Some investigators such as Nowell (1923), Blair (1970) and Giblin-Davis (1991) reported premature nutfall as a main symptom, but Maramorosh (1964) reported nuts to be normal. The dying off of the older leaves is very difficult to distinguish from normal dying off. However, the younger leaves gradually become affected as well. Sometimes, one or more older leaves may remain green while discolouration successively progresses in the younger leaves above them. Palms in a late stage are exceedingly conspicuous by their rich colouring of brown, orange, and yellow, involving all the leaves with the exception of the central tuft of young leaves. These rarely turn yellow but take on a greyish colour before collapsing. The internal symptoms are very characteristic. When the leaves are split in the median line, a red-spotted or red and yellow discolouration is revealed in the petiole, extending from 15 to 75 cm from the base outwards. In the leaves of the central column which still remain green, the internal discoloration is very intense, often extending to the developing leaves which they enclose.

The cross-section of the stem shows a red ring, 2-4 cm wide, about 3-5 cm from the periphery. A longitudinal section through the centre of the palm shows that the discoloured zone completely surrounds the base of the stem, extending upwards several feet, then breaking up into longitudinal streaks of the same colour intermingled with mustard yellow, and finally into individual spots of 1 mm or more in diameter. In later stages, discolouration extends to the soft meristematic tissue below the bud. In older palms, the red ring is also visible just below the crown and much more inward in the trunk. Discolouration is usually not as dark as in younger palms. The typical discolouration of the stem is also found in the roots where the visible effect of the disease is confined to the cortex. Seedlings usually do not show disease symptoms until stem formation has taken place.

Little Leaf symptoms, suspected to be caused by the same nematode causing RRD occurs rather commonly in oil palm in Surinam, Colombia, Venezuela and Ecuador (Rognon, personal communication) and rarely in coconut. The first visible symptom is failure of the central leaves to develop normally. They remain short and erect, and are often deformed. Descriptions by Hoof and Seinhorst (1962) are very similar to those of boron deficiency. They observed that inoculation in the wet season by dropping a nematode suspension on young leaves produced typical Little Leaf symptoms after some time. After such palms were felled 15 months later, numerous nematodes were found. Maas *et al.* (1970) found no nematodes in the stems of palms that showed red rings but he found the nematodes on deformed leaves. He suggested that these nematodes moved up through the stem to the soft tissue of less susceptible palms, without causing lethal Red Ring. The first symptoms developed four weeks after inoculation, and six to twelve weeks later these palms died. When roots are artificially infected, it may take up to sixteen weeks before the first leaf symptoms appear.

The causal agent of the disease is the nematode *Rhadinaphelenchus cocophilus* (Cobb), which is about 1 mm long. The scattered red dots in otherwise sound living tissue which mark the upper extension of the red ring in the stem and the beginning of the infestation in the leaf stalk, are initial nests usually containing a few adult nematodes together with larvae and eggs. According to Naranjo (1991), the life cycle of the nematode in inoculated immaturely harvested nuts was 9-10 days only, which is very short for phyto-pathologic nematodes. The red ring is highly infested and contains myriad active nematodes. Large numbers of nematodes are also present in the roots. In all cases, the infestation is confined to the ground tissue, the vascular bundles remaining seemingly unaffected. The nematodes also occur, to a lesser extent, intracellularly, in the main body of the discoloured tissue. Cavities teeming with nematodes are formed by the breakdown of parenchyma cells. Xylem vessels become occluded by tyloses in areas where they pass through discoloured tissue.

Injection of palms with coloured water showed that in healthy palms the colour was entirely distributed around the stem, 60 cm above the site of injection, but in diseased trees there was very little movement of the colour, indicating a reduced water uptake by diseased palms. In Surinam, in a varietal observation field, only palms of Fijian origin were affected by the disease (Kastelein 1986). External symptoms exhibited by the diseased palms were indistinguishable from those of the Hartrot-affected dwarf type variety of Fijian origin. This is not surprising, as in both cases the sieve system is blocked by the causal agent.

Transmission may take place from diseased roots to roots of a healthy palms where they are touching. Warwick and Bezerra (1991) observed that transmission through root contact occurred in seven out of fifteen plants growing around artificially inoculated plants in a greenhouse. In another experiment, each plant was contaminated when a suspension of nematode was placed on mechanically damaged roots. Chips flying away from a diseased palm being cut down, may also infect adjacent palms. But the main infection takes place through a vector, the palm weevil *Rhynchophorus palmarum*

L. These insects carry nematodes on the outside of, as well as within, their bodies. Adult weevils emerge from their cocoons in a rotted palm and disperse to other palms. According to Giblin-Davis (1991), the weevil may not be able to complete its life cycle in oil palm. Transmission takes place at oviposition of the vector that lays its eggs in leaf axils or in the stem. The weevils are often attracted to wounds or existing stem damage. But they may also bore into undamaged stems, particularly in succulent parts of the stems of vigorously growing young palms. The larvae, when feeding on diseased coconut, become infected with juvenile nematodes. Artificially infected seed nuts showed the presence of a few nematodes up to sixteen weeks after germination, but after twenty weeks no more nematodes were recovered from them.

However, not all *R. palmarum* weevils are potential vectors of the nematode. Griffith (1987) found that the genetics of the weevil indicated that these were homozygous recessive for the ability to remove the nematode. Therefore, the potential vector population was 25%. The Red Ring nematode had therefore segregated the palm weevil population into two sub-populations. In addition to limiting the size of the vector population by parasitizing, the reproductive potential of the survivors is affected through reduced fecundity. Further, many female vectors may oviposit unfertilized eggs, since mating might not be accomplished with the larger insects around. Thus, a female insect, although able to transmit the nematode may not give rise to progeny. Therefore, any further transmission depends largely on the attractiveness of the affected tree to palm weevils producing fertile eggs with the genetic requirements for suceptibility to the nematode. This replenishment of vectors comes from homozygous parents in the population. The rate of spread of the nematode will correlate with the gene frequency of the recessive allele (Griffith 1993). Environmental conditions determine the rate of the weevil population to increase, humidity being an important factor for adult weevil longevity. In addition to being a vector of RRD, palm weevils may cause severe damage to the palms themselves.

Rhina barbirostris, which is a minor pest of coconuts, also carries nematodes inside and outside its body. But it is not known whether it transmits the disease. Kastelein (1986) reported the presence of nematodes inside the weevil *Castnia daedalus* within a diseased palm. However, there are no records of this weevil transmitting the disease.

According to Giblin-Davis (1991), in Brazil *Oenocarpus distichus* Mart., growing in the jungle is an important host plant of the Red Ring nematode. In addition to this palm, Griffith (1993) mentions a series of palms as hosts of the nematode.

Control can be obtained by control of the vector. Plantation sanitation is very important in this matter. Affected palms should be destroyed as soon as possible, either by poisoning or by burning and before the young weevils spread to other trees. Traps can be used to catch adult weevils. These traps consist of containers filled with pieces of fresh coconut wood that attract the adult beetles. The bait could be poisoned by spraying with 0.1 per cent lannate. Chemical control is possible by applying insecticides such as Endrin to the leaf axils of healthy palms at monthly intervals. A combination of these methods could keep a plantation relatively free of weevils.

Burrowing Nematode

The burrowing nematode, *Radopholus similis* may cause considerable damage to coconut. It has been reported from coconut palms in Florida, Sri Lanka, India and Western Samoa. It is also known to infest and cause considerable damage to various intercrops in coconut plantations. It has caused very severe damage to pepper in Indonesia, citrus in Florida and bananas in Central America. The burrowing nematode is a migratory endoparasite, capable of spending its entire life within the roots. All stages except adult and fourth stage males are infective. The nematode is known for its sexual dimorphism. It occurs in various soil types. The nematode survives for up to six months in roots of felled coconut stumps. In host-free dry sandy soil it survives for three months and in host-free wet soil for 15 months under greenhouse conditions (Koshy *et al.* 1993).

Koshy *et al.* (1991) discussed its life cycle, epidemiology, symptomatology in coconut and control in India. Roots of Root(wilt) affected coconuts are often heavily infested with *R. similis*, but the nematode is not involved in the etiology of the disease (Sosamma and Koshy (1991a). The wide distribution of this nematode is due to transference of infected planting material, especially banana rhizomes from country to country, without observing plant protection measures. Other means of dissemination are through floods, irrigation water, farm implements and bulk transport of soil. Burrowing nematode, infested coconut seedlings do not establish easily and when they get established initially they may succumb in the subsequent dry period.

Maximum population of the nematode is found in the root zone of coconut at a distance of 100 cm from the bole of the palm and at a depth of 50-100 cm. Tender, creamy-white to orange coloured, semi-hard, main roots showing lesions and rotting, contain live populations in large numbers.

Symptoms of damage above the ground are non-specific decline such as stunting, yellowing, reduced number and size of leaves and leaflets, delay in flowering, button shedding and yield reduction. The nematode penetrates the delicate region behind the root-cap. Primary lesions also show cracking of the epidermis under stereo-microscopic examination. The nematode also attacks the plumules, leaf bases and haustoria of coconut seedlings. The only reliable method to identify an infested palm is to look for symptoms on fresh roots during the rainy season (Koshy *et al.* 1993).

Control of burrowing nematode infestation on coconut with its extensive root system is difficult. Intercropping with susceptible intercrops worsen the situation. Such intercrops should be replaced by nematode-resistant crops such as cacao, pineapple, cinnamon, clove colocasia and casava. Incorporation of green manure crops and neem and marotti (Hydnocarpus) oil cakes has a favourable effect on the build-up of predatory nematodes and on nematode parasiting fungi. While it is not possible to achieve complete control with biocontrol agents at field level, this method can be part of an integrated management programme (Koshy *et al.* 1993). Chemical control with phenamiphos or phorate at 10g. a.i. per palm gave thirty per cent control. Complete control may be obtained with soil application of phenamiphos or phorate at 25 kg a.i. per ha in infested nurseries. However, unlimited use of nematicides may cause problems of residues in coconut water and meat, and also in intercrops. Application of Aldicarp at 10 g. a.i. per palm showed residues in coconut water and meat after 45 days of application. A dip in 1000 ppm dibromochlorpropane (DBCP) for fifteen minutes is effective in controlling the nematode in seedlings. In nematode-infested areas already planted with coconuts, an integrated control, combining the various methods may be the most effective.

Sosamma and Koshi (1991b) observed that the prior establishment of VAM ameliorated the negative effect of the nematode on coconut seedlings. Among the three nematicides tested, Carbofuran, Phorate and Ebufos, Phorate had the least deleterious effect on VAM colonization.

The planting of nematode-resistant planting material, coconut palms as well as intercrops, may the most effective method to control the nematodes. Koshy *et al.* (1991) observed that all 49 genotypes screened for resistance to *R. similas* in India were found susceptible in varying intensities. The dwarf cultivars Kenthali and Klappawangi recorded the least nematode multiplication and lesion indices. Similar reactions were observed in some hybrids, such as Java Giant x Kulasekharam Yellow Dwarf (KYD), KYD x Java Giant, Java Tall x Malayan Yellow Dwarf and San Ramon x Gagabondam. Also Chowghat Orange Dwarf x West Coast Tall, and LO x Gangabondam may be added to this list (Koshy *et al.* 1993).

Diseases caused by Fungi

There are many fungi affecting coconut but only a few are of great importance, although some of the minor diseases cause severe damage in some localities. Only the most important fungal diseases will be discussed here.

Basal Stem Rot

Basal Stem Rot, Ganoderma Wilt, or Thanjavur Wilt, affects coconut palms as well as oil palm and arecanut palm. The disease occurs widespread over the world. It is the most destructive disease in Tamil Nadu, India, where it was first observed after a cyclone in 1952. Often, it occurs in badly managed groves. The disease incidence is positively correlated with mean maximum soil temperatures, and it is not correlated with minimum temperatures, rainfall or relative humidity (Bhaskaran *et al.* 1989). The disease progresses rapidly in dry areas and more slowly in wet areas. Soil water-stress may predispose the palms to infection (Nambiar and Rawther 1993). Infected palms may die within months in dry areas and survive another five to six years in areas with higher rainfall (Peries *et al.* 1975). Infection occurs primarily through root contact.

(Peries *et al.* 1975) presented a detailed description of the symptomatology of the disease. The roots are first affected and destroyed, and visual symptoms of the disease are similar to those of severe drought. The female flowers are few and poorly developed. Nuts become narrow and elongated in the immature stage, and small and distorted when mature. The husk is thicker than usual and marked with dark brown streaks. There is also considerable immature nutfall.

At the base of the stem, a characteristic reddish-brown discoloration develops, accompanied by exudation of a brown, viscous gummy substance. Initially, these bleeding patches appear on various places as parallel vertical streaks. These soon coalesce, forming a discoloured band around the trunk. According to Bhaskaran *et al.* (1989) the bleeding patches may extend to 3 m upwards as the disease progresses. Production of a dry rot of internal tissue at the base of the stem is characteristic. The bole decays rapidly, resulting in the formation of large cavities. Frequently, the palms break off at the base and falls.

Of the affected roots, the cortical region turns brown first, followed by the stele, and the roots become friable and disintegrate. As the roots in contact with the soil die back, the palm forms new roots from higher up the stem and sometimes new roots may grow out from healthy tissue through the affected tissue. According to Thirumalaiswamy *et al.* (1992), in Thanjavur palms sometimes succumb without expressing external symptoms. Palms aged 10 years and older are more susceptible to the disease than younger palms.

As causal agent of the disease, the fungus *Ganoderma lucidum* is generally mentioned, but Holliday (1980) stated that this fungus occurs in the northern temperate zone and in Central Africa above 1500 m and that its name has been misapplied in the tropics. According to him, the causal agent is *G. boninense*. *Ganoderma* spp. are saprophytes or weak parasites which can build up massive inocula on debris of woody crops; new crops then become affected. They have a wide range of hosts. The fruit body is flat bracket-shaped and is stalked, varying in size from one to 50 cm in diameter with a thickness of up to more than 10 cm. In the button stage, the colour is white but the upper side of the bracket soon becomes shiny and light to dark brown or almost black, and concentrically furrowed. The margin is generally white and so is the underside.

For control, Briton-Jones (1940) recommended digging out and burning the bole and the affected part of the stem. To prevent spreading through the roots to neighbouring palms, he recommends digging a 60 cm-deep trench around the affected palm. The holes should be left open for at least one year. Burying coconut husks in trenches around the diseased palms helps soil moisture conservation and was found useful for the management of the disease (Bhaskaran *et al.* 1990). Peries *et al.* (1975) recommend the removal of the affected area of the diseased palm and treatment of the area with a readily available fungicide, such as a 10% copper sulphate solution in water, if the palm can still be saved. Vijayaraghavan *et al.* (1987) suggested that tapping palms may lead to the release of pressure responsible for bleeding. This, in turn, may reduce the disease intensity at the initial stages of infection. Trials showed evidence of lower disease intensity due to tapping. The diseased palms gradually produced less toddy (totally 43.5 l compared to 69.6 l) with a lower sugar content than healthy palms.

Bhaskaran *et al.* (1989) reported that irrigation with fertilizer application increased disease intensity, whereas irrigation combined with Bordeaux mixture drenching checked the disease intensity considerably. Application of neem cake alone and in combination with drenching 1% Bordeaux mixture thrice at quarterly intervals was most effective in reducing disease intensity, as was soil drenching with 0.1% IBP, carboxin tridemorph or 0.05% carbendazim in combination with neem cake at 5 kg per palm. Soil drenching with 40 litres of 1% Bordeaux mixture and stem injection of Aureofungin-sol 2g + 1g of copper sulphate in 100 ml water thrice at quarterly intervals significantly reduced disease intensity and increased nut yields. According to Thirumalaiswamy *et al.* (1992), treatments with fungicides to be taken up by roots are effective only in the early stages of the disease. The treatment should be repeated once in three years. However, drenching the soil with copper solutions might lead to copper toxicity.

Trichoderma harzianum and *T. viride* were found to be antagonistic to *G.lucidum* (Bhaskaran *et al.* 1988). Neem cake application to diseased palms encouraged the saprophytic soil micro-flora, especially *Trichoderma* in coconut basins and was effective in the control of Thanjavur wilt. Other substrates can be used for multiplication of *T. viride*, such as rice bran, coir dust, sugarcane bagasse, and sorghum grain; however, neem cake is the most efficient. A number of plant extracts were tested for their effect on the growth of *G. lucidum*. Neem extract completely inhibited its growth. Banana rhizome extract and *Tephrosia purpurea* root extract gave 86% and 54% inhibition respectively.

The economy of control treatments depends on the stage of disease development. It is much more important to prevent than to control the disease incidence, by improving the growing conditions of palms. Better aeration through better weeding, improvement of drainage and correction of mineral deficiencies, combined with sanitation measures and removal of the diseased palm tissues may all contribute to the prevention or further spread of this disease. Ploughing and flood irrigation should be avoided to prevent the spread of infective propagules (Nambiar and Rawther 1993).

Grey Leaf Blight

This disease is widespread over all coconut producing countries, but is of little importance in well managed plantations. Its causal agent, *Pestalotiopsis palmarum* is a weak parasite. Symptoms are small, yellow spots surrounded by a grey margin, developing on older leaves. The centre of these spots later become greyish and spots may coalesce, giving the leaves a blighted appearance. Some varieties are more susceptible than others. Correction of unfavourable conditions controls the disease. Sea salt application or K fertilizers are often effective.

Stem Bleeding

Stem Bleeding of coconut occurs over the whole tropical zone. The disease is mostly of minor importance, but occasionally it may cause considerable damage. It also attacks oil palm and arecanut palm.

The characteristic symptom of this disease is the presence of bleeding patches on the stem. Reddish brown liquid oozes from longitudinal cracks in the stem. Patches may coalesce and form a large affected area. Old lesions cease oozing and the dark brown fluid dries up and turns black. Inside the stem, the tissues under the lesions are rotting and turn brown yellow to black. Infection may occur at any place on the stem but is usually seen on the lower part, progressing upwards. The palm gradually shows the common symptoms of weakness, yellow leaves and pencil-point stem, decrease in number of leaves, bunches and nuts. According to Chandar Rao *et al.* (1993), the extent of external lesions observed in Goa bore little relationship to the internal decay. Internal decay was characterized by disintegration of tissues into a dry, powdery mass. Affected palms at first appeared robust and healthy. Progression of the disease was rapid, the sequence of characteristic visual symptoms developing within a period of two years.

The fungus *Cerastomella paradoxa* has been suspected to be the causal agent of the disease, but artificial inoculation did not produce disease symptoms except in very few cases. Nambiar *et al.* (1986) isolated the fungi of diseased palms from various locations in India. The fungus *Thielaviopsis* sp. was consistently isolated from all locations from young lesions only. Artificial inoculation with inoculum from this fungus produced rusty brown discoloration of the bark within 2-8 weeks in all cases, followed by the oozing of a brownish liquid. *Thielaviopsis* was re-isolated from advancing margins of lesions. Mathew *et al.* (1989) developed a method for indexing the disease severity. Radhakrishnan (1990) observed that soil drenching with the systemic fungicide Calixin 0.1%

(Tridemorph) was superior over other control treatments, confirming the fungal nature of the causal agent. Nambiar and Rawther (1993) named *T. paradoxa* as the causal agent. They observed that infection by the fungus was faster on the stem during wet and cooler months when high humidity and moderate temperature prevailed. The optimum temperature for mycelial growth was found to be 28-30°C. In addition to this fungus, stem bleeding can be caused by lightning strike, and pest attack.

The fungus is a weak pathogen and penetrates into the tissue through stem damage or growth cracks, such as may develop after sudden, heavy rains following a severe dry period. Sudden heavy manuring is also suspected to cause such cracks in the stem. Burning of trash at the base of the stem damages the stem and may also lead to infection. Uexkull (1985) found a correlation between the occurrence of the disease and chlorine deficiency. Correction of the deficiency was found to prevent the disease. Also, excessive salinity with high sodium during summer was found to be associated with the disease (Nambiar and Rawther 1993). Renard and Dollet (1991) reported that the disease may also be triggered by excess nitrogen. Potty and Radhakrishnan (1978) observed that increasing doses of P fertilizers increased the disease incidence, whereas increasing N fertilizer doses reduced the disease incidence. They concluded that an unbalanced nutrient situation, disrupting the physiological system of the palm, was the primary cause of the disease. This conclusion may explain the different observations on nitrogen as a cause of the disease, the main reason being an unbalanced nutrient situation rather than lack or excess of nitrogen itself.

Control of the disease in the first place should be based on preventive measures, such as a balanced fertilizing programme, aeration of the plantation and avoiding damage of coconut stems either by men or insects. In the initial stage of the disease, the infected tissue of the palm may be chiselled out and burnt and the wounds covered with wound dressings such as oil, tar, or Bordeaux paste. Root feeding or drenching Carbendazim or Tridemorph were found effective in controlling the disease. Irrigation during the dry season may be practised to reduce prolonged moisture stress (Nambiar and Rawther 1993). The traditional practice of coconut growers in Goa is to make a 10 cm diameter hole in the stem from one side to the other, just above the bleeding site and to fill the hole with a mixture of ash, lime and cow-dung. The making of holes through the stem to cure a palm from disease was also observed by the author in Mozambique (CP 19). It is possible that Indian coconut plantation managers on the estates in Mozambique may have introduced this practice into E. Africa. Another method used in Goa is the tapping of diseased palms. However, the effectiveness of this method to control the disease may be questioned, as it does not affect the causal agent.

Lethal Bole Rot

Around 1962, outbreaks of Lethal Bole Rot (LBR) were observed in small areas near Mtwapa, Kenya and South of Dar es Salaam in Tanzania (Bock, K.R. *et al.* 1970). It mainly attacked young, about 8 year old palms, but seedlings transplanted from the nursery to the field were also highly susceptible. It has been stated that the disease had spread all over the coconut growing area in both countries. When the author was invited to inspect some diseased adult palms in Tanzania, South of Dar es Salaam in 1968, the very first palm that was cut down had a large cavity in the bole, surrounded by blackish tissue. Apparently, not only young coconuts were suffering from some kind of bole rot.

Basal stem rot was also observed in a nursery at Yandina on the Russell Islands. Descriptions of the symptomatology by Holliday (1980) and by Jackson and McKenzie (1988) are rather similar. The first visible symptoms in seedlings in the nursery were premature death of the oldest two or three leaves. White mycelium and sporophorus of a fungus were found at the base of the leaves. Younger leaves were infected successively as the fungus colonized the leaf bases, producing a brown rot. Cracks were common in these leaf bases. The outer tissues of the bole were often cancroid and had isolated rot, 1-1.5 cm deep with shallow reddish brown margins extending into the stem. The bole was seldom entirely decayed, but where this had occurred it was caused by bacterial soft rot. Seedlings may snap at the junction of the stem and nut.

In both places, *Marasmiellus cocophilus* was identified as the causal agent of the disease (Bock *et al.* 1970; Jackson and McKenzie 1988). In both regions, the disease was observed in certain periods only, the late 1960s in E. Africa and 1978-79 in the Russell Islands. Although sporophores of *M. cocophilus* continue to grow among polybagged seedlings in the field nursery at Yandina, there have been no further outbreaks of the disease since 1979. Nevertheless, some cases of diseased seedlings on other islands in the neighbourhood of the Russell Islands indicated that the disease could be seed-borne. As seed nuts attached to the palm can be affected only through the stalk or through fissures in the husk, dip treatments are used to control disease development in seednuts. Seednuts are taken directly from the tree, trimmed at the top and on three sides to expose about 75 per cent of the internal layers and then dipped in 250 ppm a.i. Antimucin plus a wetting agent for fifteen minutes. (Jackson and McKenzie 1988). The absence of new disease incidences suggests that former outbreaks have been favoured by some unknown predisposing factors. As LBR was reported to occur predominantly in young palms up to 8 years old, Schuiling and Mpunami (1992) in Tanzania evaluated 300 coconut palms varying in age from 2 to 9 years, in various stages of Lethal Disease. In 60 out of the 300 palms a rot was observed in the bole but all 60 palms exhibited very advanced leaf symptoms when cut. The incidence of bole rot became more frequent with increasing severity of leaf symptoms and rarer with increasing age of palms. Rot, dry and dark brown in colour, appeared to originate at the base of the bole, spreading radially and to a lesser extent upwards. They invariably coincided with complete collapse of the root system. It was concluded that the bole rot was clearly not primary rot as described for Lethal Bole Rot. It appeared to be a secondary symptom of Lethal Disease in young palms resulting from invasion of immature bole tissues by soil micro-organisms. Similar bole rot has been recorded in pre-bearing coconut palms affected by Lethal Yellowing in Jamaica (Eden-Green and Schuiling 1978).

Occasional outbreaks by other species of *Marasmiellus* have been reported from various countries, such as *M. inoderme* attacking seedlings in nurseries in Western Samoa, *M. semiustus* causing embryo and shoot rot in Malaysia and *M. palmivora* in nurseries in Malaysia. Such outbreaks are usually favoured by high relative humidity conditions. Control is mainly a matter of sanitation.

Bud Rot

Bud Rot is one of the most common diseases of coconut around the world, especially in the very humid regions. Usually, only sporadic cases of Bud Rot occur in plantations, but sometimes the disease may kill a few per cent of the palms each year. Serious epidemics are rare. According to Renard and Darwis (1993), there has been a resurgence of symptoms caused by *Phytophtora* more or less everywhere in the coconut growing zone. In certain regions in the Ivory Coast, up to 50 per cent of the coconuts initially planted are killed by Bud Rot, and a 20 to 25 per cent nut fall can be recorded each year; the extent of the damage depending on climatic conditions and type of planting material (Quillec *et al.* 1984).

The first symptom of Bud Rot is withering of the youngest unfolded leaf. Light brown speckles are present on the petiole bases of the youngest leaves, and on those of older leaves large yellowish to brown necrotic areas may be observed. The same is sometimes true of the peduncles of inflorescences. Apparently healthy trees in contaminated areas may show symptoms in the petiole bases alone, or also in stem and bud. The primary infection probably occurs at the level of the petioles of young leaves of ranks 1-5, without causing visible external symptoms. The pathogen develops slowly, and it is not until these leaves occupy a rank of at least 10 that penetration of the stem may occur. According to whether or not it reaches the soft tissues of the stem, two types of development may be observed:

- either the rot becomes stationary because it has penetrated into the root tissue that has already become very woody, or it penetrated too low in the bole where further development is impeded for the same reason, and the tree is unaffected;
- or the rot develops and progresses towards the centre of the root bulb and the top of the tree in the direction of the meristem.

The first typical external symptom of withering and tilting of the spear cannot be detected until the meristem is completely destroyed and the base of the spear is affected. At this stage, the disease is at an advanced stage and therefore curative treatment cannot be considered. The spear topples and hangs down between the older leaves. These older leaves remain green and retain their position for several months, which is very characteristic of this disease. According to Ohler (1984), the nuts also remain on the tree and older nuts may still mature. But Uchida *et al.* (1992) in Hawaii observed abnormal loss of small to almost mature nuts as a common early sign of the disease. Different observations may be due to different coconut varieties and different fungus species involved. The leaves fall progressively, one by one, starting with the youngest ones; the fall of the leaves extends over a period of 8-12 months until only the bare stem remains, Quillec *et al.* (1984). The evil smell of the rotting bud is not a characteristic of Bud Rot alone, as it develops in all palm diseases that cause the death of the central spear.

Sometimes the damage is restricted to premature nutfall. Nut damage, which in Indonesia is particularly severe on Yellow Dwarf (Renard and Darwis 1993), mainly concerns immature bunches, particularly during the rainy season. Pathogen propagation is either horizontal, by contact between bunches or vertical, between nuts within a given bunch. The role of rainwater and insects in the disease progress was demonstrated. The infection generally starts around the floral parts, sometimes in the equatorial part, and extends towards the apex of the nut and inside towards the shell. The disease is characterized by the presence of irregular lesions spreading from the surface of the nut to the endosperm and the nut stalk. Lesion development following artificial inoculation with *Phytophthora palmivora* is more rapid at the stalk end of the nut where tissues are softer and richer in sugars, and less towards the apex. The rate of lesion development is inversely proportional to nut age (Bennet *et al.* 1986). In young nuts, lesions penetrate the soft shell, the infested testa becoming grey and sticky. The adjacent infected endosperm is more translucent and softer than usual. In older nuts with hard shells, the fungus may penetrate through the germpore. Bennet *et al.* (1986) suggest that a diptera, *Telostylinus* sp., probably feeding on sugary exudates of young nuts which leak on the husk surface of healthy and diseased nuts, could be a vector of the fungus in Indonesia.

Studies have shown the existence of two dominant *Phytophthora* species: *P. palmivora* in Indonesia and The Philippines and *P. katsurae* in the Ivory Coast. *P. katsurae* does not occur in Asia. Both species are found in Jamaica. *P. arecae* and *P. nicotianae* are also found in Indonesia (Renard and Darwis 1993). Uchida *et al.* (1992) reported a new *Phytophthora* fruit and heart rot of Coconut in Hawaii, the pathogen of which resembled *P. katsurae*. Quillec and Renard (1984), Franqueville *et al.* (1989) and Steer and Coates-Beckford (1990) reported *P.heveae* in the Ivory Coast, Vanuatu and French Polynesia, and *P. parasitica* in Costa Rica as causal agents of the disease. Joseph and Radha (1975) found that *P. palmivora* causes dry rot in coconut prior to wet rot in later stages which is due to secondary invaders such as *Fusarium* sp. and bacteria. Steer and Coates-Beckford (1990) observed in greenhouse trials in Jamaica that both species of *Phytophthora,* as well as another isolate, *Thielaviopsis paradoxa*, caused Bud Rot only in wound-inoculated coconut seedlings, and that isolates of *P. palmivora* varied in virulence. The bacterium *Enterobacter* sp., the most frequently occurring organism in the crowns, caused tissue necrosis but inoculated seedlings did not develop Bud Rot. Studies in Jamaica (Anonymous 1989c) indicated that *T. paradoxa* always produced lesions and nutfall in laboratory tests but that its association with naturally infected fruits was very low.

Inoculations in the leaf axils with sporangial suspensions of *P. palmivora* isolated from diseased palms developed Bud Rot in 2-year-old seedlings. Subsequently, these yielded *P. palmivora* from dry rot lesions. Similarly, 5-year-old palms in the field developed Bud Rot when inoculated in young leaf axils during the wet season. On dissection of the palm, it was observed that rotting had spread further down from the site of inoculation. Epidemiological study models applied to plots affected by *Phytophthora* in the Ivory Coast suggest the existence of two propagation phases:

- an aggressive phase, during which contamination occurs from tree to tree;

- a regular phase, during which new cases appear some distance away from the initial foci (Renard and Darwis 1993).

Dissection of trees with nascent external symptoms often invisible to the untrained eye reveals the existence of an evil-smelling internal rot, already in an advanced stage of development, with a consistency of soft cheese, and purple to pale pink in colour. This rot is surrounded by a brown border, and a few brown fibres can be seen towards the base in the unaffected area. The meristem is still unaffected, but it is the poor supply to tissues above it that causes withering of the spear and youngest leaves. In apparently healthy palms when dissected, a central rot of the stem may show, totally different in appearance from advanced rot. The consistency of these tissues is still normal, and are white or slightly pink. The lower limit of the diseased tissue is more or less clearly marked by a brown border, whereas the upper limit is much more diffuse and can be detected only by rapid oxidation of neighbouring tissues when exposed at the air; the meristem is still healthy. A dissection in such an early stage shows that the rot originates from behind the growing point, in the soft tissues of the bud, and not in the spear, in the direction of the meristem. (Quillec *et al.* 1984).

The disease is favoured by conditions of high humidity, such as found on low, badly drained lands, in plantations with a very dense stand and under conditions of extensive rainfall periods. Control would, in the first place, involve measures reducing the relative humidity of the atmosphere in the plantation. Such measures might include improved drainage, wide spacing for better aeration, and adequate weed control. Affected palms should be cut down and burned as soon as possible. This deters breeding of rhinoceros beetles (*Oryctes rhinoceros* L.), which serve as carriers of the fungi spores (Abad 1985). Uchida *et al.* (1992) also emphasize preventive measures against the spread of *P. katsurae*. They recommend prompt removal of all diseased material from the plantation. Oospores of most *Phytophthora* species are capable of surviving in soil without a host plant. These thick-walled structures allow the fungus to survive in a dormant state for long periods. They suggest that the fungus is probably seed-borne as oospores in the husk. They also recommend disinfection of knives used on diseased palms. Removal of nut clusters and heavy leaf pruning of large trees has probably stimulated spread of disease.

Renard and Quillec (1984) reported that Fosetyl-Al (aluminium fosetyl) and Ridomil injected into the stem was very effective in controlling the disease. Two years after treatment had been stopped, the sanitary situation remained unchanged in the plots treated with these chemicals. Franqueville and Renard (1989) observed that Fosetyl-Al as wettable powder injected into the stem was effective, but in liquid form used in root absorption it was not. Even 43 months after treatment the percentage of nutfall due to Phytophthora was always lower on trees treated with Fosetyl-Al wettable powder than on those observed in other treatments. Stem injection with this chemical against *P. katsurae* is practised on a large scale in commercial plantations in the Ivory Coast. The Chemjet injector seems to be more effective and less traumatic for the stem than injecting after drilling a hole in the stem. Renard and Dollet (1991) reported that stem injection with Phosethyl Al and Metalaxyl offers complete protection against Bud-Rot. Although Metalaxyl is not effective against *Phytophthora*-induced nutfall, Phosethyl reduces loss of crop by at least 80%. However, such stem injection is difficult, costly and may give variable results, perhaps because different Phytopthora species are involved, or resistant strains may have developed. Chemical spraying is difficult and costly in tall palms but may be an effective method in young plantations. A relatively inexpensive fungicide such as Bordeaux mixture may be used, as long as no copper toxicity is induced.

Abad (1985) recommends removal of infected tissues including those bordering the infected lesion, and subsequent treatment with a solution of any copper-based fungicide at 3 tablespoons per gallon of water. However, Quillec *et al.* (1984) state that at the stage of the first visible symptoms the disease is already too far advanced for curative treatment.

Great variability of tolerance between varieties and hybrids has been observed. MYD passes on marked susceptibility to Bud Rot caused by *P. palmivora* to its hybrids. On the other hand, MYD introduces a source of resistance in the MAWA hybrid in zones infected with *P. katsurae*. Taffin *et al.* (1991) observed in the Ivory Coast that West African Tall proved to be resistant to nutfall, but relatively susceptible to Bud Rot. On the other hand, the MYD x WAT hybrids proved tolerant to Bud Rot, but susceptible to nutfall. The NIWA hybrid (Nias Yellow Dwarf x WAT) is more tolerant of nut fall than the MAWA hybrid, when faced with *P. katsurae*. In Asia, local ecotypes are generally more tolerant of *P. palmivora* than introduced ecotypes, although the Polynesia Tall and Rennel Tall are less severely affected than the Bali Talls in North Sumatra, Indonesia. Polynesia Tall, and Rennell Tall, to a lesser extent, perform as well in the presence of *P. katsurae* as in the presence of *P. palmivora*. The MYD is susceptible to nutfall caused by *P. palmivora* but it is tolerant to this disease when affected by *P. katsurae*. Two out of three progenies of MAWA hybrids in Ivory Coast proved to have an excellent level of tolerance to nutfall due to *P. katsurae*. The Malayan Red Dwarf is susceptible to Bud Rot caused by *P. katsurae* in the Ivory Coast. In Jamaica, MRD x Tall hybrids are more tolerant of Bud Rot than MRD. In The Philippines, only in Laguna the disease incidence was not significant (one diseased palm in five years of observation). Imported coloured cultivars were more susceptible compared to local green cultivars. Catigan, Tacunan, Tagnanan and Laguna were least susceptible to the disease (Rillo and Paloma 1989). Variation of tolerance also exists within a certain ecotype, particularly in WAT, which is surprising as the WAT is generally considered to be a rather homogeneous ecotype. It seems that coconut resistance to *Phytophthora* is more likely to be of the horizontal type than the vertical type (Franqueville *et al.* 1989).

Leaf Spot

Leaf Spot disease occurs widespread over the tropics and is commonly found on palms growing under unfavourable conditions. In India, it is widespread in areas where coconut palms are affected by the Root(wilt) disease. In those areas it is often associated with other, secondary fungi.

The disease is caused by the fungus *Drechslera halodes* = *Drechslera incurvata*. The fungus develops on the spear and sporulates as soon as the first leaf opens (Renard and Dollet 1991). Visual symptoms are the development of small, dark brown sunken necrotic spots on young foliage. At first, the spots are small, oval, and brown, enlarging and becoming pale buff in the centre, with broad dark brown margins. In cases of severe attack, the leaf edges become extensively necrotic. The lesions gradually enlarge and coalesce. No new spotting occurs on mature leaves. In advanced stages, spotting may result in drying up of the leaf (Abad 1985). On old leaves, the necrotic lesions resulting from the attack may be invaded by *Pestalotiopsis*. This fungus was often considered to be the initial parasite; which it may be when it develops following initial damage to the leaves by insects.

Trials in Papua New Guinea showed that nitrogen fertilizers increased the susceptibility of the palms, whereas both potassium and phosphorous fertilizers decreased their susceptibility. Similar results with fertilizer application were observed in Malaysia (Fagan 1985) and Tanzania (Romney 1987). Susceptibility of seedlings decreases with age. Application of sulphur does not influence the severity of the disease. Keeping the plantation well aerated helps to avoid disease attack. Removal and burning of diseased leaves is recommended.

Leaf Spot attack in the nursery can be avoided by a balanced mineral nutrition and cautious nitrogen application where this mineral is deficient. Additional doses of potassium chloride at a rate of 150 g per palm can alleviate the situation. Renard and Dollet (1991) reported weekly applications of Chlorothalonil or Mancozeb among the most effective fungicides. The best effect is obtained by applying the fungicide to the under surface of the leaves where the parasite penetrates. Systemic fungicides, in general, are not effective. To prevent disease development, in Indonesia positive results were obtained with Dithane-45, Bordeaux mixture 1%, Copper Sandox and Cobox 50%. NaCl solution at doses of 50g per l sprayed on the leaf surface delayed the development of the disease (Darwis 1991).

Renard and Dollet reported the Polynesian, Rennell and Vanuatu tall varieties, and the red and yellow dwarfs among the most sensitive cultivars, and the West African Tall and its hybrid with the

Malayan Yellow Dwarf among the most tolerant. Karkar is much more susceptible than the Markham Valley and Rennell Island varieties and several D x T hybrids. Tahiti varieties are affected all their lives and palms from Rangiroa are even more susceptible than those from Tahiti.

Leaf Blight

Reports of this disease have come from various parts of the world, such as Trinidad, Brazil, Malaysia and Sri lanka. The highest incidence of the disease is during the months with the highest temperatures and low relative humidity and rainfall, and the lowest incidence is during the cooler months and high relative humidity and rainfall (Ram (1989b).

The first symptoms are withering and drooping of the distal ends of the leaves, almost breaking away from the remainder of the leaves at a weak point, varying from 30 to 100 cm from the top, but remaining attached, hanging directly downwards as a pendulous section. A brown mark produced on the leaf stalk is particularly noticeable on the lower surface. The drooping end of the leaf, being yellow at first, finally turns brown. A cut through the rachis at the breaking point shows brown tissue. The fungus spreads into the lower part of the leaf, travelling down the rachis and the petiole, finally resulting in the yellowing of the whole leaf.

The disease is caused by *Botryodiplodia theobromae*, a weak parasite only attacking coconut palms growing under unfavourable conditions (Ram 1989). The fungus merely accelerates the death of palms having been weakened by other causes such as lack of drainage, water-stress or malnutrition. Potassium deficiency is one of the main conditions that makes the palm susceptible to attack by this fungus (Naranjo 1991). For control of the disease he recommends adequate fertilizing, irrigation during drought periods and the removal and burning of diseased leaf parts, as well as the burning of all organic material from the tree that has fallen. According to Ram (1989) chemical control is possible by injection or leaf application of a mixture of Benomyl and Carbendazim. The Brazilian Tall is highly susceptible to the disease; hybrids perform much better. The best hybrid found was PB 141 (EGD x WAT).

'Lixa Pequena'

Small verrucosis, or 'lixa pequena' was first described in Brazil in 1948. It earned its name because of the wart-like symptoms provoked by it. It is a serious disease, appearing mostly in areas that are not located near the sea shore. It is associated with 'Qeima das folhas' (leaf scorch). This pathogenic complex reduces the assimilating leaf area by around 30-50 per cent, causing considerable drop in production (Anonymous 1992a). These two diseases appear to be spreading and have also been observed in French Guiana since 1990.

Warwick *et al.* (1993b) presented the following description of the symptoms of this disease: 'The first symptoms are small, black, charcoal-like fruiting bodies formed superficially. This formation, known as stroma, is either found isolated, in lines, or in a diamond shape on the leaflets and mid-ribs. Eventually, the plant tissue around the stromata forms brown, necrotic lesions, which enlarge to about 15 x 2 cm. Numerous brown areas coalesce, the leaflets become necrotic and the whole leaf collapses.' The leaflet browning spreads from the tip to the base of the leaf in a V-shape. Rachis browning is often accompanied by a gummy excretion. Premature breakage of leaf stalks at the point of attachment to the stem also occurs, causing the bunches to hang downwards or even to break, leading to premature nutfall. (Warwick *et al.* 1991). The fungus also attacks the nuts (Resende *et al.* 1991). Greenhouse experiments carried out by Warwick *et al.* (1993b) confirmed the positive correlation between water-stress and disease development.

The causal agent of 'Lixa pequena' is *Phyllachora torrendiella = Catacauma torrendiella).* (Warwick *et al.* 1994). Chemical control of the pathogens is possible over small areas with a mixture of Benomyl and Carbendazim. However, all attempts of control over large areas in humid zones have failed. An antagonistic mycoflora, *Septofusidium elegantulum,* and *Acremonium sp.* affect the fungus in the States of Para and Paraiba, which means that the incidence of the disease can drop at certain times.

Sooty Mould
Sometimes, a large portion of the leaves is covered by a black layer of fungus growth. These fungi develop on the sweet exudation of some insects such as aphids. The fungus growth does not attack the leaves but impedes photosynthetic activity of the covered leaves. Where spraying is possible, a cheap, effective and harmless mixture of 225 g soft soap, 4.5 l of water and 9 l of kerosene may be used. The water should be heated, and the soap dissolved in the water, after which kerosene is added and stirred in the mixture until an emulsion is obtained. For spraying, the emulsion may be diluted to a 1:8 or 1:9 mixture with water. For fogging, pure emulsion is used.

Diseases caused by Bacteria
Bacterial Stripe
Bacterial stripe has been observed on seedlings in various nurseries in The Philippines. According to Abad (1985), the disease symptoms, which generally start on the older foliage, are characterized by the presence of yellowish water-soaked streaks on the leaf lamina parallel to the veins. The streaks then develop into stripes and, depending on the severity of the infection, may involve the whole length of the lamina. These stripes rapidly turn brownish and the whole leaf becomes blighted within a week. In case of severe infection, even the unopened spear leaf is infected. Sometimes the disease is fatal. Two bacteria pathogenic to coconuts, *Pseudomonas sp. and Erwinia sp.* were isolated from diseased seedlings.

Control can be achieved through timely application of the bactericide Physan 20 at a rate of 2 tablespoons per gallon of water at 2-weekly intervals. For heavily infected seedlings, removal and burning of infected foliage prior to weekly sprayings is recommended.

Diseases caused by Flagellates
Hartrot
Hartrot is the name given in Surinam to the disease causing severe damage to coconuts in Central America and the northern part of South America. In Spanish-speaking countries, the disease is called 'Marchitez', and in Trinidad it is called Cedros Wilt. The disease has been found in Surinam, Brazil, Guyana, Venezuela, Colombia, Ecuador, Trinidad and Tobago and Costa Rica.

The disease does not usually attack the palms before flowering, but occasionally may attack young palms under 2 years of age. Symptoms of the disease are quite similar to those of Lethal Yellowing. There is also some similarity to the symptoms of Red Ring. Renard (1989) sums up the symptoms of the disease, in order of importance, as follows: 'Total or partial fall of young nuts (at leaves 12, 13 and 14) and of the flowers immediately after fruit set (leaf 11). Older nuts remain attached or fall much later; browning and/or drying out of spikes on the inflorescence (at leaf 10) and premature fall of male flowers; browning of spikes on the oldest unopened inflorescence (leaf 9) and internal greyish brown coloration of the ovules. The female flower itself can sometimes turn totally brown; slight yellowing, then rapid browning of the terminal leaflets of one, then several lower leaves, developing from the tip towards the base of the leaf, from the lower leaves towards the upper leaves. This coloration may differ between varieties; browning, then rapid and generalized drying out of the foliage within 4 to 6 weeks, sometimes combined with rachis breakage on the upper leaves, along with rotting in the spear and bud causing them to topple over. The tertiary and quaternary roots quickly rot as soon as the leaves start to turn brown; cortex rot then reaches the secondary and primary roots.' Apparently, the disease attacks coconut palms from the stage of initial flowering. The incidence of the breaking of leaves is considerably enhanced by conditions of drought (Slobbe 1977).

The causal agent of the disease is a flagellate protozoa, *Phytomonas staheli* (Trypanosomatidae), one of the greatest phytopathological problems in Latin America. The flagellates are present in large numbers in sieve tubes, and are apparently confined to them. They literally plug the food conduit to growing regions in the crown, thus bringing about the death of the palm. The flagellates can also be found in the sap of the husk, the calyx and in coconut water. Flagellates were observed in nuts of varying ages of up to 11 months, but were not detectable in nuts showing advanced deterioration symptoms or in dry nuts. Flagellates were observed in healthy looking nuts and nuts with initial brown discoloration on infected trees, or fallen to the ground. As soon as brown rot was advancing, the chance of detection decreased, whereas other micro-organisms like bacteria, unicellular protozoa, yeast and fungi became dominating (Nanden-Amattaram, and Parasad-Sewkaransing 1989).

The flagellate also causes phloem necrosis in coffee, and it has also been found in oil palm in various South American countries. Formerly, the suspected host plants were lactiferous weeds, but actually indigenous palms and also other plants (Musaceae and Zingiberaceae of South America and the Caribbean) are the most suspected sources of the causal agent.

Transmission trials in oil palm by Perthuis *et al.* (1985) in Ecuador, and by Asgarali and Ramkalup (1985) in Surinam indicated the bugs *Lincus lethifer (Hemiptera Pentatomidae)*, *L. vandoesburgi* and *L. lamelliger* as potential vectors of the disease. In French Guyana, *L. croupius* was identified as a vector (Louise *et al.* 1986). It is not known whether other *Lincus* species are involved in the transmission. Kastelein (1985) observed that the flagellate *Phytomonas sp.* associated with the plant *Cecropia palmata* Willd is transmitted by another Pentatomid bug, *Edessa loxdali.* As this bug has occasionally been observed on coconut, the possibility was suggested that it might somehow be linked to Hartrot disease. The genus *Ochlerus* is responsible for Hartrot on a plantation in the Belem region, of Brazil (Renard 1989).

According to Perthuis *et al.* (1985), epidemiological data indicate that the incubation period of the disease is around 4 months. Louise *et al.* (1986) calculated the incubation period in coconut to be 6-8 months in general. The spread of the disease is rapid, often taking the form of an exponential development. Mortality rates of up to 90-100 per cent within 5 years have been observed on small plots in Surinam. Louise *et al.* (1986) observed that the disease advanced faster in Yellow Dwarfs than in Red Dwarfs. It also spread faster in MAWA than in WAT. Mariano *et al.* (1990) observed that in Brazil hybrids are more susceptible than either talls or dwarfs.

The particular biotope to which the vector seems to be linked is the 'terricolous' environment which forms inside the leaf sheaths consisting, among other things, of the remains of male flowers (Louise *et al.* 1986). The particular biotope upon which the insect seems to depend may explain why the disease in most cases appears only when the trees are 4-5 years old (Louise *et al.* 1986). The *Ochlerus* vector can also be found in colonies on the ground.

According to Renard (1989), in all situations Hart Rot foci developed in conjunction with poor plot upkeep, associated with humid zones near a river or forest. During a visit to various Latin American countries the author also observed much higher incidence of Hart Rot affected coconut palms in badly managed than in well kept plantations. According to Rognon (personal communication) this does not mean that the disease does not appear in well managed plantations, where it has also been observed. The apparently higher incidence in badly kept plantations may be due to greater weed variety and possibly also to host plants of the vector growing between the coconut palms. For control, Renard (1989) recommends a combination of two large-scale interventions to reduce vector populations: insecticidal control, and good upkeep through elimination of regrowth and clearing the approaches to plantations. He recommends the following measures:

- Ensure regular circle upkeep;
- Prevent winding weeds from invading the crown;
- Cut and windrow lower, dry leaves, which provide access to winding weeds;
- Cut, within reason, the tips of lower green leaves, touching the ground or the windrow in foci where transmission is via *Ochlerus*;
- Slash wood regrowth in plots along plantation borders;
- Lay out a laterite road or a pathway free of vegetation around plantations, to isolate the plot from surrounding forests or from rivers.

Deltamethrine at the rate of 2 g a.i. per l and lindane are effective in controlling *Lincus*. However, lindane is a dangerous insecticide because of its cumulative effects and its concentration in food chains. Against *Ochlerus*, spraying can be done of Thiodan (35% endosulfan) on to the weeds around the circle, into the axils of the lower leaves, at the base of the stem and on to the windrow as soon as the first cases of Hart Rot occur. Approximately 4 litres of a solution containing 2 ml per l of Thiodan should be sprayed per tree, 2 litres on the tree and 2 litres in the inter-row. In foci, treatment frequency should be stepped up to every 2 months.

Diseases caused by Viruses or Viroids

Cadang-Cadang

The word Cadang-cadang in The Philippines means slowly dying. This disease is responsible for killing a very large number of coconut palms in The Philippines. Potentially, it is a very dangerous disease, as its spreading to other countries with vast coconut areas could cause great losses. In 1917, the disease was already described on the island of Guam (Maramorosch 1993). The first outbreak of the disease in The Philippines occurred in 1926, on the small island of San Miguel, a few miles east of the coast of Southern Luzon. By 1960, out of the 250.000 original palms, less than 100 healthy palms were left. By 1983, diseased palms began to appear in considerable numbers on the main island of Luzon, across San Miguel, on the slopes of the Mayon volcano. The disease spread to the surrounding islands of Catanduanes, Samar, Masbate, Ticao and Burias, spreading all over the central part of Luzon, as far as Quezon. Maramorosch (1993) suggested that the possible origin of the disease may have been an imported ornamental palm of unknown origin when no plant quarantine yet existed.

The present distribution of the disease is between the latitude of Manila in the North and the island of Homonhon in the South. The spread of the disease is slow and gradual. Within a plantation, the spread is also slow and at random without definite loci of infection. The outward rate of spread within infected plots is less than 500 metres per year. But fresh sites of infection have been recorded as far as 50-100 km away from the nearest infections. The incidence between locations varies considerably. Altitude is negatively correlated with the incidence. Some sites show little expansion of the disease, other sites show active epidemics. Diseased palms are not clustered. Available data suggest that the disease may not be spread by any one specific route, but that it could be distributed by a variety of means. The viroid can be detected in the husk and in the embryo of nuts and is seed-transmitted at a low rate of about 1:300. It has also been detected in pollen (Hanold and Randles 1991-b). So far, numerous transmission tests with a great variety of insects have so far failed to indicate a vector (Pacumbaba 1985; Hanold and Randles 1991b). However, the possibility of transmission by some insect is not ruled out, either.

In The Philippines, Maramorosch (1993) found a consistent correlation between the destruction of palms owned by the Bicolanos, and the lack of infection on plantations owned by Tagalogs. Workers employed decades ago on San Miguel island were Bicolanos from Tabaco, the first locality where subsequently the disease appeared on Luzon Island. Bicolano plantation owners prefer to hire Bicolanos, whereas Tagalogs prefer mainly Tagalogs to work in their plantations. He suggested that it is most likely that the Cadang-cadang viroids were being carried from infected to healthy palms by the machetes (bolos) used by plantation workers to cut steps in the stems or when working in the crown to cut off nuts or to collect sap for toddy production. The fact that these workers walk from palm to palm on ropes connecting the palms high up in the crowns may explain why clumps of diseased palms often occur at considerable distance from other diseased palms.

The first clear symptoms indicating a palm to be affected by the disease may be the production of rounded nuts with characteristic equatorial scarifications. These nuts are smaller, and occasionally distorted. After 1-4 years, nut production comes to an end. The reduction of nut size in the early stages of the disease is due to reduced thickness of the husk and shell. Later, the kernel is also reduced.

The first symptom, although not so clearly visible from some distance, is the development of small rough, circular bright yellow tiny spots on the leaf lamina of the third or fourth leaf below the spear.

Maramorosh (1964) stated that the spots are already present on the bud-leaves and that they only become clearly visible when the leaf turns green. When the leaflets develop, the spots enlarge, becoming more and more irregular in shape and forming blotches or streaks, producing the characteristic yellow mottle. When the spots fuse with one another, they turn orange-yellow. Sometimes, this may be seen on one side of the leaflet, while the other side appears normal. Since the veins of affected leaflets seem more orange than bronze coloured, compared to those of normal leaves, the leaf colour of diseased palms often is bronze or orange-yellow. Leaves and leaflets of affected palms are also smaller and more brittle than normal, and the leaflets have a tendency to bend over or to break in the centre. Gradually, leaves take on an erect position in the crown. The stipules of the leaves of affected trees remain attached to the base of the leaves, giving them a winged appearance. Nut production ceases and leaf production slows down. This causes a gradual reduction in number of leaves until only a tuft of small, erect, yellowish-green leaves at the apex of the trunk is left.

Extrusion of inflorescences may also occur. Male flowers are dwarfed and female flowers are also smaller than usual. The inflorescences produce more buttons than usual, which usually fail to develop to maturity. Initially, the number of nuts produced increases; in later stages the palm bears male flowers only. Diseased palms have fewer primary and secondary roots than healthy palms, which may be attributed to reduced photosynthetic activity. The progress of the disease is very slow, the shortest time recorded for a palm to succumb being about five years, the average being ten, and sometimes it takes as long as 15 years before a palm is killed.

Usually, palms become naturally infected only after they have reached the age of flowering. In rare cases where young palms become infected, they are stunted and fail to produce inflorescences, although they survive well past the age of first flowering (Hanold and Randles 1991b). In 1973, two ribonucleic acids (RNA) of small molecular weight, called cc RNA^1, and cc RNA^2 were found to be uniquely associated with the Cadang-cadang disease. Both RNAs have viroid characters. The RNA properties resemble those of the viroid (small, single-stranded circular RNA molecules, about one tenth the size of a virus, which replicate when introduced into susceptible plants, at times producing a disease) groups of plant pathogens. The fact that RNAs may be considered components of the Cadang-cadang viroid was confirmed in a transmission experiment where inoculum containing the CCCVd (Coconut Cadang-Cadang Viroid) RNAs was infectious (Imperial 1985).

According to Hanold and Randles (1991a), symptomatology is unreliable for disease diagnosis. Tests for detection of CCCVd by polyacrylamide gel electrophoresis and molecular hybridization have been developed. These tests are sensitive and are definite for the viroid because they test for size, structure and nucleotide sequence. CCCVd can be detected up to about 6 months before the appearance of initial symptoms. A mobile, diagnostic laboratory has been developed making possible a rapid, cost-saving, and reliable yield diagnostic technique for Cadang-cadang disease (Rodriguez *et al.* 1989). CCCVd is the smallest known pathogen; with CTiVd (see Tinangaja disease) it is the only viroid known to affect monocotyledonous plants and the only viroid that is lethal in a host plant; and its pattern of changing molecular forms has not been reported from any other viroid (Hanold and Randles 1991). No vector of the disease has yet been identified.

Viroid-like molecules related to CCCVd have been found in some areas of the southwest Pacific, such as in oil palms in commercial plantations, in coconut palms and in various herbaceous monocotyledons. The infected oil palms show symptoms resembling those in naturally infected oil palms in The Philippines, but the coconut palms are without the typical Cadang-cadang syndrome (Hanold and Randles 1991b). No epidemic of Cadang-cadang has been reported in these countries. The orange spotting of oil palm is a well known disorder named Genetic Orange Spotting (GOS). It has been reported from both West Africa and the Malay Peninsula (Forde and Leyritz 1968, as referred to by Hanold and Randles, 1991b), and is associated with reduced productivity. However, it is transferred through the seed from parent to progeny in a pattern unlike any genetically inherited trait (Gascon and Meunier 1979, as referred to by Hanold and Randles 1991b).

Analyses of a number of GOS palms by Hanold and Randles (1991a) have demonstrated an association between the orange spotting and the detection of viroid-like nucleic acid. Hanold and

Randles (1991b) are of the opinion that until more is known about the epidemiology of these viroid-like molecules, an embargo should be placed on the uncontrolled movement of germplasm between countries. This is particularly important as the viroid infections appear to be latent and their pathogenicity in new areas is unpredictable. Embryo cultures of coconut palm and tissue culture of oil palm should be derived only from viroid-tested material. However, in addition to oil palm, other palm species successfully inoculated with CCCVd include betelnut palm *(Areca catechu)*, date palm *(Phoenix dactylifera)* and various other palm species. Many cultivars and hybrids have been tested for susceptibility by inoculation in The Philippines, but none have shown any indication of immunity to the viroid. If the viroid-like molecules are indeed identical to those identified in Cadang-cadang-affected palms, it may be expected that susceptible palms in these countries will show symptoms of the disease.

Some experts are not convinced by the work of Hanold and Randles, as they have not demonstrated the viroid nature of the molecules (which may be coconut or palm RNAs) and there is no evidence of a pathogenic effect. The relationship with Genetic Orange Spotting is at most circumstantial evidence and the call for an embargo is certainly overdone, causing a lot of problems for germplasm exchanges. Finally, Hanold and Randles recently revised steeply downwards the rate of infection observed on the original samples after modifying the probes (Rognon, personnal communication). In view of this controversy it is recommended that investigation on these suspected viroids be intensified, preferably at different laboratories.

Methods for control of Cadang-cadang are not known. Trials of control by eradication have failed, probably because the infection may have spread before visible symptoms appear. But Pacumbaba and Alfiler (1987) reported a very low rate of spread, indicating a significant reduction in disease incidence after eradication. Maramorosch (1993) stresses the possibility of disease control by cleaning the knives used, by dipping them in a solution of concentrated sodium carbonate ($NaCO_3$). It is difficult to understand why this possible control method has never been used before. The effects may become visible only after several years, because several palms may have already been infected at the start of the trial, but gradually the number of palms affected should decline, especially in an area where the disease appearance is recent. A combination of eradication and sterilization of knives might enhance the results.

Tinangaja Disease

This disease was also called the Yellow Mottle Decline of Guam, Cadang-cadang of Guam, or Guam disease. The disease was first reported in 1917, and by 1964 most coconut plantations on the island had been destroyed. Tinangaja disease shows symptoms similar to those of Cadang-cadang except that the affected trees bear spindle-shaped nuts with reduced kernel or without kernel.

It has been found that the Coconut Tinangaja Viroid (CTiVd) is very similar to that of CCCVd. With the differences in symptomatology and molecular structure, CTiVd is possibly a milder strain of CCCVd. (Rodriguez and Estioko 1987).

Foliar Decay Disease

Foliar Decay disease (FDD) was formerly also called New Hebrides disease and Vanuatu disease. The abbreviation FDMT for 'Foliar Decay induced by *Myndus taffini*' is also used, but it does not seem logical to name a disease after its vector. In addition to being confusing, this name is also too complicated. So far, the disease has only occurred in Vanuatu. It was discovered when in 1962 a programme of germplasm introduction was started. Among the first batch of imported seednuts were three dwarf varieties, Malayan Green Dwarf, Malayan Red Dwarf and Niu-Leka. Rennell Tall was also imported. In later years, various Tall varieties were added to the collection. Within eighteen months, wilt symptoms developed on the Red Dwarf and later on also on the other imported varieties. Only the Vanuatu Tall was resistant, indicating that the disease is endemic to the islands.

Calvez *et al.*, (1980) described the symptomatology on the Malayan Red Dwarf variety, presenting the most typical and marked symptoms of the disease as follows: The yellowing of the leaflets at the base of a middle leaf (generally rank 7-13) is the first visible sign of the disease. This discoloration spreads towards the tip of the leaf, also affecting nearby leaves. These leaves gradually turn brown from the tips of the leaflets, then dry up from the top to the bottom; the stalk breaks at a variable distance from the stem; the leaves at the base of the crown remaining green for some time, so that even if all middle leaves hang down along the stem, the palm has a lower stage of yellow-green leaves, which are more or less horizontal, and a tuft of erect leaves at the top. At a later stage, the lower leaves turn yellow, then brown and dry out. Only the young leaves remain. Their growth is slowed down very much, and leaf 1 is often stunted. The few leaves which remain in this state for a few weeks finally turn yellow and dry out completely. Closer observation shows that as soon as the first signs of yellowing occur, the leaflets are covered with small brown spots, round shaped and sometimes zoned. Besides, the yellowing is not uniform; in the first stage in particular, one half of the leaf least exposed to the sun remains green longer than the other half. The edges of the stalks of the yellowing leaves decay; drying starts from the base, and up to the level at which the first leaflets are inserted. Then the entire thickness of the stalk is affected, as well as the fibrous tissue around the stem. Susceptible cultivars die between 1 and 2 years after the appearance of the first symptoms.

Sometimes, particularly in rainy periods, the spathes of young inflorescences in the axils of these leaves turn brown prematurely from the top down, then rot entirely. The spathes already visible do not open. On the young inflorescence, the tips of the spikes dry up rapidly, and the very young nuts or flowers turn brown and fall. The nuts already formed develop slowly but remain fixed to the diseased tree for some time. Total drying of the crown and the death of the palm generally occur without a sign of rot in the bud. In some cases, there may be a start of external stem-rot in the zone where tissues are only somewhat woody, at the site of leaf scars left by the lowest leaves of the crown which fell prematurely. In addition to these cases, the stem presents neither external nor internal symptoms. The root system of the trees that show the first symptoms is healthy. At an advanced stage of the disease, the rootlets begin to dry up; subsequently the secondary and primary roots are affected by a dry, brown rot, starting at the tips. Towards the end, when the coconut has only a terminal leaf bunch left, 20-50% of the roots are brown at the level of the bole, and all roots rot quickly. This development indicates that the appearance of these symptoms is not due to rotting of the root system but to lack of chlorophyll assimilation causing root asphyxia. No insects were observed inside these roots.

On other varieties, the symptoms are sometimes slightly different. Discolouration of the foliage of Rennell Tall is bronzed and the first leaves to be affected are generally those at the base of the crown. Necrosis appearing on the edges of the leaf stalk begins about 50 cm from the first leaflets, and does not begin where they are inserted, as in the case of the MRD. On the convex side of the leaf stalk, the oily green spotting, normal on a healthy Rennell Tall palm, is more pronounced on a sick palm. MRD x RIT hybrids present symptoms similar to those of MRD. Palms of other varieties survive longer than MRD, which appears to be the most susceptible variety tested so far. Less susceptible palms have less leaf remission and narrowing of the stem while the bud continues to grow. The leaves fall and the stem turns green in the empty spaces left. The palm dies slowly due to lack of leaf emission. In Rennell Tall there is sometimes remission and even complete recovery.

Treatment of affected trees and seedlings with tetracycline did not show any effect, excluding mycoplasmas as a probable causal agent (Julia *et al.* 1985). Nor could fungi, bacteria, or nematodes be implicated as the pathogen of the disease. Research in Australia has demonstrated a correlation between the presence of FDD symptoms and the detection of an unusual single-stranded DNA. The DNA was detected in both experimentally and naturally infected MRD palms, and this component was disease-specific (Randles *et al.* (1987). Further research indicated the presence of 20-nm viral particles probably acting as the disease's etiological agent (Anonymous 1992a). The incubation period of the disease covers 6-13 months.

Transmission trials indicated *Myndus taffini* (Homoptera, Cixiidae) as the sole vector of the disease, which was always found in disease foci, with gradients superimposable on those of the disease (Randles *et al.* 1987). This insect lays its eggs in the roots of *Hibiscus tiliaceus*, which is a very common shrub in the country. The nymphs develop on the roots of the same plant, but the adults only seem to feed on the sap of coconuts and woodland palms (Anonymous 1992a).

The causal agent is still unknown and control methods are also unknown. Eradication of diseased palms and the control of the vector for the time being seem to be the only measures that might be tried to slow down the spread of the disease in the New Hebrides. Any movement of coconut from these islands to other countries should be banned until more is known about the disease and its control.

VTT shows a high degree of tolerance to the disease, and also the local red dwarf, VRD, has a high degree of tolerance. Unfortunately, VTT is a rather low yielder. The hybrid VRD x VTT is also very tolerant and a much higher yielder than its parents. Rennell Tall also has an appreciable level of tolerance. Other slightly susceptible or tolerant cultivars are BGD x VTT, VRD x TGT, and VRD x RIT (Julia *et al.* 1985). The average yields per ha in tons of coconut for the various varieties are: VTT - 1.5, Selected VTT - 2.5, VRD x VTT - 3.4-4, RIT x VTT - 2.8, BGD x VTT - 4.3, and MYD x VTT - 3.9 (Malosu 1987).

Diseases caused by Mycoplasma-like Organisms

Mycoplasmas are wall-free prokaryotes, discovered in 1967. They are known to be the smallest free-living organisms, capable of autonomous reproduction. Because of their small size, detection is difficult, even under an electron microscope. They belong to a class of microscopic organisms that are taxonomically intermediate between bacteria and viroids. Among the mycoplasma like organisms (MLOs), spiroplasmas play an important role in plant diseases and they are also suspected to be the causal agents of various diseases in coconut. Diseases caused by MLOs are also called yellowing diseases. Although the yellowing diseases discussed below are very similar in their symptomatology, they appear not to be caused by the same MLO. Differences in varietal resistance and epidemiology suggest that different MLOs are involved. The MLO involved with Root(wilt) disease in India seems to be very different from the African and Caribbean MLOs.

Fragments of MLO genomic DNA from Caribbean Lethal Yellowing-diseased palms have been isolated and cloned. This was the first step towards the development of practical diagnostic tests based on specific DNA probes and for the application of molecular biological techniques to increase the sensitivity, specificity and rapidity of detection of MLO. This offered the possibility to compare MLOs associated with diseases from different regions at genome level (Anonymous 1991a). Rohde *et al.* (1993) developed a more rapid method for MLO diagnosis, based on polymerase chain reaction (PCR) to amplify a sequence of the 16SrDNA gene, as developed by Ahrens and Seemüller. This PCR method for the detection of MLO DNA in coconut palm with Lethal Disease is a rapid, sensitive and non-radioactive approach. It is envisaged that this technique may become useful for epidemiological studies on Lethal Disease and for identifying pathogen-free germplasm for coconut breeding programmes.

The following description of the disease symptoms is that of Lethal Yellowingly. Where other yellowing diseases show differences, this will be indicated.

Premature nutfall is the most obvious and often the most visible symptom as far as the farmer is concerned. Initially, the pattern is usually unilateral, the nuts first falling from one side of the palm, followed by a general fall requiring 2-12 weeks to complete. Sometimes a palm will start shedding nuts, stop for two weeks, after which the fall is resumed. Usually, nuts from the middle of the crown are dropped first, followed by nuts distal and proximal to those, and finally, mature nuts. The nuts leave the calyx behind and darken at the apex within 24 hours after shedding. Sometimes the nuts show signs of decay at the base and, when cut, will show discolouration. The meat in some nuts may appear almost normal, in other nuts it may be spoiled. The water may be slimy and tasteless. However, if a nut has matured sufficiently, it will germinate, even from a diseased tree.

The first leaf symptoms are seen on the bud leaves, where pale-brown irregular water-soaked spots appear on the tips of the leaflets, progressing into the cabbage. The innermost youngest leaves begin to discolour shortly after the older leaves begin to yellow. Browning of the spear leaves, yellowing of the mature leaves, and nutfall all start within a narrow space of time, nutfall and yellowing usually being detected earlier than the browning of the spear leaves. Death of the bud occurs about halfway through the yellowing sequence. For a while, the spear remains upright but as the rotting of the base progresses, it may sag sideways, the cabbage of the palm decays, producing a very strong nauseating odour.

Discoloration of the leaves may differ between varieties and ecological conditions. In the Panama Tall variety, the leaf colour turns from green to brown-bronze, whereas in other varieties the leaf may turn to bright golden yellow. The mid-rib of some of the younger leaves may break about half way. Yellowing of the leaves, like nutfall, may also occur unilaterally. It usually starts in one or more of the oldest fronds, progressing, sometimes slowly, sometimes rapidly, towards the centre of the palm. Yellowed leaves later turn brown, desiccate and hang down. They fall readily, or are easily pulled off. The remaining leaves may turn yellow when still in an upright position until the whole crown breaks off. Infected trees usually die within three to six months after the appearance of the first symptoms, but much earlier deaths have also been observed.

In the newly opened spathes, inflorescences show a necrosis usually starting at the tip of the spikes, turning black or dark-brown and hanging limply. Discoloration continues until the whole inflorescence is blackened. Blackening of the inflorescences may also happen in the unopened spathes. The spathes finally turn brown and dry without opening. Usually, all the young spathes are affected and show blackening progressing from the tip downwards.

Roots seem to be severely affected after nutfall has set in and no new roots are produced afterwards. Dark brown patches give evidence of root discoloration, interrupting the normal red colour. Root collapse does not precede the aerial symptoms but rather parallels them. Stem abnormalities are rare. Sometimes, bole rots develop in young palms, starting at the base, spreading radically and to a lesser extent upwards. LY has been shown to require an incubation period of 4 to 9 months before visible symptoms appear.

Lethal Yellowing

Lethal Yellowing wiped out almost all tall coconuts in Jamaica. Although the devastating epidemic outbreak of Lethal Yellowing in the Caribbean occurred after the Second World War, similar diseases were reported from the Cayman Islands as early as 1834 and in Cuba, Jamaica and Haiti at the end of the nineteenth century. The disease has spread all over the Caribbean region, involving Cuba, the Bahamas, Haiti, the Dominican Republic, Mexico, Florida, and Texas. (Howard 1989). In Mexico, world's seventh largest coconut producer, in 1982 the first affected coconut palms were observed in Cancun and Isla Mujeres, from which it spread along the coast to Yucatan, Chizén Itzá and Telchac. The first cases have also been observed in the city of Mérida. It has been estimated that the disease spreads at a velocity of 50 km per yr. which may have enormous economic consequences for the country. (Diaz and Villareal, 1990).

Eskafi *et al.* (1986) observed decreased water transport in diseased palms. The uptake and xylem exudation from excised roots of diseased palms were 10% of such rates in comparable healthy palms. In addition, the rate of waterflow in excised petioles of diseased palms was about 44% of that of healthy palms. The almost complete stomatal closure, even on turgid leaves, was the earliest detectable symptom of LY. Stomata closure was coupled with a vastly increased resistance to diffusion of water vapour from leaves; which is likely to be the primary factor affecting the decreased rate of water transport in LY diseased palms. Studies by León *et al.* (1993) confirmed that coconut palms with LY had a lower stomatal conductance at noon than symptomless palms. This occurred in all fronds of the canopy. The plants progressively lost their diurnal fluctuation. Already in the very early stages of the disease, stomatal closure was spread from the base to the top of the canopy. They concluded that stomatal closure is a central event leading to the appearance of other symptoms. Results of their study are consistent with the idea that part of the mode of action of the causal agent is a hormonal imbalance which promotes physiological changes. Some of these, such as stomatal closure, may be central to the development of the disease, altering functions such as water transport and photosynthesis. Some of the changes, for instance protein loss and increased availability of amino acids (including arginine, which may be involved in the susceptibility to the disease), may favour the growth of the causal agent and further decay of the diseased palm. The disease symptoms are described in paragraph 5.6.

In 1972, Plavsic Banjac *et al.* (1972), using electron-microscope investigations found mycoplasma-like organisms (MLOs) in sections of diseased inflorescences. Studies indicated that MLOs associated with LY diseased coconut palms of Yucatan, Mexico, are genetically similar, perhaps identical, to MLOs associated with LY disease of Florida palms (Escamilla *et al.* 1991).

Lethal Yellowing has a wide range of hosts, including many palm species and one non-palm host, *Pandanus utilis* (Tsai 1988). The susceptibility of palm species varies greatly, ranging from highly susceptible to slightly susceptible. Also, between different coconut varieties there are differences in susceptibility, the Malayan Dwarf and King Coconut being highly resistant, most talls being very susceptible to moderately susceptible. Highly susceptible varieties are Indian Tall, Jamaica Tall, Vanuatu Tall, and Rangiroa Dwarf. The hybrid Malayan Dwarf x Panama Tall is also highly tolerant; crosses between MD with Rennell Tall and with Jamaica Tall are somewhat less tolerant. Most hybrids have levels of tolerance intermediate between those of the parents, but generally closer to the level of the more tolerant parent.

However, it was discovered that at three sites each in Florida and Jamaica there were abnormally high losses of Malayan Dwarf palms due to a disease with symptoms similar to, but sometimes atypical of, Lethal Yellowing (Howard *et al.* 1987). Some palms exhibited leaf-folding, which is considered a symptom of a wilt disease and not typical of LY. However, MLOs were found in all samples from palms with typical and atypical symptoms. In Florida, two of the sites were golf courses, the third was adjacent to a golf course. It was suggested that high rates of fertilization or irrigation or a combination of both could predispose palms to Lethal Yellowing susceptibility by bringing about physiological changes in the palms conductive to the survival of the pathogen, by increasing attractiveness to vectors, or by altering their feeding behaviour, or other factors. There seems to be a general tendency for the disease to spread faster under well managed palms than under palms that are stressed. From Jamaica it has been reported that varieties that had previously shown good resistance have been succumbing in higher numbers than usual. Even the mortality of putative Malayan Dwarfs at some sites was alarmingly high (Steer 1989). It was suggested that more than one strain of the pathogen that causes LY might be active on the island.

Two types of distribution have been noted in Florida and Jamaica. In the first type, one or two palms function as an infection centre and new infections appear randomly around the centre. In the second type, a jump spread is followed by a pattern of local spreads (Tsai 1988). This pattern of spread is characteristic of an airborne vector. There is strong evidence that the vector of the disease is *Myndus crudus* (Cixiidae), a grasshopper (Howard *et al.* 1983). Nymphs of the insect develop on the roots of stoloniferous grasses. Adults have a wider range of plants to feed on. They are often found on coconut leaves. Jump spreads may be the result of involuntary transport of the vector, either by strong winds or by some vehicle. It is unlikely that the disease can be spread by nuts from diseased palms. However, movement of nuts from diseased areas to healthy areas cannot be recommended for safety reasons, not to speak of other parts of the palm or even seedlings including, of course, similar material from other palm species.

Three methods of LY management are possible: treatment of the palms with antibiotics; the planting of tolerant coconut varieties; and control of the vector. Control by injection of the palms with oxytetracycline may save diseased palms if the disease is not too much advanced. Three grams per litre of water, given at four-monthly intervals is highly efficient. But continuous treatment is uneconomic unless in special circumstances such as treatment of palms planted for ornamental purposes. And in most cases, control of the vector with insecticides will be difficult and uneconomic. Eliminating the undergrowth of grasses and planting of cover-crops that do not harbour the vector is a possible measure to reduce disease pressure. A green cover with dicotyledons such as legumes, seems to be preferable. It could be combined with the planting of tolerant coconut varieties.

Lethal Disease

The first reliable reference to this disease in Tanzania was in 1905. A few early outbreaks were reported in the first two decades of this century in the Kisarawe, Rufiji and Kilwa districts. Lasting outbreaks were not reported before the 1940s when the modest early groves had been greatly expanded. But it never caused so much damage as since 1960. It occurs in almost the entire coastal coconut belt of the mainland but tends to be absent or less active in groves further inland (Schuiling *et al.* 1992a). Since 1988 it has also been observed on the island of Mafia near the Tanzanian coast, but the disease incidence on the island is very low. Affected palms on Mafia are cut and burned as soon as they are observed, which may contribute to the low incidence. The islands of Zanzibar and Pemba are still free of the disease. The disease incidence in the various districts differs widely. A few cases of the disease may be observed at all times in Kenya as well, but the economic impact is very low. The Bagamoyo disease in Tanzania (Ohler 1984) and the disease in Mozambique, called 'doença desconhecida', or 'unknown disease' may be similar to LD.

Symptoms of the disease are very similar to those of Lethal Yellowing. Symptoms in pre-bearing palms are similar, although here collapse of spear leaves and roots are early symptoms (Schuiling and Mpunami 1992). Field observations and infection rates show decreasing disease incidence with increasing age. This could suggest that the most susceptible palms in the population gradually succumb whereas the resistant ones remain. However, in resistance trials the progeny of surviving old palms showed no decrease in susceptibility. Preliminary results of resistance trials in Tanzania indicated that, distinct from Lethal Yellowing, none of the imported varieties, including dwarfs, show promising resistance to LD. Losses in hybrids are not lower than in corresponding tall parents, another indication that there is no resistance in dwarf parents. Consequently, no hybrid or variety has shown enough resistance for commercial exploitation, contrary to the situation with Lethal Yellowing in Jamaica where several dwarf varieties show excellent resistance, some tall varieties show intermediate resistance and their hybrids good, commercially acceptable resistance. In the present trials, promising resistance to LD was observed only in sub-populations of the local East African Tall (EAT), specifically a sub-population from Tanga. Lower losses in this sub-population suggest that the EAT is not uniformly susceptible to LD. In areas where LD has existed for a long time, resistance to the disease in the surviving population may have come to predominate. (Schuiling *et al.* 1992a).

The rate of spread of LD is rather similar to that of yellowing diseases in West Africa but much slower than that of Lethal Yellowing in the Caribbean region. In certain areas of Tanzania the disease is very active, whereas elsewhere it hardly advances. Even in areas of rampant disease, active centres are often relatively small and distributed in patches. They may become inactive after a certain period and then reactivated later. The mode of spread is erratic, some groves escape infection whereas others are totally destroyed. Active foci may extend as much as half a kilometre per year but generally advance far more slowly. Jump spreads are rare. The disease rarely affects less than 18 month-old palms and is not a juvenile disease. Invariably, the first cases of disease and consequently higher incidence occur along the wind-exposed borders of the field. Least affected groves are surrounded by mature stands, forest or bush. Highest incidence occurs where the groves border open spaces with uninterrupted air movement such as roads, valleys, and tidal channels. These aspects of spread strongly suggest an air-borne vector. Incidence in older palms may be related to the height of the palm, which suggests a vector that is not a strong flier. Extensive transmission trials with insects have so far failed to implicate a vector (Schuiling *et al.* 1992a).

Cape St. Paul Wilt

Cape St. Paul Wilt in Ghana was first observed at Weh, near Keta, in 1932. By 1951, the epidemic was spreading rapidly north-east of Keta and by 1970 the vast majority of palms in the Weh-Cape St. Paul and Tegbi areas had been wiped out. The spread of the disease is slow when compared to Lethal Yellowing in Jamaica.

The symptoms are very similar to those of Lethal Yellowing, but there are also some differences. In Ghana and in Togo (Kaincopé disease) the spear was often dead while many leaves still remained green, contrary to Lethal Yellowing, where the spear is often still living, although with patches of necrosis, when many leaves have already yellowed.

MLOs were observed in tissues of root-tips and in the sieve tubes of inflorescences of diseased palms (Dabek *et al.* 1976). No resistance of coconut varieties has been reported, although dwarf palms appeared to be affected only after twelve years of age. (Gianotti *et al.*, 1975). Bourdeix, in this book, reports that especially the dwarfs of Asian origin, MYD, MRD and Sri Lanka Green Dwarf, most of their hybrids and, the progenies of MLT and VTT, show a better tolerance than other palms.

Kaincopé Disease

In 1932, at the same time that Cape St. Paul Wilt was first observed in Ghana, various coconut palms near the village of Kaincopé in Togo, about 10 km east of Lomé suddenly died. For a period of more than twenty years, the disease was confined to this area, but increasing in intensity. In 1954, the disease suddenly appeared 20 km further eastward at Port Seguro, along the coast, from where it spread to various sites in the surroundings. Since 1974, the disease seems to have disappeared from the coastal region and to have spread north-west. The spread of the disease ended on riverbanks and lagoons, which might indicate a vector that cannot move over long distances. However, jump spread of the disease has also been observed. Along the coast, the spread of the disease was slow, whereas in the interior the spread was much faster. New incidences of the disease tended to increase with the onset of rains and during rains. Dabek *et al.* (1976), Dollet *et al.* (1976), and Nienhaus and Steiner (1976) all found MLOs in tissues of diseased palms.

Kribi Disease

There are various foci of Kribi disease in Cameroon. The disease was first detected in Ebodie in Cameroon, in 1937, between Kribi and the Equatorial Guinee border. Dollet *et al.* (1977) found MLOs in sieve cells in tissue taken from diseased inflorescences.

Although the spread within a focus is rapid, the general spread of the disease over the country is not rapid. For a long time the Cameroon Red Dwarf, as well as its hybrids with local talls, remained unaffected by the disease. It was not until 1975, that the first diseased CRD palms were observed. It was suspected that the CRD palms remained unaffected as they were usually planted very close to human dwellings.

Blast

Blast of coconut was discovered in Ivory Coast in 1971. It usually occurs in nurseries and young plantations, together with another disease, Dry Bud Rot. The disease is not widespread, but locally it is of economic importance.

Symptoms of the disease are identical to those of Blast in oil palm, also occurring in the same country. Blast always leads to wet rot of the spear and the roots, yellow-brown colouring of the bole and rapid drying up of the palm, starting with the oldest leaves. Losses occur mainly in the first year after planting. The disease is very rare in older plants. It was found that the age of maximum susceptibility was when the seedlings had 4 to 6 leaves, younger and older plants being more tolerant. Susceptibility and incubation time differs according to variety. The incubation time of the disease also varies with the season. It was found that the first symptoms developed about 4 weeks after transmission in July or October, whereas it takes only 10 days for the first Blast symptoms to occur after transmissions in November (Anonymous 1992). The most tolerant genotypes showed 100 per cent symptom remission, the susceptible ones only 44 per cent.

Transmission to herbaceous plants produced symptoms of wilting and yellowing in *Vinca rosea*. Electron microscope examination of these *Vinca* revealed the presence of MLOs. Results of tetracycline treatment suggest a mycoplasmic origin of the disease. Transmission trials indicated some species of the Delphacidae, *Sogotella kolophon* and *S. cubana* as the vectors of the disease (Julia and Mariau 1982).

Treatment with Temik effects about 80-90 per cent reduction in Blast attacks on oil palm, but only about 40 per cent reduction on coconut. It was suggested that uptake of the insecticide by the roots was not rapid enough in coconut. Injection of Temik or Azodrine into the husk of very young seedlings was not effective either. Shading is by far the best control method, in the nursery as well as in the field. Clean weeding in and around the nursery also helps to keep the insect population down. Suppressing grass growth by establishing a legume cover-crop such as *Pueraria* also reduced the vector population in the field.

Root(Wilt)

Root(Wilt) disease (RWD) was first reported in the State of Kerala, India, after the great flood of 1882 in three independent locations, each about 50 km apart. Since then it has spread from the original foci of infection. According to a survey conducted during 1984/85, the disease was prevalent in more or less a contiguous manner in 410000 ha in the eight southern districts of Kerala. It was also observed in a few isolated pockets in the northern districts of that state and in the bordering districts of the State of Tamil Nadu (Solomon and Pillai 1991). The intensity of the disease in the contiguous diseased tract ranged between about 1.5 per cent and 75.6 per cent. The annual loss due to the disease was estimated at about 968 million nuts. RWD is non-lethal but debilitating, and palms of all age groups are affected. It was given its name by early investigators, probably because rotten roots had been observed on diseased palms. Later, the name Wilt was considered more appropriate and these names have been combined (Rawther and Pillai 1991).

The first symptom is deterioration of the spear, indicated by whitening and softening of the leaflets. These soft leaflets are whitish-brown to pale-green. Round to rod-shaped necrotic spots appear at the margins. Usually, the leaflets are rotten at their tip margins. They have inter-veinal chlorosis followed by marginal necrosis. Chlorosis starts from the tips and progresses towards the bases of the leaflets. The general yellowing and drooping of the middle and outer whorls of leaves and flaccidity of leaflets have been indicated as the most pronounced symptoms of the disease. The flattening and bending of these leaflets and the drooping of leaves give the diseased palm a wilted appearance. At a later stage, the bending of leaflets from two sides of the mid-rib give the appearance of ribs of a human skeleton (Rajagopal 1991). But the expression of foliar symptoms varies both in frequency and in association with each other, depending on the soil types and ecological conditions. Intensity of foliar symptoms also varies according to the age of the palms, yellowing and marginal necrosis being virtually absent in palms less than ten years old (Pillai and Rawther 1991). The growing point remains unaffected. In certain cases, drying up of the spathe and necrosis of the spikes extending from the tip downwards was observed. Inflorescences may grow weaker, producing fewer buttons and nuts. Buttons and small nuts may be shed. There may be a reduction in nut size, the endosperm is thinner than usual and uneven in thickness, and when dried remains flexible. Also the husks are thinner and their fibres weaker than usual. Palms affected before flowering may not flower at all.

Rotting of roots was once considered a major symptom. However, this could not be substantiated in later studies (Solomon and Pillai 1991). Notwithstanding, the uptake of water by roots of diseased palms is much less than that of healthy palms (Rajagopal *et al.* 1986). Thomas (1988) observed that the endomycorrhizal symbiosis of coconut is adversely affected by RWD.

Electron microscopic examination of juvenile tissues showed the presence of a phloem-bound MLO, but not in all cases. It is assumed that MLOs produce metabolites that influence stomatal regulation, either their closure or opening (Rajagopal 1991). In addition to reduced water uptake, the stomatal regulation plays a key role in the ultimate expression of foliar symptoms. Contrary to stomatal closure, a characteristic feature of all other yellowing diseases, RWD palms have an abnormal stomata opening with impaired regulation leading to excessive water loss, irrespective of the time of day, season or growing conditions. By two tests, the serological test using the cross absorption technique, and stomatal resistance determination, the disease could be detected 6-20 months earlier than the actual manifestation of flaccidity symptoms (Rajagopal *et al.* 1988).

Root(Wilt)-diseased palms have higher stomatal frequency than of healthy palms. Diseased palms have consistently lower leaf-water potential than healthy palms (Mathew *et al.* 1991). Imbalance of the water economy caused by a deranged root system and impaired stomatal regulation culminates in an irreversible flaccidity symptom of leaves through changes in leaf-water potential components. The disease is not lethal but is accompanied by other diseases such as fungal leaf diseases, which may result in the death of the palm. No recovery of affected palms has been recorded yet.

In general, the spread is erratic and irregular. Jump spread also occurs, with a reach of about four kilometres from the nearest source of infection. The disease occurs in all major soil types, but the spread is faster in sandy, sandy loam, alluvial and in heavy textured soils than in laterites. The disease incidence is relatively higher in water-logged low lying areas adjacent to rivers and canals.

The lace bug *Stephanitis typica* is suspected to be the vector of the disease. Transmission trials with this insect had positive results (Mathen *et al.* 1990). The lace bugs are found colonizing in increasing numbers towards the inner leaves of the crown. The number of lace bugs on diseased palms is found to be much higher than on healthy palms. A survey showed a direct correlation between the number of insects colonizing the palms and fresh incidences of the disease (Mathen 1985). In RWD palms, the nitrogen content is higher in the middle and advanced stages of the disease compared to that of healthy trees, and the accumulation is found more towards the young and expanding leaves when the disease intensity increases (Wahid and Kamalam 1988). Accumulation of nitrogen may make the cell walls thin, which is evidenced by observations that leaflets from diseased palms were thinner and their cell constituents showed a general reduction in size. A high nitrogen content of plants with the application of nitrogen fertilizers is often correlated with a heavy incidence of a number of insects. Electron microscope studies have revealed the presence of MLOs also in a plant hopper. The vectorial role of the hopper is to be assessed through a transmission experiment (Solomon and Govindankutty 1991).

All exotic cultivars tested were found susceptible to the disease with varying degrees of intensity. However, the San Ramon variety recorded significantly lower disease incidence, followed by Guam Tall and the St Vincent and Kenya Talls. Remarkably, the local West Coast Tall was found to be the most susceptible variety, followed by Java Tall (Mathai *et al.* 1991). Chowghat Green Dwarfs (CGDs) showed maximum field tolerance (over 90%) to the disease. A breeding programme apparently involving disease-resistant CGDs and selected West Coast Talls free from disease growing in diseased areas has been initiated (Jacob and Rawther 1991). Palms treated with 3 - 6 g a.i. of oxytetracycline hydrochloride clearly indicated remission of symptoms (Pillai *et al.* 1991).

In the long term, the provision of resistant varieties seems to be the only economically justified solution to the problem. An eradication programme of diseased palms followed by surveillance was started in 1971 in some districts. By 1984, the recurrence of the disease was observed only in one village where the initial disease intensity was high, indicating that the disease could be eliminated from mildly affected areas if phytosanitary measures were adopted (Radha, K. *et al.* 1985). Integrated management practices, such as fertilizing, irrigation, proper drainage, control of other diseases such as Leaf Spot, and intercropping may substantially improve yields, indicating that in the mildly affected areas one can live with the disease (Rajagopal *et al.* 1986; Rethinam *et al.* 1991).

Tatipaka Disease

This disease was first observed in 1952 in the State of Andhra Pradesh, India. It is a slowly developing disease that debilitates the palms. Sporadically but progressively, the disease spreads at a very slow rate of 3.5 per cent in five years (Jayasankar 1991).

The disease is mostly observed in palms 20-60 years old. Young palms below 20 years of age are rarely affected. The symptoms described by Rajamannar *et al.* (1993) are as follows: 'Reduction in the number and size of roots; extensive root rot; root regeneration is greatly affected, resulting in slow decline of the diseased palm; reduction in the number and size of the leaves which become light green and turn yellowish with a green tinge; chlorotic spots develop on leaflets; in some cases leaflets of some leaves adhere together without normal spitting; the leaves bend in the middle giving a bow-like appearance; spathes are very small with few spikes; the bunches contain a mixture of normal and atrophied nuts, and in extreme cases palms become barren. The production of nuts reduces with the progress of the disease; the development of the nuts is not normal, they become round with a soft mesocarp and with the advancement of the disease the nuts become shrivelled and atrophied and do not contain water, copra or shell.'

Electron microscope examination of tissue from diseased palms indicated the presence of MLOs in addition to rod-shaped virus particles, cytoplasmic crystalline inclusion-like structures and rare inclusions looking like parallel arrays of virus particles in different samples (Rajamannar *et al.* 1993). Further investigations indicated the association of MLOs with the disease.

The dwarf cultivar, Gangabondam was noticed to be disease free, even in the areas where a very high incidence of the disease on local East Coast Talls was observed. Good management and fertilizer application helps to reduce the disease effects.

Diseases of Unknown Etiology

Ampas Tebu Disease

Ampas tebu disease was first reported in 1985, from Benkulu on the Island of Sumatra, Indonesia, affecting local tall palms. Leaves of the affected palms dry and eventually wilt, and palms may finally die. The disease seems to occur on various soil types and there is evidence of its spreading (Sitepu 1983).

Awka Wilt

The Awka area in the State of Anambra, Nigeria was the site of a severe epidemic in 1917 of what, from the description in reports, was presumably the same disease as the one causing a severe epidemic in the Awka-Onitsha area in 1951. The disease was called 'bud rot' at that time. More than 5000 diseased palms were felled in the course of the epidemic, which must have been almost all the coconut palms in that area. The palms showing disease in 1951 were probably planted on the site of the old epidemic, some time during the 1920s and 1930s. It was suggested that the disease was identical to Bronze Leaf Wilt disease of the West Indies, a name covering several diseases and physiological disorders that were later distinguished and named differently. Since 1980, the disease has also been observed in the neighbouring Bendel State (Ekpo and Ojomo 1990).

Symptoms of the disease are quite similar to those of MLO yellowing diseases. The first symptom in mature coconut palms is the premature dropping of nuts, regardless of size. Most of the nuts have a brown or black water-soaked area immediately under the calyx on the stem end. Then the leaves start yellowing, usually from the lowest leaves upwards. Thereafter, the leaves turn brown, desiccate and hang down. Such leaves fall easily and can be pulled off easily. As yellowing progresses, the bud tissues begin to rot, the necrosis of the meristematic tissues apparently beginning at the base of the central spear, and the bud dies. Finally, the top of the tree falls away, leaving a bare stem like a telephone pole.

At the nutfall stage, frequently a marked brown discolouration occurs at the terminal part of the inflorescence in the oldest unopened spadix subtended by a healthy leaf. Usually open inflorescences are necrosed and die. Unopened inflorescences are the last part of the crown turning yellow and then brown (Ekpo and Ojomo 1990). They eventually collapse due to rotting of the base. Transverse sections of the stem do not show abnormal colouring or rotting. Discolouration of the roots starts at the outer roots, while the central roots are still normal. With time, the central roots also discolour and become necrotic. Infected trees usually die within six months after the appearance of the first symptoms. The infection rate in local talls is sometimes as high as 100 per cent. According to local farmers, the local green dwarf is not affected by the disease. Young palms below 4.5 m in height are rarely diseased, but the disease incidence increases rapidly with the age of the palm. The author has not found any report on the detection of the causal agent of the disease. Descriptions of the symptoms may indicate an MLO as the causal agent, but similar symptoms have been observed in diseases not caused by MLOs. Therefore, Awka Wilt has been classified here as a disease of unknown etiology.

Dry Bud Rot

Dry Bud Rot is a juvenile disease in Ivory Coast, where it was first observed in 1972. Similar to Blast, losses occur mainly in the first year of planting, but a few cases have been observed 2 and 3 years after planting. The disease occurs in the nursery as well as in the field. According to Julia (1979), the disease is identical to stem necrosis, which caused substantial losses among nursery seedlings and in young plantings in Malaysia, Indonesia and The Philippines.

The first visible symptoms are small, square whitish spots, individual or linked, forming streaks on the first just-opened leaf or on the spear. These spots are also visible on the basis of the leaf where it has not yet turned green. Pink patches develop on the petioles. In cross-section, petioles show pink streaks, which in the early stages of the disease do not affect the vascular bundles. At this stage, the heart of the palm does not show any abnormality. At a later stage, the white spots turn brown and coalesce, spreading over the whole blade of the leaflets. On the rachis of leaf 1, and sometimes also on leaf 2, 3-10 mm long longitudinal cracks may appear, corresponding with brown, cork-like, more or less hollow zones within the rachis. Right from the first symptoms, elongation of the spear and the leaves is halted, they lose their turgescence, gradually turning to yellow and brown, the symptoms also spreading to the other leaves. Simultaneously with the leaf symptoms, cork-like spots appear on the bole, not far from the meristem. There is an abnormal thickening of the collar. Generally, the roots have a healthy appearance; sometimes, however, some brown patches may be found on the cortex, similar to those found on the bole and the leaf rachis. Finally, all the leaves dry up, and the brown rot reaches the growing point, killing the palm within one or two weeks after the appearance of the first symptoms (Renard *et al.* 1975).

The disease is suspected to be transmitted by one or several species of the *Delphacidae* (*Sogatella yubana* and *S. kolophon*) which usually multiply in grasses and bushes in very damp places, flying off in sunshine. *Hibiscus tiliaceus* is an alternative host plant (Renard and Dollet 1991).

Elimination of grasses in and around the nursery reduces the disease incidence. Shading in the nursery as well as in the field is a very effective method of reducing losses. Also, the establishment of a leguminous cover crop before transplanting seedlings to the field considerably reduces the vector population. In the nursery, monthly applications of the systemic insecticide Temik (aldicarbe 10%) at a rate of 4 g per plant reduces the disease incidence by 80-90 per cent.

The West African Tall is the most tolerant variety and the MYD x WAT is the most tolerant hybrid. It is also more tolerant than Rennell Tall (Renard and Dollet 1991). The green dwarf is very susceptible.

Fenerive Decline

Recently, a new coconut disease, named Fenerive Decline, was reported from Madagascar (Dabek 1993). The disease had already been reported in 1989, but the report had limited circulation.

The symptom syndrome of Fenerive Decline is characterized by progressive yellowing, necrosis and shedding of leaves, starting with the outer ones and gradually working inwards to the younger, centrally positioned leaves. Sometimes, the outer leaves hang down from the trunk, forming a 'skirt'. At mid-stage, palms bear virtually no mature nuts, while the remaining developing nuts are abnormally elongated. On dissection of the palm the unexpanded spear leaves show rot on the petiole basis and small, brown necrotic patches on the white pinnae; no necrosis has been observed on inflorescences and no necrosis has been observed on the apical growing point. No healthy-looking roots could be found by excavating around the bole of a mid-stage diseased palm. At an advanced stage, only a central tuft of young, recently expanded leaves remain. These show a tendency to dry to a brown necrotic colour when the leaves are still in the upright position. Finally, these leaves may break off. Tapering of the trunk of some affected palms suggest that the palms may have been in decline for at least two to three years. By electron microscopy, occasional bodies containing central, prominent congealed strands of DNA were detected in some sieve cells, and these showed some similarity in size and appearance to those reported in sieve cells of coconut palms affected by Root(wilt) in India, which were reported to be MLOs. However, the MLO nature of these structures is questioned.

Leaf Scorch

Leaf Scorch (LS) also called Coconut Withering Disease or Ceylon Wilt was first observed in 1955. Its symptomatology is somewhat similar to Root(wilt), and to some extent to Cadang-cadang.

Usually, only adult palms are affected but the disease has also been observed on young palms before the flowering stage. The clearest visible symptom is withering of the leaflets, starting at the tip and proceeding towards the base. They develop a greyish-brown colour, often preceded by yellowing. The transition from living to dead tissue seems to be relatively sudden. Withering usually starts with older leaves but occasionally middle leaves also show this symptom. In a more advanced stage, the outer margins of the lamina begin to wither and necrosis eventually reaches the mid-rib. The lamina may break away in small pieces, leaving an almost barren mid-rib. Sometimes, necrosis also may appear on young leaves. Necrosis streaks are first seen as thin, translucent, whitish or yellow lines, between 2 and 20 mm long, gradually turning into deep yellow, brown or grey. Their length increases with time and adjoining streaks may coalesce into broader streaks. They create withered patches, somewhat reminiscent of Root(wilt) disease. Flaccidity and ribbing of the leaflets are always associated with the disease. Sometimes the rachis of the leaf tips break at an advanced stage of the disease. The crown gradually becomes smaller and the stem tapers until finally the palm succumbs. Inflorescences have more buttons than usual. A reduction in the number of leaves is followed by a reduced number of nuts produced. Only one or two nuts are produced per bunch. The diameter of the nuts reduces, although their length is normal. No shedding of nuts takes place. Usually, palms die within 2-6 years, but instances of palms surviving much longer also occur.

There is little, if any spread from palm to palm but affected palms are grouped together only when soil conditions are bad. The disease seems to be linked with certain soil conditions, such as laterite soils with a hard pan near the surface and soils with a high fluctuating water table, which suggests that the disease is related to drainage problems.

Leaf Yellowing Disease

The first report of Leaf Yellowing of coconut in Indonesia in 1978, came from South Sulawesi where the disease spread and caused damage to many smallholdings. In 1989 it was reported from Poso, Central Sulawesi, as well, where it attacked a large area of smallholdings. It occurs in coconut on various soil types and ecological conditions (Sitepu and Darwis 1989).

The disease attacks young and adult palms, talls and hybrids. The typical symptom of the disease is an intensive yellowing of the leaves, starting from the tips of the leaflets and the older leaves. Eventually, all leaves except the youngest become yellow. In some cases the tips of affected leaves become necrotic. In an advanced stage, leaves and leaflets become smaller, until the whole crown is reduced and stunted. Nuts gradually become smaller. Nuts appear normal, but the copra content is small and of poor quality. Usually, yields are reduced by about 60 per cent over a period of 10 years. Some trees stop production completely (Sitepu and Darwis 1989).

Malaysian Wilt

This disease was first reported in 1928, and has been observed in various places in the country. It usually affects single palms, but groups of ten or more affected trees have also been observed.

Premature nutfall is the first visible symptom. Sometimes, distorted or small nuts may be seen on diseased palms. Immediately thereafter, the lower leaves gradually wilt and dry and the newly produced leaves gradually diminish in size, becoming stiff-looking and yellowish. The affected leaves commonly show rotten spots at the base, the rot running well into the stem tissue. Prior to falling, dead leaves often remain suspended from the apex of the stem. The mid-rib of some of the younger leaves breaks half way along, often while the leaves are still green. The crown is eventually reduced to a few small yellowish-coloured stiff leaves, which die later, leaving a bare stem. The bud becomes affected in the later stages of the disease only, showing necrosis and developing an evil smelling soft rot. The inflorescences develop necrotic spots on the spikes. A cross-section of the stem shows salmon-pink discoloured patches, intermingled with mustard-coloured tissues. These may be traced from the base of the stem to the crown, in a cylinder of gradually decreasing diameter. The pink discolouration becomes slightly less pronounced with height, until approximately 60 cm from the apex the tissues look pale apricot coloured, remaining white and healthy below the bud. The stem hardens and is difficult to cut. When a tree is cut, its break is characteristic, tearing apart the tissue that is not cut, leaving the vascular bundles standing erect and stiff to form a brush-like arrangement.

The roots show a pale pink to deep vermilion and brick red discolouration. If the discolouration is confined to the cortex, it is usually pink, whereas when occurring in the stele, it is red. The discolouration is usually more pronounced close to the bole. Affected roots become brittle, their outer tissues sloughing off easily. Some roots remain healthy until the death of the palm. Discolouration of the stem has been observed in apparently healthy palms, showing that this feature may start before nut shedding. The time between premature nutfall and the death of the palm is six months on average. No recovery has been observed.

Natuna Wilt

Natuna Wilt was first reported by local growers on the Natuna islands group, Indonesia, between Peninsular and East Malaysia in 1965. Only a few trees were affected on the western part of the island and not much attention was paid to it. However, in 1976, an outbreak caused serious problems and more than 12,000 palms were killed and another 4500 were infected by the end of that year. Since its outbreak, the disease has continued to spread to other plantations as well. Since 1976, the disease has killed more than one million productive palms (Sitepu and Darwis 1989).

The disease affects and kills young and old palms alike. Initial symptoms of the disease appear on the youngest leaves, characterized by wilting and bending at the tips, slightly changing in colour. In more advanced stages, the mid-rib breaks and the lower leaves dry up and hang down, then fall. Immature nuts start dropping at the same time as the leaves, until the tree is devoid of fruits and buttons. Fruits that still ripen on diseased palms do not show any abnormality. The bases of the young leaves show black or brown discolouration as a result of rotting. As the disease advances, the inflorescences do not open fully, or remain unopened. In certain stages, rotting develops inside and at the tip of the inflorescence. Diseased palms have more deteriorated roots than do healthy palms. The cross-section of the tree near the crown shows discolouration and the internal tissue is of a slimy nature. Apparently, the disease attacks only palms older than 25 years. It can kill a palm within a short time.

The causal agent of the disease is still unknown. Visual symptoms show similarities with Malaysian Wilt. Bacteria such as *Xanthomona* sp and *Erwinia* sp, as well as fungi such as *Thielaviopsis* sp and *Botryodiplodia* sp. are consistently present in diseased trees. Treatment with oxytetracycline appears to suppress the pathogen, which could indicate the involvement of an MLO (Sitepu and Darwis 1989).

The disease spreads quickly, the spread not showing a regular pattern. Initially, the incidence of the disease was sporadic but it spread rapidly over a wide area. New infections could be near or far away from the area of initial infection. Patches of diseased palms became a common sight in plantations on the small islands. The pattern of spread suggests an airborne micro-organism.

Socorro Wilt

Socorro Wilt (SW) has been observed in Eastern Mindanao in The Philippines since the early 1960s. It affects bearing and non-bearing palms, mostly below the age of 25, killing them within four to six months after the appearance of the first symptoms. The disease seems to be spreading. It is still confined to the island of Mindanao, especially around the town of Socorro, but it has been reported to have spread to the nearby towns of Puerto Galera and San Teodoro (Napiere 1985).

The early symptom is premature senescence of the outer whorl of leaves hanging down, through buckling or breakage at the proximal portion of the petiole. The affected leaf dries up from the tip towards the base. Younger leaves gradually turn to pale yellow and wilt, some of them breaking about half way along the mid-rib, until only the youngest leaf or leaves remaining erect. When the youngest leaf collapses, the palm is dead. Rotting of the growing point seems to be influenced only by the invasion of secondary organisms. Premature nutfall accompanies or precedes the leaf symptoms. Inflorescences fail to develop entirely. Emerging spathes turn brown and die. Some nuts that are formed during disease development are small, usually oblong and younger nuts never increase in size either. Nuts having attained maturity are likewise deformed and in some cases waterless, with dry kernels, loose, or relatively easy to detach from the shell.

The stem tissue just below the crown may be off-coloured and relatively soft. This symptom gradually diminishing at some distance downwards. In the same stem region, cracking of the pith may also be observed in some palms. Tapering of the stem or the manifestation of a dehydrated appearance of the stem just below the crown may be evident on younger palms. Most roots are decayed and blackened; however, they may not easily be broken. The parenchym tissues are discoloured, looking dehydrated. The central cylinder is also discoloured or light brown. Some roots in the advanced stage of decay may exhibit a white mycelial growth of fungus. The symptoms are somewhat similar to Malaysian Wilt and Natuna Wilt.

So far, pathological studies to implicate isolated fungi and bacteria from diseased tissues have not shown an association of any micro-organism with the disease. Sap transmission trials failed to reproduce the disease on inoculated healthy palms. Electrophoresis did not reveal the presence of virus or viroid organisms. Nor did electron microscopy, tetracycline injection or phase contrast microscopy detect the presence of any MLO, ricketsia or protozoa. The plant parasitic nematodes found in the soil could not induce sudden wilting in coconut (Concibido 1985). From an analysis of variance it appeared that coconut palms between 11 and 15 years of age were most susceptible, followed by palms of 16-20 and 21-25 years of age. Disease severity of palms aged between 6 and 10 was significantly less and palms of 1-5 years old and 26 years and older were practically free from the disease. Disease-free areas had a significantly lower incidence of rhinoceros beetle, but this can be a result of fewer dead palms presenting fewer breeding places for this insect. Dominance of the weeds, *Imperata cylindrica* and *Mikania cordata* were positively correlated with the disease severity, whereas *Paspalum conjugatum* was negatively related. Soil pH, cation exchange capacity and textural grades such as per cent sand, silt and clay were also related to the disease incidence. The results of these studies do not allow any conclusions yet (San Juan, N.C. *et al.* (1986a). Spatial distribution and severity of the disease relative to time indicate that the disease appears to increase in a jump spread pattern. Within the active focus, the disease does not spread contiguously but rather attacks one, skips several, and hits another. This pattern of spread bears some similarity to that of Lethal Yellowing. The disease tends to increase very rapidly during the early phase of the epidemic and appears to slow down towards the terminal phase. The rate of infection is much slower compared to that of Lethal Yellowing. As far as rapidity of infecting healthy palms is concerned, there is some similarity to the Cadang-cadang disease. Information obtained from the analysis of spread also adds to the evidence that SW is caused by an infectious agent and is disseminated through the air or possibly by an airborne vector (San Juan *et al.* 1986b).

Stem Necrosis

This disease was reported from Sumatra, Indonesia, in 1979 (Turner *et al.* 1979; 1980). The incidence occurred in MYD x WAT hybrid planting material imported from the Ivory Coast. Later on, there were indications that the disease occurred in similar planting material planted in Malaysia. The disease incidence occurred when plants were about 18 months old. Incidence dropped to insignificant levels when the plants had reached the age of three. Insignificant disease incidence was noted in the Nias Dwarf and none at all in the local tall palms. A large number of MYD palms succumbed to the disease during the first year in the field.

The first symptom is that the youngest opened leaf is shorter and does not open fully. At this stage the spear is either already discoloured brown, or is showing light brown necrotic spots prior to death. In the early stage, the spear is usually whitish but marked throughout with brown spots on the leaflets as well as on the mid-rib. Soon, the youngest leaf turns brown and dies, followed by the older leaves. Dieback of the leaflets starts at the tip and the discolouration is directly from green to brown. Before the first leaf dies, the unexpanded leaflets frequently show a number of brown, parallel necrotic streaks, about 1 mm wide, often positioned in a mirror image on both sides of the leaflet mid-rib. Some young leaves may show a dark grey-brown longitudinal streak on the mid-rib, about one to two mm wide, which corresponds to the position of internal necrosis. The leaf bases are rotten and the inflorescence initials are aborted. Rotting spreads towards the youngest leaves. When split, many young leaves show extensive internal tissue disorganization of the mid-rib. In many instances, the large extent of internal disorganization makes the normal healthy green colour of the leaflets look remarkable. A similar situation is found in the few inflorescences reaching anthesis. By the time the external symptoms have appeared, extensive necrosis has occurred below the bud. The oldest part of the largest lesion lies closest to the meristem. It has a purplish-pink colour and is surrounded by a yellowish zone which is typical of diseased coconut tissues. Further down the stem, the necrosis becomes more sharply defined in cross-section, usually in roughly circular patches which often appear as dark brown cores with smaller 'satellites' of lighter coloured lesions around them. There is an apparently healthy area between patches and roots. Vascular strands tend to be reddish brown for a short distance only, ahead of the necrosis, which is an indication of vascular tissue being primarily affected. The bud is the last tissue to become affected. In later stages of the disease, when stem necrosis has reached the root system junction, the root system is killed. About five months elapse between the first visible symptoms and the death of the palm. Symptoms of the disease are rather similar to those of Dry Bud Rot in Ivory Coast, and Malaysian Wilt in Malaysia. The observation of MLO's has been reported but has been questioned.

6. Vertebrate Pests

J.G. Ohler

The most important pests of coconut are rodents and insect pests. In addition to these pests, there are other animals causing damage to coconut palms or to fallen nuts, but they will not be dealt with as they are generally of local and occasional nature. Insect pests will be dealt with by Dr. D. Mariau, in the following chapter.

Rodents

Rodents have a pair of ever-growing sharp chisel-like front teeth with which they can easily cut through coconuts husk and shells. Fallen nuts with a hole in them are common in many coconut plantations.

Among the rodents, rats are the major pest of coconut. The black rat *Rattus rattus* seems to be the most widespread species. In coconut plantations they mainly live in the crowns of palms where they build their nests either in the interspaces of nuts or inside stipules, in the palm spear. They move from palm to palm through the leaves and seldom come to the ground for foraging (Bhat and Sujathna 1993). Rats gnaw holes in the nuts, mainly for coconut water. It has been suggested that ample availability of fresh water might reduce rat attack on coconuts. This might be the reason for heavier damage caused by rats on flat coral islands than on mountainous ones. The holes are generally located in the perianth end of the nut and have a diameter of about 4-5 cm. Three-to six-months-old tender nuts are preferred by these animals. In addition to tender nuts, damage is also caused to leaf stalks, unopened spathes, female flowers and mature nuts, as well as to nuts stored in the field. Rats often favour particular palms in a plantation, most of the damage being caused to a minority of the palms. Use can be made of this fact by concentrating control measures on these palms. The reason for the preference of the rats may be the taste of the nutwater or other substance. Damage by rats may be compensated to a certain extent by a reduction of physiological premature nutfall. Compensation by increased copra content of undamaged nuts is of very little economic importance. Damage increases when certain intercrops such as cocoa and cassava are grown under the coconut. Also, rice cultivation near coconut groves may cause heavy rat infestation.

To control rat damage, metal bands, about 40 cm wide may be fixed around the palm stems, sticking out like skirts to prevent the rats from climbing up to the crown. The bands are placed about 2 m above the ground to avoid damage by animals. It is also possible to fix a 0.2 mm thick transparent polythene sheet of 30 or 40 cm width around the stem (Sadakathullas and Abdul Kareem 1994). Before application, all rats should be chased from the crown and leaves touching adjacent palm leaves should be eliminated. The initial cost of this method is high, but it is very effective and long lasting, especially if non-corroding metal such as aluminium is used. However, it may be a hindrance to harvesting by climbing.

The crowns of coconut palms can be cleaned by removing all old leaves, sheaths and spadices, exposing the rats to their predators. Further effective control can be achieved by poisonous baits. For this purpose, anticoagulants such as warfarin, fumarin and coumarin or difenacoum can be used, mixed with various preferred foods of the rats. Rat food preference may be related to their earlier feeding experience. In rice growing areas rats may prefer rice to other grains. The addition of sugar may increase the attraction of these baits (Bhat and Sujathna 1989). Bhat and Sujathna (1993) recommend the use of blocks made of a mixture of rice, palm sugar, paraffin wax and poison in the ratio of 12:1:16:1, which are placed in the crown of coconut trees. In heavily infested plantations, three blocks of about 35 g are placed in the crown of every fifth palm. This baiting is repeated once or twice with intervals of about 2-3 days, each time omitting one block.

A single dose of the anticoagulant bromadiolone (0.005%) in wax cake is much more effective because the lethal dose of this poison is so small that even at the recommended concentration of 0.005% rats usually consume 3 to 6 times the lethal dose in a single feeding. Rats die 3-4 days after feeding and so they do not develop a fear for the bait. Placing 10 g bromadiolone blocks twice, with an interval of 12 days, in the same way as the other baits usually provides satisfactory control. If the rat-damage is restricted to certain trees, these trees may be selected for baiting (Bhat and Sujatha 1993).

However, the baiting of rats is not without danger. Baits placed in the crown may fall down, either by strong winds or even by the rats themselves when eating from them. These baits are very dangerous for animals or children that may be attracted by them.

Squirrels also have their nests in trees. Poison baiting against squirrels is not effective. Trapping, using single catch traps is recommended (Bhat and Sujathna 1993). Ripe coconut kernel is an ideal bait to use in the trap. Trapping is most effective in periods there are few coconuts and also only little other food available.

Burrowing rodents and porcupines may feed on fallen nuts. They may also dig underneath coconut seedlings and eat the cabbage portion. Burrowing rodents can be poisoned, using baits with zinc phosphide. Damage by porcupine can be minimized by covering the basal portion of seedlings with wire mesh (Bhat and Sujathna 1993).

7. The Insect Pests of Coconut

D. Mariau

Introduction

In the book 'Les insectes des palmiers' (Lepesme 1947), the existence of over 750 species associated with the coconut palm have already been mentioned, including not only those insects that feed on the palm, but also all their associated insects. Actually, that number would be closer to one thousand. However, only a small percentage of these insects can be of real economic importance. Nevertheless, it is not exceptional that an explosive multiplication can be observed of an insect considered to be only an occasional pest, or even that had never been observed on the palm. However, in this book only insects that cause serious damage to the coconut palm can be dealt with.

Although all parts of the palm are attacked, the leaves are the hosts of the largest number of noxious species. They are mostly leaf-eating insects that attack either the unfolded leaves of the spear, or, more commonly, the mature, fully opened leaves. These leaf-eating insects belong to the following orders: Orthoptera, Phasmida, Lepidoptera and Coleoptera. The leaves are also attacked by some sucking insects (Hemiptera). These can cause direct damage but can also be vectors of diseases. There are only a few leaf foragers (Coleoptera). Flowers and fruits in particular provide food for sucking insects (Hemiptera) and mites. In the stem, some borers may develop, belonging to the orders of Coleoptera and Lepidoptera. Finally, the roots accommodate only a few insects.

Special attention will be given to parasites and predators, natural enemies of insect pests. These insects are often difficult to observe for non-specialists because when their host population is small they have a high efficiency. These auxiliary insects are very sensitive to a large number of insecticides, and inconsiderate treatments may cause an unbalanced situation between a pest for which the treatment was not intended, and its natural enemies. To avoid such inconveniences, the pesticides, the use of which in many cases is indispensable, should be applied only when absolutely necessary, and should be employed only when a population explosion causing heavy economic damage is almost certain.

In many instances, the parasites and predators are not the only helpers in controlling the pests; other biological agents and cultural methods are available. For instance, the population explosions of the coccic *Aspidiotus destructor* are not inevitable, even in the absence of their principal enemies, represented by different species of *Coccinellidae* (ladybird beetles), when the palms grow under conditions of satisfactory water and mineral supply. Similarly, the rhinoceros beetle would not be able to find its shelters, consisting of decomposed wood, if these were covered with a cover-crop before they become attractive to the beetle. All these cultural techniques can be applied in consideration of, and in combination with, the chemical control methods to better spare the beneficial fauna, thus maintaining the best possible balance of the system.

Orthoptera

Acrididae

Locusts, eminent omnivorous insects, attack coconut only periodically, especially when favoured by particular climatic conditions (drought). Under certain conditions, locally very serious damage has been observed. This, for instance, has been the case with the *Locusta migratoria manilensis* Meyen in The Philippines (Otanes 1956), *Schistocerca gregaria* Forskal in East Africa (Lever 1969), *Tropidacris dux* Drury in Trinidad and Tobago and in Venezuela and with *Valanga nigricornis* Burmeister (CP 41) in Southeast Asia. Numerous insecticides among which are pyrethrines (deltamethrine) and organophosphates (fenitrothion), which efficiently control locusts, especially during the young stages.

Tettigoniidae

The only species of this family that merit citing are those that belong to the genus *Sexava* (treehoppers). The adults are large (5 to 6 cm). These insects are characterized by their long antennae, the length of which can be more than three times that of the body. Females have long, curved ovipositors which permit them to lay their eggs in the soil to a depth of 1 to 2 cm. The fertility is not very high (20 to 40 eggs per female) and the incubation period is long (2 to more than 3 months). The young larvae climb the coconut stems. Six instars can be distinguished with a total duration of about 130 to 180 days. Together with a pre-oviposition period of about one month, the life cycle of the insect varies from 7 to 10 months. Four species, the most important of which are *S. coriacea* Linnaeus CP 42) and *S. nubila* Stäl, occur over the various islands of the Moluccan archipelago up to Irian Jaya (New Guinea). Two egg parasites are generally very active: *Leefmansia bicolor* Waterston (Hymenoptera Encyrtidae) and *Doirania leefmansi* Waterston (Hym. Trichogrammatidae), Franssen, 1954). However, these parasites cannot prevent extremely serious outbreaks resulting in almost total defoliation, after which the insects may attack inflorescences and fruits.

Treatment with a chlorpyriphos solution (4 to 6 g per tree) on the basic part of the stem will prevent some of the population from climbing the tree. It is also possible to spray the crown with a solution of diazinon (0.3 g a.i. per l) or phosphamidon (0.5 g per l). Due to the height of the tree, such treatments are difficult. Therefore, stem injections or root absorption with monocrotophos solutions (10 to 15 g a.i. per tree), giving excellent results, may be preferred.

Various species of the related genus *Segestes*, among which *S. decoratus* Redtenbacher in Irian Jaya can occasionally cause serious damage as well.

Phasmida

Phasmidae

Various species of the genus *Graeffea* are known as locally important coconut defoliators. The most important species *G. crouani* Le Guillou is widespread in the Pacific and Melanesia. The adults generally are greenish or reddish. Females can reach a length of 12 cm; the males are noticeably shorter. The duration of their development cycle is about 6 to 7 months (O'Connor 1954).

Two species of egg parasites of the genus *Paranastatus* (Hym. Eupelmidae), originating from the Fiji islands, have been introduced to other Pacific islands with more or less success (Rapp 1989). Systemic injections into the stem or absorption through the roots of monocrotophos or dicrotophos (6 g a.i. per tree) have given excellent results.

Various other species of stick insects attack coconuts (Lever 1969). The damage caused may be important, but these species have a very localized geographic spread and their economic incidence is limited. Such is the case with *G. seychellensis* Ferrari on the Seychelles Islands, *G. lifuensis* Sharp on the Isles of Loyauté (New Caledonia), *Ophicrania leveri* Günther, principally on Savo (Solomon Islands) and *Megacrania phelaus* Westwood on the Island of Ugi (Solomon Islands).

Isoptera

It is within the family of the Termitidae that the largest number of species is found. Most frequently reported are *Microcerotermes biroi* Desnaux on the Solomon Islands (Harris 1958), various species of *Nasutitermes*, among which *N. ephrates* Holmgren in Panama and *N. novarum-hebridarum* Holmgren in the Pacific (Szent-Ivany 1956), *Allodontermes morogoroensis* Harris in East Africa, *Macrotermes bellicosus* Smeath (CP 24) (*Bellicositermes nigeriensis*) in West Africa (Mariau 1971); and two species of *Odontotermes* have been reported from India and Sri Lanka (Nirula *et al.* 1953). Damages can be serious in nurseries and on young palms. All these species principally attack coconut palms suffering from unfavourable climatic conditions or from the attack by some other insect. But this is different in the case of several species that attack coconut palms of all ages growing under normal conditions. Of these, particularly *Neotermes rainbowi* Hill (Kalotermitidae) has been reported and is considered an important pest on the Cook Islands (Hopkins 1927) and *Coptotermes curvignathus* Holmgren (Rhinotermitidae) (CP 43) (Mariau *et al.* 1992).

The control is essentially preventive, by nest destruction, as often as possible. As far as chemical control is concerned, spraying with endosulfan (0.50 to 1.0 g per l) or chlorpyriphos (1 to 2 g per l) gives very good results.

Heteroptera

Coreidae

The most important species of the Heteroptera belong to they family. That can be divided into two genera: *Amblypelta* with the most important species *A. cocophaga* China, and *Pseudotheraptus* with two species, *P. wayi* Brown in East Africa and *P. devastans* Distant in West Africa. These insects attack the flowers and young fruits, often causing their downfall. Attacks on older nuts only result in more or less serious deformations.

Adults of *A. cocophaga* are 13 to 16 mm long, according to their sex, the males being somewhat smaller with antennae as long as their bodies. In a medium stage of premature nutfall, the yield loss is about 10% (Brown 1959).

The species belonging to the genus *Pseudotheraptus* can cause much more serious damage (CP. 45 and 46). The size of the adult can be compared to that of *A. cocophaga*. The duration of the development cycle is 25 to 40 days (Way 1953). *P. devastans* Distant was reported for the first time in West Africa in the 1960s (Mariau 1969). The attacks can be extremely violent and cause the almost complete shedding of young nuts (Julia and Mariau 1978).

Chemical treatments with endosulfan or propoxur (5 g per l) are efficient against *P. devastans*. However, these treatments should be repeated frequently, as even small populations of the order of 30 bugs per hectare can cause intolerable damage. Trees colonized by weaver ants should not be treated.

The most important control method is through favouring the development of the weaver ant population of the genus *Oecophylla*: *O. smaragdina* F. in the Pacific zone and *O. longinoda* in Africa (Way 1954; Vanderplank, 1960; Julia and Mariau, 1978; and Phillips, 1956). The development of the populations of these ants are in their turn limited by antagonistic ants of the genera *Pheidole*, *Camponotus* and *Crematogaster*. Weaver ants build their nests by joining several leaflets. They give perfect protection against bug attack. Great progress has been made in the enhancement of the *O. longinoda* population in coconut groves in Tanzania. Löhr *et al.* (1993) introduced the poisonous ant bait AMDRO that was developed against fire ants in the USA. With only 3 g per tree of this bait, the populations of *Pheidole megacephala* - the greatest enemies of the weaver ant - were wiped out for a period of more than one year, providing an opportunity for weaver ants to multiply and spread to other coconut trees. The other local enemy of the weaver ant, the ant *Anoplolepis custodiens*, was controlled with three applications of permethrin to the stems of the palms, the nest entries and the major trails of the ants. This treatment was also effective for a period of more than a year. Weaver ant population development can be further favoured by intercropping with shrubs and small trees with broad leaves, such as cocoa and citrus species, much favoured by weaver ants for nesting.

Pentatomidae

Within the family of the Pentatomidae, the species *Axiagastus cambelli* Distant is wide spread in the Pacific area, but the economic importance of this insect, which sucks the flowers, is not the same everywhere. Its development cycle covers 7 to 8 months. The eggs are attacked by two egg parasites of the Hymenoptera order.

Various species of the genus *Lincus*, belonging to the Pentatomidae family, are vectors of the 'Hartrot' disease in Latin America, the causal agent of which is a *Phytomonas* (Louise *et al.* 1986).

Tingidae

The species *Stephanitis typicus* Distant of the Tingidae family is wide spread, especially within the Indo-Malaysian archipelago and The Philippines. The black-coloured adults are only 2.4 mm long. The pre-imago development takes 3 to 4 weeks. Adults and larvae suck the leaves, which turn to a bronze colour, but damage is generally rather limited. This insect has been suspected to be involved in the spread of the Cadang-cadang disease in The Philippines (Mathen 1960).

Homoptera

Aphididae

Only two species of the Aphididae family merit mention as having some incidence on coconut: *Astegopteryx nipae* Van der Goot and *Cerataphis lataniae* Boisd. The aphid *A. nipae*, with a diameter of 1.5 to 2 mm is dark green. It has been reported from Malaysia, Vanuatu, and Papua New Guinea. Its reproduction is asexual. Damage is generally not very serious, but outbreaks may occur.

C. lataniae originates from Central America and is widespread in Africa and the Pacific. The adults, with a diameter of 2 mm, are blackish and, like the green larvae, have white waxy edges. Outbreaks are possible but the direct damage is of little importance. However, the abundant honey-like excretion favours the development of sooty mould, which may cover the palm's foliage.

Coccidae

A large number of species belonging to the family of the Coccidae may be found on coconut trees. However, the greatest part of these scales, even when having a very wide geographic spread, are of little or negligible importance. Worth mentioning are *Aonidiella aurantii* Maskell, which locally has had outbreaks in Vanuatu, *Ischnaspis longirostris* Signoret and *Pinaspis buxi* Bouché, which have presented themselves notably in the Seychelles Islands. In most cases, these species are controlled by numerous chalcid hymenoptera species and predators, among which the Coccinelidae are the most active.

The transparent scale insect *Aspidiotus destructor* Signoret (CP 47), which can be found along the whole tropical belt with the exception of some Pacific islands, has or has had real economic importance. The larvae and the females have light yellowish bodies covered by a waxy semi-transparent scale, measuring less than 1 to 2 mm. They are attached to the underside of the leaflets. The eggs, numbering about 50, are laid under the scale. Young larvae crawl out from underneath, searching for a place to which they can attach themselves. In severe infestations, the scales are positioned side by side, forming a kind of crust that can cover the leaf completely. The duration of the development cycle covers about 5 weeks. The populations are generally well controlled by parasites of the Chalcidoidae family, such as the aphelinid *Aphytis chrysomphali* Mercet, an egg parasite of the same geographic distribution as its host, and even more by predators from numerous coccinellid beetles (CP 47) belonging to the following principal genera: *Scymnus, Cryptognatha, Chilocorus* (48) and *Exochomus*. Accidental introductions of *A. destructor* in various countries have been followed by generally successful adaptations of various Coccinellid species. Thus, *Lindorus lophantae* Blaisdell has been introduced in Vaté Island (Vanuatu), *Chilocorus nigritus* Fabricius in Mauritius and *Cryptognatha nodiceps* in the Fiji Islands.

The establishment of the weaver ant *Oecophylla longinoda* in the canopies of coconuts may disturb the existing balance between the scale and its predators because the ant has a very predacious nature, destroying eggs and nymphs of the ladybird beetles (Mariau and Julia 1977). Absence of predacious ladybird beetles, however, is not sufficient to trigger off an outbreak on the scale that develops on coconut palms growing under rather unfavourable agro-climatic conditions. One could eliminate the ant to favour the development of the beetles. In the early stages, the balance between the pest and its antagonists is unstable and treatment is often necessary. Spraying with dimethoate (0.4 g per l), covering the underside of the leaves well, is very effective. Other organophosphorus compounds such as parathion and monocrotophos, are equally effective.

Aleurodidae

Various species of the Aleurodidae family attack the coconut, the best known being *Aleurodicus destructor* Mackie, which is widespread in South-East Asia. The nymph, which is up to 1.5 mm long, is provided with 7 pairs of wax glands that excrete long, white rolled threads. In severe infestations, it looks as if the underside of the leaflets are covered with a layer of cotton. Outbreaks are violent but generally of short duration and localized. Important damage, however, has been reported from The Philippines (Lever 1969). Where additional treatment is necessary, the same treatment as against *Aspidiotus destructor* can be used with equal effectiveness.

Fulgoroidea

Numerous species belonging to these groups of sucking insects live on coconut foliage. They do not cause damage themselves, but some of them may transmit diseases. For instance, the fulgoroid *Myndus crudus* (Cixiidae) is known as the vector of the mycoplasma disease Lethal Yellowing in Florida, Mexico and the Caraibbean area (Howard *et al.* 1983). A species belonging to the same genus, *M. taffini* transmits the virus disease Foliar Decay in Vanuatu (Julia 1982). In Africa a disease of unknown etiology may cause serious damage to nurseries and to palms in the early field stage. This disease, called Dry Bud Rot is transmitted by two Delphacidae species: *Sogatella kolophon* Kirdaldy and *S. cubana* Crowford (Julia and Mariau 1982).

Lepidoptera

Defoliators

Numerous species belonging to this order forage on the coconut foliage. This group, having a common feeding habit, is presented in Table 10; others with a different biology are discussed separately. Chemical control of the defoliators is rather similar. The following chemical insecticides can be used generally: trichlorfon (1 to 2 g per l), carbaryl (1.25 to 2.50 g) or deltamethrine (0.75 to 1.25 g). Insecticides with a more selective action may also be recommended, such as *Bacillus thuringiensis*, as Dipel (1800 UAAK per mg) at the rate of 0.75 to 1.0 l per ha or the Bactospeine (1600 U per mg) at 0.6 kg per ha. Chitin inhibitors such as alsystin have given excellent results on some species. The critical levels at which intervention is recommended is linked to various natural factors, such as the number of caterpillars (10 to 50 per leaf) but also to the general vegetative aspect at the time of the attack, as well as the pressure exercised by the antagonistic factors (parasites and predators).

Table 10: *The defoliating Lepidoptera of coconut*

Families - genera species References	Country or geographic region	Description of stages Biological characteristics Natural enemies
Brassolidae *Brassolis sophorae* L. (Marconi, 1952)	Tropical South America	Butterfly, 70-105 mm (female larger). Forewings with large oblique orange bands. Caterpillar 80 mm, herd animal, during the day with hundreds in a nest. Cycle of 80-125 days. Nocturnal damage. Various parasites, often little active (*Spilochalcis* and Tachinids on the pupae). A *Beauveria* fungus develops on the caterpillars. Control by nest elimination.
Hesperidae *Hidari irava* Moore and Horsfield (Corbett, 1932)	Malaysia Indonesia	Greenish yellow caterpillar with a longitudinal violet line. Butterfly 50 mm. Each of the forewings has 4 large patches, of which 3 are yellow. Total cycle duration is 50 days. The caterpillars are herd animals and hide within a sheath of leaflets, held together by silk threads. There are two parasites, a hymenopter (*Apanteles bridaridis* Rohw.) and a tachinid.
Oecophoridae *Opisina arenosella* Wlk (*Nephantis serinopa* Meyrick) (Nirula, 1956) (Dharmarajn, 1962) (Cock and Perera, 1987)	India Sri Lanka	Butterfly, 25 mm. Forewings light grey with black spots. Caterpillars are rose coloured with black heads; Total cycle duration 2-2.5 months. This insect is attacked by numerous parasites, on the eggs: *Trichogramma* sp. (Hym. Trichogrammatidae); on the caterpillars: *Apanteles taragnae* Viereck (Hym. Braconidae), *Perisierola nephantidis* Muesebeck (Hym. Bethylidae), *Elasmus nephantidis* (Hym. Elasmidae), *Stomatomyia bezziana* Baranov (Dipt. Tachinidae); and on the pupae: *Trichospilus pupivora* Ferrière (Hym. Eulophidae), *Brachymeria nephantidis* Gahan (Hym. Chalcididae), *Xanthopimplas punctata* (Ichneumonidae). Their efficiency is often reduced by the presence of hyper-parasites. Heavy outbreaks.

Agonoxenidae *Agonoxena argaula* Meyrick (Hinckley, 1963) (Bradley, 1965)	Numerous Pacific islands	The development cycle covers 33-50 days. The caterpillars spin webs over the underside of the leaflets, which are scarified in bands 2 mm wide. Numerous parasites of the genera *Apanteles*, *Bracon* (Hym. Braconidae) and *Tongamyia* (Dipt. Tachinidae) attack the caterpillars, and *Brachymeria* (Chalcididae) attack the pupae. The parasites *Elachertus agonoxenae* Kerrich (Hym. Eulophidae) and *Actia painei* Crosskey (Dipt. Tachinidae) were introduced to the Fiji Islands.
Pyraustidae *Hedylepta blackburni* Butler (*Phostria*) (Zimmerman, 1958)	Hawaii	The butterfly (30 mm) has brownish forewings. The caterpillar (35 mm) is green with white lines. During their early development they are herd animals and live under tissue; the older caterpillars bring the leaflet margins together with silk thread. There is an important parasite complex: *Zaleptopygus flavoorbitalis* (Cameroon) (Hym. Braconidae), *Echthromorpha fuscator* F. (Hym. Ichneumonidae), *Frontina* sp. (Dipt. Tachinidae).
Limacodidae *Setoria nitens* Walker (Mariau, *et al*. 1991) (Soekarjoto *et al*. 1980)	Malaysia Indonesia	The female butterfly (20 mm) is slightly larger than the male. It is brown with a dark band over its forewings. The cater pillar (40 mm) varies with age from green to yellow (CP. 57), with a purple band in the middle. The body has an abundance of warts covered with itching hairs. The development cycle is 2 months, of which 1 month is for the larvae instars 8 and 9. The pupal stage is passed at ground level. The parasite complex is important: *Trichogrammatoidea thoseae* Nagaraja (Hym. Trichogrammatidae) on the eggs; *Euplectromorpha malayensis* Wilkinson (Hym. Eulophidae) on the young caterpillars, and *Spinaria spinator* Guérin (Hym. Braconidae) on the older caterpillars; *Chaetexorista javana* Brauer Bergenstamm (CP. 58) on the cocoons. But there are also several hyper-parasites. The bugs *Sycanus leucomesus* Wlk. (Reduviidae) and *Eocanthecona furcellata* (CP. 59) Woeff (Pentatomidae) are active predators of the caterpillars and a fungus (*Cordyceps*) on the pre-nymphs.
Limacodidae *Orthocraspeda catenatus* Snellen (Ponto and Mo 1950) (Mariau *et al*. 1991)	Sulawesi Papua New Guinea	The forewings are brown to ochre, the male (8-14 mm) with a darker brown border. The caterpillar at the end of its development is 15 mm long, whitish green with discontinuous black lines on its back. The development cycle is 6-7 weeks. The cocoons are fixed near to the central vein of the leaflet. It is attacked by hymenoptera: *Euplectrus* sp. (Chalcididae), *Apanteles* sp. (Braconidae), *Chrysis* sp. (Chrysididae); and by diptera, *Chaetexorista* sp. and *Bessa* sp. (Tachinidae). A virus disease limits the population.
Limacodidae *Parasa lepida* Cramer (Desmier de Chenon 1982)	India South East Asia	The butterfly measures 30-45 mm. The forewings are dark brown with a meandering brownish line. The head and thorax are yellowish green, the abdomen reddish brown. The caterpillar is 20-25 mm long, yellowish green with a light blue stripe over the middle of its back (CP. 60). The female lays 380-650 eggs. There are 7-8 larval instars. The life cycle is about 10 weeks (40 days for the caterpillars). The most important parasites are hymenoptera: *Apanteles parasae* Rohwer (Braconidae), *Trachysphyrus oxymorus* Tosquinet (Ichneumonidae), and a Tachinid dipter *Chaetexorista javana* Brauer and Bergenstamm. Coconut palms can be seriously attacked from their first year onwards.
Limacodidae *Macroplectra marnia* Moon (Menon and Pandalai 1958)	South India Sri Lanka	The butterfly is ochre-brown. The caterpillar (8-10 mm) is yellowish green on top, rose coloured at the underside. Its life cycle is 7-8 weeks The parasitism is taken care of by a hymenoptera of the Eulophidae family, *Neoplectrus maculatae* Ferriere, a fungus of the genus *Aspergillus*, and a Baculovirus of the granulose type. Damage can be serious during the dry season.
Limacodidae *Latoia pallida* Möschl *L. viridissima* Holland (Mariau *et al*. 1981) (Igbinosa 1985)	West Africa	The butterfly measures 30-40 mm. *L. pallida* is uniformly white with black veins. *L. viridissima* is green with a brown spot on the front part (CP 61). At the end of the development, the caterpillar is uniformly green with 4 rows of tufts of very itch hairs on the rear part, on *L. viridissima* these hairs are orange-red and black on *L. pallida*. The development cycle takes 3-4 months. There are several parasites (Ichneumonidae-Tachinidae) that are not very effective. A virus disease (Picornavirus) plays an important role in the population control.
Zygaenidae *Brachartona catoxantha* Hampson Merino 1938) (Lever 1953) (Van der Vecht 1950)	South East Asia	The butterfly is 15 mm in size. On the backside, the wings are brown with a yellow border. The caterpillar is 10 mm long, pale coloured with a purple median line and tubers at the sides. Up to the fifth and last instar the caterpillar feeds on the leaf tissues, with the exception of the upper epidermis, in long, discontinued bands. The life cycle is 5-6 weeks, of which 3 are for the larval stage. There is an important complex of parasites. Among the hymenoptera, various species of *Goryphus* (Ichneumonidae) are parasites of the pupae, *Apanteles artonae* Wilkinson (Braconidae) and other species, *Euplectromorpha artonae* Ferrière, (Eulophidae) are parasitized in their turn; other parasites are two important diptera of the Tachinidae: *Bessa remota* Aldrick and *Cadursia leefmansi* Baranov. A fungus of the *Beauveria* genus can be of local importance. Outbreaks are heavy and violent and occur mostly during the dry season.

Levuana iridesceus Berthune-Baker (Tothill, *et al.* 1930)	Fiji	Dark purple butterfly. Pale caterpillar with 2 purple median lines. Its biology and behaviour are very similar to that of the former species, but its feeding behaviour remains the same during the whole larval stage. The development cycle lasts 6-7 weeks. The introduction of the tachinid *Bessa remota* has resulted in considerable reduction of the populations.
Homophylotis catori Jordan (Cachan 1959)	West Africa	Similar to the two preceding species. The adult measures 12 mm. The upper side of the wings are dark brown with blue reflections. The male has comb-shaped antennae and a yellow spot on the underside of the wings. The caterpillar is golden yellow with purple longitudinal lines (CP 62). The development cycle is 45-60 days. Outbreaks are rarely widespread; on the contrary, they are often localized in the same spot. The parasite complex comprises two species of the Ichneumonidae, 1 of the Braconidae on the pupae and a chalcid. A fungus of the genus *Nemuraea* plays an important role in the population regulation. The dry season is unfavourable for its development.

Other Lepidoptera

Galleriidae

Two species with comparable behaviour: *Tirathaba complexa* Butler, reported from many Pacific islands, and *T. rufivena* Walker, mainly in Indonesia, Malaysia and The Philippines, are known. The moths have a span of about 25 mm and are uniformly greyish brown coloured with a silvery aspect. The caterpillar reaches a length of 27 mm, it is glossy brown and has long antennae projecting forward. The nymph stage develops in a silken cocoon.

The development cycle is short: 30 to 40 days, of which 15 to 20 days are for the 5 instars and 10 for the nymphal stage. The caterpillars develop on male flowers as well as on female flowers and young fruits into which they bore, so they fall to the ground. Damage is often limited, considering that an important percentage of the flowers and fruits are shed in anycare. The caterpillars are attacked by various parasites, among which: *Apanteles tirathabae* Wilkinson (Hym. Braconidae), *Devorgilla palmaris* Wilkinson (Hym. Ichneumonidae) and *Argyrophylax basifulva* Bezzi (Diptera Tachinidae) have been successfully introduced from Java to the Fiji Islands (Lever 1969). Sometimes a chemical treatment with trichlorfon or carbaryl is required.

Phycitidae

The adult of *Hyalospila ptychis* Dyar has a span of 14 mm; its forewings are white except at the base which is brownish with reddish violet designs. The white caterpillar measures 15 mm. The development cycle covers about one month (Bondar 1940a). The damage is comparable to that of *Tirathaba*. Authors have different opinions about its economic importance in the north-east of Brazil.

Castniidae

Castnia daedalus Cramer. The moth has a span of 110 to 180 mm; it is dark brown with a whitish bar across the forewing. The caterpillar reaches a length of 10 cm and is cream coloured; the pupa is enclosed in a cocoon made of stem fibres coated together. The caterpillars attack the base of the petioles and the stem, through which they bore tunnels that can reach a length of up to 1 metre. The insect can be found throughout the Amazonian basin. The coconut palms becomes very weakened by its attack, and serious production losses are suffered. The natural control factors seem to be rather weak. Chemical control is inevitable. Applications of trichlorfon (1.5 g per l) give good results, provided that the whole crown is well drenched (about 5 litres per tree) (Huguenot and Vera 1981).

Pyralidae

Until recently, insects of the genus *Sufetula* have been reported only from oil palm, where caterpillars under the soil surface destroy the root extremities that start growing out from the base of the stem. The attack takes place either before the root has penetrated into the soil or when it has grown only a few centimetres. Beyond this depth the caterpillar cannot penetrate and the root extremity is out of its reach. This is not the case in peat soils, in which these caterpillars can move more easily. In Indonesia, such attacks on coconut by the species *Sufetula sunidesalis* Walker have been observed, which before had been observed only in oil palm (Desmier de Chenon 1975). In this particular situation, the caterpillars attack the root extremities at all levels (CP 40). After the attack, a scar develops on the root, followed by the emission at that place of several secondary roots, which in their turn are attacked. The primary roots become stumped and cannot develop any more, resulting in severe retardation in the growth of the palm and loss of production. Spraying with endosulfan or ethylchlorpyriphos reduces the population considerably. However, the required frequency of the treatment is unknown.

Coleoptera

Lymexilonidae

The wood-boring beetle *Melittomma insulare* Fairmaire can be grouped within this small family, which, in certain situations, can be an important pest of coconut in Madagascar and especially in the Seychelles Islands (Vesey-Fitzgerald 1941). The long and slender adults are 6 to 18 mm long. The body is dark, with 4 lighter dorsal spots on the head, the thorax and the outer wings. The males are smaller than the females. They do not feed, and live a few days only. The eggs are laid in groups of up to more than one hundred, in crevices in the stem. The larva, which reaches a length of 2 cm at its full development, has a strongly chitinized head and thorax. This is also the case with its last, hoof-shaped segment. The larvae develop over a period of one year (Brown 1954), preferably within the central part of the stem, where sometimes up to 200 specimen can be counted. The pupal stage is passed in a cell excavated in the wood. The main damage is caused to the basal part of the stem (up to 60 cm). Young trees are killed and the stems of older trees can be completely excavated, breaking off with the slightest gust of wind. Chemical control is particularly effective with insecticides that have a high gas pressure. This is the only way to reach the insects in the galleries.

Chrysomelidae Hispinae

According to their behaviour, the insects of this family can be divided into two groups: leaf miners and those that forage on young, undeveloped leaves. The damage done by the leaf miners is mainly caused by larvae that bore tunnels within the foliar tissue. Adults, which gnaw the surface of the underside of the leaves, generally cause less serious damage. Various species belonging to the genus *Promecotheca* attack coconut in south-east Asia and in the Pacific. The adults are about 8 to 10 mm long. They live on the underside of the leaves, where they lay their eggs and feed. The larvae are yellow-coloured, legless and with a flattened facies. The complete development cycle covers about 3 months.

Promecotheca cumingi Baly is regarded as an important pest in The Philippines and Singapore (Lever 1951) but it has also been reported from other countries, such as Malaysia and Indonesia. It is generally well controlled by various parasites: *Pediobius parvulus* Ferrière (Hym. Eulophidae) and *Sympiesis javanica* Ferrière, which attack the larvae, while *Achrysocharis promecothecae* Ferrière parasitizes the eggs.

Promecotheca coeruleipennis Blanchard occurs in many Pacific Islands: Fidji, Tonga, Samoa, etc. (Taylor 1937). Like the former species, *P. coeruleipennis* is normally controlled by the egg parasite *Oligosita utilis* Kowalski (Trichogrammatidae) and the parasite of the larvae, *Elasmus hispidarum* Ferrière Elasmidae (Hymenoptera). The omnivorous mite *Pyemotes ventricosus* Newport can destroy almost all larvae and nymphs, which has caused important imbalances between the pest and its other parasites, resulting in even greater pest infestation the following year. This required the introduction of *Pediobius parvulus*, which also feeds on *P. cumingi* and which can survive the time span between one generation of its host and the next. This species has been equally successfully introduced in New Britain (Papua New Guinea) for the control of outbreaks of *Promecotheca papuana* Csiki (Gressit 1959), where the local egg parasite *Closterocerus splendens* Kowalski and the parasites of the larvae, *Eurytoma promecothecae* Ferrière and *Apleutropis labori* Girault did not give sufficient control. On Vanuatu *P. opacicollis* Gestro is mainly controlled by the egg parasite *Oligosita utilis* Kowalski (Kowalski 1917; Risbec 1937).

Also on Madagascar, some chrysomelid leaf miners have been observed. The genus *Coelaenomenodera* comprises 41 species, 34 of which are from Madagascar. *C. perrieri* Fairmaire CP 44) and *Balyana mariaui* Berti and Desmier de Chenon develop on coconuts (Mariau 1974). These two species are well controlled at the egg stage by two species of Trichogrammatidae, *Oligosita robusta* and *O. minuta* Viggiani, as well as in the larval stage by *Chrysonotomyia* sp., *Sympiesis aburiana* Waterson, *Cotterellia* sp., and *Pediobius* sp., all belonging to the family of the Eulophidae. The West and Central African species *C. minuta* Uhman, a very serious pest of oil palm, has been observed on coconut only occasionally.

Other species, in all stages of their development, live between the leaflets of leaves not yet unfolded. *Brontispa longissima* Gestro is widespread in south east Asia and the Pacific. Larvae and adults feed on the leaflet tissue in narrow bands (CP 49). In cases of severe attack, the palms may become completely defoliated. This type of damage is very harmful to the coconut palm, particularly when the emission of new leaves is rather slow, especially when the palm is young, or when it grows under unfavourable agronomic conditions. The life cycle differs according to the situation and also according to the various authors. In Indonesia, the complete cycle covers 5 to 7 weeks (Kalshoven 1957). Adults are slender and long, measuring 8 to 12 mm. The colour varies between localities. Thus, the reddish brown types dominate in Java, whereas the almost completely black ones are found in the Solomon Islands and Papua New Guinea. Some authors attach special importance to the geographic races. The parasite composition includes various species, as much for the eggs (the Trichogrammatidae chalcids *Haeckeliana brontispae* Ferrière and *Trichogrammatoidea nana* Zehntner, and Encyrtidae *Oencyrtus* sp.), as for the larvae (the Eulophidae *Tetrastichus brontispae* Ferrière).

A related species, *B. mariana* Spaeth caused substantial damage in Micronesia in the 1930s. The introduction of *H. brontispae* and *T. brontispae* established a better balance between the populations. Another related insect, *Plesispa reichei* Chapuis, lives on the Indo-Malaysian Peninsula. Its head and thorax are brown-orange and its outer wings are black. Its development cycle covers about 1.5 to 2 months (Corbett 1923).

Among the undeveloped leaf-eating Chrysomelidae, two species from Madagascar and one from Brazil can be mentioned. The genus *Gestronella* (CP 50) has two species affecting coconut, *G. centrolineata* Fairmaire and *G. lugubris* Fairmaire. The former is 7.6 mm long and is light brown, its outer wings being traversed by 8 lines of yellowish brown points, grouped in pairs. The latter is larger, the females being 12.8 mm long and the males 8.8 mm. They are uniformly black with dark blue reflections. The larvae have a flattened head; the thorax and abdominal segments, divided by deep furrows laterally, are extended by small appendices, and the last abdominal segment has two curved extensions. The development cycle covers about 5 months (Appert 1974). *G. centrolineata* can cause serious damage to young coconuts, whereas *G. lugubris* can seriously defoliate adult palms, especially those that grow slowly.

In Brazil, young coconut trees during the first 2 or 3 years of life may be seriously attacked by a chrysomelid of the sub-family of the Cassidinae, *Coraliomela brunnea* Thunb. (CP. 55). The red adult is 25 mm long. Like the former species, its larvae develop between the leaflets of leaves not yet unfolded that may be consumed through their total thickness. The complete development cycle covers 264 days (Ferreira and Morin 1986). Due to three egg parasites, one *Tetraticus* and two *Closterocerus* species, control can reach about 60%. Nevertheless, chemical treatments are often required in nurseries and young plantations.

Curculionidae

Ten species of the genus *Rhynchophorus* are known, of which the greatest part are pests of cultivated palms. Adults are large beetles, 30 to 50 mm in length, mostly dark coloured, generally with longitudinal brightly coloured bands (CP. 52). Adults, which are very good fliers, lay their eggs on wounds of various origins, on the mature stem as well as in the crown. Larvae penetrate into the living tissues of the palm, where the insect completes its development cycle in about 3 months. The nymphal stage is passed within a cocoon of vegetal debris made by the larva at the end of its development, when it reaches a length of 40 to 50 mm. The larva is whitish, legless, with a swollen median; the head is armed with strong mandibles. The most important species are *R. ferrugineus* Oliver, occurring on the entire Indian and Indo-Malaysian continent, *R. phoenicis* Fabricius in Africa and *R. palmarum* L. in Latin America. Adults of the latter are the main vector of the Red Ring disease, of which the causal agent is the nematode *Rhadinaphelenchus cocophilus*.

Apart from the tachinid dipter *Paratheresia menezesi* Townsend, which attacks the larvae of *R. palmarum* (Moura *et al.* 1993), no natural enemies of this insect (which is well protected within its tunnel) are known. Control of the Rhynchophorae in the first place is preventive, by avoiding making wounds in the stem in which the insect can lay its eggs, and by eliminating all existing and potential breeding places. Attraction to the wounds is not visual, but by smell. This characteristic is used to attract adult insects to traps containing pieces of coconut wood or other vegetal matter, which are renewed periodically (Morin *et al.* 1986). Application of an attractive pheromone excreted by the male insects increases the efficiency of the traps considerably (Rochat *et al.* 1991). Using a vegetal bait with the attractant, 25 times more *R. phoenicis* were captured in the traps than with vegetal matter only (Rochat *et al.* 1993). Comparable results have been obtained with other species. The chemical composition of the attractants is different for each species.

Contrary to the Rhynchophorae, the species of the genus *Rhinostomus* (CP. 53) do not need wounds in the stem to lay their eggs, which are often deposited on small uneven parts of the palm. The South American species, *R. barbirostis* Fabricius attacks living coconut trees. The black adult is about 30 to 40 mm long, not counting its long, very hairy rostrum. Attacks may vary between one coconut palm and another. On some of them very great quantities of eggs can be observed, up to more than 100 on a band 10 cm wide (Mariau, not published). The stem is bored through in all directions, causing a general weakening of the tree, and breakage in case of gusty winds. Adults are nocturnal. In the daytime they may be found in the grass at the foot of the tree or, more often, in the axils of old leaves. The most serious attacks by these insects have been reported from Brazil, especially from the State of Sergipe. The African species, *R. afzelii* Fahraeus has only been observed on dead or withering palms.

Also in Brazil, between the States of Bahia and Paraiba, coconuts are being attacked by *Homalinotus coriaceus* Gyllenhall (Bondar 1940b). The adult is black and stocky, and is about 2 to 2.5 cm long. Eggs are laid at the base of the inflorescence. If the inflorescence is still young, the damage results in rapid, total rot. More often, tunnels are bored in the periphery of the rachis of the inflorescence. This weakens the rachis, which may break when it carries many fruits. Otherwise, the nuts may fall off as a result of insufficient nourishment. The pupal stage is passed in a small excavation made by the larva in the surface of the stem at the base of the rachis. Dwarf coconuts are much more susceptible than talls.

Another beetle that attacks Brazilian coconuts is *Amerrhinus ynca* Sahlb. Adults are yellow with black points. They lay their eggs in the stems of young palm trees. In cases of severe attack these palms yellow and break. The presence of the parasite *Paratheresia menezesi* (Diptera Tachnidae) is generally insufficient for controlling serious outbreaks (Moura *et al.* 1994).

Dynastidae

Thirty-nine species of the genus *Oryctes* have been registered (Bedford 1980) but only some of them have a real impact on the development of palm trees, and more particularly on coconuts. Due to its geographic spread, the species *Oryctes rhinoceros* L. is the most important (CP. 54). To three other important species in Africa and Madagascar, other species of related genera with a very similar biology to that of *Oryctes* can be added. Without doubt, *Oryctes* is the most studied pest of coconut (Gressit 1953). In 1980 Bedford had already registered about 200 biographical references, of which the majority deal with *O. rhinoceros*. In the meantime, perhaps another 50 have be added.

The adult of *O. rhinoceros* is about 35 to 50 mm long; it is dark brown and carries a horn on its head which is bent backwards. The horn is more developed on males than on females. Its distribution is very widespread and reaches from India to most of the Pacific islands, including all south-east Asian countries. From the axil of a young leaf, the adult bores a tunnel through the spear. When opened, the affected leaves show a V-formed cut. In severe and repeated attacks (the tunnels may reach a length of 1 metre), defoliation may be serious, even in adult palms. The palms are rarely killed, except under conditions that are particularly favourable to the insect. In young palms, however, this is not the case, as the tunnels often reach the tissues near the meristem of the growing point, causing serious deformations of the stem and, if the meristem has been affected, resulting in the death of the palm.

The larvae live in decomposed wood and in compost, where they complete their development through three stages. Their cycle covers about 3 months. The larvae are curved and are white-greyish, with a head equipped with powerful, hardened jaws. The pupal stage, which takes about 20 days, is passed in a hole cut in the wood, or within a fibrous cover. The adult displaces itself only at the end of the day, during the first hours of the night.

In Africa, *O. monoceros* Ol. is the dominant species; *O. boas F.* occurs much less frequently. This is due to the latter only completing its development in compost, and not in decomposed wood (Mariau 1967); thus it can only be found near the villages.

Eleven species of Oryctes have been registered in Madagascar, of which two may have some economic influence on coconut: *O. simiar* Coquerel and *O. pyrrhus* Burmeister. The nature of the damage caused by the former are comparable to that described above, but the species from Madagascar can lay its eggs in much lower leaf axils, because of which the insect can reach the growing point much easier, making it much more dangerous (Mariau, not published). *O. pyrrhus* attacks only young palms. The adult bores a hole in the soil at a few centimetres' distance from the basis of the tree, which it penetrates after having perforated the husk of the nut.

Species belonging to other genera have a biology and especially a feeding habit very similar that of the *Oryctes* species. The most important ones include *Scapanes australis* Boisduval in Papua New Guinea, The Philippines and the Solomon Islands (Bedford 1976), the attacks of which are often followed by those of the *Rhynchophorus*. *Augosoma centaurus* Fabricius is found in Africa. These large insects can move in solid swarms. In one month, in a plantation of 75 hectares, 138,000 insects, or an average of 13 per coconut palm could be collected (Venard-Combes and Mariau 1983). Also *Xylotrupes gideon* L. (CP. 55) may be mentioned, which is widespread over Micronesia (Bedford 1975), and *Strategus aloeus* L. which feeds in a similar way to *O. pyrrhus* from Madagascar (Hurpin and Mariau 1966).

Control of the Scarabeidae, and especially of the *Oryctes* species, is above all preventive. Control becomes very difficult when breeding places are not removed.

Rhinoceros beetles breed in moist, decomposing organic matter such as rotting wood, stems of dead palms, felled logs and stumps of palms and other trees. They can also breed in compost heaps as long as these have enough mass not to dry out quickly. Compost heaps can be used as traps, either by spraying them with insecticides (BHC 0.01% or Aldrin 0.01%), or by turning them over regularly and killing all visible adults and larvae. Split pieces of coconut stem laid down with their flat side downwards can also be used as traps. Beetles tunnel their way to the inside of the logs through the soil. Turning over the logs twice a week and killing the larvae is a very effective way of control. The effectiveness of traps can be improved by treating them with attractants, the most effective of which is ethyl dihydrochrysanthemumate (Chrislure). Regular inspection of these traps is labour-intensive and the high cost of chemical attractants is a disadvantage. Removing tunnelling beetles with a hooked wire is an old traditional method.

At the stage of land preparation it is recommended that all possible future breeding places be planted with a cover crop such as *Pueraria phaseloides*. This method has proved itself as well in plantations established on forest lands (Julia and Mariau 1976a), as in replantings (Mariau and Calvez 1973), as long as the wood or the palm stems are covered within one year after felling. In replantings, the same results were obtained with *O. rhinoceros*. (Wood, 1969). As the beetles are also attracted by the silhouette of the young palms, interplanting young coconuts with high intercrops, such as maize and bananas, may reduce beetle attack.

There are only few natural enemies of *Oryctes*, e.g. the parasitic wasp *Scolia ruficornis* Fabricius (Scollidae) and the predator *Platymeris rhadamanthus* (Reduvidae) (Hemiptera) Gerst, introduced from Africa and Madagascar into the Pacific area. Micro-biological control, which has been the subject of an abundant literature, has proved itself much more promising, either by the use of the fungus *Metharhizium anisopliae* Metschnikoff (Swan 1974) or, above all, by the use of *Baculovirus oryctes* brought into prominence in Malaysia (Hüger 1966). Although this virus is active against *O. rhinoceros* and can play an important role in reducing its populations, this is not the case with the African *O. monoceros* that has been shown to be much more tolerant. For some species, the use of various attractants can be a method of control. Thus, *A. centaurus* and *O. rhinoceros* can, in certain conditions, become strongly attracted by ultraviolet-enriched light. Smelling baits treated with ethyl chrysantemate can be used against *O. monoceros* (Julia and Mariau 1976b).

Acarina

Eriophyidae

Eriophyes guerreronis Keifer belongs to this family. The adult, which is no longer than 255 microns, lives under the protection of the floral parts of the fruits, where it grows so rapidly that its development cycle covers no more than 10 days, while it feeds on the developing tissues (Mariau 1977). After being attacked, the fruit's development is disturbed (CP. 56), resulting in a smaller size and, consequently a more or less reduced kernel content. When the fruits are not prematurely shed, the production decline may be about 10 to 20% on average, and sometimes more when agro-climatic conditions are unfavourable. This mite was reported for the first time in the early 1960s in the State of Guerrero in Mexico, and a few years later in all Latin American countries, in the Caribbean and in Africa.

The intensity of the attack varies with environmental conditions, but also with varieties and hybrids (Mariau 1986). For instance, West African Tall is very sensitive, whereas Malaysian Tall is very tolerant. The hybrid PB 121 (West African Tall x Malayan Yellow Dwarf is less sensitive than its parents. Chemical control (chinomethionate - monocrotophos) may be practised under very special conditions, such as in seed gardens, but cannot be recommended for plantations due to the necessary frequency of application and height of the trees. Micro-biological control, particularly with the fungus *Hirsutella thompsonii* has not yet yielded very positive results, at least not in the field. Another Eriophyidae, *Colomerus novahebridensis* Keifer, whose distribution area is S.E. Asia and the Pacific behaves in the same manner as *E. guerreronis*, but causes much less damage.

Tetranichinidae

These mites are widely spread over coconut leaves. They belong to the genera *Tetranychus, Oligonychus* and *Eutetranychus*. However, the mites live in balance with their environment and no outbreak has been mentioned in the literature.

Tenuipalpidae

This is not the case with *Raoiella indica* Hirst. Serious attacks were reported on young coconut palms in Mauritius (Moutia 1956).

Conclusion

The group of the most important insect pests consists of numerous species of defoliating caterpillars. However, in most locally restricted cases attacks by other insects may also have important economic consequences. For instance, the various *Sexava* species have caused serious defoliations of coconut in the Molucca Islands; *Oryctes* can kill many young palms, and in South America many plantations have been destroyed by Red Ring Disease, the causal agent of which is transmitted by the *Rhynchophorus* weevil.

Coconut is often planted almost continuously over large areas, increasing the danger of insect population explosions. Many of these pest populations can develop very inconspicuously, due to the presence of parasites and predators, and then suddenly explode very strongly. The planter is therefore surprised and cannot avoid serious damage. Although he may well recognize the principal pests in his region, he should always be very alert.

Over vast areas, in many cases chemical control is inevitable, requiring aerial treatments, the only method to control such an explosion rapidly. However, wherever possible, biological control methods will be preferred, which preferably should be applied preventively so as to create unfavourable environmental conditions for the explosion of one or other insect population. It has been observed that application of inadequate cultural practices could have serious consequences for the development of a pest. For example, it is possible that the passage of a tractor in a young coconut plantation can cause serious attacks by *Rhynchophorus* weevils. This insect can oviposit near slight damage to the leaf axils caused by tearing as a result of tractor wheels rolling over leaves trailing on the ground. Thus, preventive control requires no more than some knowledge of the pests and regular supervision of the plantation.

PART II. PLANTING MATERIAL AND PLANTATION MANAGEMENT

8. Selection and Breeding

R. Bourdeix

Introduction

The genetic improvement of cultivated plants is a strategic sector on which important social and economic interests are based. On the other hand, this substantial discipline in general requires considerable material and scientific investment, programmed for continuity over many years.

In a perennial plant such as the coconut palm, the constraints connected with its biology increase the cost of scientific progress and aggravate the consequences of possible errors. In fact, a genetic experiment frequently covers an area of eight hectares for a minimum period of twelve years. Consequently, coconut research not only needs high investment but also great functional stability. To obtain convincing results in the field of varietal improvement, a research station should be operational for at least twenty years. Finally, on the human level, coconut genetic research requires patience proof against anything and it requires a certain stoicism: mostly, a researcher analyses the trials planted by his predecessor, and establishes experiments that will be analysed by his successors.

Scientific research on coconut started in India in 1916 (Harries 1978), but many studies have been interrupted by world wars or by the economic crisis in 1929. With the exception of the work of Patel (1938), genetic improvement trials were not resumed until after 1950. Actually, about twenty stations, spread all over the tropical zone, study the improvement of coconut. However, the scope of the programmes and their budgets differ widely.

After a brief review of the specific character of the plant and the objectives of its improvement, two successive chapters will present the genetic resources of the coconut and selection methods that have been applied to it.

Coconut Breeding, Constraints and Advantages

The coconut palm is a cumbrous plant, both in terms of space and time. Planting densities generally range between 100 to 250 trees per hectare, according to varieties and cultural practices. The time between planting and flowering varies from one year, for most precocious dwarfs, to seven years for certain tall ecotypes. It takes a further year to get mature seeds from female flowers; and yet one more year for raising seednuts in the nursery before planting.

Its low rate of multiplication is another limiting factor. In natural conditions, most cultivars produce less than 100 nuts per palm per year. With hand controlled pollination, the yield is even less: a mother tree generally not producing more than 20 fruits per year. The fruits have no dormancy, hampering storage of seednuts. Only about 65% of the seednuts will give plants available for field transfer. These factors considerably limit the choice between crossing plans. The absence of natural vegetative propagation is another constraint. However, research on *in vitro* multiplication should provide an alternative within a few years.

On the other hand, the plant has some advantageous characteristics. Its continued production all through the year allows for a balanced planning of the breeding programme. Hand pollination work is easy because of the large size of the inflorescence and the flowers. No inter - or intra-ecotype sterility has ever been observed. Its perennial nature permits enduring conservation of successive generations and living collections.

Selection Criteria

Plant breeding objectives are fairly complex. They result from a compromise between various consumption and cultural habits, processing techniques and knowledge of the plant. This complexity becomes even greater when the studied plant has multiple uses. In the case of coconut, there is no unique ideal type integrating its numerous and different uses.

Fruit characteristics

The fruit characteristics determine the variable criteria for selection, according to the region of cultivation or the type of use of the nuts. Traditionaly, large fruits, although weaker (thin shelled) are appreciated in certain regions in Asia and the Pacific. Exporters of fruits to be sold per piece prefer medium sized, shock resistant (thick shelled), late germinating nuts.

It is estimated that in many countries about one-third of the production is consumed as waternuts. Some cultivars have sweeter and more aromatic water than others. For this purpose, the number of fruits produced per hectare is important, as long as their water content is sufficient to satisfy its consumers. In some countries, such as India and Sri Lanka, the coir is an important commercial product and also coir dust is increasingly used. For these purposes, large, long fruits with a large proportion of husks are well suited.

Another use of coconut is the production of toddy and sugar. For this purpose, trees are selected which produce a high number of long, flexible inflorescences, producing good toddy yields. Certain hybrids seem to be suitable for this type of exploitation (Jeganathan, 1974).

Mechanized processing of fruits is little developed. The greatest part of the operations is done manually, such as dehusking, shelling and peeling. In the case of manual processing, operating costs rise with the number of fruits treated: large fruits, which reduce operation costs per equal volume of nuts are preferred. As mechanization takes over, the thickness of the meat becomes a more important criterion: the thicker the meat, the less percentage of loss is caused by mechanical peeling. The facility by which the meat separates from the shell can also be taken into consideration: in this aspect, considerable differences between cultivars seem to exist (Noel, personal communication). During the last thirty years, new products have been developed from the fresh kernel, such as desiccated coconut, coconut milk, powders and creams that keep the flavour of the coconut kernel. In view of its declining importance on the vegetable oil market, coconut should gradually change its role from that of an oil crop, to that of a fruit tree.

Pest and disease resistance

Returns often depend on phytopathologic conditions. The success of genetic control of pests and diseases is still moderate. In Ivory Coast and Indonesia, sources of tolerance to *Phytophthora* have been discovered. Vanuatu produces hybrids that are tolerant to the Coconut Foliar Decay Virus (CFDV). In Jamaica, a hybrid suffering little from Lethal Yellowing has been widely distributed. However, various lethal diseases still cause serious problems. Little or no tolerance to the lethal yellowing diseases in Ghana and Tanzania, to Cadang-cadang disease in The Philippines, to Red Ring in the Caribbean region and Latin America, and to Root(Wilt) disease in India, has yet been identified. In certain regions, insects and mites practically reduce production to nil. Research on genetic resistance to pests and diseases thus has an important priority. Disease resistance has been dealt with in this book's chapter on diseases (Chapter 5).

Adaptation to the environment

Most coconut plantations in the world are smallholdings, where traditional cultural practices are applied. They are often planted on poor sandy soils where other crops do not grow well. Phytosanitary control and fertilizing are seldom practised. For these farms, cultivars have to be selected that can develop and produce under unfavourable agronomic and climatic conditions. Situations of seasonal water-stress are frequent and drought tolerance should be included as a selection criterion. However, the cultivars to be distributed must clearly also respond to improved cultural techniques. Distribution of planting material that reacts only a little to fertilizing condemns smallholders to under-development. Such mistakes have been made by large-scale distribution of strong tall coconuts of which production remained moderate, whatever the farmers' efforts for improvement.

Genetic Diversification

Introduction

There is a spectacular morphologic diversity in coconut that is particularly expressed in its fruits, by colour, size, and shape (CP 25). However, it is probably on the level of the genetic structure and variability of the species that knowledge of this crop is most fragmentary. Its bibliography does not mention proximate species with which coconuts are interfertile, even partially. *Cocos nucifera* is thus a genetically isolated taxonomic unit.

There are two types of coconut, dwarfs and talls. The world coconut population contains no more than 5 per cent dwarfs, but these can be found all over the intertropical zone. The origin and genetic determinism of dwarfism are still not well known. Apart from their little height, most of the dwarfs show a combination of common characteristics: autogamic preference (talls are allogamic), small size of their organs, precocity, and rapid emission of inflorescences. Because of the last two characteristics, dwarfs play an important role in genetic improvement programmes.

Distribution of the coconut is a result of the fruits floating in the sea being transported by the currents and, later, fruits being carried during human travels and migrations (Child 1974; Harries 1978). The fruit size, and the fact that they are produced in low numbers, have influenced the diversity of the species. The fruits, distributed by a capricious sea or transported by sailors, were probably introduced in small numbers from place to place. The history of coconut distribution is probably a series of founder effects that have induced considerable genetic random drift. The human role played in selection has also been important, but is difficult to quantify. Actually, practically no spontaneous populations exist any more: almost all coconuts on this planet were planted by people.

A controversy exists about the terminology to be used in the description of the various coconut populations. The term 'cultivar' seems to be quite suitable, as almost all coconuts have been planted and are cultivated by man. However, to distinguish traditional cultivars from more recent ones, the terms 'ecotype' and 'hybrid' are also used. The ecotype is defined by a group of individuals from the same environment showing morphologic similarities. The hybrid, in its widest sense is defined as a cross between two structures belonging to different ecotypes. The term 'structure' here means a population, a family, or an individual.

Surveys, Conservation and Exchanges of Germplasm

Generally, the objective of a survey is to collect living material with the widest possible variability. However, sometimes a survey aims at the collection of material with determinated characteristics: ecological adaptation, phenologic characteristics, or resistance to certain diseases.

Survey methods

Assuming, after Pernes (1984) that the financial problems are resolved, which is a survey in itself but of quite another nature, two main types of surveys may be distinguished. Continuous surveys are generally conducted over a limited area around a research station by researchers living in the surveyed country. Spot surveys consist of the exploration of a pre-defined region within a relatively short time. When a survey is not continuous, it is important that the various points brought up by Harlan (1973) be solved: administrative limitation of the area to be surveyed, government authorization, and agreement with local authorities. This is followed by logistical problems, common to the two types of surveys: the choice of personnel, research equipment and means of transport adapted to the country where the survey is to take place.

Particular constraints on coconut

Until recently, coconut surveys were faced with two constraints linked to the biology of the plant. The first is the large size of the fruits; a sample of a hundred fruits often weighs more than 150 kg. The volume of the fruits considerably restricts the number of samples that can be transported, or leads to a reduction of the effectiveness of the samples. Another constraint is the structure of the seed. The coconut is classified within the species of the recalcitrant seeds (Roberts *et al.* 1984): a high water content, unable to withstand much dehydration, and a rapid loss of germination capacity. With numerous vegetable species, the humidity of the environment induces germination. In the case of coconut, water is already available within the fruit. Most cultivars have no dormancy period; the seeds start to sprout one to three months after reaching maturity. Moreover, the coconut seed is relatively sensitive to cold.

Due to these characteristics, numerous samples of coconut varieties have been lost partly or totally because conditions did not allow for sufficient pre-sampling, or because ships transporting the fruits passed through zones that were too cold, or because the duration of transport and customs clearance exceeded the survival time of the seeds. For these reasons, in all research stations some coconut accessions can be found that are represented by numbers that are too low.

Contribution of new technologies

The application of new technologies makes it possible to get around some of the problems mentioned above. Many trials were conducted on *in vitro* cultivation of zygotic coconut embryos. The growing of plantlets from mature embryos has been fairly quickly achieved by the various teams that have worked on it (Cutters and Wilson 1954; Abrahams and Thomas 1962; and Fisher and Tsai 1978) A standard growth medium, such as that of Murashige and Skoog (1962) is enough for obtaining leaved shoots (Iyer 1981).

It has been more difficult to obtain plantlets with a balanced development of roots and aerial parts. The first methods used treatments with auxines and alternation of liquid and jelly-like growth media (Guzman 1970; Guzman *et al.* 1971). Finally, Assy Bah *et al.* (1989) showed that the root system obtained by growing the embryos only in a jellylike medium containing the Murashige and Skoog solution, 2 g per l active carbon and 60 g per l of saccharose favoured the later development of the plantlets under natural conditions. The weaning and transplantation in the field of plantlets obtained from embryos have been difficult steps to overcome. Assy Bah *et al.* (1989) started the first trials with large-scale weaning. These trials provided 38 to 75% of plantlets that could be transferred to the field. Actually, according to Assy Bah (personal communication) 95% of the embryos grow to be plants and weaning losses are about 10%. Complementary agronomic studies are still required to improve growth in the nurseries, which is still too slow compared to that of seedlings.

Interest in this new technology is linked to problems of transfer and conservation of vegetative material. In the case of a survey, collection of embryos cultivated *in vitro* permits a considerable reduction in the volume of collected material: 100 excised embryos rarely exceed a weight of 20 g, compared with some 150 kg for this number of seednuts. This technique also considerably reduces the risk of transmitting diseases (Blake 1989). With financial support from the International Board for Plant Genetic Resources (IBPGR), Assy Bah *et al.* (1989) developed a portable laboratory that permits the collection of the embryos directly in the field. Removal of the embryo can be done at the same time as an analysis of the fruit components, providing interesting data on the genetic variability (Harries, personal communication).

On the other hand, a growth medium permits the storage of embryos for six months at environmental temperature without notable loss of germination capacity (Assy Bah and Engelmann 1993). This short-term storage solves the problems involved with the lack of dormancy of seednuts. Finally, freezing mature or immature embryos in liquid nitrogen permits long term conservation of accessions (Assy Bah and Engelmann 1992a and 1992b). Recent developments of micro-informatica also offer possibilities. Collecting and analysing data directly in the field permits a rapid estimation of the variability and possible choice between the populations to be sampled.

Finally, in future it will be interesting to dispose of an additional method of diversity analysis based on the molecular structure of the embryos. A fraction of the material collected during the survey can be sacrificed for genome analyses aimed at direct quantification of the genetic variability collected.

Choice of populations to be sampled

According to Pernes (1984) 'Good collection programmes are carried out in two stages: a first exploration and preliminary harvest as a basis for studies that will permit better planning of the second, more systematic campaign. An example of such a programme is the recent survey in Mexico (Zizumbo *et al.* 1993). Fruit analyses have been made first at 47 localities all along the Atlantic and Pacific coasts and in the narrowest part of the country (the States of Oaxaca and Veracruz). Thereafter, the collection has been carried out in only 19 localities, 90% on the Pacific coast, where the greatest variability can be found. However, available budgets seldom permit such organization. Mostly, the survey team passes only once through each place. Planning of the itinerary has to based on data found in literature. This preliminary research should be as complete as possible and cover various fields, such as ethnobotany, ecology, agronomy and cultural systems, phytopathology and genetics.

The flora, old and modern, often contain useful information. Unfortunately, the size of the vegetative parts of the coconut has never allowed for conservation in a herbarium, which has interesting implications for other cultivated species. Contacts with local organizations indirectly concerned with coconut cultivation are often also valuable: agricultural research organizations, associations and co-operatives of planters, development institutions and botanical gardens.

In the case of continuous coconut planting covering several hundred kilometres, sampling at regular intervals can be planned. In a collection project in Indonesia, Meunier (1986) considered systematic sampling on various coastal regions every 50 or 100 km. However, this method cannot be systematized because the density and the genetic richness of the coconut plantation often differ from place to place. The data collected directly in the field during the survey sometimes lead to interesting reorientations. Because of the perennial nature of coconut, populations or individuals with original characteristics are generally known by the farmers.

Intra-population sampling methods

Two options can be considered for sampling a population: the selection of progenitors, generally according to production criteria; or random sampling of the population. In view of its cost, a survey should cover the largest possible spectrum. In particular, genetic resistance to pests and diseases always has a high priority. The most productive coconuts are not necessarily the ones that possess genes that confes desease resistance.

On the other hand, the method of sampling seednuts often limits the importance of intrapopulation selection. In most cases the samples result from open pollination. Even when a rigorous selection has been made, half of the inherited genes have not been contolled. In Tall populations, the highest-yielding coconuts have a tendency to more pronounced autogamy. When selecting with emphasis on productivity, an undetermined, but sometimes significant number of self-pollinations will be obtained. The resulting congeniality could affect the later evaluation of the accessions at the research station. It is therefore, better to avoid rigorous selection based on production criteria, and sample a large number of trees. If, notwithstanding this, a selection based on production criteria is made, the following precautions should be observed:

- selection within a homogeneous plantation;
- elimination of border trees and those near open spaces, which are favoured by the absence of competition;
- yield recording during an even number of years for the observation of possible alternate bearing. A tree full of nuts in one year may yield practically nothing the following year;
- avoidance of selection for yield components alone when genetic correlations between the various components are unknown. In particular, a selection based on weight of copra per nut may in some cases cause a decline in total copra yield per tree and per year (Bourdeix 1988a);
- recording of selection targets and criteria, which are fundamental data.

Even if all precautions have been respected, the efficiency of the selection remains limited. A simulation made in the Ivory Coast based on the data of a survey has shown that by using one parent tree per 20 on the basis of copra weight per tree, the yield of the progeny under conditions of open pollination did not improve more than 6.4%.[1]

Sample numbers
The number of trees to be sampled in a population also is important. There exists an intra-population genetic variability in the case of preferably allogamous ecotypes. Various trials have shown that the exploitation of this variability can result in an important genetic progress (Bourdeix 1989). In an article on the survey of palmae, Meunier (1976) has proposed a random choice of 60 to 100 trees in each population, which should be marked on a map containing precise details about the environment. Within the framework of the Indonesian project, a collection of 300 nuts from a minimum of 50 trees was recommended for the tall ecotypes. For the autogamous dwarf types, this number may be reduced to 150 (Meunier 1986). In this case, it concerns a minimum number, allowing for the conservation of a collection on a single site. These figures allow for losses, in the germination bed and in the nursery, of about 40%.

Where embryos are excised, the sample number to be used should allow for expected losses by weaning and transplantation to the field. Moreover, part of the embryos will probably be sacrificed to phytosanitary control (detection of viroids) or to molecular biology studies. The collected accessions should be kept at the various research stations, either in a living condition or frozen within liquid nitrogen. The collection of an average of six embryos from each of 100 parent trees per population seems to be a satisfactory sampling method, provided the trees are sufficiently loaded. Otherwise, the number of parents to be sampled should be increased. This figure may seem excessive. However, a certain part of the collected embryos will be used for international exchange. In this way, for each sample collected, the institution responsible will receive one or more accessions derived from other collections. International exchanges are an excellent means to increase the value of a coconut survey.

Recording of survey data
Management and standardization of data on the surveyed populations and the collection of vegetal material are major priorities. No worldwide survey of coconut germplasm has yet been made. Identical planting material has frequently been referred to under different local names and, the same name used for different material. In 1978, the IBPGR started with a minimal list of descriptors for coconut germplasm. This list was revised and improved in 1992 after the International Workshop on Coconut Genetic Resources at Cipanas, Indonesia, from 8 to 11 October 1991. This list follows the model established by the IBPGR for cultivated plants and it is composed of five sections, of which the first, called 'passport', concerns data collected during the survey. The 'passport' consists of two parts. The first, called 'Accession data' contains identifiers of the accession: number of the accession, name of the cultivar, date of acquisition, type of maintenance, etc. The second part, called 'Collection data' contains geographical and ethnobotanical data recorded during the survey, as well as the applied sampling technique.

Under the auspices of the IBPGR, an international database was set up by the Centre de Coopération et Recherche Agronomique pour le Développement - Département des Cultures Pérennes (CIRAD-CP), permitting the identification of coconut material kept in the various national collections. According to Perry (1992), this database will have multiple functions:

- to assess the current status of conservation and characterization of the genetic resources in all participating collections;
- to provide an indication of duplication (including intentional security duplication) of material between collections;
- to provide an indication of gaps that may exist in the representation of geographical provenance, area of origin, or phenotypic/genotypic variability inherent in the collection - and assess the regeneration requirement at an international level.

To complete this database, attempts were afterwards made to establish a 'passport' for existing collections. This experience has shown that certain important headings often remained empty, because the necessary information had not been kept in writing. For instance, for certain accessions it is not known how many parents were sampled, nor how these parents were selected. Rigorous application of the methodology proposed by the IBPGR will thus avoid the loss of valuable information. In addition to bearing their local name, the ecotypes should bear a name in an internationally used language. These names generally include the type (dwarf or tall) and a geographical or cultural reference. For ecotypes with a homogeneous colour, the colour is generally also mentioned.

The first inventory of coconut ecotypes was made by Chomcallow in 1978 and was transmitted by a circular letter to the various national research institutions. To each ecotype, a code of three letters was attributed. This nomenclature has recently evolved. The international codes for coconut cultivars count four letters and two numbers. The last letter must be 'T' for a tall ecotype or 'D' for a dwarf ecotype. Numbers are used only to describe subpopulations. This can be illustrated by some examples:

- the Malayan Yellow Dwarf is coded as MYD;
- the Tagnanan Tall (Philippines) was first coded as TAG. To be in agreement with the new rules, it has been changed to TAGT;
- the Bali Yellow Dwarf (Indonesia) was previously coded as AYD (because the abbreviation BYD corresponds to Brazil Yellow Dwarf). For easier understanding, it has been changed to BAYD;
- two similar looking sub-populations of Panama Tall from two distict sites (Aguadulce and Monagre) are coded as PNT01 and PNT02.

As the international cultivar list evolved only very recently, this chapter has been written using the previous codification by Chomchalow. In order to facilitate reference, Annexe 1 provides an updated and revised list, including both classifications.

Transfer and exchange of vegetal material

For transfer and exchange operations, the choice of partner depends on the available vegetative material, his decision to export the material, and the phytosanitary conditions of the region where the material is located.

Phytosanitary rules

Import and export of vegetative material should respect the international phytosanitary rules. These are to be published in the next edition of the *Guidelines for Safe Movement of Coconut Germplasm*, under the auspices of the IBPGR:

- Germplasm should be collected from palms that appear healthy;
- Germplasm should not be moved from sites at which diseases of unknown etiology occur;
- Germplasm should preferably be moved as embryo culture or pollen;
- Palms from which pollen, embryos and seedlings will be moved, should be indexed for Cadang-cadang and other viroids, and for Coconut Foliar Decay Virus (CFDV) where appropriate (Frison 1992).

Certain authors recommend a systematic indexing of all accessions, even in regions where the diseases mentioned above are unknown. However, actual indexing, based on molecular biology still require some improvements. Thus, in Sri Lanka, 10 coconut trees were recently tested for the Cadang-Cadang viroid (CCVd). CCVd viroid-like sequences were found in 8 of the 10 palms, although the Cadang-Cadang disease does not occur in Sri Lanka (Randles and Hanold 1992). Coconut seed can be imported from regions of origin that are free of diseases of unknown etiology or diseases caused by mycoplasma-like organisms and phytomonas (Frison 1992). The numerous seed imports made in the past seem to have been made with caution. In fact, no recorded example of a disease that has been transmitted by coconut seednuts from one region to another seems to exist. By providing a correct inventory of the germplasm available in each region, the database of the Coconut Genetic Resources Network (COGENT) permits an optimal and secure transfer of planting material. For instance, the

Vanuatu Tall is available free of disease in the Ivory Coast, whereas this material can no longer be exported from its original country.[2] Similarly, Jamaica Tall free of Lethal Yellowing disease has been planted in Nicaragua. Coconut quarantine centres should still be improved. Various organizational types can be envisaged:

- quarantine centres for embryos and plantlets grown *in vitro*. Here, embryos can be tested for the presence of diseases, and possibly used for genome analysis for determination of their diversity, and finally the *in vitro* cultivated plantlets or embryos can be forwarded in liquid nitrogen. Such a centre could be placed outside the production regions;
- short term quarantine centres for coconut seednuts. Here the seednuts can be received and checked for diseases. Six to 8-month-old seedlings can be forwarded to their destinction. Such centres can be placed outside the production regions where local environmental conditions permit coconut growing. The high costs involved in forwarding the seedlings are a disadvantage of this system;
- long-term coconut seednut centre. Reception of seednuts and planting of seedlings followed by observation during an interval of one generation. The seednuts to be delivered in this manner will be the progeny of the original accessions. This centre should be located within a large producing country free of serious coconut diseases. Such an organization exists in India; The CPCRI has used one of the Andaman Islands for testing various accessions collected in the Pacific in 1981 (Bhaskara Rao *et al.* 1993).

Germplasm transfer techniques

Sampling methods and the cultivation of obtained embryos have already been discussed above. For a more technical approach, the papers of Ashburner *et al.* (1991) and Assy Bah *et al.* (1992) present detailed descriptions of experimental set-up. Sending, transfer and reception of seednuts should be done with some caution. In the Ivory Coast, the floral bracts are removed before sending, and seednuts are submitted to an insecticide and fungicide treatment,[3] dried for 24 hours and bagged (IRHO CIRAD-CP 1980). With the use of a permeable cloth, like jute, it might be possible to treat the whole bag, when already closed and filled with seednuts.

Whitehead (1968) formulated a fumigation method for coconut seednuts: an exposure to hydrogen cyanide (up to 3.5 g per m^3 per hr) or methyl bromide (16.5 g per m^3 per 2hrs) for a short period is not harmful to coconut seed. Shipments from Tahiti to Jamaica were fumigated with methyl bromide for 1 hour, after treatment with dieldrin (4%) and thorough insecticide dusting. (According to Whitehead DDT powder 2% can be used, or aldrin 2.5%, or BHC 0.25%, or other suitable insecticides.) All consignments were again inspected on arrival and fumigated with the same dosage of hydrogen cyanide or methyl bromide. However, the methyl bromide treatment may be risky when doses and duration of treatment are not properly controlled (IRHO CIRAD-CP 1980).

Taffin and Wuidart (1981) have described the precautions to be taken on the arrival of coconut seednuts that have made a long journey:

- rapid rehydration of the seednuts;
- no cutting of husks to avoid cutting the germ;
- placing germinated seednuts under shade;
- insecticide and fungicide treatments right from transfer to the nursery, and every ten days for three months (Recommended treatment: Organil 66 (40g commercial product per 15 l of water) and Sevin or Carbaryl at 20 g commercial product per 15 l of water.

Under natural conditions, ripe seednuts with coloured epidermis germinate within seven to eight weeks. Le Saint *et al.* (1989) stored such seednuts in watertight packing. They found that such hybrid seednuts could be kept for four months without affecting the germination capacity. So far, this type of conditioning has not been used on a large scale because of its high cost and because of recent progress in the transport industry.

Conservation of Planting Material

Conservation in the field

This expensive method has the advantage that it permits an evaluation, and that the material will be readily available for selection programmes. Thus, before a planned genetic trial is planted out, a period of about three years will pass, as follows:

- six to twelve months for the planning and carrying out hand pollinations;
- one year for the development of the flowers into ripe fruits;
- one year in the seed bed and the nursery.

Under normal conditions, collection plots are renewed every 20 to 30 years, according to the cultivars. When Talls are over 15 metres high, it will be difficult to hand pollinate the palms under satisfactory safety conditions. However, phytopathologic or climatic problems may considerably reduce life expectation of the accessions. Thus, two typhoons recently decimated the collections established in Jamaica and in the Albay region of The Philippines (Persley 1992).

Cryoconservation of zygotic embryos

This method consists of dehydrating the mature zygotic embryos and keeping them in liquid nitrogen (Assy Bah and Engelmann 1992a; 1992b). This method permits protected storage of collections for a considerable time. But the use of accessions stored in this way will induce considerable delays. Before a genetic trial using planting material derived from embryos can be established, a period of about 12 years will pass, as follows:

- minimum of nine years between the stage of a frozen embryo and the stage of a flowering parent (in the case of tall ecotypes);
- six to twelve months for carrying out hand pollination;
- one year for the development of fruits;
- one year in the seedbed and the nursery.

Thus, establishing a population from frozen embryos implies long term planning.

Cryoconservation of pollen

Pollen is dehydrated to about 5% humidity and subsequently stored in liquid nitrogen. Such storage does not cause any important loss in germination capacity. So far, it has not been possible to determine the maximum storage period for pollen in liquid nitrogen, but theoretically its conservation is unlimited.

This intermediate method permits a rapid use of the genetic resources: the use of frozen pollen does not involve any additional delay when compared to fresh pollen. However, this type of conservation cannot entirely substitute conservation in the field. For a cross, at least one of the parents should grow in the field.

Actually, the exchanges of vegetative material between research institutes are of a mixed type: dehydrated pollen that permits immediate use of new vegetative material, and *in vitro* cultivated embryos that will provide useful material to be used within a period of eight to ten years. Pollen is sent at environmental temperature and stored in the freezer after reception. This method involves organizational problems, as this pollen should be used within a period of six months, after which it will lose its germination capacity. Without doubt, in future transfer and storage of pollen will be done in a frozen condition in liquid nitrogen.

Balance: surveys conducted and collections established

Due to biological constraints involved in coconut, only a few surveys were conducted before 1980. One of the exceptions is the work carried out by Whitehead, who made a large collection in the

Pacific for the Coconut Industry Board in Jamaica. The regional influence of this collection, however, is small due to the prevailing Lethal Yellowing disease and typhoons sweeping the country.

The CIRAD-CP (ex IRHO) played a pioneer role in this respect, by way of its important collections of vegetative material in the Ivory Coast and Vanuatu. In the Ivory Coast, planting material has been imported from 1955-60 onwards, in large numbers, such as have been achieved rarely. Some examples can be given:

- Rangiroa Tall, Tahiti Tall, Rennel Island Tall and Solomon Tall, planted in 1968 in numbers per entry of over 400 trees;
- Vanuatu Tall, with 450 trees planted in 1970;
- Sri Lanka Tall and Pumilla Green Dwarf, introduced in 1972 and 1975, of which 275 and 194 trees were planted, respectively;
- Tagnanan Tall imported from The Philippines in 1975 and 1978, totalling 800 trees.

Situated in a region without serious diseases and benefiting from favourable climatic conditions, the Ivory Coast collection has played an important international role. All research centres involved in coconut genetic improvement possess at least some entries reproduced in the Ivory Coast. One of the recent actions involved the transfer of numerous accessions to Ghana and Tanzania to test their susceptability to lethal yellowing diseases. In 1978, consultation by IBPGR on coconut genetic resources has showed an urgent need for collecting coconut germplasm for breeding. In addition, attention was drawn to substantial replanting programmes involving the risk of germplasm losses. Finally, lack of co-operation and co-ordination was observed between the various research centres. Following these findings, IBPGR supported various collection operations in the 1980s:

- in Indonesia, germplasm was collected in eleven provinces by the Coconut Research Institute in Manado under the supervision of the National Plant Genetic Resources Committee. The collected populations are kept at Mapanget (North Sulawesi) - 30 accessions; Bone-Bone (South Sulawesi) - 50 accessions; and Pakuwon (West Java) - 34 accessions (Darwis and Luntungan 1992);
- a grant was allocated to The Philippine Coconut Authorities for the establishment of a genetic resources centre. This collection contains 22 dwarf types and 53 tall types, with a minimum number of 90 trees per entry (Santos 1990);
- the Central Plantation Crop Institute of India was commissioned to collect the germplasm in the Pacific region. An important collection was established in 1981 on one of the Andaman Islands, using 100 seednuts per accession (Bhaskara Rao and Koshi 1981);
- *In situ* evaluations were made and collections were organized in Mexico (Zizumbo *et al.* 1993).

In the Pacific, in the PDICC (Production and Dissemination of Improved Coconut Cultivars) programme, financed by the European Development Fund and France, eight countries of the Asian Pacific Coconut Community (APCC) region are participating (Fiji, Kiribati, Papua New Guinea, Solomon Islands, Tonga, Tuvalu, Vanuatu and Western Samoa). Within the framework of this project, an inventory and collection of genetic resources have been undertaken since 1989. It should lead to a morphometric description of each ecotype in accordance with the criteria recommended by the IBPRG.

In Papua New Guinea a large survey, covering the major coconut regions was conducted by the Central Plantation Crops Research Institute (CPCRI), funded by the Australian Centre for International Agricultural Research (ACIAR). Its results are being analysed. An initial collection of 59 local populations (and also 6 foreign ecotypes) are being established near Madang. Simultaneously, hybrids of local talls and dwarfs are being tested. A great effort has been made in the collection of coconut germplasm. Within fifteen years, the volume of these collections has been multiplied by four or five. However, it is not sufficient merely to collect. The collections should be maintained, evaluated and disseminated. The maintenance costs often surpass the budget capacities of the national research centres. International support should therefore be provided in this field.

It is very difficult to answer the question of what proportion of the germplasm already collected will be useful in the future. Of some coconut growing regions, such as Latin America, Malaysia, and Madagascar, only a small part has been surveyed. In other very rich regions, like Indonesia, certainly numerous interesting ecotypes have not yet been identified. In relation to the world coconut population, the proportion of coconut palms kept in collection represents less than one in 30,000. This calculation is without doubt simplistic; however, the size of this figure, and the fact that the world coconut population is largely built up of traditional, empirically selected cultivars, indicates that much still needs to be done on the collection of the genetic diversity of the coconut. However, there is an alternative not yet exploited: the dynamic conservation of germplasm in the agricultural regions where variability is great, such as in the Indian Ocean, the Pacific, Indonesia, and The Philippines. In particular, certain islands could play the role of agro-ecosystems reserved for the conservation of coconut. To avoid genetic erosion, the use of hybrids should be forbidden, in exchange for compensation to farmers. Certain exogenous ecotypes could even be introduced in these islands to create really dynamic conservations of genetic resources. However, the economic aspect of such a conservation method remains to be studied, especially in comparison with the creation of living collections at research stations.

Diversity Analysis Methods

Evaluation in the field

For a proper understanding of the plant it is necessary to underline the sensitivity of the coconut to environmental variations. The aspect of a tree at a certain moment is usually not representative of its genetic value. Various phenomena coincide in the creation of the situation:

Variations resulting from the heterogeneity of the environment

There is a classic variation linked to heterogeneity of the field: gradients or zones of soil fertility or water availability. Minimal differences in plantation techniques may have considerable consequences. An extreme case has been reported from regions subject to floods, where a difference of 20 cm in planting depth could explain wide differences of yield (Meunier, personal communication).

Production rhythms

The growth of coconut palm is controlled by various rhythms, depending on internal or external factors. Although it may produce continuously, yields are often is irregularly distributed over the year. In Ivory Coast, 55 to 70% of the fruits are harvested during the first semester. Biannual cycles are superimposed on this variation. These phenomena of alteration are rather frequent in young trees or following droughts. They can completely change the general aspect of these trees, as is shown in photographs CP 3 and CP 4, showing the same palm in a year of high yield and in a year of lowh yield, respectively. There are other, slower rhythms (of three to four years) that particularly affect the number of female flowers and fruit set. These rhythms depend on climatological events, age and the reserves of the palms (Bourdeix *et al.* 1994). Figure 4 illustrates this phenomenon on a Cameroon Red Dwarf population, planted in 1974. Between 1987 and mid-1990 an increase in fruit set was observed. This phenomenon was probably caused by the severe drought in 1985 and the improvement of climatic conditions thereafter. The unproductive phase, during which the stem diameter and root-disc are formed, appears to be determining. A severe drought, a phytopathologic event, or even bad agronomic management may have consequences for the plantation's whole future. Even if conditions improve afterwards, it seems that coconuts suffering from stress at a young age will never catch up completely.

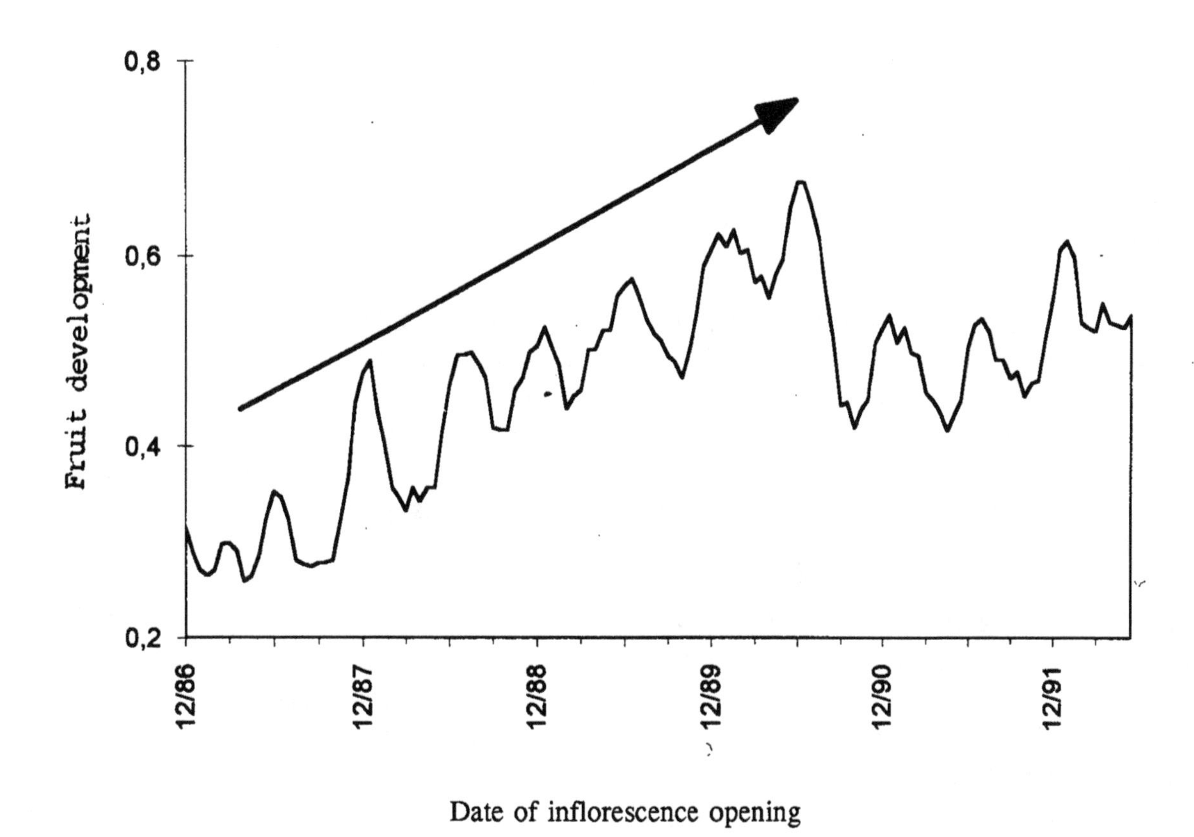

Figure 4: *Trend in the part of total flowers developed into fruits three months after pollination in a Cameroon Red Dwarf population planted in 1974*

Conclusion

For a proper evaluation, vegetative and yield characteristics of a sufficient number of trees, should be measured and over a period of several years. Nuce de Lamothe and Wuidart (1982) have emphasized that it is difficult to conduct such a study outside a research station. Multilocal tests are an ideal but rarely realized solution, due to the costs involved. However, this type of test is sometimes conducted in nearby localities, such as at various stations belonging to a national network. In such a case, such plantations should be set up over a period of time with, for instance, a one-year interval. This precaution permits the integration of the climatic variations during the unproductive stage.

From the time a tree is used as a female parent in a hand-pollinated crossing, it becomes difficult to evaluate its proper value. The depressive effect of bagging induces perturbation in the production rhythms. Breeders are often faced with a difficult choice: the immediate use of a new accession in the crossing programme, or determining in an exact manner the characteristics of this accession. A partial solution may be the use of young populations as male parents only, as pollen collection does not influence production.

Nuce de Lamothe and Wuidart (1982) proposed a detailed protocol of observation on the characteristics of vegetative development, flowering and production of the coconut. This work was refined in 1992 under the auspices of the IBPGR by the International Steering Committee of Coconut Genetic Resources Network (COGENT). A list of descriptors was published in 1992 (IBPGR 1992).

Electrophoresis

This method of genetic variability evaluation is barely developed for coconut. The initial study, undertaken with pollen has involved nine enzyme systems (Benoit and Ghesquiere 1984). After several technical difficulties, only four systems were used to compare eight ecotypes:

- Malayan Yellow Dwarf (MYD);
- Cameroon Red Dwarf (CRD);
- Pumilla Green Dwarf (PGD);
- Niu-Leka Dwarf (NLA);
- West African Tall (WAT);
- Malayan Tall (MLT);
- Tahiti Tall (TAT);
- Vanuatu Tall (VTT).

The eight ecotypes showed a weak enzyme polymorphism: few polymorph loci are observed per system, and never more than two alleles per locus. The intra-ecotype variability is weak for autogamous dwarfs, stronger for the Niu-Leka Dwarf and the talls, with the exception of the West African Tall, which is monomorph for the four enzyme systems tested. The weak enzyme polymorphism of coconut contrasts with the morphologic diversity within the species and suggests that the marked phenotypic differences could hide rather homologous genetic structures. The apparent absence of variability in WAT is possibly due to successive founder effects that have led to a high level of consanguinity.[4] The sampling of this study is limited, at the level of the number of enzyme systems studied as well as at that of the ecotypes analysed. Together with the low variability observed, this sampling limits the importance of the analysis of the genetic distance. As is emphasized by the authors of this study, the fact that the WAT and the autogamous dwarfs appear to be close in genetic distance is probably an artefact. On the other hand, all the observations made elsewhere are in contradiction with this result.

Within the framework of an ACIAR project, an Australian team recently started the study of 20 enzyme systems that group about 30 loci (White and Knox 1988). According to Meunier (1992) the conclusions of this study will be of notable equivalence, particularly where it concerns the low enzyme polymorphism as shown by electrophoresis.

Another study, on 35 different coconut populations, was recently undertaken in Indonesia (Asmono *et al.* 1993). Of six enzymes analysed by starch gel electrophoresis, five showed variation in the isoenzyme banding pattern: peroxidase, esterase, acid phosphatase, endopeptidase and glutamate oxaloacetate transaminase. Many populations had more than one banding pattern. Genetic diversity averages were generally less for dwarf than for tall types. A genetic similarity analysis indicated that Nias Yellow Dwarf and Malaysian Yellow Dwarf were quite similar. The genetic distance between dwarf and tall populations on this basis was considerable.

Foliar polyphenol analysis

Analysis of the polymorphism of leaf polyphenol using high performance liquid chromatography provided an original approach to the study of genetic diversity in numerous vegetal species (Hory 1989). The first analysis on coconut has involved the measurement of 16 sufficiently individualized peaks or major items of chromatographic information, each corresponding to a molecule or a few molecules of strong structural affinity. A more detailed description of the experimental protocol can be found in the publication of Jay *et al.* (1989). From 32 ecotypes, 171 trees were sampled in the collection of the Marc Delorme Station in the Ivory Coast. The data have been worked out with multivariate analysis. Most of the results presented here are derived from discriminant analysis.

Analysis according to the Dwarf or Tall criteria

This analysis has shown a clear distinction between dwarfs and talls. Only 19 out of 171 individual trees show atypical behaviour. Certain tall trees of various ecotypes behave like dwarfs: AGT, MLT, RGT, TAG, RIT, TAT, WAT, PNTO1; one NLA tree behaves like a tall. Most dwarfs present common characteristics that clearly distinguish them from talls; the morphologic and polyphenolic analyses are in agreement on this aspect.

Analysis according to ecotype

The second analysis consists of a discriminant analysis per ecotype. This study on the individual data favours the differences between ecotypes at the expense of intra-ecotype variability. Figure 5 presents the averages per ecotype of this new analysis. Classification within this analysis is not based on geographic groups; the image obtained, however, permits such an interpretation. Five groups are recognized for the classification of the collection of the Marc Delorme station: Pacific, Far East, Indian Ocean, Africa and America, the latter represented by one ecotype only. Among the tall ecotypes, the representation permits the determination of three distinct groups corresponding to the Pacific, the Far East and Africa. The ecotypes of the Indian Ocean may be divided between the African and the Far East groups. This intermediate position can be explained geographically. The Comoros archipelago, although situated in the Indian Ocean, is closer to Mozambique than to India, partly explaining the proximity between CMT and MZT. On the other hand, the Andaman Islands, from where the Andaman Giant Tall (AGT) originates, are relatively close to Thailand, the gate to the Far East. The variable and intermediate characteristics of the coconuts from the Indian Ocean suggest that the dissemination has followed a path extending through the Far East, India and Africa.

Although the division between dwarf and tall is not a binary classification, dwarfs can almost all be found on the left half of figure 5. The centre and the right part of the figure contain only tall ecotypes originating from Africa and the Indian Ocean, suggesting that introduction of dwarfs into Africa was recent. Certain points strengthen the historical hypothesis precisely. The Ghana Yellow Dwarf (GYD) and Malayan Yellow Dwarf (MYD) are very close, confirming the old hypothesis that the yellow dwarf was introduced from Malaysia into Africa during the time of the British colonial rule.
The Cameroon Red Dwarf (CRD) is isolated between the Pacific ecotypes. This dwarf was introduced into Cameroon from the Caroline Islands during the 19th century (Nuce de Lamothe and Rognon 1977). On the discriminant analysis, all green dwarfs appear to be grouped: Pumilla (PGD) from Sri Lanka, Thailand (THD), The Philippines (PIL, CAT) and Equatorial Guinea (EGD). It is known that the latter, although collected from Africa, came from Brazil (Nuce de Lamothe and Rognon 1977).

The analysis indicates that the Brazil Green Dwarf originates from the Far East. The Sri Lanka Green Dwarf (PGD) probably comes from an introduction from The Philippines or Thailand. The eccentric position of the Tahiti Red Dwarf (TRD), when compared to other Pacific ecotypes, remains to be explained. This red dwarf might not have originated in Polynesia, but might have been imported from the Far East.

Another analysis, not presented here, gives surprising results. The geographic origin of tall ecotypes has been used as a discriminating criterion; dwarfs do not feature in the calculation but are represented as supplementary individuals. This option was used to characterize the various dwarf ecotypes in function of the origin of the tall ecotypes. The picture obtained (plan 1,2) confirms the conclusions mentioned earlier about the dissemination of tall ecotypes from the Far East to Africa. The superposition of dwarf ecotypes shows that these cover an important region: all tested green dwarfs and the Polynesian Red Dwarf appear to be linked to Far Eastern origins, whereas other dwarfs, including those of Malaysia, are included within the Pacific region. This division of dwarfs into two groups, which does not alter any information presented earlier, still needs to be confirmed by a supplementary analysis.

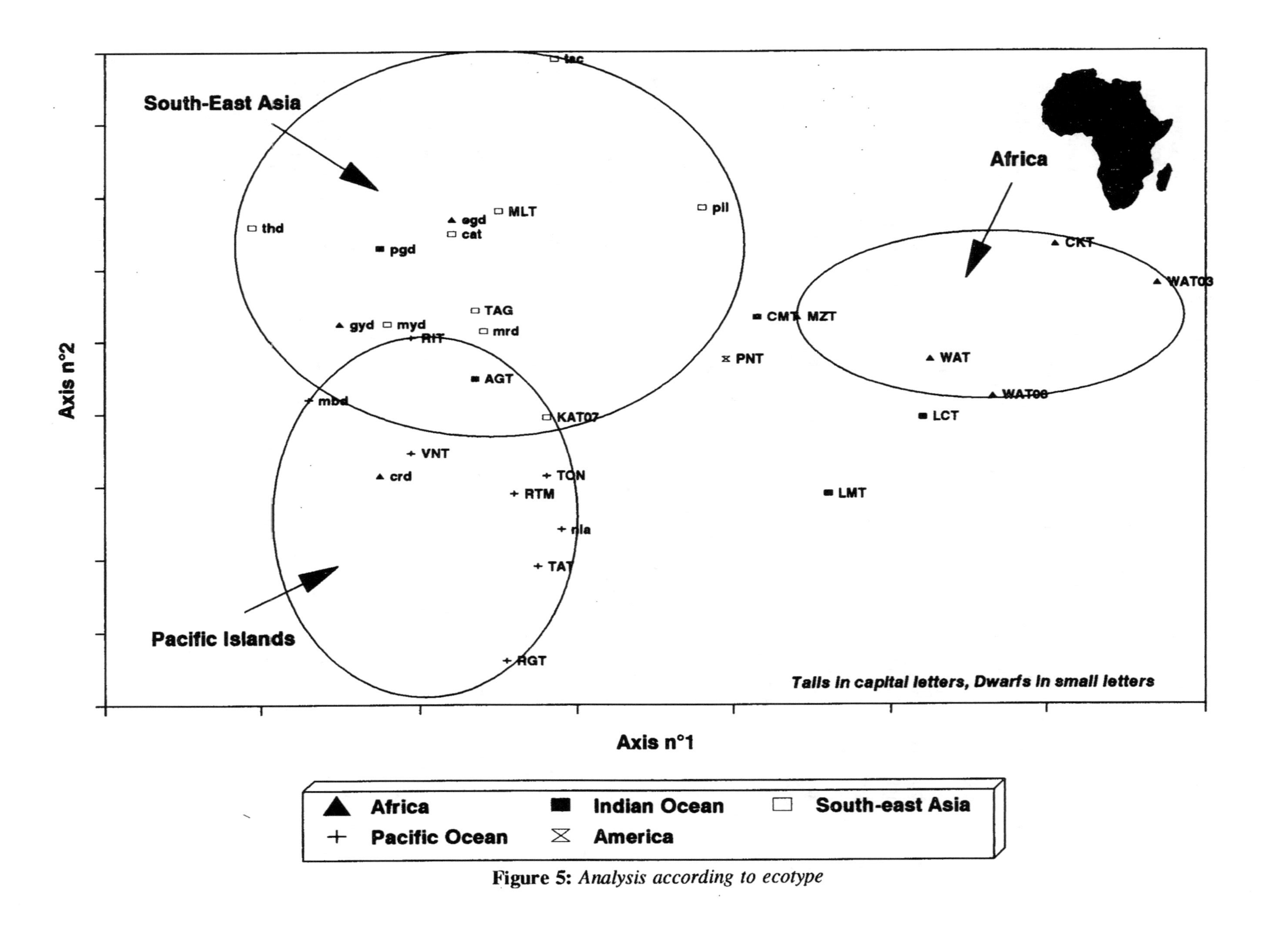

Figure 5: *Analysis according to ecotype*

Conclusion
The foliar polyphenol analysis permits an original and suitable approach to the genetic variability of the coconut. A polyphenolic variability exists between the different ecotypes and, in varying degrees according to the way of reproduction, within each ecotype. The picture of variability confirms the earlier hypotheses that the dispersal of the coconut followed a path passing through the Far East, the Indian Ocean and Africa. The analysis confirms that dwarfs were introduced into Africa during the time of colonial rule. Some dwarfs came from the Far East, others from the Pacific. The study does not permit an exact determination of the origin and history of dwarfism, but shows that variability of dwarf ecotypes is important, although less so than that of talls.

The coconut is a perennial plant with a long life cycle. Morphophenologic evaluations often demand measurements carried out over many years. One of the advantages of polyphenolic polymorphism is its rapidity: within the framework of a survey, the preliminary polyphenolic polymorphism study should permit a considerable increase in the effectiveness of the collection. Trials have been conducted to confirm the stability of polyphenolic profiles of the same material within different environments. This study, conducted with some control varieties grown simultaneously in Indonesia, Vanuatu and the Ivory Coast, has shown certain limits of the method. The relative foliar polyphenol contents seem to be rather sensitive to environmental conditions. It will be difficult to use this technique to compare populations belonging to collections planted in various countries.

Molecular biology
International funds were recently allocated to projects involved in genome analysis of coconut. Therefore, it is useful to discuss briefly the possibilities offered by this technique.

In the short and medium term, molecular biology will be particularly useful for comprehending and structuring genetic diversity. In particular, one of the major preoccupations of breeders is to establish the variability partitions permitting a better orientation for selecting crosses to be tested (for instance, a classification of the population in groups of combining ability). Genome analysis techniques make it possible to get rid of environmental variations that often limit the efficiency of other diversity analysis methods. According to Persley (1992), the techniques of genome analysis make it possible to identify involuntary duplications in planting material collections. These techniques will also serve in the certification of the legitimacy of certain crossings; breeders in fact are sometimes faced with atypical cases raising doubts. In the longer term, the establishment of a genetic map will make it possible better to understand important characteristics such as disease and drought resistance, fruit quality and probably yield. Finally, in future, transformation techniques will make it possible to introduce genes originating from other living organisms into the coconut genome.

Description of the main ecotypes
It is difficult to make a precise inventory of coconut cultivars maintained in collections, and even more difficult to propose a synthetic view of the various ecotypes existing all over the world. Bringing up to date the database of the Coconut Genetic Resources Network should make it possible to establish a more precise overview in the near future. Rather than presenting a more complete overview, the descriptions proposed here are limited to the 50 ecotypes that make up the Marc Delorme Station (Ivory Coast) collection. For more than 25 years, numerous observations have been made on this collection which have been the subject of various publications.

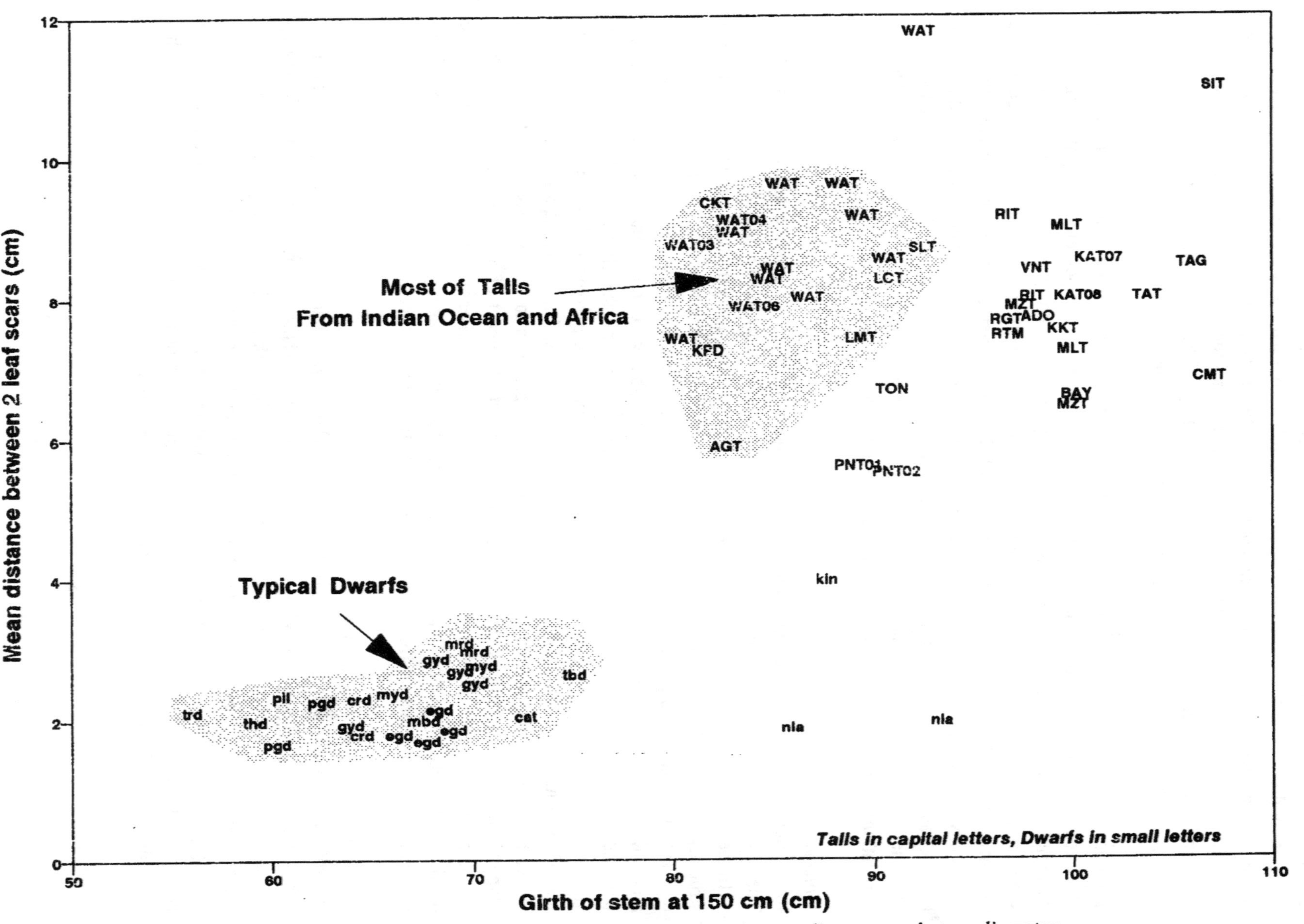

Figure 6: *Relationship between the leaf scars distance and stem diameter*

Another initiative will be the publication of a handbook on coconut cultivars. The first edition will give a precise and illustrated description of 40 ecotypes and 20 hybrids derived from material available at the Marc Delorme Research Station. Each cultivar will be described on a conductsheet classifying its origin, its reference code, its principal morphological and production characteristics, its behaviour *vis-à-vis* the various diseases and its possible use in development and genetic improvement programmes throughout the world. These sheets will be illustrated by photographs taken according to precise criteria, permitting visual identification of the various cultivars. Finally, material kept or created in other research centres may complete this first edition.

Stem, leaves and inflorescences

In Ivory Coast, the morphological characteristics of the stems, leaves and inflorescences were observed according to methods proposed by Nucé de Lamothe and Wuidart (1982). These characteristics are to a large extent represented in the IBPGR descriptor list. Observations on the stem are generally made at the age of ten. Measurement of the leaves and inflorescences are semestrial for a minimum period of three consecutive years.

In Annexe 2, tables summarizing the observations made on 2189 trees of the collection are presented. Certain populations were analysed several times. Sometimes, the same accession has been planted on different sites or different dates and sometimes the original accession and its progeny have been analysed. Finally, controls of the collection, planted systematically with the new introductions, have been the subject of several analyses. This repetition of observations, although laborious and costly, is particularly useful to evaluate environmental effects.

Stem

In regions where typhoons prevail, resistance of the stem may be an important selection criterion. Measurements on stem development taken in Ivory Coast are the following:

- girth of the bole at a height of 20 cm above ground level (G 20);
- girth of the stem at a height of 150 cm above ground level (G 150);
- number of leaf scars between 100 and 200 cm above ground level for the talls, and between 100 and 150 cm above ground level for the dwarfs.[5] This parameter has been converted into the mean distance between two leaf scars;
- height of the stem.

In general, tall ecotypes have strong stems. The girths of the largest boles measure over two metres. These can be found in the Pacific ecotypes, such as the Rennel Island Tall (RIT) and the Solomon Island Tall (SIT), the Vanuatu Tall (VTT) and the Tahiti Tall (TAT). Also the Comoros Tall (CMT) has a very large bole. Among the talls, the thinnest stems can be found in Africa and India, with girths of about 140 cm at 20 cm above ground level. The example of the West African Tall (WAT) shows that the girth of the bole may vary from 140 to 177 cm, according to environmental conditions.

Boles of dwarf ecotypes are barely developed, with the exception of two ecotypes, the Niu Leka Dwarf (NLA) and the Kinabalan Green Dwarf (KIN), with G 20 values of over 115 cm. The other dwarfs have girths of 50-80 cm (Pumila Green Dwarf from Sri Lanka and Catigan Green Dwarf and Tacunan from The Philippines respectively). With Malayan Yellow and Red Dwarfs, the presence or absence of boles depends on the growing conditions. Under very favourable conditions, a slight basic swelling may be observed. On the other hand, under bad conditions, the stem is often thinner at the base than at 150 cm above the ground (Figure 1).

Figure 6 shows the relationship between the mean distance between two leaf scars and the stem diameter at 150 cm above soil level of the various ecotypes (the ecotype codes are presented in annexe 1). The majority of dwarfs clearly differ from talls: having a thinner stem and a small internodal distance. Among atypical dwarfs, NLA shows an internodal distance of dwarf type and a girth of tall type; KIN appears to be intermediate between dwarf and tall. KIN is characterized by greater variation coefficients for the characteristics G 20 and mean difference between leaf scars. This population probably contains a mixture of dwarf and tall genes in a situation of segregation. The population of KIN introduced in Ivory Coast through open pollinated seednuts may be illegitimate. However, this population comes from a PCA research station and was introduced at the same time as other green dwarfs that have been shown to be homogeneous and of good legitimacy. Comparable distributions of stem girth may be observed in D x T and T x D hybrids.

The majority of African Tall and Indian Ocean ecotypes are localized within a narrow region on the map, corresponding to a relatively thin stem (78 to 95 cm girth). The Comoro Talls (CMT) and Mozambique Talls (MZT) are exceptions, with wider stems. These two ecotypes probably result from an introgression between populations of different geographical origins; the leaf polyphenol analysis suggests characteristics that are intermediate between African and south east Asian types.

The height of the palm is a characteristic influenced by environmental variations. The most striking example is that of the Equatorial Green Dwarf (EGD). In two different plots at the station, the height of the palms at the age of eight years varied between 77 cm and 140 cm. Height is also influenced by competition between palms. In one plot, where WAT and SIT were planted in 1968, the latter showed a very strong growth and has created an unfavourable situation for the WAT. In the WAT quite an abnormally large distance between leaf scars may be observed, which is clearly visible in Figure 6 (the WAT positioned right at the top of the figure). Subjected to competition, the WAT has reacted by etiolation, increasing internodal distance for increased growth in height.

Stem growth facilitates a distinction between dwarfs and talls. At the age of 10, dwarfs studied measured between 114 and 205 cm. The atypical dwarf KIN is characterized by considerable growth, with 239 cm at the age of eight. At the age of ten, talls studied were between 372 and 647 cm high.

Leaf

Leaf dimensions influence photosynthesis assimilation capacity and canopy size of palms. The latter particularly determines optimal planting densities. Measurements on the leaves taken in Ivory Coast are the following:

- petiole length;
- rachis length;
- number of leaflets on one side of the leaf;
- length of the middle leaflet;
- leaflet width.

According to ecotype, the total leaf length varies between 3.5 m (EGD) and 6.5 m (Karkar Tall from New Guinea). Typical dwarfs are clearly different from talls. All dwarfs studied, with the exception of KIN, have a leaf length below 470 cm. All talls have a leaf length above 530 cm. The relative proportion of the rachis compared to the whole leaf varies little between the populations (72 to 77%).

The influence of the environment on leaf length can be seen partly in the behaviour of the WAT. According to plantations, for this ecotype, the mean leaf length varies between 5.4 and 5.9 m. Within the EGD, substantial differences in leaf length (from 3.5 m to 4.3 m) may be observed between palms growing in different plots. This difference is due to starting the measurement in one plot at a much younger age of the palms. Or, the length of the leaves increases with age until maturity (8-10 years). In one EGD plot, leaf length increased from 2.8 m at the age of 5 to 4.1 m at the age of 10.

Figure 7 shows the relationship between the width of the leaflets and the number of leaflets on one leaf side. Dwarfs are characterized by a lower number of leaflets with a smaller surface. KIN for these parameters shows the characteristics of a tall. NLA has a number of leaflets similar to those for a tall, but a total leaflet surface that is closer to that of dwarfs (wide but short leaflets). Among the tall ecotypes, with these two parameters the origin corresponding to the great geographical regions cannot be identified. Several WATs show large measures. The great leaf surfaces observed in particular are due to the great width of the WAT leaflets.

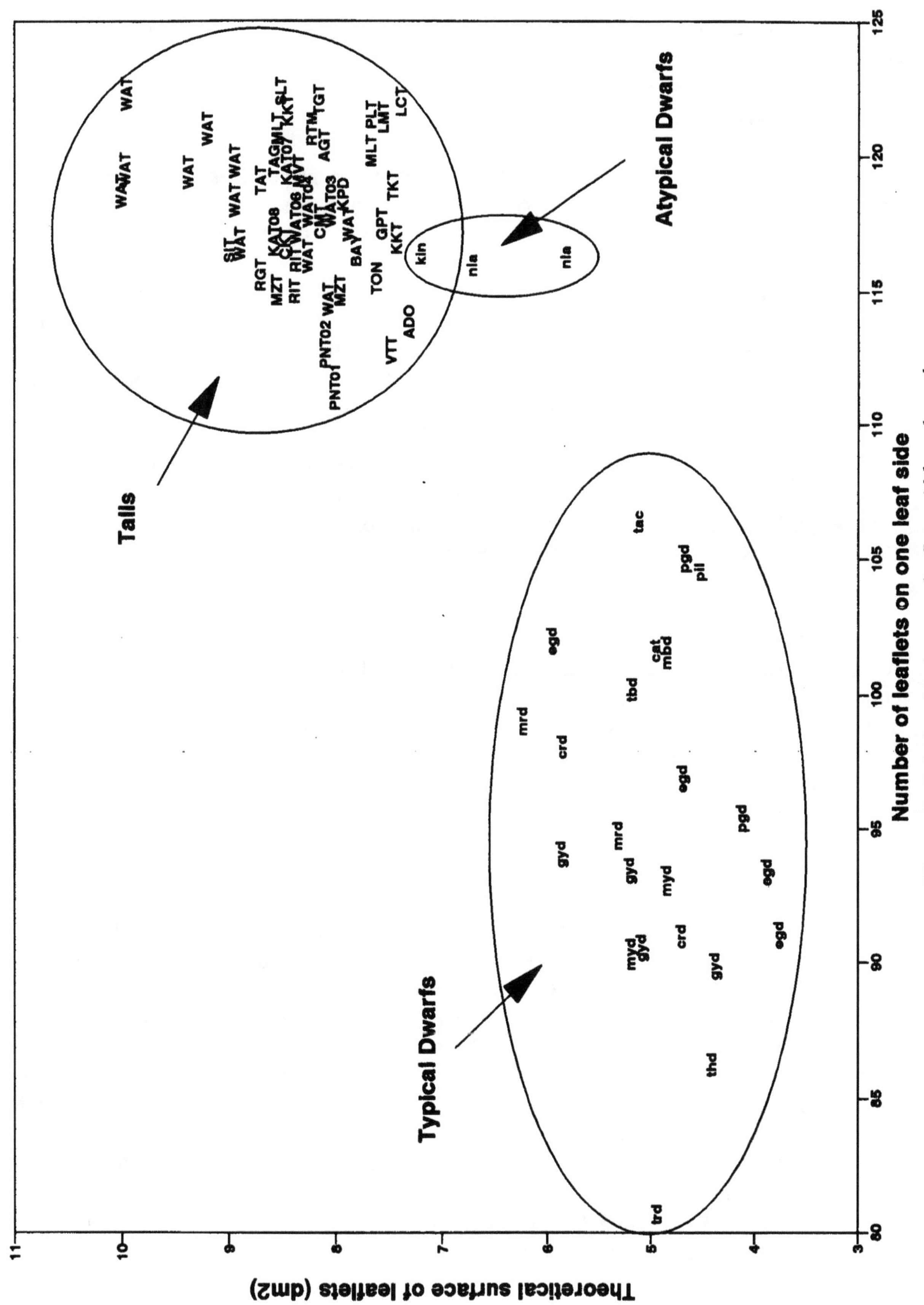

Figure 7: Relationship between leaflets width and number

Inflorescence
The following measurements were taken on inflorescences in the Ivory Coast:

- stalk length;
- length of the inflorescence axis;
- length of the spikes;
- distance between the point of attachment of the spike on the axis and the point of attachment of the first female flower on the spike.

Inflorescence shapes may differ substantially, as is shown in photographs CP 5 to CP 8. The proportion of the axis compared to the total inflorescence length (stalk plus axis) is variable: about 30% for brown dwarfs (MBD and TBD) and MZT, and up to 45% for Malayan dwarfs (GYD, MYD, MRD, EGD).

Figure 8 presents the number of spikes per inflorescence related to the length of the inflorescence stalk. This parameter should probably be integrated into the selection criteria for coconut. In fact, in the more productive hybrids certain bunches abort before maturity, due to their excessive weight. This phenomenon may occur particularly at a young age. A short inflorescence stalk limits the abortion problems and provides better support of the bunches by leaves. Dwarfs are characterized by a short inflorescence stalk and a low number of spikes. The atypical dwarf NLA is an exception with a greater number of spikes than in all other ecotypes. The Malayan Dwarf types and the PGD have the shortest stalks. Among the talls, the RIT and the CMT have the longest stalks. TAT has the highest number of spikes per inflorescence. On the other land, certain talls from Indonesia (TGT, TKT) and from New Guinea (GPT, KKT) are characterized by the shortness of their inflorescences.

Bunch and fruit numbers
Tables 11 and 12 present the production data for the accessions and progenies kept in Ivory Coast. About 12,000 talls and 4000 dwarfs have been observed individually during periods that sometimes covered a period of 18 years after planting. The photographs CP 9 to CP 12 show various bunch shapes. The Malayan Dwarf types and the WAT represent about 40% of the total number of trees observed. There are two reasons for this priority. Some of these ecotypes have been systematically used as control references in the collections. On the other hand, various dwarf ecotypes have been used in comparison trials, accession progenies obtained either by selfing or by intercrossing.

Dwarf ecotypes
In the Ivory Coast, dwarf ecotypes start producing between 3 and 5 years after planting, much depending on environmental conditions, as is shown by the example of the red and yellow dwarfs. However, it should be noted that certain dwarfs of Indonesian origin, not yet introduced into Ivory Coast, distinguish themselves by a much greater precocity. For example, when planted under good conditions, the Salak Green Dwarf (SKD) flowers on average one year after planting (CP 23), and starts producing from the second year onwards (Noiret, personal communication). Compared to other dwarfs, the Niu Leka Dwarf is an exception, as always with a rather late production start (7-8 years).

For a comparison of the yield of the various ecotypes at a young age, reference should be made to a communal production period. This period here is fixed at 3-8 years. Even if the Marc Delorme Station has relatively homogeneous environment conditions, the environmental effect remains important. An extreme example may be mentioned: The MYD planted in 1981 annually produced 10.8 bunches and 64.3 fruits between 3-8 years on average. The same ecotype, planted in 1982 in another site produced only 7.4 bunches and 20.4 fruits per year.

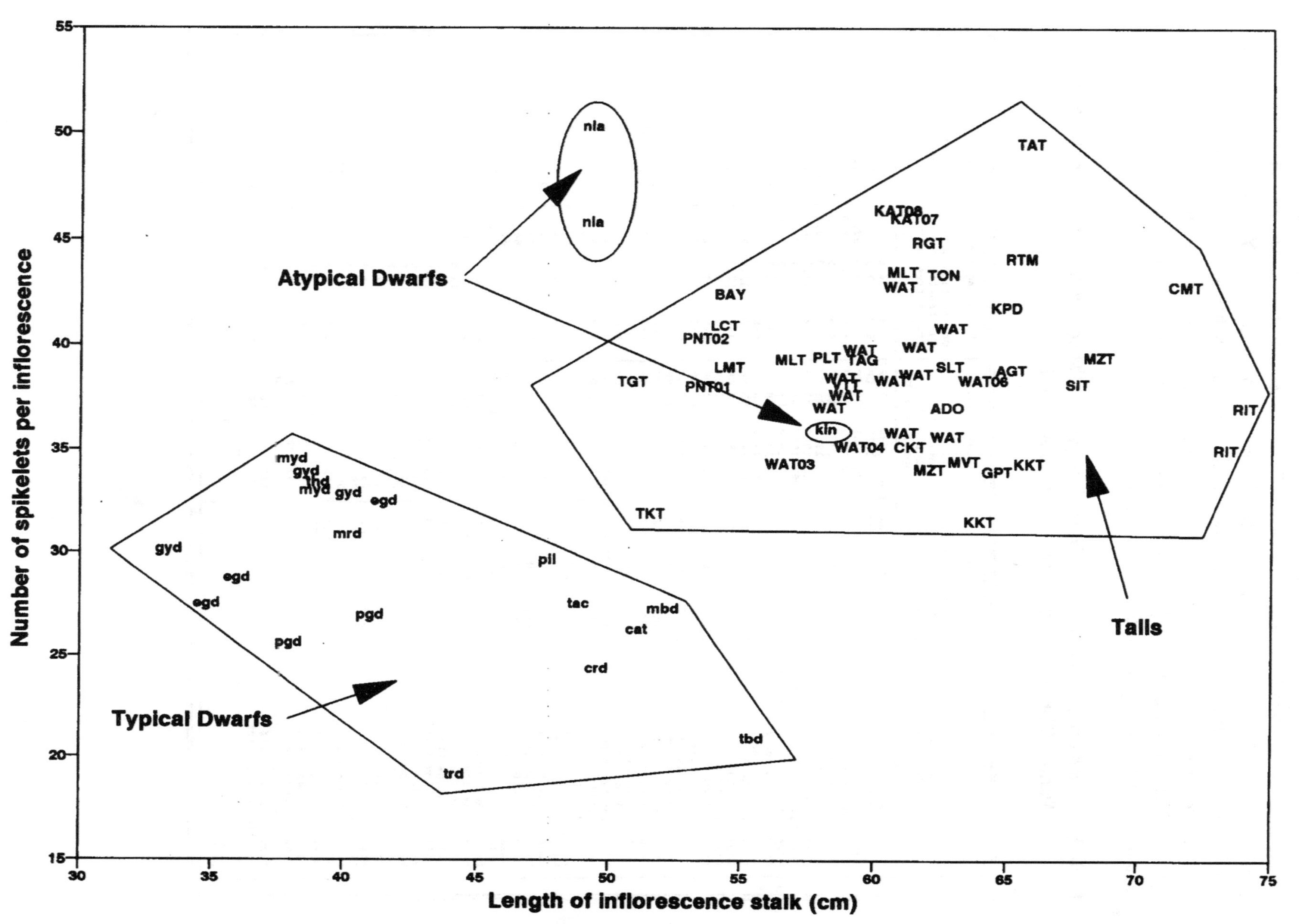

Figure 8: *Relationship between number of spikes and length of the inflorescence stalk*

For the period of 3-8 years of age, a dwarf plantation (good genotype and favourable environmental conditions) annually produces 10 or 11 bunches and 50 to 70 fruits. The least productive ecotypes are Niu Leka Dwarf, Tahiti Red, Kinabalan Green and Aromatic Green, with less than 25 fruits per year. On the other hand, Pumilla Green Dwarf and Pilipog Green Dwarf are classified as producing the greatest number of fruits per tree, followed by Ghana Yellow Dwarf and MYD. When mature, a well performing plantation of dwarfs annually produces 14 to 16 bunches and 80 to 130 fruits per tree. Classification according to production of bunches and fruits is more or less the same as at a young age, with higher numbers for the Madang Brown Dwarf. The ecotypes followed until 17 years of age have not shown any production decline due to age.

Table 11: *Bunch and fruit production of dwarf ecotypes in the collection of the Marc Delorme Station. (For new international codes see Annexe I)*

International Cultivar name	Codes International, French, Country.			Localization Plot, Planting Date, Number of trees, Propag. method'				Young Age Observation period, Bunches/tree/year, Fruits/tree/year.			Adult Age Observation period, Bunches/tree/year, Fruits/tree/year.		
Aromatic Green Dwarf	ARO	NVP07	THA	092	1980	15	OP	5-8	4,6	9,5	9-10	11,5	46,6
Cambodian Green Dwarf	KGD	NVC	KHM	111	1970	143	OP	6-8	7,1	38,8	9-17	11,4	47,5
Cameroon Red Dwarf	CRD	NRC	CMR	132	1981	98	SF	4-8	9,2	58,5	9-12	9,4	74,4
Cameroon Red Dwarf	CRD	NRC	CMR	132	1981	135	CP	4-8	9,2	58,0	9-12	9,4	73,3
Cameroon Red Dwarf	CRD	NRC	CMR	132	1982	112	SF	4-8	8,5	53,2	9-11	10,4	80,1
Cameroon Red Dwarf	CRD	NRC	CMR	132	1982	75	CP	4-8	8,2	50,8	9-11	9,9	72,7
Catigan Green Dwarf	CAT	NVP02	PHL	092	1980	71	OP	5-8	11,3	34,4	9-13	14,3	70,4
Equatorial Guinea Green Dwarf	EGD	NVE	GIN	092	1978	60	CP	3-8	6,7	29,1	9-15	14,1	78,2
Equatorial Guinea Green Dwarf	EGD	NVE	GIN	092	1980	53	CP	5-8	10,2	39,6	9-13	11,9	60,1
Equatorial Guinea Green Dwarf	EGD	NVE	GIN	132	1981	68	SF	4-8	11,7	53,8	9-12	11,3	57,2
Equatorial Guinea Green Dwarf	EGD	NVE	GIN	132	1981	60	CP	4-8	11,7	54,1	9-12	11,7	62,7
Equatorial Guinea Green Dwarf	EGD	NVE	GIN	132	1982	52	SF	4-8	12,3	53,4	9-11	11,5	72,8
Equatorial Guinea Green Dwarf	EGD	NVE	GIN	132	1982	72	CP	4-8	12,2	54,5	9-11	11,2	73,9
Ghana Yellow Dwarf	GYD	NJG	GHA	092	1978	60	OP	3-8	9,3	44,1	9-15	15,2	94,3
Ghana Yellow Dwarf	GYD	NJG	GHA	092	1979	30	OP	3-8	9,8	46,7	9-14	14,0	91,3
Ghana Yellow Dwarf	GYD	NJG	GHA	142	1982	60	OP	4-8	11,8	64,4	9-12	15,1	111,8
Ghana Yellow Dwarf	GYD	NJG	GHA	132	1981	90	OP	4-8	11,4	62,0	9-12	14,8	107,3
Ghana Yellow Dwarf	GYD	NJG	GHA	132	1981	63	SF	4-8	13,6	84,2	9-12	14,5	100,5
Ghana Yellow Dwarf	GYD	NJG	GHA	132	1981	60	CP	4-8	12,7	69,0	9-11	13,2	87,6
Ghana Yellow Dwarf	GYD	NJG	GHA	132	1982	30	OP	4-8	12,5	69,1	9-11	13,0	86,8
Ghana Yellow Dwarf	GYD	NJG	GHA	132	1982	147	SF	4-8	11,7	63,4	9-11	14,1	95,6
Ghana Yellow Dwarf	GYD	NJG	GHA	132	1982	150	CP	4-8	12,9	68,4	9-11	9,2	49,7
Kinabalan Green Dwarf	KIN	NVP06	PHL	142	1982	60	OP	4-8	9,4	14,4	9-11	10,7	21,5
Madang Brown Dwarf	MBD	NBN	PNG	092	1979	75	OP	3-8	10,3	49,3	9-14	13,5	90,0
Malayan Green Dwarf	MGD	NVM	MYS	142	1983	66	CP	4-8	10,3	53,7	9-10	10,7	51,3
Malayan Green Dwarf	MGD	NVM	MYS	142	1983	60	SF	4-8	9,6	51,8	9-10	11,2	48,2
Malayan Green Dwarf	MGD	NVM	MYS	142	1984	54	CP	3-8	10,4	56,7	9-9	10,1	59,4
Malayan Green Dwarf	MGD	NVM	MYS	142	1984	42	SF	3-8	9,4	52,9	9-9	11,0	58,4
Malayan Red Dwarf	MRD	NRM	MYS	112	1972	52	OP	5-8	10,5	56,3	9-17	14,5	87,5
Malayan Red Dwarf	MRD	NRM	MYS	132	1981	40	SF	4-8	12,8	64,5	9-12	13,6	74,6
Malayan Red Dwarf	MRD	NRM	MYS	132	1981	90	CP	4-8	12,7	64,8	9-12	14,0	74,6
Malayan Red Dwarf	MRD	NRM	MYS	132	1982	110	SF	4-8	12,1	59,0	9-11	13,9	77,8
Malayan Red Dwarf	MRD	NRM	MYS	132	1982	120	CP	4-8	12,1	56,6	9-11	13,9	81,2
Malayan Yellow Dwarf	MYD	NJM	MYS	112	1972	54	OP	5-8	9,4	48,7	9-17	13,0	72,6
Malayan Yellow Dwarf	MYD	NJM	MYS	092	1982	23	OP	4-8	12,9	77,1	9-12	13,9	92,2
Malayan Yellow Dwarf	MYD	NJM	MYS	132	1985	90	OP	4-8	8,8	24,5	9-11	9,4	36,3
Malayan Yellow Dwarf	MYD	NJM	MYS	132	1981	45	CP	4-8	12,5	65,6	9-11	13,2	95,8
Malayan Yellow Dwarf	MYD	NJM	MYS	132	1982	174	SF	4-8	12,4	66,3	9-11	12,8	89,3
Malayan Yellow Dwarf	MYD	NJM	MYS	132	1982	165	CP	3-8	8,9	50,5	-	-	-
Niu Leka Dwarf	NLA	NNL	FJI	092	1978	66	CP	8-8	7,8	17,9	9-15	10,9	24,7
Niu Leka Dwarf	NLA	NNL	FJI	092	1979	39	CP	7-8	9,1	16,5	9-14	10,2	24,4
Niu Leka Dwarf	NLA	NNL	FJI	092	1980	33	CP	7-8	5,4	6,1	9-13	8,1	16,1
Pilipog Green Dwarf	PIL	NVP05	PHL	092	1980	15	OP	5-8	11,1	52,1	9-13	13,2	67,9
Pilipog Green Dwarf	PIL	NVP05	PHL	092	1982	42	OP	4-8	10,5	43,5	9-11	10,4	33,1
Pumilla Green Dwarf	PGD	NVS	LKA	112	1972	44	OP	5-8	8,7	72,1	9-17	11,5	88,6
Pumilla Green Dwarf	PGD	NVS	LKA	092	1978	150	OP	3-8	8,3	61,2	9-15	14,9	124,6
Tacunan Green Dwarf	TAC	NVP03	PHL	092	1980	30	OP	5-8	9,3	22,8	9-13	11,4	44,5
Tacunan Green Dwarf	TAC	NVP03	PHL	092	1982	31	OP	4-8	9,5	32,0	9-11	8,2	33,0
Tahitian Red Dwarf	TRD	NRY	PYF	092	1978	90	OP	5-8	5,0	31,1	9-15	9,8	71,8
Ternate Brown Dwarf	TBD	NBO	IDN	132	1985	90	OP	3-8	8,0	35,2	-	-	-
Thailand Green Dwarf	THD	NVT	THA	092	1978	150	OP	3-8	8,3	25,0	9-15	15,3	65,9
Total						3804							
Mean						75		3-8	8,1	40,2	9-X	11,7	66.7

Table 12: *Bunch and fruit production of tall ecotypes in the collection at the Marc Delorme station*

International Cultivar name	Codes: International,	French,	Country.	Localization: Plot,	Planting Date,	Number of trees,	Propagation method	Young Age: Observation period,	Bunches/tree/year,	Fruits/tree/year.	Adult Age: Observation period,	Bunches/tree/year,	Fruits/tree/year.
Andaman Ordinary Tall	ADO	GND02	IND	M63	1982	150	CP	5-8	11,0	67,1	9-11	13,1	96,7
Andaman Giant Tall	AGT	GND03	IND	142	1982	125	CP	6-8	6,3	23,2	9-11	7,5	34,2
Baybay Tall	BAY	GPH04	PHL	142	1982	100	OP	8	9,6	36,7	9-11	8,7	43,3
Cambodia Battambang Tall	KAT09	GCB09	KHM	101	1970	175	OP	6-8	5,7	22,9	9-15	10,1	35,3
Cambodia Koh Rong Tall	KAT10	GCB10	KHM	101	1970	175	OP	6-8	6,3	26,2	9-15	10,7	41,6
Cambodia Ktis Battambang Tall	KAT11	GCB11	KHM	111	1971	19	OP	6-8	8,3	31,8	9-16	11,3	35,0
Cambodia Ream Tall	KAT07	GCB07	KHM	101	1970	200	OP	6-8	6,8	29,5	9-17	11,3	48,8
Cambodia Sre Cham Tall	KAT08	GCB08	KHM	101	1970	125	OP	6-8	6,2	28,0	9-15	11,1	45,6
Cameroon Kribi Tall	CKT	GCA	CMR	M63	1982	150	CP	5-8	11,0	58,0	9-11	12,7	76,1
Comoro Moheli Tall	CMT	GCO	COM	111	1972	363	OP	7-8	6,2	32,1	9-17	11,3	57,4
Kappadam Tall	KPD	GND05	IND	142	1982	42	CP	6-8	10,3	38,6	9-11	9,7	40,9
Karkar Tall	KKT	GNG01	PNG	102	1975	75	OP	7-8	2,7	10,9	9-18	10,0	48,0
Laccadives Micro Tall	LMT	GND07	IND	101	1979	125	OP	6-8	7,8	81,9	9-14	10,4	94,7
Laccadives Ordinary Tall	LCT	GND08	IND	101	1978	150	OP	7-8	6,6	47,2	9-15	11,3	76,4
Malayan Tall	MLT	GML	MYS	M63	1982	100	CP	7-8	11,4	30,4	9-11	9,5	34,1
Mozambique Tall	MZT	GMZ	MOZ	101	1970	225	OP	6-8	3,6	19,7	9-15	11,6	70,0
Mozambique Tall	MZT	GMZ	MOZ	M63	1981	150	CP	6-8	9,9	43,6	9-12	12,3	64,2
Panama Aguadulce Tall	PNT01	GPA01	PAN	M63	1980	150	OP	7-8	4,9	6,5	9-13	10,9	30,9
Panama Monagre Tall	PNT02	GPA02	PAN	M63	1980	150	OP	7-8	6,5	7,5	9-13	11,6	30,8
Rangiroa Tall	RGT	GPY02	PYF	081	1968	512	OP	6-8	5,2	21,3	9-14	9,6	41,1
Rangiroa Tall	RGT	GPY02	PYF	091	1969	525	OP	6-8	5,6	19,7	9-18	10,5	46,3
Rennell Island Tall	RIT	GRL	SLB	081	1968	463	OP	6-8	8,0	46,8	9-18	10,4	48,2
Rennell Island Tall	RIT	GRL	SLB	111	1970	83	OP	6-8	6,3	33,8	9-17	11,9	61,4
Rennell Island Tall	RIT	GRL	SLB	M61	1980	296	CP	7-8	11,9	51,8	9-13	13,1	64,1
Rotuman Tall	RTM	GRT	FJI	101	1970	75	OP	6-8	6,4	33,3	9-17	11,3	59,3
Solomon Island Tall	SIT	GSN	SLB	081	1968	400	OP	6-8	9,0	72,9	9-18	12,1	104,9
Sri Lanka Tall	SLT	GSL	LKA	112	1972	275	OP	7-8	7,0	32,1	9-17	12,5	63,6
Tagnanan Tall	TAG	GTN	PHL	102	1974	400	OP	6-8	4,5	14,9	9-19	11,4	45,9
Tagnanan Tall	TAG	GTN	PHL	083	1978	400	OP	6-8	3,4	5,5	9-13	10,7	21,6
Tahitian Tall	TAT	GPY01	PYF	091	1969	550	OP	6-8	7,1	34,0	9-18	12,2	68,0
Tahitian Tall	TAT	GPY01	PYF	111	1970	83	CP	6-8	5,4	26,7	9-17	12,2	63,2
Tahitian Tall	TAT	GPY01	PYF	M62	1980	843	CP	7-8	7,7	21,0	9-13	12,2	56,4
Thailand Sawi Tall	THT01	GTH01	THA	060	1968	113	OP	6-8	8,1	42,4	9-18	12,8	73,0
Tonga tall	TON	GTG	TON	101	1970	100	OP	6-8	8,7	47,8	9-17	12,7	64,8
Vanuatu Tall	VTT	GVT	VUT	111	1970	450	OP	6-8	7,4	55,1	9-18	13,0	95,9
West African Tall	WAT	GOA	CIV	081	1968	225	OP	6-8	6,4	35,3	9-18	10,7	72,0
West African Tall	WAT	GOA	CIV	091	1969	525	OP	6-8	8,9	45,0	9-18	12,9	82,0
West African Tall	WAT	GOA	CIV	101	1970	225	CP	6-8	7,2	34,3	9-17	11,6	69,9
West African Tall	WAT	GOA	CIV	111	1970	84	CP	6-8	6,4	38,2	9-18	12,1	85,3
West African Tall	WAT	GOA	CIV	111	1972	144	CP	7-8	9,2	47,5	9-17	11,9	74,5
West African Tall	WAT	GOA	CIV	112	1972	168	CP	7-8	8,8	41,3	9-17	12,2	59,4
West African Tall	WAT	GOA	CIV	102	1974	175	CP	6-8	6,9	30,8	9-19	11,2	49,6
West African Tall	WAT	GOA	CIV	102	1975	45	CP	7-8	8,9	33,5	9-18	11,5	53,7
West African Tall	WAT	GOA	CIV	083	1978	211	CP	6-8	10,1	41,1	9-13	13,3	63,5
West African Tall	WAT	GOA	CIV	M63	1980	75	CP	7-8	12,8	34,6	9-13	13,1	68,0
West African Tall	WAT	GOA	CIV	035	1980	1690	CP	7-8	13,2	44,3	9-13	13,3	63,0
West African Tall	WAT	GOA	CIV	142	1982	50	OP	6-8	11,8	61,1	9-11	10,4	63,2
West African Tall pop. Akabo	WAT03	GOA03	CIV	M63	1982	150	CP	6-8	11,5	41,0	9-11	11,8	52,5
West African Tall pop. Mensah	WAT04	GOA04	CIV	M63	1982	150	CP	6-8	11,1	36,6	9-11	11,6	48,9
West African Tall pop. Ouidah	WAT06	GOA06	BEN	M63	1982	150	CP	6-8	12,2	42,9	9-11	11,5	51,9
Total						12384							
Mean						248		3-8	3.6	16.8	9-X	11.5	58.8

Table 13: *Fruit composition of ecotypes in the collection at the Marc Delorme Station*

International Cultivar name	Codes: International,	French,	Country.	Sampling size (g): Trees sampled	Total number of sampled fruits	Mean weight of fruit component: Whole fruit	Husk	Shell	Water	Kernel	Copra	Oil	Ratio copra/ fruit without water
Andaman Giant Tall	AGT	GND03	IND	13	129	1474	607	217	233	418	249	163	0,20
Andaman Ordinary Tall	ADO	GND02	IND	73	1234	990	368	158	138	327	193	123	0,23
Aromatic Green Dwarf	ARO	NVP07	THA	9	140	637	205	103	132	198	116	72	0,23
Baybay Tall	BAY	GPH04	PHL	50	421	1263	334	191	282	457	254	—	0,26
Cambodia Battambang Tall	KAT09	GCB09	KHM	42	490	1679	502	248	417	511	284	179	0,23
Cambodia Koh Rong Tall	KAT10	GCB10	KHM	28	500	1400	404	214	350	433	249	158	0,24
Cambodia Ream Tall	KAT07	GCB07	KHM	49	1349	1699	508	252	410	531	293	186	0,23
Cambodia Sre Cham Tall	KAT08	GCB08	KHM	32	305	1786	513	267	441	565	303	193	0,23
Cameroon Kribi Tall	CKT	GCA	CMR	34	711	859	366	148	78	268	172	140	0,22
Cameroon Red Dwarf	CRD	NRC	CMR	108	3086	757	215	134	141	267	143	103	0,23
Catigan Green Dwarf	CAT	NVP02	PHL	51	1897	823	302	123	114	284	165	90	0,23
Comoro Moheli Tall	CMT	GCO	COM	51	1289	1280	483	193	201	402	254	159	0,24
Equat. Guinea Green Dwarf	EGD	NVE	GIN	115	4007	773	257	131	102	283	162	107	0,24
Ghana Yellow Dwarf	GYD	NJG	GHA	221	7537	699	231	100	124	244	117	76	0,20
Kappadam Tall	KPD	GND05	IND	7	80	1524	638	201	240	446	262	173	0,20
Karkar Tall	KKT	GNG01	PNG	49	1029	1219	426	188	224	382	222	142	0,22
Kinabalan Green Dwarf	KIN	NVP06	PHL	51	1142	1552	368	256	396	533	315	200	0,27
Laccadives Micro Tall	LMT	GND07	IND	52	1080	554	244	93	36	183	118	74	0,23
Laccadives ordinary Tall	LCT	GND08	IND	51	1195	668	300	109	53	206	134	80	0,22
Madang Brown Dwarf	MBD	NBN	PNG	67	2142	499	240	72	35	152	76	51	0,16
Malayan Green Dwarf	MGD	NVM	MYS	101	3051	736	240	115	124	258	131	82	0,21
Malayan Red Dwarf	MRD	NRM	MYS	88	3394	890	312	129	159	291	144	90	0,20
Malayan Tall	MLT	GML	MYS	92	1008	1582	522	225	351	484	269	177	0,22
Malayan Yellow Dwarf	MYD	NJM	MYS	100	1936	619	188	91	107	233	112	—	0,22
Mozambique Tall	MZT	GMZ	MOZ	112	3210	1104	462	183	138	322	203	138	0,21
Niu Leka Dwarf	NLA	NNL	FJI	32	656	1021	438	149	119	316	192	113	0,21
Panama Aguadulce Tall	PNT01	GPA01	PAN	50	391	1528	471	226	345	486	266	173	0,22
Panama Monagre Tall	PNT02	GPA02	PAN	50	423	1467	451	217	328	472	262	166	0,23
Pilipog Green Dwarf	PIL	NVP05	PHL	30	981	555	206	78	79	192	119	77	0,25

Pumilla Green Dwarf	PGD	NVS	LKA	91	3109	413	206	57	30	120	78	50	0,20
Rangiroa Tall	RGT	GPY02	PYF	51	1349	1389	492	208	204	485	268	173	0,23
Rennell Island Tall	RIT	GRL	SLB	186	5375	1681	476	252	386	568	301	194	0,23
Rotuman Tall	RTM	GRT	FJI	50	2040	1435	469	218	239	509	299	195	0,25
Solomon Island Tall	SIT	GSN	SLB	68	2527	1050	359	166	169	355	203	132	0,23
Sri Lanka Tall	SLT	GSL	LKA	49	1456	1349	522	205	186	436	269	171	0,23
Tacunan Green Dwarf	TAC	NVP03	PHL	27	798	1019	345	159	177	338	206	109	0,24
Tagnanan Tall	TAG	GTN	PHL	86	1587	1521	405	236	364	516	280	159	0,24
Tahitian Red Dwarf	TRD	NRY	PYF	40	1064	252	99	51	17	85	45	25	0,19
Tahitian Tall	TAT	GPY01	PYF	260	7365	1208	422	191	170	425	252	167	0,24
Ternate brown Dwarf	TBD	NBO	IDN	50	415	666	258	103	65	241	140	—	0,23
Thailand Green Dwarf	THD	NVT	THA	79	2566	739	222	126	130	262	124	75	0,20
Thailand Sawi Tall	THT01	GTH01	THA	92	2772	1854	570	274	454	555	299	190	0,21
Tonga tall	TON	GTG	TON	43	1837	1254	383	202	213	458	270	176	0,26
Vanuatu Tall	VTT	GVT	VUT	67	1653	916	284	150	156	327	194	124	0,26
West African Tall (mixed)	WAT	GOA	CIV	203	4675	1051	448	162	110	331	203	—	0,22
West Afr. Tall Agriculture	WAT01	GOA01	CIV	150	2581	1068	452	164	122	329	198	—	0,21
West Afr. Tall Benin 2	WAT05	GOA05	CIV	118	2569	1166	543	173	118	332	205	135	0,20
West Afr. Tall pop. Akabo	WAT03	GOA03	CIV	140	3433	1036	446	162	105	323	200	130	0,21
West Afr. Tall pop. Mensah	WAT04	GOA04	CIV	395	9057	986	421	160	92	313	188	125	0,21
West Afr. Tall pop. Ouidah	WAT06	GOA06	CIV	84	976	1136	516	164	119	336	209	136	0,21
Total				4137	104017								
Mean				82	2080	1097	383	168	190	356	204	119	0,22

Tall ecotypes

At the Marc Delorme Station, tall ecotypes start producing between five and eight years of age. Environmental conditions may influence the beginning of production by a period of one year. For comparison of the production of the various tall ecotypes, reference should be made to a constant production period. The period at this station has been fixed at 3-8 years (to maintain comparison with the dwarfs). A young, good tall plant produces 5 to 7 bunches and 25 to 45 fruits per tree annually. The bunch and fruit production of young talls is a little less than half of that of dwarfs, on average.

A mature, well performing tall plant annually produces 13 or 14 bunches and 80 to 100 fruits per tree; However, these figures correspond to the best results obtained at research stations, and should not be extrapolated to field conditions. On average, the fruit production of mature talls is about nine-tenths that of dwarfs. The effect of the environment on the bunch and fruit production at maturity is well illustrated by the behaviour of WAT on two neighbouring plots:

- 85.3 fruits and 12.1 bunches on average between 9-18 years on one plot, planted in 1970;
- 49.7 fruits and 11.6 bunches on average between 9-16 years on another plot, planted in 1974.

Ecotypes producing the highest number of fruits are those that produce the smallest fruits: Solomon Tall, Vanuatu Tall, Lacadives Micro Tall, etc. Ecotypes producing the lowest number of fruits are those that produce the biggest, round fruits: Panama Tall, Malayan Tall, Cambodia Battambang Tall. No ecotype monitored until 18 years of age has shown a decline in yield.

Fruit composition

In Ivory Coast the fruit analysis data, processed for about thirty years, permit the establishment of an overview based on about fifty different geographic origins (accessions and descendants of accessions). The sampling and analysis methods are those described by Meunier *et al.* (1977), and by Wuidart and Rognon (1978): one fruit analysis per harvest round (monthly for dwarfs, bimonthly for talls) on 50 trees for a period of 6 years whenever possible.

Table 13 shows the results of these fruit analyses. The data presented are averages per ecotype, often combining plantations of the same planting material on various plots. Twenty-five nuts from each of the 4.137 analysed trees were sampled on average. Only 9% of these trees were characterized by less than ten fruits. Some populations are still under observation.

Coconut fruit characteristics vary with climate and soil conditions. For instance, different progenies of Ghana Yellow Dwarf (GYD) from the same parent trees have been planted on different dates and on different plots on the Marc Delorme Research Station; these progenies have a mean copra per nut content varying between 106 and 147 g.[7] For one particular tree, the average fruit weight may differ with time, according to the nature of the pollen, the age of the tree and its production rhythm. Periods of heavy production are often accompanied by lower average fruit weight (CP 15). Environmental variations in general barely influence fruit weight compared to other production components, such as number of bunches and number of fruits per bunch. Figure 9 shows the copra percentage of fruits without free water (CFWW). The CFWW ratio is often used for the characterization of the fruit composition. The free water weight is eliminated as it fluctuates strongly with environmental conditions. The water content of the kernel shows little genetical variation; a similar ratio based on the weight of the kernel instead of copra would be equally valuable. Data presented in Figure 9 are the averages per plot and per year of plantation. Various dwarf ecotypes and the Lacadives Micro Tall (LMT) of Indian origin have the lightest fruits. The Thailand Tall, Cambodian Tall, Rennell Tall, Panama Tall and Malayan Tall bear the heaviest fruits. A Philippine green dwarf (KIN) also produces a rather heavy fruit for a dwarf. It is interesting to observe the individual CFWW values. The individual palms with the highest CFWW reach values of 30%. These are found mostly among the Pacific talls: TAT, VTT, RGT, TON, RTM, and among The Philippine ecotypes: PIL, KIN, TAG, BAY. Some belong to the Indian type, LMT and AGT. There is no clear relationship between the circumference of the fruit and its copra content. VTT and LMT are small fruit types, TAG and TAT are larger fruit types; all of them have a good composition. The lowest individual CFWW values are 13%, which can be found among the WAT, the MZT, the KKT tall and the MBD dwarf from New Guinea, and to a lesser extent in the PGD of Sri Lanka.

Flower biology

The study of the flower biology of coconut showed a variability in the reproduction, dividing the ecotypes into groups of allogamous and autogamous palms.

Particular phenotypes

As with all plants, particular phenotypes appear in relatively low frequencies, such as coconuts with sweet, edible husks, Makapuno nuts, Spicata trees, and palms with their leaflets remaining attached to each other. None of these particular phenotypes have any important impact on genetic improvement programmes. The only relatively rare, but widely used characteristic is dwarfism, which represents, without including hybrids, less than 5% of the world coconut population.

Conclusion

Overview of progress

Since 1980, considerable progress has been made with genetic diversity of the coconut. New techniques have been developed. The *in vitro* culture of zygotic embryos facilitates the collection of new accessions and safeguards germplasm exchanges. The study of lethal coconut diseases has reduced contamination risks, although certain detection techniques still have to be completed. Freezing embryos and pollen in liquid nitrogen provides a new alternative for germplasm conservation.

The methodologies for variability analysis have been diversified. The analysis of the enzyme polymorphism by electrophoresis has given indications about coconut dissemination, but it has not led to a structuration of the variability. Foliar polyphenol analysis has given some precise results on the genetic distance between ecotypes; it remains difficult to compare populations planted in different ecological conditions with this method. The next stage will consist of the application of genome analysis techniques to coconut: these will permit elimination of variations linked to the environment, reducing the precision of other methods. In the more or less long term, they may provide a wide range of applications. The international community has agreed there is an urgency to collect coconut germplasm threatened by the increasing danger of lethal diseases and by extension of hybrid plantations. The volume collected has been multiplied by four or five in less than 15 years. However, certain culti-vation regions have still been insufficiently surveyed. In addition, it is not enough merely to collect material; it is also necessary to maintain, evaluate and disseminate the collections. An effort to obtain financial aid should therefore be made in this field of activity. Finally, other conservation methods of germplasm resources should be developed: embryo cryo-banks, and perhaps dynamic conservation of germplasm on islands that would play the role of specially reserved ecosystems.

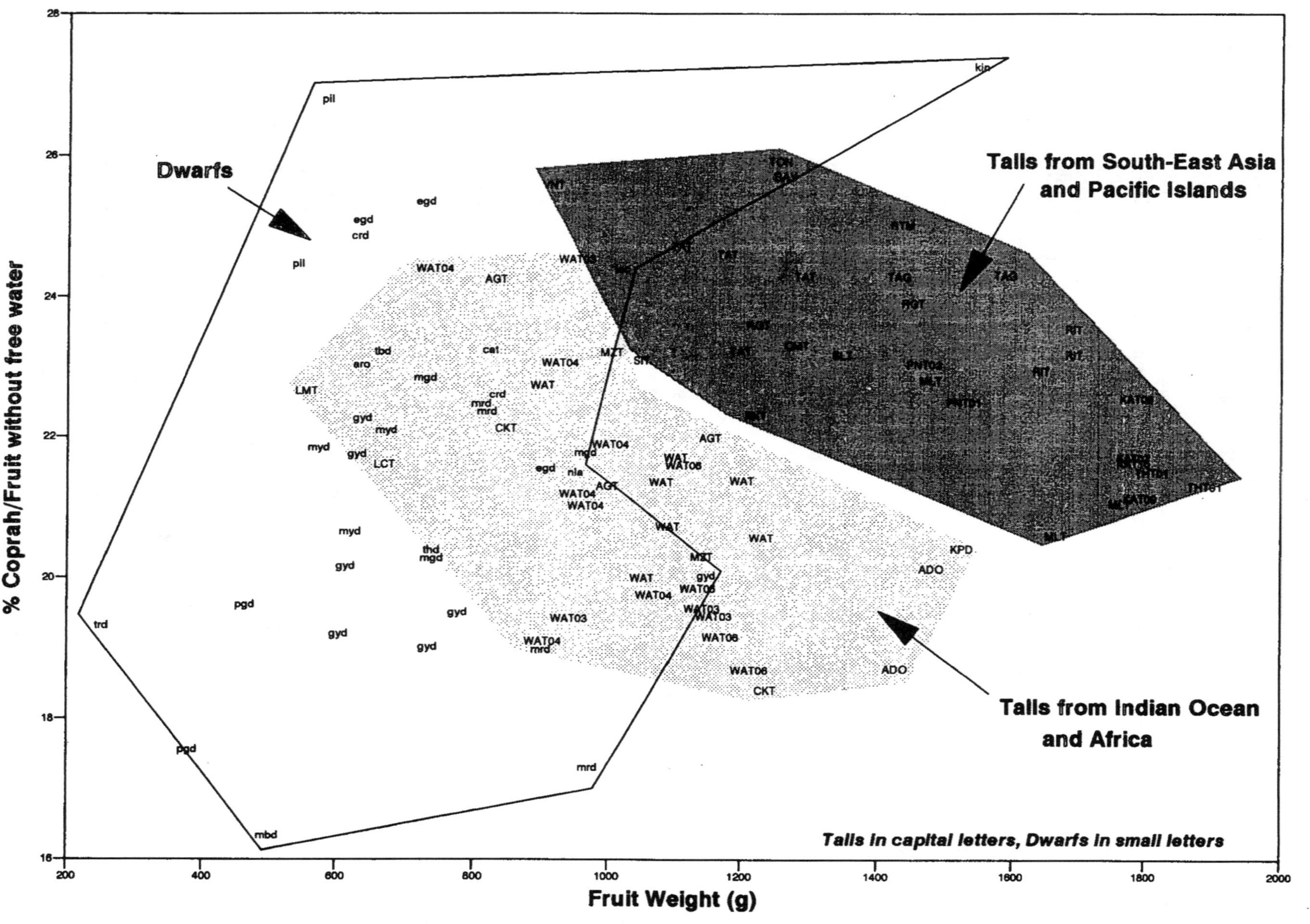

Figure 9: *Copra percentage of fruits without free water*

In 1991, the IBPGR established a coconut genetic resources network. This network brings together representatives of coconut producing countries, institutions involved in coconut research and various donors. Its objective is to promote research activities and facilitate information exchange of coconut genetic resources. One of the first recommendations of the network has been the establishment of an international database to make an inventory and a description of material kept in collections. The CIRAD-CP (France) has been charged with the establishment of this database which, in December 1993, already contained 502 accessions. The standardization, initiated by the IBPGR has been followed by publication in 1992 of an international list of coconut descriptors.

Coconut diversity

Appraisal of the relative characteristics of ecotypes is related to the extension of the observed variability. Among the talls studied, the analysis of the fruit composition permits the distinction of two groups:

- an Indo-African group, whose fruits are characterized by a low copra content and a thick husk;[8]
- a group from S.E. Asia and the Pacific, the fruits of which are characterized by a low husk percentage.

The Indo-African ecotypes correspond well with the Niu kafa type described by Harries (1978). Ecotypes from S.E. Asia and the Pacific correspond more with the Niu vai type. According to this model, the coconut should have been disseminated naturally from Asia across the Indian Ocean towards the East African Coast. The foliar polyphenolic analysis has accurately indicated the origin of the African ecotypes; the coconut has been disseminated from the Far East to Africa, all the time undergoing changes and genetic impoverishment. However, the path of dissemination is not entirely regular. Certain Indian Ocean types, such as talls from the Comoros and from Mozambique show some characteristics that are closer to the Far East type than to the Indo-African type. However, the polyphenol analysis has made it possible to separate coconuts from S.E. Asia from those from the Pacific.

Dwarf ecotypes show a combination of characteristics that are entirely different from those of the talls: reduced size of its organs, precocity, rapid emission of bunches and preferential autogamy. The Niu Leka dwarf is exceptional in its allogamous nature and late productivity. Other ecotypes show so many common points that it is difficult to evaluate the variability of the group. Nevertheless, as far as the fruit composition is concerned, the variability of dwarfs is far from negligible. On the other hand, the foliar polyphenol analysis indicates that certain dwarfs should be of Far Eastern origin and others from the Pacific.

The origin and genetic determinism of dwarfism are still unknown. D x T crossings generally show an intermediate growth between that of their parents. Crossings with the atypical Niu Leka dwarf are an exception and behave more like their dwarf parent. Never has a complementarity of recessive alleles been observed: the crossing of two dwarf ecotypes has never produced a tall. These results led to the formulation of two hypotheses on the origin of dwarfism:

- if dwarfism originates from only one unique 'selection centre', and has been disseminated by mankind, it should be recognized that introgressions with tall types have caused a final diversification of dwarf ecotypes;
- according to the second hypothesis, various forms of dwarfism have appeared independently and in different places, and were conserved and disseminated by mankind.

The problem of the centre of origin

Neither the morphometric analyses, nor the studies on enzymatic and polyphenolic polymorphism were able to indicate the centre of origin of coconut. The ways of dissemination of the coconut could explain the difficulty of localizing its centre of origin. Two mechanisms have played a role in coconut dissemination: the nuts floating along with sea currents and, later, human travels and migrations. These mechanisms have modelled the genetic structure of the populations.

The natural dissemination - its multiple founder effects inducing genetic drift - should lead to a mosaic of clearly differentiated ecotypes, showing variable but high levels of consanguinity, and distributed according to geographic gradients.

The human action is inseparably linked with the natural dissemination: the world coconut population consists almost entirely of plantations. Omnipresent mankind harvests, plants and selects. Certain selected forms keep the ability of spontaneous dissemination. A dynamic equilibrium is created, including successive founder effects, multiple introductions, primary and secondary selections, and introgressions between natural and selected forms. All intermediate stages between the pure line and the population in its true sense are hidden under the term 'ecotype'.

Genetic Improvement Methods

As has been mentioned in the general introduction to this chapter, a coconut breeder often analyses trials established by his predecessor and establishes trials for his successors.

In various countries, numerous research years have been lost as a result various accidents: fires, floods, revolutions, turnover of personnel or simply lack of funds leading to complete programme discontinuation. Therefore, it is of major importance to ensure that the data collected at the research stations will be available and safely kept for many years. These data should be duplicated systematically in two geographically different places. These may be two different national institutes, or a national institution co-operating with a specialized international research institute. There should be a unitary identification key for each coconut tree in a research station. This key is generally composed as follows:

- a code for the experimental station;
- a code number for the planting plot at the station (often three figures);
- a code number for the planting line within the plot (generally two figures);
- a code number for the tree within the line (two figures);
- a code corresponding to the year of planting.

In this way, even if a plantation has been cut down and replanted again, the unitary identification will avoid confusion.

Reproduction Techniques

The basis of plant breeding consists of creating new genetic structures by means of crossing. In the case of coconut, various reproduction techniques may be used. Here again, taking into account the time required for the trials, recording the operations is of fundamental importance. Each coconut tree used in a crossing should be identified by a parent number, attributed once and for all in a unique way. In this way, progressively a catalogue of parents can be made containing at least the following information:

- parent number;
- identification key, described above;
- ancestors of the parent tree.

Similarly, each progeny should be numbered. According to the type of cross studied, the progeny number should be attributed to various genetic structures. It could be:

- a cross between two populations (open or hand pollination);
- a cross between an individual palm and a population;
- a cross between two individual palms;
- selfing of an individual palm.

Open pollination

This method, the most simple, is natural pollination. Pollination of coconut is effected both by wind and insects. In Sri Lanka, coconut pollen was found at 180 m from the emission source (Manthriratne 1971). Within a plantation the probability of pollination between two trees probably decreases as a function of the distance between these trees. Wind plays an important role, as it can transport pollen as well as insects. No study has yet been able to quantify these phenomena exactly.

Among the ecotypes classified as preferentially allogamous, self-pollination may occur unexpectedly. The rate of autogamy thus fluctuates according to several parameters. Inflorescence production over the year is not regular. In India, a study on talls has revealed that in April 75% of the trees showed overlapping between phases, allowing for selfing, whereas in November no overlapping took place at all (Patel 1938). In Ivory Coast, a more rapid emission of inflorescences can be observed at the beginning of the dry season. During this period, the autogamy rate of the population increases.

In addition, not all individuals in a population have the same rate of autogamy. High yielders produce more inflorescences and may have higher selfing rates. Competition between pollen can also increase selfing rates. MYD, pollinated with a mixture of its own pollen and that of a tall, preferentially produces dwarf seeds (Sangare 1981). Without doubt, similar phenomena can also be observed among some talls.

Controlled open pollination

This method, essentially used for the production of hybrid seednuts, consists of emasculation of inflorescences to produce seeds resulting from a controlled cross. This idea was first launched by Tammes (1955), who proposed to apply to coconut the technique used in maize, consisting of interplanting two ecotypes to use one as the male and the other as the female parent.

In 1961, Liyanage described a similar technique, but adapted to intra-population crosses. It consists of selecting male and female parent trees within a plantation. All trees that are not used as male parents are emasculated. Seednuts will be collected only from those trees selected as female parents and previously emasculated. This technique was developed for the production of hybrid seednuts (Nucé de Lamothe and Rognon 1972a). It is still being used, for instance on Sri Lanka, where a large seed garden has been interplanted with PGD and SLT for the production of the hybrid CRIC65 (Liyanage *et al.* 1988). CP 16 shows an operator emasculating a tree in this seedgarden. According to Nuce de Lamothe and Wuidart (1992), 'the cost of a seedgarden varies largely with country and site.... This often is twice as high as the investment required for the establishment of an industrial plantation'.

Seedgarden isolation

The seedgarden has to be bordered by a barrier to avoid contamination by unwanted pollen from outside. In most cases this barrier is free of coconuts and planted with forest vegetation or another perennial crop, such as rubber or oil palm. However, the barrier can also be planted with coconut under the following conditions:

- the planting material is the same as the male parent used in the seedgarden;
- the planting material is another variety but is emasculated regularly.

The width of the barrier must be sufficient to guarantee isolation. Following the experimental results obtained with dwarf palms, Nucé de Lamothe and Wuidart (1992) recommend a barrier, 400 to 500 m wide, or more, according to local conditions.

Proportion of pollinators
The proportion of pollinators in the seedgarden should be sufficient to assure good pollen distribution. Nucé de Lamothe and Rognon (1972a) recommend planting 20% pollinators. They admit that this proportion is probably excessive, but it allows for a selection of pollinators based on their intrinsic value: certain trees showing unfavourable characteristics can be eliminated without affecting pollination. A final proportion of 5 to 10% of male parent trees seems to be sufficient (Ohler 1984).

In Sri Lanka, in a seedgarden without pollinators, parthenocarpic bunches have been observed, resulting from a lack of pollination (CP 13). Only the husk develops and cracks under the influence of internal constraints. In this case it concerned a local and temporary problem, as in the neighbourhood a plot of pollinators started flowering.

Emasculation
In a well isolated seedgarden, the quality of emasculation determines the rate of legitimacy of seednuts. Timing of emasculation should be adapted to the floral biology of each cultivar. If the latter is an autogamous dwarf, emasculation is carried out two days before the natural opening of the spathe (Wuidart and Rognon 1981). Emasculation generally comprises four operations:

- elimination of the spathe that protects the inflorescence. This operation facilitates the work of the pollinator;
- cutting off the part of the spikes on which male flowers grow. The spikes should never be cut close to the last female flower, but about 3-5 cm above this point;
- elimination of the two male flowers situated at the axil of each female flower, and those on the remaining part of the spike;
- subsequently, the eliminated male flowers, which can still be a source of contamination, should be done away with. A simple solution is to bury them at the foot of the palm.

The errors most often made are delayed emasculation after the natural opening of the spathe, and leaving some male flowers after emasculation. Such errors cause selfings. Experience has shown that a system of regular control is indispensable. Slack control is rapidly followed by an increased rate of illegitimate seednuts. In properly carried out operations, this rate does not exceed 5 to 7%. In certain seedgardens large-scale abortion has sometimes been observed, due to internal necrosis of pollinated flowers. The cause is the cutting away of spikes too close to the female flowers, the necrosis developing from the cut towards the place where female flowers are growing. This problem has been resolved by cutting the spikes 5 to 6 cm above the female flowers and removing the remaining male flowers by hand (Wuidart, personal communication).

Assisted pollination
For seednut production this method, developed by Nucé de Lamothe and Rognon in 1972b, is currently used in most coconut-producing countries for seednut production. A recent series of publications (Nucé de Lamothe and Wuidart 1992; Rognon and Bourgoing 1992 and Wuidart and Rognon 1993) presented detailed descriptions of this method. It consists of bringing controlled exogenous pollen on to emasculated inflorescences. The seedgarden is planted only with one or more cultivars used as female parents. The pollen is brought from an exterior source, which can be thousands of kilometres away. The consideration presented earlier on the isolation of the seedgarden,[9] the emasculation technique and the control of seednut quality, can also be subject to this technique. The operations concerning the pollen and the pollination are more specific.

Preparation and storage of pollen
The pollen is collected from the selected male cultivar. It is not necessary to bag the inflorescences to avoid contamination by nearby trees of another variety. The proportion of foreign pollen that could be present on an inflorescence in full anthesis is considered negligible. When the pollination is carried out

to produce crosses for genetic experiments, it is better to bag the inflorescence (see the following section on hand pollination).

In Indonesia, inflorescences are being treated to increase the quantity and the quality of the harvested pollen. This treatment consists of the application of a mixture of naphthalene acetic acid 400 ppm and ethrel 100 ppm, the day after the natural opening of the spathe (Darwis 1991). This method does not appear to be applied on a large scale. Its economic value needs to be confirmed insofar as pollen harvesting is really a limiting factor with coconut. A pollen harvester visits 150 palms per day and collects 20 to 30 kg of fresh male flowers. The processing of these flowers consists of crushing, drying and sieving. Subsequently, the pollen is used for a cross, after verifying that it has not lost its germinating capacity.

When pollen is to be used within a period of 10 to 15 days, drying to a humidity content of 10-12% is enough. For conservation of pollen for a period of several months, supplemental drying and vacuum storage is required (Rognon and Bourgoing 1992). Pollen treated in this manner has a humidity content of less than 4-5%. The viability, expressed by the germination rate in a sweet gelatinous medium, is generally is about 40%. Cameroon Red Dwarf pollen prepared by the Rognon/Bourgoing method and frozen for a period of seven years has kept a viability above 30% (Santos, personal communication). However, it is better to use the pollen within about six months of preservation.

Preliminary experiments conducted in Ivory Coast have shown that freezing of dehydrated pollen in liquid nitrogen (-196°C) does not cause a decline in germination rate. Long-term conservation in liquid nitrogen has interesting implications. Taking into account the evolution of improvement methodologies, the pollen from improved parent trees will be a limiting factor. Due to commercial constraints, the production of seednuts is not always continuous. With the possibility of storing the pollen of improved parent trees, the loss of a precious material can be avoided.

Pollination

A homogeneous mixture of 5% pollen and 95% talcum powder is prepared daily by the operator responsible for 1200 to 1500 trees. This mixture is applied to the emasculated inflorescence each day during the receptive period of the female flowers. The spraying system may consist of a washbottle, a flexible rubber tube attached along a bamboo or a light metallic (aluminium) rigid tube (CP 19).

The total quantity of pollen applied to each inflorescence depends on the duration of the receptive phase of the female flowers of the cultivar used. For an autogamous dwarf with a long female phase (10-12 days), this quantity is about 0.4-0.5 g per inflorescence. For a tall with a short female phase, this quantity is reduced by one half.

Hand pollination

In a coconut genetic experiment, the legitimacy of crosses is an indispensable priority. In fact, a genetic experiment must be monitored for at least twelve years; what a waste of time and investment if the observations are made on illegitimate trees!

The principle of hand pollination focuses on emasculation of the inflorescence followed by isolation of the female flowers by means of material impermeable of pollen but permeable to air. When the flowers become receptive, selected pollen are introduced for pollination. The procedure covers three different steps:

- collection and preparation of the pollen;
- isolation and the pollination of the female flowers;
- harvesting and marking of fruits.

Collection and preparation of pollen

Isolation of the inflorescence for the collection of pollen is not always done systematically. Certain breeders are of the opinion that contamination by undesirable pollen must be negligible in view of the quantity of pollen obtained. However, Sangare (1981) showed the existence of pollen competition phenomena that may favour the contaminating pollen. Therefore, isolation of the inflorescence is necessary and should be carried out at least eight days before collecting spikes (Nucé de Lamothe *et al.* 1980).

Nucé de Lamothe *et al.* (1980) have also emphasized that the pollen should be constantly isolated from the environment. A detailed system has been proposed: application of a thick canvas bag for pollen harvesting, use of a sterilized isolation box, disinfection with alcohol before transfer, water-permeable but pollen-impermeable containers for drying in the oven. These precautions are sometimes considered excessive, but they are justified if in a conditioning laboratory pollen from different origins is treated simultaneously, which increases the contamination risk substantially.

Isolation and pollination of female flowers

The first step involves isolation of flowers with material permeable to air but not to pollen (of thick canvas). Two technical options may be considered:

- individual isolation of each female flower;
- isolation of the entire inflorescence.

Individual isolation of female flowers was developed in Jamaica (Harries 1967). Small cylinders of polyurethane foam surround the female flowers and remain in place due to their elasticity. A foam plug inserted into the cylinder protects the stigmas. It is necessary to open the cylinder to verify if the female flower is receptive. When this is not the case, the plug is replaced and pollination is postponed. Contamination is possible when the cylinder was opened. The technique is improved by leaving some flowers uncovered. These control flowers, eliminated later, serve only to investigate whether the covered flowers are receptive and should be pollinated.

Bagging the entire inflorescence is the most widely used method. The material for the isolation bag (thick canvas) should meet some precise criteria: permeable for air, but not for pollen, sufficiently strong to stand three weeks of severe tropical conditions. Nucé de Lamothe *et al.* (1980) have presented a detailed description of the quality of the bag and the way to place it, taking the floral biology of each ecotype and the period of pollen viability into account. During pollination, contacts between the fertilized isolated flower and the exterior environment are reduced as much as possible to prevent contamination. The use of transparent windows permits the observation of flower maturity without opening the bag. A mixture of pollen and talcum powder is blown in with the use of a washbottle. After disinfection with alcohol, its spout is introduced through a small hole in one of the windows. Before and after dusting with pollen, the small hole is covered with tape. This method has sometimes been considered to be exaggeratedly careful. However, experience has shown that, even when respecting the rules, some illegitimates are always obtained. The errors most often made are inadequate emasculation of the inflorescences and wrong numbering.

Harvesting and marking the fruits

Recording pollinations requires special attention. Low yields of crosses require a great number of operations-multiplying the risk of errors. A unique number is given to each pollination. By this number, recorded in a register or computer file, the exact pedigree of each seednut is registered. The seednuts are harvested one year after pollination. The development of the seedling in the nursery requires another year. It is therefore necessary to have a marking system that can survive severe climatic conditions. On the tree, the fruits are marked with a felt-tip pen in indelible black ink. At harvest, an engraved aluminum tape is attached with a metal wire to each fruit, carrying the same number of the artificial pollination. In the nursery, another aluminum tape with the same number is attached to a leaf of the seedling. After planting, removal of the pollination numbers provides a final check on the pedigree of the plant. This control almost always shows that 1 or 2 per cent of the plants have lost their identification during transfer to the field. These plants are replaced by others of known pedigree.

Seednut production and harvest

Dwarfs, often used as female parents in seedgardens, are subject to phenomena of alternating production. In certain periods, they can produce an excessive number of fruits per bunch and 'suffer' to enable these fruits to reach maturity. The harvested seednuts are therefore small and have little water. Mostly, when placed in a well managed germination bed in a well equipped nursery, these seednuts can develop satisfactorily. In Ivory Coast, it was tried to intensify the alternating production mechanisms by periodically castrating the trees in the seedgarden. This castration induces a catching-up reaction which permits keeping the seed production within a certain period and optimizing the use of pollen from improved parents. Pollen is often an important limiting factor. However, this method should be used with caution and should not be used without having conducted complementary trials. In fact, in certain periods, the seednuts produced are too small, with too little water. These nuts germinate badly, in spite of good care in the nursery. (Bourdeix *et al*, 1991 and 1994).

If small seednuts have to travel, or when they are distributed directly to the farmers, germination problems may appear. In certain Philippine seedgardens, the number of fruits per bunch is limited to 6-8. The surplus fruits are eliminated when they are 3 months old. This method guarantees good development of the remaining fruits and avoids germination problems (Santos, personal communication). For proper germination, seednuts should not be collected before they are 11 months old. A simple technique is not to harvest seednuts from a bunch that does not have at least one splashing, apparently dry nut (brown epidermis). Seednuts harvested too early run the risk of developing abnormal shoots. These shoots often have rumpled leaf-tips, similar to the symptoms of boron deficiency.

Seedling control

In seednuts derived from a seedgarden with yellow or red dwarfs as female parents, the germination shoot colour allows for recognition of the illegitimate seedlings. The colour seems to be controlled by two independent genes, each having two alleles (Bourdeix 1988b). The brown or green shoots are considered hybrids, whereas the yellow or red shoots, depending on the colour of the mother tree, are considered selfings (Nucé de Lamothe and Rognon 1972a). However, this system indicates too many selfings, as red and yellow coding alleles are found with low frequencies in most of the tall populations. When the female parent's colour in the seedgarden is green or brown, the germination speed may sometimes identify the illegitimate seedlings (Wuidart and Rognon 1993).

In vitro cultivation

Somatic embryogenesis

Coconut breeders are of the opinion that a vegetative multiplication technique would represent decisive progress. Particularly, such a process would resolve most of the problems resulting from the low rate of multiplication, which is one of the main limiting factors in the breeding and seednut production.

After sometimes more than ten years of research, three teams have obtained the regeneration of plants derived from somatic tissue cultivated *in vitro* (Branton and Blake 1983; Raju *et al.* 1984; Buffard-Morel *et al.* 1988).[10] However, obtaining some plantlets does not mean that the method is operational. Until 1991, the results were not reproducible. An important step forward was made recently with the reproducible production of in-vitro-plants from five different parent trees (Verdeil *et al.* 1992). The outline of the experimental process used was as follows:

- callogenesis from fragments of somatic tissue. Various types of tissue have been used: foliar tissues (Pannetier and Buffard-Morel 1982); inflorescences (Eeuwens 1976), and root points (Branton and Blake 1983a). This stage leads to the formation of disorganized calli;
- maintenance and cultivation of isolated calli and embryogenesis induction;
- embryo maturation, followed by conversion;
- transfer of normal shoots to a weaning bed for 3-4 weeks, to the pre-nursery for 3-4 months, to the nursery for 8-10 months, and finally to the field.

The example of the oil palm shows that a long period can pass between the first laboratory success and the distribution of clones to the growers. Complementary research is necessary for large-scale use of the process. In particular, one must be sure that passing through the callus stage does not provoke soma-clonal variations but ensures a true reproduction of the cloned genotype. In the case of coconuts, these research activities will probably take many years.

Haplo-diploidizing

With *in vitro* haplo-diploidizing, homozygotic plants can be obtained rapidly, and may be used either for further reproduction (as parent) or for the creation of stabilized variations within a selection cycle. This method is particularly useful for a crop with a long life cycle, like coconut, for which the production of 95% homozygotic material requires at least 25 years (four generations).

A thesis at the University of Paris-Sud was devoted to research into a method for obtaining coconut haploids *in vitro* (Montfort 1984). Two methods were considered: cultivation of female gametophytes (gynogenesis) and male gametophytes (androgenesis). The first method did not yield results due to the morphologic structure of the ovary, which is large and its cultivation causes problems of bulkiness and disinfection. The ovary is attached to the placenta and its dissection has been shown to be difficult. With the androgenetic method, embryos have been developed to increasingly advanced stages: globular, pear-shaped, development of a characteristic haustorium, presence of a root point and of a leaf design. To date it has not been possible to develop plantlets. Recent progress in the field of *in vitro* somatic embryogenesis should make it possible to complete this process.

Discussion

Risks of open pollination

The major inconvenience of open pollination in coconut is the unknown rate of selfing, which fluctuates according to numerous parameters. From a genetic point of view, this rate is important because selfing often induces inbreeding depression (among the majority of tall cultivars). The only advantage of open pollination is its simplicity; it is therefore the last solution to be considered. It should be avoided as far as possible in genetic experiments.

Comparison of seednut production methods

Techniques of seednut production have been properly worked out but should be followed systematically and precisely. Even if these techniques are under control, they remain relatively difficult due to the nature of the plant. One hectare of seedgarden (205 trees in the case of dwarfs) will produce seednuts for only 50 to 60 hectares of hybrids (Rognon and Bourgoing 1992), whereas one hectare of oil palm may produce sufficiently for the annual planting of about 1500 hectares of hybrids. By contrast, in the case of oil palm, the collection of pollen is often an important limiting factor.

The main difference between the system of assisted pollination and that of controlled open pollination lies in pollen management. In the first case it is produced separately and brought to a seedgarden consisting only of mother trees. In the second case, pollinators and mother trees are interplanted within the same seedgarden. The major advantage of assisted pollination is its great flexibility of use. It facilitates the production of various types of hybrids on demand, by changing the pollen applied. In fact, due to the absence of pollinators in the seedgarden, seed production per hectare is also slightly higher. However, these pollinators must be planted on a separate plot of land, also occupying a non-negligible area of land. The advantages of controlled open pollination are its simplicity and its low labour requirement. After planting, the only operation required is emasculation of the female parent. However, the original composition of the plantation determines the type of hybrid to be produced. Thus, a seedgarden with a mixed stand of Pumila Green Dwarf and Sri Lanka Tall can only produce the hybrid CRIC65 (PGD x SLT).

These two techniques appear to be more complementary than exclusive. After planting a dwarf x tall seedgarden, using the controlled open pollination technique, dwarfs in general flower two to three years earlier than talls. During this intermediate period, pollen for the production of hybrid seednuts has to be brought from outside. On the other hand, if so required, it will always be possible to change a controlled natural pollination system into an assisted pollination system.

Finally, the choice between one of these systems of seednut production should depend mainly on economic factors: the cost of pollinators and pollen collection, and the share of labour in the seednut production cost. One line of research not yet exploited would be the detection of male sterility recessive genes. Their use in seedgardens would render emasculation unnecessary and reduce seednut production cost.

Hand pollination techniques

Hand pollination techniques could be improved. In certain cases, precautions taken to avoid contamination are insufficient; in other cases, the reliability of the system is realised at the cost of low productivity and prohibitively high costs. The prescriptions from Whitehead (1963) and Nucé de Lamothe *et al.* (1980) are increasingly regarded as international standards. The pollen conditioning in flasks sealed in vacuum, the use of transparent windows to avoid unnecessary opening of the bag, and dusting with a washbottle with a mixture of talcum and pollen, are technical options increasingly being used.

One of the points of disparity concerns the material from which the isolation bags are made. Bags used in Africa have been developed for oil palm and later adapted to coconut. They are made of very reliable material (synthetic green tissue CD 72). This tissue resists about ten successive pollinations, or eight months of exposure to precarious weather conditions, without losing its impermeability to pollen.

Preliminary trials were conducted to try to explain the reduction in nut yield induced by bagging. The majority of the fruits abort before the age of three months. Neither the heat, nor the light nor the humidity within the bag seem to be directly responsible. One hypothesis is that bagging with the synthetic CD 72 tissue provokes a retention of carbon dioxide: this accumulation of carbon dioxide would be harmful to the oxygen requirement of the inflorescence and the pollen. This hypothesis is still to be confirmed by experiment.

Mass Selection

The majority of world coconuts are derived from mass selection, informally done by growers. At the end of the 19th century, large plantations were established by planting fruits imported from a region known for its production (Ziller 1962). In most cases, seednuts were selected according to their own characteristics: some preferred large and heavy fruits (Zuninga *et al.* 1969), others medium-sized fruits, preferably roundish (Apacible 1968). The genetic structure of the coconut populations was modified by successive selection of fruit characteristics. The system of selection of the best trees within the best plots began to be applied more recently. In this period, all research stations involved with coconut breeding have used this mass selection system.

The beginning of a coconut selection programme is characterized by a period of at least ten years during which no result will be obtained. It is therefore necessary to distribute only material expected to possess genetic advantage or, in other words, the seednuts collected from the locally highest producers.

In most cases, the selection criteria include the yield of copra per tree or one of its components such as number of fruits produced or nut copra content; various authors have included resistance to disease (which can be the main limiting factor), drought tolerance, or certain vegetative and reproductive characteristics. Thus, there are three variants of mass selection, differing according to the reproduction system used: mass selection using open pollination, selfing, or intercrossing. Mass selection using open pollination has been practised most. The advantage of the method is its simplicity: seednuts are collected from trees which show attractive characteristics at a certain moment or over a certain period. Progenies resulting from open pollination are the basis of an improved population, which will then undergo other selection cycles.

The other variants of mass selection consist either of selfing the selected trees, or intercrossing them. These methods, more cumbersome because they require hand pollination, were less often used. After a brief summing up of the theoretical bases determining their efficiency, a summary of the experimental results is presented for each of the three variants, which may provide some lessons.

Mass selection using open pollination

Where the male parent is not being controlled, breeding involves only half of the additive genetic effects. Some improvements may be passed along where the plot is isolated from any other source of pollen. It is possible to emasculate or to cut down all trees showing unfavourable characteristics. In this way, from the very first generation, open pollination takes place only between selected trees.

History

In the case of coconut, the efficiency of mass selection of mother trees based on production characteristics has been the subject of controversy for some time. Although a large number of this type of selection programme have been conducted all over the world, only a few experiments have a statistical structure that permits a real judgement of the efficiency of the method.

The experiments started in 1937, with a statistical design known as the 'Ceylon Latin Square', compared four combinations:

- with mother tree selection, or without;
- with nursery culling of seedling, or without.

Twenty years later, it was proven that although culling in the nursery may result in about 10% improvement, mass selection had been totally ineffective. Harland (1957) concluded that the high-yielding characteristic could not be transmitted and proposed various other breeding methods, considered better adapted to coconut. However, a detailed examination of the experimental protocols (Sakai 1960) reduced his credibility. The high and low yielding trees had been selected in two different provinces and did not belong to the same population. In addition, the period of observation of the mother trees had been too short and certain trees, classified as weak producers, some years later appeared to be rather good yielders.

This observation induced renewed interest in mass selection, which was supported by the first publication on heritability estimations in coconut. In 1961, Liyanage and Sakai, following calculations based on the variance analysis of the progeny of nine exceptionally yielding coconut palms, concluded that mass selection based on copra content per nut and yield ought to be very effective. That same year, a study on the progeny of 37 high-yielding trees (Ninan *et al.* 1961) showed the absence of any relationship between the value of the mother tree and the characteristics correlated with yielding capacity of the progenies in the nursery, such as collar girth and number of leaves. The authors concluded that the phenotype of an individual tree is no adequate indication of its genotype.

These two studies used a basically asymmetrical sample, as it concerned the best trees of the original population. The response to the selection of the sample cannot be extrapolated to the totality of the population. The first study which involved a complete population was carried out with the SLT ecotype (Liyanage 1967). It shows that the selection of the 5% best trees out of 104 led to an increase of 14.4% in yield of the progeny. However, the strictness of this selection considerably limits the potential seednut production: in most cases, the annual mean yield of the selected individual trees does not exceed 200 fruits per tree. Culling in the nursery eliminates about 40% of the seedlings. Under these conditions, with a selection rate of 5%, with one hectare of 'seedgarden' an area of six hectares can be planted annually. This low multiplication coefficient is incompatible with rational management of a production organization, postponing the seednut production for one generation.

The division into five yielding classes of a Philippine tall population (Santos *et al.* 1980) clearly showed the absence of a correlation between the yield of the parent trees and that of their progeny. After heritability calculations, based on various trials, Meunier *et al.* (1984) observed that in a certain number of cases the yield components (number of fruits and copra per nut) taken separately, present a good heritability. On the other hand, this is not the case with their totality, copra per tree, due to the strong negative correlation between number of fruits and copra per fruit.

A recent study (Bourdeix 1988a) made it possible to simulate a cycle of mass selection using open pollination. The population studied was a plantation of 400 WAT. Of each parent, between 2 and 5 open pollination descendants were planted totally at random and evaluated. The simulation results, presented in Table 14 reveal the existence of significant genetic correlations. Selection based on one yield component modifies the value of the population for the others, hence a strict selection on copra per nut, which is the most heritable characteristic, simultaneously induces a loss in the number of fruits and in yield. Selection based on the number of fruits increases the yield in a way practically equivalent to selection based directly on yield itself; but it also provokes a strong decline in copra per nut. Only selection based directly on yield enables a simultaneous gain in all three parameters: number of fruits, copra per nut, and yield. The response to the selection remains low compared to the estimates obtained by Liyanage.

In the studied WAT population, the number of fruit appears to be the predominant genetic yield component. Although the yield is obtained by multiplying the number of fruits by the nut copra yield, the following example shows that selection aimed at an increase in copra per nut may reduce yield. It should be stressed that the networks of genetic correlations vary from one population to another, and the result could have been different with another ecotype. However, if the objective is to increase copra yield per tree, the selection based on copra per nut (or fruit size) is a method to be considered with caution.

Table 14: *Results of mass selection simulations on the yield and its components of a West African Tall (WAT) population*

Selection according to copra per tree			
Percentage of trees chosen	Progress on copra per tree %	Progress on copra per nut %	Progress on number of nuts %
5 %	4.9	7.0	2.4
10 %	2.7	3.7	1.6
15 %	2.9	3.9	1.5
25 %	3.7	3.4	0.0
50 %	2.9	2.8	-0.1

Selection according to number of nuts			
Percentage of trees chosen	Progress on number of nuts %	Progress on copra per tree %	Progress on copra per nut %
5 %	8.9	6.4	-1.9
10 %	5.5	3.6	-1.3
15 %	6.0	4.5	-1.0
25 %	3.2	1.4	-1.4
50 %	3.2	2.1	0.9

Selection according to copra per nut			
Percentage of trees chosen	Progress on copra per nut %	Progress on copra per tree %	Progress on number of nuts %
5 %	13.0	-3.4	-13.5
10 %	8.1	-2.1	-9.7
15 %	6.2	-2.3	-7.8
25 %	4.3	-0.2	-4.3
50 %	2.5	0.8	-1.3

The influence of the reproduction systems

The study of mass selection effects on yields, using open pollination, is characterized by a number of divergent results. The differences originate in the reproduction characteristics of the tall ecotypes.

Although the latter are preferentially allogamous, natural selfing is sometimes possible. The rate of selfing increases with the rhythm of inflorescence production. This rhythm depends on the individual vigour of the tree and on climatic conditions. When selecting well-performing trees, one can select trees with a higher tendency towards selfing. Consequently, their progeny suffers from an inbreeding handicap.[11] On the other hand, the rhythm of inflorescence emission also varies between seasons. Selection results could differ according to the season within which seednuts are harvested.

A method of varying performance

To end with mass selection using open pollination, it may be concluded that this method may sometimes give genetic progress: the best result obtained consisting of a 14.4% gain in the first cycle for a selection of 5% of the trees. Results are variable, even contradictory, possibly due to the absence of control of the reproduction system. Although response to the selection of mother trees exists, yield increases are not rapid enough (Liyanage 1972). The radical selection necessary for obtaining progress substantially limits seednut production capacity.

Mass selection using selfing

Obtaining pure coconut lines remains a long-term prospect which, according to Charles (1961) 'would discourage the most ardent'. The four generations required to create 95% homozygous structures represent a period of 25 to 60 years, depending on the method of parent evaluation. Such a long period implies the distribution of partly heterozygotic seednuts.

Compared to open pollination, mass selection using selfing maximizes the short time use of the genetic additive effects. Nevertheless, it does not permit genetic combination of individuals. As it is not very likely that one of the trees has all the best genes right from the beginning, the chances of finding the most favourable combinations are therefore rather slim. Inbreeding depression and homeostatic loss linked to homozygosity could also limit the effectiveness of the method.

History

Studies on coconut usually do not deal with more than one selfed generation (S1). Generally, in talls this induces inbreeding depression. A comparison of the progenies of S1 and those of open pollination (OP) of 18 tall ecotypes has demonstrated that in most cases selfing induces reduction of vigour, numbers of fruit produced and copra content per nut (Satyabalan and Lakshmanachar 1960). The intensity of the depression varies between ecotypes, which is not surprising as in coconut various types of reproduction exist. The spread of coconut has been accompanied by founder effects that have created naturally inbred isolates and induced genetic drift. Possibly, ecotypes that can withstand inbreeding well are those that have already undergone natural inbreeding. Such ecotypes would have a restricted variability and fewer possibilities of within-population improvement. Studying the cytologic behaviour of the S1 of the same ecotypes, Nambiar *et al.* (1970) showed that selfing in most cases induced an increase in chromosomic aberrations and in pollen sterility. For three tall ecotypes this phenomenon was accompanied by a reduction in chiasma frequency, which corresponds to a decrease in recombination frequency.

Satyabalan and Lakshmanachar (1960), when analysing the S1 and OP progenies of 14 high-yielding tall coconuts on the west coast of India, found that there was an inbreeding depression of 22.2% in the number of fruits produced. It seems that there are differences in performance between parents, as in certain progenies the S1 are one-third better than the OP, and in others the S1 produced one-third less than the OP. Liyanaga (1969) studied the S1 and OP progenies of 17 non-selected SLT palms. With the data produced, the mean inbreeding depression values and their intervals may be calculated at the 95% confidence level.

o Leaf number at 48 months: 13.8% ± 6.0;
o Flowering precocity: 8.5% ± 6.4.

The first genetic experiment conducted at the Marc Delorme Research Station compared, among other treatments, the S1 and OP progenies of 50 WAT. For all parameters observed, except for nut copra content, the averages of the selfings were inferior to those of the crosses. Inbreeding depression particularly affects copra per tree (19.3%), the number of fruits (15.4%), the number of bunches (11.7%) and flowering precocity. The fact that it does not affect the nut copra content could be explained by the triploid nature of the tissue; the flow of genes brought by the pollen restores a heterozygosity which limits the inbreeding depression. The strong heritability of the nut copra content may also have played a role.

In 1991, Sukumara Nair and Balkrishnan were the first to describe two successive generations of selfing. Their results, presented in Table 15, seem to indicate a strong inbreeding depression for nut yield. The first selfing generation produced on average 64% less than its parents. The difference between the first and the second generation of selfing is much less, not more than 4%. However, environmental effects probably introduce some divergence: the following generations were planted at intervals of more than twenty years, without controls permitting an exact comparison between following generations.

Homeostatic loss

Ohler (1984) suggested that the depressive effect on yields induced by selfing could be compensated by greater production uniformity. Genotypic homogenization should lead to greater phenotype homogeneity. However, in certain cases it can be observed that the homeostasis of pure lines of allogamous plants is much less than that of heterozygous plants, and that they are as variable as populations (Demarly 1977). For coconut, where autogamous, allogamous and intermediate ecotypes coexist, the situation is undoubtedly much more complex.

Table 15: *Yield performance (nuts per palm per year) of Indian West Coast Tall, grandparents and self progenies, (from Sukumaran Nair and Balkrishnan, 1991)*

Motherpalms	Family groups					
	1/109	1/174	VI/4	1/129	VIII/27	1/109
1. Grandparents (mean of 10 years, 1920-29)	125.90 (100%)	117.20 (100%)	105.90 (100%)	104.60 (100%)	117.50 (100%)	116.17 (100%)
2. First generation selfs (S1) (mean of 10 years, 1950-59)	48.43 (38.5%)	59.57 (50.8%)	48.65 (45.9%)	48.57 (46.4%)	46.98 (40.0%)	36.18 (38.5%)
3. Second generation selfs (S2) (mean of 10 years, 1976-85)	34.84 (27.7%)	49.91 (42.6%)	39.96 (37.7%)	44.60 (42.6%)	39.85 (33.9%)	34.86 (30.0%)

Autogamous dwarf ecotypes, due to their homozygosity are usually more homogeneous than talls. In a plot planted with the same dwarf ecotype, the estimated mean production over several years varies little between trees. Nevertheless, dwarfs are considered less hardy and less drought-resistant (Ziller 1962). In addition, the seasonal production fluctuation is much greater among dwarfs than among talls. It is difficult to determine if these characteristics can be attributed to dwarfism or to the genetic organization resulting from autogamy. If the second hypothesis is correct, the low homeostasis linked to homozygous structures would express itself by a loss of hardiness and a more marked production alternation. An estimate of the variation coefficents of progenies of 14 talls (Satyabalan and Lakshmanachar 1960) shows that for the number of nuts, selfings are at least as variable as open pollinated palms. In this case, selfing has induced a production decline without reducing the phenotypic variability.

A method unadapted to varietal creation
In most tall ecotypes an inbreeding depression is observed reflecting the existence of a heterosis effect. Loss of vigour induced by selfing reduces mass selection effectiveness substantially. Rapid decrease in variability reduces the probability of favourable combinations without apparently improving the homogeneity of production. In addition, it is impossible to assure large-scale seednut production by using hand pollination. The only solution would be to plant selfs of each selected tree in seedgardens, isolated from each other, which would mean a certain delay of one generation. Rather than using this generation for multiplication of trees selected for their phenotypes, it would be better to select trees on the basis of the value of their progeny. Mass selection using selfing is therefore a method that is not well adapted to coconut.

Mass selection using intercrossing
The system involves the selection of parent trees on the basis of their phenotypic performances and intercrossing them. Various crossing programmes may be used, such as independent pairs and factorial crosses. They are essentially justified by their progeny analysis. In the case of mass selection only one is being considered: the one by which a group of mother trees is pollinated by a mixture of pollen from various pollinators.

Under open pollination conditions, each mother tree transmits one-half of its additive genetic effects to its progeny, the other half being derived from the population. With intercrossing, this second part of the additive effect is 'enriched', as it comes from a selected sub-population. Forty to fifty mother trees can be pollinated with pollen from a single parent tree, permitting much stricter selection among father trees. Twenty per cent of the population could be retained as mother trees and a selection rate of less than 1% could be used for the male parents. This shows that intercrossing is the most effective reproduction method within the context of mass selection. In case a change from one cycle to another requires the use of hand pollination, it is possible to produce seednuts in the usual manner, by controlled open pollination or assisted pollination. For coconut, mass selection using intercrossing was, so it seems, never evaluated systematically. It has been used occasionally, but without controls permitting determination of its effect. Some recent experiments seem to be better conceived. However, they are still too recent to produce results.

Conclusion
This study makes it possible to classify the three mass selection variants by order of increasing efficiency: selection using selfing, open pollination and intercrossing. The first two methods are of very limited importance. Selfing induces a yield decline without increasing production homogeneity appreciably. Open pollination leads to variable results, without doubt due to absence of pollen origin control. In the most favourable cases, the drastic selection required to obtain an improvement reduces potential seednut production considerably. One generation of multiplication is inevitable. Therefore, it is better to use the available generation for the evaluation of parents by the performance of their progeny. The only advantage of mass selection using open pollination is its simplicity. Mass selection using intercrossing appears to be more effective, as it allows for a strict selection of pollinators, while retaining a large potential seednut production. However, there is no experimental result that permits an assessment of the genetic progress that may be realized. The various mass selection variants have been presented as complete methods where the final target was seed production. However, even in the case of selection based on progeny performance, parent trees have often been selected first on the basis of their phenotype; mass selection is an integrated component in most coconut improvement programmes.

Consideration may also be given to using mass selection to improve two populations separately, with the objective of later creating hybrids. In other words, this comes back to carrying out selection based on the intrinsic value of the populations, with the objective of improving their crossing value. In general, it appears that in plants whose improvement is recent, correlations between intrinsic value and crossing value remain low. The example of oil palm shows that selection based on yield, separately within two populations may lead to counter-selection of hybrid performance (Gasçon *et al.* 1966; Meunier personal communication). On the contrary, Moll *et al.* (1971) have shown that, with maize, a double within-population selection was as effective as a reciprocal selection based on progeny test in the short term. Therefore, this method involves a certain amount of risk, which is difficult to quantify for coconut.

Within-Population Selection based on Progeny Test

As with mass selections, the methods described below have in common that, according to the authors, the selected population was initially limited to only one ecotype.

Selection based on half-sib families

The 'prepotency' conception was introduced in coconut by Harland (1957) to indicate trees of which the open pollination progenies show superiority. He proposed the collection of seednuts from such prepotent trees. This method corresponds with starting a recurrent selection on half-sib families.

In 1967, Liyanage published a study on the analysis of the open pollination progeny of 104 SLT. The low number of each progeny (9 half-sibs) seems insufficient to safeguard against divergence due to sampling environmental effects. However, this study permits the assessment of the effects of selection based on open pollinated progeny. The six best families among the 104 show a production 32% higher than average. This genetic gain, although possibly overestimated, can be compared to the progress that could have been obtained by mass selection. Selection of the six highest yielding parents induces only 14% yield increase in the progeny. This study underlines another limit of mass selection; to evaluate the effectiveness of this method, recourse to progeny tests is indispensable. But once these tests have been realized, it is much better to select the parents on the basis of the progeny value than on the basis of their intrinsic value. Selection on the basis of half-sib families proves to be more effective in the long term, the more so as it does not exclude a preliminary phenotypic choice.

A similar study was undertaken in Indonesia. In 1926 and 1927, 100 Mapanget parents were evaluated and selected in the villages around the research station of Manado, Balitka (Tammes 1955). Open pollination seednuts were collected and a trial was established, using 1400 trees, combining 43 half-sib families. Unfortunately, the results of this well-planned study were lost during the Second World War (Liyanage *et al.* 1986).

Critical analysis

The use of open pollination remains open to question. Testing the parents as mother trees is inconvenient. Selection by families allows for the selection of only a limited number of mother trees, insufficient for large-scale seednut production. An alternative would be to test the parents as pollinators. Each of these should be crossed by hand pollination with a population of mother trees. To avoid the loss of valuable genotypes, each parent tested for its value in crossing should be multiplied by selfing. Seednut production could then be assured by assisted pollination, using the pollen collected from the sclfings of the best parents.

Selection of full-sib families

Harland (1957) proposed to cross two high yielders, the best progenies afterwards to be reproduced by hand pollination. This method exploits within-ecotype heterosis, a phenomenon clearly indicated by the existence of inbreeding depression. In Indonesia, a programme was undertaken with 43 half-sib families of Mapanget Tall produced by Tammes in 1927. In this experiment, parents were selected and crossed, sometimes within and sometimes between families. About fifty progenies were planted between 1957 and 1959 with an initial number of fifty trees each. Extremely promising yields, of the order of 40 kg of copra per tree per year were obtained. At the time of the measurements, in 1975-1979, 59% of the trees were dead, clearly favouring the yield of the remaining trees. Converted to a real density, the best progeny tested reaches 3.05 tons of copra per hectare. However, this trial has made it possible to identify four élite families which were used for the production of seednuts.

Critical analysis
Hand pollination, cumbersome and of low productivity, cannot be used economically for seednut production, as it generally produces forty to fifty plants per pair per year. This explains why this method, whatever its efficiency in terms of genetic progress, was never developed for direct use. An improvement in this method would be the selfing of each parent tree, immediately after the establishment of the trial. Therefore, once the interesting pairs have been identified, crossings can be produced by using these selfings. The seednut production capacity, even if improved, remains insufficient, at least in the first generation. The selfings are of low productivity and require a substantial area, due to the necessary isolation between seed gardens.

Conclusion
The two methods described by Harland (1957) have a low seednut production capacity. Nevertheless, selection by half-sib families, by testing parent trees as pollinators and not as mother trees is without doubt the best method thus far described. Results obtained by Liyanage with open pollination of half-sib families suggest that important genetic progress is possible. However, breeding with half-sib families uses only part of the additive genetic effects. On the other hand, if the initial population consists of one single ecotype, there is a risk that the variability may reduce the progress that can be realized. Hence, creation of a population combining various ecotypes needs several generations, and delays the improvement programme considerably. From these two examples it can be seen that all methods based on within-population hybridization may be used. But, as one can see in the following paragraphs, why concentrate on within-ecotype heterosis when the hybrid vigour between ecotypes can give immediate genetic progress? This strategy is justified only in special cases, for instance when one disposes of only one ecotype resistant to a certain lethal disease.

Selection of Hybrid Type
Although the first hand pollinations were done in India in 1920, the first hybridization programme between ecotypes is attributed to Marechal who, in 1926, crossed the Malayan Red Dwarf with the Niu Leka Dwarf of the Fiji Islands. Unfortunately, his work did not survive the 1929 economic crisis and the pedigree of the hybrids was lost (Child 1975). Patel (1938) made the first crossings between talls and dwarfs, with the objective of combining the hardiness of the first with the precocity of the second. These hybrids, although they were planted under unfavourable conditions, proved more precocious and more productive than their tall parents (Bhaskaran and Leela 1978). Between 1940 and 1960 numerous hybrid tests were conducted, comprising D x T and T x T crossings with local ecotypes. These trials generally involved low numbers of trees. Most results showed superiority of hybrids over local talls. However, hybrids were sometimes considered too close to their dwarf parents, as they were alternate bearing, subjected to bunch abortion and sensitive to drought. In some cases, legitimacy of these hybrids may need confirmation.

For a long time these studies have remained essentially theoretical as the absence of a viable seednut production technique hampered the distribution of the hybrids. Some countries abandoned left this path of research, which did not appear to lead to practical applications. Development of viable seednut production techniques opened the way for the distribution of hybrids and made it possible to produce legitimate seednuts at relatively low cost.

Methodology of hybrid tests
The principle of the first phase of hybrid selection seems to be similar in all research programmes. Crosses between accessions are made and compared to a control, generally the local tall cultivar. The term 'hybrid' is used in a wider sense to indicate crosses between populations.[12] However, according to Nucé de Lamothe (1993), 'a great majority of stations for their coconut improvement programme don't have more than some, more or less well planned trials of first generation hybrids derived from not always well known populations that have been crossed with each other essentially because they were available'. In general, uniformity of methodologies is desirable for better comparison of results.

Determination and publication of precise and reproducible techniques is a prerequisite for progress in this field. In particular, the importance of hand pollination, assuring the legitimacy of crosses, should again be stressed.

Choice of parents

Parents used for crosses are usually selected on phenotype, based on yielding performance or disease resistance. In some rare cases parents are randomly selected after elimination of trees with an atypical phenotype. The strictness of selection and the number of parents kept for crosses differ according to programmes. In particular, the number of parents varies according to availability. For instance, the Rennell Island Tall entry in the various collections is represented by numbers varying between three and more than one thousand trees. In Ivory Coast, the number of each ecotype actually kept is between 24 and 48 trees, according to its use as male or as female. Such samples that generally result from a not very strict phenotypic selection remain representative and make it possible to conduct relatively short-term pollination programmes (IDEFOR/DPO 1992).

Number of crosses

According to the research programmes, the numbers used for evaluation of a cross varies from about ten trees to over one hundred and fifty. Natural tall coconut populations show genetic variability. None of the trees has the same genotype nor the same value in the cross. For instance, the Rennell Island Tall parents have been crossed individually with the Malayan Red Dwarf. The progenies compared are each derived from a different father tree. Their mean productivity between 4-11 years varies from 2.4 to 4.3 tons of copra per hectare (IDEFOR/DPO 1992). All individual trees of a cross between populations do not have the same genotype. The genetic variability of a hybrid is at least equal to that of its two parents. Marshall and Brown (1973) estimated that 60 to 100 trees could represent a natural population of allogamous plants. For coconut, and in the case of a hybrid between ecotypes, a minimum number of 100 to 120 trees seems necessary for a proper evaluation.[13]

Experimental design

Conducting a genetic experiment requires the use of a control and a statistical design. The control may be an ecotype used locally, or a widely distributed hybrid of known characteristics. Differences in height and width of canopy explain why the hybrids of dwarfs, talls and dwarfs x talls are usually planted separately. In addition to pure mathematical and statistical considerations, one phenomenon should be taken into consideration: the competition effect. In particular, the various hybrids planted in a comparison trial rarely have the same growth. The competition for light develops at the age of nine or ten years, and may influence the evaluation of the final production. Because of competition risks, the experimental designs frequently used will be blocs, complete or incomplete, consisting of large elementary plots: for instance five lines of five trees or four lines of five trees. Such a system permits comparison of the trees at the inside of the plots (surrounded by the same planting material) with the trees at the outside (in contact with different planting material and susceptible to competition).

Recording of production data

Usually, production is distinguished between young age, two to eight years after planting, and adult age from 9 to 12 years and older. The number of years of observation should preferably be even, to reduce the effects of biannual alternation. Trials are harvested monthly (dwarfs) or bi-monthly (talls and dwarfs x talls). The number of bunches and fruits from each tree are recorded separately. Meunier *et al.* (1977) have analysed all fruits of two coconut populations produced during two years. This has made it possible to determine the minimal sampling that can be used to assess the mean composition of the fruits of a tree, or of a group of trees:

- to characterize a tree, four fruits should be analysed every two months over a period of two or three years;
- to characterize a population or a progeny, one fruit each of 50 trees should be analysed per bi-monthly harvest over a period of 6 years;
- in the case of an experimental design in blocs per elementary plot, one fruit per tree per harvest-round over four years.

Wuidart and Rognon (1978) proposed a method to determine the copra content per nut based on the dry matter of a representative sample of endosperm taken from the equatorial part of the nuts. The same sample may also be used for the determination of the oil content.

Review of hybridization programme between ecotypes

Around the world, about 400 hybrids have been created in national research programmes. For the time being, less than ten of these have been tested internationally in view of their performance under various ecological conditions.

Ivory Coast

The genetic improvement scheme used in Ivory Coast is based on the constitution of a collection containing numerous ecotypes; detection of the best crosses between ecotypes; followed by an improvement of these crosses using individual combining ability tests (Nucé de Lamothe 1970; Gascon and Nucé de Lamothe 1976). This scheme was recently changed into two improvement procedures, T x T and D x T, for both of them using the method of recurrent reciprocal selection (Bourdeix *et al.* 1990; 1991a; 1991b).

Table 16: *Compared yield of the first performing hybrids identified in Ivory Coast and the West African Tall control*

Production and percentage of WAT control							
Trial	Trial n°3	Trial n°8	Trial n°5	Trial n°5	Trial n°5	Trial n°11	Trial n°11
Period of observation	9-12 yr	9-13 yr	9-15 yr	9-15 yr	9-15 yr	9-14 yr	9-14 yr
Hybrids tested	WATxRIT PB 213	WATxVTT PB 214	MYDxWAT PB 121	MRDxTAT PB 132	MYDxTAT PB 122	NJMxRIT PB 123	CRDxWAT PB 111
Earlyness of flowering (months)	51 (91%)	57 (90%)	51 (84%)	50 (82%)	52 (85%)	40 (75%)	44 (83%)
Number of bunches per tree	12.7 (109%)	13.5 (118%)	14.8 (123%)	14.2 (118%)	14.2 (118%)	16.5 (120%	15.5 (113%)
Number of fruits per tree	102 (134%)	100 (238%)	109 (188%)	101 (174%)	110 (190%)	123 (164%)	132 (176%)
Copra per nut (g)	311 (138%)	212 (89%)	247 (105%)	282 (120%)	253 (108%)	289 (128%)	240 (107%)
Copra per ha (tons)	4.26 (186%)	3.15 (235%)	3.67 (197%)	3.88 (209%)	3.80 (204%)	4.80 (207%)	4.35 (188%)

Detection of the best hybrids between ecotypes

Between 1965 and 1993, 123 hybrids between ecotypes were tested in Ivory Coast (without the reciprocal crosses or the complex hybrids). These experiments cover 132 hectares, *i.e.* more than 1 ha per tested combination. Figure 10 recapitulates the total of the crosses between ecotypes created at the Marc Delorme Station.

The initial trials used a population of WAT derived from two cycles of mass selection as control. The tests have made it possible to compare 35 hybrid combinations, involving 13 parental ecotypes. Table 16 presents the production results of the best hybrids. All hybrids, except five, have shown a significantly higher yield than the WAT control; none of the hybrids was inferior to the control. The 35 hybrids showed a yield on average 66% higher than that of the control, and four of the hybrids produced more than twice the yield of the control.[14]

Trials planted between 1976 and 1992 use PB 121 (MYD x WAT) as hybrid control (CP 20) (Nucé de Lamothe and Benard 1985). These trials have made it possible to discover new hybrids that yield more at the adult stage and are more resistant to *Phytophthora* than the PB 121 hybrid. In fact, the controls used are the improved PB 121 hybrid (Bourdeix *et al.* 1992), and the improved PB 213 hybrid (WAT x RIT).

In Figure 10, more empty intersections than occupied intersections can be observed. It is practically impossible to test all hybrids. The 123 between-ecotype combinations created in almost 30 years of improvements represent only 26% of the possible hybrids (without the reciprocal crossings). A drastic choice between combinations to be tested is inevitable. At the start of the improvement programme, a deliberate choice was essential, taking various criteria into consideration, such as the apparent genetic distance between the two parents, complementarity of their characteristics and their yield level. Progressively, some better known populations were used as testers. Two types of trial are distinguished - basic trials and complementary trials.

In the case of the improvement of T x T hybrids, the basic trials consist of crossing each new entry with two complementary tall testers of well known characteristics (see the paragraph on the evolution of selection strategies). In Ivory Coast, the WAT and RIT populations (or parent trees selected from these two populations) are used as testers.

In the case of the improvement of D x T hybrids, the basic trials consist of crossing each new entry with one single tester of well-known characteristics. Thus, all talls are crossed with MYD and all dwarfs are crossed with a tall (in the beginning WAT was used as a tester but for various reasons the hybrid WAT x RIT is currently preferred).

Complementary trials have more complex crossing programmes. Their objective is to improve the knowledge of the way of transmission of their characteristics, and also to test certain intuitions, such as, for instance, when it is expected that crossing two specific entries could be interesting because of their complementarity. These trials should be planned to provide a maximum of genetic information; their crossing plans are complete or incomplete balanced factorial design (Figure 11).

Improvement of the best hybrids

A second improvement phase has been running at the Marc Delorme Research Station since 1970 (Gascon and Nuce de Lamothe 1976). It consists of the separate improvement of the best performing hybrids. For a description of this method it is better to use an example. The good performance of the hybrid PB 122 (MYD x TAT) has stimulated its further improvement. Forty-five TAT parent trees were selected and individually crossed with the same MYD population. The 45 progenies thus obtained have thereafter been planted in comparison trials. The tested units are half-sib families, each resulting from a different TAT male parent and the MYD population (female parent).

Theoretically, improvement of a between-ecotype hybrid comprises two kinds of trials, presenting complementary crossing plans: trees of one of the ecotypes are individually crossed with a group of trees of the other ecotype, and vice versa. However, in practice, the crossing plan has been simplified because the two ecotypes involved showed unequal levels of variability. Thus, to improve D x T hybrids, the combining ability of numerous tall parents has been tested, without using the reciprocal test; in fact most of the dwarfs of the collection are autogamous and appear to be an almost pure line. For the experiments, lattice designs were used (4x4 or 5x5), enabling the comparison of 15 to 24 progenies plus one control.

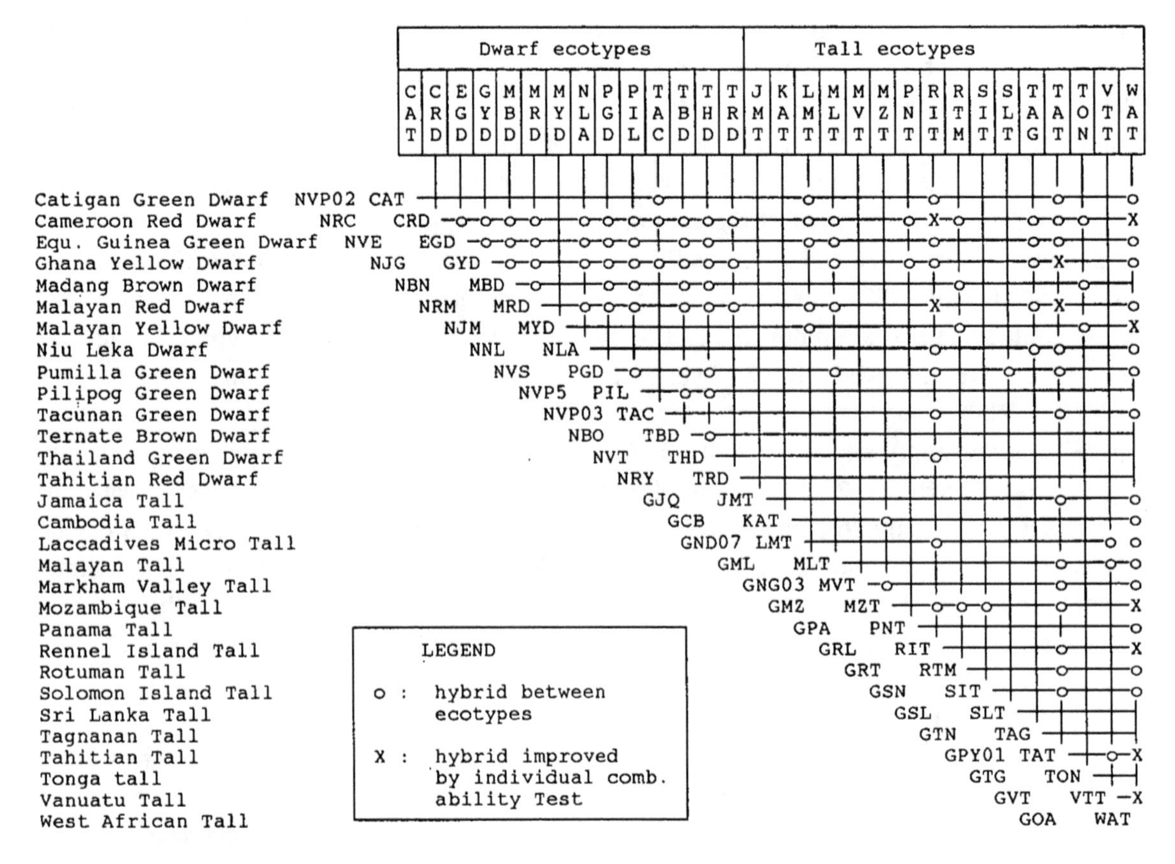

Figure 10: *Overview of hybrids tested by the Marc Delorme Research Centre (Ivory Coast)*

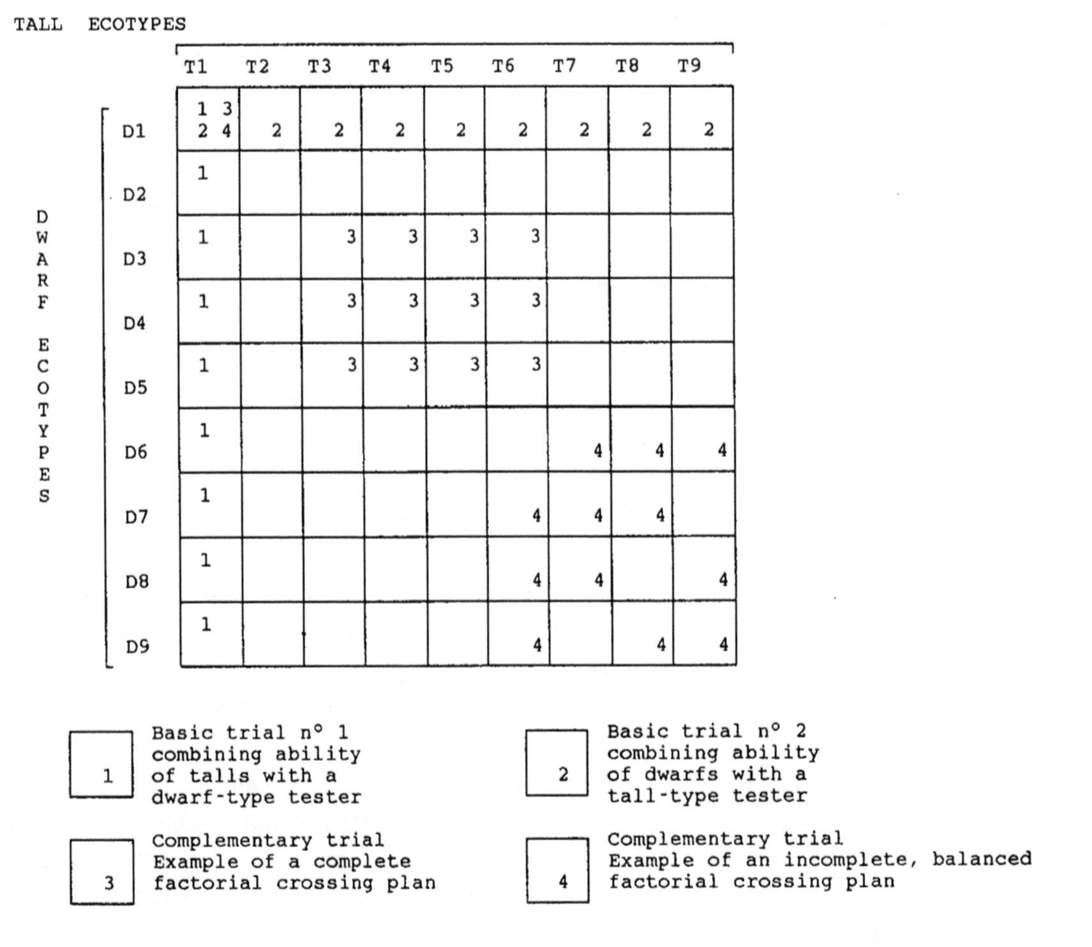

Figure 11: *Illustration of basic trials and complementary trials in the case of improvement of the DxT hybrids*

The individually tested male parents are selfed to assure their conservation and their multiplication. These selfings will provide the pollen needed for the seednut production. Initial results show that the selection of 7 to 8% of the best families may give 15 to 30% genetic progress, according to the trials (Bourdeix *et al.* 1989).[15] An exploitable variability exists within the natural populations. The analysis of these trials has led to some conclusions that may influence improvement strategies:

- o the phenotypic parent selection is sometimes effective, but cannot replace the progeny test. Certain high-yielding parents have medium values in crosses, and vice versa. The progeny test, although costly, cannot be excluded;
- o The highest-yielding progenies in terms of copra per hectare do not show a higher number of bunches nor a greater weight of copra per fruit than the average. The realized genetic progress appears essentially to be due to an improvement in the number of fruits per bunch. However, in some cases, the percentage of copra from the fruit without water is slightly improved;
- o Some parents always seem to behave well in crossings, either when they are crossed with WAT or with various dwarf testers. These parents are particularly interesting for the continuation of the improvement programme.

Creation of complex hybrids
Creation of complex hybrids (three or four ways) has been undertaken in Ivory Coast since 1976. This programme essentially aims at evaluating the genetic variability of this type of crossings and at discovering exceptional individuals. These will be reproduced vegetatively as soon as the cloning system becomes operational. The combinations tested, covering 23 hectares, are the following:

GYD x (WAT x RIT);
LMT x (WAT x RIT);
GYD x (WAT x TAT);
(WAT x TAT) x RIT;
(MYD x WAT) x (EGD x RIT);
(MRD x RIT) x (WAT x TAT);
(CRD x MYD) x (WAT x RIT);
(CRD x RIT) x (EGD x WAT);
(MRD x MYD) x (WAT x TAT);
(MRD x WAT) x (TAT x VTT).

India
According to Nair *et al.* (1991), a total of 86 crosses have been tested in India, including 31 D x T hybrids (the dwarf to be used either as male or female), 51 T x T hybrids and 4 D x D hybrids. But excluding reciprocal crosses and complex hybrids, the total number reduces to 75, as shown in Figure 12. The experimental tree numbers are low in general, rarely exceeding 20 trees per cross. This option, coming from too little area available to the research station, is the main limit of Indian research design in matters of coconut breeding. However, the results clearly show the superiority of the hybrids over the tall control (West Coast Tall), as shown in Table 17. At maturity, the best of these hybrids produce over 50% more than the control, with a production level of 3-3.5 t of copra per ha. In India the Root(wilt) disease prevails, to which no cultivar has yet shown any tolerance. Some individuals derived from local populations seem to perform well in seriously attacked regions and are being used in an intercrossing programme (Jacob *et al.* 1991).

The annual number of coconut seednuts needed in India is estimated at 15 million (Nair *et al.* 1991). But the actual seedgardens do not produce more than 700,000 hybrid seednuts annually (Muliyar and Rethinam 1991). Additional plantations, established before 1988, under the most favourable conditions cannot reach a production higher than 1.5 million. An important effort still has to be made in the field of seednut production in India.

Table 17: *Yields of some hybrids created in India (Nair et al. (1991)*

Material tested	number	fruit production 18 a 21 years		annual copra yield per tree (Kg)	
		average per tree	percentage of control	average per tree	Percentage of control
Chowgat Orange Dwarf x West Coast Tall	11	99.8	136 *	20.8	152 *
West Coast Tall x Chowgat Orange Dwarf	11	88.4	120 *	17.5	128 *
West Coast Tall x Gagabondam	13	65.7	89 *	12.4	90 *
Laccadives Ordinary Tall x Chowgat Orange Dwarf	11	99.3	135 *	19.4	142 *
Laccadives Ordinary Tall x Gagabondam	10	101.5	138 *	19.8	145 *
West Coast Tall (control)	18	73.5	-	13.7	-

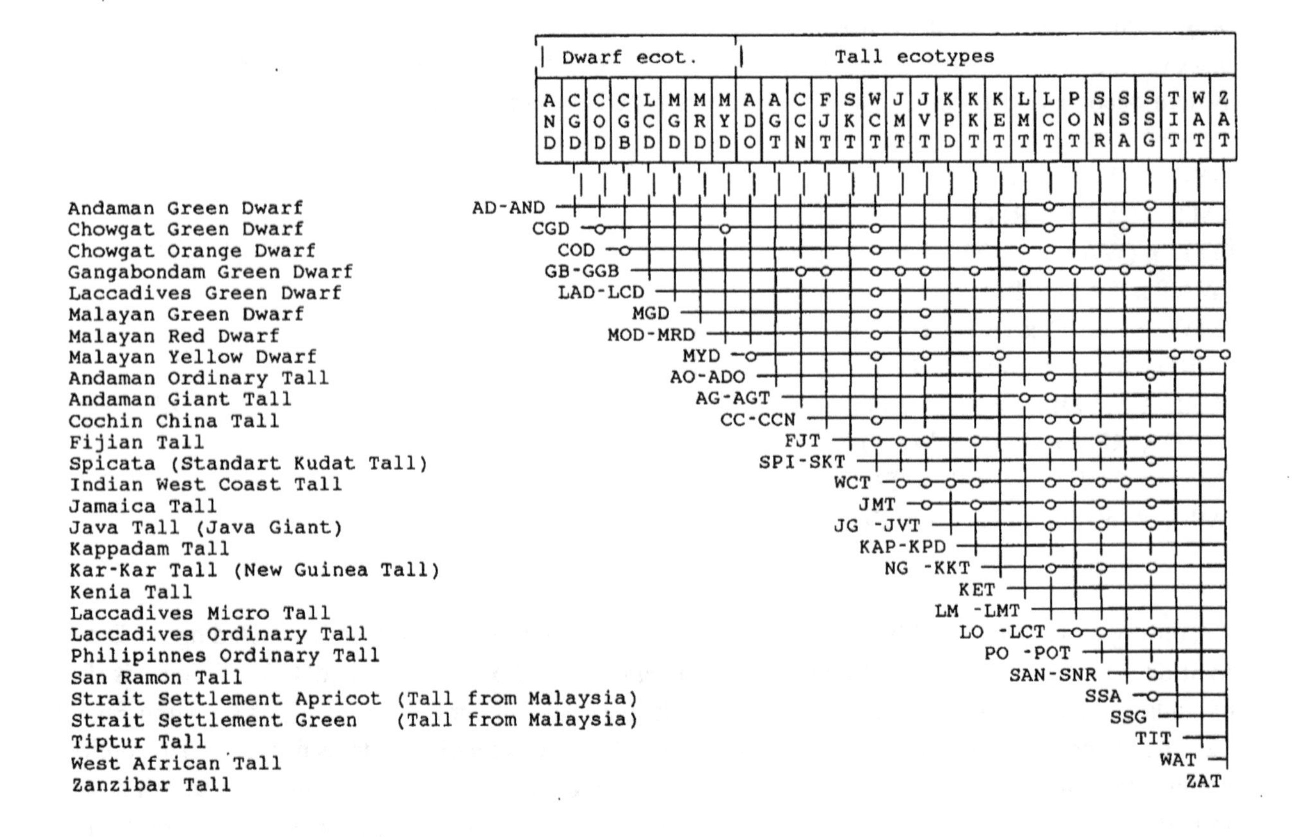

Figure 12: *Overview of hybrids tested by Central Plantation Crops Research Institute in India (NAIR, M.K.- personal communication)*

Indonesia

Between 1977 and 1993, at least 52 hybrid combinations were planted in Indonesia. The majority of these combinations were realized by the Coconut Research Institute of Indonesia (BALITKA), but some private enterprises have also undertaken research programmes, particularly the Multi-Agro Corporation (MAC) and P.T. Riau Satki United Plantations (RSUP). Figure 13 shows the crosses tested in these three organizations.

Table 18: *Results of trial no. 4, Balitka Pakuwon Station, Indonesia (IRHO document 2191, 1989)*

	Young Age 5-8 years		7-10 Years	Adult Age 9-11 years	
	Nuts per tree per year	Copra per hectare (Tons)	Copra per nut (Grams)	Nuts per tree per year	Copra per hectare (Tons)
Nias Yellow Dwarf x Tenga Tall (Khina 1)	52 (176 %)	1.76 (241 %)	263 (119 %)	74 (163 %)	2.82 (+216 %)
Nias Yellow Dwarf x Bali Tall (Khina 2)	37 (128 %)	1.48 (213 %)	306 (134 %)	62 (149 %)	2.74 (232 %)
Nias Yellow Dwarf x Palu Tall (Khina 3)	48 (166 %)	1.77 (242 %)	287 (129 %)	74 (166 %)	3.16 (241 %)
Tenga Tall	9	0.34	290	32	1.32
Bali Tall	8	0.34	307	24	1.07
Palu Tall	8	0.32	293	30	1.28
Nias Yellow Dwarf	50	0.96	151	59	1.29

The percentages in brackets correspond with the value of the hybrid compared to the average of its two parents.

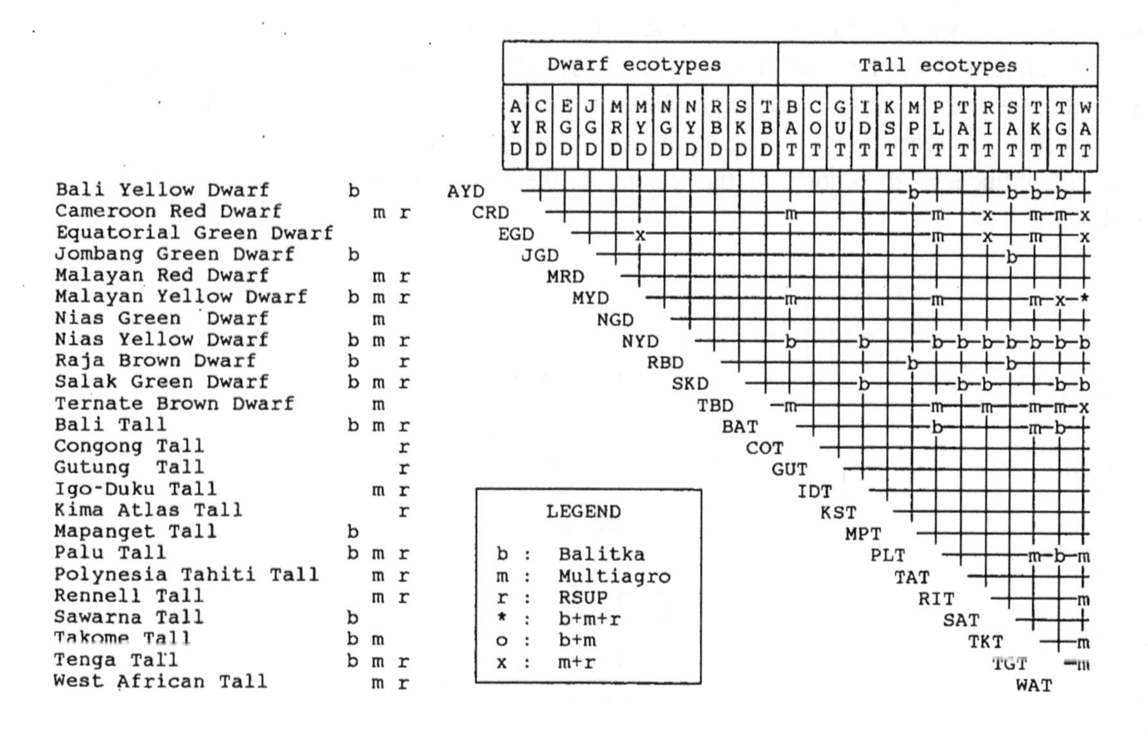

Figure 13: *Overview of hybrids tested by Balitka Research Center (except Pakuwon Station), MULTIAGRO and RSUP (Indonesia)*

Trials realized by BALITKA between 1977 and 1984 used the Fisher block design with 12 to 16 trees per cross, which seems insufficient for a proper evaluation of the crosses between populations. Trials planted after 1984 used higher numbers, generally more than 80 per cross.

In spite of their small size, some of the trials planted before 1984 yielded interesting results. In particular, trial no. 4 at Pakuwon Station may be the only trial in the world comparing hybrids and their parental ecotypes within the same experimental design. Such a set-up makes it possible to measure the heterosis effect, defined here as the ratio between the production of the hybrid and the average production of its two parents.

These productions are presented as percentages in Table 18. The hybrids produced more than double (about 230%) the average of their two parents. This performance in the first place is a result of higher nut production (about 60%) but also of higher copra content per nut (about 30%). However, these calculations should be regarded with caution; the mixture of populations with different height (talls and dwarfs) may produce diverging results due to competition.

In the most humid region of Indonesia, *Phytophthora* causes serious damage to coconut (bud rot causing the palm's death). The hybrid PB 121 (MYD x WAT), largely used in development programmes, has proved susceptible to this disease. A study conducted in 1991 showed the high tolerance of hybrids with Polynesian Tall and certain local talls, which has led to partial reorientation of seednut production. After 1985, the three hybrids tested in the trial described above were distributed by Balitka using the production of three seedgardens, situated at Paniki (100 ha), Pakuwon (100 ha) and Paya Gajah (120 ha). Numerous private enterprises also established seedgardens. Those from the two enterprises mentioned above (MAC and RSUP) cover areas of 210 and 402 ha respectively (IRHO CIRAD-CP 1992).

Jamaica

The main problem of coconut production in Jamaica is Lethal Yellowing disease. The genetic improvement programme focuses mainly on varietal resistance against this disease. Thirty-two Hybrids between ecotypes have been realized (Been 1992). The experimental designs used are complete blocs with intermediate numbers (30 to 40 trees per tested combination). The planning of the crossing programme in Jamaica is original and modern. The accessions of the collection are systematically crossed with three testers: Panama Tall (PNT), Malayan Yellow Dwarf and Jamaica Tall (Figure 14). An identical system has recently been adopted at other research centres, particularly in Ivory Coast. The disease has influenced the choice of the testers, as the MYD and PT have some tolerance to Lethal Yellowing.

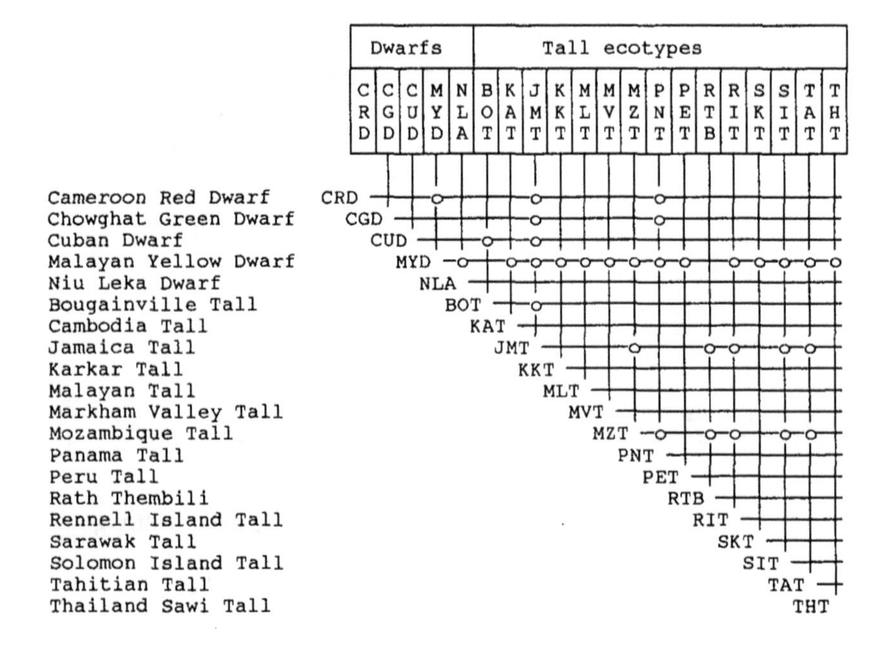

Figure 14: *Overview of hybrids tested by the Coconut Industry Board (Jamaica)*

Unfortunately, cyclone 'Gilbert' in 1988 destroyed about 45% of Jamaican coconut and seriously reduced the collections and genetic trials. Some trials conducted in private plantations seem to have been monitored irregularly due to logistical and budgetary problems. Damage caused by the disease led to the planting of the MYD, planting material which appeared totally resistant (Harries 1971). This planting material has been shown to be little adapted to marginal conditions and is susceptible to other coconut pests (Smith 1970). Finally, a tolerant hybrid has been distributed, the MAYPAN. It is a cross between MRD and MYD as female parents and PNT as the male parent (CIB 1973; Harries and Romney, 1974). About 110,000 hybrid seednuts and 35,000 dwarf seednuts were distributed in 1993.

Malaysia

Malaysian coconut covers more than 300,000 ha, but its area is being progressively lost to oil palm and cocoa. Hybrid tests were established by the Highlands Research Unit (HRU) and the United Plantation Berhad (UPB) company. The HRU has tested the crosses of MYD with the following three talls: WAT, Malayan Tall (MLT), and RIT (Odi Link Hoak 1981). Between 1973 and 1982 UPB planted the following combinations (Chan 1981):

- crosses of WAT with the three talls RIT, MLT and TAG;
- crosses of four talls RIT, MLT, TAG and WAT with MRD and MYD.

Between 1975 and 1984, UPB produced several million seednuts or PB 121 hybrid plants. The Federal Land Development Authority (FELDA) also produced seednuts, but on a smaller scale. However, since 1985, following the adoption of a new government policy discouraging new coconut planting, the demand for and the production of seednuts have decreased substantially.

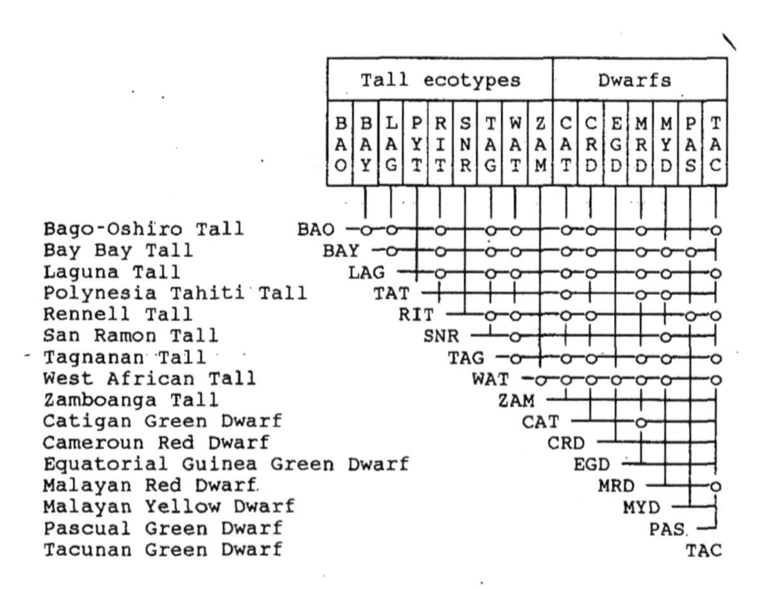

Figure 15: *Overview of hybrids tested by The Philippine Coconut Authority*

The Philippines

In The Philippines, the most important coconut improvement programme is that of The Philippine Coconut Authority (PCA). At least 52 different hybrid combinations were tested between 1974 and 1993. The experimental design comprised Fisher blocs, often with more than 80 trees each. Several controls were used, such as various local talls and the PB 121. Figure 15 shows the tested crosses. The Philippines also has a large network of performance trials localized in nine places within the country, each comprising eleven different cultivars, such as tall ecotypes, D x T and T x T hybrids.

The PCA programme is characterized by a complete test of a relatively low number of parental populations. Excluding the reciprocal crosses, more than 50% of the 120 possible crosses were made from 16 parental populations cited.

As far as productivity is concerned, local and imported hybrids showed superiority over local talls. Some local hybrids reached a production level at least as high as that of PB 121 (Table 19). Two severe drought periods in 1981 and 1983 are largely responsible for the low production level of the trial presented (Santos, personal communication). These figures do not reflect the real potential of hybrids created in The Philippines. Table 19 presents a much better result, obtained with the same material within the more favourable ecological conditions in Ivory Coast.

Table 19: *Yield comparison of hybrids in The Philippines (Zamboanga Research Centre, annual report, 1989) and in Ivory Coast*

Country	Trial number	Crosses tested	Trees per Cross	Nuts per hectare per Year	Copra per hectare per Year (T)
Philippines	PHGC04			5-12 years	5-12 years
	1979	TAG Control	96	3394	0.99
		CAT x WAT	96	6981	1.45
		CAT x TAT	96	6596	1.60
		CAT x TAG	96	5612	1.49
		MRD x TAG	96	6904	1.86
Philippines	PHGC03			5-11 years	5-11 years
	1978	MYD x WAT control	144	9173	1.76
		CAT x LAG	144	7689	1.94
		CAT x TAG	144	6438	1.84
		CAT x BAO	144	6109	1.68
Ivory Coast	PBGC18			5-13 years	5-13 years
	1978	MYD x WAT control	144	18612	4.22
		MYD x TAG	144	16281	4.38
		MRD x TAG	144	14930	4.34
		EGD x TAG	144	12481	3.59
		CRD x TAG	144	14035	3.68

An original trial was established in 1992. It compares 15 possible combinations of the following six tall ecotypes: BAO, BAY, LAG, TAG, RIT, and WAT (Santos, personal communication). Within each population, 26 parents obtained through mass selection were selected. One hundred trees per cross have been planted in a totally random design. This trial appears to be interesting for the following reasons:

- its diallel design permits good measurement of genetic parameters (heredities and genetic correlations);
- use of the WAT (from Ivory Coast and Benin) and RIT (Pacific) makes it possible to establish some links with several other improvement programmes;
- finally, this trial is the first attempt to create a synthetic variety in coconut. The single hybrids created will be used to produce a second generation. This second one will be evaluated and planted in isolated seedgardens. The open pollinated seednuts produced by these seedgardens (the third generation) will form the synthetic variety to be distributed among farmers.[16]

The Philippines need for selected planting material is considerable, of the order of 5 million seednuts annually. The BUGSUK seedgarden made it possible to plant about 35,000 ha of the PB 121 hybrid between 1980 and 1983 (Santos 1990). In 1986, the PCA initiated the production of seednuts of the hybrid PCA 15-1 (CAT x LAG). The production potential was not very great, about 240,000 seednuts per year. Production was suspended in 1989, permitting the multiplication of the female dwarfs, to plant seedgardens at Aroman, Carmen, and Cotabato. The latter, which covers 303 ha, comprises various dwarf types (CAT, MRD and TAC). It will have a production potential of about 3.5 million seednuts annually, widening the range of hybrid distribution.

Sri Lanka

The relatively low number of hybrids tested in Sri Lanka results from a strict government phytosanitary policy. According to Liyanage *et al.* (1988), five hybrids of the D x T type were created between 1949 and 1980 (excluding reciprocal crosses): PGD x SLT; Sri Lanka Yellow (CYD) x SLT;

Sri Lanka Red Dwarf (ARD) x SLT; PGD x Rath Thembili (RTB); and PGD x San Ramon Tall sub-population (SNR). Two cultivars were distributed, CRIC 60, derived from a cross between selected SLT parents and CRIC 65, from a cross between PGD and selected SLT parents. Table 20 presents some evaluations of these cultivars. Exchanges of planting material between Sri Lanka and Ivory Coast permitted comparison between varieties distributed in both countries. In all cases, CRIC 65 yielded more than the tall cultivar CRIC 60. However, Liyanage *et al.* (1988) mentioned that after 12 years the differences between talls and hybrids become smaller and disappear. On the other hand, the hybrid CRIC 65 would be less adapted to regions with a severe water deficit (Peries 1993).[17] Difference between talls and hybrids are clearly more pronounced in Ivory Coast. This phenomenon is undoubtedly due to better climatic conditions in Ivory Coast. At a young age, the CRIC 65 hybrid produces twice as much as in Sri Lanka.

In 1982, a new series of trials was started to compare some of the above cultivars in various places. Two additional combinations were made: crosses between a sub-population of San Ramon Tall with SLT and CYD. These trials use the complete bloc design with considerable numbers of trees per hybrid (80-100). Initial results of these trials in two localities confirm the superiority at a young age of CRIC 65 over CRIC 60. Over the first four harvest years, the hybrid CRI 65 yielded 39% (Bandirippuwa estate) and 116% (Thammenna estate) more nuts than the cultivar CRIC 60 on average (calculated by Peries 1994). The copra weight per nut of the two cultivars is practically equivalent (194 g and 192 g respectively at Bandirippuwa estate and 200 g and 201 g respectively at Thammenna estate).

The seedgarden of Ambakele was established and interplanted with the cultivars SLT and PGD according to a controlled open pollination design. Sometimes it produces CRIC 60 seednuts (harvested from the talls) and sometimes CRIC 65 seednuts (harvested from the emasculated dwarfs). The improved seednut production is still small compared to the national need. From 1983 to 1990 the CRI produced about 11 million seednuts, which comprised 79% talls, 18% of the improved tall CRIC 60 and only 3% hybrids (calculated by Peries 1994).

Table 20: *Comparative yields of varieties from Sri Lanka and Ivory Coast CRIC60, CRIC65, WAT and PB121*

	SRI LANKA			
	(Liyanage, 1961)		(Manthriratne, 1971)	
	SLTxSLT CRIC60	SLTxPGD CRIC65	SLTxSLT CRIC60	SLTxPGD CRIC65
Young age	4-8 years	4-8 years	5-8 years	5-8 years
Kg copra/ tree/year	5.9	10.8	—	—
Adult age	9-12 years	9-12 years	9-12 years	9-12 years
Kg copra/ tree/year	27.5	30.4	7.5	11.5

	IVORY COAST					
	WATxWAT Plot 112	SLTxSLT CRIC60 Plot 112	WATxWAT Trial n°5	MYDxWAT PB121 Trial n°5	MYDxWAT PB121 Trial n°24	SLTxPGD CRIC65 Trial n°24
Young age	4-8 years	4 years	4-8 years	4-8 years	4-8 years	4-8 years
Kg copra/ tree/year	5.4	5.9	5.9	19.0	22.8	21.6
Adult age	9-17 years	9-17 years	9-15 years	9-15 years	9-11 years	9-11 years
Kg copra/ tree/year	11.9	16.5	13.7	27.0	32.6	29.1

Tanzania
Tanzania has the largest coconut area of Africa. However, the crop suffers from unfavourable conditions, such as Lethal disease, the *Pseudotheraptus* bug, droughts and old age.

The 'National Coconut Development Programme' was founded in 1979 to revitalize the coconut sector with GTZ and IBRD funds. A programme of varietal tests, established by IRHO, was initiated in 1980. According to Mkumbo and Kullaya (1994), 22 hybrids and 23 ecotypes were tested on four different sites of different disease and climatic pressures: Kifumangao, Chambezi, Sotele and Pongwe (Figure 16). According to Schuiling *et al.* (1992), most of the cultivars are represented in each site by lines of 10-11 trees, with 5-6 repetitions.

All cultivars tested in Tanzania were attacked by Lethal disease with a lethality varying from 14 to 65% (Schuiling 1992). In an attempt to identify local sources of resistance, 29 tall accessions have recently been collected in Tanzania and Kenya and have been planted in performance fields. One East African Tall (EAT) entry, coded LBS has aroused some interest. On the Pongwe site, the LBS population has lost 14% of the trees, but its tolerance level is still to be confirmed. On the same site, the Cameroon Red Dwarf (CRD) has lost only 7% of the trees. According to these results, these two ecotypes could be classified as rather tolerant. However, CRD has also been tested at Kifumangao, where the disease pressure is highest, and where the disease has already killed 90% of the trees.

Since 1981, two seedgardens with an area of 166 ha have been established on the Islands of Mafia and Zanzibar, disease-free at that time. In 1989 about 550,000 hybrid seednuts were produced (Sangare and N'cho 1990). From 1990, the seednut production was reduced. In some areas, the hybrids have confirmed their superiority over the local talls. However, in other areas, the serious drought of 1987 appears to have caused greater damage to the hybrids than to the EAT.

The definition of drought resistance characteristics requires a precise quantification and clear objectives. The example of the oil palm may be brought to mind (Meunier, personal communication). At the station of Pobé in Benin, a progeny had been considered to be susceptible to water-stress. The slightest droughts provoked massive abortions. But during one abnormally severe drought, numerous trees of the high-yielding progenies were killed whereas, of the progeny considered sensitive to drought, not one palm died, because all the fruits had been aborted early. This hyper-susceptibility is a resistance mechanism to severe water-stress.

A negative correlation is frequently found between productivity and drought resistance. In the case cited above, the question could be raised as to what would be the best selection: a cultivar that loses part of its fruits every year, or a higher-yielding cultivar running the risk of being killed off periodically? It is not easy to answer this question. It depends on multiple socio-economic factors, such as the social consequences of a transitional penury, frequency of severe droughts, availability of vegetative material, cost of planting, etc.

Cameroon Red Dwarf
Equator Green Dwarf
Malayan Red Dwarf
Malayan Yellow Dwarf
Pumilla Green Dwarf
East African Tall
Malayan Tall
Rennell Island Tall
Sri Lanka Tall
Tahitian Tall
Vanuatu Tall
West African Tall

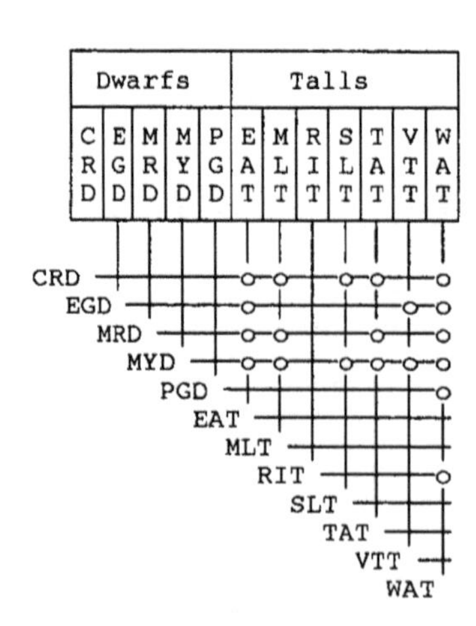

Figure 16: *Overview of hybrids tested by the National Coconut Development Programme (Tanzania)*[18]

All growers will one day have to choose between traditional low-yielding varieties adapted to their environment and modern varieties of uncertain adaptation. Decisions taken too early - not always made by the farmers - have sometimes had tragic consequences. However, from experience it is known that after various trials, the suitable varieties will ultimately be planted. In the case of Tanzania, it would be possible to identify new hybrids[19] more resistant to drought than the imported crosses, and higher-yielding than EAT.

Thailand

A programme of genetic improvement was started by the Chumphon Agricultural Research Centre at the station of Sawi in 1960. According to Anupap Thiracul *et al.* (1992), the following hybrids between ecotypes are actually being tested:
MYD x Maphrao Klang Thailand Tall (KLG);
MYD x Maphrao Thailand Tall (MPE);
MYD x WAT;
Pak Chok Thailand Tall (PCK) x MPE;
SLT x MPE;
WAT x MPE and WAT x RIT.

However, according to various papers, such as Maliwan Rattanapruk *et al.* (1985), other combinations were also tested between 1982 and 1983:

Polynesia Tahiti Tall (TAT) x WAT;
MRD x TAT;
MRD x WAT;
MYD x PCK;
Thailand Green Dwarf (THD) x MPE;
MRD x KLG;
Thalai Roi Thailand Tall (TLR) x Kaloke Thailand Tall (KLK).

The experimental designs used were Fisher blocs with 4 to 5 repetitions, with 20 to 30 trees per elementary plot. Some trials have been repeated in various places. Initial results of a trial planted in 1975 (Maliwan Rattanapruk *et al.* 1985) show good performance of MYD x WAT (PB 121) and WAT x MPE when compared to Thailand Tall (Table 21). After this, the three-wayhybrid creation was undertaken. This programme essentially aims at evaluating the genetic variability of the three-wayhybrids and determining whether these hybrids may be distributed to farmers. The detection of outstanding individuals that could be multiplied once cloning becomes operational is without doubt another objective. The following combinations were tested:

(WAT x RIT) x Selected MPE;
(MYD x WAT) x Selected MPE;
(MYD x TAT) x Selected MPE;
(MRD x RIT) x Selected MPE;
(MYD x MPE) x RIT.

Thailand's seedgardens produce annually at least 250,000 MYD x WAT and MRD x WAT seednuts under the names of SAWI 1 and SAWI 2 respectively. Fruits of the PB 121 are sometimes considered too small by growers. Hence, distribution of a hybrid with large fruits was started in 1984. A seedgarden was established to deliver seednuts of MPE x WAT (Maliwan Rattanapruk *et al.* 1985). In 1994, the production of this hybrid called 'Chumphon Hybrid 60' remained limited to 20,000 seeds per year; however, the production potential should reach 170,000 seednuts in future years (Chulapan

Petchipiroon, personal communication). This is without doubt the first use of a T x T hybrid in a development programme.

Table 21: *Hybrids and Thailand Tall. Copra yield per ha during the 5th to the 8th year (Maliwan Rattanapruk et al., 1985)*

	Year 5 (Kg)	Year 6 (Kg)	Year 7 (Kg)	Year 8 (Kg)	Mean 5-8 Years (Kg)
MPE control	10.6	92.2	266.8	859.6	307.3
TAT x WAT	0	164.6	794.0	1744.0	675.7
MPE x WAT	49.4	410.5	1198.0	2131.6	947.4
MRD x TAT	172.6	566.6	996.2	1861.0	899.1
MYD x WAT	289.7	1105.8	1608.6	2533.4	1384.4

Vanuatu and Pacific Region

The Vanuatu Agronomic Research and Training Centre (VARTC) plays an important regional role in coconut cultivation. In its conception, the genetic improvement programme is similar to that of Ivory Coast. The only notable difference is the number of trees of the tested crosses, which is seldom higher than 50 in Vanuatu, due to lack of available land.

Between 1974 and 1993, the Saraoutou Research Station tested 65 between-population hybrids (excluding the reciprocal crosses and the complex hybrids). The centre is also the initiator and co-ordinator of a project called 'Production and Dissemination of Improved Coconut Cultivars' which since 1989 has combined eight APCC countries: Fiji, Kiribati, Papua New Guinea, Solomon Islands, Tonga, Tuvalu, Vanuatu and Western Samoa. In this project, financed by the European Development Fund and co-financed by France, 18 hybrids have already been tested (to the end of 1993). Figure 17 shows the total of the crosses made. In Vanuatu hybrids are tested for their yielding capacity and for their tolerance to Foliar Decay, the major constraint on hybrid use. The local tall (VTT) is totally resistant. Calvez *et al.*, (1985) classified ecotypes and hybrids according to their susceptibility to this disease.

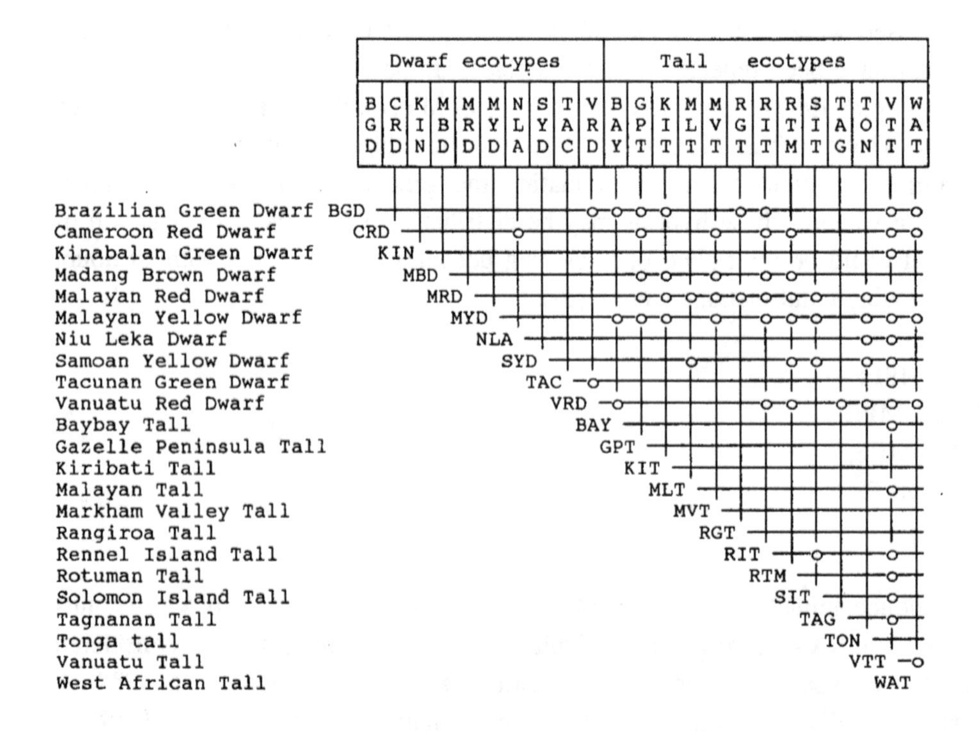

Figure 17: *Panorama of hybrids tested by the Vanuatu Research Center*

These tests identified one hybrid, the cross between Vanuatu Red Dwarf (VRD) and Vanuatu Tall (VTT), showing a good tolerance, and clearly higher-yielding than VTT. Some growers, however, do not appreciate the small size of its fruits. Improvement of the VRD x VTT hybrid was undertaken after individual combining ability tests. Table 22 shows the initial results, which are encouraging: a selection of the best progenies could improve the yields by 15-30%. The hybrid VTT x RIT, although less tolerant, has also been improved by this method.

In 1992, the centre produced about 110,000 seednuts and plants, 50% VTT and 50% VRD x VTT. In view of the low demand, this production is sufficient. Smallholders often prefer planting seednuts free of charge. The number of seednuts required for the maintenance of the Vanuatu coconut area was estimated at 600,000 per year (Calvez *et al.* 1985).

Other genetic improvement programmes were started in the Pacific region. In Fiji, a hybrid programme was re-started in 1970. However, the experiments have been planted with low tree numbers (of the order of 12 trees per cross, according to Vernon *et al.* 1975) and on soils barely suitable for this crop. A 384 ha research and development centre was established in 1986. From 1995 onwards, a seedgarden will deliver about 250,000 MRD x RIT and MRD x RTT hybrid seednuts per year. In Tuvalu, Western Samoa and on the Solomon Islands,[20] seedgardens also produce hybrids between Malaysian dwarfs and RIT (Duhamel 1993).

Other Programmes

In Brazil, coconut hybridization work involves a few imported hybrids only, planted before 1984. No hybrid using the local ecotypes has yet been tested. In 1984, the 'Centro de Pesquisa Agropecuária dos Tabuleiros Costeiros' (CPATC), belonging to the Empresa Brasileira de Pesquisa Agricola (EMBRAPA), established comparative trials with seven hybrids imported from Ivory Coast, compared to a local tall control. The tested combinations include:

CRD x WAT;
CRD x RIT;
MRD x TAT;
MYD x WAT;
MYD x TAT;
WAT x RIT;
BGD x WAT.

The hybrids have clearly shown to be higher yielders than local talls. In addition, the hybrid BGD x WAT was shown to be less susceptible to Leaf Blight (*Botryodiplodia theobromae* P.) than the Brazil Tall and the hybrid WAT x RIT, and is resistant to drought (Warwick *et al.* 1991).

Table 22: *First results of the improvement of the hybrid VRD x VTT in Vanuatu (CARFV 1993)*

Crosses tested	Nuts per tree per year	Copra per nut (g)	Copra per hectare (t) (x 0.95)
	6-7 years	4-5 years	
Mean of 16 progenies VRDxVTT (half-sib families)	147	151	3.4
	6-7 Years	4-5 Years	
Mean of best two progenies VRDxVTT (half-sib families)	168	169	4.3

The FRUTOP company also tested three of the hybrids mentioned above, in the region of Fortaleza in 1983. Notwithstanding unfavourable agronomic conditions, the trials showed the superiority of these hybrids over Local Tall. During the first four years, the hybrids yielded an average of 40 fruits per tree against only 3 for the Brazil Tall (CIRAD-CP 1993). In an agreement between the SOCOCO company and CIRAD-CP, a seedgarden for the production of hybrid seednuts has been established. This seedgarden, planted between 1982 and 1987 includes 40 ha of dwarfs and 10 ha of talls. As the planting programmes are rather small, the use of this seedgarden has been limited.

In Costa Rica, no genetic improvement programme has been started as yet, but seednut production has been undertaken by the SACRAC Ltd company (Illingworth 1991). The seedgarden, totalling 20 ha produces hybrid crosses between Malaysian Dwarfs and Panama Tall (PNT).

In Mexico, seedgardens (30 ha) and performance trials were established in 1978 and 1979 by the 'Impulsora Guerrerense del Cocotero' (IGC). The experiments compared three imported hybrids and a local one:

CRD x WAT;
MYD x WAT;
MRD x WAT;
MYD x Mexican Tall (Pacific Coast).

Various problems, partly related to land tenure, have resulted in the abandonment of the project. Two government institutions recently initiated coconut research and development projects (Morin, personal communication). The 'Centro de Investigación Cientifica de Yucatan' (CICY) tests local accessions recently collected by Zizumbo *et al.* (1993) against Lethal Yellowing disease and *Phytophthora*, and tries to rehabilitate planting material introduced from Ivory Coast in the 1970s, such as WAT, RIT, TAT, MYD, MRD, and CRD. The 'Instituto Nacional de Investigación Forestal, Agricola e Pecuaria' (INIFAP) has replanted a seedgarden with yellow dwarfs originating from India and Ivory Coast, that would permit an annual production of about 300,000 hybrid seednuts from 1995 onwards. Some hybrids from crosses between yellow dwarfs and local talls are being tested.

In Guyana, a collection of planting material consisting of some ecotypes and some hybrids were planted in 1978 and 1986 in Saut Sabbat and Combi. The main objective of this programme was testing tolerance against Hartrot, a lethal disease.

In the Dominican Republic, some seedgardens were planted between 1985 and 1987 by the LAVADOR SA company. However, the area actually used for seednut production covers only 5 ha of planted MYD in a perfectly isolated area near Santo Domingo. Hybrids produced are crosses between the yellow dwarf and a local tall called Criollo Tall (Wuidart, personal communication).

In Cameroon, some comparative trials were established in 1974 and 1980. Most of these trials were abandoned, due to pest attack and a lethal disease (Kribi). A seedgarden of about 30 ha of red and yellow dwarfs, established in 1975, has been exploited irregularly.

In Ghana, coconut is affected by the lethal Cape Saint-Paul disease. A rehabilitation project was started in the Western Region in 1981. Seven performance trials, established between 1981 and 1983 studied 21 hybrids and 6 different ecotypes, mostly originating from Ivory Coast (Nipah and Dery 1994). During the same period, two seedgardens totalling 26 ha were planted in a disease-free region. A better tolerance is observed in the pure dwarfs from Asian origin: MYD, MRD, and Sri Lanka Green Dwarf, most of their hybrids and of the progenies of MLT and VTT (Sangare *et al.* 1993).

In Nigeria, a limited genetic improvement programme was undertaken by the Nigerian Institute of Oil Palm Research (NIFOR). The cross between the Nigerian Tall and the Nigerian Green Dwarf proved to be significantly higher-yielding than the tall parent: 45 fruits per tree at 8 years, against only 8 fruits for the tall. The precocity was also improved by two years (NIFOR 1989). Other hybrids, using MYD as their female parent have also been tested, and initial results seem to show that the new hybrids are superior to the Green Nigerian Dwarf x Nigerian Tall. For the time being, seednut production activities remain limited.

In Vietnam, 14 hybrids were tested in a genetic improvement programme started in 1983. The strategy adopted in Vietnam is a good example for countries wishing to start an improvement programme, making use of international experience. Vietnam, at the beginning of its programme imported the best cultivars produced by various countries, such as Ivory Coast, Fiji, Philippines, and

Sri Lanka. Seven exotic hybrids were compared to two local hybrids and various Vietnam Talls. Seedgardens with a total area of 120 ha were established for hybrid seednut production (Long 1993). Actual production consists mainly of PB 121. This will be diversified when the results of the on-going trials become known.

In Papua New Guinea, a coconut research programme was started in 1987. According to Ovasuru (1992), the following hybrids were tested:

MRD x Madang Brown Dwarf (MBD);
MRD x RIT;
MRD x Gazelle Peninsula Tall (GPT);
MRD x WAT;
MRD x MLT;
MRD x Solomon Island Tall (SIT);
MYD x WAT.

In 1994, it was planned to establish a bloc of 179 ha, comprising collections, genetic trials and seedgardens. In particular, 18 local and imported tall populations will be systematically crossed with three dwarf testers: MYD, MRD and MBD. Among the main problems of Papua New Guinea are the insect pest attacks (*Oryctes rhinoceros, Scapanes australis* and *Rhynchophorus bilineatus*) killing large numbers of coconut palms. The hybrids MRD x SIT and MRD x RIT are susceptible, with a mortality of over 50% in some areas (Lee Heng Lye and Jerry 1985, cited by Ovasuru, 1992). Hence, research for genetic resistance against these insects is a top priority.

Genetic analysis of yield and its components

Genetic analysis, described in the paragraph below, refers to progenies derived from hand pollination (known parents). For analyses of experiments with open pollination (male parent unknown) referrence is made to the paragraph on mass selections. In Ivory Coast, some trials have a factorial or diallel crossing plan which permitted genetic analysis (Meunier *et al.* 1984; Bourdeix 1989). The number of parental ecotypes involved is always low (never more than nine) and the hybrids tested are crosses between populations, and not between individuals. Two examples are given by trials no. 5 and no. 8 at the Marc Delorme Station. Trial 5 is a factorial design: three tall cultivars (WAT, RIT and TAT) have been crossed with three dwarfs (MYD, MRD, and EGD), thus nine combinations in total. Trial 8 corresponds with a half-diallel between four tall cultivars (WAT, TAT, MLT and VTT). Tables 23 and 24 present the analyses for yield and its components of the two trials. In trial 5 the principal effects are significant but the interactions are not (with the exception of nut copra content). In trial 8 the interaction effects are significant for the number of nuts and yield, but are small when compared to the principal effects. Such conclusions can be drawn from the analyses of the majority of the studied trials. The relative performances of the various hybrids essentially depends on the parent ecotypes used. Some particular (WAT, RIT and VTT) always seem to perform better than others (MLT and MZT) irrespective of the female parent.

This additive behaviour is not incompatible with a strong heterosis effect. A comparison between the average production level of hybrids and that of WAT control and parent ecotypes in the collection shows that their production is significantly higher on average: about 60% for copra per tree when considering 21 ecotypes and 42 hybrids in Ivory Coast (Bourdeix 1989).

Finally, there is no clear connection between production of parent ecotypes (copra per tree) and that of their resulting hybrids. For instance, in Ivory Coast the Mozambique Tall (MZT) produces better than the WAT. But the WAT hybrids are superior to those of MZT.

Table 23: *Genetic analyses of yields and components of trial no. 5 (Bourdeix 1989)*

	ddl	Number of bunches		Number of fruits		Copra per nut		Copra per tree	
		C.M	F	C.M	F	C.M	F	C.M	F
Dwarfs	2	0.7	4.5 *	1262.3	17.4 *	3432.8	81.0 *	18.3	3.6 *
Talls	2	2.0	13.3 *	720.3	9.9 *	1742.4	41.1 *	18.8	3.7 *
Interaction	4	0.3	1.9	33.8	0.5	265.1	6.3 *	5.8	1.2
Residual	45	0.2		72.8		24.4		5.0	

* test significant at level of 5%

Among the hypotheses explaining these results, the following may be stated: the majority of ecotypes show inbreeding depression at variable degrees, resulting from two factors:

- as regards the ecotype: successive founder effects would have caused allelic losses, - as regards individuals: partial selfing is possible within all ecotypes whose reproduction system was studied (Sangare et al. 1978);
- in addition, the progeny of a tree is often planted in a group and close to that tree, favouring crossing between related palms.

Inbreeding causes a yield decline in tall coconuts. Hybridization between ecotypes would produce non-inbred, hence higher-yielding trees. Within the hybrid population (in the absence of inbreeding) the relative performances could in the first place be explained by genetic additive effects. Although there may be other explanations for these heterosis phenomena (Demarly 1977; Mac Key 1974), the first hypothesis seems to be supported by results obtained from improving the best hybrids. For that purpose, at the Marc Delorme Research Station some parents have been crossed individually with various testers. Fifteen RIT parents have been crossed with populations of WAT, CRD and MRD; Various TAT parents have been crossed with WAT, MYD and MRD. Cross-checks between trials show that the same parents are always the best, whatever tester used. This result, which is very important for further improvements shows that parents with very good combining ability have been identified.

Table 24: *Genetic analyses of trial no. 8 of yield and components (Bourdeix 1989)*

Sources of variation	ddl	Number of bunches		Number of nuts		Copra per nut		Copra per tree	
		C.M.	F	C.M.	F	C.M.	F	C.M.	F
Principal effects	3	12.7	50.1 *	5806.0	208.3 *	19257.3	208.2 *	179.2	99.8 *
Interaction	2	0.2	0.9	418.5	15.0 *	295.5	3.2	6.5	3.6 *
Bloc	5	0.5	2.3	3651.0	131.0 *	418.3	4.5 *	110.1	61.3 *
Residual	25	0.3		27.9		92.5		1.8	

* test significant at level of 5%

Evolution of Selection Strategies

Dwarfism utilization strategy

There exists a controversy about which hybrid type to recommend: D x D, D x T or T x T (Harries 1991). According to Ohler (1984), the attention of breeders and growers was too much focussed on the D x T type. Almost all hybrids planted throughout the world consist of this type, although some similarly productive T x T hybrids are also known (Nucé de Lamothe and Benard 1985a). One of the most important questions is how dwarfism must be integrated into the selection scheme. In particular, what type should be the base of genetic improvement: D x D, D x T, T x T, or intermediate forms?

Intermediate forms

It is not a priori evident that a scheme separating dwarfs and talls is an ideal solution. For instance, two D x T hybrids could be intercrossed (F2) to produce situations of dwarfism segregation. This could permit the introduction of favourable dwarf characteristics into populations of much wider genetic variability. In coconuts, the factor determining dwarfism remains unknown. F2 progenies of D X T hybrids are heterogenous in tallness. Such material is difficult to use in a plantation, as competition between trees would have a generally depressive effect. To produce a cultivar that could be distributed, a supplementary stage of selection for tallness would be necessary. Such a fixation selection is costly and would delay the improvement of other characteristics. In addition, various selection methods, *e.g.* reciprocals, are based on a division of the variability in sub-groups. These are kept in reproductive isolation and are each improved as a function of the other. In general, the groups formed in this manner combine well with each other: they are genetically divergent, or present complementary characteristics. The importance of groups selected for their complementarity is based on the pre-existent hybrid vigour: right from the start, the between-groups hybrids show such superiority that it would require several generations of breeding with another method to reach a similar level.[21]

A complementarity exists between dwarfs and talls. Dwarfs are a group entirely apart, with specific characteristics and a good ability to combining with talls. This complementarity should be exploited. These arguments are a reason not to create dwarfism segregation situations for sexual multiplication in the first stage of a selection scheme. On the other hand, intercrossing between D x T hybrids is fully justified in view of clonal selection. Once it is possible to multiply a genotype, the problems of heterogeneity in tallness could be avoided by planting monoclonal plots.[22]

Dwarf x Dwarf hybrids

The greater the share of dwarfism in a hybrid, the better its precocity and the more the smaller canopy will permit high-density planting. In Ivory Coast, the hybrid MYD x MRD has produced 3.8 tons of copra per ha. (Le Saint *et al.* 1987). In the same trial, the MYD used as a control produced one ton less. The production level of the hybrid MRD x MYD in this trial is comparable to that of the best D x T hybrids.[23] However, in general dwarfs are not very hardy and have little drought resistance (Ziller 1962). Possibly, these shortcomings are transmitted partly to the hybrid. In addition, on the scale of world coconut dwarfism remains a marginal phenomenon. Variability of dwarfs is inferior to that of talls. The long-term potentialities of dwarf hybrids are therefore probably limited. The precocity and rapid emission of bunches of most dwarf ecotypes are of major importance. D x D hybrids will be created for improvement needs, even if their direct distribution is not a first priority (Bourdeix *et al.* 1991a).

Dwarf x Tall or Tall x Tall?

The pre-eminence of the D x T hybrid type throughout the world is essentially related to its precocity. However, the differences are far from sufficient to eliminate the T x T hybrid type. In a recent article, a comparison was made between the world's most distributed D x T hybrid, PB 121 and the improved WAT x RIT tall x tall hybrid (Bourdeix *et al.* 1995). In cumulated yields, the latter reaches the level of PB 121 after nine years, and it subsequently produces one ton more per ha per year. The economic impact of precocity depends on various factors, according to production conditions. If the need is to recover planting costs as soon as possible, the D x T hybrid should be considered. When a higher profitability in the long term is to be preferred, with the consequence of needing at least one additional year to pay off the investment cost, the T x T hybrids should be selected.[24] Precocity and productivity are not the only criteria determining the choice of planting material. In particular, the price of seednuts should be taken into account. In the case of D x T hybrids, the use of dwarf mother trees is easier. Precocity of the dwarfs, their slow growth and the existence of markers for the detection of illegitimate seedlings are practical advantages of great economic importance. T x T seednuts by nature have a higher production cost. However, once cloning becomes operational, this difference will disappear. There is another argument in favour of the T x T hybrids: the fruits collected from the D x T hybrids cannot be used as seednuts (segregation of dwarfism genes). Unfortunately, experience has shown that when development programmes are abandoned, such useless seednuts are largely used. In the case of T x T cultivars, the nuts collected in the plantations can still be used as seednuts, although their yielding capacity will be lower than that of the hybrid. A recent point of view is to recommend T x T hybrids for intercropping, as their rapid growth would permit better development of the lower vegetation layers. In the trial described above, the growth difference between T x T and D x T hybrids remained less than 50 cm for the first 11 years (Bourdeix *et al.* 1995). Such a small difference does not actually justify adoption of a special cultivation system.

Conclusion

Considering a scheme limited to D x T hybrids would oppose two populations very unequal in variability. Neither the improvement potentialities, nor the evolution of the characteristics desired by the users would permit the elimination of the T x T type. A diversification in types of cultivars will certainly develop in the next decade.

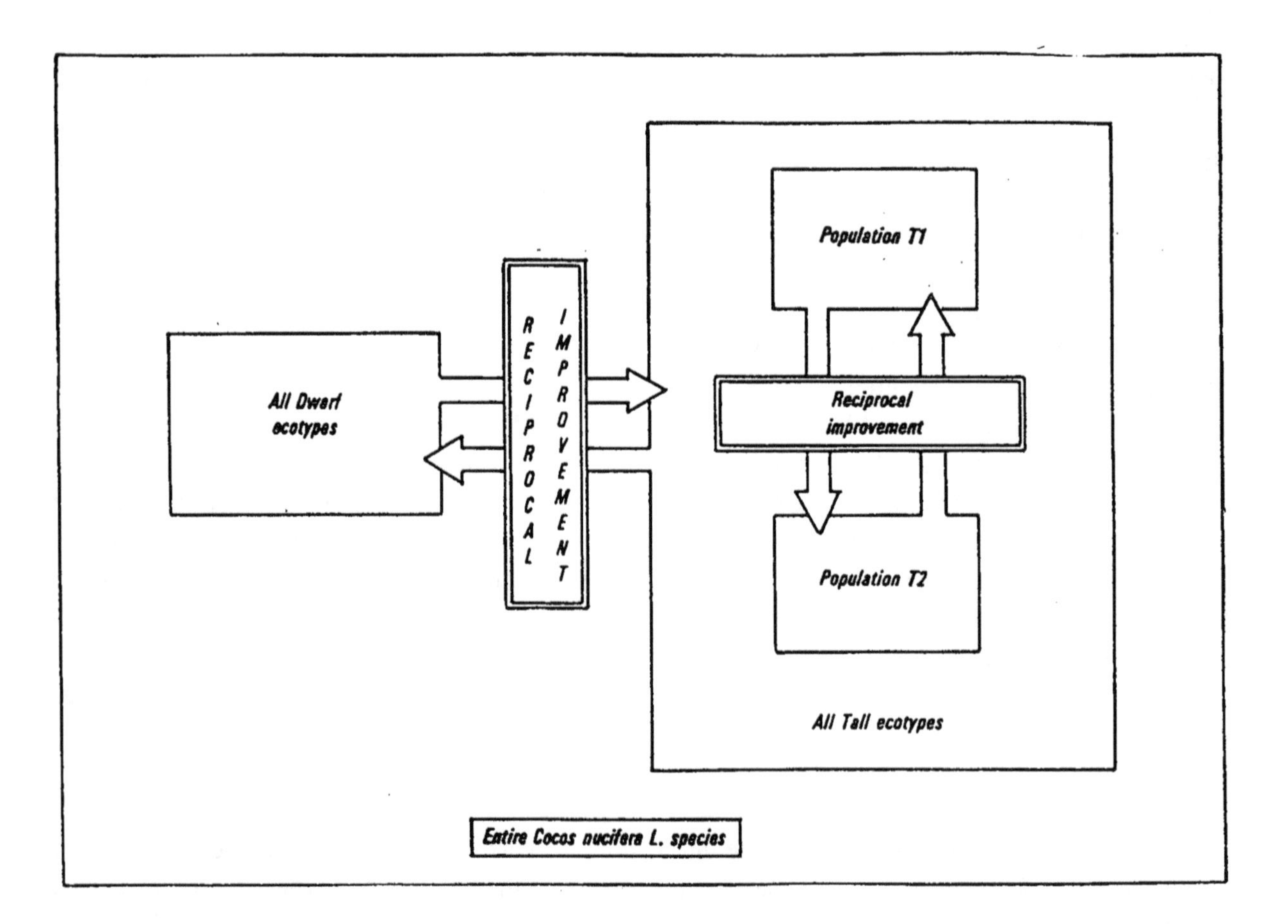

Figure 18: *Schematic presentation of the two selection axes, D x D and T x T*

Management of selection procedures

It may be concluded that at least two selection procedures should be maintained, one of the D x T type, the other of the T x T type. The question is which selection strategy could permit the optimal exploitation of these two procedures. Harries (1991) proposed the production of three-way-hybrids, using hybrid female parents (D x D, D x T or T x T). Some plantations would then be of mixed character - that of fruit production or seednut production, according to the opportunities. This evolution seems inevitable and almost automatic. With the progress of improvement programmes, successive generations will become available, and the crosses will become more and more complex. Some varieties in the year 2000 may have genes originating from three or four different accessions, or even more. In any case, the question remains what the nature of parent trees will be and how they should be combined to obtain the best possible progenies.

Within the scheme of CIRAD-CP, a balance of genetic programmes and trials has made it possible to propose new orientations for coconut genetic improvement (Bourdeix *et al.* 1990; 1991a; 1991b). It does not concern a revision of earlier recommended methods, but rather a generalization of principles that are already being applied. The scheme is based on the method of reciprocal recurrent selection (RRS): the creation of populations kept in reproductive isolation and improving with respect to each other (Comstock *et al.* 1949). Two major selection procedures have been described, improvement of the T x T and the D x T hybrids. Figure 10 shows a synthetic illustration of this method. The conception of the procedure of the D x T selection is relatively simple.Two rather different groups, dwarfs and talls have complementarities and good reciprocal combination ability. The constraints related to the biology of the plant, as well as other genetic factors, have led to the choice of the use of the RRS method for half-sib families.

The conception of the T x T procedure has proved more difficult. The greatest part of coconut variability is within the talls. Some crosses show a strong heterosis effect. Attempts have been made to structure this variability. These have led to divisions based on morphologic characteristics (Harries 1978; N'Cho *et al.*, 1993) or the analyses of foliar polyphenols (Jay *et al.* 1989). However, actual knowledge is still insufficient to evaluate the suitability of these divisions for the genetic improvement plan, particularly that of the identification of complementarity of the groups and their combination ability. Available data do not permit a definite conclusion. The method kept for the improvement of the T x T hybrids takes this ascertained ignorance into account. It consists of the artificial creation of two populations, starting with two 'founding' ecotypes specifically selected for this purpose. These two populations are then improved, each *vis-à-vis* the other, in RRS on half-sib families.

The selection of the founding ecotypes should take into consideration the constraints on each research project. These may vary from the availability of planting material to the phytopathologic situation. For practical reasons, it would be desirable if that at least one of the ecotypes was a local one. In Vanuatu, the local tall is resistant to Foliar Decay, the main limiting factor of the crop. In this case, this ecotype should necessarily be kept as founder. Some results obtained by various research centres show that the SLT and TAG are also potentially interesting candidates.

In Ivory Coast, an analysis of the genetic trials has led to the selection of RIT and WAT as founding types. These two types combine well, as well with each other as with other dwarf and tall ecotypes. In addition, individual parents with an excellent combining ability have been identified within the two ecotypes (Bourdeix *et al.* 1989; 1992). The use of these parents as testers has numerous advantages.

Clonal selection

Cloning does not lead to new variability. Once the process becomes operational, there will be a great temptation to clone, wherever, whatever high-yielding coconut, supposing to dispose of the best cultivar in the world. It is clear that this is not a good method. To obtain convincing results, cloning should be integrated within a general improvement programme. Taking into account the time required for field trials, the development of the method must be anticipated, and an adapted selection strategy should be selected today. In this way, the necessary ortets would be available at the right time.

Various clonal uses may be considered, such as conservation of ortets, reproduction of accessions, multiplication of male-sterile trees, varietal exit.... Especially the latter possibility (production and distribution of clones) will be explained here.

Selection of clonal ortets in genetic trials
Initially, the means for preparing a clonal selection will be limited. Candidates for cloning will be selected within the trials of the general breeding scheme (devized for making varieties by sexual propagation). The method used will probably be analogous to that used for oil palm (Meunier *et al.* 1988):

- selection of ortets within the best progenies of the best individual palms. This stage can be started before cloning has become operational;
- multilocal testing of these clones to identify and distribute the best among them.

In Ivory Coast, ortets of the clones will be selected preferably in the progenies of best parents individually tested for combining ability. In general, the value of the clones will be the higher when:

- the level of the original cross is high;
- the within-the-cross variability is high;
- the selected characteristics will be heritable;[25]
- the number of clones tested is high.

Crosses specifically designed for cloning
Some crosses will be specifically devoted to clonal exploitation. Their objective is rapidly to accumulate certain favourable genes that are scattered in several populations (disease-resistance, precocity, productivity) within a single genome.

It seems to be particularly interesting to create segregation situations of dwarfism by intercrossing D x T hybrids. This type of crossing has been implemented in Ivory Coast (Meunier *et al.* 1984) and planned in The Philippines (Santos 1990). It is hoped to obtain very variable progenies, within which some trees will have the precocity, the low slow growth of dwarfs, and an excellent production level. This method can be systematized at each selection cycle, without spending too much on it. Each cross to be made starts with four ecotypes, two talls (T1 and T2) and two dwarfs (D1 and D2), according to the following model: (D1 x T1) x (D2 x T2)

In the absence of an epistasis effect, the value of a double hybrid is equal to the average of the single hybrids of which it is composed. To obtain the best double hybrids, the parent should be selected in such a way that the average of the four following hybrids will be as high as possible: (D1 x T2); (D2 x T1); (D1 x D2); (T1 x T2).

Conclusion
It seems to be agreed now that hybrids are the varietal type best adapted to coconut cultivation (Santos *et al.* 1986; Shuhaimi Shamsudin *et al.* 1990; Harries 1991; Manciot and Sivan 1988; Nair *et al.* 1991; and Nucé de Lamothe *et al.* 1991). Hybrids, higher-yielding than local talls, have been identified in practically all countries with a research organization. Distribution of these new varieties is still insufficient. The hybrids represent less than 15 per cent of the coconuts planted during the past ten years. Smallholders, representing almost 95% of the world coconut plantations, are too often satisfied with unselected seednuts free of charge.

How can this insufficient use of research results be explained? The causes are multiple. Some are not coconut-specific; their analysis would go beyond the scope of this chapter. According to Nucé de Lamothe (1993) 'the bi- and multilateral aid programmes have contributed to the research effort on coconut. However, they have almost always lacked a global and long term vision on these problems. This has had some fatal effects, the beneficiary sometimes finding himself completely isolated and in a more difficult situation at the end of the project than at the beginning because of the sudden stop of the financing and the disappearance of the best educated staff'.

Numerous coconut seedgardens have been established throughout the world, but only a low percentage of these have finally achieved a regular production. Seednut production requires regular and meticulous monitoring. Even if the techniques have been mastered, they remain difficult, due to the nature of the plant. The low spread of the hybrids thus seems to be related to certain characteristics of the coconut: its perennial nature, its large canopy and above all its low multiplication rate. The cost of the seednuts and young plants reduces their use, except when a development project delivers them free of charge.

The most important problem of coconut cultivation is not finding a variety that is even more productive than the hybrids, nor a comparison of the hybrids created in the different research centres, but above all to make sure that the varieties already developed are produced and planted. An enormous effort of extension and training must be made to pursuade small farmers to invest in these new varieties. The development should be planned in such a way that the exclusive use of one unique type will be avoided, as it would cause an impoverishment of genetic resources and would be liable to phytopathologic risks.

After all, the potentialities of coconut productivity improvement are huge. Selection of the best between-population hybrids has made it possible to double the yield within one generation. In Ivory Coast, improvement of some of these hybrids has resulted in new progress of about 15-30%, according to the trials. This is not surprising in a crop whose the improvement begun only recently and among which a multitude of allogamous, autogamous and intermediate forms coexist. The success of genetic control of pests and diseases is still rather moderate. In Ivory Coast and Indonesia, sources of tolerance to *Phytophthora* have been detected; Vanuatu produces hybrids tolerant to the Foliar Decay virus (CFDV); a hybrid showing good tolerance of Lethal Yellowing has been distributed in Jamaica. On the other hand, no tolerance has yet been found against the yellowing diseases in Ghana and Tanzania, nor to the Cadang-cadang in The Philippines or to Red Ring in the Caribbean region and South America.

Practically all improvement programmes have, more or less recently, been oriented towards the exploitation of hybrid vigour. However, this option is far from restrictive; it even demands to be often redefined. In fact, very different methodologies can still be considered, such as the comparison of progenies of populations, individuals, half-sib families or sib families; the structuration of the variability to optimize the genetic gain; the means by which to anticipate the completion of cloning technique. The answer to these questions will determine the future success of the improvement programmes.

The methodological options described in this chapter were defined after much deliberation within the CIRAD Tree Crops Department. They can be summarized in T x T and D x T, according to the method of recurrent reciprocal selection in half-sib families. These are actually applied in Ivory Coast within the framework of co-operation with the Department of Oil Crops of the Forest Institute (IDEFOR/DPO).

Notes

1. The surveyed population was observed during three campaigns, from 1951 to 1954 (Bourdeix, 1988a). The selection without doubt would have been more efficient if, as mentioned above, the production of the parent trees had been recorded during an even number of years (in fact: four campaigns).

2. In Vanuatu, a lethal disease prevails (Coconut Foliar Decay Virus). As detection tests are not yet operational, no vegetative material can be exported from the country. The Vanuatu Tall is the only known ecotype with excellent resistance. It was introduced into the Ivory Coast in 1969. This was done after a thorough investigation of a large, disease free area. After 25 years no phytopathologic problem has yet presented itself.

3. The nuts are immersed for three minutes in a solution containing: 0.7 g commercial product per l Azodrine (55.2% monocrotophos against mites; 5 g commercial product l Dithane M45 (80%) mancozebe). This product may be replaced by Organil 66 (16% manebe and 64% carbathene) at the same concentration.

4. This apparent absence of variability of WAT is problematic as this ecotype had been sampled in a large region, including various plantations in Ivory Coast and Benin. On the other hand, the following text shows that certain tests of progenies reveal important differences between WAT parents.

5. The international descriptor used for this characteristic is slightly different: it speaks of the distance between 10 leaf scars from one metre above ground-level upwards. This measurement, after all, seems insufficient. In fact, in the case of dwarf ecotypes, which often emit more than 15 leaves annually, the measurement relates to less than eight months of a tree's life. Stability of the measurement may be reduced by seasonal rhythms, biannual alternating phenomena, or possible climatic events. For a more reliable measurement it would be better to count at least 30 leaf scars, corresponding to about two successive years of a tree's life.

6. OP: open pollination, female parent identified, male parent not identified
   CP: controlled pollination. Both parents identified
   SF: self-cross. Same parent as male and female.

7. There is a real problem in the varietal definition among dwarfs of the "Malaysian" type, to which the Ghana Yellow Dwarf belongs. These dwarfs can be found almost anywhere and show phenotypic similarity. It is often difficult to decide about the genotypic or environmental origin of the observed differences. Liyanage and Luntungan (1978) compared various Yellow Dwarfs of a presumed Malaysian origin, and observed notable differences in the level of copra content per nut. The scope of these variations ranges from 99 grams (MYD in Sri Lanka) to 188 grams (NYD) in Indonesia, similar to the Malayan type. The authors came to the conclusion that the differences were mainly genotypic. Taking into account that data of Liyanage and Luntungan (1978) are collected from various continents and from regions with diffferent ecologic conditions, it continues to be difficult to be certain about the environmental or genetic origins of the observed differenc-es. Only a comparison of different materials in one and the same place or by molecular biological techniques would make this possible.

8. However, such coconuts have been described from Australia and the Philippines (Buckley and Harries, 1984). These ecotypes are not represented in the collection of the Marc Delorme station, and may bring the proposed classification into question again.

9. However, in the case of assisted pollination, the barrier around the seedgarden cannot consist of coconut trees. Even coconut trees of the same variety as the pollinator initially used are not suitable, as the type of pollen for different crossing programmes may be changed.

10. In the Philippines, a team has also obtained the regeneration of a plantlet derived from callus developed by zygotic embryos (Nguyen *et al.*, 1992).

11. Among tall coconuts, selfing in most cases induces an inbreeding depression. The progenies obtained in this way are generally less productive than those derived from inter-crossing. This phenomenon is dealt with in the following chapter.

12. The nomenclature usually used for hybrids may be illustrated by an example: MRD x TAG indicates a sample of MRD female parents crossed in mixture with a group of TAG male parents.

13. Crosses between autogamous dwarf ecotypes are a case apart. In fact, certain dwarfs have undergone numerous founding effects and are very homozygous. An F_1 between two of these dwarfs can be considered equal to a cross between two pure lines: its genetic structure is homogeneous, all the individual palms resulting from the cross are of the same genotype. In such a case, an experimental number of about twenty trees per cross may be sufficient.

14. In Ivory Coast, the difference between the local tall and the hybrids is frequently more important than in other countries: certain hybrids produce more than twice the WAT yield, although, for instance in India and Sri Lanka hybrids rarely yield more than 150% of the yield of the local tall. Two phenomena may explain these differ-ences: 1) compared to certain Indian and Sri Lankan talls, WAT is a lower yielder. Results obtained at the Marc Delorme Station confirm this difference; 2) the agronomic conditions in Ivory Coast

are better, and favour the expression of differences between planting material. For instance, the hybrid SLT x PGD in Ivory Coast produces more than twice as its reciprocal CRIC65 in Sri Lanka, although the direction of the cross is not important (Manthiratne, 1971; Bourdeix and N'Cho, 1992).

15. The improvement of the PB 121 hybrid (started in 1978) has recently been the subject of a publication (Bourdeix et al., 1992). In the described trial, 15 male WAT parents were crossed individually with a population of female MYD. The hybrid progenies of the best WAT parents are significantly better than the PB 121 commonly distributed among farmers. In terms of copra per tree, the difference is 19% for young palms (4-8 years) and 15% for adult palms, during which period the progenies produced over 4.8 t/ha of copra while the PB 121 control reached only 4.2 t/ha. Two out of three progenies show an excellent level of tolerance to premature nutfall caused by ***Phytophthora katsurae***. The use of the "second generation" seednuts will certainly develop increasingly in the near future. Finally, from a genetic point of view, the result has questioned a preconceived idea. The relatively homogeneous aspect of the WAT populations invited the attribution of a low genetic variability to them. However, the trial shows that within the populations an exploitable variability exists for yield as well as for *Phytophthora* tolerance.

16. This project is original, but runs a risk by using open pollination for seednut production. The percentage of selfing is not controlled and varies with the season. During some periods, the seednuts may have a reduced value due to inbreeding depression resulting from selfing.

17. In Ivory Coast, several trials conducted for as long as 15 years have not shown any relative decline in the performance of hybrids compared to the local tall. On the other hand, problems encountered in Sri Lanka with the hybrids in dry areas may be of agronomic nature. The seednuts are distributed directly to the farmers. Those collected from the PGD are small, poor in reserves and water and require better nursery treatment than tall seednuts.

18. This figure contains 20 crosses only, although Mkumbo and Kulaya (1994) counted 22. We could not find the two missing ones. Possibly they concern reciprocal crossings not shown in the figure.

19. These may be of the T x T type with the possible use of a local ecotype.

20. For some decades, the Department of Agriculture of the Solomon Islands carried out experimental work on the plantation of Lever Pacific Plantations Pty Ltd, particularly at Yandina on Russell Island (Child, 1974). However, cyclones have seriously affected the experiments and a fire has destroyed many data that unfortunately had not been duplicated.

21. These successive generations, in addition the time they require, would also induce a loss of favourable alleles by genetic drift.

22. If all trees on one plot are of the same genotype, their cross will be homogeneous. It should be noted that such an option does not imply the establishment of large plantations of only one clone. Such a structure would be fragile. But a plantation consisting of a mosaic of monoclonal plots of different genetic origin could be considered.

23. On the other hand, various crosses with the Niu Leka Dwarf (Fiji) have not yielded the expected results. In Ivory Coast, this allogamous dwarf is a late producer and it transmits its lateness and its low production to its progeny (IRHO-CIRAD, 1989). About forty other dwarf hybrids have recently been planted in Ivory Coast. However, these trials are not yet in the productive stage.

24. In fact, this alternative is not limited to the choice between the two types D x T and T x T. Various D x T crosses (for instance MYD x TAG and MRD x TAG) behave much like the improved WAT x RIT hybrid: yielding lower than PB 121 during the first years, followed by higher yields at maturity (IDE-FOR/DPO, 1992).

25. Certain methods of environmental effect analysis, such as the one of polishing (Baudouin *et al.*, 1987) permit an increase in the heritability indices.

ANNEXE I: International cultivar list

Old codes refers to Chomchallow 1987, new codes were established by Bourdeix and Baudouin 1995. See legend for an explanation of symbols and country codes

Code old	Code new	Code french	Country	International cultivar name	Synonym	Remark
ADO	**ADOT**	GND02	IND	Andaman Ordinary Tall	=AO	
	AET		IDN	Aertembaga Tall	=DAG! or DAE	
AGA	**AGAT**		PHL	Agta Tall		Blackish appearance of immature fruit, due to premature drying and cracking of epiderm
AGD	**AGDT**		PHL	Aguinaldo Tall		
AGT	**AGT**	GND03	IND	Andaman Giant Tall	=AG	
	AIT		IDN	Amahai Tall	=DAI	Novarianto et al., 1994;
	ANGD		IND	Andaman Green Dwarf	=AD,=AND,=AGD	Nair M.K (Pers. Communic.)
ARO	**AROD**	NVP07	THA	Aromatic Green Dwarf	=Nam Hom	
AYD				Bali Yellow Dwarf		See BAYD
AYR	**AYR**		IND	Ayiramkachi Tall	=AYK; ~LMT?	
BAG	**BAGD**		PHL	Baguer Green Dwarf		
BAN	**BAND**		PHL	Banga Green Dwarf		
BAO	**BAOT**		PHL	Bago-Oshiro Tall	=Hijo Tall, =Romano Giant	
BAT	**BAT**		IDN	Bali Tall		Large nuts (see Ohler 1984)
	BAYD		IDN	Bali Yellow Dwarf	=AYD	
BAY	**BAYT**	GPH04	PHL	Baybay Tall	<-LAG	Related to Laguna
BDR	**BDRT**		LKA	Bodiri Tall	~Maphraeo Phuang	
BGD	**BGD**	NVB	BRA	Brazilian Green Dwarf		
	BGT		IDN	Banyuwangi Tall	=DBG	Novarianto et al., 1994; Aromatic water
BIB	**BIB**		GUM	Bibola Tall		Extinct ?
	BIT		VNM	Bi Tall	=Bung Tall	Vo Van Long 1994
	BJT		IDN	Beji Tall	=DBJ, <BAT	Sub population of Bali Tall
BLI	**BLI**		TTO	Blanchisseuse	misspelled as Blanch Issues	according to Harries; Type not stated
BLT			SLB	Bellona Tall		location in Rennell Island, see RIT
BNG	**BNGT**		PHL	Benigan Tall		
	BOT		PNG	Bougainville Tall		
	BRD		BRA	Brazilian Red Dwarf		
	BRT	GBR	BRA	Brazilian Tall		Sub populations : BRT01 Praia Do Forte; BRT02 Merepe BRT03 Sao Jose do Mipibu
BTC	**BTCT**		PAL	Bertachel		A Tall according to Harries. There is a "Partagel" in Yap
BUS	**BUSD**		PHL	New Buswang Green Dwarf		
	BYD		BRA	Brazilian Yellow Dwarf		
CAN	**CANT**		GUM	Cannon		
CAT	**CATD**	NVP02	PHL	Catigan Green Dwarf	=Rabara,<Bilaka	Bilaka is a type of philipino gren dwarf including also TAC
CCN	**CCNT**		VNM	Cochin China Tall	CC	This name is rather vague : there are several tall populations in the southern part of Viet Nam (formerly cochinchina). It refers to an accession of 26 palms introduced to India in 1940.
	CGD		IND	Chowghat Green Dwarf	~PDG?	Very similar to PGD

Code old	Code new	Code french	Country	International cultivar name	Synonym	Remark
CHA	**CHA**		GUM	Chaca		Type not stated
CIT	**CIT**		SLB	Christmas Island Tall		
	CKT	GCA	CMR	Cameroon Kribi Tall		
	CLT		PNG	Papua New Guinea Central Tall		Faure, 1994
	CMT	GCO	COM	Comoro Moheli Tall		
CNO	**CNOD**		PHL	Coconino Green Dwarf	>Mangipod, Dahili, (Mamareng ?)	Pugai is not a coconino according to Copeland
COD	**COD**		IND	Chowghat Orange Dwarf		
	COT		IDN	Concong Tall	DCG	
CRD	**CRD**	NRC	CMR	Cameroon Red Dwarf		
	CUD		CUB	Cuban Dwarf		
CUL	**CULT**		PHL	Culaman Tall		
CYD	**CYD**		LKA	Sri Lanka Yellow Dwarf	Ceylon Yellow Dwarf	
DAG	**DAGT**		GUM	Dagua Tall		
	DAUT		VNM	Dau Tall		Vo Van Long, 1994
	DIT		LKA	Dikiri Tall	~Macapuno	Peries, 1994; see MAC
	EAT		TZA	East African Tall		
EGD	**EGD**	NVE	GNQ	Equatorial Guinea Green Dwarf	~Brazilian Green Dwarf	
	ELT		PNG	East Sepik Tall		Faure, 1994
EMD			PAL	Emadech		See TOBD
	EOD		VNM	Eo Brown Dwarf		Vo Van Long, 1994
ERI			PAL	Eriiech		See TOGD
FAI	**FAID**		THA	Fai Orange Dwarf	=Maphrao Fai,<Mu Si Som	An orange dwarf according to Chulapan Petchpiroon
	FIT		PNG	Nuguria Tall		Faure, 1994
FJT	**FJT**		FJI	Fijian Tall		
FMD	**FMD**		SLB	Fiami Dwarf		Colour ?
GAL	**GALD**		PHL	Galas Green Dwarf		
GAT	**GAT**		PHL	Gatusan Tall		
GBT			SLB	Gilbertese Tall		See KIT (Gilbert Isl.=Kiribati)
GGB	**GGBD**		IND	Gangabondam Green Dwarf	=GB	
	GGD		PNG	Papua New Guinea Green Dwarf	=PGD!	Faure, 1994
	GIT		VNM	Giay Tall		Vo Van Long, 1994
GPT	**GPT**	GNG04	PNG	Gazelle Peninsula Tall	=Gazelle Tall, GLT	
GTB	**GTBT**		LKA	Gon Thembili Tall		Lyanage
	GUT		IDN	Guntung Tall	=DGU	
GYD	**GYD**	NJG	GHA	Ghana Yellow Dwarf	~Malayan Yellow Dwarf	
HLG	**HLG**		THA	Hua Ling		Comprises tall as well as Dwarfs (brown and green), The shape of the nut recalls a monkey's face (Chulapan Petchpiroon pers. comm. 1995)
	IBT		IDN	Igo Bulan Tall	=DIB	Novarianto et al., 1994From Ternate, only 1 palm available at Mapanget !
	IDT		IDN	Igo-Duku Tall	=DID	Shape of the nuts elongated/oblong
	IIT		IDN	Ilo-Ilo Tall	=DII	Novarianto et al., 1994
	IRD		PNG	Iokea Red Dwarf		Faure, 1994
	JGD		IDN	Jombang Green Dwarf	=GHJ	

Code old	new	french	Country	International cultivar name	Synonym	Remark
JMT	**JMT**	GJQ	JAM	Jamaica Tall	Tres Picos, Alto Atlantico	
	JPT		IDN	Jepara Tall	=DJP	Novarianto et al., 1994
JVT	**JVT**		IDN	Java Tall	=Java Giant, JG	
	JYD		IDN	Jombang Yellow Dwarf	=GKJ	Novarianto et al., 1994
KAP	**KAPD**		PHL	Kapatagan Dwarf		
KAT	**KAT**	GCB	KHM	Cambodia Tall		Includes sub populations: KAT02: Tuk Sap KAT03: Kampot KAT04: Kopal Tani KAT05: Kompong Trach KAT07: Ream KAT08: Sre Cham KAT09: Battambanb KAT10: Koh Rong KAT11: Ktis Battambang
	KBT		IDN	Kulaba Tall	=DKB	Novarianto et al., 1994
KCK	**KCKT**		THA	Kon Chuk Tall		Pointed nut
	KGD	NVC	KHM	Cambodian Green Dwarf		
KIN	**KIND**	NVP06	PHL	Kinabalan Green Dwarf		
	KIT	GKI	KIR	Kiribati Tall	=GBT	
KKD	**KKD**		SLB	Kukum Dwarf		
KKT	**KKT**	GNG01	PNG	Karkar Tall	=Kar Kar, Kar-Kar, NG	
KLG	**KLGT**		THA	Maphrao Klang Thailand Tall	<Thailand Tall	medium sized nut. Klang = name of a farmer
KLK	**KLKT**		THA	Kalok Thailand Tall	<Thailand Tall	Very large nuts
	KLYT		IDN	Kalasey Tall	=DKY	Novarianto et al., 1994; avoid KYT!
	KMT		LKA	Kamandala Tall		Peries, 1994; large nuts
	KNT		IDN	Kinabuhutan Tall	=DKN	Novarianto et al., 1994
KPD	**KPDT**	GND05	IND	Kappadam Tall	=KAP	
KRD	**KRD**		SLB	Kira Kira Red Dwarf		
KRJ			IDN	Kelapa Raja (Tall)		Unknown. Raja is supposet to be a dwarf
	KST		IDN	Kima Atas Tall	<-Mapanget; =DKA,KAT!	Mapanget population improved at Kima Atas (not Kima Atlas)
	KUT		IDN	Kairatu Tall	=DKU	Novarianto et al., 1994
KWG	**KWG**		MYS	Kalpawangi		Type ?
	KWT		PNG	Kiwai Tall		Faure, 1994
KYT	**KYT**		KEN	Kenya Tall		
LAG	**LAGT**		PHL	Laguna Tall	Galimba; Limbajon; Malapon	Synonyms refer to local languages
LAP	**LAPT**		GUM	Lapugua		
LCD	**LCD**		IND	Laccadives Green Dwarf	=LD, LAD	
LCT	**LCT**	GND08	IND	Laccadives Ordinary Tall	=Lakshadweep Ordinary, LO, LKO	
LFT	**LFT**		NCL	Lifou Tall		
LKM			IND	Lakshadweep Micro		See LMT
LKO						See LCT
LMT	**LMT**	GND07	IND	Laccadives Micro Tall	=Lakshadweep Micro, LM, LKM	
LON	**LONT**		PHL	Loong Tall		
	LPT		IDN	Lubuk Pakam Tall	=DLP	Novarianto et al., 1994

Code old	new	french	Country	International cultivar name	Synonym	Remark
MAC	**MACT**		PHL	Macapuno Tall		Endosperm with jelly-like consistance. Many other names for this rather ubiquitous character
	MAD		IDN	Malabar Dwarf	=DMA	Novarianto et al., 1994
MAG	**MAGD**		PHL	Magtuod Green Dwarf		
MAK	**MAKD**		PHL	Makilala Green Dwarf		
	MAT		PNG	Manus Tall	=MLT!	Faure, 1994: Avoid MLT!
MBD	**MBD**	NBN	PNG	Madang Brown Dwarf		
	MBT		PNG	Milne Bay Tall		Faure, 1994
MDV				Maldiviana		Not a cultivar, refers to the classification of Narayana & John 1949, a subclass of *nana*
MGD	**MGD**	NVM	MYS	Malayan Green Dwarf		NVM
MGP	**MGPD**		PHL	Mangipod Green Dwarf	=Ipod	
MLT	**MLT**	GML	MYS	Malayan Tall		GML
MND	**MNDT**		PAL	Minado Tall		
MPE	**MPET**		THA	Maphraeo Thailand Tall	>Yai, Klang, Kalok	comprises several populations/cultivars, see also THT
	MPT		IDN	Mapanget Tall	=DMT	
MRD	**MRD**	NRM	MYS	Malayan Red Dwarf	=RMD	
MSI	**MSID**		THA	Mu Si		Thai (Dwarf)
MSK	**MSKD**		THA	Mu Si Khieo		Thai (green Dwarf)
MSL	**MSLD**		THA	Mu Si Luang	>Nali Ke	Thai (yellow dwarf)
MSS	**MSSD**		THA	Mu Si Som	=Mu Si Daeng, >Maphraeo Fai	Thai (Orange Dwarf)
MVT	**MVT**	GNG03	PNG	Markham Valley Tall	=GMV; Markham Tall	
	MWT		IDN	Marinsow Tall	=DMW	Novarianto et al., 1994
MYD	**MYD**	NJM	MYS	Malayan Yellow Dwarf		
MZT	**MZT**	GMZ	MOZ	Mozambique Tall		
NAG	**NAGT**		MRI	Niyug Agaga Tall		
NAR	**NART**		COK	Nu Araketa		
NAT	**NATD**		TON	Niu Ati Red Dwarf	~Niu New Guinea in WSM	According to Harries
	NCT	GNC	NCL	New Caledonia Tall		
NDG	**NDG**		MRI	Niyug Dagua		Type ? Not "Niyng"
NFS	**NFSD**		NIU	Niu Fisi Red Dwarf		See Harries
NGA	**NGAD**		MYS	Nyiur Gading	=Nior Gading	Malaysian Dwarf, any colour
NGD	**NGD**		IDN	Nias Green Dwarf	=GNH	
NGU	**NGU**		NIU	Niu Gau	~Navassi;~Niu Hiata	Type ?
NID	**NID**		NGA	Nigerian Dwarf		Colour ?
NIM	**NIM**		THA	Nim	~Navassi	Type ?
NIT	**NIT**		NGA	Nigerian Tall		
NJK	**NJKT**		MSI	Ni Jok Tall		
NKE	**NKED**		THA	Nali-ke Dwarf	=Nar-Ri-Kay (?), <Mu Si Luang	
NKF	**NKFT**		TON	Niu Kafa	~Niu Afa, Niu 'afa	Means "rope", large husk
NKI	**NKIT**		NIU	Niu Kini Tall		
NKL	**NKLT**		NIU	Niu Kula Tall	~Macapuno?	Jelly endosperm? See Firman, Jackson & Guarino Several countries (Niue, Tonga, Tuvalu)
NKM	**NKMD**		THA	Nok Khum Green Dwarf	=Nok Koom	
NKN	**NKNT**		MRI	Niyug Kunon		

Code old	new	french	Country	International cultivar name	Synonym	Remark
NKU	**NKUT**		COK	Nu Kura Tall	~Macapuno?	See Firman, Jackson & Guarino 1986 ; Related to Niu Kula?????
NLA	**NLAD**	NNL	FJI	Niu Leka Dwarf	=Fidji Dwarf, Niu Le'a; =NLKD ?	A cross pollinating dwarf, may be green or brown
NLB	**NLBT**		MRI	Niyug Laipuga Tall		
NLK	**NLKD**		NIU	Niu Leku Green Dwarf	=NLAD?	
NLL	**NLLT**		TON	Niu Matakula Tall		ref. in the Philippines. Base of fruit red
	NLT		PNG	Namatanai Tall		Faure, 1994
NME	**NMET**		TON	Niu Mea Tall		
NMG	**NMGT**		COK	Nu Mangaro Tall		
NMJ	**NMJ***		MSI	Ni Mej		Type : Tall or brown Dwarf ?
NML	**NMLT**		NIU	Niu Malua Tall		
NMM	**NMMD**		MRI	Niyug Mogmog Dwarf		
NMR	**NMR***		MSI	Ni Maro		Type : Tall or Dwarf ?
NMU	**NMU**		MSI	Ni Mur		Type : Tall or Dwarf ?
NMV	**NMVT**		TON	Niu Mealava Tall		Red ring at calyx
NPL	**NPLT**		NIU	Niu Pulu Tall		
NRK	**NRKT**		MSI	Ni Rik Tall		
NSM	**SMOT**		MRI	Niyug Samoa		
NTE	**NTE**		NIU	Niu Tea	=Nu Tea	To be checked: a Tall according to Harries, a Malayan Dwarf (yellow or red) for Efu... Several places (Niue, Tonga)
NUL	**NULT**		TON	Niu 'uli		
NUR	**NUR**		COK	Nu Uri	?= NULT ????	Type : Tall or Dwarf ?
NUT	**NUT**		TON	Niu 'utongau	~Navasi	Type: Tall or Dwarf ?
NVI	**NVIT**		TON	Niu Vai Tall		Vai=water
	NVT		LKA	Navasi Tall		Peries 1994; Edible husk; Several cultivars have a similar phenotype
NYD	**NYD**		IDN	Nias Yellow Dwarf	=GKN, ~MYD?	
NYL	**NYL**		MSI	Ni Yalu		Check: a Tall or a yellow Dwarf?
OCT	**OCT**		PAL	Ongchutel		
OIL	**OILT**		PAL	Oilol		
	OLT		PNG	Oro Tall		Faure, 1994
	PARD		PNG	Papua New Guinea Red Dwarf	=PRD!;=PAD	Faure, 1994
PAS	**PASD**		PHL	Pascual Green Dwarf		
	PBT		IDN	Parit Baru Tall	=DPB	Novarianto et al., 1994
PCK	**PCKT**		THA	Pak Chok Thailand Tall		
PDL	**PDLD**		PHL	Philippine Dailig	<Coconino; =Dahili?	Was Dalig in Chomchallow's list.
	PDT		IDN	Pandu Tall	=DPD	Novarianto et al., 1994
	PELT		IDN	Pelimpaan Tall	=DPE	Novarianto et al., 1994 (avoid PET!)
	PET		PER	Peru Tall		
PGD	**PGD**	NVS	LKA	Pumilla Green Dwarf	=Sri Lanka or Ceylon G. D.	Very similar to CGD
PGI	**PGID**		PHL	Pugai Green Dwarf	=Pilipog*	(* according to Copeland)
	PGLT		IDN	Pungkol Tall	=DPL	Novarianto et al., 1994 (avoid PLT!)
PIL	**PILD**	NVP05	PHL	Pilipog Green Dwarf	=Pugai*	(* according to Copeland)
PKW	**PKW**		THA	Pluak Wan		
PLN	**PLNT**		PHL	Philippine Lono Tall		
	PMT		IDN	Peniraman Tall	=DPM	Novarianto et al., 1994

Code old	new	french	Country	International cultivar name	Synonym	Remark
PNT	**PNT**	GPA	PAN	Panama Tall	=Choco; > Redondo, Pacific Tall	Avoid "San Blas" which refers to the "Tres Picos" type. Among the existing sub-populations : PNT01: Aguadulce PNT02: Monagre
	PPT		LKA	Pora Pol Tall		Peries 1994: Thick shell
PPW	**PPWT**		PHL	Philippine Palawan		
	PRD		TZA	Pemba Red Dwarf		From Zanzibar
	PRT		IDN	Pangandaran Tall	=DPR	Novarianto et al., 1994
	PSNT		IDN	Paslaten Tall	=DPN	Novarianto et al., 1994 (avoid PNT!)
PTI	**PTID**		THA	Pathiu Green Dwarf		
PUA	**PUAT**		THA	Phuang Tall	=Maphrao Phuang	Large number of nuts/bunch
PUR	**PURT**		PHL	Puringkitan Tall		Noted for high number of nuts
	PUT	GDO03	IDN	Palu Tall	=DPU,(=PLT)	
	PYD		PNG	Papua New Guinea Yellow Dwarf		Faure, 1994
	RARD		PNG	Rabaul Red Dwarf	=RRD	Faure 1994. Avoid RRD!
	RBD		IDN	Raja Brown Dwarf	<Ternate Brown Dwarf ?	Brown-red colour. Origin : Ternate, found in Java
RGT	**RGT**	GPY02	PYF	Rangiroa Tall		
RIT	**RIT**	GRL	SLB	Rennell Island Tall	=Rennell Tall, (=Bellona Tall)	
RRD	**RRD**		PYF	Rangiroa Red Dwarf		
RTB	**RTB**		LKA	King Coconut	=Rath Thembili	Semi tall, autogamous
RTM	**RTMT**		FJI	Rotuman Tall		
	RUT		IDN	Riau Tall	=DRU	Novarianto et al., 1994
SAL	**SALT**		PHL	Salambuyan Tall		
	SAT		IDN	Sawarna Tall	=DSA	
SCT	**SCT**		SYC	Seychelles Tall		
	SDT		IDN	Sungei Daun Tall	=DSD	Novarianto et al., 1994
	SET		IDN	Sea Tall	=DSE	Novarianto et al., 1994
	SFIT		IDN	Sofifi Tall	=DSI	Novarianto et al., 1994 (avoid SIT!)
SGD	**SGD**		WSM	Samoan Green Dwarf		
SIT	**SIT**		SLB	Solomon Island Tall		
	SKD		IDN	Salak Green Dwarf	=Kalimantan; KGD, GSK	
	SKT		MYS	Sarawak Tall		
	SLRD		LKA	Sri Lanka Red Dwarf	=Ceylon Red Dwarf, (=ARD)	Was ARD
SLT	**SLT**	GSL	LKA	Sri Lanka Tall	=Ceylon Tall	
SMO	**SMOT**		WSM	Samoan Tall		
SMT	**SMT**		WSM	Samatau Tall	<-Samoan Tall	More uniform than SMOT
	SNID		PHL	San Isidro Green Dwarf		
SNO	**SNOD**		PHL	Sto. Nino Green Dwarf		
SNR	**SNRT**		PHL	San Ramon Tall	SAN	
	SOD		IDN	Sagerat Orange Dwarf	=GSO or GOS	Novarianto et al., 1994
SOX	**SOX**		THA	So		Type ?
SPI	**SPIT**		PHL	Spicata Tall	Many (See Child 1974)	A botanical character (Branchless inflorescence)
	SRD		IND	Spicata Red Dwarf		
	SRT		IDN	Sungei Rasau Tall	=DSR	Novarianto et al., 1994
	STD		IDN	Sri Tanjung Dwarf	=GST	Novarianto et al., 1994

Code old	new	french	Country	International cultivar name	Synonym	Remark
STK	**STKT**		IND	Standard Kudat Tall	<SPIT	
STV	**STVT**		TTO	St. Vincent Tall	~Jamaica Tall ?	(In India)
SUD	**SUD**		SUR	Surinam Dwarf		(in India)
SUT	**SUT**		SUR	Surinam Tall		(In India)
SYD	**SYD**	NJS	WSM	Samoan Yellow Dwarf		Maybe related to MYD
	TAAT		VNM	Ta Tall		Vo Van Long. 1994
TAC	**TACD**	NVP03	PHL	Tacunan Green Dwarf	=Rabanuel; < Bilaka;Linkoranay	
TAG	**TAGT**	GTN	PHL	Tagnanan Tall		
TAL	**TALD**		PHL	Talisay Green Dwarf		There seems also to be a Talisay Tall
TAT	**TAT**	GPY01	PYF	Tahitian Tall	Polynesia Tall n°1; PYT1	
TBD	**TBD**	NBO	IDN	Ternate brown Dwarf	>Raja Brown ?	
TBL	**TBLD**		PHL	Tambolilid Green Dwarf	~Niu Leka ?	
TGD	**TGD**		PYF	Tahitian Green Dwarf		
	TGKD		IDN	Trenggalek Dwarf	=GTK	Novarianto et al., 1994(avoid TKT!)
	TGT	GDO02	IDN	Tenga Tall	=DTA	
	THD	NVT	THA	Thailand Green Dwarf	<Tung Khled, Pathiu ?	
THT	**THT**	GTH	THA	Thailand Tall		Includes sub-populations THT01: Sawi THT04: Ko Samui
	TIT		IND	Tiptur Tall	Karnataka: TT	Nair M.K. pers. Communic.
TKD	**TKD**		THA	Thung Khled Green Dwarf		
	TKT	GDO01	IDN	Takome Tall	=DTE; DTE	
TLB	**TLB**		THA	Tha-le Ba		Type ? High number of fruits
TLR	**TLRT**		THA	Thalai Roi Thailand Tall	~Phuang Tall	
TOB	**TOBD**		PAL	Tobi Emadech Brown Dwarf	=EMD (Emadech)	
TOG	**TOGD**		PAL	Tobi Eriiech Green Dwarf	=ERI (Eriiech)	
TON	**TONT**	GTG	TON	Tonga tall		
TOP	**TOP**		GUM	Topalau		Type ?
TPK	**TPKT**		PHL	Tampakan Tall		
	TRD	NRY	PYF	Tahitian Red Dwarf		
	TST		IDN	Talise Tall	=DTS	Novarianto et al., 1994; # Talisay in the Philippines
	TTD		IDN	Tebing Tinggi Dwarf	=GTT	Novarianto et al., 1994, Not "Tening"
	TTT		IDN	Tontalete Tall	=DTT	Novarianto et al., 1994
TUP	**TUPD**		PHL	Tupi Green Dwarf		
TUR	**TUR**		PAL	Turang		Type: D or T?
	TYD		VNM	Tam Quan Yellow Dwarf		Vo Van Long. 1994
VEN	**VENT**		PHL	Venus Tall		
VIC	**VICT**		PHL	La Victoria		Tall or Dwarf ?
VKT	**VKT**		SLB	Vanikoro Tall		
	VLT		PNG	Vailala Tall		Faure, 1994
VRD	**VRD**	NRV	VUT	Vanuatu Red Dwarf		

Code old	Code new	Code french	Country	International cultivar name	Synonym	Remark
VNT	**VTT**	GVT	VUT	Vanuatu Tall	=VNT	Includes these sub-populations: VTT01: Surrenda, VTT02: Leroux, VTT03: Banks, VTT05: Surrenda, VTT06: Port Olry, VTT07: Tanna, VTT08: Torres, VTT09: Tanna.
WAT	**WAT**	GOA	CIV	West African Tall		Includes these sub-populations WAT02: Agriculture, WAT03: Akabo, WAT04: Mensah, WAT06: Ouidah.
WCT	**WCT**		IND	Indian West Coast Tall		
	WNT		PNG	West New Britain Tall		Faure, 1994
	WST		IDN	Wusa Tall		Novarianto et al., 1994
	XGD		VNM	Xiem Green Dwarf		Vo Van Long, 1994
YAI	**YAIT**		THA	Maphrao Yai	<MPET	
	YIT		YAP	Yap Island Tall		
ZAM	**ZAMT**		PHL	Zamboanga Tall		
ZAT	**ZAT**		TZA	Zanzibar Tall		

Legend:
-The abbreviations which are followed by a "!" should be avoided because they may lead to confusions (ex: Kima Atas should be "KST" and not "KAT!" because "KAT" is the Cambodia Tall.
-The symbols before the synonyms means :

symbol	meaning	Example
=	Is identical to	Laccadives Micro Tall = Lakshadweep Micro
~	is similar to (but not necessarily identical)	Nim ~ Navassi (both have an edible husk, but are found in different countries)
<	Is included in	Beji Tall < Bali Tall (sub population)
>	Includes	Bali Tall includes
<-	Is derived from	Baybay Tall <- Laguna

List of country codes (using ALPHA-3 ISO code)
* not in ISO code

BEN	BENIN
BRA	BRAZIL
CIV	CAMEROON
CMR	CAMEROON
COM	COMOROS
COK	COOK IS
CUB	CUBA
FJI	FIJI
GHA	GHANA
GNQ	EQUATORIAL GUINEA
GUM	GUAM
IDN	INDONESIA
IND	INDIA
JAM	JAMAICA
KEN	KENYA
KHM	KAMPUCHEA
KIR	KIRIBATI
LKA	SRI LANKA
MOZ	MOZAMBIQUE
MRI*	MARIANA ISLANDS
MSI*	MARSHALL ISLANDS
MYS	MALAYSIA
NCL	NEW CALEDONIA
NGA	NIGERIA
NIU	NIUE
PAL*	PALAU ISLAND
PAN	PANAMA
PER	PERU
PHL	PHILIPPINES
PNG	PAPUA NEW GUINEA
PYF	FRENCH POLYNESIA
SLB	SOLOMON IS
SUR	SURINAME
SYC	SEYCHELLES
THA	THAILAND
TON	TONGA
TTO	TRINIDAD TOBAGO
TZA	TANZANIA
VNM	VIET NAM
VUT	VANUATU
WSM	SAMOA
YAP*	YAP ISLAND

ANNEXE II: Measures of stems, leaves and inflorescences of the ecotypes at the Marc Delorme Station collection

LEGEND

C20 - Circumference of the bole at 20 cm above the ground
C150 - Circumference of the stem at 150 cm above the ground
CF - Number of leaf scars between 100 and 200 cm above the ground for Tall ecotypes and between 1 and 1.50 m above the ground for Dwarf ecotypes
LP - Length of petiole
LR - Length of rachis
NBF - Number of leaflets on one side of the leaf
LF - Length of leaflet
LGF - Width of leaflet
ILP - Length of bunch stalk
ILA - Length of inflorescense axis
NBE - Number of spikelets
ILE - Length of spikelet
ILPRF - Distance between the point of implantation of the spikelet on the axis and the point of implantation of the first female flower on the spikelet

Measures of stems, leaves and inflorescences of the ecotypes of the Marc Delorme Station collection.
Average measures of Tall ecotypes;

International and French codes, Plot, Year of planting and number of trees Observed	High (cm) and date of measure	C20 (cm)	C150 (cm)	CF	LP (cm)	LR (cm)	NBF	LF (cm)	LGF (cm)	ILP (cm)	ILA (cm)	NBE	ILE (cm)	ILPRF (cm)
ADO GND03 142 82 30	358.0 (8)	176.3	98.3	12.8	157.0	448.8	120.4	126.9	6.5	65.1	39.7	38.5	47.3	9.2
AGT GND02 M63 82 30	411.0 (10)	142.5	82.7	17.0	143.7	403.6	113.9	114.2	6.4	62.7	37.0	36.1	42.4	8.3
BAY GPH04 142 82 30	289.0 (8)	200.6	100.2	15.4	153.3	443.6	116.5	124.0	6.3	54.5	40.1	42.4	41.7	7.6
CKT GCA M63 82 30	551.0 (10)	148.7	82.2	10.7	147.3	429.1	116.8	128.2	6.6	61.3	41.5	35.6	47.0	8.6
CMT GCO 111 72 30	428.0 (10)	210.5	106.9	14.6	154.6	448.1	117.6	132.2	6.2	71.7	42.9	42.8	48.3	9.1
GPT GNG04 142 84 30	254.0 (8)	—	—	—	175.4	453.6	117.1	128.0	5.9	63.1	36.7	32.9	47.2	9.5
KAT07 GCB07 101 70 30	515.0 (12)	196.8	101.3	11.7	164.0	468.5	119.8	130.9	6.4	61.4	48.6	46.5	48.8	8.5
KAT08 GCB08 101 70 30	582.0 (12)	185.9	100.3	12.5	165.6	467.4	117.2	131.9	6.5	60.8	46.5	46.4	46.0	8.0
KKT GNG01 102 75 30	421.0 (10)	179.0	99.5	13.3	174.3	460.8	117.0	121.7	6.1	63.9	39.6	31.5	50.2	9.6
KKT GNG01 142 84 30	220.0 (8)	—	—	—	171.8	473.6	121.1	132.2	6.3	65.4	39.3	34.3	48.9	8.5
KPD GND05 142 82 30	319.0 (8)	151.3	81.8	13.8	155.7	440.0	117.5	125.2	6.4	64.9	44.6	41.8	47.7	9.1
LCT GND08 101 78 30	367.0 (9)	180.1	90.8	12.1	148.0	427.2	122.0	121.1	6.1	54.3	38.9	40.9	45.4	8.9
LMT GND07 101 79 30	422.0 (10)	174.2	89.4	13.5	143.2	423.8	121.4	123.7	6.1	54.5	37.3	38.6	46.0	8.8
MLT GML M63 82 30	569.0 (10)	185.0	99.7	11.1	168.3	455.1	120.2	121.8	6.3	56.8	38.3	39.3	42..8	7.1
MLT GML S22 56 30	970.0 (17)	177.0	100.0	13.8	150.0	161.0	121.0	135.0	6.3	61.0	46.0	43.0	—	—
MVT GNG03 142 84 30	211.0 (8)	—	—	—	173.6	466.2	119.1	133.0	6.3	61.8	36.6	33.1	48.8	8.7
MZT GMZ 101 70 30	—	168.0	100.0	15.5	127.0	404.0	115.0	134.0	6.0	62.0	30.0	34.0	—	—
MZT GMZ M63 81 30	590.0 (12)	189.9	97.4	12.7	161.0	446.7	115.0	136.5	6.3	68.4	39.2	39.4	45.4	8.2
PNT01 GPA01 M63 80 30	410.0 (10)	161.4	89.3	17.9	149.1	410.7	111.2	123.3	6.5	53.7	40.4	38.3	43.8	7.7
PNT02 GPA02 M63 80 30	377.0 (10)	165.7	91.1	18.0	151.9	420.8	113.4	122.7	6.6	53.6	37.8	40.8	42.7	6.7
PLT GDO03 142 84 30	261.0 (8)	—	—	—	160.9	466.2	121.4	125.2	6.0	58.4	37.5	38.5	48.1	7.3
RGT GPY02 091 69 30	591.0 (12)	193.5	96.6	13.4	147.6	448.4	115.6	134.3	6.4	61.9	42.4	44.9	55.5	9.2
RIT GRL 060 68 30	—	225.0	98.0	12.5	162.0	403.0	115.0	131.0	6.4	74.0	53.9	37.0	—	—
RIT GRL 081 68 30	764.0 (12)	221.7	96.8	10.9	163.0	422.4	116.5	127.3	6.6	73.3	48.4	35.0	55.8	11.3
RTM GRT 101 70 30	584.0 (11)	193.3	96.8	13.4	156.6	423.0	121.0	121.4	6.8	65.5	46.3	44.1	53.3	9.0
SIT GSN 081 68 30	884.0 (12)	217.4	107.1	9.1	170.0	453.9	116.5	137.3	6.5	67.6	43.5	38.1	55.9	10.5
SLT GSL 112 72 30	516.0 (12)	184.1	92.5	11.5	149.0	451.4	122.4	133.0	6.4	62.8	38.0	39.0	48.4	9.3
TAG GTN 102 74 30	642.0 (12)	187.8	106.0	11.8	153.8	470.0	119.8	133.9	6.4	59.5	41.5	39.3	48.1	7.6
TAT GPY01 091 69 30	683.0 (12)	209.6	103.7	12.5	158.4	442.4	119.1	129.3	6.7	65.8	45.9	49.5	51.7	8.3
TKT GDO01 142 84 30	238.0 (8)	—	—	—	172.8	455.3	119.0	123.1	5.9	49.9	36.5	31.1	46.6	8.7
TGT GDO02 142 84 30	46.0 (6)	—	—	—	161.9	463.4	122.5	127.1	6.4	48.7	35.5	37.0	43.5	6.9
TON GTG 101 70 30	515.0 (11)	174.7	90.9	15.0	150.6	399.5	115.5	120.9	6.3	62.5	40.7	43.4	46.7	7.6
VTT GVT 111 70 30	535.0 (10)	211.4	98.2	11.9	135.5	398.6	112.8	121.9	6.1	58.9	39.7	38.2	48.6	9.0
WAT GOA 081 68 30	718.0 (12)	180.9	92.3	8.5	139.2	446.1	122.3	136.5	7.3	62.8	41.7	40.8	46.0	8.2
WAT GOA 091 69 30	627.0 (12)	177.3	90.8	11.7	147.0	440.1	119.5	136.8	7.3	60.9	42.1	42.8	45.5	8.9
WAT GOA 101 70 30	570.0 (11)	168.2	88.5	10.4	149.6	440.8	118.7	139.0	7.2	59.4	40.2	39.8	46.2	8.0
WAT GOA 102 74 29	484.0 (11)	163.4	85.6	10.4	143.9	437.9	119.8	131.4	6.8	58.3	39.7	37.4	46.1	8.0
WAT GOA 102 75 28	606.0 (12)	162.5	80.5	13.5	138.6	401.8	114.7	122.3	6.6	61.0	38.0	35.8	43.4	8.7
WAT GOA 111 70 30	647.0 (10)	163.8	89.5	10.9	141.5	435.7	121.0	133.3	6.9	58.7	39.1	38.4	45.7	8.3
WAT GOA 111 72 30	447.0 (10)	170.9	86.8	12.5	141.4	430.5	118.9	131.4	6.8	61.5	39.1	38.6	47.0	8.4
WAT GOA 112 72 30	418.0 (10)	173.1	85.3	12.1	142.5	441.8	119.4	133.8	7.0	60.6	40.1	38.3	46.4	8.7
WAT GOA 142 84 30	256.0 (8)	—	—	—	139.2	412.7	116.6	125.6	6.3	63.7	36.6	34.1	43.4	7.9
WAT GOA M63 80 30	462.0 (10)	140.1	84..8	12.1	141.1	423.9	116.8	127.0	7.0	61.6	45.0	39.9	46.9	8.4
WAT GOA M63 82 30	378.0 (8)	147.0	83.1	11.2	148.2	424.4	116.3	126.0	6.6	58.9	39.5	37.6	45.1	8.2
WAT03 GOA03 M63 82 30	498.0 (10)	143.0	80.9	11.4	143.1	420.2	117.9	124.3	6.5	56.8	37.6	34.3	43.2	8.3
WAT04 GOA04 M63 82 30	531.0 (10)	147.5	83.5	11.2	145.0	418.6	118.3	125.1	6.6	59.4	39.1	35.1	44.7	8.5
WAT06 GOA06 M63 82 29	526.0 (10)	152.4	84.1	12.7	154.6	413.9	117.4	125.4	6.7	64.0	40.9	38.3	44.6	8.8

Measures of stems, leaves and inflorescences: Variation coefficients of Tall ecotypes.

International and French codes, Plot, Year of planting and number of trees Observed	High (cm) and date of measure	C20 (cm)	C150 (cm)	CF	LP (cm)	LR (cm)	NBF	LF (cm)	LGF (cm)	ILP (cm)	ILA (cm)	NBE	ILE (cm)	ILPRF (cm)
ADO GND03 142 82 30	15.6 (8)	9.1	10.7	19.6	8.2	6.6	4.6	11.6	8.1	9.9	13.2	12.2	10.2	13.4
AGT GND02 M63 82 30	14.4 (10)	12.5	6.7	22.5	7.8	5.6	5.3	5.2	6.7	12.3	11.6	12.2	8.5	16.7
BAY GPH04 142 82 30	16.7 (8)	9.6	5.8	15.4	6.9	8.3	4.7	8.3	6.2	13.1	8.8	11.5	12.5	16.8
CKT GCA M63 82 30	14.8 (10)	7.7	5.4	16.3	6.0	5.8	3.6	6.4	5.1	7.5	8.6	11.9	11.1	14.1
CMT GCO 111 72 30	21.0 (10)	9.9	11.6	21.9	9.7	7.1	4.9	7.3	8.0	10.6	15.2	15.0	1.3	15.7
GPT GNG04 142 84 30	23.7 (8)	—	—	—	6.5	6.5	5.9	6.3	6.3	17.4	16.5	12.1	12.8	14.7
KAT07 GCB07 101 70 30	14.0 (12)	11.0	8.1	13.8	6.9	5.4	5.3	7.4	9.0	11.6	12.6	16.2	9.5	16.9
KAT08 GCB08 101 70 30	11.0 (12)	14.1	8.9	14.5	6.9	7.0	5.4	6.5	8.2	13.4	11.4	13.0	7.4	12.8
KKT GNG01 102 75 30	27.0 (10)	10.0	9.5	18.7	4.6	6.5	3.9	8.3	7.2	8.8	12.6	12.3	9.6	15.4
KKT GNG01 142 84 30	21.4 (8)	—	—	—	10.1	6.8	5.9	5.7	7.6	11.3	15.5	13.3	14.0	16.9
KPD GND05 142 82 30	18.5 (8)	10.2	6.6	24.3	8.3	6.4	4.5	8.1	9.3	9.9	11.5	10.5	9.5	18.0
LCT GND08 101 78 30	17.0 (9)	8.4	9.8	12.7	7.6	5.6	4.6	6.0	8.2	12.6	10.0	11.9	13.6	13.3
LMT GND07 101 79 30	16.9 (10)	8.6	10.1	22.3	8.2	5.6	4.9	7.7	7.3	12.5	10.4	18.1	9.0	12.9
MLT GML M63 82 30	15.1 (10)	7.4	6.7	14.4	7.4	6.8	4.9	6.0	5.7	10.6	9.9	14.1	8.9	13.3
MLT GML S22 56 30	10.1 (17)	12.8	8.2	14.5	8.5	6.1	8.0	9.3	10.3	18.7	14.7	14.7	—	—
MVT GNG03 142 84 30	27.2 (8)	—	—	—	7.4	4.4	4.9	6.9	5.5	17.2	15.4	18.2	7.4	14.9
MZT GMZ 101 70 30	—	10.2	10.1	22.9	8.1	6.1	5.7	11.4	13.0	15.9	27.5	23.2	—	—
MZT GMZ M63 81 30	13.3 (12)	8.7	8.4	13.4	8.3	6.1	4.6	6.5	6.4	12.0	11.8	16.2	11.6	15.6
PNT01 GPA01 M63 80 30	13.4 (10)	8.6	6.2	16.6	6.0	4.9	4.4	4.6	5.9	9.1	9.3	12.1	8.6	9.0
PNT02 GPA02 M63 80 30	14.7 (10)	7.7	5.9	14.8	4.9	3.8	3.2	5.6	5.6	14.3	11.4	10.1	11.0	17.5
PLT GDO03 142 84 30	23.5 (8)	—	—	—	6.7	7.5	4.7	7.9	7.7	12.3	18.6	18.1	11.1	13.5
RGT GPY02 091 69 30	14.0 (12)	13.4	12.6	21.2	8.5	6.2	5.9	8.2	9.8	14.7	14.8	18.6	3.9	13.1
RIT GRL 060 68 30	—	18.0	8.0	16.0	7.0	6.0	6.0	6.0	11.0	10.0	9.0	5.0	—	—
RIT GRL 081 68 30	12.6 (12)	12.6	8.1	11.6	5.5	3.7	4.5	5.3	4.9	8.1	10.0	12.5	7.3	12.6
RTM GRT 101 70 30	18.0 (11)	12.3	6.9	16.3	7.0	6.1	5.0	7.2	7.7	10.1	10.7	17.0	9.6	12.3
SIT GSN 081 68 30	9.0 (12)	10.0	8.0	9.6	6.3	4.7	4.1	5.2	8.2	10.9	9.1	13.0	1.5	14.5
SLT GSL 112 72 30	19.0 (12)	9.1	12.0	18.0	8.0	6.5	3.8	8.4	4.8	11.3	10.1	11.3	2.5	13.4
TAG GTN 102 74 30	15.0 (12)	6.5	7.1	9.6	7.1	5.5	4.2	6.7	7.3	11.2	13.0	14.1	8.7	14.3
TAT GPY01 091 69 30	15.0 (12)	9.2	8.6	15.5	7.3	8.2	5.9	7.9	8.5	13.8	12.6	10.6	1.5	12.4
TKT GDO01 142 84 30	25.1 (8)	—	—	—	6.4	6.3	4.9	5.8	6.0	23.0	23.3	18.7	11.2	21.2
TGT GDO02 142 84 30	46.7 (6)	—	—	—	5.7	7.4	6.5	7.7	6.8	23.9	16.1	13.4	14.0	23.3
TON GTG 101 70 30	14.0 (11)	8.6	7.6	15.6	7.0	4.9	5.3	6.0	7.1	10.2	10.0	9.7	11.0	15.4
VTT GVT 111 70 30	16.3 (10)	15.1	9.8	22.1	8.8	6.0	5.9	7.4	11.7	17.5	12.2	13.0	16.5	24.2
WAT GOA 081 68 30	13.0 (12)	14.2	6.8	10.6	5.8	4.9	3.2	4.8	5.5	7.2	11.3	10.6	1.3	14.2
WAT GOA 091 69 30	13.0 (12)	10.4	10.0	19.0	6.2	6.5	4.3	4.8	6.8	9.5	8.7	12.7	2.9	13.1
WAT GOA 101 70 30	12.0 (11)	11.1	8.3	10.9	6.0	4.9	3.6	6.1	7.5	11.4	13.1	10.0	8.4	14.1
WAT GOA 102 74 29	12.0 (11)	9.4	9.8	12.7	8.5	8.3	6.0	8.0	6.5	10.1	13.8	15.9	10.1	12.4
WAT GOA 102 75 28	15.0 (12)	7.4	8.0	16.4	5.6	7.2	4.7	7.9	7.6	11.0	11.8	9.6	17.9	12.3
WAT GOA 111 70 30	22.0 (10)	10.3	7.8	23.3	7.2	5.7	3.3	6.4	6.3	10.9	13.7	10.3	3.9	12.7
WAT GOA 111 72 30	22.0 (10)	12.0	9.7	29.0	9.8	6.8	5.1	9.0	8.7	11.4	12.9	11.5	2.9	14.1
WAT GOA 112 72 30	14.0 (10)	11.6	10.4	17.0	8.1	5.1	5.1	6.7	7.9	9.8	12.5	13.0	1.8	15.1
WAT GOA 142 84 30	22.8 (8)	—	—	—	7.4	5.6	5.8	5.5	7.1	9.2	11.8	11.1	9.2	18.0
WAT GOA M63 80 30	7.0 (10)	7.0	6.1	11.4	5.8	4.3	3.4	6.2	4.5	6.8	7.9	8.6	9.0	14.6
WAT GOA M63 82 30	15.9 (8)	7.1	5.9	12.5	6.0	5.2	4.8	5.9	4.5	10.8	10.6	11.3	8.3	13.7
WAT03 GOA03 M63 82 30	13.8 (10)	7.0	5.3	11.9	4.6	4.4	3.9	6.3	3.8	8.1	7.4	11.4	7.5	11.5
WAT04 GOA04 M63 82 30	11.4 (10)	8.5	7.2	14.0	5.6	4.6	4.0	5.2	4.0	8.3	8.4	10.4	6.6	11.2
WAT06 GOA06 M63 82 29	14.0 (10)	10.5	6.3	20.7	7.4	4.6	4.7	5.0	5.3	8.1	12.6	13.9	8.2	12.6
Mean	16.9	10.2	8.2	16.6	7.1	5.9	4.8	6.9	7.2	11.9	12.4	13.1	8.5	14.7

Measures of stems, leaves and inflorescences. Averages of Dwarf ecotypes.

International and French codes, Plot, Year of planting and number of trees Observed	High (cm) and date of measure	C20 (cm)	C150 (cm)	CF	LP (cm)	LR (cm)	NBF	LF (cm)	LGF (cm)	ILP (cm)	ILA (cm)	NBE	ILE (cm)	ILPRF (cm)
CAT NVP02 092 80 30	149.0 (9)	77.4	72.6	24.3	106.6	333.7	101.6	107.6	4.6	51.0	29.0	26.3	40.3	8.1
CRD NRC 132 82 30	146.0 (8)	59.0	64.3	21.6	102.8	278.7	90.9	89.0	5.3	49.5	27.4	24.4	39.6	8.4
CRD NRC S20 56 30	—	61.0	65.0	25.9	100.0	305.0	98.0	101.0	5.8	45.3	30.4	—	—	—
EGD NVE 092 78 30	118.0 (10)	60.1	66.3	28.0	89.0	256.1	91.0	87.6	4.3	34.8	25.9	27.6	34.1	7.2
EGD NVE 092 80 31	77.0 (8)	62.2	67.7	29.4	91.8	270.4	93.4	90.2	4.3	35.9	24.9	28.8	32.6	7.2
EGD NVE 132 82 30	140.0 (8)	68.4	68.3	23.1	100.3	301.4	96.9	102.1	4.6	41.5	29.9	32.5	36.3	7.8
EGD NVE S20 58 30	—	68.0	69.0	26.7	103.0	329.0	102.0	124.0	4.8	38.7	30.4	—	—	—
GYD NJG 092 78 30	199.0 (10)	61.2	63.9	25.6	98.5	285.4	89.8	91.2	4.8	33.3	24.9	30.2	32.9	5.9
GYD NJG 132 82 30	201.0 (8)	70.4	69.3	18.5	106.4	308.4	90.5	99.9	5.1	38.5	27.7	33.8	34.2	6.3
GYD NJG 142 82 30	194.0 (8)	73.9	68.1	17.4	108.3	314.1	93.4	100.1	5.2	40.1	28.3	32.9	34.1	5.8
GYD NJG S20 55 30	—	78.0	70.0	19.7	101.0	342.0	94.0	115.0	5.1	35.9	30.3	—	—	—
KIN NVP06 142 82 30	239.0 (8)	130.1	87.7	12.6	132.5	412.2	116.4	120.0	6.0	58.2	35.7	36.6	41.5	7.2
MBD NBN 092 79 30	152.0 (10)	56.9	67.5	25.0	117.7	320.2	101.5	104.4	4.7	52.0	25.2	27.3	36.9	7.4
MRD NRM 132 82 30	200.0 (8)	77.0	69.3	16.4	114.0	337.7	94.7	110.8	4.8	40.1	31.4	30.9	36.7	7.1
MRD NRM S20 59 30	—	74.0	70.0	16.7	108.0	359.0	99.0	130.0	4.8	34.6	32.2	—	—	—
MYD NJM 112 72 30	208.0 (10)	67.0	66.0	21.0	108.0	305.0	93.0	99.0	4.9	38.0	30.0	34.0	37.0	—
MYD NJM 132 82 30	205.0 (8)	69.4	70.4	17.9	108.0	307.8	90.3	99.9	5.2	38.9	27.8	33.0	34.8	6.2
NLA NNL 092 78 31	120.0 (10)	118.7	93.4	25.3	105.0	287.1	116.0	95.9	7.0	49.3	35.2	50.3	35.4	5.9
NLA NNL 092 82 30	114.0 (10)	108.0	86.0	26.5	99.6	285.0	116.3	93.9	6.2	49.3	33.9	45.8	34.3	5.3
PGD NVS 092 78 31	156.0 (10)	51.5	60.2	30.0	95.7	273.2	95.4	91.5	4.5	37.9	25.2	25.7	33.2	6.9
PGD NVS 112 72 30	140.0 (9)	64.0	63.0	22.0	104.0	319.0	105.0	102.0	4.6	41.0	27.0	27.0	37.0	—
PIL NVP05 092 82 30	113.0 (9)	57.5	60.4	21.4	93.6	311.1	104.5	103.1	4.4	47.6	27.9	29.7	33.8	7.2
TAC NVP03 092 82 30	121.0 (10)	84.9	—	—	108.1	330.7	106.4	100.4	5.1	48.8	32.8	27.6	31.0	6.9
TBD NBO 132 85 30	54.0 (5)	61.5	75.0	18.9	120.4	318.2	100.2	106.8	4.7	52.7	20.1	18.9	38.2	7.7
THD NVT 092 78 30	182.0 (10)	63.5	59.1	25.2	104.7	295.1	86.2	93.8	4.7	39.0	29.9	33.2	33.9	6.3
TRD NRY 092 78 30	131.0 (10)	62.2	56.0	23.7	102.7	281.0	80.7	105.0	4.7	44.2	23.2	19.2	35.2	10.1

Measures of stems, leaves and inflorescences: variation coefficients of Dwarf ecotypes.

International and French codes, Plot, Year of planting and number of trees Observed	High (cm) and date of measure	C20 (cm)	C150 (cm)	CF	LP (cm)	LR (cm)	NBF	LF (cm)	LGF (cm)	ILP (cm)	ILA (cm)	NBE	ILE (cm)	ILPRF (cm)
CAT NVP02 092 80 30	15.0 (9)	9.0	3.2	11.9	5.5	3.8	2.3	4.0	3.4	5.4	6.9	7.8	5.9	7.7
CRD NRC 132 82 30	12.8 (8)	8.0	3.0	7.5	4.6	4.3	2.6	3.6	3.5	5.9	7.5	8.6	4.9	5.5
CRD NRC S20 56 30	—	6.4	5.7	12.0	5.3	5.6	2.5	8.2	9.8	11.2	15.4	—	—	—
EGD NVE 092 78 30	13.2 (10)	5.1	2.6	8.7	4.3	5.9	3.1	5.5	3.3	6.8	6.6	5.6	5.0	5.2
EGD NVE 092 80 31	17.2 (8)	3.9	2.1	8.5	4.8	6.0	3.2	4.2	3.2	6.9	6.6	7.2	6.4	9.3
EGD NVE 132 82 30	11.9 (8)	4.9	2.3	5.5	4.2	5.2	2.8	3.6	3.7	3.9	6.1	5.6	4.0	5.8
EGD NVE S20 58 30	—	3.8	3.4	11.3	10.2	4.3	2.6	4.7	4.8	11.4	9.7	—	—	—
GYD NJG 092 78 30	15.6 (10)	8.2	3.4	14.5	4.4	3.7	1.7	4.1	3.4	5.0	6.9	8.4	4.4	6.2
GYD NJG 132 82 30	9.0 (8)	5.5	4.2	6.6	3.0	4.2	3.4	3.4	4.2	5.3	4.7	5.9	4.5	6.2
GYD NJG 142 82 30	10.7 (8)	7.0	3.5	11.9	3.5	5.8	2.7	4.3	3.8	6.9	6.0	6.6	5.0	8.1
GYD NJG S20 55 30	—	6.2	3.2	8.5	8.2	4.2	3.9	9.5	7.5	15.5	15.6	—	—	—
KIN NVP06 142 82 30	15.7 (8)	22.7	7.8	26.5	8.6	5.3	4.4	5.6	5.6	9.3	8.3	12.1	15.2	14.5
MBD NBN 092 79 30	14.9 (10)	6.2	3.9	7.5	5.1	4.1	2.1	3.9	4.1	4.2	9.0	9.7	3.4	5.2
MRD NRM 132 82 30	12.9 (8)	6.1	2.6	6.9	3.4	4.0	2.4	3.7	3.9	4.2	5.1	5.2	5.8	6.6
MRD NRM S20 59 30	—	5.6	3.8	12.8	5.8	3.0	3.1	4.8	5.3	12.0	11.5	—	—	—
MYD NJM 112 72 30	16.0 (10)	10.0	6.0	13.0	8.0	6.0	4.0	8.0	5.0	15.0	10.0	9.0	8.0	—
MYD NJM 132 82 30	6.0 (8)	6.0	4.0	7.7	5.0	6.1	3.1	5.7	4.0	5.9	8.1	8.0	5.6	8.9
NLA NNL 092 78 31	23.0 (10)	18.0	5.4	17.6	6.1	7.0	6.0	7.3	8.5	7.3	10.7	15.4	8.1	9.6
NLA NNL 092 82 30	22.9 (10)	12.2	8.0	13.6	8.6	6.6	5.2	7.5	8.9	10.6	11.0	13.8	8.7	12.5
PGD NVS 092 78 31	8.3 (10)	8.0	3.2	9.0	3.3	2.8	2.2	3.1	4.4	4.2	4.9	5.9	7.1	7.0
PGD NVS 112 72 30	14.0 (9)	7.0	3.0	12.0	5.0	4.0	2.0	6.0	7.0	12.0	10.0	5.0	13.0	—
PIL NVP05 092 82 30	17.4 (9)	9.7	4.8	12.9	7.1	6.9	2.3	6.1	3.2	7.0	9.1	10.7	7.2	9.1
TAC NVP03 092 82 30	21.9 (10)	15.1	—	—	8.1	6.6	3.3	5.4	4.3	6.3	13.2	12.3	8.3	11.3
TBD NBO 132 85 30	41.1 (5)	7.8	—	15.8	8.2	8.4	5.4	7.8	6.0	12.1	11.2	11.4	10.2	10.7
THD NVT 092 78 30	9.7 (10)	7.3	3.8	6.8	3.5	3.6	2.2	2.4	2.6	5.7	4.9	5.2	3.4	5.2
TRD NRY 092 78 30	23.7 (10)	10.9	5.3	21.5	6.4	7.0	2.3	7.6	5.9	7.9	9.6	10.4	6.7	7.3
Mean(except atypical Dwarfs NLA and KIN)	15.3	7.3	3.7	10.6	5.5	5.0	2.8	5.2	4.6	7.9	8.6	7.8	6.3	7.4

9. Cultural Practices

J.G. Ohler

Plantation Lay-out

Before preparing the lay-out of a plantation, a decision should be taken on the type of coconut farming that will be adopted. In this consideration some important factors may play a role, such as:

- selection of farm type: monoculture or multiple cropping;
- selection of planting material: talls, hybrids or dwarfs;
- planting system: square, rectangular or triangular;
- crop combination: monocropping, mixed cropping, intercropping, or mixed farming;
- type of intercrops: annuals or perennials, and their degree of shade tolerance.

Water availability, soil texture and depth are other factors playing an important role in the decision on the planting system and crop combination. The decision on farm type may be the most important for the entire farming period, with long-lasting effects, as coconut palms have a very long productive life, especially if well managed.

The lay-out of a new plantation should be planned before clearing the land, in case wind-rowing of the debris is to be carried out. For this purpose, the land will have to be lined first. To reap maximum benefit from sunlight on flat land, lining should be carried out in a N-S and E-W direction. When planting in hedge systems the wide lanes should run in the E-W direction, the narrow lanes N-S. On slopes, lining may be carried out on the contour, permitting erosion control measures such as contour bunds, etc. After clearing and burning the debris, the land has to be staked before a drainage and a road system can be laid out. Stakes of about 1.5-2 m should be used to mark the planting sites. A wooden frame may be used to mark the size of the planting holes. This may be done with the stake in the middle or with the stake in one corner.

Drainage and road arrangements should be made before planting, so as not to interfere with the seedlings after planting. Drainage is necessary in low areas and in depressions without natural water outflow. On flat lands along the coast, drainage canals running towards the sea may have a gate on the sea side provided with a hanging door. When the tide is high, the seawater will push the door against the gate, closing the passage for the water; when the tide is low, the accumulating rainwater on the inside will push the door open, allowing the water to flow out. No general recommendations on drainage systems can be made as these depend very much on local climate, soil and topography conditions. The drainage and road systems must be compatible.

Clearing

Clearing Forest Land

Clearing forest land for new coconut plantations does not differ from that for any other crop. Felling and de-stumping will be necessary to permit planting and subsequent treatments. After wind-rowing and drying, the burning of the vegetable debris is almost inevitable because any other means of clearing the land will probably not be economic. The burning should be complete, so that no obstacles may prevent the following activities. Burning of vegetable debris will leave much valuable ash, containing important plant nutrients on the topsoil. Wherever possible, this ash should be incorporated into the soil, particularly on sloping land, before the first heavy rains of the rainy season flush it away. Where the debris will be burnt before planting coconuts, the windrows may be arranged on the future coconut lines, thus making use of the ash as a fertilizer for seedlings. This system may be used on poor soils where farmers have no means of applying chemical fertilizers to the planting hole. Some topsoil may be scraped into the planting hole.

Where quick establishment of a covercrop is desirable, the opposite may be done: placing the windrows between the future coconut lines. Leguminous covercrops react especially strongly to fertilizing with ash. Thus, the planted covercrops will grow out quickly and spread over the inter-row area. In this case, fertilizers will have to be mixed with the soil in the planting hole to give the seedlings a good start. On sloping land, clearing may be done in contour strips. The usual anti-erosion measures such as bunding and ridging must be taken. When heavy machinery is used for clearing and other operations, these operations should be carried out in the dry season so as to avoid damage to the structure of the topsoil. The year after the cleared strips have been planted and protected, the remaining land may be cleared and planted. The steepness of the slopes determine the system to be used. As soon as soil humidity permits, a fast-growing covercrop should be established on the land, for protection against erosion, for better water penetration, for uptake of soluble plant nutrients, and for soil temperature regulation. Leguminous covercrops which root on internodes are to be preferred, as they are very effective against erosion. Covercrops should be sown at a heavy sowing rate for quick cover establishment.

Clearing Savannah Land

Where shrubs and trees are scarce, they may be cut and removed by hand. On large areas, they may be removed by bulldozers or by two tractors dragging a heavy chain between them, sweeping away all woody vegetation. After removing bushes and trees, the land can be ploughed and harrowed. On smallholder plots, this may be done by hoeing. Some tropical grasses, such as *Imperata cylindrica* are very hard to eradicate, as they will sprout again from their underground stolons. Control of this grass will be dealt with in the section on weeding.

Clearing old Coconut Plantations.

Palms may be felled by cutting into the root mass from different sides. This can be done by axe or by the corner of a bulldozer blade, after which the tree can be pushed over. When the palm falls, the bole and the rest of the root-mass will emerge from the ground. The felled palm is left on the ground for several weeks and the crown is removed only when completely dried out, evaporation by the leaves helping to dry out the stem. Removal of the spear (a delicacy!) for consumption is worth waiting a little longer for the stem to dry out.

For cutting up coconut stems, saws and chains of special material are to be preferred, as the wood is very hard and blunts a normal saw in a relatively short time. This process is discussed in the chapter on wood processing. Where the wood is to be used as timber, its value may largely compensate for the cost of removal of stems. Where coconut wood has a value and can be processed, cutting down palms should be done in accordance with the processing capacity of the saw mill(s). Where the wood is not used as timber, the elimination of old coconut stems is not easy. The stems are too heavy to be swept in windrows, except by bulldozers. The wood is very moist and takes a long time to dry. Where the stems are to be left in the plantation, between the newly planted trees, they will have to be covered by a fast growing covercrop to prevent the breeding of rhinoceros beetles in the decaying wood. Other alternatives are either removal or burning the stems.

If the palms are to be burned, the stem can be cut into pieces of about 2 m long, which are split into halves. These halves can be stacked with the round sides upwards to facilitate run-off of rainwater. As soon as the wood has dried, dry leaves and coconut shell may be placed in the centre of the stack and lighted. Large numbers of stems can better be burned in a central place, burning many stems at a time. Large pits, about 4 x 4 m wide and 1.50 m deep are dug out by hand or tractor. About one pit is required per 4 ha. The stem portions are stacked near the pit. The bottom of the pit is covered with a 30 cm layer of dry palm leaves, shells and husks. The shells, particularly give great heat, facilitating the initial burning of the stems. The stem pieces are stacked in the pit to a height of at least 1 m above ground level, and the mass is set on fire. Stem pieces are thrown into the fire at regular intervals. In this way, about 2 ha of coconut palms can be burned per day, the fire being maintained for two days. It requires a team of six to carry out this operation. As the boles do not burn properly, they may be left in the lane and be destroyed when they start decomposing. Sometimes, the boles are placed on top of the stacked stem parts, but they do not burn easily. Such heavy work can only be done by means of a drag line, using a pair of steel grips instead of a grab bucket.

Where rhinoceros beetles are no problem, the palms may be poisoned by introducing an organic arsenic product, monosodic methylarsonate (MSMA) through a hole in the stem. The chemical is used in a solution of at least 500 g per l. This chemical is poisonous and the application must be done carefully. It has a toxicity of around 1.8 g per kg (oral test). The hole, with a diameter of 2 cm, is made as high as possible above the bole, at least 20 cm deep, slanting down at an angle of 45° (Biberson and Duhamel 1987). Immediately after drilling the hole, 50 ml of the solution is poured in without spilling. The hole must be plugged immediately afterwards. Plugging can be done with clay or pegs made from the rachis of fallen leaves. A mixture of MSMA and paraquat can also be used, at a rate of 85 cc with a 1:1 dilution with water. In Jamaica, 'Coopers Cattle Dip' which is used to control ticks on cattle and other livestock, has also been used for poisoning coconut palms.

The best time for poisoning is in the dry season, when palms absorb the liquid more rapidly. Usually the drying out of the lower leaves starts within two days after application, and the entire crown dries out within a few weeks. After another two weeks, the crown may topple over. The stem remains and rots gradually. This method is very economic compared to cutting and burning, but if rhinoceros beetles breed in the stems, they may damage the new plantation heavily.

The Nursery

The Lay-out

The lay-out and organization of a nursery depends very much on its size. Government nurseries servicing a coconut-growing region need access roads for lorries and/or tractors and trailers. Nurseries should preferably be located near the main road. They may have modern sprinkler systems and a specialized labour force. A small nursery of a private plantation may have only a few beds and transportation of seedlings may be done manually or by using an animal-drawn cart. Water availability is of paramount importance. The best available land should be selected for the nursery. The soils should preferably be light and seedlings should be watered when rainfall is inadequate. The soil should be properly worked to a depth of about 40-50 cm and well manured with organic as well as with inorganic manure. The organic material improves the soil structure and water-holding capacity, but it may also attract termites. In that case, dusting or spraying of insecticides will be necessary. Nurseries should be fenced to keep out stray animals. Where seedlings are to be planted in polybags in a continuous production, soil must be available in sufficient quantity to fill the bags.

With a spacing of the seedlings at 60 cm distance in a triangular planting system, about 32,000 seedlings can be placed per hectare of planted land. The use of about one-third of the area for access roads, service areas, etc., leaves room for about 20,000 seedlings per ha. The number of nuts to be planted in the nursery depends very much on the planting material and the expected percentage of seedlings to be culled out. The number of seedlings to be produced must also allow for filling in where some planted seedlings fail in the field, which may happen due to careless handling, drought, termites or invading grazing animals.

Seednuts

In the nursery, the basis is prepared for about 60 years of production, hence it is of major importance to use the best available planting material and to manage the seedlings as well as possible. Wherever possible, seeds obtained through artificial pollination or from a special seedgarden consisting of selected trees only should be used. The average improvement that may be obtained from the planting of seednuts from open - pollinated selected mother-trees is small. For seednut selection see also the chapter on Selection and Breeding by Dr. Bourdeix (Chapter 8).

Seednuts from the oldest bunch germinate and sprout better than those from younger bunches or fallen nuts. It is recommended not to use seednuts younger than those from the second oldest bunch. The time required for germination and sprouting is determined by genetic factors, but it is also slightly influenced by the season. Earliness of germination may differ during the same months at the same latitude in different countries. Differences occur in seedling growth due to differences in size of seednuts within the same cultivar, but disappear within a period of about six months. Apparently, there is a depression of the net assimilation rate in seedlings from large nuts, so that the total dry matter supplied by assimilation as well as by the endosperm is equal for seedlings from larger and smaller nuts. Under conditions of water-stress and reduced leaf assimilation, the rate of transfer of dry matter from the endosperm to the plant has greater influence, and under such conditions nut size might influence growth directly (Foale 1968b). Bourdeix, in the chapter on Selection and Breeding, points to the fact that alternate bearing coconut trees in years of high yields may produce smaller nuts with little water which may have germination problems if not well tended.

Storage of seednuts 'to break the dormancy period' and economize on seedbed expenses is a common practice (one m^3 of nuts contains approximately 250 nuts, depending on ecotype). However, coconut seednuts do not undergo a dormancy period (Wuidart 1981a) and the differences between sprouting times is a result of differences in germination speed. Stored nuts kept in moist conditions may sprout earlier after planting than nuts planted directly, but the total time between harvesting and sprouting might be almost the same. Storing nuts reduces the occupation of the germination bed. According to Wuidart (1981a), storing periods vary with the germination speed of the variety or cross to be planted:

- rapid germination (dwarf type), 10 days
- average germination (hybrids and some talls), 15 days
- slow germination (tall type), 21 days.

In commercial seednut production, involving shipping seednuts over large distances, lack of dormancy is a problem. Non-chlorophyllous seedlings may break or become dehydrated before they are placed in a seedbed. Saint *et al.* (1989) found that a temperature of 15°C slowed down the germination process without affecting germination capacity. They also found that it is possible, within certain limits, to delay germination simply by packing the seednuts in sealed opaque bags, made from a compound of 50 g/m^2 Kraft paper + 12 micrometer aluminum sheet + 80 g/m^2 polyethylene. The permeability characteristics of this material is under 2 cm^3 per m^2 per 24 hours for oxygen. Germination inhibition can be maintained throughout storage for periods of up to 8 months. The seednut maturity stage affects this phenomenon significantly. Seednuts harvested when still coloured never germinated whereas there were a few cases of germination in the dry nuts. For dry nuts, the germination percentage appeared to be unconnected with storage time. This most probably resulted from the advanced degree of maturity of certain nuts whose germination had entered an irreversible, but invisible phase when packed. None the less, seedling development inside the bag was slowed down considerably, since after 8 months it did not exceed 5 cm, a length which is reached within a few days in a normal atmosphere. The nut maturity stage and length of storage prove to be of utmost importance in this type of operation. A longer time in sealed packages led to a drop in germination capacity for both types of seed, but the drop was sooner and steeper for dry nuts. Storage within the range of 2-7 months did not seem to affect germination speed of coloured nuts, and germs appear closer together in time. Nuts harvested when dry had a tendency to germinate slower with time. Humidity on which fungi develop is seen on the nuts and the inner wall of the sealed bag when opened. Trials showed that ditane at a rate of 27 g of commercial product per bag of 30 nuts gave 70% of fungus control. Soaking treatments were ineffective.

Seednuts stored for 6 weeks in normal, aerated synthetic bags usually used for nut transport germinated within two months after having been packed in sealed bags. The germination percentage increased substantially during storage, levelling out at 5-6 months. After this period, conditions suitable for transfer to the seedbed do not significantly improve the final germination rate. Control batches sown after harvesting revealed that after 4 months in storage, dry nuts germinated only up to about 35% of their original capacity, on average, whereas less mature nuts reached satisfactory levels of 80%.

When storage did not exceed 4 months, the performance of the control appeared perfectly satisfactory. The proportion of healthy plants was close to, or better than, that of sealed packages, provided the seednuts were handled carefully during storage, which is not always the case during shipment. Beyond 4 months of storage, drying out of seedlings led to losses of 40-90%, especially in batches of nuts harvested when dry. The germination percentage of nuts stored in a room decreases with time. Remison and Mgbeze (1988) in Nigeria observed that germination decreased with storage from 55.6% for non-stored nuts to 8.3% for nuts stored for three months at ambient temperature in the laboratory. Nuts planted without previous storage sprouted from 12 weeks after harvest onwards, and nuts stored for one month had rapid sprouting from 14 weeks onwards. In contrast, storage for 2-3 months reduced sprouting and most of the nuts stored for 3 months lost their viability. It goes without saying that temperatures, and humidity in the storage environment may have a considerable influence on germination and sprouting percentages within a certain period. The subsequent vegetative growth of the seedlings was similarly affected by storage. Seedlings from unstored nuts had relatively bigger seedlings, as reflected in plant height, girth and number of leaves, while those from stored nuts had a comparative disadvantage in size.

Foale (1991d) harvested coconuts at an early stage of maturity and dehusked them within one week. From earlier trials he reported that there was no loss or damage to seedlings grown from dehusked nuts, provided they were adequately protected. In this trial, the dehusked nuts were exposed in sheltered, dry storage for nil, two, or four weeks, and then germinated in a vapour-saturated atmosphere. He found that earliness of germination was improved in proportion to the duration of the exposure. Setting the time scale at '0' on the day dehusking was performed, two weeks prior exposure raised germination in the first week to 30%, which was the level unexposed nuts reached after 3 weeks. Two more weeks exposure raised germination to 60% in the first week. However, by week 5 all treatments had reached about the same level. Exposure for longer than 4 weeks was harmful, probably due to the drying out of the embryo through the germpore cover. Complete dehusking of nuts permits the observation of germpore opening, which may be important in seednut selection. There is such variation between coconuts in husk thickness and toughness due to variation in water content, that true germination performance is masked by sprout emergence when no dehusking is performed. Also, the true growth performance of the sprout is masked. The less homogeneous the planting material, the more important such practice may be. Also, Kumar and Pillai (1990) observed in a field trial that dehusking of seednuts was advantageous for early germination and that it had no adverse effects on seedling vigour and the production of quality seedlings. This may be true for very well managed nurseries and plantations where seedlings are adequately manured. It may be useful in selection trials or when nuts have to be transported, reducing their volume and weight. But it cannot be recommended for smallholder practice, as it increases the risks of damage during handling and of drying out, and it increases labour costs. In addition, the husk functions as an important source of minerals. Under conditions of poor soil and low fertilizer use, dehusking of seednuts might affect seedling development.

Germination Bed, Seed Bed or Pre-nursery Bed

Germination beds are used to obtain rapid, grouped germination and healthy, well formed sprouts (Wuidart 1981a). Germination beds should be as close to the nursery beds as possible to reduce transport and the risk of drying out during transplanting. Placing nuts in a germination bed before planting out in the nursery has several advantages. In the germination bed, nuts can be planted much closer than in the nursery bed, saving space, water and labour. Before placing the nuts in the germination bed, the soil should be weeded thouroughly. The soil should be friable and well drained to facilitate the lifting of the seedlings for transplantation to the nursery, and to reduce the risk of termite attack. Where termites or other insects pose a problem, they should be controlled chemically. Spacing could be 5 cm between nuts in the row and about 20 cm between rows, using a density of about 16 seednuts per m^2. Seedling selection in the germination bed, based on early sprouting may considerably reduce the number of seedlings to be moved to the nursery beds. However, this difference will decrease with increasing homogeneity of the planting material.

The period between planting and sprouting depends very much on the variety and may range from about 4 to 6 months, the Niu vai type nuts and dwarf nuts sprouting earlier than Niu kafa type nuts. Early germinating and sprouting of seednuts are related to early bearing and high-yielding. According to Wickramaratne *et al.* (1987), seednuts from selected palms should not be selected or rejected on the basis of size, quantity of nut water, or shape, as it has been found that none of the variables measured bear a strong enough relationship with germination to use them as a basis for more efficient seednut selection. Only immature, empty, exceptionally small or outsized nuts should be rejected, and all others used as seednuts. Water content may not be a criterion when it depends on nut size, but when the water fills only a small part of the nut cavity, the haustorium may fail to make contact and germination may be hampered.

From the results of a trial relating sprouting time to yield performance, carried out in India, Satyabalan (1990) concluded that all early germinating seedlings need not necessarily turn out to be high yielders, and that early germination cannot be considered an important criterion for selection of seedlings of the tall coconut. However, in this trial, nuts harvested over a period of 5 months were all planted on the same date, leaving out the possibility of germination during storage of the early harvested nuts. Not surprisingly, the first harvested nuts showed a very high sprouting percentage after 1-2 months, this percentage gradually decreasing in the batches of nuts harvested in the subsequent months.

Although irrigation of the germination bed should not be excessive, it should be enough to keep the husks and the soil moist. A rather common practice to improve sprouting is the slicing off of a part of the husk ridge positioned on top at planting, at the stalk end of the nut above the germination eye. Removal of the exocarp facilitates penetration of water through the husk and the moisture will reach the sprout earlier, providing optimum moisture conditions in the husk for germination and sprouting. However, where a gain of a few days is not very important and where rainfall is regular and irrigation of the seedbed is adequate, the husks will remain moist and soft, and slicing the nuts is not necessary; it only increases the cost of seedling production. But this practice will be all the more beneficial when the husk is dry and tough and where rainfall conditions are inadequate. However, slicing the husk will also cause rapid drying out if the nut is not watered regularly. Therefore, such practices should be recommended only for well-managed nurseries. The labour required for the slicing is about one man-day per 1000 seednuts (Peries 1984). Sometimes a slice of husk is also removed on the opposite side of the nut, permitting easier penetration of the roots through the husk into the soil. Borah (1991) observed that in a nursery under rain-fed conditions, partial removal of the husk recorded the maximum height, girth and number of leaves, followed by complete removal of the husk and control. He suggested that the reason was a faster and maximum absorption of water, thus permitting the nuts to germinate earlier. It should be kept in mind that early visibility of the sprout is not caused by earlier germination alone, but also by earlier emergence due to a thinner husk layer it has to penetrate. However, if the nuts in the germination bed are watered regularly, the husk around the eye will be moist and soft, and the husk at the underside of the nut will be soft and will not prevent roots from penetrating into the soil. When seednuts have been stored for some time, the slicing of the husk involves the risk of damaging the sprouts of early-germinating nuts. The economy of husk slicing in adequately watered nurseries may be questioned, more so as the decomposing husk functions as an organic manure for the seedling.

Germination beds should be narrow enough to permit easy reach of the nuts in the middle position when watering, or when the nuts are to be lifted out for transplanting. In places where good drainage poses a problem, beds should be raised with additional soil to avoid waterlogging conditions. The number of rows in a bed depends on the size of the nuts, but about eight should be the maximum for dwarf nuts and 5-6 for talls. The beds should not be too long either, to permit reasonable movement from one bed to another. The nuts are placed side by side in furrows about 15 cm deep, so that when filling the furrow again, about one-third of the nut remains uncovered. The paths between beds should be wide enough to permit watering and free movement. Usually they are made 50-60 cm wide. Where heavy rainfall occurs, part of the soil to cover the nuts may be taken from the path, facilitating drainage of rainwater, provided there is an outflow for the water.

Laying the nuts with the soft eye towards the upperside might help the sprout to emerge faster, as it would be closer to the surface. Placing the nuts with the soft eye closer to the lower surface may ensure that the soft eye, and therefore the embryo, remains in contact with the nut water for a longer period, which might enhance germination. Since there is no method to determine the position of the soft eye of the seednut, it is not possible to position it in the nursery such that early sprouting is facilitated. Practically, the best planting position of the nuts is horizontal, irrespective of which side is lowermost. This will increase the speed of sprouting and reduce nursery costs considerably (Wickramaratne and Padmasiri 1986). Although sprouting of the nuts may be somewhat later if the germinating eye is in a low position, this has no relation with germinating earliness. The time difference between the emergence of sprouts from nuts placed to the germinating eye in a lower or higher position may be negligible. As the roots grow rather quickly, the position of the soft eye in relation to contact with the soil will be of only minor importance, provided the nuts are adequately watered.

In some countries, nuts are planted in a vertical position, permitting the roots to grow through the husk for a longer time, an advantage at transplanting. But this can be done only with nuts that have enough water to fill the entire or almost the entire cavity, for the haustorium not to lose its contact with the nut water. Results of a large-scale trial conducted by Wuidart and Nucé de Lamothe (1981) showed that vertically placed nuts (germpore on top) germinated later than nuts placed horizontally (both with the husk sliced at the top position). The differences at 50% germination were 14 days for MYD; for WAT the difference was 5 days at 50% germination, but 22 days at 80% germination. In the case of RLT, the difference was very small. The results confirmed earlier observations that MYD germinates very rapidly, WAT slowly. MYD, RLT and WAT take 51, 81, and 126 days respectively to attain 50% germination. It was concluded that late germination was influenced by two factors:

- husk moisture, increased by slicing the husk on top;
- the nut's water content and the distance separating the haustorium from this water. This factor varied with variety: a 3-4 weeks delay for fruits containing little water (WAT) and having oblong nuts resulting in a relatively large distance between the haustorium and the water; 1-2 weeks delay for MYD, the fruits of which contain more water and have a round nut; and a 1-week delay for fruits of RLT with a large volume of water, and nuts of which the equatorial diameter exceeds the polar diameter-resulting in a relatively short distance between the water and the haustorium.

The differences between varieties are thus determined mainly by the length of time the haustorium takes to reach the water in the nut's cavity. In this trial, horizontally placed nuts did not develop better than vertically placed nuts. It even seemed that the latter developed better, although they started off slower at the root level. It was concluded that the vertical position is at least as good as the horizontal position.

In a trial conducted by Remison and Mgbeze (1988) it was observed that horizontally placed nuts had a higher germination percentage than vertically placed nuts. In a trial conducted under rain-fed conditions, Borah (1991) observed that horizontal planting resulted in maximum height, girth and number of leaves when compared to vertical planting. He also observed that seednuts that floated vertically upright in water were more vigorous than those floating horizontally. The strong and stout seedlings from such nuts resulted from early germination and less exposure to drought. Possibly, the cavity of vertically floating nuts is better filled with water than those floating horizontally. The coconut water may be absorbed completely within a period of about 5 months and by this time about half the kernel may still be left in the nut, available to the developing seedling. After about 18 months, the endosperm is completely exhausted or decomposed.

Covering the nuts with grass or straw should not be recommended, as it may attract termites and hamper the observation of sprouting. It also lowers soil temperature, delaying sprouting. Only where water availability is not sufficient for irrigation may covering for protection against drying out be an advantage. Germination beds attacked by termites or other soil-borne insects may be treated with BHC or Endrin. When all the nuts in a seedbed have been harvested and planted at about the same time, seedling selection based on early germination is one of the most important measures to guarantee the use of the best available planting material. Early sprouting is related to high leaf production, early flowering and high-yielding capacity. Depending on the number of available nuts in the seedbeds and the nuts required for planting, the percentage to be selected for planting may differ. Where heterogeneous planting material is used, this percentage should not be much higher than 50%. Where homogeneous planting material is used, this percentage can be as high as 70%, depending on the quality of the planting material. The selection is done to guarantee the use of quite outstanding planting material. For commercial seedling production, a proper recording system should be used, containing all the required data of the nuts in each seedbed.

The nuts should be taken out of the germination bed to the nursery as soon as possible, to avoid the shock transplanting. In light soils, seedlings may be lifted out of the germination bed with an iron hook. In heavier soils, rooted seedlings may be lifted out with a shovel, to avoid damaging the sprout. Seedlings should never be lifted by the sprout as this may damage their attachment to the nut. The roots are clipped off close to the nut. The sprouted nuts are planted in the nursery without delay to prevent their drying out. When sprouting becomes visible, one or two roots may have emerged from the husk.

The Nursery Bed

In general, spacing of about 60 cm is used in the nursery. When the seedlings are planted too close together, they will become etiolated and lanky. Wide spacing produces vigorous seedlings, but requires a larger area of nursery beds, needing more weeding, irrigation, etc.

The lay-out of the nursery is adapted to the irrigation system used. Usually the beds are long and narrow. To facilitate access to the seedlings, these are sometimes planted in double rows, the spacing within the row about 15 cm and between rows about 45 cm. Seednuts in a single row should face the same direction. In the adjacent row they should be placed in the opposite direction and should be alternated in position. Shading of the nursery is not necessary, provided the seedlings do not suffer from water-stress. Mulching is often practised in the nurseries, as it prevents the topsoil from drying out, at the same time reducing weed growth. Fertilizing the nursery is essential, in spite of the fact that the seedlings receive a nutrient supply from the endosperm during the first few months after germination. As soon as the roots penetrate into the soil, an additional nutrient supply becomes available. Culling of abnormal, or slowly developing seedlings may start from the second leaf stage onwards. The girth at the collar and number of leaves are important selection criteria. This second selection will again increase the average quality of the planting material.

Polybags

Most modern nurseries use polybags for planting seednuts (CP 27). For smallholder nurseries, the use of polybags may be too expensive. The use of polybags involves several great disadvantages, such as:

- the large volume of good soil that has to be collected to fill the bags;
- the bulk which has to be transported from the nursery to the field. Where large numbers are involved, the availability of sufficient means of transportion is required. Smallholders could carry one or two seedlings on a bicycle, or ten to twenty on a oxen-drawn cart, if available;

The advantages of polybag use are:

- little or no root damage at transplanting and therefore little or no shock at transplanting;
- less restriction in time for transplanting;

- o less weeding;
- o homogeneous fertilizer doses per bag resulting in better seedling development;
- o better rooting in the field;
- o safer handling of seedlings;
- o easier culling of undesired types.

Bags of various sizes are used. The bags are made of 0.2 mm thick black polyethylene, resistant to ultraviolet rays, 20/100 mm thick. In Ivory Coast, bags of 40x40 cm are used. In Malaysia bags measuring 60 x 40 cm were preferred. Where seedlings are to remain for a relatively long time in the bags, large bags are more suitable. The lower half of the bags has 4-5 mm-wide holes, the lowest row at about 5 cm above the ground. The number of holes should be sufficient to drain the bags, but not so numerous as to involve the risk of the soil drying out. Important factors determining the choice of the bag size will be the availability of good soil to fill the bags and the cost of labour. The bags are two-thirds filled with the best available soil and placed on a levelled bed at a spacing of about 60 cm in a triangular planting system. A 40x40 cm bag filled in this way contains about 10 l of soil, weighing about 16-18 kg. Fertilizers and/or manure should be used wherever necessary, in quantities depending on soil fertility. Where no basic fertilizer doses have been applied, a mixture of urea, bicalcium phosphate, potassium chloride, and kieserite in a ratio of 1:2:2:1 may be applied monthly.

Monthly doses per plant are:

- o 1-2 months 15 g;
- o 3-5 months 30 g;
- o > 5 months 37.5 g.

A germinated nut is placed after in each bag, which it is filled up to about 1 cm below the rim, but never higher than the collar of the seedling. For polybags, placing the nuts in a vertical position has some advantages, especially for large fruits, the polar diameter of which exceeds that of the nursery bag:

- o the seedling is better centered in the bag;
- o carriage is better, reducing the risk of uprooting the plant during handling and transport, which is very important when the distance between the nursery and the field is large, or when the planting site is hard to reach and the bags have to be carried by hand;
- o the rooting of the plant is possible better after transplanting;
- o planting can be done at a suitable depth without burying the collar.

The bed is thoroughly weeded before the bags are placed in position. The bags are placed with the sprout always on the same side. After placing the bags, weeds between them are kept down. With the increasing size of the seedlings, spacing between bags is also increased. In a trial conducted in Nigeria, Iremiren (1986) found no difference in growth parameters between seedlings raised in polybags and seedlings raised in the soil. However, the small size of the bags, only 40 x 40 cm, may have been the reason why polybag seedlings did not show any advantage. Growth parameters were also similar at spacing treatments of 30 x 30, 45 x 45, 60 x 60 and 75 x 75 cm, except for height and leaf area index, which decreased significantly with wider spacings. Wuidart (1981b) recommends as a general rule the following spacings:

- o 0- 6 months 60x 60 cm;
- o 6- 9 months 80x 80 cm;
- o 9-12 months 100x100 cm.

The seedlings should be watered regularly, depending on climatic conditions. The polybags can also be placed on a sheet of black plastic, thus preventing weed growth and possible penetration of coconut roots into the soil. Where blast and dry rot are important nursery diseases, the plastic sheet impeding grass growth will also impede the development of the disease vectors in the nursery.

Regular watering is very important. Wuidart (1981b) suggests the following quantities of water every other day:

- 0-2 months 8 mm;
- 2-4 months 10 mm;
- 4-6 months 12 mm;
- > 6 months 15 mm.

From 6 months onwards, the requirement will be about 75 m^3 of water per day per ha so that the hourly discharge of water should be about 10 m^3 per ha. The seedlings should be inspected regularly for disease or pest attack, and when attacked they should be treated accordingly. Seedlings in polybags can stay in the nursery for longer compared to seedlings planted in the soil. Dwarfs develop slower than talls and can stay longer in the nursery. There is no general rule, but it can be accepted that a plant fit for planting will measure 1.20 m (from the nut to the youngest leaf unfurled in a normal position) and be 20 cm in girth (Wuidart 1981b). Older seedlings become 'pot bound', suffering from a check of growth. At transplanting, large seedlings will suffer more from shock and may need to be staked so as not to be blown over. On the eve of their transplantation to the field, the seedling should be watered abundantly. All handling of the seedling should be by the bag, not by the collar as they would be lifted out of the bag. If roots have grown through the bag, they should be cut before moving the bag (Wuidart 1981b). Only healthy, well developed seedlings should be planted in the field. Off types should be destroyed.

Planting

Planting Holes

Following the preparation of the drainage system and the roads, the planting holes are dug, preferably a few months before planting, except in sandy soils as the holes may cave in. The dimensions of these holes depend on soil conditions and the depth of the water table. Recommendations on size vary substantially and will also depend on farm type. On large estates where large numbers of seedlings will be planted, digging large holes may be uneconomic. In such cases, especially on light soils, the holes may be only slightly bigger than the polybag containing the seedling. In heavy soils, and especially in soils with a high content of lateritic gravel, large holes, subsequently filled with light soil are an advantage for palm development. In practical, this may be possible only on smallholdings. In Jamaica, experiments showed no advantage in digging holes any larger than to accommodate the seedling and a good supply of topsoil, but this depends very much on soil conditions. Coconut husks may be placed at the bottom of the planting hole. They contain a fair amount of nutrients and of water.

In India, coconuts are sometimes planted in holes as deep as 1.5 m, particularly in areas with a long dry season and in soils with a deep water-table. As the coconut stems grow, these holes are gradually filled with organic waste and soil. Care should be taken to keep the soil level below the growing point, otherwise the palm may be killed. Therefore, the planting hole should be wide enough to permit keeping the soil away from the seedling collar. In this way, the bole is nearer to the water-table where the soil will not dry out so quickly, and the rooting surface of the palm is much larger than that of palms planted at shallow depths. Ramanathan (1987) found that in the dry climate of Karnataka (India), planting at depths of 60 and 90 cm gave better results than at depths of 0 and 30 cm. The best results in earliness and cumulative yield were observed at a planting depth of 60 cm. The development and total number of roots were much better at this depth as compared to surface planting. Deep planting provides resistance to cyclones and induces earlier flowering and higher yields. Results of trials have shown that planting at 60 and 90 cm depths resulted in 10 months earlier bearing and increased annual yields by 10 and 7 nuts per palm respectively. It also increased the root numbers by 100% and 104% respectively, as compared to the control. Trewren (1991) reports that on Kiritimati (Christmass Island) it has been found that coconuts under conditions of low and irregular rainfall can be successfully established by digging planting holes down to the water-table (up to 2.5 m deep), refilling to 30 cm above the water-table with topsoil and planting 3-6 month old seedlings.

Usually, seedlings are transplanted to the field when they have about three or four leaves. Planting at a later stage increases the transplanting shock, with the exception of seedlings raised in polybags. Seedlings selected for planting should be taken out of the nursery bed with the use of a shovel, cutting all the roots before lifting it. All further handling should be done carefully to avoid damage. Seedlings should not be lifted by the sprout. This is particularly important for dwarf seedlings, as the sprout can become easily detached from the nut. Thampan (1981) reported that transplanting even after a month's delay after lifting out the seedlings may not adversely affect their performance after transplanting. But this may depend much on the relative humidity of the air. Normally, such risks should be avoided. As a rule, no more seedlings should be lifted than can be planted within one day. The lifted-out seedlings should be kept in the shade as much as possible until planted. Exceptions can be made when seedlings have to be transported over long distances. The transplanting age should be determined beforehand to allow effective planning of the planting schedule in the nursery and in the field.

The best time for planting seedlings in the field is at the beginning of the rainy season, after the rains have started to fall regularly. Seedlings planted late in the rainy season may not be able to develop a root system large enough to survive the dry season. Seedlings in polybags should be watered the day before transplanting, to keep the soil firm at transplanting. Handling of the seedling should be done by means of the plastic bag. Holding the seedling by the collar when attached to the bag heavy with wet soil is dangerous, as the plant may break away from the nut. Before placing the bag in the hole, the two lower corners of the bag are cut with a sharp knife and the bottom is slit. After planting the bag in the hole, the hole is partly filled with soil and the sides of the bag are cut and the bag is removed. Including the digging of the plant holes, one man can plant 60-80 seedlings per day. Having dug the holes earlier, one man should be able to plant about 150 seedlings per day. This, of course, depends very much on the size of the bag and the hole. Mulching around the young palm may keep the soil moist and may reduce weed growth. Where available, coconut husks could be used for this purpose.

Where termites may be expected to cause damage (CP 24), the planting hole should be treated with an appropriate insecticide (BHC, Endrin) before the seedling is planted. The planting hole may be filled with topsoil from the surrounding area, mixed with the recommended quantity of fertilizer or manure. Organic manuring is very beneficial to the seedlings, due to its great waterholding capacity and the micro-nutrients that will become available with its decomposition. However, in practice it will be difficult to collect sufficient organic manure, especially if large numbers of seedlings have to be planted. A useful organic product, often available, is coconut husks. Two layers of husks, concave side up, can be placed at the bottom of the hole. Weeding the area around the seedling is very important to avoid competition for water and smothering of seedlings by weeds. The circle to be kept free of weeds should be about 1.50 m in diameter. These circles should be widened with the development of the seedling. Optimal treatment of seedlings will pay dividends throughout the long life of the coconut palm. Adequate fertilizing and weeding and irrigation where needed may accelerate the onset of flowering by one or more years.

Planting coconut in peat soils

Increasing pressure on land often pushes coconut cultivation to less suitable areas. Peat soils pose very specific problems to coconut cultivation. In Southeast Asia, especially in Sumatra and Kalimantan, Indonesia, but also in Malaysia, large areas of coconut have been planted in such soils. Such lands can be used only if they can be drained and if they are at least 2 m above sea level, to avoid tidal flooding (Bourgoing 1989). The peat soil should preferably be no deeper than 1.50 m in order to provide anchorage for the coconut roots in the subsoil. However, coconut can also be planted on deeper peat soils. Ochs *et al.* (1992) describe the planting of coconut on peat soils of 4 m deep or more, where coconut roots cannot reach the clay-subsoil. Forest should be cleared at least two years, and shrubs one year, before planting. A drainage system should be established as soon as possible, at least 7 months before planting. Planting should not be done when heavy rainfall in combination with high tides may result in drenched or flooded soil conditions. Planting rows must run parallel to the drains. Collection drains are made about 6 months before planting, at a spacing of about 500 m. Secondary drains are made in every 4th inter-row, at least 2 weeks before planting (Bourgoing 1989). Ochs *et al.* (1992) used tertiary canals every 16th row, at a spacing of 112 m. Tertiary canals are dug before windrowing and compacting to prevent damage to the drenched soil by heavy machinery and to facilitate their movement. Planting is done in an equilateral triangle system, 8 m in the row and 7 m between rows. The drainage system aims at keeping the water-table sufficiently low to prevent asphyxiation of the coconut root system. A mean depth of 70 cm below the surface of the compacted peat was considered adequate, although the real depth, due to imperfections of the system varies between 40 and 100 cm (+ 30 cm). To keep the water level as stable as possible, primary canals are closed off with an overflow gate, to ensure a constant water level, independent of tides. The secondary canals are 4 m wide and 3 m deep and they are fitted with adjustable gates where they run into the primary canals, to act as barrages and sluice gates. Primary and secondary canals are used as access routes to the plantation, using small boats or water scooters. For plantation supervision, motorbikes are used, which can easily use the narrow, compacted tracks alongside the canals. In the tertiary canals, gates are constructed of planks propped up with peat, fixed at 70 cm, to keep water within the plots in case the water level in the secondary canals drops too much by accident or prolonged drought. Soil compacting is performed by a D6-type bulldozer dragging a roller behind it, about 3 m wide and 1.90 m in diameter, weighing 10 tonnes, which can be made heavier with water or concrete weights, if necessary. Its surface has 168 'sheep feet'. A 3 m wide strip is compacted along the planting rows, *i.e.* about 45% of the total plot area.

Before preparing the planting hole, about 1 m^2 is dug out to a depth of about 10 cm, also removing roots and other debris. Subsequently, the soil in this pre-hole is compacted with a wooden hammer to a depth of about 25 cm. A wooden planting implement is used, made of solid wood for complete rigidity, with a handle section about 1.30 m long terminating in a 30 cm diameter cylinder of solid wood tapering to a point. The cylinder section is as long as the height of the soil cylinder in the polybag. The planting hole has to be 7-8 cm wider than the soil cylinder of the polybag. The planting hole is widened by circular movements with the planting instrument, compacting the sides of the hole as well, and providing a more solid soil condition for the seedling. When the seedling is placed in the hole, the top of the seednut should be about 5 cm beneath the surface of the pre-hole. Seedlings older than 6-7 months should be planted somewhat deeper, about 1 cm for each additional month in the nursery. The planting hole is filled with soil mixed with about 350 g triple super-phosphate. This soil also has to be compacted (Bourgoing 1989). Ochs *et al.* (1992) found that planting at a depth leaving 15 cm of compacted peat above the level of the collar base provided the best growth. Of course, the collar is not to be covered with the peat soil.

Covercrops should be planted as soon as possible to suppress grasses and other weeds. Where flooding may occur, the creeping covercrops could be planted on small mounds in the inter-row. Creeping covercrops that develop rapidly on peat soils are *Pueraria javanica* and *Calopogonium mucunoides* (Bourgoing 1989). Ochs *et al.* (1992) observed that in their area, *P. javanica* did not appear to fix atmospheric nitrogen, the roots being virtually without nodules and the leaves having a pale green colour. Fertilizing of the covercrop with 500 kg per ha of a mixture of urea, copper sulphate and iron sulphate gave good results, but even then, the green manures needed regular upkeep to prevent domination by weeds. However, the maintenance of a creeping covercrop is too expensive. It is better to permit the development of the natural fern on the non-compacted inter-rows, keeping the coconut circle and a 2 m wide strip along the row clear of ferns to enable the workers to move easily around the plantation. Manual weeding is very costly. The herbicide round-up, applied twice a year at a rate of 2 l per ha (300 l of solution per ha) costs less than five annual slashing rounds. It is recommended to use this herbicide from the fourth year onwards.

Planting Density

Most coconuts in commercial plantings have been planted at spacings varying between 8 x 8 m to 9 x 9 m, either in a triangular, a rectangular or a square system. Dwarfs are planted at spacings of 6.5 x 6.5 m or 7 x 7 m. The planting density is based on the selection of the farming system to be adopted.

Where wide inter-rows are preferred, such as in mixed farming systems, single or double hedge planting systems have sometimes been used, in which the spacing within the row is narrow and between the rows wide. Planting on the triangle is done mostly in coconut monocropping with the objective of maximum copra yield per ha. Planting on the square or rectangular is sometimes used in monocropping, but mostly when coconut is intercropped or undergrazed, especially where cultivation has been mechanized. For intercropping, planting in avenues is even better, as it will provide space and light for intercrops throughout the entire coconut growing period. Planting density influences individual palm development and yield as well as yield, per hectare, maintenance and harvesting costs.

A rapid estimate of the planting density when planting in equilateral triangles can be made by using the formula presented by Duhamel (1987):

$$D = \frac{11547.}{d^2}$$

In this formula 'd' represents the distance between palms. The formula has been obtained from the calculation: density = $\frac{100}{d} \times \frac{100}{e}$; [e = $\frac{\sqrt{3}d}{d^2}$ = 0.866 d].

Influence of density on girth and leaf number

Generally, there is little influence of density on stem girth and number of leaves produced. Pau and Chan (1985) noted a slight reduction in the number of leaves produced at high densities, but the differences were not significant.

Influence on leaf length and leaf area

Pau and Chan (1985) reported a slight but insignificant increase in the leaf area of leaf 14 with increasing density. The trend of leaf growth suggested that petiole length is likely to show a greater etiolation effect than rachis length. At a density of 185 palms per ha, the drip circle of the palms was 5.11 m and the maximum overlap between leaves at eight years from field planting was about a quarter of the rachis length. The leaf pinnae per leaf and the width of the middle pinnae were similar in all density treatments. The mean number of leaves per palm was 30.

Influence on stem size

Stem size increases with planting density.

Influence on flowering and fruit set

The age of first flowering is not much affected by planting density within a certain range. Pau and Chan (1985) observed first spathe appearance in MAWA palms two years after field planting, or 32 months after transplanting from the nursery to the polybag. Six months later, 45% of the palms had started flowering and a further 3 months later, 90% of the palms had produced at least one inflorescence. There was no difference between treatments in age of first flowering.

Density trials in Jamaica (Anonymous 1990a) showed that spacing significantly affected the number of female flowers per inflorescence and the fruit set. The number of female flowers per inflorescence decreased with increasing density, whereas percentage fruit set increased with planting density.

Influence on nut and copra production

From trials conducted in various countries, an adverse effect of high density planting on nut production per palm was generally reported. However, due to the higher number of palms per ha, the total number of nuts produced per ha was higher at higher densities. Pau and Chan (1985) in Malaysia planted coconut at four different densities in equilateral triangle planting systems. Nut yields related to planting density in this trial are presented in Table 26.

Table 26: *Nut production per palm at four different planting densities*

palms per ha	(nuts per palm) age of palms (year)					total
	5th	6th	7th	8th	9th	
136	86	122	153	139	162	662
160	92	115	142	132	156	637
185	93	112	137	122	148	614
210	84	100	127	112	129	554

The decreasing nut production per palm with increasing planting density is clearly demonstrated by this table. However, the nut production per ha shows an opposite trend, increasing from 21,971 nuts for 136 palms per ha to 27,200 nuts for 210 palms per ha. Assuming a conversion rate of 6,000 nuts per ton of copra, copra yield would have increased from 3.53 t per ha to 4.53 t per ha, a difference of 1 t per ha. However, copra content per nut might have been affected at high planting densities, due to reduced total photosynthesis under conditions of reduced light availability for each palm. It is not excluded either, that with increasing age, the highest density would not continue to produce the highest yields.

It was calculated that for monocropping the optimum density would be about 180 palms per ha. Nucé de Lamothe (1990) reported that results of large-scale research programmes in The Philippines and Indonesia have shown that no significant differences were observed in production per hectare within a range of 115 to 180 trees per ha. By low density planting, a larger soil volume is available to each palm, which is one of the reasons for high yields per palm. Fewer palms per ha also means less production per ha of total biomass, and less fertilizer may be needed to increase nut production because less nutritious elements will be used for the production of biomass. Fewer palms to be climbed for harvesting also reduces the harvesting cost. But wider spacing favours weed growth, increasing weeding cost.

Optimum density for coconut monoculture differs for different ecological conditions. In Ivory Coast, optimum density for MAWA hybrid was found to be around 150 palms per ha, the limiting factor mainly being water deficit. Smith (1972) in Jamaica, concluded that the optimum spacings for talls, hybrids and dwarfs were 7.9, 6.7, and 5.5 m, corresponding with densities of 185, 257 and 381 palms per ha, respectively. He also observed that the curve representing the yields per ha with increasing density is flat topped, indicating that under the conditions in Jamaica there is a rather wide range where increasing densities produce almost equal yields. He concluded that where coconuts are grown as a monocrop, planting can be carried out at densities at the higher limit of this curve. The heavy shade would reduce weeding costs considerably. Losses of palms due to lightning, etc. would not reduce yields as long as the number of palms correspond with the flat top of the curve. High densities will also produce high initial yields. On the other hand, factors such as fungus disease may be a reason for adopting a spacing below optimum, especially in very humid areas.

Low density planting
As the yield curve for different planting densities is rather flat between densities of 100-180 palms per ha, a wider spacing is preferred where multiple cropping is planned. It will provide more land, light and water for intercrops. Planting in a square or rectangular system will facilitate mechanized cultivation and crop management. In climates with long dry seasons, spacing at 9 x 9 in a triangular system, or 10 x 10 in a square system will not give very different yields, but the latter may improve light conditions for the intercrop considerably. The selection of the coconut variety to be planted is also of importance when intercropping is considered. When high-yielding hybrids are preferred, tall x tall hybrids may be selected instead of dwarf x tall hybrids. Nuce de Lamothe *et al.* (1991) suggest that in such a situation the high-yielding WAT x RIT hybrid (PB 213) might be selected, which gives just as high yields per hectare when it is planted in 10 m triangles as when it is planted at 9 or 8 m spacings. The PB 213 is distinguished by more rapid growth and a greater number of leaves than the WAT. The length of the leaves would justify planting this material at lower densities. When a palm dies in a low-density planting system, gap-filling will be necessary.

Hedge row or double hedge row planting
The hedge row system is based on narrow spacing between palms in the row, and large spacing between rows (CP 65). The double hedge row system is based on alternation of a narrow spacing between two rows with a very wide spacing. All dead leaves and other organic debris can be collected within the row, and the inter-row be kept clean for cultivation. The narrower the spacing in the row, the more the palms will tend to bend outwards, reducing the effect of wide inter-row spacing. With a hedge system, intercropping is possible during the entire life-time of the palms, with a reduction of space when the palms are young, and low leaves may hamper cultivation of other crops close to the palm. The system to be chosen depends very much on the crop(s) planned for intercropping.

Replanting Old Coconut Plantations.
In most countries, not much land may be available to plant new coconut groves to increase national coconut production. On the other hand, the area of old and senile coconut plantations may be relatively large, and replanting of the old stand might increase coconut yields considerably, if done properly. When planning the substitution of an old coconut grove by new palms, the main question is whether or not to cut down all the old palms before planting the new seedlings, or to underplant with seedlings first, and subsequently remove the old palms. Actually, most smallholders will be reluctant to cut down palm trees still bearing a few fruit. They will wait until the palm is felled by lightning or by senility. Smallholders in Sri Lanka generally attach more importance to a continuous flow of income from the existing palm stock, although it is lower than the expected revenue from the new planting (de Silva, S. 1988). The decline in coconut production in Sri Lanka may be attributed to the reluctance of growers to remove senile palms whose useful life has long passed (Silva and Tisdell 1986). There may be no replanting system by which the farmers' income from coconut will not show a decline over a certain period. But the decline and the period should be reduced as much as possible. When the replanting is accompanied by a system of improved management, this is certainly possible. The immediate planting of intercrops between the young palm trees can help the farmer to overcome this problem.

Estates may prefer total clearing for organizational reasons and to avoid problems with late-removal of high trees between seedlings, although the gradual removal of old palms is more economic. Clearing and cleaning the plantation can all be completed in one season and the new trees will grow up in full sunlight, and all trees grow in equal conditions. Pai *et al.* (1984) compared three different replanting systems: Felling-Replanting (FeRe), Poisoning-Replanting (PoRe) and Replanting-Poisoning (RePo). In the FeRe system, stems were cut down with a chain saw three months after poisoning. As the coconut was intercropped with cocoa, the cut at the base of the stem was done at the appropriate position so that the stem fell down along the coconut row. Where necessary, ropes were used to guide the stem down. To facilitate replanting and movement in the field, the felled stems were cut into manageable pieces and stacked along the coconut row. The cost of this system was about two-and-a-half times that of other systems, due to the high labour cost for precise felling, cutting the stem into pieces and stacking. Labour requirement was three chain saw operators per ha, and 18 man-days for stacking. Another significant expenditure was the collection of rhino beetles to contain the damage to seedlings. Beetle infestation, favoured by the additional breeding sites in the decaying stems, posed a major problem.

In the PoRe system, palms were left to rot after poisoning, the debris of fallen leaves and nuts was stacked along the palm rows. Stems took a long time to decay, the lower half of 70% of the palms were still standing three years later, making grub collection difficult and increasing the beetle problem. In the RePo system, hybrid seedlings were planted along the coconut rows some six months before the old stand was poisoned, providing a grace period of eight months during which no beetle collection was necessary. The fallen crowns did not cause severe damage to the coconut seedlings. During the six months there was only little shading effect by the old palms. Also in this system the beetle problem was serious. With better beetle control system, such as the use of poisoned baits with attractants, combined with the use of the baculovirus, and overgrowing the stems on the soil with rapid growing covercrops, the beetle problem could have been reduced.

In the three described systems above, all old palms were removed within the period of one year. A great disadvantage of such systems, however, especially where no intercrop is growing, is the lack of income during a period of about 5-8 years which the farmer has to overcome, depending on the use of hybrid or tall planting material. Therefore, complete clearing is almost impossible for smallholders who cannot do without a regular income from their coconuts. Of course, when all palms are removed and new palms are planted, some intercrops may be grown, which could provide some income for a few years, provided there was a market for the products. But the income from these intercrops may not always be enough to compensate for the loss of the income from coconuts. The most difficult period is just before the first harvest from the young palms, because the overhanging leaves of the young palms will make intercropping impossible before the first coconut yields are produced.

Gradual removal of old palms can be done in blocks, which has the same (dis)advantages as total clearing, only on a smaller scale. If, for instance, each year one fifth of the plantation is replanted, as was the system in The Philippines (Pordesimo and Noble 1990), the farmer's income from coconuts would be very much reduced after three years of cutting down his palms. Moreover, within each block some reasonably producing palms may have been cut down, while in the remaining area some palms may grow that produce almost no nuts. Replanting systems that continue to provide some income for farmers during the transition period are more acceptable to smallholders. Underplanting is the most obvious way of rejuvenating smallholders' coconut groves. Yields per ha and per palm of improved planting material under conditions of adequate management will soon be higher than yields from old, senile palms, even if these young palms have not yet reached full productivity. Shortening the transition period by replanting with early-yielding hybrids could solve part of the problem.

When senile palms are to be substituted by new, high-yielding planting material, timing of the operation is the most important factor. Underplanting the old palms with young seedlings and thinning out only when these seedlings start producing, may give higher total yields when these young palms come into production. Such results were obtained by Romney (1987) in Tanzania in a climate with a severe dry season and a rainfall of 1100 mm. MAWA hybrids under full shade of 43 palms per ha produced 5 nuts per tree, 5 years after planting in the field, against 26 nuts per tree for similar seedlings growing in the open. However, during these 5 years the talls had yielded 30 nuts per palm per year. In such a system, intercropping will be impossible. In this case, the original density of 43 palms per ha was very low. Underplanting in high density stands of old palms, stem diameter, and production capacity of the young palms, may be affected by growing in too much shade. After removal of the old palms these young palms might never reach the production capacity level of palms that have been grown in the open field.

Gradual thinning of the old stand is another alternative. Young coconuts can stand some shade during the first years of growth without affecting their quality. They can be planted all at the same time in the lanes between the rows of old palms. Where the old plantation has been planted irregularly, all old palms growing within 2 m distance of a planting hole should be eliminated. Thinning out of the old palms can be done either before planting the young palms or one year after planting. General recommendations cannot be put forward, as results will depend very much on the density of the old plantation and the condition of the palms. According to Darwis (1991), the best system is the felling of 50% of the old palms after the first year and the remainder after the third year. With this system, tried out in Indonesia, the development of replants was better than in the clean cutting system. The seedlings had a larger average girth, carried more female flowers and produced 48 nuts per palm against 42 nuts for replants in the clear cutting plot. The original density of the plantation was not mentioned. However, with this system, when planting hybrids the total old stand will have been removed one year before new palms start producing and farmers will face serious income reduction, as also intercropping at this stage is almost impossible.

Wherever possible, a system should be used involving overlapping of the first yields of the young palms with the last yields of the old palms, to achieve a continuous, although temporarily somewhat reduced, coconut production. When thinning, the least productive palms should be removed first. The old plantation can be selectively reduced to about 50-60% of the original stand within two years (depending on the original density), removal of the remainder of the old palms may be undertaken at a slower rate, so that farmers may still have some of their highest yielding old palms in production when the new palms start yielding. The remaining adult palms will benefit from better clearing of the undergrowth which has to be done for the planting of the young trees, and from less competition for light, and probably also for water and minerals. Where leguminous covercrops are to be planted, this will also be to the advantage of the remaining old palms. Individual yields of the remaining old palms may increase after reduction of their density, thus compensating for some part for their reduction in number.

Foale (1968) conducted a trial with four palm thinning rates in a plantation with an original density of 171 palms per ha. The last palms were felled in year 6. He observed that the yields of the young palms were reduced compared to those of palms growing in the open. However, from the ninth year onwards this difference was insignificant. The combined yield of copra from the old palms and the young palms amounted to 16,900 kg per ha compared to 8,349 kg per ha in the young plantations where all the old palms had been poisoned before planting the young seedlings. Using a simulation model, Prodesimo and Nobel (1990) came to the following conclusions: iIntercropping substantially increases the farm's profitability and significantly reduces the period over which net farm income declines to below subsistence level. Gradual thinning and replanting at a rate of 20% is not the best rate, whether intercropping is practised or not. Strip replanting is best if monoculture is being practised, but gradual thinning and replanting at a rate of 10% per year is best if intercropping is practised. Based on the yield data for Laguna (Philippines), the benefits of applying inorganic fertilizer did not outweigh application costs. As was stated, the latter conclusion has local value only and depends on a series of factors, such as soil fertility, crops planted, prices of basic materials and products, etc.

Planting in the lanes of the old plantation provides little choice for other inter-row spacings that might be required when other varieties or hybrids are to be planted or when certain intercropping systems are to be used. Proper planning is essential before starting. Such rejuvenation systems demand proper management and great care when cutting down palms to avoid damage to the young palms. Intercropping in such systems is more complicated. Removal of old palms requires considerable labour. Under conditions where family labour cannot be economically employed elsewhere, this should not be a major constraint. Under conditions of labour shortage and lack of funds for hiring labour, gradual thinning will be the most logical choice. Where the coconut wood has a market, it may be sold to dealers who might even do the removal against a certain deduction of the price. Income from wood might also provide funds for hired labour and the acquisition of proper equipment, such as chain saws, which in their turn would reduce labour cost.

Sometimes, soils of old coconut groves have been so exhausted by continuous exploitation without fertilizer use that a new plantation on the same land develops very badly. Exhaustion is sometimes so severe that fertility restoration by chemical fertilizers is almost impossible. Pomier *et al.* (1986) and Ochs *et al.* (1993) discuss such problems caused by overcropping areas with high population densities in Ivory Coast. Continuous cropping plus removal of the husks, leaves and other vegetative waste for fuel resulted in a drop in topsoil organic matter content from 3 per cent under the original forest to less than 0.5 per cent. Experimental attempts to replant failed, even with the use of high doses of fertilizer. Apparently the mineral exchange capacity in these coarse sands had been reduced so much by the reduction of its organic matter content that fertilizer applications were of little use. In particular, nitrogen deficiency was difficult to correct with mineral fertilizers. It was impossible to obtain over 1.5 tons of copra with hybrid material under conditions where cost-effectiveness was marginal. In any case the required investments were incompatible with the means of the smallholders involved.

Using conventional legume covercrops was also impossible, since they would have been unable to develop in such an environment. Only with the use of leguminous trees with deep and hardy root systems, such as *Acacia auriculiformis, A. mangium, Albizia falcata* and *Casuarina equisetifolia*, were improvements obtained. One out of every three coconut rows was replaced by a double row of leguminous trees. Dupuy and N'Guessan Kanga (1991) obtained comparable results with *Acacia crassicarpa*. Such trees, used as intercrops, restore fertility of the topsoil. In addition to enriching the soil with nitrogen and organic matter, these trees can also provide fodder for cattle and firewood for farmers' households or for sale on the local market. It must be kept in mind that on shallow soils, where leguminous trees and coconut would compete within the same soil layer, results might be quite different, especially in the long run.

Weeding

Weeds compete with palms for moisture, minerals and light. In mature plantations the main competition is for water. Minerals taken up by herbaceous weeds will return to the soil by decomposition of organic mass fallen to the ground. In some cases, slashing may have a depressive effect on coconut yields. This can be attributed to the adverse effects on palm nutrition, particularly that of nitrogen. Possibly, a large portion of the nitrogen being released through the decomposition and mineralization of the organic material was either taken up by the remaining vegetation or made unavailable to the palm by temporary fixation by soil micro-organisms, or lost by leaching. High weeds, even herbaceous weeds hamper cultural practices in the plantation. Often, under such conditions, fallen nuts are not found by their owner. They become visible after germinating, some months later.

The availability of water and minerals determines the effect of competition between weed and crop. In very humid areas, competition for water may be non-existent, but competition for minerals may be severe. On the other hand, a higher root density in the soil in a non-weeded condition would reduce soil mineral losses due to leaching. In such areas, selected undergrowth such as a leguminous covercrop is to be preferred to weeds. On the other hand, in areas with a marked dry season, weeding may be essential to obtain satisfactory coconut yields. In this case, weeding should be performed before the end of the rainy season, to avoid rapid soil water depletion and drying out of the soil early in the dry season. In Ivory Coast it was observed that once the nitrogen supply was ensured through fertilization, moisture stress was the major limiting factor. Regular slashing and clean weeding treatments in particular were superior to a water-demanding leguminous cover crop.

Direct weed control can be obtained either manually, mechanically or chemically. Intercropping is an indirect method of weed control. The method to be selected depends on local conditions, such as wages, availability of implements and materials, access to the land, farm size and management level, and cost involved. Although complete weed control may give better immediate response of coconuts than regular slashing, clean weeding involves the danger of erosion and also affects soil quality. Soil temperatures rise, the soil organic matter content decreases gradually, reducing microbiological life in the soil, altogether decreasing soil fertility.

Young seedlings transplanted into the field should be protected against overgrowing by weeds or covercrops. Ringweeding in a circle, widening with the age of the palm, is a common practice. The initial diameter of the weeded circles is usually 1-1.50 m, increasing to about 3 m in subsequent years. In very dry climates wide circles are to be preferred, to reduce water competition as much as possible. Weeding within a radius of 4 m is almost similar to clean weeding. Ringweeding has usually to be done several times a year. Saving on the care for the seedlings is bad economy, because the lower yields resulting from retarded flowering and less sturdy development of the young palm in many cases will not be compensated by savings on weeding and other cultivation practices. When weeding the circle with a hoe, no topsoil should be scraped off and thrown outside the circle, as is so often done. In addition to removing the topsoil, deepening of the circle may also affect the root system, particularly in soils with a high water-table. Weeds should be cut just under ground level and left to dry and function as a mulch. Weeding frequency should impede production of seeds which later would germinate within the circle. When the palms are still young, the inter-rows do not need intensive weeding, but if high weeds grow around the weeding circle, these should be slashed regularly in order not to shade the seedling. Woody shrubs and trees in the inter-rows, however, should be eradicated. They impede free movement in the plantation, and hamper the reaping of fallen nuts and control against theft.

On clay soils, the movement of machines in the plantation when the soil is still moist may compact the soil and affect yields. Practices such as disk-harrowing may damage the root system, especially where the roots grow very close to the surface, as is the case when the soil water-table is high. In Sri Lanka it was found that a simple tractor-drawn rotavator with steel blades revolving at very high speed, driven by the power take-off was the cheapest weeding method. It can even control *Imperata cylindrica* by repeated slashing, preventing flowering and retarding its growth due to depletion of nutrition reserves, finally resulting in its death. The capacity of this implement is about 2 ha per 8-hour day. Chemical weed control is often very effective and beneficial. It permits treatment over large areas at the optimum time, and damage to the palm roots is avoided. It can also be carried out when soil conditions are unsuitable for mechanical or manual weeding. One man with a sprayer can circle-spray about 300-500 palms per day, depending on the distance he has to walk in order to fetch water. Spraying around seedlings should be done with great care, so as not to hit the palm-leaves. Before spraying, weeds may be slashed and allowed to regrow a little. Thus, the nozzles of the sprayer may be kept low, to avoid drifting of the herbicide. Regrowing lush weeds are at a stage of enhanced susceptibility to herbicides. In general, rainfall immediately prior to or after the application of a contact herbicide is a disadvantage as moisture on the leaves dilutes the chemical or washes it off. Systemic herbicides can be very effective, as they may be taken up by the leaves as well as by the roots.

Coconut is rather susceptible to a wide range of herbicides, even to those not harmful to other monocotyledons, such as the phenoxyalkyl (or 'hormone') herbicides 2,4-D; 2,4, 5-T, and NCPA. The amine form of 2,4-D (DMA-6) is less phytotoxic than the volatile ester formulation, and can be used without affecting coconut seedlings, provided normal precautions are taken to prevent the chemical from contacting the seedlings. Under the same conditions, paraquat, diuron, terbacil, linuron, and atrazine can be used. One of the most suitable herbicides is glyphosate (Roundup), which has low toxicity to coconuts and a wide range of toxicity to weeds. The addition of ammonium sulphate to the spray mixture greatly enhances the efficiency of this herbicide (Romney 1991). The choice of herbicide also depends on which weeds are growing in the area.

Imperata cylindrica L. is one of the toughest grass weeds in the tropics. It is a very competitive weed and it is suspected of exudating substances from the roots that are noxious to other plants. It grows very well on less fertile soils and in conditions of bad drainage. It forms long rhizomes, from which it sprouts. It also multiplies by seed. Pieces of rhizomes, when left in the soil after digging out the plant may sprout again quickly and the weed may again cover the area if not regularly controlled. Repeated digging is a very costly operation, and in coconut plantations can seriously damage coconut roots. Dalapon is very effective against *Imperata cylindrica*, but due to its phytotoxic effect on coconut it should be used before planting only. This weed can also be controlled very effectively with

Roundup (glyphosate), a systemic herbicide that is absorbed only by the leaves of plants, not by their roots. After the *Imperata* has succumbed, creeping legumes can be sown, which can suffocate any surviving sprout and also other weeds (Rognon *et al.* 1984a; 1984b). Heavy shading by healthy, vigorous coconut palms also affects the growth of this weed. Thus, a combined treatment with herbicides, leguminous covercrops and fertilizing is a very effective combination of weed control methods that may also increase yields of the coconuts considerably.

Sasidharan *et al.* (1991) observed that chemical weed control, using paraquat, 2,4-D sodium salt or dalapon, did not influence the mycorrhizal colonization in the soil. Incorporation of 2,4-D sodium salt stimulated the VAM intensity in the root and rhizosphere soil of weeds and coconuts. Mechanical weed control generally reduced the VAM colonization in weeds and coconuts.

Bourgoing and Boutin (1987) developed an inexpensive and simple device to keep down the grasses, using a light wooden roller without blades, heavy enough to break the grass-stems so that they cannot completely straighten themselves again. The roller is 1.50 m long and can be drawn by one animal. Passing four times along each inter-row, one hectare can be treated per day. In addition, *Pueraria* is sown in a well prepared strip along the coconut rows. By the time the creepers start to overgrow the inter-row, about 6 weeks later, the *Imperata* may again be up to half a metre high. The grass is then rolled for a second time. The roller does not affect the *Pueraria* very much and does not slow down its development. Subsequently, the centre part of the inter-row is rolled twice, when the *Pueraria* advances over the area. The method may also be used to rehabilitate an area where, due to lack of management, *Pueraria* is being overgrown by *Imperata*. Rolling the area at intervals of three weeks may solve the problem. Depending on the regrowth of the grass, these intervals may be extended later. After some time, the *Pueraria* is capable of controlling the grass. The best time for rolling is during the rainy season, when young *Imperata* shoots bend easily and when *Pueraria* development is quickest and its damage by the roller is least. Simple, cheap rollers can be made locally. They can be drawn by animals or light tractors. This method is very suitable for smallholders. When drawn by two persons, about 800 - 1000 m^2 can be rolled per day.

Cover Crops

Covercrops, especially leguminous creepers, can to a great extent help suppress natural weed growth. The advantages of replacing the weeds by covercrops are usually easier control of the covercrops compared to a mixture of weeds and, in the case of legumes, their nitrogen fixation capacity in the root-nodules, reducing cost of fertilizing. A covercrop also shades the soil and keeps the soil temperature at a lower and more stable level, and it prevents very high rates of mineralization compared to clean weeding. It also protects the land against the impact of heavy rain which can cause the start of sheet erosion, mobilizing the soil particles. Creeping covers which root on the internodes form a network of rooted stems in the soil that will act as a barrier against run-off of water and soil. Denser vegetation under coconut palms also improves the soil's microflora. Organic matter produced by the covercrops favours microbiological activity in the soil, improving the N and P nutrition of the palms.

Depth of the root system of covercrops is very important. Legumes usually root much deeper than grasses, thus drawing their nutrient and water supply from soil layers different from coconut feeder roots. Through the deep roots, minerals are absorbed by the plant and are returned to the topsoil when leaves fall off or when the plant is incorporated into the soil. A disadvantage of lushly growing covercrops is that fallen nuts may be hard to find. In such cases, going over the cover crop with a light disc harrow may solve the problem without completely eliminating the covercrop. In smaller plots, this may be done by slashing. In climates with a severe dry season, covercrops may delay the development of young coconut palms, as was observed in Ivory Coast by Frémond and Brunin (1966) and Pomier and Taffin (1982). Compared to bare soil, palms growing in soils planted with cover crops came into bearing about one year later. Drought resistance is sometimes mentioned as a desirable character for a covercrop. However, drought susceptibility may be rather an advantage, as the covercrop will wither and dry instead of competing with the coconut during the dry season. When such crops are prolific seed producers, such as *Calopogonium mucunoides*, new cover will be quickly established after the first rains. Desirable characters in covercrops are:

- easy multiplication;
- quick growth and establishment;
- shade tolerance;
- tolerance to regular slashing;
- deep rooting;
- non-susceptibility to pests and diseases of coconut;
- easy to eradicate when required;
- easy to handle (no spines);
- rooting on the internodes (for creepers).

In new plantations, covercrops should be planted as early as possible, even before the seedlings are planted. When sowing covercrops by broadcasting, the land should be ploughed and harrowed before sowing. When covercrops are sown in strips or pockets, these sites should also be made weed-free before sowing. Vigorously growing covercrops may take up a considerable amount of nutrients and initial fertilizing of covercrops is no luxury on poor soils. Tropical soils are often poor in available P, and are more or less acid. Application of rock phosphate can solve both problems and is especially favourable for legumes. Fertilizing covercrops may seem an exaggeration, but it may be one of the most rational ways of fertilizer application. When planting cover crops in a row at 1 m from the palm row, where the palm root system is very dense, and applying fertilizer along the row of cover crops, not only the covercrops but also the coconuts will profit from the applied minerals. The total root mass of coconut and covercrop is greater than that of coconut alone. In addition, in a combination of coconuts with cover crops, leaching is less and total mineral uptake is greater. Part of the nutrients absorbed by the covercrop will be recycled and become available to the coconut when the covercrop is incorporated into the soil, or by natural leaf fall from the covercrop to the ground if the cover is a permanent one. Acid soils should be limed.

Some legumes nodulate easily; others need inoculation if they have not grown on the same land before. Wherever available, effective rhizobia inoculates should be applied. Where such inoculum is not available, but where the legume to be planted is growing somewhere in the neighbourhood, soil from this site can be dug out and the seeds to be planted mixed with a slurry of this soil and water and then dried, so that they are coated with inoculated soil.

Leguminous creepers are often winders as well. These winders can climb into high weeds, smothering them. Some of these winding legumes, such as velvet bean or *Stizolobium aterrimum* are so aggressive that they can climb metres high around the stem of a coconut palm. When using such covercrops, regular control is necessary. With a stick with a hook attached to its end, such winding vines may easily be pulled back into the inter-row. In areas with a long dry season, these covercrops may be slashed or rolled with a roller fitted with some blades, to cut down the vegetation and reduce evapotranspiration. With the first rains, they may grow out again, or, as in the case of some covercrops such as *C. mucunoides*, the seeds may germinate and a new cover will rapidly develop again. Each set of ecological conditions may be suitable for different covercrops. *Centrosema pubescens* and *Calopogonium caeruleum* (perennial calopogonium) are fairly drought resistant. *C. pubescens* grows well on sandy soils and can survive a dry period of three months. *Psophocarpus palustris* is also quite drought resistant. It can be sown alone or in a mixture, whereas *C. caeruleum* should be sown alone. The latter is usually propagated by cuttings due to the high cost of the seeds (Bourgoing 1990). *Pueraria javanica,* which has the most substantial leaf development, requires a well distributed rainfall pattern to survive. *C. mucunoides* is even more sensitive to drought and may completely die off after a severe dry season. Flemingia (*Moghania macrophylla*) is a bushy legume with a well developed root system and is very useful against soil erosion. It is drought-resistant and tolerates shade. (Bourgoing 1990). Often, a mixture of covercrops is used, so in case one of the crops suffers from disease or pest attack, the others may take its place. A combination of the three creepers (winders) *C. mucunoides, C. pubescens* and *P. javanica* is a much-used system. The first crop is a quick starter, the latter two are slower but may finally become dominant. On very sandy soils, where the establishment of covercrops is hampered by rapid drying out of the soil, burying coconut husks may considerably improve soil moisture availability and help the establishment of cover crops.

Some shrub legumes, such as *Crotalaria* spp. (sunhemp) and *Tephrosia* spp. may also yield very good results. However, they have the disadvantage of becoming woody, which may hamper other culture practices and the movement of labour through the plantation. The performance of a certain cover crop may be very good in one place and unsatisfactory in another. This is the result of ecological conditions that differ from place to place. Thus, for each area the best performing cover crop has to be found by experience.

The seed requirement for the mixture of Calopogonium, Centrosema and Pueraria is 2 kg of each species per ha. For better development, these seeds may be mixed with triple super-phosphate (TSP) in a ratio of 2 parts of TSP to 1 part of seed. The seeds of the three legumes are first mixed, subsequently TSP is added and the blend is carefully mixed again (Bourgoing 1990). However, fertilizer requirements depend on soil conditions. Flemingia seed requirements for sowing at the base of terraces is 2 kg per ha, for sowing every inter-row on flat land 3 kg per ha and for sloping land 4 kg per ha.

When cover crops are incorporated into the soil, they function as a green manure. Thomas and Shantaram (1984) observed that green manure incorporation into the soil caused a considerable increase in the populations of different various and non-specific groups of micro-organisms due to the availability of easily decomposable sources of energy and carbon. The population of asymbiotic nitrogen bacteria and phosphate solubilizers increased considerably. Dehydrogenase, phosphatase and urease activities also increased. The degree of response due to incorporating *Mimosa invisa* into the soil was lower compared to *Pueraria phaseoloides* and *Calopogonium mucunoides*. This was possibly due to the woody nature of the plant residue decomposing only slowly.

Khan *et al.* (1990) in India estimated that *M. invisa* grown in interspaces could yield about 3280 kg of dry matter per ha and contribute 75 kg of N, 6 kg of P_2O_5 and 38 kg of K_2O per ha if incorporated into the soils. Vijayaraghavan and Ramachandran (1989) recorded a weight of 12.8 kg green leaf matter per basin produced by *Desmodium tortuosum*, and 7.59 kg and 7.46 kg by cowpea (*Vigna unguiculata*) and sunhemp (*Crotalaria juncea*) respectively. Thomas and Shantaram (1984) measured 19.4, 17.0 and 14.7 kg of green matter per basin for *P. phaseoloides, M. invisa* and *C. mucunoides* respectively. Such yields depend very much on soil, climate and other cultural practices such as fertilization. It must also be remembered that, with the exception of nitrogen, all minerals are taken up from the soil. The cover crops enrich the soil with N only when legumes are planted. When edible products are harvested from the cover crop, such as from some *Vigna or Phaseolus species*, minerals are removed from the land and the beneficial effect of these cover crops on soil fertility will be less than when no products are taken out. Table 27 presents some characteristics of some much-used cover crops.

Table 27: *Some characteristics of some much used cover crops*

Covercrop	Germination rate	Growth rate	Soil cover and weed control capacity	Resistance to drought	Resistance to parasites	Resistance to grazing	Resistance to shading
C. pubescens	rapid	average	poor	very good	quite good	average	average
C. caeruleus	slow	slow	good	good	very good	very good	good
P. palustris	average	average	quite good	quite good	very good	quite good	average
C. mucunoides	average	rapid	average	poor	quite good	average	poor
P. javanica	slow	average	very good	average	quite poor	average	poor
M. macrophylla	slow	slow	quite good	good	good	very good	good

Source: Bourgoing (1990).

Moisture Conservation

The success of coconut growing depends a great deal on adequate water supply throughout the year. Where rainfall is not well spread throughout the year, and where there are no additional soil water supplies, soil water conservation is a very important cultural practice. Soil water conservation not only provides the palm with water for a longer time, but also with nutrients dissolved in the water. Adequate weed control may reduce soil-water losses during the dry season.

There are various methods of moisture conservation, such as:

- contour planting, contour tillage, and trenching on sloping land;
- mulching;
- increasing of the soil organic matter content;
- covercropping.

Contour planting on sloping land, combined with contour tillage, ridging or bunding of the soil not only prevents the water from rushing down the slope, it also improves water penetration into the soil. This effect can be increased by digging short trenches in the soil, placed in such a way that all run-off water between bunds or ridges is collected. With the run-off water, soil particles moved by the water are also collected in the trenches. Such trenches or pits thus have a double function of erosion control and water conservation. They can be filled or partly filled with coconut husks or with coir dust, which will contribute to the improvement of soil fertility. Coir dust has the ability to absorb and retain ten times its weight of water (Liyanage, M. de S. 1988). Eventually, when the trenches or pits are filling up with organic matter and soil, new trenches can be dug in different places. Where rhinoceros beetles pose a problem, coconut husks or coir dust placed in the trenches may become a breeding place for the beetles. In that case, they can be used as traps by applying insecticides to the organic material. To prevent beetles from breeding, the coir may also be covered by a soil layer, as described below. On flat land, pits or trenches can be filled with husks or coir dust almost to soil surface level. Liyanaga M. de S. (1988) recommends alternate layers of 8 cm coir dust and 5 cm soil. In gravelly soils, he observed a 20% increase in the number of nuts and a 15% increase in copra yield as a result of burying coir dust.

Mulching the weeded circles or the basins around the stem keeps the soil temperature lower and reduces evaporation. In addition to this effect, mineralization of organic material will supply nutrients to the most intensively rooted zone around the palm. Where rhinoceros beetles pose a problem, layers of mulch should not be so thick that beetles can use them as breeding places. Mulching can be done with slashed weeds from the inter-row, coconut husks and coir dust, or any organic material that can be found in and around the plantation. Das *et al.* (1991) reported a 50% yield increase after incorporating coconut husks as a mulch to the basin of coconut palms. The effect of the husks lasted for 6 years. Coir dust can also be used for mulching. Its value as a mulching material is shown by the results of a trial conducted by Uthaiah *et al.* (1989) in the State of Karnataka, India, where annual rainfall is high, 3000 mm, but where also a prolonged drought prevails during the months of November to May. Without protective measures, a large part of the coconut seedlings is lost due to drought. The trial included various mulching materials, such as coir dust, rice husks, black polythene and two systems of watering. Table 28 shows the results of this trial, ten months after planting. The soil moisture content was measured before the pre-monsoon showers. Coir dust and rice husks yielded very good results, which was attributed to the higher soil moisture content and lower soil temperature which would have induced a better growth of the root system. Another advantage of the use of coir dust is that the coir industry does not need to dump it somewhere - a great problem in many places. Varadan *et al.* (1990), observed that soil temperature under polythene is higher than under other mulches. Results of an experiment carried out in Indonesia showed that mulching with *Gliricidia maculata* was better than with coconut leaves, sedge-grass, or coconut husk. Total numbers of main roots, rootlets, and respiratory roots were larger in mulched coconut seedlings than in seedlings that were not mulched (Darwis 1991).

Soil organic matter content may be increased by incorporating decomposed mulching material into the soil and by regular discing of weeds, green manures and cover crops. Tillage will also reduce evaporation by capillary water movement, and the organic matter will improve soil conditions. In addition to its great waterholding capacity, organic matter improves the soil structure and soil microbiological activities and it reduces evapotranspiration. Planting leguminous cover crops is recommended for the supply of nitrogen that otherwise may become deficient when soil nitrogen is used for the decomposition of the organic material.

Perennial intercrops provide shade and also change the micro-climate at soil level. The leaves falling from these crops may provide additional mulch.

Table 28: *Effect of mulches on the survival of coconut seedlings and on soil moisture*

	Survival out of 30 no.	Survival out of 30 %	Soil moisture %
Coir dust, 10 cm thick	26	86.7	4.41
Rice husk, 10 cm thick	25	83.3	3.87
Black polythene 400 gauge	17	56.7	2.37
Black polythene 600 gauge	15	50.0	2.57
Earthen pitcher, 10 l/week	23	76.7	3.75
Pot watering, 10 l/week	21	70.0	2.81
Control	16	53	2.18

Cutting off old coconut leaves to reduce evapotranspiration may even be considered. Rao (1989) observed that during the dry season in Kerala, India, some farmers cut off mature lower leaves of the canopy to increase the transpiration and stimulate the growth of younger leaves. They must have found that this method has some advantages. To a certain extent this may be a confirmation of the findings of Magat and Habana (1991), who found that yields were not significantly affected by leaf pruning, leaving as few as 13 leaves in the crown only. Although their trial lasted for only one year, and the effects during the following years of pruning at that level could have been negative, there may be a level of pruning at which some young nuts may come to maturity at the expense of some old leaves.

Irrigation

Fresh Water Irrigation

As water availability is one of the most important requirements of the coconut palm for the production of nuts, irrigation in dry areas or in regions with a marked dry season is essential to yield improvement. Due to pressure on land, coconut has been increasingly planted in less suitable areas with long dry seasons. Occasionally, some of these areas may be subject to severe droughts, such as in East Africa. In India, in the States of Tamil Nadu and Karnataka, coconut is predominantly an irrigated crop (George *et al.* 1991).

When establishing new plantations, the water availability should be studied seriously to select the right type of coconut, adapted to the environmental conditions. In dry areas where coconuts have been grown for a long time, local talls may have been subject to a natural selection and have acquired some degree of drought tolerance, such as the East African Tall in Tanzania. In general, talls are more tolerant of drought than hybrids, which in their turn are more tolerant than dwarfs. Under conditions of regular water supply, D x T hybrids may produce much higher yields than talls, but during droughts hybrids may suffer much more, resulting in serious yield losses for one or two years. Some palms may even succumb to the drought. According to Daniel *et al.* (1991), in Ivory Coast PB 121 hybrid showed good drought resistance, as opposed to the WAT parent's susceptibility. This might be explained by a higher number of secondary and tertiary roots in PB 121 compared to WAT. However, although PB 121 may be more tolerant to drought than WAT, this does not mean that it is highly tolerant compared to other cultivars. The author has observed PB 121 hybrids suffering severely from drought in Tanzania and in Eastern Indonesia. The impression was obtained that heavily bearing palms suffer more from drought than less productive palms, confirming the observations of Bourdeix (1994) discussed in Part I of this book. Magat *et al.* (1988b) reported that in The Philippines these hybrids did poorly under conditions of uneven distribution of rainfall.

Depending on the severity of these droughts, high yields in good years may not compensate for losses during dry years. As a result, yearly income from the coconut plantation will vary considerably, which for smallholders is very undesirable. On the other hand, if irrigation water is available in such areas, very high yields may be obtained, compensating largely for the irrigation costs. This means that in the first case drought tolerant coconut cultivars may be preferred that respond well to improved management techniques, and in the second case hybrids would be the best choice.

Rajagopal *et al.* (1989) observed in India that West Coast Tall (WCT) coconut palms are subject to severe stress when soil water deficiency (SWD) exceeds a threshold of 115 mm. They suggest that this can be taken as the critical level for initiating irrigation, and planning water application schedules. The total biomass response to SWD implies that coconut palms are highly sensitive to water deficit conditions in terms of overall photosynthetic activity, CO_2 assimilation, dry matter production and partitioning. At an irrigation/cumulative pan evaporation (I/E) level of 0.5, the palms experienced severe moisture stress, resulting in greater stomatal resistance (111%), epicuticular wax content (32%), reduced transpiration rate (10%), leaf-water potential (68%) and reproductive dry matter production (22%), as compared with well-watered palms. Based on the relationship between soil water deficit and the stomatal resistance, the critical soil water deficit for irrigation scheduling was deduced to be 110 mm. However, vegetative dry matter production was reduced at much lower soil water deficits than this value. No significant difference in the stomatal resistance was found at irrigation levels equal to the rate of pan evaporation of 1.0. Indications were also found that irrigation could mitigate only soil, but not atmospheric, drought. This confirms the findings of Chaillard *et al.* (1983) that coconuts close their stomata, thus reducing their photosynthetic activity, when the relative humidity is too low, even though the water supply is adequate.

Rajagopalan *et al.* (1991) observed that at the Pilicode research station in India irrigation of T x GB hybrids at the I/E level of 1.00 was significantly superior to irrigation at I/E levels of 0.50 and 0.75. At I/E level 100, maximum collar girth was recorded and there was a significant increase in the number of functional leaves. At this level also, the maximum percentage of flowering was recorded. This percentage correlated with collar girth and number of functional leaves and total number of leaves. Mathew *et al.* (1993) observed the threshold at which West Coast Tall palms maintained normal water relations was at Irrigation/Cumulative Pan Evaporation (I/CPE) ratio of 1.02, corresponding well with the I/CPE ratio of 1.0 observed by Rajagopal *et al.* (1989).

Various different irrigation systems are possible. The water may be obtained from rivers, lakes or man-made reservoirs, with or without the use of a pump, and brought to the plantation via canals, ditches or pipes. In other cases ground water may be pumped up. Pipes in continuous use may interfere with other cultural practices in the plantation. Chaillard *et al.* (1983) reported severe damage to high density PVC pipes in Ivory Coast, caused by the Gambian rat *Cricetomys gambianus*. Such pipes should be buried in order to avoid rodent attack. The water may be applied in rills parallel to the row of palms, or it can be collected in a basin around the palm, or be applied through a system of drip irrigation. In the later case usually four drippers are placed around the palm. Where water availability is low, drip irrigation is to be preferred, as much less water is needed. Local conditions determine the choice of the irrigation system. Investment and operation costs of the various systems vary widely and no general recommendations can be given. Investment cost may be prohibitive, even if the applied system could be very profitable, due to the simple fact that the necessary finances are not available.

The management capacities of the farmers also play an important role in the successful exploitation of an irrigation system.

A very simple method of irrigating young palms practised by smallholders in India is described by Mahindapala (1987). In this method, two large (20 litre) earthenware pots are buried about 0.75 m away on either side of the seedling. The surface of the pot facing away from the seedling is made impervious by the application of tar or other suitable paint to prevent unnecessary water loss. The pots are filled with water, which will seep through the sides. The rate of seepage depends on the porosity of the pot. In dry sandy soils, water had to be replenished every 6 or 7 days. This system can be compared to drip irrigation, although the water supply will gradually reduce after each filling of the pot, as the water level in the pot drops, reducing pressure and area of seepage. Liyanaga and Mathes (1989), in the Eastern Province of Sri lanka tried out this method with several different volumes of water and different intervals of application, using 9-month-old palms. The soil belongs to the group of regosols, with 93.5% sand, 2.0% silt and 4.5% clay with very rapid infiltration and drainage, and a high water table. The seedlings received a total rainfall of 1067 mm during the first three months after planting, which was followed by a drought period of 10 months during which total rainfall was only 324 mm. This period was followed by six months of heavy rain, and a period of eight months in which months with ample rainfall alternated with dry months. Surprisingly, there was no significant response to irrigation during the first year after planting, although some height increase was observed during the second 6-month period. However, the response to irrigation was considerably higher thereafter. After 30 months, in the non-irrigated plots 48% of the seedlings had died, the surviving seedlings showing retarded growth. The first response was shown for girth. However, after the first year after planting all the characters showed marked response to irrigation treatment. The lowest response was given by the application of 20 l once every two weeks, the highest by the application of 40 l twice a week, which was the highest dose given. However, the difference between this treatment and 20 l applied twice a week was not large.

Percolation of the water through the soil should be studied properly. Most absorbent roots of the palms are usually found in the first 0.5 to 1.0 m of soil. To avoid water losses, water should be applied at such a rate that percolation does not go beyond this rooted zone. Chaillard *et al.* (1983) found that coconut palms concentrate their roots in and around the wet bulbs around the irrigation rills. For the 0-70 cm surface layer, they found a weight of root orders II, III, and IV near the rill that was three times greater than that corresponding to the inter-row without an irrigation rill. Applying irrigation water in four 1-hour periods per day did not improve the water use coefficient due to reduced percolation. This theoretical advantage of splitting was out-done by reduction in the volume of wet bulb around the irrigation rill that was not completely filled by this method. They also found that water applied at the end of the day is less well used by the trees. For systems working day and night applying the water in rills, they recommend that the rills should be made as long, narrow and shallow as possible; the dose of water should be split into two daily applications, allowing minimum filling of the rills (sufficient amount of water applied during each period); losses of efficiency due to night irrigation should be reduced by applying only half the amount. However, in very sandy soils, the higher infiltration speed limits the spreading of the water in the rills, and an increase in the number of distribution points will be an advantage.

Irrigation by flooding should be done only under conditions of ample water availability. It can also be done only on flat land. The advantages are the low technical skills required and low investment costs. Disadvantages are the difficulty in controlling of the volume of water applied, increased leaching of nutrients and discontinuity of water application. Irrigation in basins around the stem has about the same advantages and disadvantages as irrigation by flooding, but on a lower scale. Mulching of the basins with organic material reduces loss of water by evaporation. Water can be brought to the basin through a canal system, or through permanently buried pipes from which water is manually distributed from the take-off points by use of a hose.

Sprinkler irrigation has some advantages on sloping land, as the risk of water runoff is small. It requires a complicated piping system and investment costs are high. Risk of damage to pipes and sprinklers is considerable. Water loss through evaporation is higher. In a system with permanently fixed sprinklers, supplied by underground pipes, the wet circle will be always the same, which may result in more intensive rooting and more efficient use of water. Water doses can be controlled by the length of sprinkling times. Equal distribution of water over the circle will be more difficult with increasing radius of the circle.

Drip irrigation also requires a permanent piping system. The advantages of this system are reduced water doses, good control of water volume applied, and continuous water supply to the palm without excessive percolation. Drip irrigation is also low labour and energy intensive. By using this system, the soil surface layer remains dry, there will be much less weed development and weeding costs will be reduced. However, the pattern of wetting front in the soil will be different for different soils, due to variation in soil texture, permeability, presence or absence of impermeable layers, quantity of water applied per irrigation, discharge rate of the emitter and the initial moisture content of the soil (Dhanapal *et al.* 1995). Results from a trial conducted at Kasaragod, India, showed that the vertical and horizontal movement of water was directly related to the quantity of water applied, and the same was true for volume of root zone wetted in the coconut palm basin. Sub-surface placement (in the root zone) of emitters covered a greater volume of the basin and maintained higher moisture content than the surface-placed emitter. However, it may lead to loss of water due to deep drainage beyond the root zone. Varadan and Chandran (1991) compared the effects of drip irrigation, using 15, 30 and 45 l per day per palm with basin irrigation at 600 l per week per palm. The drip method at 15 l per day per palm was on par with control, whereas the other two drip treatments were on par with 600 l per week per palm basin irrigation. During the first two years of the trial, drip irrigation at 30 l per day per palm gave the best results, but in the third year 600 l per week per palm was superior. The average increase of nut yield over the 3 year period of the trial was highest for the 30 l treatment. Evaporation losses are minimal. Investment costs are high and good management skill is required. Fertilizers may be mixed with irrigation water, increasing their efficiency, especially in irrigation systems with an adequately controlled water supply. For drip irrigation, special soluble fertilizers and filters are required (Mahindapala 1987).

When irrigation is applied, the total exploitation of the coconut land is increased. Leaf and nut production will increase and only a certain proportion of these products will be returned to the land. This means that more mineral nutrients will be exported from the soil. Nutrient losses through leaching through the soil may also increase, depending on the irrigation system and management level. This means that an irrigated system requires higher fertilizer applications to be sustainable, increasing investment and operation costs. Thus, such systems are viable only in situations where all factors for maintaining the system are available. Bhaskaran and Leela (1978) observed that palms in the high yield groups failed to show parallel response to irrigation, as observed in the low yield groups. It was suggested that the yielding capacity of the high-yielding palms was already fully exploited even under normal conditions, and when irrigated nutrients became a limiting factor for further yield increases. However, in addition to nutrient limitation, it may also reflect a difference in root system development between high-yielding and low-yielding palms. The palms with a better developed root system suffering less from drought than the others.

An initial reaction of a mature coconut palm to irrigation in the dry season is reduced premature nutfall. Drooping leaves may recover their turgor, but yellowed and withered leaves will not recover. Reduced premature nutfall may result in higher nut yields within one year following the start of irrigation. Reduced water-stress also influences flower production at initiation, resulting in a secondary increase in nut production about two years after starting irrigating the palms. Chaillard *et al.* (1983) observed that, with irrigation, yields of the hybrid PB 121 growing in the coastal area of Benin could be increased from the usual average of 1.3 t copra per ha to about 5,5 t copra per ha. Daniel *et al.* (1991) in Ivory Coast, observed that the hybrids can produce yields of around 5 t of copra per ha, even with an annual water deficit of around 350 mm. They calculated that for PB 121 hybrids with a theoretical potential of 7 t of copra per ha per year (equivalent to about 250 nuts per tree), each additional 100 mm of water deficit leads to a reduction in yield of about 700 kg of copra per ha, *i.e.*

as an initial estimate, a 10% reduction in potential for every additional 100 mm of annual water deficit.

Mathew *et al.* (1993) conducted a 4-year-long study on water requirements of 10-year-old coconut palms, and the effect of irrigation during the 5-months dry season in Kerala, India. They observed that irrigation did not influence yields of nuts during the first two years of experimentation, which they designated as a transitional phase. The palms responded significantly to irrigation during the subsequent two years, which were designated as the influence phase. The little influence during the transitional phase was attributed to the effect of water-stress on inflorescence primordia during the pre-treatment period. The average increase in annual nut yields per palm over pre-treatment yield, due to these treatments ranged from 46 to 69 (68-113%) during the first year of the influence phase, and from 47 to 67 (68-104%) in the second. The highest yields obtained in the experiment were about 130 nuts per palm per year. According to Blaak (personal communication), averages of 180 nuts per palm per year have been obtained with West Coast Tall on a commercial scale with irrigation and adequate nutrient supplies in Kerala, India. Net returns to irrigation can be considerably increased by multi-cropping, as more products will benefit from the water. Especially in combinations of crops with root systems exploring other soil depths than coconut, the practical use of the water will be increased as well, and less water will be lost through percolation.

Salt Water Irrigation

Coconut is a plant of halophytic tendency with good tolerance to salt in the soil. Plants of this type usually need high concentrations of electrolytes in their cells to grow and equilibrate their water-balance. In this context, manifestations of Cl deficiency can be interpreted as the inability of a plant to maintain its water potential at sufficiently low values due to the deficiency in monovalent anions (Ollagnier *et al.* 1983). There are numerous examples of the beneficial influence of Cl application on the health of coconut trees growing in conditions of low Cl availability.

However, seawater is not always applied for its Cl content, but also in cases of serious drought. Coconut palms can stand salt concentrations of up to about 1.0 per cent. Repeated irrigation with seawater involves a risk of too high salt accumulation in the soil. In a trial conducted in Brazil, Costa *et al.* (1986) found that when irrigating seedlings in polybags with increasing doses of salt water mixed with irrigation water, the number of tertiary roots decreased, whereas the seedlings treated with the highest doses of salt had fewer primary roots, and these were thicker than primary roots in other treatments. Therefore, total root weight between treatments showed no difference. After irrigation with fresh water, the palms recovered. For repeated irrigation with seawater, soils should be friable with good internal drainage. Rainfall during the wet season should be sufficient to wash out excessive salt from the soil. In addition to similar effects as with fresh water irrigation, seawater contains some important nutrient minerals, such as Cl and Mg, which may have a positive effect on the palm yields. Remison *et al.*, 1988) in Nigeria treated seedlings in polybags filled with 15 kg soil with fortnightly applications of 0, 2, 4, 6, 8, 10, and 12 g salt. The chemical analysis after twelve months showed that Na and Cl content of the leaves had increased appreciably, while N, K, Ca and to a lesser extent, P decreased with salinity. Antagonistic effects occurred in the soil between Na and K, and Ca and P. Soil pH, N, Mg and Ca were not affected by salt application. Compared to the initial nutrient content in the soil before experimentation, N and P decreased drastically after experimentation. In a similar trial (Remison and Iremiren 1990) it was observed that seedlings responded to salt at low rates of about 2 g by increases in girth, height and leaf number. Higher levels were tolerated but did not significantly increase any growth parameter. Leaf splitting was not affected either. Generally there was an increase in dry matter at levels of 2-6 g salt. In one experiment, dry matter of leaves and total dry matter increased by 48% and 39% respectively when salt was applied at the rate of 6 g per plant. In the second experiment, the highest increase in total dry matter was 32% at 2 g salt level. At the highest concentration of salt, some seedling mortality was recorded.

Fertilizing

Any soil under agricultural exploitation will sooner or later need fertilizer application to replace the minerals that have been exported by the removed harvested products and by leaching. Mineral export can be kept as low as possible by returning as much waste material as possible to the plantation. In coconut plantations this means recycling husks, shells and leaves. Also, in the stem of the palm many nutrients are stored that will become available only after many years, provided the stem is burned or buried in the plantation. Where stems of felled palms are removed, these nutrients are lost as well.

Modern coconut management, aiming at high yields and the commercialization of more coconut products such as shells, husks, and also the wood, means a much increased removal of mineral nutrients from the coconut land which has to be replenished annually. In the case of intercropping or mixed farming, more products are removed from the land, and the export of nutrients is even higher than when growing coconut in monoculture. Coconut growing in rather depleted soils often responds remarkably to the application of fertilizer. Perera (1989) in Sri Lanka estimated an average return of 50% on investment for the application of fertilizers on smallholder coconuts.

An investigation conducted in some districts in Kerala State, India revealed that only 6.51% of the farmers cultivated hybrids along with other cultivars. Twenty-four per cent of the farmers adopted correct spacing for planting. The majority of farmers did not apply fertilizer at all, whereas about 30 per cent applied fertilizers only at a low level. Only about 3 per cent of the farmers applied fertilizers in split applications. Green manures and cover crops were planted by only 3 per cent of the farmers. The practice of husk burial was not adopted by any of the farmers in the study area. Unawareness of the recommended technology, lack of conviction in recommendations, and lack of sufficient capital were found to be the major constraints facing the adoption of such techniques (Bastine *et al.* 1991). The most acceptable way of getting farmers to use fertilizers on their coconut land might be through the cultivation of intercrops which give a rapid return on investment. The coconut may benefit from the fertilizer as well, and part of the minerals will be returned to the soil by leaves of the intercrop falling to the soil or by incorporation of the crop residue into the soil after harvesting the intercrop.

Smallholders often claim that 'fertilizing is too expensive for coconuts'. In this case 'too expensive' does not refer to a negative return on investments but to the financial position of the farmer, who cannot afford to invest considerable sums of money and wait for about 2-3 years for the first returns. Sometimes, as a result of a subsidy, fertilizers are available at reduced prices. However, such programmes are not very successful if farmers do not have the means to buy fertilizers even at reduced prices. Ochs *et al.* (1993) stressed that small rates of fertilizer application lead to a relatively higher yield increases than higher rates. Using smaller application rates than the optimum leads to lower yields, but also relatively lower fertilizer costs, thereby improving the cost-effectiveness of the operation. Planning for higher income through fertilizer application of smallholder coconut farms with low capital availability should aim at modest results, gradually increasing fertilizer doses when incomes rise. It is this aspect of coconut growing that merits much more attention from extension services, credit institutions and development planners.

Application of modern techniques as developed by research could multiply coconut production. However, production figures show that actual increases are only very small and mostly due to new plantings and area increase. For many, this is due to lack of fertilizer application. Ouvrier and Taffin (1985a) estimated that less than 1% of the world's coconut groves are regularly fertilized. Planting new and highly productive material to replace old coconut groves makes very little sense if no fertilizers are applied, as the new palms will require more nutrients than the old ones and the soils may be expected to be depleted if no fertilizers have been used on the old plantation. As long as no investments are made in the application of fertilizers and other improvements, all research efforts to develop high-yielding cultivars and new techniques that can considerably increase production will have little impact. Higher pressure on land might become beneficial to coconut production when intercrops are incorporated into the Coconut Based Farming System (CBS) and when these intercrops are fertilized. Ditablan and Astete (1986) reported on an intercropping trial conducted in The Philippines, in which coconut production increased more than twofold within a period of seven years. A substantial increase was observed in coconut production, especially in the last years of the experiment, which was attributed to the impact of improved cultivation and more absorption of nutrients from the soil. There are three different ways of improving nutrient availability in the soil:

- green manuring;
- application of organic manures;
- application of chemical fertilizers.

Organic fertilizing has some advantages over chemical fertilizing, because:

- it involves macro- as well as micro-elements;
- it improves soil texture;
- it improves soil moisture-holding capacity;
- it improves conditions for soil micro-biological life;
- it improves the soil's cation exchange capacity.

All these factors are very important for the regulation of mineral supply to the plant. Therefore, the use of manure and/or compost, or just green leaves cut off from some source outside the coconut grove, should always be given preference. Detection of deficiencies of minor elements is often rather difficult, especially if the deficiency is combined with a deficiency of one or more of the major elements. When visible, the deficiencies are usually already in a rather advanced stage. Correction of mineral deficiencies with chemicals is often very problematic in remote areas, as the materials are not available, nor the institutions to supply them. For large plantations and in areas densely planted with coconut, organic material other than recycled products may be hard to obtain. In most cases, therefore additional chemical fertilizers may have to be applied to obtain high yields.

Organic Manuring

Green manuring

For cover crops and green manuring the same crops are frequently used. Cover crops as a cultural practice to reduce weed growth and evaporation from the soil have already been discussed. Green manures, which have the ability to fix nitrogen in symbiosis with soil micro-organisms will mainly add nitrogen to the soil in addition to the organic matter. In an experiment conducted in Ivory Coast by Pomier and Taffin (1982b) in soil where nitrogen was the essential production factor, yields in all plots declined after N fertilizer treatments were discontinued, except in plots covered with legume cover crops. The N-content of the soils where leguminous cover crops were planted was much higher than in other plots. In some soils, *Rhizobium*-inoculation and seed pelleting can enhance the nodulation and N fixation of green manures. Thomas and Ghai observed significantly more response in nodulation behaviour, dry matter yield and nitrogen concentration when pelleting was done on *Rhizobium* inoculated seeds sown in an acidic coconut soil. The difference in response due to the use of different pelleting materials (rock phosphate, calcium carbonate, charcoal, and dolomite) was not significant.

If the legumes produce edible products, such as beans and pods, harvesting these products will also remove that part of the nitrogen used by the plant to produce these beans and pods. As already mentioned, legumes may have very deep roots by which they can take up minerals that have already leached to depths below the dense root zone of coconut palms. These minerals will be returned to the topsoil when the green manure is incorporated into the soil. However, this cannot be regarded as additional fertilizer. It is rather a recycling of minerals that otherwise would have been leached out and lost. Often, soils are so poor that green manures also grow very slowly, and are barely beneficial. In such a case, the growth of the green manure crops could be improved by applying chemical fertilizers. In plantations with poor soils it can often be observed that green manure crops grow much better in places where some organic debris has been burned before planting the legume. The coconuts would benefit from the fertilizers. Alternate growing of green manures and annual intercrops, using chemical fertilizers, may be an effective system of maintaining soil fertility.

The dense mass of vegetation, especially where a mixture with winding cover crops are planted can hinder movement in the plantation and hamper nut collection. Fernando (1989) describes a low-cost roller that can be made at village level. It consists of a rounded tree trunk of 50 cm diameter and 1 m in length, in which six lengthwise saw cuts have been made at regular spacing. Steel blades are inserted into these cuts. The roller can be drawn by a pair of oxen. The capacity of the rolling equipment is about 0.8-0.9 ha per day, without exhausting the animals or the operator. Such a roller crushes the cover crop without cutting and uprooting it, such as is the risk with disk harrows, which might affect the re-growth of the cover crop. With such a roller, the cover crop is not incorporated into the soil but functions as a mulch on to of the soil. The roller developed by Bourdeix and Boutin has no blades and is claimed to be effective as well.

Manuring with other organic materials

When fertilizing with organic material, recycling of plantation material is the most obvious choice. Dehusking the coconuts and leaving the husks in the field are generally recommended practices. The husks have a high mineral content, 50 husks contain about 0.5 kg potash (Liyanage, L.V.K. 1987), equal to about 1 kg muriate of potash. The husks should be placed with their convex side upwards, in circles around the stem of the palm. Ochs *et al.* (1993) draw attention to the fact that in an intensive cropping situation, husks may represent the equivalent of 200 to 300 kg KCl per ha per yr. This should not be overlooked when calculating the profitability of alternative industrial uses of husks (fibres, energy), which should compensate financially for the corresponding loss in fertilizer. Different authors mention different nutrient contents of the husks, especially for the main components N, K and Mg (Table 29). This may be due to differences between varieties and between the soils in which the coconuts were growing.

Husks can be placed around the palm as a mulch, but they can also be buried. When buried, they will improve soil conditions, and the soil waterholding capacity will be increased. Husks can absorb and retain about six times their own dry weight of water. It takes about 3-4 years for husks to decompose. Nunez *et al.* (1991) experimented with the cellulolytic fungus *Trichoderma* as an agent to accelerate the decomposition of husks. Treated and untreated husks were buried in the soil. Distinct areas of decomposition in the treated husks were noted two months after inoculation with the fungus. In these areas, fibres were reduced to tiny pieces. The original soil K content was 190 ppm. After two months, soil with untreated husks and soil with treated husks contained 517-593 ppm and 700-747 ppm extractable K respectively. Leaving the husks in the field may attract termites. Ouvrier and Taffin (1985b) observed that when leaving mounds of husks in the plantations, the mounds most colonized by termites were also those whose mineral elements evolved most rapidly.

Another important recycling product is the coir pith or coir dust, representing about 45% of the coconut husk. It contains about 25-30% lignin and 33% cellulose. The word 'coir' is derived from 'kayar' in the Malayalam language. The quality of coir pith depends on the fibre extraction process. Coir pith obtained after mechanical fibre extraction is richer in nutrients compared to coir pith obtained after retting, as it loses much of its original mineral content during this process. The coir pith obtained from the fully mature and older nuts contains a higher amount of lignin and cellulose and fewer water-soluble salts compared to younger nuts. A great variation of C:N ratios have been reported, ranging from 58:1 to 112:1 (Savithri and Kahn 1994). Often, in the neighbourhood of processing centres where large numbers of coconut are husked, huge amounts of coir pith can be found, lying waste. The tannins that ooze from the dump yards during the rainy season are considered to create environmental pollution problems. Sometimes it is set on fire, but it burns very slowly, due to its high moisture content. However, coir pith is an excellent product to be returned to the soil, either as a mulch or as an organic fertilizer.

Coir pith can be applied as a mulch, used as litter (after drying) in stables or poultry pens before applying it to the soil, or applied directly to the soil. In a trial conducted in India, a coir pith mulch,

10 cm thick around seedlings resulted in a higher seedling survival and higher soil moisture content than other treatments, such as mulching with rice husk or black polythene, and earthen pitch and pot watering at a rate of 10 per week (Utaiah *et al.* 1989). Using coir pith as a litter will enrich the material with minerals and reduce the C:N ratio, improving its quality as a manure. Due to its fibrous and loose nature, incorporation of coir pith into the soil improves soil's physical properties and water-holding capacity considerably.

When incorporating coir pith into the soil, decomposition for which soil nitrogen will be used may result in temporary nitrogen deficiency of the palm. Reduction of the C:N ratio before applying the coir pith to the soil is recommended. The effect of blending retted coir dust with fertilizers was studied in the laboratory by Nambiar *et al.* (1988). They observed that available nitrogen decreased progressively with time of incubation in all treatments. The organic carbon content showed a similar trend as that of available nitrogen. The application of NPK fertilizer generally favoured accumulation of humic acid in the soil. There were no marked changes in ammoniacal nitrogen during the incubation period. The NPK treatment increased the nitrate N content whereas in the NPK + coir dust treatment the increase was marginal, suggesting a nitrification inhibition property of coir dust. Joshi *et al.* (1985) observed in laboratory trials that production of total mineralized nitrogen was consistently lowest when urea was blended with retted coir dust (1:1). Unretted coir dust blended with urea (1:1) produced lower mineralized nitrogen than urea alone. Other blendings with lower proportions of retted and unretted coir pith (9:1) showed intermediate behaviour. Results indicate the usefulness of blending urea with coir dust for controlled and gradual release of urea nitrogen. This finding is important for soils prone to heavy leaching losses, where controlled release of N can be achieved in the root zone of perennial crops. Nagarajan *et al.* (1985) treated 1 t coir pith with *Pleurotus sajor-caju* (an edible basidiomycetous fungus) and 5 kg urea and kept the mixture as a heap to decompose in an open yard. After 30 days, a drastic reduction in the lignin content from 30% to 4.8% was observed. The cellulose content had reduced from 26.5% to 10.1%. The C:N ratio was reduced from 112:1 to 24:1, which is comparable to farmyard manure and compost when applied to the soil.

Table 29: *The mineral content of coconut husks presented by various authors (N-Mg in percent; Mn-Cu in ppm): A = Nagarajan et al. (1985); B = Bopaiah (1991) and C = Ouvrier and Taffin (1985b)*

	N	P	K	Ca	Mg	Mn	Zn	Cu
A	0.26	0.01	0.78	0.40	0.36	12.5	7.5	3.1
B	0.27	0.15	1.71	0.21	0.14			
C-fibres	0.19	0.011	1.343	0.065	0.057			
C-parenchyma	0.25	0.011	1.794	0.127	0.061			

For composting coir pith, an area of 3x5 m may be selected in a shady place. Coir pith is spread out over the area in a 2 cm thick layer (about 100 kg of coir pith), over which one bottle of spawn (300 g) of *Pleurotus sajor-caju* is uniformly spread. Then the layer is covered with another 2 cm thick layer of coir pith, over which 1 kg of urea is uniformly spread. This process can be repeated until the heap reaches a height of about 1 metre. For each ton of coir, 5 kg of urea and 5 spawn bottles are required. The moisture content in the heap is maintained at about 200% by sprinkling water (Savithri and Khan, 1994). In a trial conducted in The Philippines with several fungi, including *Pleurotus sajor-caju*, it was observed that *Phanerochaete chryosporium* UPCC 4003 was the most effective fungus for degrading the lignocellulose components of coir dust at optimum conditions (Uyenco and Ochoa 1984). However, the degradation process was carried on with minimal nitrogen concentration.

Trichoderma sp. is among the native microflora observed in the raw coir pith. An additional advantage associated with *Trichoderma* sp. is its ability as a biocontrol agent of soilborne fungal diseases, such as *Ganoderma*, the causal agent of Basal Stem Rot (Savithri and Khan 1994). The use of stable manure is well known and does not need special discussion. However, as long as stable manure is produced from vegetation growing under the coconut palms only, it does not add anything new to the soil; rather it is a system of recycling nutrient elements in an organic form. Minerals taken up by animal bodies are lost to the land, as is the case with the nitrogen that may evaporate during production and application of stable manure. Waste materials can sometimes be obtained from agro-industries, such as rice bran, straw, or the residue from beer breweries. Such materials, if available at a reasonable price, can be composted in the plantation, or spread out in stables, to be mixed with the animal excrement.

Organic material for the manuring of coconut can also be obtained by interplanting coconut with other trees or shrubs that can be pruned regularly. In a mixed cropping trial with *Gliricidia sepium* and *Leucaena leucocephala* in Sri Lanka, these trees were planted in double rows in the coconut avenues at a spacing of 2 x 0.9 m. They yielded 7-10 tons and 12-16 tons of green matter per ha respectively and 14-20 t per ha of fresh firewood during the first and second years (Liyanage, L.V.K. *et al.* 1987). Harvesting started one year after planting the trees by lopping the plants at one metre above the ground at 3-month-intervals. No data on resulting coconut yields were presented. In some countries, coconuts are manured with leaves and branches of other trees growing outside the plantation. This is possible only on very small holdings, as the practice is labour intensive. For large plantations it may be very difficult to obtain sufficient quantities of such material from outside the plantation.

Gunasekera (1989) recommends fertilizing with organic material combined with a system of root rejuvenation. A trench could be opened 30 cm from the bole, forming one-quarter of a circle. This trench could be filled with organic material and covered with soil to prevent the breeding of rhinoceros beetles. After two years another quarter of the circle may be filled, etc. This method induces new root formation from the bole. In these experiments using 30 kg of *Gliricida* or *Leucena* leaves to fill the trenches, after two years of treatment nut yield increases of 15-20 nuts per palm per year to 50-60 nuts per palm per year were observed after two years of treatment.

Chemical Fertilizers

Most cultivated lands need regular supplies of mineral nutrients, especially after having been cultivated for some time. When planting or replanting coconut on land previously cultivated, fertilizer application to the seedlings is very important to guarantee good development of the bole, which is very important for the productivity of the tree, as it increases the rooting surface. With good nutrition, the stem will also attain its maximum width. Nutrient deficiencies at a later stage may cause a narrowing of the stem, which may decrease the palm's productivity. Fertilizing after such a period may again allow the stem to attain its original width, but not in the narrow section, as the stem has no cambium. Such narrow stem sections may also be caused by drought and other unfavourable growing conditions, and show something of the palm's history.

Adequate fertilizing of young palms lowers the pre-bearing age. Depending on soil conditions, adequately fertilized young palms may start bearing one ore more years earlier than palms growing without fertilizers. Nelliat (1972) recommended the application of one-tenth of the adult palm dose of fertilizers about three months after planting in the field, one-third in the second year, two-thirds in the third year and the full dose in the fourth year. Teoh *et al.* (1985) in Malaysia observed that in young PB 121 hybrids growing on coastal clay soils, from 6 months after field planting onwards, potassium had been taken up in the highest amount, followed by nitrogen, Ca, Mg and lastly P. Unfortunately Cl was not included in the study. Cecil and Khan (1993) observed that in general the effect of N on vegetative growth of young palms was maximum, followed by K, whereas P showed favourable interaction with N and K.

It goes without saying that high-yielding cultivars need more mineral nutrients for the production of a greater number of nuts than low-yielding palms. On the other hand, high-yielding varieties can also respond much more strongly to fertilizer application. However, according to Ooi (1987), there is no evidence available to justify the belief that hybrids would do worse than local varieties under low management conditions. Results of a fertilizer experiment in Thailand suggested that, in the absence of fertilizer, hybrids did as badly as local varieties, but no worse. Results of a trial in Kerala, India, showed that the hybrid Chowgat Orange Dwarf x West Coast Tall was superior over West Coast Tall, whether or not-fertilizers were applied (Green 1991). However, the duration of the trial was not mentioned and the possibility exists that after a number of years (depending on soil fertility) in the non fertilizer plots the greater depletion of minerals by the hybrid might affect its superiority, especially if husks are removed from the field with harvesting. Ochs *et al.* (1991) stress the fact that the economically optimum level of fertilizer application is much lower than the rate of fertilizers required for optimum yield. In the benefit:cost ratio, the price of copra plays an important role.

When recommending fertilizer application, it should be verified whether the planting material is of adequate quality and not too old and senile to respond economically to fertilizer application. Where this is doubtful, improvement might better be achieved through renovation of the plantation and building up soil fertility with fertilizers and green manures. Ouvrier (1984b) observed that in coconut planted on sandy soil in Ivory Coast the total amount of dry matter produced was directly linked to the dose of potassium applied, provided a sufficient amount of kieserite was applied to correct the magnesium deficiency induced. KCL increased the proportion of albumen in the nut significantly. KCL had no effect on the K and Cl contents in the albumen, but only on contents in the husk and shell. The effect was considerably greater for the husk. K and CL contents in the husks increased with the doses of KCL applied. Increase in leaf K content corresponded to identical increases in the husk. There was a linear relationship between K applied and K removed, and consequently little unnecessary consumption. If the husks were returned to the field, the amount of potassium exported per ton of copra varied little. Nitrogen, phosphorus, magnesium and sulphur were found mainly in the albumen, and the amount exported per ton of copra remained constant, regardless of the manuring rate. The shell contained no phosphorus, calcium or magnesium.

For mature coconut palms, the amount of minerals consumed by the palm may give an indication of the fertilizer requirement. The amounts of minerals removed per ton of copra harvested, as presented by various authors, differ to some extent. These differences can be due to ecological differences, scarcity of nutrient elements in the soil, luxury feeding by the palms in case of abundance, varietal differences in nutrient uptake and, last but not least, varietal differences in the composition of the nuts (percentage husk, shell and albumen). Table 30 presents the nutrient export by unhusked nuts reported from three different studies conducted in Ivory Coast:

Table 30: *Nutrient export by unhusked nuts*

Element	A	B	C
N	16.1	21.2	18.9
P	2.2	3.2	3.9
K	28.8	41.0	36.4
Ca	1.3	2.6	5.2
Mg	2.2	3.6	1.1
Na	3.0	2.2	-
Cl	18.7	21.2	-
S	1.3	1.8	-

Source of data: Ochs *et al.* (1993)

A = Mineral exports by unhusked nuts, calculated per ton of copra of 12-year-old PB 121 hybrids, based on a yield of 6.7 t per ha year (kg per ha);

B = Mineral exports by unhusked nuts, calculated per ton of copra of 6-year-old PB 121 hybrids, based on a yield of 5 t per ha year (kg per ha);

C = Mineral exports by unhusked nuts, calculated per ton of copra of adult West Coast Tall in India (kg per ha).

The husk is responsible for the greatest proportion of mineral exports of PB 121, 67% of potassium and 85% of chlorides. For the total export of minerals, the nutrients taken up by the stem should be added, which for 6-year-old palms were (in kg per ha): N-48; P-6; K-54; Ca-11; Mg-9, Na-5; Cl-28; and S-2. It was assumed that in the sixth year PB 121, under conditions of this study, could be considered functioning stably (Ochs *et al.* 1993). These figures show that, for coconuts chlorine may be regarded as a macro-element, whereas the importance of P is at the same level as Mg. The importance of leaving husks and leaves in the field is evident. The husks contain almost one-half of the total potassium and chlorine. The leaves contain 75 per cent of total calcium and 53 per cent of total magnesium. These figures also clearly indicate that when population pressure on land increases, increased removal of husks and leaves for the use of fuel and roofing material may have serious consequences for coconut land fertility. If the minerals removed by these products are not replaced by fertilizers, yields may soon decline sharply.

It is clear that nutrient removal by coconuts depends on soil conditions as well as on coconut variety and yielding capacity, and no general recommendations for fertilizing coconut plantations can be given. The figures presented above can be used only as guide-lines. When deficiency symptoms become visible, the palms may have suffered from a shortage for a long time and the plantation has been producing at sub-optimal level. Monitoring of mineral contents in the palm is therefore imperative for good management. The three main systems used to determine fertilizer needs are leaf analysis, soil analysis and fertilizer experiments. Leaf analysis is a rather accurate system to determine the nutritional situation of the plant. Soil analysis not only determines the soil availability of certain nutrients, but it also provides information on other chemical soil characteristics, such as pH and nutrient fixation. Also other soil characteristics, such as cation exchange capacity, texture and depth, may all be important to determine the system of fertilizer application and the doses to be used. Fertilizer experiments provide the final answer of plant reaction to certain fertilizer applications, the influence on the nutrient levels in leaves, the interactions between the various nutrients, and the potential yield responses within local ecological conditions. Fertilizer experiments also give information on costs and on the most economic level of fertilizer use.

Leaf analysis is based on certain sampling techniques that should be generally accepted and used to compare results obtained in different places. The system most used is the one developed by IRHO. The following description of this sampling method is based on an article by Taffin and Rognon (1991). The best time for sampling is at the start of the dry season. After rainfall, at least 36 hours should pass before sampling, as nutrients may have leached from the leaves. Sampled trees may be marked, to be sampled again later to investigate the results of treatments applied on the basis of sampling results. As leaves of different ages have varying mineral contents, only one leaf position is used. For young seedlings, leaf no. 4 is used; as soon as it is available, leaf no. 9 is used for somewhat older palms and for fully grown palms leaf no 14 is used. The position of the corresponding spathe indicates the turning direction of the leaf spiral. If its position is slightly to the left of the leaf, the spiral turns left, if it is on the right, the spiral turns right. Leaf no. 1 is the youngest leaf just detached from the central spear, Usually, leaf no. 9 supports the largest unopened spathe and leaf no. 14 generally supports an inflorescence with nuts the size of a fist. Without cutting the leaf, 6 leaflets are cut, 3 on either side, from the central part of the lamina. Of each leaflet, only the central 10 cm portion is used. The edges of each leaflet (about 2 mm) and the central vein are removed. Of each leaflet fragment, one side is used for analysis, the other is kept as a duplicate. The leaflets are cleaned with cotton wool and distilled water, then carefully dried. Duplicates are wrapped and stored in a dry place. Samples should be adequately labelled, giving all necessary information on date, site, and conditions, as well as on the condition of the plot at the time of sampling. After labelling, the samples are dried in an oven at 70-80°C for about 10 hours. Where no oven is available, a 250 W electric bulb can be used for drying. The sample batches can be sent in sealed plastic bags to the laboratory for analysis.

The sample covers 25-30 trees. For large areas, one sample is used for every 50 or 100 ha. In smallholder areas, one sample is taken per holding. In large plantations, for immature trees (requiring very close mineral nutrition monitoring) uniform 50 ha sectors are marked out, based on the soil map if possible. For mature trees, one sample should be taken per sector of 50 - 100 ha. In small plantations, trees are randomly selected; in large plantations a system is often used to facilitate follow-up work. Abnormal trees should be excluded. Jayasekara (1988) recommends sampling 2-3 palms from each one-acre-block to eliminate variability.

Leaf nutrient contents, and consequently deficiency levels, fluctuate between seasons, and also depend on soil characteristics such as pH, nutrient contents, texture, and humidity. Therefore, results of leaf analyses should preferably be compared with those of fertilizer trials. Assuming that leaf analysis methods cannot be applied universally in different parts of the world with different soil and climatic conditions, Jose *et al.* (1991) studied the nutrient contents in the leaves of WCT palms in Kerala State, India. They found that the correlation between the nitrogen content and the potassium content of the 10th leaf and yield were higher than for other leaves. They concluded that the 10th leaf will best reflect the nitrogen and potassium status of the palm in relation to yield. The phosphorus content of this leaf also correlated significantly with yield. However, using different methods for different regions may lead to confusion, and the adoption of one uniform system is to be preferred. If necessary, critical levels for fertilizer application can be adapted to locality.

Table 31: *Critical level of mineral elements in leaf rank 14*

element	critical level %	critical level ppm
N	2.0; 2.2*	
P	0.12	
K	0.80-1.0; 1.20-1.40*	
Cl	0.50-0.60	
Mg	0.24; 0.20*	
Ca	0.30-0.40	
S	0.15-0.20	
Fe		50**
Mn		30**
B		10-13
Cu		2

* = for hybrids
** = tentative

The content of a certain element in leaf 14, below which the application of this element as a fertilizer may result in an economical yield improvement, is called the critical level of this element. The critical levels in leaf rank 14 for the various elements are presented in Table 31. These values are indicative only. Variations may occur under different climatic and/or soil conditions or between different varieties. Ochs *et al.* (1993) reported that leaf contents for mineral nutrients differed very significantly in hybrid comparative trials with uniform fertilization. They recommend that this should be taken into account to verify whether these differences are worth exploiting, to determine fertilizers and critical levels specific to different ecotypes and hybrid varieties, and even to examine whether it would eventually be possible to find planting materials that are less demanding of inputs.

Chemical fertilizer application

Chemical fertilizer should be applied either after the heavy rains have passed, or early in the rainy season, preferably some months before the heavy rains will fall. For young palms, a split application is recommended, but when rainfall is heavy, and in adult plantations, split application may be considered to reduce leaching of fertilizers. Anyhow, it is almost inevitable that part of the fertilizers will be lost through leaching, especially in soils with a low cation exchange capacity. Surprisingly, Ochs *et al.* (1993) found that on deep sandy soils in Ivory Coast results did not reveal the slightest difference between single applications or three split applications of fertilizers. They concluded that these deep sandy soils (5-10% clay) are capable of slowing the downward movement of fertilizer solutions sufficiently for the very deep coconut root system in such soils to be able to absorb them, even with a single application per year.

It may be expected that a larger portion of fertilizers will be taken up where root density is higher. As the root system of the coconut palm spreads out from the stem in widening circles, it is logical that with increasing distance from the stem, root density becomes lower. Therefore, it is recommended to apply fertilizers within a circle of about 2 m from the stem. Avilan *et al.* (1984) in Venezuela observed that irrespective of cultivars and plant age, most roots were met in a circular area from the stem to half the radius of the rim of the canopy. Anilkumar and Wahid (1988), using labelled P as a fertilizer, observed that in a laterite soil consisting of 57.5% sand, 16.5% silt and 26.0% clay, 82.5% of the root activity occurred within a radius of 2 m around the stem.

However, root system development depends very much on soil type and in lighter soils roots will grow out further than in heavy soils. Root development is also influenced by cultural practices. Fertilizer application stimulates root growth. Pomier and Bonneau (1987) found that the effect of fertilizer application on the root system was about the same, whatever the distance between the roots and the stem, *i.e.* both outside and inside the circle of application, meaning that the relation between fertilizer and distance from the stem was not significant. This shows that mineral fertilization encourages overall development of the root system without a significant increase in the weight of the roots immediately below the place where fertilizer is applied. Due to a better root system development, and the spread of roots over a wider area by fertilizing, a larger volume of soil may be exploited by the roots, which in turn may reduce water-stress during dry periods.

The use of coated fertilizers for slow release of nutrients may be recommended, especially in soils with low cation exchange capacity. However, such fertilizers are not always available and are more expensive. In Tuvalu (Anonymous 1986a) it was observed that 0.5 kg of potassium fertilizer applied in biodegradable bags produced the same level of foliar K as 1.5 kg broadcast, indicating a more efficient use of fertilizer.

The time required for a yield increase as a response to fertilizer application is considerable. During the first year after fertilizing, leaves that were yellowish due to a deficiency may turn green. Mature leaves will remain longer on the tree. As a result of a better condition of the palm, premature nutfall may be reduced in the second half of the year after fertilizing, resulting in a slight increase in yield in the second year after fertilizer application. As a result of more healthy leaves with an increased photosynthetic capacity, fruit set will improve during the second year and a considerably higher yield will be obtained in the third year. Gradually, the number of female flowers will increase and fruit set will improve and the full response to fertilizer treatment may be obtained from the fourth year onwards.

Nitrogen

Nitrogen application usually has a beneficial effect on the development of young palms. Nitrogen deficiency may occur in intensively grazed plantations without adequate fertilizer application, or otherwise exhausted lands. In an adequately managed plantation, underplanted with leguminous cover crops, additional nitrogen supply is hardly necessary.

Ouvrier (1990) distinguished two groups of elements in 89-month-old palms: those for which the stem and the crown have roughly the same mass, *i.e.* N, K, and Cl, and the others for which the leaves represent over 50% of mineral mass. Braconnier *et al.* (1992), analysing four-year-old coconut trees, 3.5 months after applying isotopically labelled nitrogen, found that total nitrogen distribution in the plant is related to the distribution of dry matter. Thus, the leaves contained 66% of total nitrogen of aerial parts. Percentages of bunches and stem were 24 and 14 respectively. Differences between these findings are probably due to age differences between groups. The highest labelled N percentages were recorded for the spear, bud and green spathes, indicating that developing organs are considerable sinks of nitrogen. By contrast, stipules and dry spathes, which did not grow and had low physiological activity, had low-labelled N percentages. However, the percentages in the stem were surprisingly high, although growth of this organ was nil. The same situation was found in old leaves. Apparently N fertilizer was distributed in all leaves, indicating that labelled N in different organs has a rather rapid turnover. These observations confirm the existence of an influx and afflux of nitrogen in each leaf, whatever its rank. The proportion of nitrogen in coconut seems to be in continuous flux. Labelled nitrogen was distributed in all parts of the coconut, except in stipules and dry spathes, indicating N distribution throughout the tree. Little nitrogen fertilizer was distributed into mature bunches, indicating that N nutrition of the albumen comes mainly from the husk and shell. Thus, nitrogen fertilization would influence production only 8 to 10 months after fertilizer application. The critical level of N in a leaf of rank 14 can be fixed at 1.80-2.0 per cent of dry matter for tall varieties and 2.2 for the hybrid PB-121.

In a trial with different N fertilizers conducted on sandy soils, Manciot *et al.* (1980) observed that ammonitrate mostly increased the weight of the palms, the foliage benefiting most. Urea, on the contrary acted very strongly on root development, particularly on root number, not on root diameter, whereas ammonium nitrate did not increase the fresh weight of the plant. Tarigans (1991) found that spraying urea on coconut seedlings does not affect stem girth or leaf splitting. The effect on nut number can only be observed with correct nutrition, which may be depressed when the plant's nitrogen contents increase (Pomier and Benard 1988). It is generally experienced that when nitrogen considerably increases the number of nuts produced, it very significantly lowers the copra content of the nuts (Manciot *et al.* 1980). Anilkumar and Wahid (1989) observed that the application of ammoniacal fertilizer increases the availability of Mn. Ballad (1991) observed that N fertilizer did not show any effect on leaf P, Ca, and Mg content. A trial in Jamaica showed that N had a negative effect on foliar K and Na content, and a positive effect on female flowers per inflorescence, fruit set, nuts per bunch and per palm, and bunches per palm (Anonymous 1990b).

In case of nitrogen deficiency, heavy doses of nitrogen fertilizer may result in rapid improvement of leaf colour and photosynthetic capacity, giving earlier yield responses than with other elements. Table 32 presents results of a fertilizer experiment on a heavy soil (Porto Bello) and on a minerally poor sandy soil (Barra), both deficient in nitrogen and potash, carried out by the author in the plantations of the Companhia do Boror in Mozambique. Fertilizers were broadcast. Treatments started in May 1955. Fertilizers included sulphate of ammonia, superphosphate, and KCL (50%). The amounts of each fertilizer used were: 1955 - 2.5 kg; 1956 - 2 x 2.5 kg; 1957 - 1.5 kg. CL deficiency was not likely, as the plantations were near to the sea with tidal seawater occasionally coming in. (A = yield during June 1955 - June 1956; B = yield during June 1956 - June 1957.)

After initial application of fertilizer, in all N treatments on both soils, palms reacted strongly within one year, the canopy colour changing from light, yellowish green to dark green. The reactions to the other elements, when applied without N, were rather insignificant. Reactions in the third year were much stronger, and the treatments NK and NPK showed beneficial reactions to K. Unfortunately, the data on this third year are no longer available. The late response to K may have been influenced by reduced exchangeable K in the soil as a result of the application of ammonium sulphate, as was found by Anilkumar and Wahid (1989), indicated in the section on potassium (below).

Also, Manciot *et al.* (1980) reported that only after N levels in the palms had normalized, did a potassium deficiency become apparent. Davis *et al.* (1985) applied 500 g N to coconuts growing in a rich volcanic soil deficient in nitrogen and with a low water-holding capacity. Significant nut yield increases were observed only in the third year. N in combination with P, K, Mg, and micro-nutrients showed less yield improvement than N alone. Difference in reaction to N application when compared to the trial in Mozambique was probably caused by differences in soil conditions and the severity of N deficiency. In Mozambique, where coconut was growing on flat low land, the applied N might have been available within reach of the roots in the soil water, whereas the palms growing in the second trial were growing on a slope, the soil water flowing downwards through the soil, the N applied to the palm being available in the top soil only during the wet season, and the fertilizer reaching deeper soil layers being transported beyond the palm's reach.

Nitrogen fertilizers should be applied at the beginning of the rainy season. In India it was found that under conditions of heavy rainfall as much as 70% of the applied nitrogen may be lost through leaching. Under such conditions, slow release nitrogenous fertilizers are more efficient compared to straight nitrogenous fertilizers. Planting of leguminous cover crops may be very beneficial under such conditions. When legumes are being planted under coconuts on a poor soil, it might be recommended to apply a dressing of N fertilizer first, to speed up the improvement of the plantation. With legumes alone, it takes longer than with chemical fertilizers alone.

Phosphorus

Although phosphorus is an indispensable nutrient for coconut, the extracted quantities mentioned above indicate that it is not the most important element in coconut nutrition as far as quantity is concerned. Phosphorus is rarely a limiting factor for coconut production. There are no characteristic visual symptoms of P deficiency, apart from slowing down of growth and shortening of fronds (Manciot *et al.* 1980). Sometimes it is even stated that coconut does not respond to phosphate fertilizers. However, situations where soil P levels are too low for adequate coconut production can occur. The critical P level in leaf 14 is fixed at 0.120 per cent.

Table 32: *Yield response to fertilizer treatment in Mozambique (nuts per palm)*

Treatment	Heavy soil		Sandy soil	
	A	B	A	B
N	16	41	11	20
P	19	17	14	13
K	21	24	10	12
NP	23	34	10	19
NK	16	39	19	23
PK	24	22	12	20
NPK	13	29	11	25
Control	29	23	12	13

Although significant relationships may be found between soil-available P and leaf P, leaf P values are not always related to yield. According to Wahid (1984), in many experiments P levels of none of the leaf ranks correlated significantly with yield in coconut. Sometimes available P in the soil has accumulated as a result of repeated P fertilizer treatments over a number of years. Kahn *et al.* (1983; 1985a) concluded that utilization of built-up reserves in the soil is the most ideal and economic way of management of coconut groves, suggesting that P application might be skipped for a number of years in situations where soil-available P is around 20 to 25 ppm in the main rooting zone in the coconut basin. This is well below the level of 120 ppm, above which in Ivory Coast P manuring had no more effect on leaf P level or yield (Manciot *et al.* 1980). They observed that plant P levels were not influenced even when soil-available P levels drained to 59 ppm at 0-30 cm depth and 10 ppm at 30-60 cm depth, and also when the soil available P increased considerably. The yield variations were never statistically significant. They also observed that even after continuous P fertilizing over 28 years, the foliar P levels did not reach the critical level of 0.12 per cent. Cecil and Khan (1993) found that in India, after a period of 22 years of phosphorus application, withholding of further phosphatic fertilizers for a period of 14 years did not affect the yield and nutrition of coconut palms. VA mycorrhizal infection and the population of P solubilizing bacteria was considerably higher in the root zone of palms where P application was withheld, compared to palms receiving P fertilizers. This suggested the beneficial association of micro-organisms in supplying P from accumulated soil reserves.

Khan *et al.* (1985b) conducted an experiment with different forms of P fertilizer on coconut- growing in a lateritic gravelly soil with clay loam structure with a pH of 5.2-6.4. As a result of a 6- year fertilizer application, the soil-available P increased considerably, regardless of the P-carrier used, and in accordance with the levels applied. Nitrophosphate was found to increase available P in the 0-30 cm depth of profile at the higher levels of application to a higher degree than ammonium phosphate. Where lower P levels were applied (160 g per palm) with nitrophosphate and rock phosphate, significantly higher quantities of P remained in the 0-30 cm depth. There was slight mobility of P to lower depths. Regardless of the P-carrier, the mobility of P increased with the increase in the level of P application. It is suggested that when the larger part of the roots is found below 30 cm depth, and as P moves to the roots mainly by diffusion, a more rational approach to P fertilization of tree crops would be by way of deeper placement. Also Anilkumar and Wahid (1989) observed that continued application of superphosphate over a long period can enhance the movement of P into deeper layers even in acid soil. In this respect, the behaviour of P in the soil is more or less similar to that of S. But superphosphate dressings over the years did not improve P uptake. A similar observation was made on the foliar S level of the palms receiving ammonium sulphate.

Differences observed in various studies on soil P levels and P fertilizing and the effects on leaf P levels are probably due to differences between soil types. In Sri Lanka most coconut soils are P-deficient, It is the only country where consistently significant effects on nut and copra yield have been obtained by P fertilizer application (Loganathan *et al.* 1984). P deficiency in the soil is confirmed by the low leaf P-levels in palms. In India, adult palms have been only occasionally found to benefit much from annual P applications. Fertilizer experiments have shown that the P requirements are low, and responses slow and inconsistent. However, deficiency of P in the soil retards growth and delays flowering and the ripening of the nuts. P application has been found to increase girth at collar, number of leaves and rate of leaf production in seedlings (Khan *et al.* 1985a).

Romney (1987) reported a strong response to P application of PB-121 and EAT seedlings growing in Tanzania on red-brown sandy loam soils very deficient in P. Within 5-12 months, the growth of the seedlings of these varieties increased by 20 and 15 per cent respectively, compared to control. By the age of 4.2 years, East African Tall (EAT) palms receiving P had more leaves; for hybrids, P increased flowering, number of bunches and fruit set. Hybrids tended to respond to the higher level of P application by higher foliar P levels, and they continued to respond up to a level of 0.135% compared to 0.11% for EAT. Application of P raised foliar N and Mg. In Jamaica it was observed that P had a positive effect on tree growth, number of inflorescenses produced, and on the number of nuts per bunch (Anonymous 1990). Phosphate fertilizer application over long periods may reduce soil pH, especially where ammonium phosphate is used. This may increase the availability of Mn in the soil. However, prolonged use of this fertilizer may lead to the erosion of Mn fertility even from very low soil depth through excessive dissolution and leaching (Anilkumar and Wahid 1989). In view of the low P requirements of coconut, rock phosphate might be the best P fertilizer for coconut.

Potassium

The causes of potassium deficiency are essentially pedological, as soils rarely contain the large amounts of potassium required by the coconut. Only on soils of volcanic origin, with high exchangeable K, is the need for correction insignificant (Ollivier 1993b). However, with time all soils will become deficient without fertilizer application, especially if high K-consuming intercrops, such as cassava are planted under coconuts. On potassium-deficient soils, the effects of potassium fertilizers on coconut are notable in all the production factors, such as number of inflorescences, number of bunches, number of nuts per bunch, total nut production and also copra percentage of the nut. But it also has a positive influence on the growth of immature palms. Improved growth in early years results in increased production at maturity, and the effect of potassium deficiency in young palms cannot be completely corrected by K fertilizer application at maturity. Potassium fertilizer should be applied annually, which is more effective than a double dose every two years (Manciot *et al.* 1980). The critical level of potassium is about 0.8-1.0 per cent of dry matter for tall varieties. For the PB 121 hybrid, the level is between 1.20 and 1.40% (Ollivier 1993b). Deficiency symptoms are not clearly visible until leaf potassium content drops below 0.5%, *i.e.* once trees are already suffering from severe deficiency (Ollivier 1993b).

Anilkumar and Wahid (1989) observed that application of muriate of potash increased available K and organic carbon in the soil. The effect on soil-exchangeable K, however, was N-dependent, as revealed by the significant N x K interaction. They found that higher rates of ammonium sulphate application led to reduced exchangeable K. This effect could be due to the removal of K^+ ions from the exchange sites by NH_4^+ ions and/or by H^+ ions generated in the nitrification process. However, in general there was a build-up of exchangeable K in muriate of potash-receiving plots. The accumulation was observed in lower layers down to 75-100 cm. They also observed that the application of superphosphate over the years reduced the exchangeable K content of the soil significantly. This effect could perhaps be due to the influence of Ca present in the fertilizer material on replacing K from the exchange sites in the soil. In Jamaica it was observed (Anonymous 1990) that K fertilizing had a negative effect on foliar N, Mg and Na levels.

Ouvrier (1984b; 1987) found that the application of KCL to hybrid coconut increased the K and Cl contents in the husk, not in the albumen. Application of potassium decreased the exported amount of N, P and S. The best correlations between leaf content and nut content were found for potassium, chlorine and magnesium. Any increase in leaf K content corresponded with an identical increase in the husk. The linear relationship observed between potassium applied and potassium removed, indicated little unnecessary consumption. For this reason, the theoretical savings made by returning the husks to the soil also increase with the quantity of fertilizer applied. It was also observed that without dehusking in the field, tall coconuts exported more potassium than the hybrids per ton of copra produced. This may result from the greater proportion of the husk in the fruits of tall palms. Due to K:Mg antagonism, KCL application had a depressive effect on the Mg content of copra. It slightly reduced the quantities of N, P and K. With calcium, a depressive effect of potassium and magnesium fertilizer was observed. Application of kieserite increased Mg contents in the husk and albumen. The shell does not contain P, Ca or Mg. In fertilizer trials in Tuvalu (Anonymous 1986a) it was observed that potassium had an antagonistic effect on boron.

Chlorine

Plant analyses have shown that Cl is an essential element in coconut nutrition. Cl deficiency results in a decline in the number of nuts per tree, and even more in copra per nut. The use of sea salt as a fertilizer for coconut has been a common practice for a long time in several coconut countries, such as Indonesia, India and Sri Lanka, without anyone knowing the essential benefit of this treatment. Ollagnier and Ochs (1971) observed in Colombia that application of KCl led to considerable coconut yield increases and leaf Cl levels, without altering the leaf K level very much. The yield increases could therefore be attributed to Cl rather than to K.

Magat *et al.* (1986; 1988a) observed a linear increase in nut production, copra weight per nut, and copra yield per palm with increasing rates of NaCl application. Lcaf Na and Cl concentrates also increased significantly, whereas other elements remained unaffected. They concluded that expensive KCl could be substituted by inexpensive NaCl. On the basis of experimental results, Ollagnier *et al.* (1983) proposed the adoption of a critical level of 0.25 per cent in leaf 14. They also suggest that Cl is probably of more importance in precocious and fast-growing material such as D x T hybrids than in talls. Magat *et al.* (1988c) suggested that the average optimal level of leaf Cl in leaf 14 could be 0.55%, and the critical level 0.30%. Actually, the critical level is considered to be between 0.50 and 0.60 per cent (Wuidart, personal communication).

In a trial conducted on inland-upland soil in The Philippines, using different sources of Cl, yield of coconut, especially in terms of copra per nut and per palm was significantly increased by Cl fertilization (Magat and Padrones 1984). Increases amounted to 3.84, 3.31 and 2.95 tons copra per ha per year due to NH_4Cl, KCl and NaCl fertilization respectively. Margate and Magat (1988) observed that Cl application to mother trees significantly influenced earliness and total percentage of germination of seednuts, girth, and leaf production of seedlings, indicating that the growth of seedlings at the nursery stage probably depends on the nutrition of the mother palms. Magat *et al.* (1991) observed clear evidence of positive residual effects of Cl fertilizers for 3-5 years after fertilization of either KCl, NaCl or NH4Cl on the yield of coconuts growing on a loamy clay soil with clay subsoil in The Philippines. During a period of five years after termination of the fertilizer applications, yields remained higher in former Cl plots when compared to former control plots.

Calcium

The quantity of Ca removed by the coconut palm is much higher than that of P. The function of calcium is mainly concerned with proper growth and functioning of stem and leaves rather than with the palm's productivity of nuts (Cecil 1991). A limited survey of different ecological conditions in Africa and Asia showed that coconuts can adapt to a range of calcium leaf contents of 0.20% - 0.50% without adverse consequences. Manciot *et al.* (1980) concluded that a Ca level of 0.30-0.40% in leaf 14 is satisfactory. According to Wuidart (personal communication) Ca deficiency in Ivory Coast was observed only in dwarf seedlings in the nursery, but never in talls or hybrids. Yellow and red dwarfs appear to be the most susceptible varieties. In adult palms the critical level varies between 0.40 and 0.60% Coconuts planted on soils of coral origin even show leaf calcium contents that can be as high as 0.90%. On such soils with an extremely high calcium content, serious nutritional disorders may occur, such as deficiencies in K, Mg and Fe. Calcium deficiency does not occur frequently. The deficiency observed in the Dabou seedgarden in Ivory Coast was the first known case of a situation where calcium nutrition had become very insufficient. Leaf contents in these palms were very low, less than 0.1% (Dufour *et al.* 1984). It was also observed that different coconut varieties may have different calcium contents, and calcium contents of the same variety may differ greatly under different conditions, without the trees showing apparent symptoms. In general, heavy liming is not necessary. Nevertheless, regular additions of Ca through Ca-bearing fertilizers, such as rock phosphate and superphosphate or other liming materials, may be adopted for supplying the Ca requirement of the palm (Cecil 1991). Anilkumar and Wahid (1989) observed that the Ca content of the leaves increased steadily with increasing levels of applied superphosphate. A tendency for the exchangeable Ca status of the soil to improve at the cost of exchangeable K in plots receiving superphosphate was also observed. Romney (1987) reported that liming reduced N and P levels significantly.

Magnesium

Mg deficiency is a rather common phenomenon in coconut plantations, particularly on acid sandy soils. Low soil Mg content may be responsible for this, but continuous heavy K fertilizer applications may also induce Mg deficiency. Dwarf coconuts are more sensitive to magnesium deficiency than talls and hybrids. The critical magnesium level for a rank 14 leaf on an adult palm is fixed at 0.24 per cent of dry matter for tall coconuts and 0.20 per cent for the PB-121 hybrid (Ollivier 1993a).

Habana *et al.* (1987) studied the influence of various sources of Mg on yields, using dolomite (Mg Co_3 - 18% MgO), Epsom salts ($MgSO_4$ - 16% MgO) and kieserite ($MgSO_4$ - 24% MgO). During the first three years, Mg fertilizers did not increase nut production per palm significantly. However, a significant increase in nuts produced by kieserite treatment was noted during the fourth and fifth years. This was probably due to sulphur, because the leaf S content that was originally below the critical level was corrected by the application of kieserite. Epsom salts showed consistent improvement of copra per nut from the second to the fifth year of the experiment, and also dolomite improved copra yield per nut during the fourth and fifth years. Correlation analysis revealed that nut and copra per tree were positively related with leaf S, whereas copra per nut was positively related with Mg. Epsom salts dissolves rapidly in water, kieserite is slower and dolomite dissolves very slowly. Dolomite can be applied as a basic treatment in the same way as rock phosphate is being used. For rapid correction of deficiencies, Epsom salts and kieserite are to be preferred. Further deficiency can be prevented by the occasional use of magnesium limestone. It is very important to maintain a proper balance between Mg and K supply, which can be checked by leaf analysis.

Sulphur

According to Taysum (1981), sulphur is often considered to be a 'secondary nutrient' alongside calcium and magnesium. However, it is extremely important to coconut. Not only does it influence the overall growth of the palm but also its ability to form healthy, well-filled fruits. On the many islands of the Pacific, coconut is fertilized with doses of elemental sulphur, ranging from 100 to 500 g per tree per year. He observed sulphur deficiency in various coconut estates in Mozambique, which was enhanced by a change in the fertilizer policy. As long as the estates were treated with ammonium sulphate, deficiency symptoms were not apparent. However, as soon as urea was used as N fertilizer, leaf colour turned yellow-orange, with numerous fungal lesions. He concluded that the application of 'nitrogen only' was strongly contra-indicated. This was confirmed in Madagascar, where it was found that imbalanced N:S ratio may result in serious nutritional disturbances. In the sulphur-deficient Markham Valley in Papua New Guinea, an examination of the N:S ratio indicated a linear correlation between this ratio and the growth of seedlings as measured by height.

Manciot *et al.* (1980) observed a deficiency below the leaf S content of 0.13 and they fixed the critical level of sulphur content at 0.15 - 0.20 per cent. Magat *et al.* (1988a), in The Philippines, found that the critical and optimum levels of leaf-S in leaf 14 were 0.12% and 0.19% respectively. Silva *et al.* (1985) stated that for coconut the sixth leaf was the most sensitive for factors involving S treatments. S concentrations in the leaf tissue were negatively correlated with applied levels of sulphur. The total content of sulphur per nut as well as the sulphur concentration in the fresh kernel followed identical patterns. In both tissues the minimal concentration of sulphur was associated with a relatively higher level of sulphur supply. On the other hand, yields of fruits and copra increased with treatments. They assumed that a resumption of the supply of S probably generated a rapid increase in growth and production, causing in effect a diluting effect on levels in the plant tissues. The response curve was concave, suggesting that when growth is normalized, tissue nutrient levels respond positively to treatment. Sulphur application reduced nut size, but this was more than compensated for by an overall increase in nut yield. Southern (1967b) observed that two years after application of 0.9 kg sulphur the copra quality improved from category 0 (extremely rubbery) to category 4 (normal), whereas the oil content increased from 40 to 65%. Magat *et al.* (1988d) fertilized S-deficient coconuts with gypsum ($CaSO_4$). They observed significant yield increases in number of nuts, as well as in copra per nut and copra per tree. The response in nut per tree was quadratic, whereas that in copra yield was linear. Leaf-S showed a quadratic response. They concluded that the critical S level in the leaf is 0.13% and the optimum level is 0.19%. The latter is within the range of 0.15% - 0.20% indicated by Manciot *et al.* (1980) as critical. The differences observed in copra per nut, compared to the study by Silva *et al.* (1985) can probably be explained by quantities of S application in relation to the initial state of S deficiency.

Sulphur being associated with nitrogen in protein composition, its critical level also depends on the nitrogen nutrition of the palm. The N/S ratio should be between 10.1 and 13.1. In Indonesia, results of recent studies on young palms growing on peat soil indicate values of 0.16 and 0.17 for sulphur and 1.87 and 1.86 for nitrogen in leaves 4 and 9 respectively (Wuidart 1994a) The N/S relationship may explain the slightly different critical levels presented by different authors.

Zinc

Successful use of zinc sulphate, at a rate of 200 g per palm, was reported by Vijayaraghavan *et al.* (1988) in the coastal belt of the State of Tamil Nadu, India, where leaf samples had indicated a deficiency. Yield increases were as high as 69.5%. Zinc levels increased significantly by means of P or Mg application, whereas in the case of N or K application, they were lowered. Barr (1993) in Kiribati observed an antagonism between Fe and Zn. There was also an indication that N reduces Zn assimilation.

Manganese
Manganese deficiency may occur on alkaline soils. However, it is difficult to define a critical level for this element. It has been shown that 30 ppm Mn in leaf 4 of 2-year-old palms was insufficient, and that manganese applications should be associated with iron. Manganese sulphate applications should be accompanied by Fe sulphate applications. Southern and Dick (1967) observed that manganese tended to accumulate in older leaves. Leaf values for young leaves were fairly similar, differences in manganese content for older leaves were very large. Anilkumar and Wahid (1989) observed that the application of ammonial fertilizers increases the availability of Mn in the soil due to lowering of pH, but prolonged use of such fertilizer leads to erosion of Mn fertility through excessive dissolution and leaching.

Iron
According to Manciot *et al.* (1980), the diagnosis of iron deficiency by leaf analysis is tricky, as it had not been possible to define the critical level in the leaf with sufficient precision. Consequently, coconuts on poor soil can show deficiency symptoms when the iron level in leaf 14 is 45 ppm of dry matter, whereas other palms growing in rich soil are still green at 30 ppm. Nevertheless, visible signs of ferrous deficiency (CP 30 and 33) can be expected to appear at values below 50 ppm. Trewren (1987) reported that on coralline soils in Tuvalu, for a tree with a stem of over two metres high, 30g ferrous sulphate appeared to be a safe level of application. Injection of ferrous sulphate to iron-deficient seedlings had a striking effect on their re-greening and weight increase. Results of a trial indicated that stem injection up to 20 g per tree should be repeated in alternate years, whereas at the 30 g level every third year would be adequate. However, it was found that the uptake after re-application was considerably reduced, probably due to dead tissue surrounding the injection hole. This is a serious drawback, since a new hole will have to be drilled each year and after five or six injections the base of the trunk would weaken severely. Therefore, new techniques of Fe application have to be developed. It was also found that injections of small quantities of ferrous sulphate into tree trunks had a positive influence on the assimilation of nitrogen and sulphur, and a negative effect on the levels of phosphorus, sulphur, calcium, magnesium, sodium and manganese in the palm.

Boron
Manciot *et al.* (1980) reported varietal differences in sensitivity to boron deficiency, Malayan dwarfs being more sensitive than talls, and the hybrids having intermediate reactions. Jayasekara and Loganathan (1988) observed boron deficiency in young palms growing on a clay-loam soil in Sri Lanka. Leaf analyses from the third leaf showed a boron content of 5.9-7.5 ppm in the early stages and 3.4-5.6 in the advanced stages against values of 7.6-10.0 for healthy palms. Based on additional observations on older palms they concluded that B-concentration of 8-10 ppm can be suggested as the critical leaf nutrient concentration range for young coconut palms, below which B deficiency is likely to occur. According to Wuidart (personal communication), the critical level for B in leaf 14 is between 10 and 13 ppm. Boron deficiency can be corrected by the application of borax (sodium tetraborate). Borax is usually applied at rates of 30-50 g per palm. Jayasekara and Loganathan (1988) observed that palms in an advanced stage of deficiency failed to recover after the application of sodium tetraborate. Other palms recovered completely within 8 months after treatment with 28-56 g of sodium tetraborate dissolved in 8 l of water and added to the soil within a radius of 1 m from the stem. As explained earlier, the addition of potassium may increase boron availability in soil with medium or low pH levels. Ohler (1984) reported the appearance of B deficiency symptoms in some coconut-growing areas; yet the application of borax did not change their appearance. In Tanzania, the application of potassium fertilizers in some cases cured such palms. This may have been due to the formation of potassium tetraborate, making the boron that was fixed in the soil available to the palms.

Copper
It is not possible to make value judgements about copper nutrition in young palms based on leaf contents if the coconuts are deficient, with reduced leaf mass varying with the deficiency stage. It is only possible to compare palms with a slight deficiency or none at all, with crowns of equivalent, or almost equivalent leaf mass. Early and substantial applications of copper sulphate can considerably attenuate, but not totally stop, fall in Cu content. Once this stage has been reached, the content gradually rises as the trees grow older, reaching values of 4-5 ppm by the time the palm starts bearing. For older palms it is assumed that the critical level is around 2 ppm Cu in leaves 4, 9, and 14, and the optimum level is reached at 3 ppm. On peat soils, copper should be applied as early as possible. In trials, an application of 2 g per seedling has been sufficient; 5 g per tree caused slightly phytotoxic effects. Deficiency trials showed that it was possible to apply up to 150g of copper sulphate to the planting hole without risk of phytotoxicity on young coconuts, but an initial dose of 20 g was enough to protect the coconut for at least 2 years. For security, and according to the leaf copper content, additional and higher doses of copper sulphate can be given later. Leaf copper contents in commercial plantations gradually increase with age, indicating that copper deficiency can affect young coconuts during the first two years after planting, but becomes increasingly unlikely thereafter (Ochs and Bonneau 1988).

Heavy metals
Concentrations of various mineral nutrients in coconut can be influenced by the presence of heavy metal solutions in the rooting medium, as was shown by trials carried out by Biddappa *et al.* (1988) and Bidappa and Bopaiah (1989). This is a fact to be taken into consideration when analysing nutrient concentrations in the various parts of the coconut. It is recommended to check possible abnormal concentrations of heavy metals in the soil when studying the results of leaf analyses.

Harvesting
Ripe nuts tend to dry up. Their skin turns brown and the volume of nut water declines. Such nuts, when shaken, make a splashing sound. According to Shivashankar (1991), coconuts can be harvested at 11 months when still green, at which stage the oil synthesis is almost completed. Jayasuriya and Perera (1985) observed that dry matter accumulation ceases after 11 months, but according to Thampan (1981) maximum oil content is reached when the nuts are 12 months old, the reduction in the percentage of oil in 11, 10 and 9-month-old nuts being 5, 15 and 33% respectively. Differences in observations may have been due to differences in varieties observed. There is no appreciable difference in the yield of coir fibre between 10 to 12-month-old nuts, but fibres of younger nuts are more pliable and are lighter coloured. According to Taffin and Ouvrier (1985), hybrids, or rapidly germinating varieties should be harvested more often. Dwarf varieties which sprout in 45-60 days should be harvested every month, PB-121 hybrids, which sprout in 70-90 days are harvested every two months, whereas WAT, which sprouts in 100-150 days can be harvested at three-monthly intervals. They recommend harvesting of bunches when two or three of the nuts have turned brown. Harvesting is still one of the weak spots in coconut cultivation, because mechanization is barely possible for practical as well as for financial reasons. Access to trees with machinery is often impossible.

A survey conducted in a replanting project with hybrid coconuts in Indonesia indicated that all farmers used hired labour to harvest coconut through a piece rate system. Under this system, hired labour earned as much as 50% of the copra value (Amrizal *et al.* 1989).

Reaping
Some coconut varieties keep their nuts attached to the bunch for a very long time, sometimes even up to their germination. Such trees should be harvested by climbing or other means. However, many coconut varieties drop their nuts to the ground when mature and turned brown. Reaping the nuts can be a good solution if it can be done at short intervals. However, this method also has some disadvantages.

The greatest disadvantage of letting nuts fall to the ground is facilitating theft, which is often a very serious problem, especially in densely populated areas. Most coconut farms are not fenced and cannot be fenced because they are crossed by paths or even roads linking houses and villages. Fencing would not prevent all theft; it would only make it more difficult. Another disadvantage would be the tendency to leave the nuts in the field too long, risking the start of germination because of humidity. Germinated nuts have a lower kernel content and are unsuitable for desiccated coconut or coconut cream, as the kernel and the oil have begun to deteriorate due to biochemical changes that have set in (Ranasinghe 1995). Thirdly, leaving fallen nuts in the field would facilitate pest attack, such as by rats and other rodents. Some nuts may also get lost when the undergrowth is tall, as is often the case in smallholders fields. Sprouted nuts may be left in badly managed plantations, developing into trees and causing increased density of the plantation.

Climbing

The number of skilled labourers available to climb coconut palms becomes increasingly smaller and the labour cost for this hard work is rising. Cutting steps in the stem of the palm for easier climbing is very bad management. It affects the palm's production potential and may attract borers. In some countries, trees are climbed without any accessories, in others ankle-rings, waist-rings or hand ropes are used to provide the climber with more support. The number of palms that can be climbed per day also varies substantially from place to place, according to the height of the palms and the ability of the climbers.

Various ingenious devices have been invented to enable non-skilled workers, particularly breeders, to climb coconut palms. However, they have never been put into large-scale practice. In India, a climbing device was invented that helps a man climb even the tallest coconut tree in 2 minutes. The climbing kit consists of 2 iron frames that can be attached to opposite sides of the tree trunk by an adjustable belt of wire rope encircling the tree. Each main frame has a sliding sub-frame with a foot-rest at the bottom and a handle linked by a locking mechanism. (Anonymous 1985b).

The number of nuts a climber can harvest per day varies considerably. It depends on the height of the trees, their productivity, the intervals between climbing rounds, the number of bunches harvested per round, and the ability of the climbers. Taffin and Ouvrier (1985) report that for a coconut grove producing 12-15,000 nuts per ha per year, from trees 7-10 m high, 2,500 nuts per climber per day is easily attainable. Ohler (1984) observed 400-800 nuts per day in a grove with trees of 15-20 m high, yielding about 5-6,000 nuts per ha per year and harvested at monthly intervals. Pomier (1984) observed a number of 3,600-4,200 nuts per day, but did not indicate the conditions under which this high number of nuts was harvested. A better way of comparison is the number of palms that can be climbed per day. Nair and Menon (1991) mention a number of about 40 palms per day in India, which is about the same number as the author witnessed in Mozambique. After the nuts have been cut off and fallen to the ground, they are collected, stacked and counted. When whole bunches are cut, some nuts may detach and roll away, but most remain attached to the bunch, facilitating collection and stacking. In Mozambique, strings of the husk were cut loose with a knife, one end of the string remaining attached to the husk, and with these strings the nuts were bound together in groups of five, facilitating counting and transportation to the vehicle. Usually, when in the tree canopy, the climber will also remove dry leaves and spathes or any other dry matter. When in need of money, or to avoid the risk of fallen nuts being stolen, farmers may tend to harvest immature nuts. If harvested too green, such nuts produce rubbery copra which will fetch a lower price and which is unsuitable for desiccated copra. There are various ways by which the climbing problems can be overcome. Unfortunately, such solutions often bring other problems. The following practices could be considered:

Harvesting with Poles

Harvesting nuts, especially whole bunches, with the use of long poles with a sickle-shaped knife attached to it (CP 18) may be the best solution, although the use of this method will be limited to a certain height of trees. The poles are usually made from bamboo, but according to Taffin and Ouvrier

(1985) also raphia palm rachis is used. Thampan (1981) reported that with climbing, a harvester can hardly cover 50 trees per day, whereas with a pole he can cover at least 250 trees. Frémond *et al.* (1966) indicate 150-500 trees per day, according to the height of the tree. Nair and Menon (1991) refered to 80-100 palms that can be harvested per man-day with a pole, against 40 palms by a climber. It is evident that the height of the tree is an important factor. The higher the tree, the less the difference between harvesting with poles and by climbing. Very high trees can no longer be harvested with a pole because it is impossible to maintain the equilibrium of the long pole.

Harvesting with Monkeys

Although monkeys are used in Indonesia, Malaysia and Thailand, it is not as easy and cheap as it may appear. Monkeys *(Macacus nemestrina)* need to be very well trained. Their daily capacity can be considerable. Up to 1000 nuts per morning have been recorded. However, their maintenance may be rather expensive, otherwise it might be expected that monkeys would be used more often.

Harvesting Intervals

Under conditions of well-distributed rainfall, a new inflorescence will be produced about every month, and an old bunch of nuts will be ready for harvesting. Where nuts are within easy reach, a harvesting interval of one month is recommended as it reduces the risk of theft or losing nuts in the undergrowth after falling from the tree. However, when trees grow tall, the farmer prefers to harvest as many nuts at a time as possible, to save labour effort and costs. He will therefore increase the harvesting intervals, harvesting two, and sometimes three bunches per harvest. This possibility depends on how long the coconut variety retains its nuts on the tree. Thus, one could harvest a bunch with several brown nuts plus one older bunch (all nuts brown) and one younger bunch (all nuts green). This method involves the risk that sprouting may begin in the oldest bunch, especially during humid weather or that too young bunches will be harvested. However, the advantages over a one-month interval are great and harvesting at two or three-month intervals is widely practised.

Harvesting slow-growing Trees

Planting slow-growing trees, such as dwarfs, would postpone the harvesting problem for some time, but finally dwarf nuts also grow out of reach. Moreover, there are many disadvantages connected with the cultivation of dwarf coconuts with respect to marketing possibilities and intercropping or mixed farming. Dwarfs are also much less hardy and are easily affected by drought. Theft would also pose a greater problem with dwarfs than with talls or hybrids. The growth of hybrids is slower than that of talls; however, the difference is negligible.

Toddy and Sugar Production

Although the coconut palm is commonly regarded as an oil crop, it can also be used as a sugar producer. Sap from young inflorescences is called toddy. In India, this sap is called 'Neera, in Sri Lanka 'Raa' and in The Philippines 'Tuba' (Ranasinghe, personnal communication). It has a rather high sucrose content. In Indonesia, analyses showed that the average sugar content varied between 12% and 14.4% (Darwis 1991). Naim and Husin (1984) reported a sugar content of 17% from Malaysia. This shows that there is great variation in sugar content between cultivars. The sap can be obtained by tapping inflorescences. Tapping is performed in most coconut-growing countries, either for the production of fermented sap as an alcoholic beverage, or for the production of palm sugar. The best time to start tapping is prior to the splitting of the inner bract and the emergence of the spikes from the spathe. Sap yield is much lower if the tapping starts after the spadices have burst open.

During the first week, the spathe is tied firmly over its entire length with strong string or with fibrous strands stripped off from petioles of young leaves, so as to prevent premature splitting of the inner bract due to expansion of the spadix. Twice a day, the outer surface of the spathe is then gently beaten all around with a wooden mallet. The tapping process is essentially an art, and results therefore depend upon the skill of the tapper. The technique consists of carefully bruising and rupturing the tender tissues of the floral branch by gently hammering and pounding the spathe. (Ranasinghe personnal communication). Special care is taken not to reduce the flower buds inside the spathe to a pulp, in which case the spadix becomes useless. After about three days, about 5 cm of the apical tissues is sliced off. During preparation, the spathe is bent downwards, so that later the sap can be collected in a container. To facilitate the bending of the spathe, the sheath covering the inflorescence may be split at the base. The bending is a delicate operation, as too much force would result in breaking the rachis tissues (Levang 1988). From the first day of tapping onwards, after beating, a very thin slice of the apical tissue is cut off (about 2 mm thick). This slicing is performed twice daily; beating is done in the morning only. This continues until the sap starts oozing out.

A spadix tapped prematurely may burst open and become discoloured, the discoloration starting at the cut portion. The same may happen to a spadix tapped when over-mature. In both cases toddy production may be much reduced because there will be a delay in the commencement of the toddy flow and a rapid spadix consumption, as each day the discoloured portion has to be cut off. Beating is stopped once the sap starts dripping. Santiago (1989) observed that the interval from the beginning of the tapping to the dripping of the toddy varied much. It differed from tree to tree and from spadix to spadix in the same tree. The commencement of the flow takes place as early as 5 days and as late as 32 days after the tip of the spadix had been cut off for the first time.

When the sap starts flowing, a container is placed under the dripping spadix. Santiago (1989) observed that the sap flow gradually increases and may reach a peak after 3-5 weeks, the peak may then endure for 1-3 weeks, after which the flow declines. The flow may continue for about a month until the length of the spathe is reduced to a length of about 10-15 cm. The longest productive period observed was 74 days. This may also depend on the spadix length. Palms with relatively low nut yields also have shorter productivity of the spadix than high yielders. The flow of toddy dwindles when the spadix is damaged by rats or insects, etc., and also when the spathe covering the inflorescence is removed partially or totally after it has cracked open. When the cut-end of the spadix touches the liquid in the container, the toddy flow may reduce as well. To prevent sap fermentation after slicing, the fresh cut may be brushed with some lime, or some lime water may be put in the container. Some farmers add some dried latex from mangosteen bark to the lime water (Naim and Husin 1984). The most effective harvesting cycle is twice a day. More harvests are not profitable in relation to the additional work, and fewer harvesters involve the risk of breaking the spathe under the excessive weight of the container, and they may also provoke other problems such as sap fermentation and drying of cuts. If, for any reason the spadix is not sliced for a period of two days, a type of healing latex exudes from the wound which impedes the sap flow. Two weeks of tapping will then be required to recover normal sap production. Sometimes a closed spadix has to be abandoned. One or two days' rest for the harvester may lead to a production loss of 15-20 days (Levang 1988). It is therefore important that tapping be performed by a team, whose members can substitute each other in case of sickness, etc.

About three weeks before the end of tapping, preparation of another spadix may begin. Naturally, the growing conditions of the palm and its vigour play an important role in toddy production. Tapping may be continued for periods of one year or longer. Contrary to a common belief, continuous tapping of coconut inflorescences does not affect the palm. After stopping the tapping, nut yields may show a temporary increase compared to the normal nut yield of the palm. According to Mathes (1984), these higher nut yields after tapping are due to a much higher number of female flowers being produced per spadix after tapping, with a similar percentage of fruit set compared to non-tapped trees. Toddy yields differ with palm variety, palm vigour and season. According to Ranasinghe (1995), the average yield of sap per day with 8 months tapping per year in Sri Lanka is 1500 ml. Santiago (1989) observed that sap flow is closely associated with the leaves' water content, suggesting an influence on sap flow by the internal water condition of the trees. The toddy yield decreased strongly after the relative water content of the leaves declined below 85%. It would be interesting to know if sugar content also varies with season and sap flow. Consequently, toddy yield was adversely affected by low rainfall, particularly when the soil water reserves were low and palms suffered from water-stress. As the flow of toddy decreases with increasing transpiration, high temperatures and low relative humidity have similar effects. Toddy yield was not significantly correlated with photosynthetically active radiation. It would be interesting to know if the sugar content of the sap correlates with photosynthetic activity. Sap production is relatively high at night, due to reduced transpiration and increased sap pressure. It was also observed that toddy flow had an inverse relationship with the total soluble solids content of the toddy.

Talls yield much more toddy than dwarf palms. A good tall palm may yield about 2 l per day, and hybrids even more. At a sugar content of about 15%, this would mean a sugar yield of 300 - 400 g per palm per day. Levang (1988) reports yields recorded in Indonesia of 300-600 g of raw sugar per palm per day, with an average (28 trees) of 438 g from coconuts growing under very favourable conditions. At 150 palms per ha, this would represent a production of 66 kg per ha per day, or almost 24 t per ha per yr, which is very high. If such yields are possible under very favourable conditions, yields of 12 tons of sugar per ha per year appear to be within reach of well-managed plantations under average growing conditions.

When a cluster of palms or an entire grove is tapped, ropes can be tied between palms, enabling the tapper to walk from palm to palm. The sap container can be emptied into a bucket and this bucket can be lowered to the ground with a rope, where an assistant may collect it. In this way, one tapper can treat 90 palms per day, whereas only about 25-30 palms may be treated if palms have to be climbed one by one; but in that case, no assistant is needed. With this system, one ha of coconut plantation used for toddy production would require about 2 tappers throughout the year, plus 2 helpers on the ground. Levang (1988) reported that a coconut tappers' daily remuneration was about two to three times as high as the wage of a common agricultural labour, referring to tappers who climbed each individual tree twice a day. Freshness of the sap is the main factor determining the quality of palm sugar. Fresh sap has a pH of 6 and any reduction in pH is indicative of deterioration. When the brix of the sap is 17 and pH 6, the sugar will set nicely and have a brownish glossy appearance and be 'grainy' when broken. Good coconut palm sugars can be obtained by gradually boiling fresh coconut sap to a temperature of 118-120°C and then cooling it. This sugar should have a moisture content of 8%. In the original packing of dried coconut leaves, deformation of the sugars starts after one month. The sugar can be hygienically bagged in oriented polypropylene laminated with polyethylene or polypropylene or thermoform container for a period of six months with slight loss of moisture (Naim and Husin 1984). Toddy production and the manufacture of sugar and other products can be a welcome source of income for smallholders, especially in situations where copra prices are low and marketing is difficult, and where sugar is being imported. It is recommended that sap yields and sap sugar contents of more cultivars be measured in different seasons in comparison trials. Also, the correlation between sap yield and sugar content and their nut yield should be studied.

PART III. COCONUT-BASED FARMING SYSTEMS

10. General Aspects

J.G. Ohler

Population increases in many parts of the world, the reduction of available arable land, loss of land due to erosion, and a decrease in soil fertility due to inadequate management, require better use of the land still available through more intensive farming and higher outputs through improved management systems. Multiple cropping is such a management system, which makes much better use of available resources such as land and water, often improving soil fertility and microclimate. As a matter of fact, a coconut-based farming system (CBFS) is an almost ideal example of agroforestry (CP 22 and 63).

Multiple cropping is already rather common in various countries. In Tonga, almost all other crops are planted under coconuts and the area of coconut represents about 80 per cent of all arable land (Ooi 1987). In India, Sri Lanka, The Philippines and Indonesia, intercropping in coconuts is a widely used practice. But even in these countries there is still wide scope for increasing the intercropping area under coconuts, or for intensifying the systems used. Although intercropping may offer many advantages, it also poses problems, often related to the farmer's management capacity, ecological conditions, and the locality of his farm. A sample survey of intercropping on coconut lands in Sri Lanka revealed some constraints of the system (Liyanage M. de S. *et al.* 1986) The seven most important problems that were faced by the farmers were, in order of importance:

- drought;
- lack of funds;
- price instability;
- lack of technical know how;
- problems of timely availability of labour;
- availability of planting materials;
- theft.

On average, each farmer faced at least three of these problems, lack of know-how and funds and non-availability of planting material being the more acute problems faced by smallholders, whereas drought, price instability and theft were reported as general problems affecting all categories of holding sizes. In addition, the marketing of perishable seasonal crops (fruits) and crops produced in bulk (tubers) could also pose a serious problem. In The Philippines, Arboleda *et al.* (1986) reported that the main problems faced by farmers were as follows:

- inadequate funds;
- delayed release of funds;
- high cost and/or absence of agro-inputs;
- difficulty in marketing;
- lack of processing and storage facilities.

At a meeting of a UNDP/FAO Working Group on Coconut-Based Farming Systems (CBFS) in Thailand in September 1989 (Anonymous 1989d), the following constraints on CBFS were identified:

1. Marketing:
- lack of assured markets;
- substantial price fluctuations in coconut and intercrops and their products.

2. Credit:
- lack of collateral, making it difficult for farmers to obtain credit for their investment;
- too high interest rates for farmers.

3. Information:
- a big gap in technology transfer from researchers to extension workers and to farmers; inadequate adaptive technology.

4. Inputs:
- inputs are generally not available on time;
- even if they are, they are too expensive.

5. Farm size:
- some farms are too small for economic operation (less than 1 ha), often caused by fragmentation every new generation.
- some farms are too large (the optimum size varying from country to country).

6. Land tenure:
- the problem of land lease still prevails, *e.g.* short-term leases, limitation of freedom to operate the farm, etc.

7. Farmers' education:
- poor level of education of farmers makes it difficult for them to adopt new technologies; some even resist changes in their old-fashioned practices.

8. Storage:
- a lack of storage facilities at village level for perishable products.

9. Transportation:
- inadequate transportation facilities and other infrastructure at village level, including feeder roads and pick-up trucks;
- too high transportation costs.

To this list, the socio-economic factors involved, as mentioned by Opio in Chapter 12 may be added. Together, they could be used as a check list for coconut development schemes. Nevertheless, some of the points mentioned need some comment. For instance, when the costs of certain items are considered too high, this does not always mean that prices are excessive, but they are still too high for farmers who lack the money to pay for them, sometimes even at subsidized prices. The restriction of farm size suitable for intercropping does not appear to be realistic, as intercropping systems can be adapted to any size. The number of problems involved with multiple cropping and better management may well explain why overall coconut production per ha is changing so very little, despite the many coconut smallholder development projects that have been established in many countries over the years. Coconut-based farming systems can be divided into three main activities:

- mixed cropping, where other crops replace a certain number of the coconut palms;
- intercropping, which includes cultivation of other crops under coconut palms;
- mixed farming, which is a combination of coconut growing and animal husbandry on the same land.

All three systems aim at a higher return from the land, expressed either in income or in products for home consumption. In remote areas, coconuts and intercrops can only be grown for home consumption and the local market unless there is scope for a processing industry that may reduce the volume of the coconut products and enhance their value so they can be economically exported to larger markets.

Mixed Cropping

Mixed cropping with coconut is usually done with other trees, which are planted in sites where coconut palms would otherwise have been planted. It is mostly done in home gardens around smallholders houses in an irregular way, seldom in a regular planting system. A new system of replanting old coconut groves in a mixed planting, with N-fixing trees on the coastal sands of Ivory Coast was rather successful (Taffin *et al.* 1981). It involved two legumes, *Acacia mangium, A. auriculiformis,* and *Casuarina equisetifolia.* The N-fixing trees received 140 g of a mixture of bicalcium phosphate and potassium chloride, applied in four split applications. The coconuts were fertilized normally. The N-fixing trees had to be protected against sea spray. After about five years the N-fixing trees were cut back to a height of 1.50 m to reduce competition with the coconuts (especially for light) and to harvest firewood for the farmers. *A. auriculiformis* produced about 49 m^3 of wood per ha, the other two species about 22 m^3. However, the total biomass produced by *A. mangium* was greater that that of the other species. The wood was used and the small branches and leaves were left on the soil, ensuring that over 80% of the potential total nitrogen from N-fixing trees was recycled, *i.e.* 457 kg per ha of intercropped planting in the case of *A. mangium*, 398 kg for *A. auriculiformis*, and 227 kg for C. *equisetifolia*. Coconut development was highly satisfactory. Ochs *et al.* (1993) reported that in the fifth year of the trial the nitrogen content in leaf 14 for the various treatments was:

o coconut monocrop	1.79%
o coconut + casuarina	2.11%
o coconut + A. mangium	2.17%
o coconut + A. auriculiformis	2.24%

The sustainability of the system is not yet known. *Casuarina* planted in double rows between the seashore and the coconuts were effective as windbreaks and reduced the effect of seawater sprays on the young coconut palms. Such systems are possible with other trees as well, although fruit trees could not be lopped systematically to reduce their competition with coconut.

Intercropping

Intercropping can be done with annuals and perennials, or a combination of such crops. On very small farms, annual food crops for domestic consumption or for local markets prevail. On larger farms, industrial crops are also grown under coconuts, including shrubs and small trees. Cocoa, coffee and bananas are among the most favoured crops grown under coconut, but crop combinations vary much with ecological conditions, marketing possibilities, etc. The main differences between weeds and useful crops under coconuts are the treatments applied to the intercrops, control of their growth, and the financial income obtained from their products.

Generally, coconut intercropping may show the following characteristics, as described by Tarigans and Darwis (1989) for intercropping in Java, Indonesia:

- o the crops are tangled - it is not clear which one is the main crop and which one the secondary crop;
- o total crop density is too high, exceeding the area's capacity;
- o a proper cropping pattern has not been employed;
- o farm maintenance, fertilizer application, pest and disease control have hardly been provided;
- o improved varieties have not been used.

Badly managed intercropping results in soil exhaustion, poor coconut palms and poor intercrops, ending in neglect and abandonment. This can have serious consequences in countries where a large part of the arable land is planted with coconuts. Or, as expressed by Nair (1986): 'The crux of the problem is the level of management.' Systematic growing of crop combinations requires a higher level management than for a single coconut crop, *e.g.* adequate weeding, collection of fallen leaves in the palm rows, pest control, regular harvesting, recycling of waste products, and fertilizer use. In particular, recycling of waste products and the use of fertilizers is of paramount importance to prevent soil mineral depletion. The choice of intercrops depends on ecological conditions and - where grown on a commercial basis - on marketing possibilities and prices. When farmers lack the knowledge, skill and/or the required investment funds, they may rent out their land for intercropping to a commercial enterprise, the management of the coconut and the income from this crop remaining with the farmer, provided the enterpreneur applies good management and maintains or improves soil fertility level. In addition to benefiting from the rent, the farmer will also benefit from a better upkeep of his plantation and from residues of fertilizers applied to the intercrop. This system is more suitable for annual intercrops or short-term perennials such as pineapples or bananas than to long-term crops.

In Ivory Coast, where population density is high due to the proximity to large cities, resulting in high demands for firewood, all coconut waste such as husks and leaves have been used as fuel for many years. Copra yields, which originally amounted to about 700-1000 kg per ha (without fertilizer), have gradually decreased to 100-300 kg per ha per yr. This yield reduction was essentially due to fertility reduction (Pomier *et al.* 1986). If coconuts and intercrops are adequately fertilized, intercropping is not harmful to coconut production. On the contrary, it may increase the coconut production where water availability is not a problem. On the other hand, intensive recycling of mineral fertilizers through a combination of coconuts and intercrops could lead to an accumulation of certain elements, which could disturb the soil mineral nutrient balance. Multicropping has several advantages over monocropping. It can not only provide a higher income for farmers, but the system is much more balanced and better adapted to sudden changes that might affect one of the crops when grown as a monocrop. If the price of a monocrop is low, or if the monocrop suffers from some pest or disease, there is not much that a farmer can do than cut his losses, whereas in a multicropping system, the other crops may compensate for a certain part of the losses. Technically, as well as financially, a multicropping system is more flexible and the total return from the land is often higher. However, environmental conditions may impede the establishment of such an intensive farming system.

Multiple cropping not only results in a more efficient use of land, but also of the farmer's labour potential, particularly on very small holdings, where part of the labour potential may be idle. In a monocrop, the main labour requirement may be in peaks, such as during harvesting time, whereas during other periods labour requirement may be low. Multicropping may flatten the labour requirement curve and may also provide more labour opportunities. According to Tarigans (1989), the main reasons for the wide acceptance of intercropping by coconut farmers in Indonesia were the higher net returns and the provision of productive work for family labour. Farms with available idle labour might opt for labour-intensive high-value crops. In larger holdings, multiple cropping may generate additional employment. According to Nair (1979), under Indian conditions the annual labour requirement for one hectare of monoculture of coconut is about 150 man-days. For multiple cropping this may increase to about 350 man-days and for mixed farming it may go up to about 1000 man-days for stable-fed milk cows. Ranatunga *et al.* (1988) in Sri Lanka estimated that a farm with two adult workers will have about 500 man-days available per year. Intercropping would soon require the hiring of labour. Comparison of man-day requirements between countries is difficult due to different labour efficiency and different cultural practices. In India, Nelliat and Krishna (1976) measured the additional labour input for some intercrops, as presented in Table 33.

Compared to short grass or weeds, intercropping will improve the micro-climate in the plantation. The existence of one or more other canopies closer to the ground will increase the shade on the land, reduce the movement of air by convection and reduce ventilation, thus acting as a buffer against drastic changes in the eco-climate. It will stabilize soil temperature, slow down mineralization processes of organic material, improve the soil's organic matter content, recycle more mineral nutrients through plant residues, improve soil structure, and improve conditions for soil micro-organisms. The existence of more root systems in the soil, providing a higher total rooting density at different depths, will reduce mineral losses through leaching and will also improve soil structure. Soil nutrient uptake will increase when more than one crop is grown on the same land, and these nutrients should be compensated for by fertilizing. However, the efficiency of fertilizer use will increase with the increased root volume of the crop combination, especially when intercrops also exploit soil layers deeper than the coconut roots.

Table 33: *Additional labour requirements (man-days per ha) for various intercrops*

Crop	Man-days
Sweet potatoes (Ipomoea batatas)	56
Yam (Dioscorea sp.)	64
Chinese potatoes (Coleus sp.)	92
Cassava (Manihot utilissima)	93
Ginger (Zingiber officinale)	108
Turmeric (Curcuma sp.)	108
Elephant yam (Amorphophalus campanulatus)	123
Banana (Musa sp.)	170

Thus, after initially higher fertilizer expenses per ha for crop establishment, the effect of the applied fertilizers will increase and the amount of fertilizers required annually may decrease due to the higher recycling percentage. Increased shading of the soil may also reduce weeding costs and facilitate the control of such hardy weeds as *Imperata cylindrica.* The improved management may also increase the coconut leaf number and size and thus again produce more shade, which in its turn will again suppress more weeds. Another beneficial effect of increased soil coverage by intercrops is its protection of the land against erosion.

However, smallholders usually lack the capital for the fertilizers necessary to give all crops a good start. Cropping systems propagated by development schemes should be within the financial and managerial capacity of smallholders. To reduce initial investment, the change to more intensive cropping might be started on part of the holding only, thus reducing risk, which is an important factor hampering management improvement. With the partly improved enterprise, income and management experience will increase, facilitating a gradual increase in the intensively cropped area. Blaak (1986) emphasizes the adoption of low cash input systems for smallholders. Such a system might be based on widely spaced coconuts. The number of nuts per palm will be increased by more light and soil volume availability, partly compensating for the reduced number of palms planted. A key role could be played by mycorrhizas in combination with leguminous cover crops inoculated with rhizobium bacteria. Through increased P availability to the legume by myccorhiza activity, the legume will grow better. In its turn, the legume in symbiosis with the rhizobium will produce more N. Both elements N and P will finally become available through the organic material falling to, or being worked into, the ground. Small doses of P applied to the system, about 20-40 kg per ha might enhance the P availability in the system and improve the N-fixation in the roots, without affecting the mycorrhizas. Not only may legumes benefit from the myccorhiza activity, other crops also live in symbiosis with myccorhizzas. Dense vegetation, increased organic material content of the soil and a more favourable microclimate in turn will favour the myccorhiza population development. Such a system could considerably reduce N and P fertilizer expenses. However, doses of fertilizer must always be sufficient to prevent soil exhaustion. It should again be emphasized that low to moderate fertilizer application may give relatively higher returns than high doses. Thus the system could be gradually built up.

As coconuts and intercrops will compete for light, water, and minerals, spacing of coconuts is an important factor. In densely planted coconut groves, the choice of intercrops will be restricted and only shade-tolerant crops can be grown. In very densely planted groves a certain number of palms may be cut down (the lowest producers!), but it is often very difficult to convince a farmer of the need to cut fruit trees, even if they produce only a few fruits. When new groves are to be planted, wide spacings are recommended, allowing for a much wider range of intercrops from which to chose. Where sun-loving annual intercrops are to be grown, coconuts can be planted in E-W rows 15 m apart or more. Where shade-tolerant crops such as cocoa are grown, the coconut palms may be planted in N-S rows, 9 to 12 m apart. The within-the-row spacing could be maintained at 8-9 m. With close spacing, the palms may bend outwards over the lanes, thus again increasing their shading effect. Opio (this volume) states that hedge-planting in Tonga, with more palms per ha than with the conventional spacing results in higher coconut yields and allows for early grazing under coconuts.

Coconut can be intercropped when palms are still young and leaves are still growing upright. Once the leaves start hanging down to the ground, about three on four years after planting, intercropping is no longer possible. During these first years, intercrops can be grown in almost full light. Zakra *et al.* (1986) observed that in the Middle Ivory Coast, young coconuts developed faster under intercropping conditions, in particular with weeded plants such as yam or cassava, whereas with the cover crop *Pueraria,* development slowed down, probably due to excessive water consumption. Intercropping may be resumed when enough light penetrates the canopy to allow for the growing of another crop. The age of the palms at which intercropping can be resumed depends very much on the spacing of coconut trees, the variety of coconuts grown, and the shade tolerance of the intercrop. When palms grow up, slant rays of sunshine will add to light coming in between the leaves. When palms grow old, after about 50 years, a gradual reduction of the canopies may occur and more light will penetrate. But in a well managed plantation, growing under favourable conditions, this stage may come many years later. In a trial carried out in India in a coconut plantation planted at a spacing of 8 x 8 m, soybeans yielded 342 kg per ha under 10-year-old palms, 610 kg per ha under 40-year-old palms, and 750 kg per ha under 60-year-old palms (Lourduraj *et al.* 1992). Where the inter-rows are very wide, intercropping could be continuous, the available area with sufficient light for crop growth reducing and expanding again with the growth of the palms. Ecological conditions influence palm development substantially; therefore, only local conditions can determine at what age of the palm trees intercropping is possible, or not. Light transmission differs between locations under the palms and is more intense in the centre of a square than near the stem of the palm. Barile and Sangalang (1990) in The Philippines measured the light intensity reduction under nine different coconut varieties planted at 7.5 x 7.5 m in a triangular pattern. They found that average light intensity reduction at 7.00-8.00 a.m, 11.30 a.m.-12.30 p.m. and 4.00-5.00 was 71.46%, 4.93% and 53.93% respectively. The sunlight reduction in the Laguna variety was significantly higher than for all other eight cultivars, due to such characteristics as leaf length, number of leaflets and length of petioles. They recommend that in the development or selection of cultivars suitable for intercropping, those with fewer leaflets, and shorter leaves and petioles should be given consideration. However, the yield capacity of these palms should also be kept in mind.

Light interception by coconuts, and consequently the photosynthetically active radiation (PAR) transmitted through the canopies, also depends on the condition of the palms. Moss (1992) observed that fractional interception of light was found to vary considerably between months, and to be associated with leaf shedding caused by dry season water-stress. Palms to which potassium fertilizer was applied intercepted more light and replaced leaf area lost during the dry season more rapidly than those that had not been fertilized. Palms planted at a higher density intercepted more light and carried more leaves per palm, but intercepted less light per leaf than those planted at a lower density. Increased palm population density increases the total light interception of the crop but reduces interception by individual palms, despite increased leaf numbers. For monocropping of coconut, increased total yield would compensate for a reduced light efficiency index (LEI). For intercropping, wide palm spacings are normal as this system maintains LEI for the coconuts while allowing more unintercepted light to reach underplanted crops. LEI of intercrops may in turn be influenced by light intensity and genetic factors. Moss concluded that with more information about LEI it would be possible to determine optimal spatial arrangements for coconuts and intercrops. Studying the efficiency of crops under shade is therefore an additional area of investigation that may be useful in determining optimal management of multi-storey cropping systems. If low light intensity is limiting yield, the potential response to fertilizer application will also be limited, so that measuring light may be a useful basis for nutrient management of intercrops.

Falling leaves may damage intercrops. This is another reason not to plant too close to the trees. Usually a spacing of 1½-2 m between intercrops and coconut stems is maintained. In small-holdings, intercrops may be grown all over the area between the palms, on larger estates they are mostly grown in one direction in the lanes between the rows, to allow for movement in the plantation. The yield of an intercrop at a distance of 2 m from the palm may be only 60% of that from the same crop growing in the centre of the lane. With very wide spacings, these differences may reduce. Fallen leaves are collected in the rows, between the palm-trees. Competition for water will occur and its effects will increase with lower water availability. In areas where the coconut root system is confined to a circle of a few metres around the stem only (depending on soil conditions and management), competition for water will be less fierce than in areas where the coconut palms have extended root systems that grow all over the area, such as in light soils and shallow soils. The presence of coconut roots all over the area does not mean that multiple cropping is impossible.

Water availability depends on rainfall and irrigation, the water-holding capacity of the soil, soil depth and depth of the soil water table, evaporation from the soil, and transpiration through the vegetation. Increased shading of the soil through intercropping, and the natural mulch from fallen leaves of intercrops, may stabilize soil temperature and considerably reduce the loss of water by evaporation. Nair (1979) found that the daily evaporation under a combination of cocoa and coconut was only about 40 per cent of that from an open area. However, water consumption through transpiration could increase when intercrops are grown under coconuts. But in the latter case, the water is used and not lost. There are indications that the total water consumption of an intercropping mixture may not differ very much from that of a monoculture. Total available water and minerals determine the result of the enterprise. In Sri Lanka, experimental evidence showed that there would be no severe competition for soil moisture between coconuts and the intercrops if the annual rainfall is over 1900 mm (Liyanage *et al.* 1986). But even some competition for water might be acceptable, as long as the combined income from the farming system is higher than that from coconut as a monocrop. In areas with a severe dry season, intercrops that can be harvested at the end of the rainy season might be preferred. The best crop combinations will be found by experience.

Changes in the ecosystem under the coconut palms caused by intercropping may have positive as well as negative effects on pest and disease development in a plantation, such as is the case with pigeon pea (*Cajanus cajan*), which can favour the development of the coconut leaf miner *(Prometotheca cumingii)* population (Abad 1983). A dense vegetation may provide shelter for insect pests and hamper their chemical control. But it may also create conditions that are more favourable to their parasites, thus creating a better natural balance, Crops that harbour the same pests or diseases that also attack coconuts should be avoided. Where coconut suffers from Bud Rot, caused by *Phytophtora palmivora*, it is not advisable to intercrop with cocoa, which is very susceptible to this fungus. In addition to thc combination with cocoa, Abad (1983) mentions some other combinations that may cause pest problems, such as intercropping with maize, rice, cassava, pineapple or sweet potato, which may attract rats. A dense undergrowth may also favour rats and other rodents. A coconut - squash combination might favour the development of scale insect populations *(Aspidiotus destructor)*. It is also suspected that diseased crops such as banana, tomato, cucurbits, crucifers and others may be inoculum sources of the leaf stripe disease of young coconut palms, caused by the *Pseudomonas/-Erwinia* complex. On the other hand, crops that harbour enemies of coconut pests have an advantage. Such is the case with certain low (fruit) trees that harbour weaving ants that forage in the coconut crowns, predating on insect pests and their eggs. Intercropping with papaya may favour the development of spider mite populations. Where weeds may be the host plants of the causal agents of certain diseases such as yellowing diseases and Hartrot; elimination of these weeds by growing intercrops may reduce disease incidence in coconuts. Additional farming practices such as ploughing or harrowing may reduce soil-born pests such as the slug caterpillar *Setoria nitens* which pupates in the ground.

A wide variety of annual and perennial crops can be grown under coconuts, their profitability depending on environmental and marketing conditions. When intercropping is practised for home consumption only, no market research is needed. Intercropping systems are site-specific and must be designed to meet local soil fertility conditions, rainfall characteristics, farmers' resources and management skills, working habits, land tenure systems, market situation and economic infrastructure. Introduction of new coconut planting material with different growing characteristics, such as hybrids, may enforce a change of an already established intercropping system. General recommendations on which crops to grow cannot be given as each set of ecological, economic and marketing conditions allows a certain range of crops to perform satisfactorily.

On smallholdings, annual food crops are the most frequently planted intercrops, cassava being one of the most important. But maize can also be found as an intercrop, and even sugar-cane is sometimes grown under coconuts. Chilli peppers often do very well and pineapples can also be a very remunerative intercrop. Pepper and vanilla are planted to climb up the stem of the palm. At the Plantation Crops Research Station in Kasaragod, India, a combined system was developed, consisting of food crops, grass for cattle, and some commercial crops by which ½ ha of irrigated and intercropped coconut can sustain a family, provided all waste material is recycled. In this case organic waste was composted in a tank, the released methane gas being used for household cooking and lighting.

Perennial intercrops, such as cocoa, coffee, banana, small fruit trees, clove, and other tree spices will show a positive cash flow only after a number of years. Such crops, requiring more capital and time, but less labour once planted, will be more appropriate for larger holdings, whereas the smallholder may be more interested in direct returns, especially when fertilizers are being used. For intercrops that produce less in the shade of coconuts, less fertilizer is needed than when these crops are growing in monoculture without shade.

Cocoa is one of the best combinations with coconut where the soil is loamy and fertile and the dry season not too long. In this combination, the highest income is often derived from cocoa and not from coconuts. Salam and Sreekumar (1990) describe a small farm in Kerala, India, supporting a family of seven. The farming system is a mixture of mixed farming, mixed cropping and intercropping, with 60 adult coconuts, 12 arecanut palms on which pepper was climbing, 3 jackfruit trees, 2 mango trees, 2 lime trees, 2 breadfruit trees and 1 tamarind tree and some other small fruit trees, intercropped with guinea grass and a number of vegetable crops. The farm supported one milk cow and a heifer. The topmost canopy was formed by coconuts, yielding 140 nuts per palm per year, the second layer by areacanuts and fruit trees, the third by banana and cassava, and the fourth layer consisted of other tuber crops, vegetables and guinea grass. These coconuts were growing in basins and were irrigated with the use of a 3 H.P. electric pump. Cattle dung and other organic material was applied in the basin around the coconuts. Cattle were stable-fed with guinea grass from the farm. The farmer also possessed 10 chickens, feeding on the land. There were also 5 beehives. The entire labour requirement was met by the family. This private farm (not at a research station) is a very good example of what good management can achieve on a small piece of land.

Leguminous shrubs or trees can also be grown between the palm rows, producing vegetable material either for manuring or mulching of the coconut tree, and firewood for the household. In Sri Lanka good results have been obtained with such a system, using *Gliricidia sepium* and *Leucaena leucocephala* (Liyanage 1993). Yields of 6-8 t per ha of fresh foliage per year have been obtained. The foliage of both trees, with a nitrogen content of about 4%, is an excellent green manure for coconut palms. Incorporating 30-50 kg of fresh prunings into the soil to a depth of 15-20 cm meets the entire nitrogen requirement of the palm. The decomposed foliage more than doubled the water-holding capacity and the organic carbon content of degraded ultisols reduced the bulk density and improved earthworm activity.

In a plantation, yielding 30-35 nuts per palm per yr, incorporation of 30 kg of *G. sepium* loppings into quarter-circle trenches dug around each palm increased nut yields by 12%. The fresh green matter may also be used as fodder. Other trees showing promising results were *Calliandra spp.* and *Acacia auriculiformis*. In another trial, *L. leucocephala* planted in double rows between the palm trees produced about 60 m^3 per ha of fresh wood after three years of growth, while copra yield increased by 8 per cent.

For mixed farming the reader is referred to Chapter 14.

11. The Smallholder and Coconut-based Farming Systems

S.G. Reynolds

The small farm is a major feature of land tenure systems in the developing world (Anonymous 1986). According to Williams (1976) the smallholder cultivator and his family make up two-thirds or more of the world population, but the great majority of small farmers have small land holdings which proportionately represent only a limited percentage of the total land area (Table 34). The amount of land which qualifies a farmer to be classified as a smallholder appears to vary from <1 ha to approximately 5 ha. In south-east Asia the average size of small farms is about 1-2 ha. (Devendra 1993). Although coconut is often thought of as a large-scale plantation crop, most of the world's production comes from small farms (Pordesimo and Noble 1990), with McDowell and Hildebrand (1980), estimating that 90% of all coconuts are grown by smallholders. In The Philippines 91% of coconut growers are smallholders (<5 ha) according to Aguilar and Benard (1991). In Indonesia the figure is 97% (Darwis 1988) and in India 98% of the 5 million coconut holdings are smaller than 2 ha (Punchihewa 1990), with the average coconut holding <0.41 ha. In Malaysia 93% of the coconut area is farmed by smallholders (Yusof and Rejab 1988) with an average land holding of approximately 2.0 ha (and a range from 1.1 ha in Kelantan to 4.7 ha in Selangor). In Sri Lanka smallholdings account for over 90% of the total area under coconut (Anonymous 1983) and according to Liyanage *et al.* (1989) 87% of the holdings were <2.02 ha with 58% <0.8 ha in size. However, in the main growing area of the coconut triangle, 35% of the estates were > 8 ha in size (Liyanage and Dassanayake 1988). In Thailand > 80% of farms were <2.4 ha in size (Rungrueng 1988); in Western Samoa > 80% of coconut production is based on smallholdings (Opio 1989) while in Papua New Guinea the plantation sector is bigger and the smallholder percentage drops to 62% (Ovasuru 1988).

Table 34: *Proportion and size of holdings by region*

Region	Smallholdings as % of all holdings	Smallholdings as % of total area	Average size of holdings (ha)
Africa	66.0	22.4	1.0
Far East	71.1	21.7	0.7
Latin America	67.2	4.6	2.7
Near East	54.6	9.8	1.8

Source: Bavappa and Jacob (1982) from World Census of Agriculture, 1970.

Economics of the smallholder coconut operation

Due to the limited size of most smallholdings, the small number of trees, the limited labour requirements of the coconut crop, and the limited returns provided, most families cannot survive on the incomes from the coconut crop. Aguilar and Benard (1991) suggest that the problem of low income in the smallholder coconut production sector can be attributed to several interacting factors, some of which are within his control, and some beyond:

- declining and unstable prices of coconut products;
- declining productivity of coconut trees due to senility and/or non-adoption of recommended coconut management practices;
- underutilization of coconut farms due to tenure problems;
- absence, of or ineffective, support services or credit and their combinations.

In Western Samoa, coconut, based on returns per man-day, failed to meet the cash return needs of the extended family (Burgess 1981). The only alternatives were to practise intercropping and to ensure that the majority of family labour had off-farm employment. For a family intercropping a 3 ha coconut holding with cocoa, pineapple, vegetables, etc., total net revenue was maximized by leaving all secondary labour in wage employment, and only at 8 ha (0.47 ha per capita) could all available labour be absorbed in on-farm tasks and provide the income needed. In Malaysia, it was observed that, in common with many coconut growing countries, coconut smallholders were plagued by poverty (Bin Mohd Kamil and Bin Ahmad 1978). In 1975, out of a total of 34,400 smallholdings, 17,500 or 51% fell below the poverty line. Details of on-farm and off-farm income are shown in Table 35, while the proportion of on-farm income from coconuts is shown in Table 36.

Table 35: *Farm/off-farm income per ha (Bin Mohd Kamil and Bin Ahmad, 1978)*

Locality	Av. farm size (ha)	Total gross Income ($)	Farm income		Income from off-farm	
			$	%	$	%
Hutan Melintang	1.2	2,573	844	33	1,729	67
Bagan Datoh	1.6	2,523	1,910	76	618	24
Teluk Baru	1.1	2,930	1,748	60	1,182	40

The survey showed that the labour resource was underutilized and it was recommended that gradual replacement of existing trees by new high-yielding coconuts be undertaken at planting densities that would allow various intercropping systems to be established. Mahendranathan (1976) indicated that keeping one crossbred dairy cow for milking and disposal of female calves either as replacements or male calves for beef would give a gross income per ha of about $100 per month, compared with only about $10 per month from beef. It was also shown that total gross income per ha from coconuts intercropped with cocoa was $3180, with coconuts providing $480 and cocoa $2700, demonstrating the importance of intercropping (Anonymous 1978b).

Table 36: *Gross farm income per ha by locality and crop (Bin Mohd Kamil and Bin Ahmad, 1978)*

Crop	Locality					
	Hutan Melintang		Bagan Datoh		Teluk Baru	
	$	%	$	%	$	%
Coconut	542	64	787	41	822	47
Cocoa	264	31	1,024	54	831	48
Banana	38	5	99	5	95	5
Total	844	100	1,910	100	1,748	100

Various surveys in The Philippines have demonstrated that the coconut farmer (where coconuts are monocropped) is productively employed for only about 27% of the available time, leaving over 200 mandays or 73% of his time for off-farm employment or other activities such as intercropping (Ontolan 1988), and livestock raising, etc. to bring in additional income and reduce his dependence on copra. Data in Table 37 show that the income from coconuts in fact represents only part of the total income of smallholders and, except for Quezon Province, is usually <50%. The rest comes from regular off-farm employment, seasonal labouring, rice farming, livestock and poultry raising, seasonal cropping, orchards, vegetable raising and general intercropping. According to Villegas (1991), income from coconut monocropping averages only P 2000 per ha per yr (and according to Calub 1989, just over P 2000), which is considerably lower than the poverty threshold level and therefore each household maintains livestock, fruit trees, vegetables and other crops in patches around the house for home consumption and cash income.

9 Bunches with fruits of different colour, size and shape *(R. Bourdeix)*

10 Bunches with fruits of different colour, size and shape *(R. Bourdeix)*

11 Bunches with fruits of different colour, size and shape *(R. Bourdeix)*

12 Bunches with fruits of different colour, size and shape *(R. Bourdeix)*

13 Heavy production with low average fruit weight *(R. Bourdeix)*

14 Inflorescence almost fully covered with female lowers *(J.G. Ohler)*

15 Parthenocarpic bunch resulting from lack of pollination *(R. Bourdeix)*

16 Operator emasculating a mother tree in seedgarden *(R. Bourdeix)*

17 Assisted pollination *(R. Bourdeix)*

18 Harvesting with bamboo poles with knife attached *(J.G. Ohler)*

19 Stem with hole cut through for curing disease *(J.G. Ohler)*

20 PB 121 hybrid with heavy load of fruits *(R. Bourdeix)*

21 Bunches with long main rachis *(J.G. Ohler)*

22 Smallholding with mixed cropping *(J.G. Ohler)*

23 Dwarf variety 'Salak' with first inflorescence emerging from below soil level *(J.G. Ohler)*

24 Damage by termite *(Macrotermes bellicosus)* attack on planted seednut *(D. Mariau)*

25 Morphologic diversity of coconut fruits *(R. Bourdeix)*

26 Pollination by male flowers of the next inflourescence *(R. Bourdeix)*

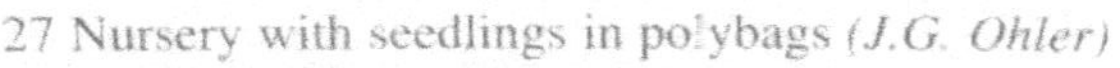

27 Nursery with seedlings in polybags *(J.G. Ohler)*

28 Young trees suffering from drought *(J.G. Ohler)*

29 Sulphur deficiency symptoms *(G. Benard)*

30 Iron deficiency symptoms *(X. Bonneau)*

31 Boron deficiency symptoms *(CIRAD-CP)*

32 Potassium deficiency symptoms *(J.G. Ohler)*

33 Iron deficiency symptoms *(G. de Taffin)*

35 Potassium deficiency symptoms *(CIRAD-CP)*

34 Sulphur deficiency symptoms *(CIRAD-CP)*

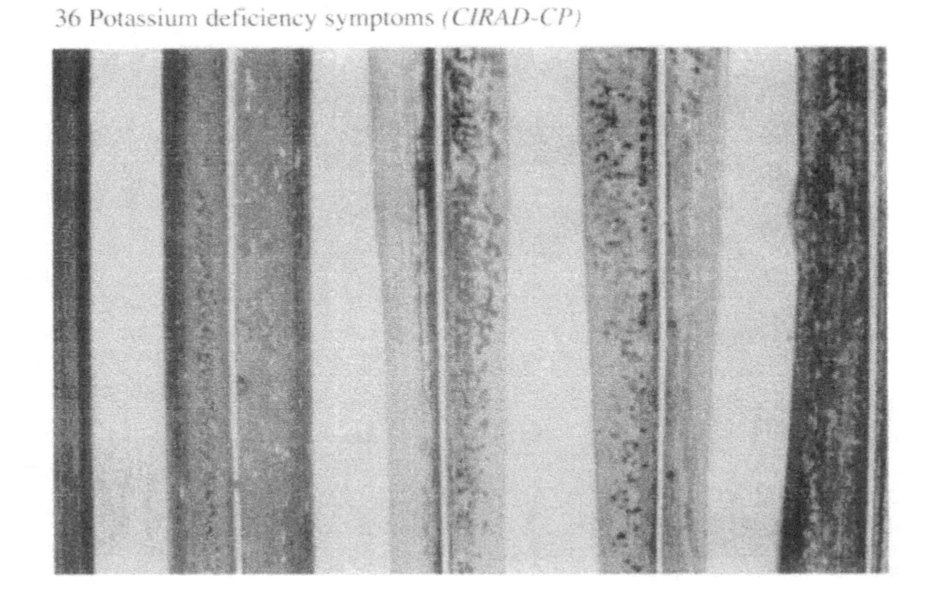

36 Potassium deficiency symptoms *(CIRAD-CP)*

37 Copper deficiency symptoms *(X. Bonneau)*

38 Copper deficiency symptoms *(X. Bonneau)*

39 Leaflets with nitrogen deficiency symptoms *(CIRAD-CP)*

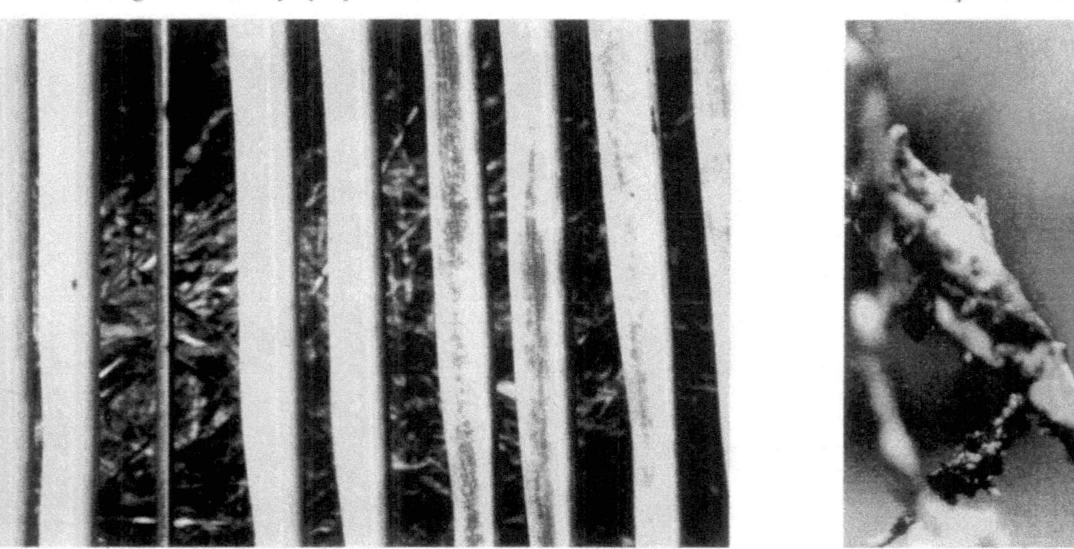

40 Root tips attacked by *Sufetula sunidesalis (R. Desmier de Chenon)*

1 Branched coconut stem *(R. Behrens)*

2 Spiralling coconut stem *(J.G. Ohler)*

3 Alternate bearing coconut palm *(R. Bourdeix)*

4 Alternate bearing coconut palm *(R. Bourdeix)*

5 Inflorescences with varying spike lengths and varying spike and nut numbers *(R. Bourdeix)*

6 Inflorescences with varying spike lengths and varying spike and nut numbers *(R. Bourdeix)*

7 Inflorescences with varying spike lengths and varying spike and nut numbers *(R. Bourdeix)*

8 Inflorescences with varying spike lengths and varying spike and nut numbers *(R. Bourdeix)*

73 Rice huller made of coconut wood in Zanzibar (*J.G. Ohler*)

74 A coconut grater made in Zanzibar (*J.G. Ohler*)

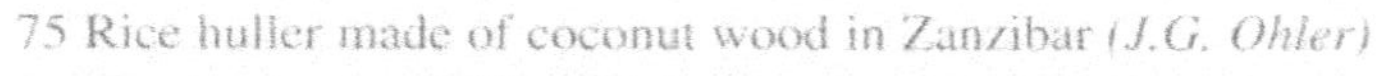

75 Rice huller made of coconut wood in Zanzibar (*J.G. Ohler*)

76 Sawing coconut stems (*J.G. Ohler*)

77 Sawing coconut stems *(J.G. Ohler)*

78 Scooping out coconut meat *(J.G. Ohler)*

79 Roof tile made of coir fibre and cement in Zanzibar *(J.G. Ohler)*

80 Cupboard made of coconut wood in Zanzibar *(J.G. Ohler)*

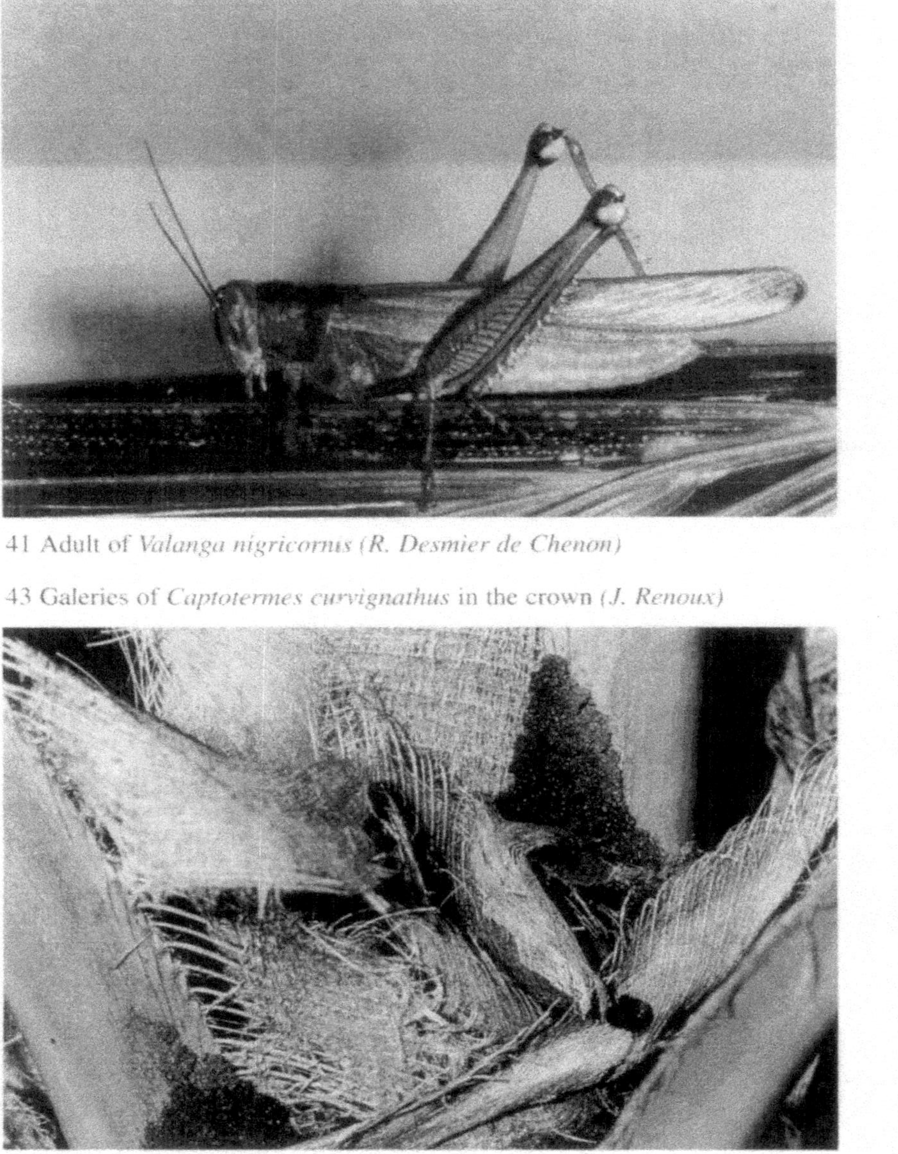

41 Adult of *Valanga nigricornis* (*R. Desmier de Chenon*)

43 Galeries of *Captotermes curvignathus* in the crown (*J. Renoux*)

42 Female of *Sexava coriacea* (*R. Desmier de Chenon*)

44 *Balyana mariaui* (top) and *Coelaenomenodera perrieri* (*D. Mariau*)

45 Coconuts damaged by *Pseudotheraptus wayi* attack *(J.G. Ohler)*

47 *Aspidiotus destructor* with some coccinelid beetles *(J.P. Morin)*

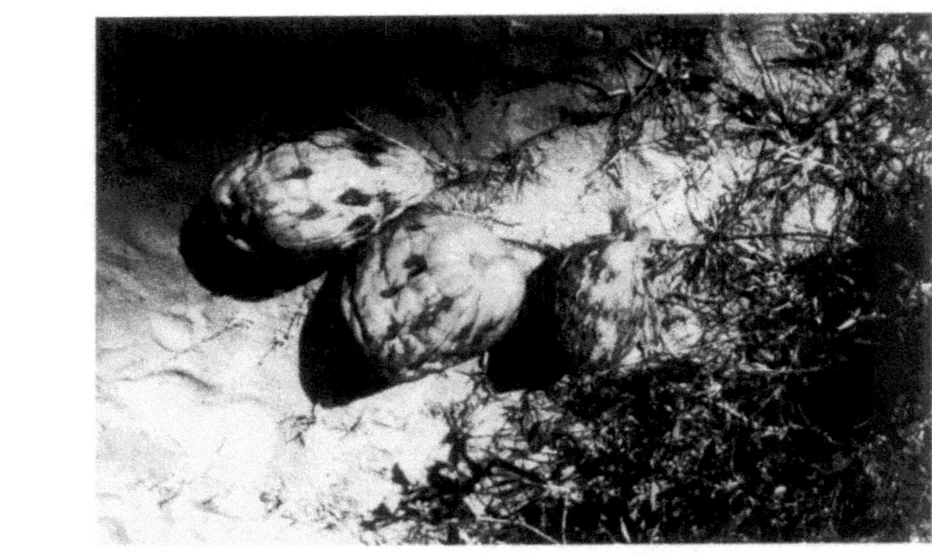

46 Coconuts damaged by *Pseudotheraptus wayi* attack *(J.G. Ohler)*

48 *Coccinellidae*: *chilococus* spp. (l.); *Cheilomenes* spp. (r.) *(D. Mariau)*

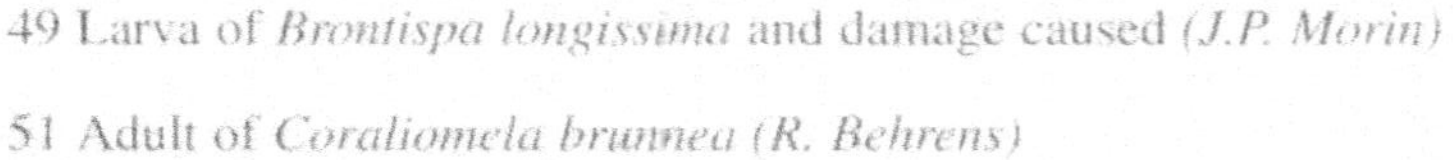

49 Larva of *Brontispa longissima* and damage caused (*J.P. Morin*)

50 Adults of *Gestronella* sp. (*D. Mariau*)

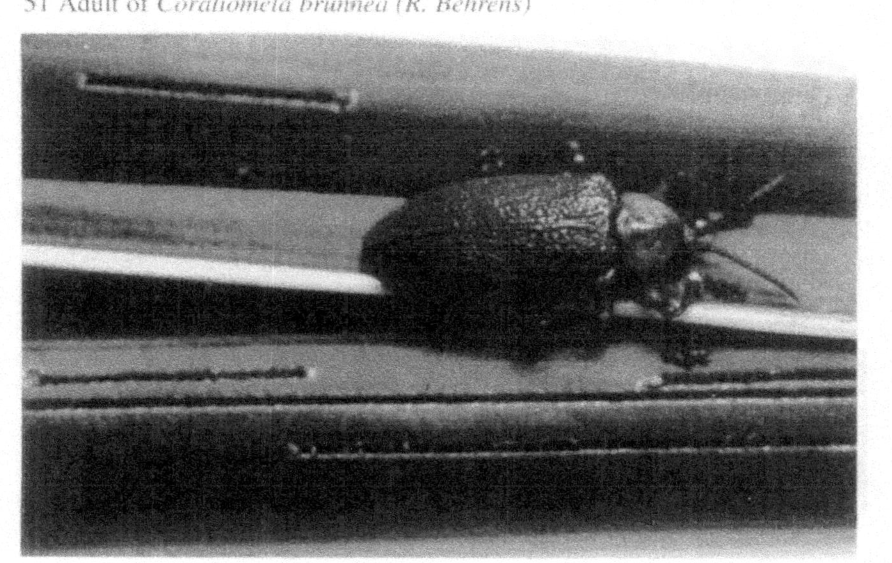

51 Adult of *Coraliomela brunnea* (*R. Behrens*)

52 Adult of *Rhynchophorus vulneratus* (*R. Desmier de Chenon*)

53 Stem damaged by *Rhinostomus barbirostris* (J.P. Morin)

54 Male of *Oryctes rhinoceros* (R. Desmier de Chenon)

55 Male of *Xylotropus gideon* (R. Desmier de Chenon)

56 YD nuts: 1 healthy, 3 attacked by *Eriophyes guerreronis* (D. Mariau)

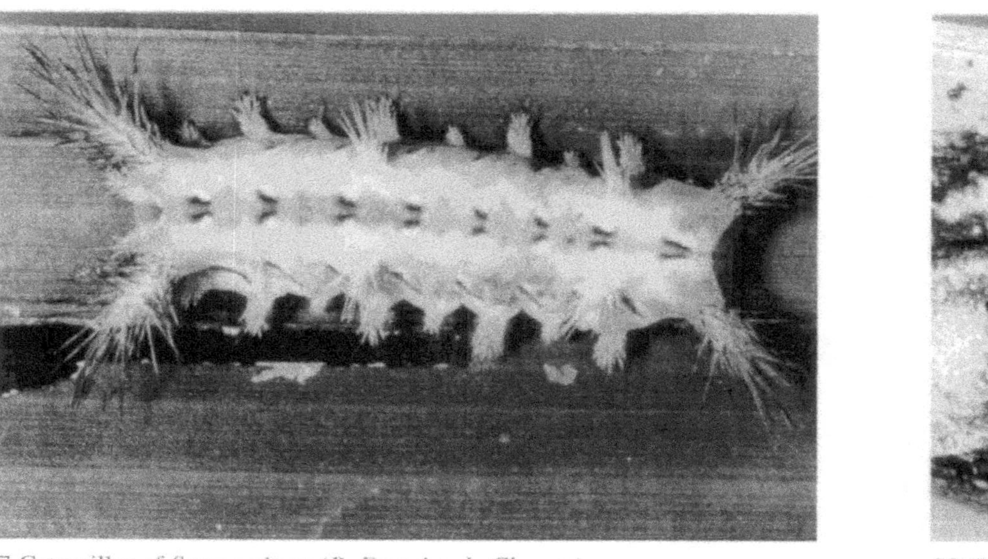

57 Caterpillar of *Setora nitens* (R. Desmier de Chenon)

58 *Chaetexorista javanica* (R. Desmier de Chenon)

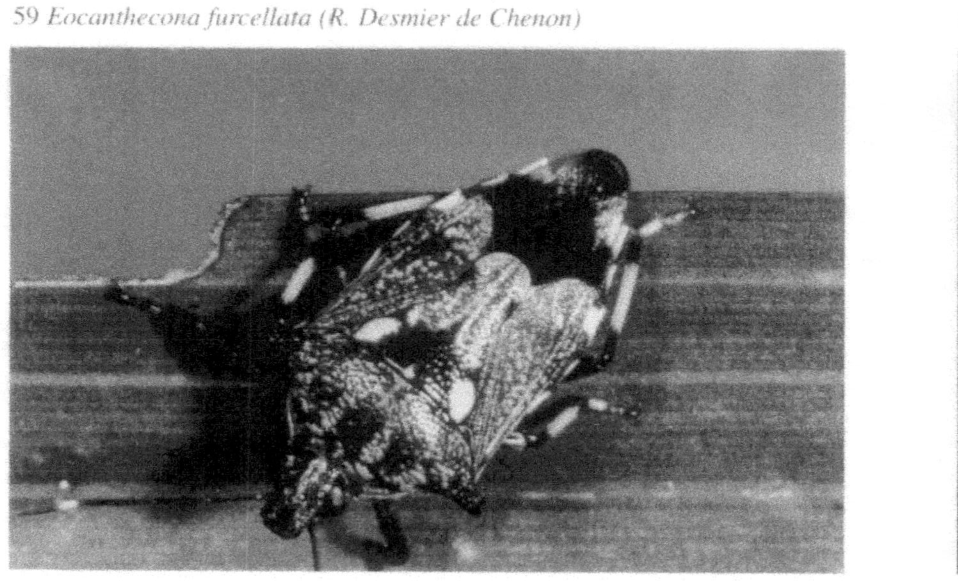

59 *Eocanthecona furcellata* (R. Desmier de Chenon)

60 Caterpillar of *Parasa lepida* (R. Desmier de Chenon)

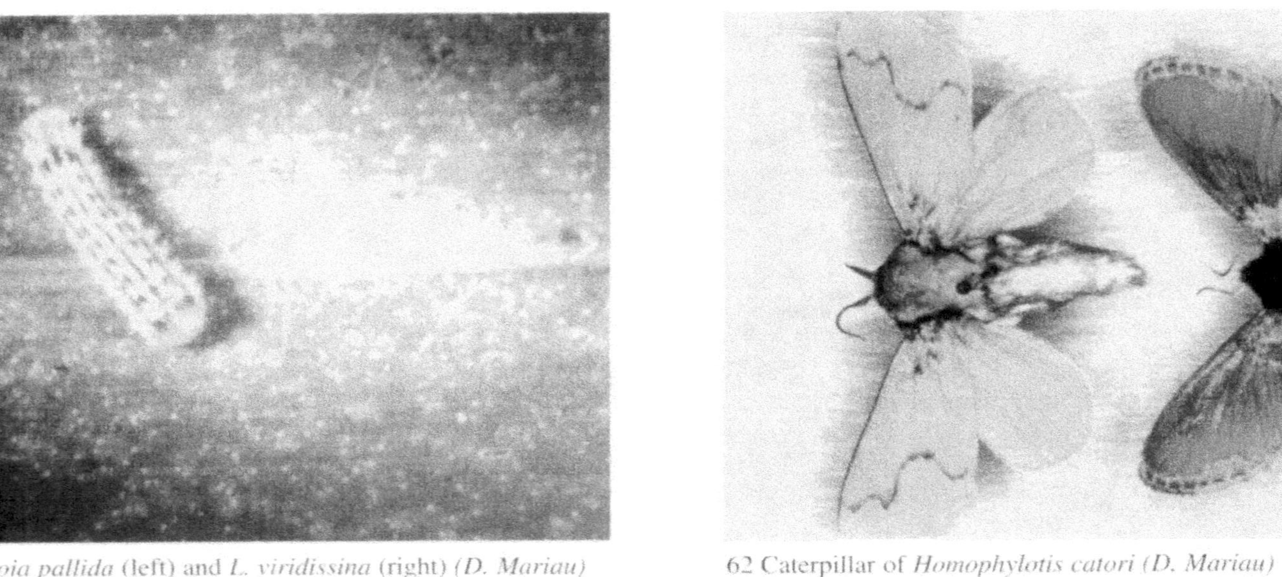

61 Butterflies of *Latoia pallida* (left) and *L. viridissina* (right) *(D. Mariau)*

62 Caterpillar of *Homophylotis catori (D. Mariau)*

63 Combined intercropping and interplanting in old plantation *(J.G. Ohler)*

64 Cattle as 'sweepers' under coconut *(S. Reynolds)*

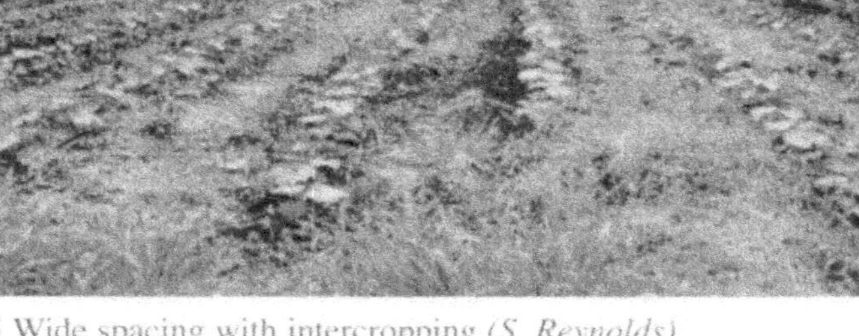

65 Wide spacing with intercropping (*S. Reynolds*)

66 Cattle grazing under coconut (*S. Reynolds*)

67 Buffalo grass (*S. Reynolds*)

68 Well-grazed Buffalo grass (*S. Reynolds*)

25 Morphologic diversity of coconut fruits *(R. Bourdeix)*

26 Pollination by male flowers of the next inflourescence *(R. Bourdeix)*

27 Nursery with seedlings in polybags *(J.G. Ohler)*

28 Young trees suffering from drought *(J.G. Ohler)*

33 Iron deficiency symptoms *(G. de Taffin)*

34 Sulphur deficiency symptoms *(CIRAD-CP)*

35 Potassium deficiency symptoms *(CIRAD-CP)*

36 Potassium deficiency symptoms *(CIRAD-CP)*

37 Copper deficiency symptoms *(X. Bonneau)*

38 Copper deficiency symptoms *(X. Bonneau)*

39 Leaflets with nitrogen deficiency symptoms *(CIRAD-CP)*

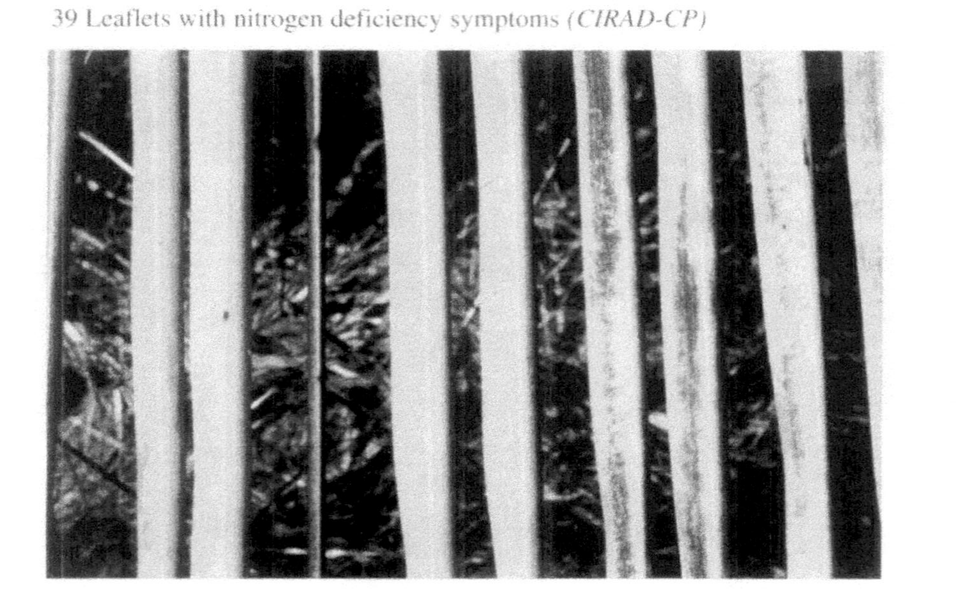

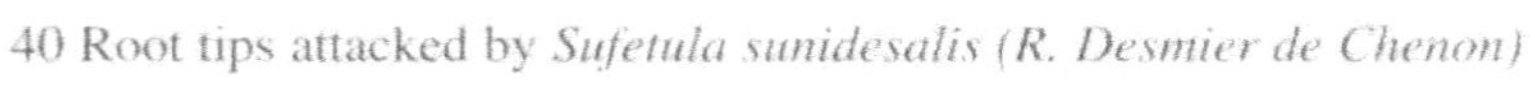

40 Root tips attacked by *Sufetula sunidesalis (R. Desmier de Chenon)*

73 Rice huller made of coconut wood in Zanzibar *(J.G. Ohler)*

74 A coconut grater made in Zanzibar *(J.G. Ohler)*

75 Rice huller made of coconut wood in Zanzibar *(J.G. Ohler)*

76 Sawing coconut stems *(J.G. Ohler)*

41 Adult of *Valanga nigricornis* (*R. Desmier de Chenon*)

42 Female of *Sexava coriacea* (*R. Desmier de Chenon*)

43 Galeries of *Captotermes curvignathus* in the crown (*J. Renoux*)

44 *Balyana mariaui* (top) and *Coelaenomenodera perrieri* (*D. Mariau*)

45 Coconuts damaged by *Pseudotheraptus wayi* attack *(J.G. Ohler)*

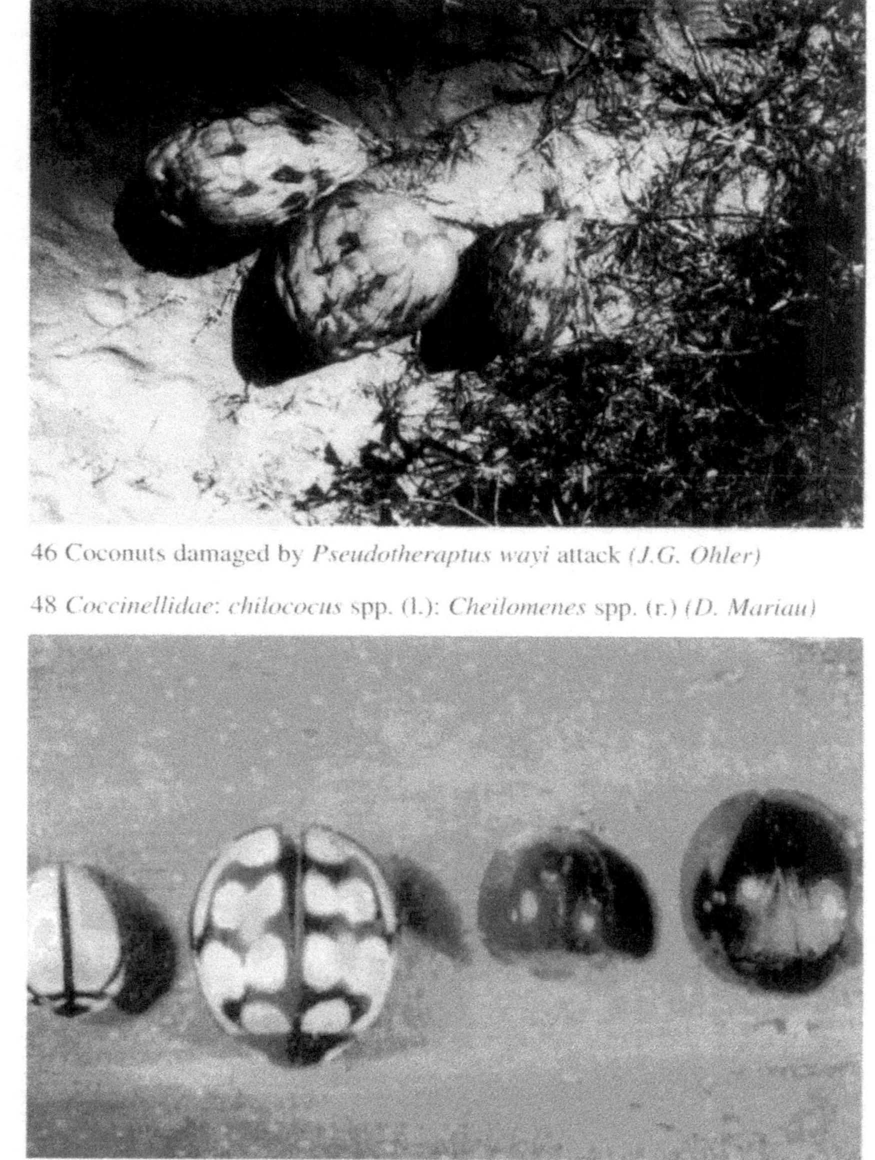

46 Coconuts damaged by *Pseudotheraptus wayi* attack *(J.G. Ohler)*

48 *Coccinellidae*: *chilococus* spp. (l.); *Cheilomenes* spp. (r.) *(D. Mariau)*

47 *Aspidiotus destructor* with some coccinelid beetles *(J.P. Morin)*

49 Larva of *Brontispa longissima* and damage caused (*J.P. Morin*)

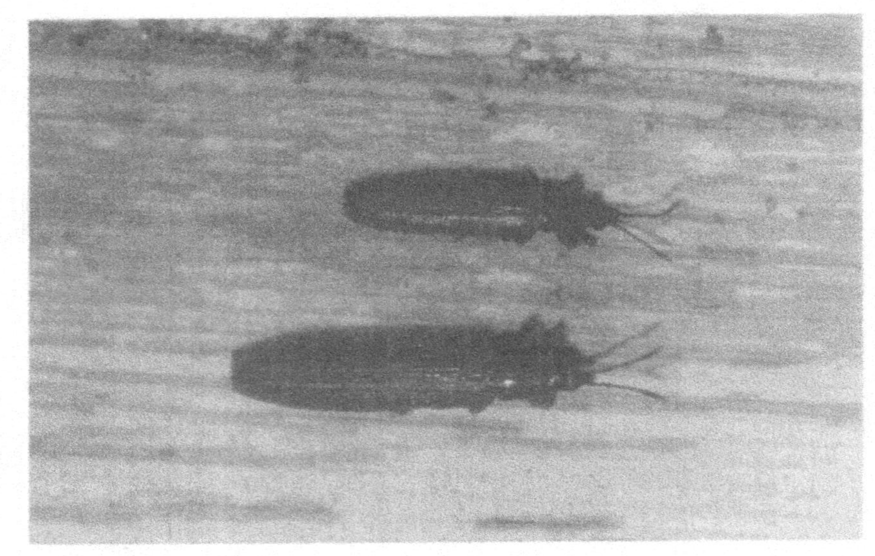

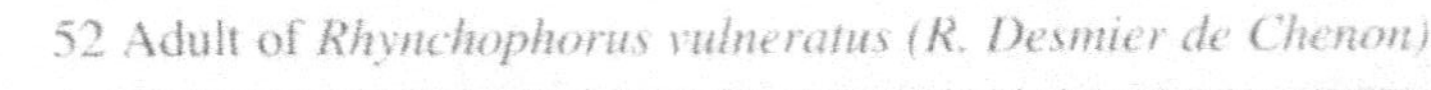

50 Adults of *Gestronella* sp. (*D. Mariau*)

51 Adult of *Coraliomela brunnea* (*R. Behrens*)

52 Adult of *Rhynchophorus vulneratus* (*R. Desmier de Chenon*)

53 Stem damaged by *Rhinostomus barbirostris* (J.P. Morin)

54 Male of *Oryctes rhinoceros* (R. Desmier de Chenon)

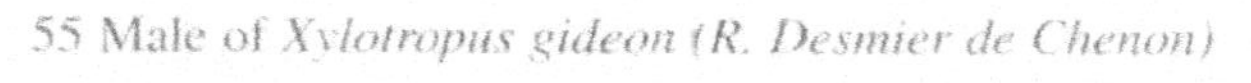

55 Male of *Xylotropus gideon* (R. Desmier de Chenon)

56 YD nuts: 1 healthy, 3 attacked by *Eriophyes guerreronis* (D. Mariau)

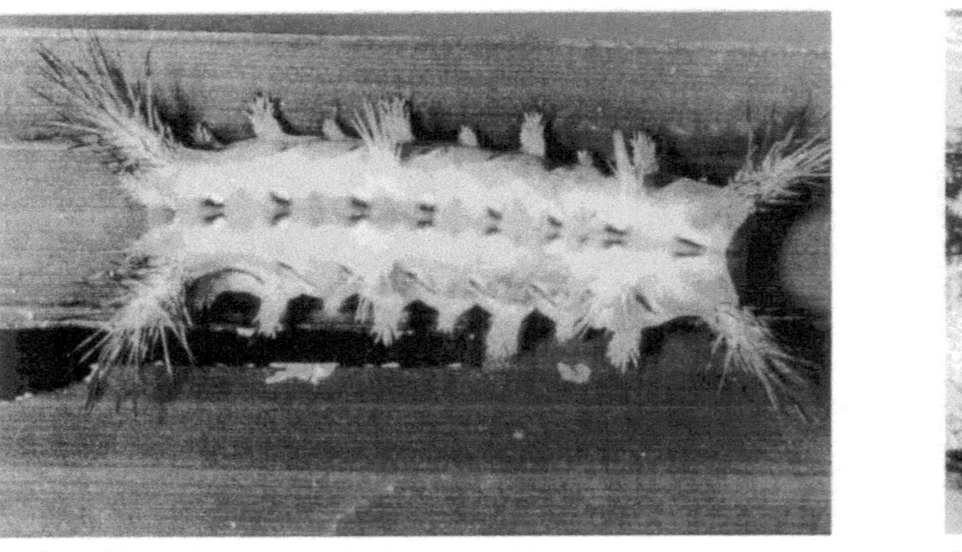

57 Caterpillar of *Setora nitens* (*R. Desmier de Chenon*)

58 *Chaetexorista javanica* (*R. Desmier de Chenon*)

59 *Eocanthecona furcellata* (*R. Desmier de Chenon*)

60 Caterpillar of *Parasa lepida* (*R. Desmier de Chenon*)

65 Wide spacing with intercropping (*S. Reynolds*)

66 Cattle grazing under coconut (*S. Reynolds*)

67 Buffalo grass (*S. Reynolds*)

68 Well-grazed Buffalo grass (*S. Reynolds*)

69 *Leucaena leucocephala* (*S. Reynolds*)

70 *Brachiaria humidicola* (*S. Reynolds*)

71 A young stand of Elephant grass (*S. Reynolds*)

72 Friesians grazing under coconuts (*S. Reynolds*)

69 *Leucaena leucocephala* (*S. Reynolds*)

70 *Brachiaria humidicola* (*S. Reynolds*)

71 A young stand of Elephant grass (*S. Reynolds*)

72 Friesians grazing under coconuts (*S. Reynolds*)

Table 37: *Dependency on coconut income (Anonymous 1982a)*

Area	Proportion of coconut income to total income (%)
Quezon	54.3
Batangas	27.7
Laguna	37.5
QBL	46.7

It is clear that the coconut smallholder must have other sources of food and income beyond those obtained from his coconuts. In many areas he has already found ways of providing these; in other areas there is considerable scope for new intercrops and integration of crops and livestock beneath the coconut canopy. In several places research is in progress to identify integrated crop-livestock systems or coconut-based farming systems (some of which would require the farmer to thin his often dense coconut stands) to maximize productivity and subsequent returns to the smallholder.

12. Coconut-based Farming Systems in the Pacific

Fred Opio

Introduction

Farming systems are fundamental to agriculture. All farmers practise some kind of system where inputs and associated materials are transformed into output. A farming system is an interactive practice involving inputs and the environment in which crops and livestock are produced. The production alternatives can take the form of a single intercrop, a mixture of crops, or a crop/livestock combination which are compatible with each with other and other environmental factors. One of the most common farming systems practised by coconut growing traditional farmers is the coconut-based farming System (CBFS). This is a multiple cropping or crop/livestock production system aimed at maximizing and/or complementing the benefits that can be derived from land under coconut.

CBFS can take many forms. In Asia and the Pacific, the traditional smallholder practice is to intercrop older coconuts with other, short-term crops in some arbitrary rotational sequence. To this extent, over 60 per cent of smallholders practise CBFS. But in the plantation sector, where long-term crops such as bananas, coffee, cocoa, lanzones, vanilla, kava, etc. are involved, continuous cropping is commonly practised (APCC 1990); Burgess 1981; Gomez and Gomez 1980; Fernando *et al.* 1984).

Morphological Characteristics of Coconut

The morphological characteristics of the coconut palm and the conventionally adopted spacing (ranging from 7x7 to 10x10m) associated with the coconut root system which normally clusters within 2 metres of the stem allow open spaces for further cultivation and/or grazing. At the much used spacing of 9x9m, at least 81 m^2 is actually allocated to each palm and yet the effective root area per palm is only 12.5m^2 or 15.4% of the available space, thus leaving 68.5 m^2 (84.6%) of spacing underutilized. Furthermore, as established by Nelliat *et al.* (1974), the top 30 cm of soil is generally devoid of functional roots and 86% of the coconut roots are found between 30 and 130 cm depth This suggests that coconut is by nature suited to intercropping. The unique leaf canopy permits a large part of the solar energy to be transmitted through it. The percentage of light transmitted depends on the age of the palms, ranging from 20% under 10-20-year-old palms, to 50% in plantations of 40 years and over (Figure 19). In older plantations light transmission increases substantially, providing ideal conditions for CBFS (Figure 21).

Light interception and the competitive ability of intercrops are believed to increase the adaptability of plants to partial shading. In fact, other crops are known to yield better under shade. These plants develop thinner leaves so that their chloroplasts become larger and richer in chlorophyll (Abilay 1983; Carandang, 1977; Paner, 1975). However, many researchers have warned that shading reduces nitrate reductase activity and nodulation in legumes (IRETA, 1988). They suggest that legumes grown under shade may have to be liberally fertilized with nitrogen compound.

Land-use Intensification

Coconut is one of the most widely grown tree crops in the tropics, occupying about 20-30 per cent of the total cultivated area. With its economic life span of 60 years or more, it occupies the land for a long time. As a monocrop, its economic life in some places averages only about 40 years, depending on growing conditions. (Burgess 1981; Opio 1990a). Beyond this period, the productivity of the land under coconut diminishes, so that it is necessary to diversify the land use or to replant coconut.

At the conventional spacings of 7x7 to 10x10 m, various trials with annuals and perennials have proven that CBFS is feasible. In Africa, Asia and the Pacific, where intercropping practices are common, various species of annuals have been grown, most of which have given up to 60% increases in their yields, as well as that of coconut, compared to the same grown area of monocrops (Creencia 1978; Cuavas 1975; Opio 1992). There are many common annuals and perennials recommended for

CBFS. Many are shade-loving (coffee, cocoa, pepper) or shade-tolerant or adapted to partial shade. Conversely, there are other non-shade-tolerant crop species whose yields would decrease if planted under coconut. Using productivity as a land value measure suggests that coconut land generates a very low return per unit area compared to other crop land. At 1988 copra prices, Opio (1992) estimated that land under coconut monocrop would yield a gross margin of less than US$ 220 per hectare per year, compared to coconut/cocoa or coconut/coffee, which would yield about US$ 620 and US$ 485 respectively. This suggests that the opportunity cost of land under coconut is relatively high, considering further that coconut often occupies prime land and does remains so for a long time.

A measure of land use intensification commonly used to evaluate effective land use under mixed cropping is the land efficiency ratio (LER). This can be expressed as:

$$\text{LER} = \sum_{j=1}^{n} \frac{(X_j)}{Y_j}$$

where: X_j = yield of crop j in a crop mixture
Y_j = yield of crop j in monocrop
j = 1..........n crops.

Studies conducted in many APCC countries indicated that the LER index was around 0.30 (Burgess 1981; Cosico, 1983; APCC, 1987), suggesting that less than 40 per cent of the total coconut land was under CBFS, of which smallholders accounted for more than 50 per cent (Table 38). The low LER suggests that coconut monocropping is an inefficient land use system.

Economic Aspects

Over 80 per cent of coconut farmers are smallholders. The average farm size among this group is less than five (5) hectares (APCC 1990; Fernando *et al.* 1984; Gomez and Orozco 1979). The high proportion of smallholders cultivating coconut is probably related to the low labour requirement associated with coconut production. Opio (1992) found that in the Pacific, one adult labour unit is sufficient to man about 3.7 hectares of coconut plantation. The average labour requirement per hectare is 290 man-hours compared to 200 in The Philippines. And yet, the household labour supply is about 3.1 and 2.7 adult labour units respectively. This represents approximately 4236 and 5772 man-hours available per household, respectively. Thus, there is ample labour available for any intensive CBFS practice.

CBFS is economic. The rationale behind the conventional wisdom is that the same land can profitably be used to produce other crops so that the productivity of the land is increased. It is also regarded as a simple but effective diversification strategy for smallholders who are often dependent on few cash crops. Proponents of CBFS argue that the system is a 'self-providing' requirement for smallholders, whose long-term planning horizon is limited and who have limited resource endowments (Jodha 1981; Opio 1986; 1990a). In effect, it is a simple but effective risk minimizing strategy suited for the uninsured farming practices adopted by many smallholders in the tropics.

Intercropping coconut land is also profitable. Experiments with various annuals and perennials have been found to give high returns. But the profitability of CBFS depends not only on yield levels, but also on other economic factors, such as costs of production, prices, other market conditions and various socio-economic factors. Yields of intercrops are known to vary with species, different climatic conditions and location. Similarly, net returns are known to vary dramatically with cost of production and producer prices. Increases in net returns from CBFS is not a function of intercrops only, but also of increases in coconut yields. Increase in nut yields under CBFS has been attributed to improved husbandry practices. Liyanage *et al.* (1984) found that intercropping coconut increased nut yield. Similar were observed results in The Philippines. They all concluded that improved husbandry practices and complementary interaction between coconuts and intercrops may have been the main factors.

Table 38: *Proportion of coconut land under CBFS in APCC countries (APCC 1990); Burgess (1981); Cosico (1983)*

	Total coconut area (ha)	Percentage intercropped smallholder	overall
Fiji	66,630	34.2	23.0
India	1,209,400	NA	NA
Indonesia	3,182,000	NA	NA
Malaysia	321,000	NA	NA
Papua New Guinea	241,000	78.9	70.0
Philippines	3,261,473	57.0	43.0
Solomon Islands	63,000	NA	NA
Sri Lanka	419,000	NA	NA
Thailand	415,000	NA	NA
Tonga	24,940	91.3	85.1
Vanuatu	93,691	53.1	39.2
Western Samoa	48,000	46.0	33.0

Socio-economic factors which influence CBFS include, among other things, farmers' receptivity to coconut production, traditional and cultural phenomena, farming systems and interplay of market mechanisms. These factors, when put together, make the practice of CBFS one of the most intricate farming systems. Despite its low returns, many smallholders regard coconut as the tree of life which is hard to abandon (Ohler 1984). Even at the less productive age of over 60, traditional coconut farmers are still unwilling to cut down their palms. It is within this context that CBFS is considered to be the most desirable practice to ensure that farmers' efforts are well rewarded.

Coconut Husbandry

Coconut cultivation is less labour-intensive than that of many other perennials, except in the first year of establishment. A total of 336 man-hours is required for various establishment activities in the first year. In subsequent years, the annual labour input requirement averages about 120 man-hours. Other major activities requiring labour throughout the crop life-span include; weed control, fertilizing, harvesting and processing (copra-making). Opio (1990a) estimated that one adult labour unit can manage about 3.7 hectares of coconut plantation per year and concluded that this is perhaps the reason why mean plot sizes under smallholder coconut farming in the Pacific are over 3.0 hectares.

Similarly, coconut is not capital intensive. Most of the high input costs are associated with establishment activities in the first year. In Asia and the Pacific, establishment costs average about US$ 80 and US$ 100 per hectare respectively. In subsequent years annual operating costs are less than US$ 20 per hectare. The major fixed cost is the construction of a hot-air copra drier commonly used in The Philippines, Malaysia, India, Sri Lanka and parts of the South Pacific. The estimated cost of constructing a modestly sized hot-air drier would be about US$ 500 (Opio 1990a). General maintenance of the drier, requiring replacement of drums and wire netting once every five years, costs about US$ 10. Due to the high initial costs involved, many smallholders prefer to sell fresh copra.

Cash-flows for Monocrop Coconut

Coconut monocropping is not remunerative. In the South Pacific, Burgess (1981) and Opio (1990a) analysed coconut returns and found that the accumulative discounted cash-flows at 10 per cent balanced negative cash-flows in years seven and five for local talls and hybrids respectively (Table 39). Over a 40-year period, the average annual returns per hectare from local talls was about US$ 55.

The undiscounted cash-flows for the first 40 years of productive life-span based on local tall and hybrid yield streams, gave four years of negative cash-flow for both types. Undiscounted cash- flows were found to vary with yield fluctuations and market prices, but generally did not give negative returns within the first 40 years, except for the first five years. Interestingly, in years of very depressed producer prices, there was a tendency among farmers to neglect their plantation temporarily.

Technological adoption, improved husbandry practices and the introduction of hybrids are expected to enhance the productivity of coconut plantations and increase farmers' income. However, input increase results in cost increase, which may reduce the expected net benefits. Thus, the increase in returns to factors of production, especially labour may be marginal. In many South Pacific countries, the estimated undiscounted net return to labour, per hour averaged US$ 0.90 and US$ 3.51 for local talls and hybrids respectively. But when discounted (at 10%), the net return to labour of US$ 0.14 per hour for local tall is significantly lower than the minimum wage of many Pacific Island countries. In The Philippines, estimated returns to labour are more reasonable, perhaps due to low rural wages.

Table 39: *(Opio 1990a) Discounted cash-flows (10%) for local talls (LT) and hybrids in Western Samoa (1988 Prices)*

Monocrops		Intercrops (Local Talls (LT)				LT/Rotational Grazing and intercropping with		
Year	LT	Hybrid	LT/Cocoa	LT/Taro	LT/Kava	LT/Cattle	Taro	Kava
1	- 173.77	- 173.77	- 340.00	+ 2623.81	- 1040.59	- 173.77	+ 2623.81	- 1040.59
2	- 198.56	- 198.56	- 447.85	+ 4387.45	- 1103.40	- 198.56	+ 4387.45	- 1103.40
3	- 216.59	- 216.59	- 508.70	+ 6657.17	+ 2578.79	- 216.59	+ 6657.17	- 2578.59
4	- 232.99	- 232.99	- 407.69	+ 8116.09	+ 6629.06	- 232.99	+ 8116.09	+ 6629.06
5	- 210.76	- 240.26	- 20.85	+ 8132.11	+11590.10	- 210.76	+ 8132.11	+11590.10
6	- 165.12	- 72.29	+ 398.43	- 8183.59	+11634.73	- 165.70	+ 8183.59	+11634.73
7	- 70.29	+ 222.66	+ 827.64	+ 8278.42	+11730.56	- 70.29	+ 8278.42	+11730.56
8	+ 14.79	+ 546.12	+1216.70	+ 8363.51	+11815.66	+ 14.21	+ 8363.51	+11815.66
9	+ 112.51	+ 890.90	+1590.76	+ 8461.22	+11913.37	+ 112.51	+ 8461.22	+11913.37
10	+ 183.39	+1215.83	+2002.59	+ 8532.24	+11984.25	+ 183.39	+ 8532.24	+11984.25
11	+ 292.74	+1552.30	+2433.28	+ 8641.59	+12093.60	+ 292.74	+ 8641.59	+12093.60
12	+ 403.63	+1888.40	+2836.29	+ 8752.48	+12204.63	+ 403.63	+ 8752.48	+12204.63
13	+ 504.43	+2193.93	+3022.65	+ 8853.28	+12305.29	+ 504.43	+ 8853.28	+12305.29
14	+ 596.07	+2471.70	+3535.71	+ 8944.92	+12396.93	+ 596.07	+ 8944.92	+12396.93
15	+ 665.94	+2711.91	+3833.52	- 9014.87	+12466.80	+ 665.94	+ 9014.87	+12466.80
16	+ 740.63	+2941.47	+4107.73	+ 9089.56	+12541.49	+ 740.63	+ 9089.56	+12541.49
17	+ 808.53	+3150.16	+4357.02	+ 9157.46	+12609.39	+ 868.53	+ 9157.46	+12609.39
18	+ 870.25	+3339.87	+4583.64	+ 9219.19	+12671.11	+ 870.25	+ 9219.19	+12671.11
19	+ 926.37	+3512.34	+4789.66	+ 9275.30	+12727.23	+ 926.37	+ 9275.30	-12727.23
20	+ 960.48	+3661.50	+4955.22	+ 9762.08	+12619.61	+ 830.77	+ 9612.55	+12619.61
21	+ 998.42	+3804.03	+5117.05	+10091.91	+12651.33	+ 970.39	+ 9942.37	+12651.33
22	+1032.92	+3933.61	+5231.59	+10126.40	+13290.85	+1097.31	+ 9921.59	+13290.85
23	+1064.28	+3051.41	+5335.72	+10157.76	+13987.14	+1212.70	+10036.97	+13987.14
24	+1092.78	+4158.50	+5430.38	+10186.27	+14823.18	+1317.60	+10141.87	+14823.18
25	+1121.27	+4251.11	+5522.27	+10495.83	+14851.67	+1415.53	+10451.43	+14718.67
26	+1151.48	+4339.48	+5607.15	+10707.28	+14881.87	+1508.86	+10622.87	+14812.00
27	+1178.94	+4420.07	+5674.79	+10734.73	+14909.33	+1593.71	+10656.01	+14896.86
28	+1203.90	+4493.22	+5736.29	+10759.70	+14868.18	+1670.85	+10733.15	+14855.70
29	+1226.59	+4559.71	+5792.19	+10782.39	+14887.97	+1740.98	+10803.27	+14875.49
30	+1243.59	+4617.22	+5837.52	+10973.92	+15188.88	+1797.21	+10994.80	+15176.40
31	+1261.73	+4672.17	+5883.10	+11104.58	+11515.69	+1854.54	+11125.46	+15503.22
32	+1278.21	+4722.13	+5924.53	+11121.06	+15908.90	+1906.66	+11120.63	+15896.42
33	+1293.19	+4767.54	+5962.20	+11136.05	+15923.88	+1954.04	+11168.01	+15892.03
34	+1306.81	+4808.83	+5996.44	+11149.67	+15937.11	+1997.11	+11211.08	+15935.10
35	+1314.04	+4844.54	+6027.94	+11264.92	+15944.73	+2031.11	+11326.33	+15969.10
36	+1322.27	+4878.66	+6057.09	+11342.71	+15922.11	+2063.68	+11404.12	+15946.48
37	+1329.75	+4909.68	+6083.59	+11349.9	+15928.24	+2093.29	+11398.09	+15952.61
38	+1336.55	+4937.88	+6107.69	+11356.46	+16067.49	+2120.20	+11424.75	+16091.86
39	+1342.73	+4963.52	+6129.59	+11362.41	+16218.38	+2144.67	+11448.99	+16241.75
40	+1348.35	+4906.02	+6149.50	+11435.09	+16399.92	+1266.92	+11521.67	+16424.11
Return to labour per mhr (WS$)	0.28	0.77	0.52	1.21	1.80	0.44	1.18	1.77

Exchange rate (1988): US$1.00 = WS$2.25; Minimum wage/hr (1988) = WS$1.25

Coconut-based Farming Systems with Annuals

Low economic returns from coconut monocropping suggest that returns could somehow be increased if coconuts were intercroppped. Intercropping with annuals or short-term crops provides such an alternative. In The Philippines and other parts of Asia, where land ownership under share tenancy is practised, intercropping with annuals is considered more appropriate, since most landowners do not allow planting of perennials between coconut. In newly established coconut plantations, intercropped annuals may provide quick cash returns to provide some income when the palms are not yet bearing. In flat land and gentle slopes certain species of annuals can provide good ground cover, minimizing erosion and improving moisture retention. However, the major drawback of annuals as intercrops is their need for regular inter-cultivation, which is labour-intensive. Besides, most annuals are adversely affected by shade.

Surveys of major coconut-producing countries have revealed that a number of annuals can perform well under coconut (APCC 1990). Table 40 shows the degree of shade tolerance of common annuals recommended for intercropping. The popularity of some of these annuals depends on geographical location and ethnic preference. In the Pacific, taro (*Colocasia esculenta*) is very popular, while in The Philippines gabi is more popular.

Table 40: *Common annuals grown under coconut (Opio 1992; Carandang, 1977)*

Annuals	Shade tolerance	Performance under given climatic conditions		
		Wet	Moderately wet	Dry
Arrowroot (*Maranta arundacea*)	M	VH	H	L
Beans (*Phaseolus sp.*)	L	H	VH	L
Bele (*Hibiscus manihot*)	H	H	VH	M
Cassava (*Manihot esculenta*)	M	VH	VH	H
Chilli (hot) (*Capsicum spp.*)	VH	VH	H	M
Chinese cabbage (*Brassica oleracea*)	M	H	H	M
Cow peas (*Vigna unguiculata*)	L	L	H	M
Egg plant (*Solanum melanongea.*)	M	VH	H	L
Ginger (*Zingiber officinale*)	H	VH	H	M
Groundnut	M	M	H	L
Kenaf (*Hibiscus cannabinus*)	L	VH	H	L
Maize (*Zea mays*)	L	H	H	L
Mustard (*Brassica sp.*)	M	M	H	M
Onion (*Allium cepa*)	L	H	H	M
Pepper (black) (Piper nigrum)	VH	VH	H	M
Pigeon pea (*Cajanus cajan*)	M	VH	H	M
Rice (*Oryza sativa*)	L	VH	H	L
Sunflower (*Helianthus annuus*)	L	M	H	L
Sweet potato (*Ipomoea batatas*)	M	VH	H	M
Taro (*Colocasia esculenta*)	H	VH	H	M
Yam (*Dioscorea spp.*)	H	VH	H	M

VH = Very high; H = high
M = Moderate; L = Low

Agronomic Implications

Screening trials suggest that yields of various annuals are reduced when grown under partial coconut shade. In The Philippines, Creencia (1978) reported that a number of annuals, including arrowroot, sunflower, groundnut and eggplant, gave 30-40% lower yields compared with yields obtained in open fields. Abilay (1983) attributed low yields of peanut to shading, which encourages vegetative growth at the expense of pod production. In areas with high rainfall, frequent cloudiness aggravates shading by coconut leaves, resulting in lower yields. Gomez and Gomez (1983) compared yields of some annual crops in shaded and unshaded conditions and found that there was significant reduction in yield (Table 41). This would, however, depend on the age of the palms and the degree of light penetration (Nelliat *et al.* 1974). There are also many examples of annuals performing satisfactorily under coconuts.

Table 41: *Shade effect and yield reduction of some annuals under coconut (Gomez and Gomez 1980)*

Crop	Yield (t per ha) unshaded	shaded	Yield reduction (%)
Maize	1.85	1.70	8
Sorghum	2.78	2.12	24
Soybean	1.02	0.71	30
Bush bean	2.24	1.04	54
Climbing bean	2.52	1.06	58
Mung bean	0.83	0.31	63
Sweet potato	16.16	2.62	84

Carandang (1975) found that gabi, ramie and chillipepper gave higher yields under coconut than in the open. He concluded that there is a need to select shade-tolerant crops for intercropping. However, of more significant importance is the increase in nut yields when coconut is intercropped. Carcallas and Aparra (1983) found that there was a significant increase in nut yields when palms were intercropped, ring weeded and fertilized. Increased nut yields are believed to be attributed mainly to fertilization and inter-row cultivation. However, leguminous annuals could contribute to increased nut yields through their nitrogen fixation in the roots. Also, Prudente *et al.* (1979) suggested that intercropping with annuals does not only increase profitability, but is beneficial to the palms as well. The proportional increase in yields in terms of copra per nut and copra per tree are attributed to appropriate intercropping (Table 42).

Table 42: *Coconut yield increase with intercropping (Prudente et al. (1979)*

Coconut + intercrops	Average no. of nuts per tree per year	Copra per nut (g)	Copra per tree (kg)
Coconut (mono)	77	213	16.4
Coconut + cucumber	86	225	19.4
Coconut + black pepper	82	205	16.8
Coconut + gabi	90	219	19.7
Coconut + camote + mungo	88	227	20.0

Economic Implications

Economic analysis of selected annuals in various countries shows that annuals under coconuts can give high net returns. In The Philippines, nine species of annuals were found to give high profits and in many South Pacific countries, intercropping with taro, yams, sweet potatoes, watermelon, pumpkins and a variety of vegetables, is highly profitable. This suggests that although other annuals may give high yields, high production costs make them less profitable. For example, in The Philippines, where ginger gives a high gross return, high production costs result in a relatively low net return.

Table 43: *Undiscounted returns from common annual intercrops from selected countries (US$) (Philippines Coconut Authority 1984; Opio, 1992)*

	Fiji (1988)			Tonga (1983)			Philippines		
	gross	net	mhr	gross	net	mhr	gross	net	mhr
LT/Corn	376	176	2.67	NA	NA	NA	170	76	0.87
LT/African daisy	NA	NA	NA	NA	NA	NA	1867	1771	4.32
LT/Ginger									
LT/Mustard	2786	1987	5.75	NA	NA	NA	572	135	0.72
LT/Sw. potato	NA	NA	NA	NA	NA	NA	1867	1094	3.96
LT/Taro	5800	4395	0.96	6400	4425	1.47	311	174	0.71
LT/Yam	1018	663	0.67	2240	1436	5.53	NA	NA	NA
	3150	2816	1.06	4997	4632	1.86	NA	NA	NA
Minimum wages			0.44			1.00			NA

LT = Local Tall, mhr = per man per hour, NA = not applicable

Since profitability of intercrops depends on yield levels and cost of production, net returns from intercrops may be increased if yields are increased and the overall production costs are minimized. In The Philippines, Felizardo (1982) found that yield of maize increased substantially when fertilized. Fertilizing maize with NPK increased its yield more than five times. In Western Samoa, Fernando *et al.* (1984) reported that crop rotation including legumes increased taro yields by two to three times. Increased yields accompanied by high farm gate prices can result in high returns, but where cost of production is very high, profit margins would be reduced. Opio (1990b) noted that adoption of appropriate crop rotation does not only result in high net returns, but it is a more cost-effective and efficient resource use. In Tonga, Opio (1992) found that yields of root crops improved where appropriate crop rotation patterns were adopted, which was also labour saving.

Table 43 shows returns from different annual intercrops in a few selected countries. It indicates that for some annuals, coconut intercropping gives high net returns, but the degree of profitability varies with the type of intercrop and location. In The Philippines, low net returns for many annual intercrops suggest that farm gate prices are relatively low compared to Fiji and Tonga. However, there are only marginal differences in return to labour, which suggests that the effort of farmers in general are well rewarded.

Coconut-based Farming Systems with Perennials

The ecological conditions to which coconut is adapted are also suitable for growing a variety of fruit and plantation crops. The most common fruit and plantation crops are avocado, bananas, breadfruit, cocoa, coffee, guava, kava, pepper, pineapple, some citrus varieties, tea and vanilla. Many of these crops are either shadetolerant or shadeloving. Although many perennial crop species are known to grow well under coconut, research on their compatibility and economic viability has been conducted on only a few of them, among which cacao may be the most important. In Asia and the Pacific, more work has been carried out on coconut intercropping with annuals than with perennials.

Effect of Perennial Intercrops on Coconut

Evaluation of some perennial intercrops suggests that growing cocoa, coffee, pineapple and bananas has no adverse effect on coconut yields. In fact, perennial intercrops, such as cocoa, benefit coconut. Nair (1977) reported that intercropping coconut with cocoa improved coconut yields by 95 per cent. Cocoa is a self-mulching crop. It has large leaves which it sheds periodically, providing good mulching material, conserving moisture and increasing soil organic matter and microbial activity, such as nitrogen-fixing and phosphate-solubilizing processes and IAA-production by *aspergillus flavus* and *A. fumingatus* in the rhizosphere. Intercropping with cocoa is one of the most beneficial practices to coconut. Nevertheless, there are other promising perennial intercrops observed in many coconut-producing countries, such as bananas (*Musa spp.*), black pepper (*Piper nigrum sp.*), coffee (*Coffea spp.*), kava (*Piper methysticum*) mainly in the Pacific, pineapple (*Ananas sativus*), avocado (*Persea americana*) and vanilla (*Vanilla planifolia.*).

Economic Implications

Intercropping with perennials is popular on large-scale plantations. Perennials are particularly suited to intercropping with coconut because once they reach maturity they continue to provide a steady flow of income with little maintenance requirements. This is also considered important under smallholder production systems where resources are limited. For instance, where the supply of family labour is limited, great pressure will be exerted on the household to allocate labour to the production of food crops and other more profitable cash enterprises at the expense of coconut production.

In Tonga, Opio (1992) evaluated the viability of the coconut/vanilla crop combination and found it to be one of the most profitable perennial intercropping systems. The coconut/cocoa combination was found to be marginally profitable in Fiji and Tonga, and uneconomic in Western Samoa. While in Fiji and Tonga, the return to labour from coconut/cocoa was slightly higher than the minimum wages in the respective countries, in Western Samoa it was lower than the minimum wage (Table 44). In The Philippines coconut/coffee and coconut/bananas were more profitable than other perennial intercrops.

The most important economic aspect of coconut intercropping with perennials is the provision of a steady cash-flow once the intercrop(s) come(s) to fruition. Opio (1990b) found that once a coconut/perennial intercrop combination reaches maturity, it continues to provide a positive cash-flow for the remainder of its economic productive life-span. Such steady streams of cash-flow provide much needed internal capital in-flow and cash insurance for farmers. Where intercrops are major food crops, they provide food security and/or marketable surpluses. To this end, coconut intercropping with perennials can constitute the basis for coconut intercropping in the long run.

Table 44: *Annual returns of common coconut/perennial crop intercrops (US$) (Philippines Coconut Authority 1984; Opio 1992)*

Intercrops	FIJI			Tonga			Western Samoa			Philippines		
	Gross	Net	Mhr	Gross	Net	Mhr	Gross	Net	Mhr	Gross	Net	Mhr
LT/bananas	NA	NA	NA	1937	830	1.78	NA	NA	NA	1995	1007	1.78
LT/cocoa	361	303	0.41	448	391	0.73	343	283	0.23	276	243	0.66
LT/coffee	NA	NA	NA	NA	NA	NA	NA	NA	NA	2418	1839	2.89
LT/orange	NA	NA	NA	NA	NA	NA	NA	NA	NA	260	199	0.56
LT/papaya	NA	NA	NA	NA	NA	NA	NA	NA	NA	1592	698	0.97
LT/vanilla	NA	NA	NA	5726	4307	2.97	Na	NA	NA	NA	NA	NA
Minimum wages			0.44			1.00			0.41			NA

LT = Local Tall
NA = not applicable
Mhr = per man per hour

Coconut-based Multi-storey Cropping System

A multi-storey cropping system is a more complex CBFS, developed to accommodate two or more intercrops of different heights, canopy patterns and rooting systems, to maximize the use of available sunlight, nutrients, moisture and land area under coconut. The fundamental objective is to increase the productivity of coconut land. The coconut palm serves as the 'top floor', whereas perennials such as coffee, bananas, papaya etc., form the mid-storey crops, and short-growing crops such as taro, vegetables, pineapples etc., form the ground floor.

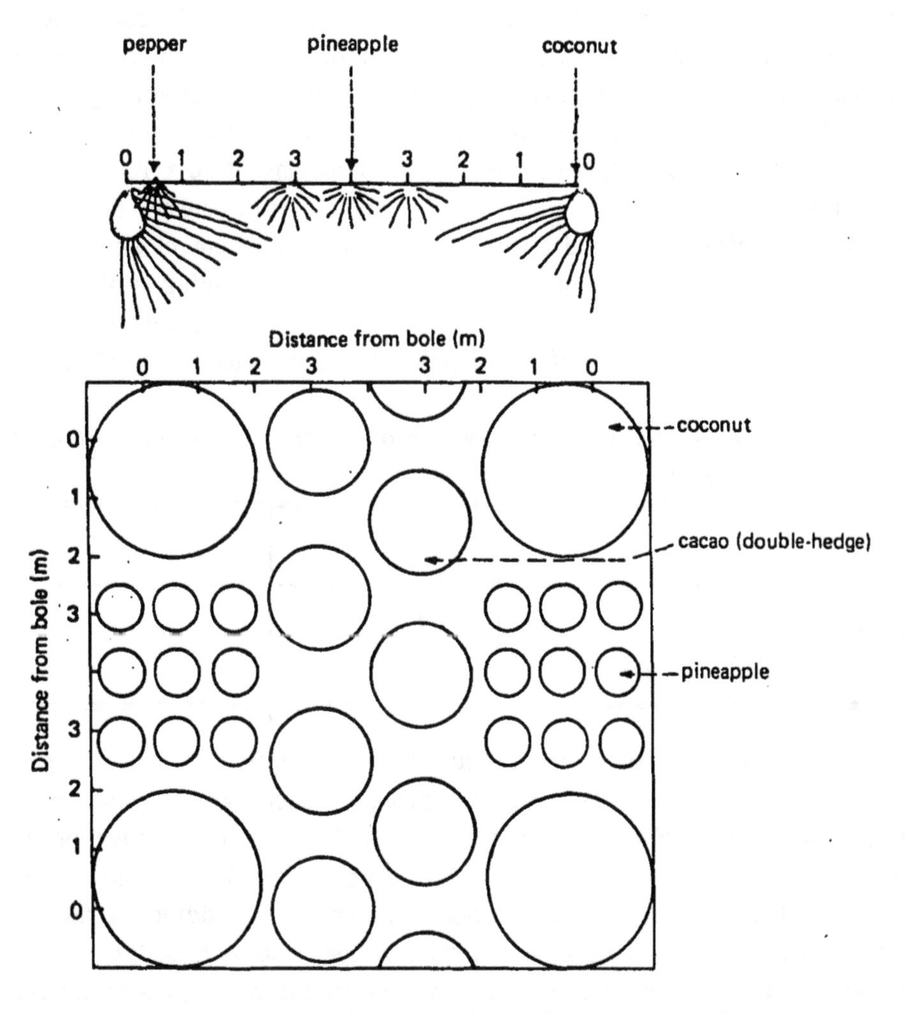

Figure 19: *Root pattern in a crop combination of a multy-storey cropping system (Nelliat et al. (1974)*

Multi-storey cropping systems require more management skills, labour and other inputs than most systems. In The Philippines, India, Indonesia, Sri Lanka and many Pacific Island countries, multi-storey cropping systems are widely practised. In Tonga, as many as 13 different crop species are grown under coconut (Havea 1991). Research is currently underway to identify the most profitable crop combinations which will satisfy various technical requirements. Nelliat *et al.* (1974) argued that the basic principle related to multi-storey cropping systems is crop compatibility, combining different crop heights and rooting systems. They found that in India, a crop combination of coconut + pepper + cocoa or cinnamon + pineapple met the general criteria of a multi-storey cropping system. While coconut formed the top floor, pepper grown along the coconut trunk formed the second floor, cinnamon grown to 2 or more metres high formed the first floor, and pineapple with its shallow root systems represented the groundfloor. Figure 19 shows the root pattern of a multi-storey cropping system. In Tonga more than 10 crops are reported to be regularly grown. Crop compatibility and crop combinations can be questioned. Crop mixing within coconut plantations does not necessarily represent multi-storey cropping systems. In fact, Tongan farmers largely practise crop mixing, although studies undertaken to evaluate the performance of some of the plantations confirm that some farmers in fact practise multi-storey cropping systems unknowingly. Nair and Varghise (1976) found that at Kasaragod, India, a coconut + black pepper + cocoa + pineapple combination was the most promising. Havea (1991) reported that in Tonga, coconut + breadfruit + vanilla + pineapple was the most popular multi-storey cropping system. It appears that the suitability of a multi-storey crop combination depends on the farmers' perception and other technical/management requirements.

Agronomic Aspects

Multi-storey cropping systems require high-level management. At the conventional spacings of 7 x 7 m to 9 x 9 m, compatible crop mixtures may be introduced, but each crop species has to be individually managed. For instance, Cuavas (1975) found that one of the promising crop mixtures in a multi-storey cropping system was coconut + pineapple + papaya + coffee + jackfruit. The cropping sequence involved planting pineapple under coconut first, at a spacing of 30 x 100 cm, followed by two rows of papaya spaced at 3 x 3 m within the inter-rows and jackfruit between the coconut. Then within the rows of papaya, coffee is planted at the centre of each coconut block of four palms. The early-maturing papaya is harvested at the end of the first year, continuing until the end of the third year. Pineapple is harvested in the second year and allowed to ratoon thereafter. By the time papaya and pineapple have been abandoned, coffee and jackfruit are expected to provide continuous return flows.

Table 45: *Effect of multi-storey cropping systems on coconut yields in The Philippines (Margate et al. 1979a); Philippine Coconut Authority 1979)*

Crop combination	Nuts per palm (accum. 27 harvests)	Copra per nut (g)	Copra per palm (kg)
Cococnut monocrop	194	269	52.0
Coconut/black pepper	189	276	52.3
Coconut/black pepper/cacao	238	274	65.2
Coconut/black pepper/pineapple	258	280	72.2

A properly selected compatible crop mixture can be beneficial to coconut palms. Margate *et al.* (1979a) found that coconut + black pepper + pineapple increased nut yields per palm and the copra content also improved (Table 45). This was attributed to inter-row cultivation, regular weeding and fertilization of intercrops. Margate *et al.* (1979a) claimed that chlorine and calcium in coconut leaves increased probably due to regular fertilization of intercrops. In addition to the benefits to coconut, it was noted that the yields of pineapple and black pepper also increased, compared to monocropping. About 2600 fruits per hectare of pineapple were harvested from a ratoon crop, but no record on pepper yield was presented.

Economic Implications

Although multi-storey cropping is one of the most effective land-use intensification systems that can increase the productivity of coconut land, its labour demand is high. Opio (1990a) observed that, while regular maintenance of a coconut monocrop requires only about 120 man-hours per hectare per year, multi-storey cropping system labour demand, although varying with the type of crop combination, is generally high. Nair and Varghise (1976) reported that in one of their most promising crop combinations (coconut + black pepper + cocoa + pineapple) about 2160 man-hours (360 man- days) per year were required. At this level of labour demand, a hectare of multi-storey intercrop would require at least two full-time adult labour units. This is further complicated by the diversity in harvesting time.

It is apparent that multi-storey coconut intercropping is more labour intensive than other coconut-based farming systems. However, the return to labour per man hour may in some cases be lower than the minimum wage. Lack of information on the minimum wage in The Philippines has made such evaluation impossible. None the less, the economic return from a multi-storey coconut intercropping enterprise makes it viable and perhaps one of the most profitable coconut-based farming systems. In a feasibility study in The Philippines, Pablo (1983) estimated that a multi-storey crop combination of coconut + sitao + ampalaya + squash + papaya + cacao gave the highest net return and coconut + ampalaya + upo + patola + squash + bananas + cocoa ranked second (Table 46).

Table 46: *Estimated annual net return per ha of multi-storey coconut/intercrop in The Philippines (US$). (Pablo 1983)*

Ranked/Crop combination	Labour mhr per ha per year	Gross return per ha	Net return per ha	Return per mhr
1. Coconut/sitao/ampalaya/ squash/papaya/cacao	2218	746	448	0.20
2. Coconut/black pepper/ cacao/pineapple	2160	745	411	0.19
3. Coconut/ampalaya/upo/patola/ squash/bananas/cacao	2239	745	372	0.17
4. Coconut/pineapple/gabi/ cacao	1946	582	337	0.17
5. Coconut/pineapple/sitao/ egg plant/ubi/gabi/ sweet potato/guava	2254	608	284	0.13
6. Coconut/pineapple/patola/ upo/sitao/chico	1992	502	259	0.13
7. Coconut/sitao/green maize/ bananas/lanzones	1719	391	213	0.12

mhr = man-hour

Relay Cropping Systems

Population pressure and land tenure systems tended to influence cultivation systems. As the size of the household increases, the land available per household member diminishes, calling for a more intensive and more diverse use of family land. In Tonga, where a land allotment system is practised, each household is restricted to less than 3.5 hectares (8.25 acres). For its food requirements, the household has to diversify its production as much as possible. To achieve this, the system of relay cropping has been adopted by many farmers. This cropping system involves rotations of a variety of mixed crops within one hectare of land subdivided into various blocks of land. Figure 20 shows the within-block crop mixture and the rotational pattern followed within the unit area. These crop mixtures are not necessarily based on a multi-storey system but on a traditional pattern and degree of compatibility.

In Tonga, since farm households cultivate food gardens of about 400-600 m^2 per crop, blocks are subdivided accordingly. Plant population per crop depends on the nature of the crop and its use. Staples, which are also highly profitable crops, usually dominate these blocks. For instance, staples such as sweet potatoes and cassava are grown in over 80 per cent of the blocks, followed by taro (*Xanthosoma spp.*) and yams (*Dioscorea alata*), which are grown in more than 60 per cent of the blocks. The estimated crop ratio with respect to coconut is 1:3 (Burgess 1981).

Agronomic Aspects

Relay cropping is based on traditional cropping practices where the land is continuously cropped for 2 to 3 years, after which it is left to rejuvenate for some time. In the first year a given compatible crop mixture is introduced, including a short fallow grazing block. In successive years, rotational cropping within the blocks is adopted each year while maintaining the same crop mixture. The main fallow period is introduced after year 3, while retaining the multi-storeyed block. Due to limited land, the fallow period in many South Pacific does not generally exceed 5 years. During the fallow period livestock are generally introduced (Figure 20). At the end of the fallow period the same process is repeated. Opio (1992) found that in Tonga, the most popular relay cropping pattern during the first year of rotation was as follows:

Block 1 - coconut + yam + giant taro + plantain + sweet potato + pumpkin + water melon + vegetables
Block 2 - coconut + vanilla + pineapple or kava
Block 3 - a short fallow grazing under coconut
Block 4 - coconut + sweet potato + taro + cassava

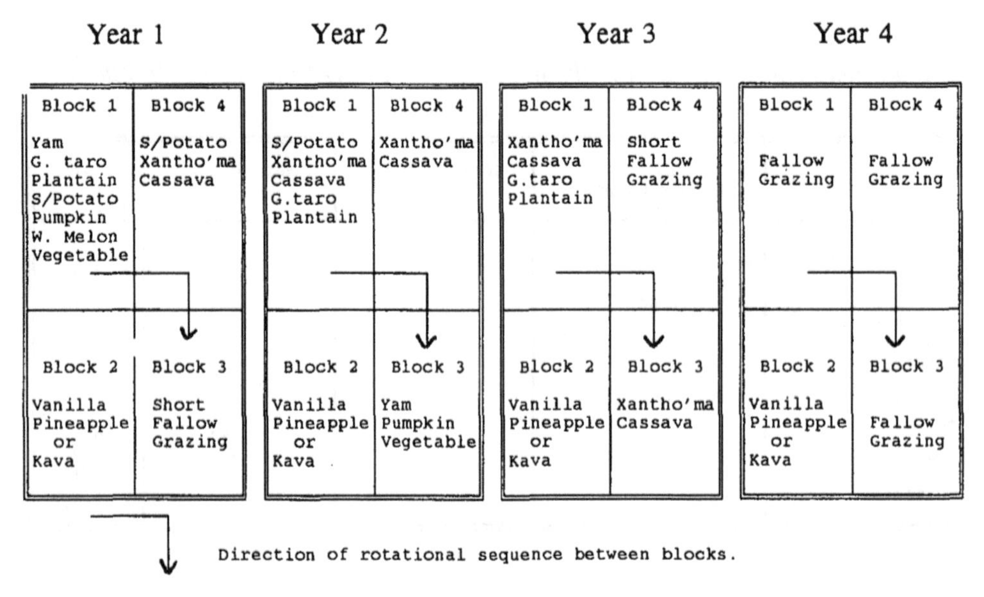

Figure 20: *Rotational sequence for coconut intercopping in Tonga*

In the second year the blocks are merely rotated, while retaining giant taro and plantain in block 1 and coconut + vanilla + pineapple or kava in block 2 for multi-storey cropping. Taro and cassava are introduced in block 4 instead of short fallow grazing, which suggests that there are alternative grazing lands. Where there is no alternative grazing land, short fallow grazing is retained while the rest of the rotational pattern is retained with the same crop mixture as above. Changes in the crop mixture is dependent on crop compatibility and viability. However, in many cases traditional staples are preferred, irrespective of their viability.

Economic Implications

Economic analysis of on-farm selected annuals commonly used in the Pacific show that relay cropping systems under coconuts are viable. Opio (1992) found that a number of compatible crop combinations in Tonga gave high returns. The combination of coconut + potatoes + water melon gave the highest net return per hectare (Table 47).

Table 47: *Economic returns to relay cropping enterprises under coconut in Tonga (Havea 1991; Opio 1992)*

Relay cropping enterprise	mhr per ha per yr	Gross return per ha per yr (US$)	Net return per ha per yr (US$)	Return per mhr per yr (US$)
Coconut monocrop	119	78	57	0.48
Coconut/cattle	231	155	127	0.55
Coconut/xanthosoma/cassava	2.209	2.167	1.436	0.65
Coconut/head cabbage/capsicum	5.079	7.041	3.709	0.73
Coconut/pumpkin/watermelon	5.564	10.881	7.665	1.38
Coconut/sweet potato/ groundnut	5.640	11.080	7.952	1.41
Coconut/sweet potato/ water melon	5.057	11.520	8.242	1.63
Coconut/yam	2.490	5.600	4.632	1.86
Coconut/vanilla	1.450	4.336	4.307	2.97

Ranking relay cropping enterprises under coconut on the basis of return to the farmer's effort indicates that the coconut + vanilla crop combination is more rewarding than other practices. However, on the basis of economic returns, a combination of coconut + sweet potato + water melon provides the highest gross margin on a per-hectare basis. Similarly, coconut + sweet potato + groundnut and coconut + pumpkin + water melon combinations also yielded high gross margins on a per-hectare basis. Returns from the above combinations are over US$ 7,000 per ha, against less than US$ 4,000 per ha from other combinations. But, apart from coconut + cattle, the returns from any relay cropping enterprise combination is found to be more remunerative than most coconut-based systems.

Although relay cropping systems under coconut are viable and more profitable, they are generally very labour intensive. In most cases, compatible crop combinations include one or more labour-intensive crop. It can be argued that intercropping minimizes resource demands. This is mainly true for land and capital, and to a limited extent for labour. High labour demand is generally required for specialized activities pertaining to a given crop species in the intercrop combination (*e.g.* planting, thinning, pruning, harvesting, marketing, etc.). These specialized activities generally account for more than 50 per cent of the total labour requirement under coconut-based farming systems. Havea (1991) estimated the individual intercrop labour requirements and found that in the relay cropping system involving coconut + sweet potato + water melon, the labour demand for coconut alone accounted for just 2.5 per cent of the total, compared to 46.2 and 51.3 per cent for sweet potato and water melon respectively. The high labour demand for crop combinations is in fact the reason behind relay cropping. It encourages block subdivisions that enable the household to manage the land with limited labour supply.

Coconut and Livestock Systems

A coconut/livestock farming system is common practice in large-scale coconut operations, where livestock, mainly cattle and goats, are grazed under coconut supported by improved or good pasture. A more recent development in the Pacific is the system of rotational grazing/intercropping where cattle or goats are kept under coconut for some time, followed by 2 or 3 years of continuous cropping, after which this process is repeated.

Coconut Intergrazing/Rotational Intercropping System

Land shortage, due to population pressure and land ownership systems in many coconut growing countries, has encouraged intensive land-use systems. The system of coconut intergrazing and rotational intercropping has been adopted on old coconut plantations (over 20 years old) in many Pacific Island countries. It involves 2 to 3 years of continuous intercropping of coconut land, followed by three or more fallow periods. During the fallow period, as the land is allowed to rejuvenate, livestock are introduced. Short-term improved pastures are sometimes used but, generally, natural pastures are used to minimize costs and labour requirements. The system is found to be beneficial to coconut, short-term intercrops and livestock, as the fertility of the soil is increased through nutrient recycling, adoption of appropriate intercrops for rotation (generally including a legume) and livestock manure. Growing root crops in rotation with legumes under coconut for 2-3 years is common practice in the Pacific.

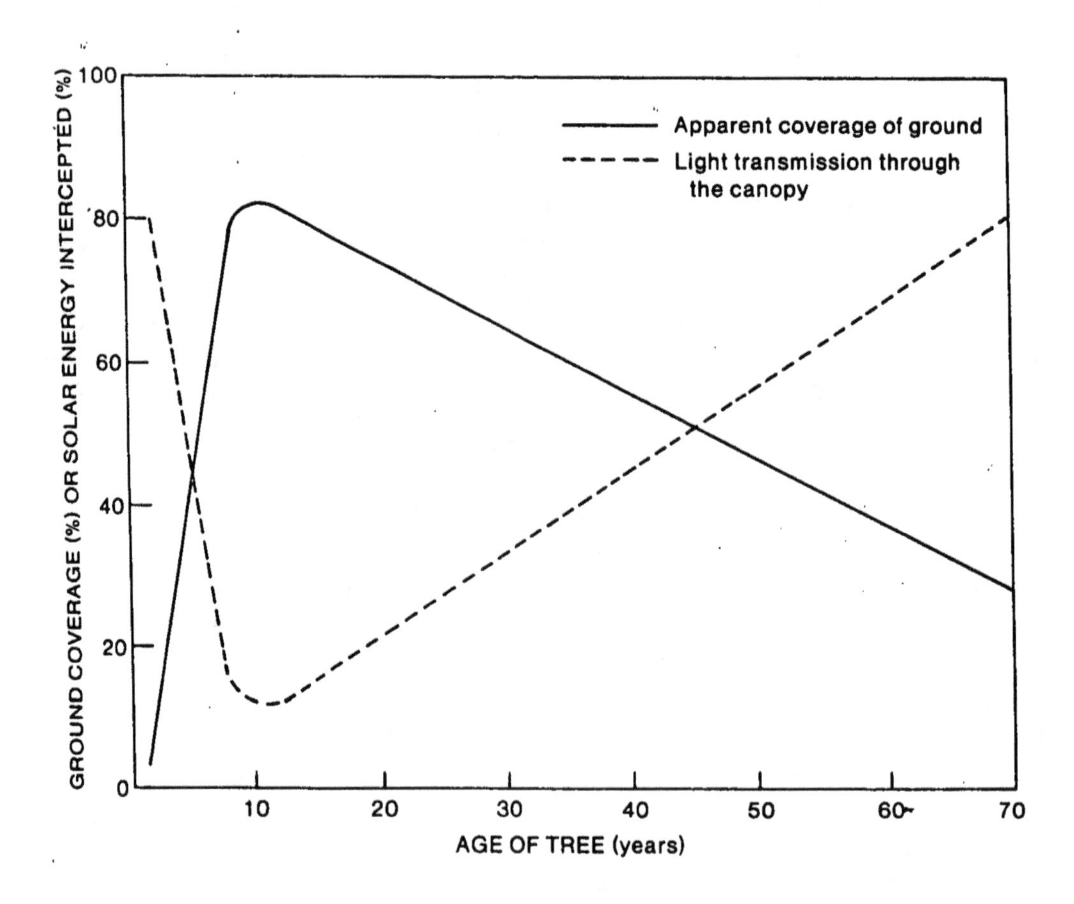

Figure 21: *Apparent coverage of ground by coconut canopies, as affected by age of the tree (Nelliat et al. 1974)*

The critical factor influencing coconut intergrazing and rotational intercropping is light. The degree of light penetration varies with the age of the coconut. As the coconut grows older, more light penetrates, thus allowing the development of undergrowth. Nelliat *et al.* (1974) observed that light transmission through the coconut canopy increases with age (Figure 21). Beyond year 30, light transmission is adequate to allow intercropping and pasture establishment. Niar (1979) and Opio (1986) claimed that even between 17 and 20 years, it is possible for pasture to be developed under coconut and subsequently grazed (Figure 22).

Reynolds, referring to Wilson and Ludlow (1991) points to the fact that the curve presented by Nelliat *et al.* (1974) significantly overestimates the minimum light levels achieved. Effective intercropping with annuals is possible from year 20, depending on whether the crop species is shade-tolerant, shade-loving, or not. Taro (*Xanthosoma spp*), a shade-tolerant aroid, was found to perform well in low light intensity (IRETA 1988). However, to minimize the problem of light penetration, the system of wide spacing of hedge planting has been adopted in many countries. An evaluation of wide spacing adopted in Tonga indicated that hedge planting can accommodate more plants per hectare compared to the convential spacing of 9 x 9 (Opio 1992). The coconut population increase does not only increase coconut yields by more than 20 per cent but also allows for continuous intercropping and/or grazing. Preliminary on-farm data suggest that the system provides good prospects for any

CBFS. Unfortunately, it was not possible to provide a realistic economic evaluation of hedge planting since it is still at the early stage of on-farm adoption in Tonga.

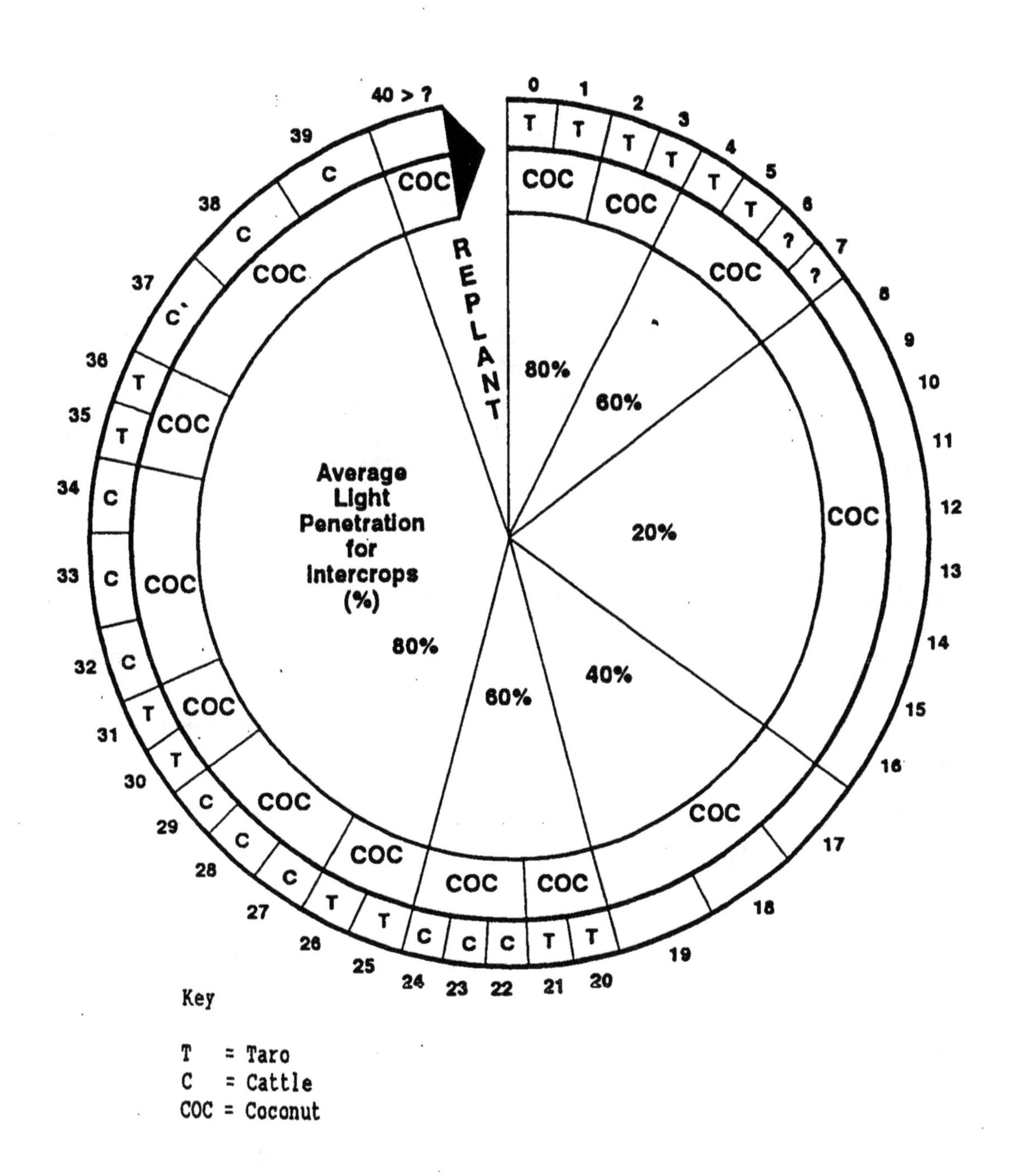

Figure 22: *Intercropping rotational grazing module for coconut plantation*

Economic Implications

Comparative economic analyses of coconut-based farming systems suggest that coconut/livestock farm ventures in most cases provide the lowest economic returns. In some cases, such ventures are not viable, but many farmers still find them convenient and fitting within traditional farming practices. Economic evaluation of coconut-based farming systems indicates that the viability of a coconut/cattle venture is marginal compared to either coconut/annual or coconut/perennial intercrops. But the system of coconut/intergrazing and rotational intercropping enhances the viability of coconut/livestock systems. In the South Pacific, Opio (1990b) in a comparative benefit-cost analysis of different CBFS, found that coconut/cattle was marginally more cost effective than coconut monocropping (Table 48).

Socio-Economic Aspects of Coconut-based Farming Systems

In spite of the compelling reasons favouring coconut-based farming systems, there are several fundamental problems that make it difficult to adopt CBFS. Coconut farmers have a choice to make insofar as they can decide on what enterprise to introduce on their farms, but the adverse conditions and the social norms that often govern coconut farmers make difficult it often for them to introduce even the most viable enterprise options. Conservatism of traditional small-scale farmers, the unique social setting of the rural environment and the rather harsh conditions under which farmers have to operate have limited their scope to adopt CBFS. Land tenure problems, off-farm employment opportunities, farm size, labour (family) supply, capital, product prices, input costs, and marketing infrastructure, which are associated with farming, also affect farmers' attitudes to CBFS. For a particular farming system to be acceptable it must fit within the socio-cultural, political, institutional environment of a society. Independent studies by Gomez (1982) in The Philippines and Opio (1988) in Western Samoa on coconut farmers' welfare and attitudes to coconut technologies, indicated that over 80 per cent of coconut farmers were smallholders with low education levels. Excluding coconut land, the average household size was over six persons operating on less than one hectare of land for their livelihood, with less than US$ 100 per capita income. The income from coconut represented about 34 per cent of the total household income.

Table 48: *Comparative analyses of coconut-based farming systems in the South Pacific (at 10% discount rate)*

Type of CBFS	Sum of net present value (US$)	Annuity (US$)	Return to mhr. (US$)	Benefit-cost ratio
LT Coconut	1.354	138	0.20	3.87
LT Coconut/cattle	2.173	222	0.44	4.59
LT Coconut/cocoa	6.158	630	0.52	4.73
Hybrid coconut	4.989	510	0.97	11.22
LT Coconut/taro/cattle	11.527	1.179	1.18	7.29
LT Coconut/taro	11.440	1.170	1.21	8.34
LT Coconut/kava/cattle	16.429	1.680	1.77	9.53

Source: Opio (1990b); LT = Local Tall.

Coconut farmers appear to have a fatalistic attitude towards coconut farming because they believe that only God can control the problems that beset coconut farming. Under traditional systems, family and kinship ties govern most of what is done. When relatives insist on it, farmers will most likely adopt new techniques or introduce new crops. This is particularly common in many South Pacific islands. Opio (1988) found that in Western Samoa 77.4 per cent of coconut farmers were not willing to adopt new technologies of coconut farming because, influenced by traditional beliefs, they believed that there were many risks associated with it. In The Philippines, more than 72.9 per cent of the farmers felt the same. Apparently, farmers are influenced in their decision-making by traditional norms.

Conclusions

Coconut-based farming systems are adopted by many small-scale farmers as a self-sustaining and risk-minimizing strategy. The rationale is that the productivity of the coconut land can be increased. It minimizes resource demand, especially under traditional farming practices where levels of resource endowments are low and where the farmers' planning horizon is limited. In large-scale farm operations, CBFS is also adopted because it provides an efficient resource allocation strategy and minimizes input costs. Since many traditional farmers are unwilling to cut down and replace over-aged palms, coconut tends to occupy land for a long time at the expense of alternative production. In this regard, coconut land can be better utilized adopting a CBFS. In addition, given that coconut monocrop is not very remunerative, it is important that an integrated coconut farming system is adopted to make coconut production worth while. The adoption of CBFS encourages improved husbandry practices, increases the productivity of coconut land, and enhances the viability of coconut ventures. The steady net cash-flow from CBFS acts as an insurance to farmers and provides a genuine reward for the farmer's effort.

13. Coconut-based Cropping Systems in India and Sri Lanka

Prafula K. Das

Introduction

Coconut holdings in India and Sri Lanka are by and large very small (less than 2 ha) and when these units are monocropped they do not provide adequate income to meet the multiple needs of the dependent families. Raising a number of crops in the interspaces of coconut stands is therefore an age-old practice in densely populated coconut areas in these two countries. But the wrong choice of intercrop species and unscientific method of cultivation may lead to intense mutual competition, resulting in low returns from such gardens (Bavappa and Jacob 1982). Evidence is, however, accumulating through farming systems research proving that scientifically adopted coconut-based multiple cropping systems are highly productive and rewarding (Nair 1979).

There is a wide variety of crops that could be raised in the interspaces of coconut stands. These will not all be discussed here; only some of the more important cropping systems are briefly dealt with in this chapter. It is not possible to present a detailed economic analysis of each of the case- studies since some information on various inputs used for the trials were not always clearly mentioned by the earlier authors. It may be kept in mind that yield and income figures mentioned in the case of different intercrops or cropping systems are merely indicating their suitability or otherwise, because for each intercrop or system widely varying yields and income may be obtained in different sites with different growing and marketing conditions (Ohler 1984). Both the inputs and outputs of intercrops shown in this chapter refer only to per ha gross coconut area, and not to net area of these crops. The returns expected in Indian Rupees (Rs) and equivalent US Dollars ($) refer to the corresponding period of the studies and not to the current currency values.

Root and Tuber Crops

In India and Sri Lanka, as in any other coconut-growing country in the world, root crops and tubers are among the most favoured intercrops under coconut stands (Ohler 1984). These include cassava (*Manihot utilissima*), amorphophallus (*Amorphophallus campanulatus*), greater yam (*Dioscorea alata*), lesser yam (*Dioscorea esculenta*), colocasia or cocoyam (*Xanthosoma sagitifolium*), sweet potato (*Ipomoea batatus*), and Chinese potato (*Coleus parviflorus*). These crops are popular among smallholders because they partially meet their food requirements and require very little cash input. In addition, they are less subject to the vagaries of weather and market fluctuations (Das 1991a).

Table 49: *Recommended cultivation methods of root and tuber intercrops*

	Method of planting	Spacing	Fertilizer dose (kg per ha)		
			N	P_2O_5	K_2O
Cassava	Mounds	1.0 x 1.0 m	80	80	80
Amorphophallus	Pits	1.0 x 1.0 m	65	50	80
Greater yam	Pits	1.0 x 1.0 m	65	50	80
Lesser yam	Pits	0.7 x 0.7 m	50	50	65
Sweet potato	Ridges and furrows	0.5 x 0.2 m	55	30	55
Chinese potato	Beds	0.2 m between plants	45	45	70

These root and tuber crops are planted in April-May following the onset of pre-monsoon rains. While sweet potato takes 4 months, Chinese potato takes 6-7 months and others take 9-10 months until harvest. Cassava is propagated by stem, sweet potato by vine cuttings and amorphophallus and yam are propagated by cut pieces of tubers. Table 49 shows the method of planting, spacing and the recommended dose of fertilizer for each of these crops. To increase the efficiency of fertilizer use, it has to be applied in two or three splits. Kannan and Nambiar (1976) reported that when cassava had been cultivated at the Coconut Research Station (CRS), Pilicode (Kerala) in the interspaces of 50-year-old coconut stands for six years, an average yield of 15.4 t of tuber per ha was realized from the intercrop. Productivity of coconuts also increased by 5.1 nuts per palm per year after the introduction of intercropping. This coconut + cassava combination could therefore provide an additional net profit of Rs. 2047 (US$ 227) at 1975 prices. In a six-year trial of a coconut + colocasia cropping system at CRS, Pilicode, the intercrop gave an average yield of 6.2 t per ha and the yield of coconuts increased by 13.7 nuts per palm per year over its monocropping stage. The total additional net profit from this combination was estimated at Rs. 2178 (US$ 268) at 1975 prices.

The Central Plantations Crops Research Institute (CPCRI), Kasaragod (Kerala) conducted a five-year field trial of coconut-based root crops and tuber intercropping systems during the 1970s. Table 50 presents the average yield of these intercrops.

Table 50: *Average yield of root and tuber intercrops in coconut stands at Kasaragod during the 1973-74 to 1977-78 crop seasons (Hegde et al. 1990)*

	Variety	Yield intercrop t per ha
Cassava	H-165	14.82
Amorphophallus	Local	13.46
Greater yam	Local	13.61
Lesser yam	Local	9.26
Chinese potato	Local	7.32
Sweet potato	H-42	8.38

These yield levels are comparable with the productivity levels when they are grown as monocrop in the open. The study further revealed that these intercrops did not have any adverse effect on coconut yield. At CPCRI, Kasaragod, a field trial was conducted during the 1970s to study the effect of growing cassava and amorphophallus as intercrops in coconut gardens with and without rotation. The result of this study, presented in Table 51, implied that the root crops and tubers should be raised in rotation rather than growing the same intercrop year after year in a particular coconut plot (Varghese *et al.* 1978).

Table 51: *Yields of cassava and amorphophallus under coconuts with and without rotation at Kasaragod during 1972-77*

Intercropping method	Average yield (t per ha).
Cassava without rotation	4.5
Cassava alternated with amorphophallus	6.3
Amorphophallus without rotation	6.4
Amorphophallus alternated with cassava	11.0

Varghese *et al.*, (1978) also reported that when in intercropping systems the root crops alone were fertilized, and the 65-year-old palms standing in the field were not fertilized, productivity of these palms was reduced by 16 nuts per palm per annum, but where both coconuts and intercrops were fertilized the yield of the palms increased by 2.4 nuts per palm per year compared to the fertilized monocropped palms. This implied that both coconuts and intercrops should be manured according to their nutrient requirements. According to Ohler (1984), most root crops are heavy feeders on potassium. Severe potassium deficiency symptoms in coconuts may therefore be observed rather commonly where cassava and sweet potato are grown in consecutive years. Hence, adequate fertilizing of both coconuts and intercrops is imperative.

At Kasaragod, amorphophallus under coconuts had yielded 12.2. t per ha. The employment generation of this cropping system, compared to coconut monocrop, was 131 additional labour days per ha. The total net income from the coconut + amorphophallus combination was estimated at Rs. 12,014 (US$ 1500) per ha as against Rs. 5500 (US$ 688) from coconut monocrop at 1978 prices (Das 1991b). The mean yield of greater yam as an intercrop under coconuts at Kasaragod was 10.3 t per ha and the net profit from this tuber alone was Rs. 2323 (US$ 290). However, since the yield from coconuts had increased from 60.8 nuts in monocropping to 69.8 nuts per palm per year in the intercropping field, the total additional net profit from their combination at 1978 prices was Rs. 7823 (US$ 975) per ha. This system also provided additional gainful employment of 76 labour-days over coconut monocropping.

In Sri Lanka, a number of field trials were conducted at the Coconut Research Institute (CRI), Lunuwila and at the Regional Agricultural Research Centre (RARC), Makandura (NWP). Liyanage and Dassanayake (1993) reported that cassava and cocoyam (*Xanthosoma spp.*) were identified as promising intercrops among root and tuber crop groups for wet and intermediate zones, and are popular among the coconut growers because of their ease of management and ready market. Table 52 presents the performance of root and tuber crops under coconuts in Sri Lanka. Under the group of rhizomatous species, ginger (*Zingiber officinale*) and turmeric (*Curcuma domestica*) are commonly grown as intercrops under coconuts in South India and wet zones of Sri Lanka. Coconut gardens, with their characteristic partial shade cover, provide an ideal environment for the growth of these rhizomatous species. The presence of an assured market is also in their favour when intercrops have to be chosen. Ginger and turmeric are planted in April - May in raised beds of 4-7 m length, 1-1.2 m width and 30 cm height in between rows of coconut palms. Compost or farm-yard manure at the rate of 5-6 t per ha is worked into the soil while preparing the beds. The details of planting materials, spacing and fertilizer doses are presented in Table 53.

Table 52: *Yield performance of root and tuber intercrops under coconuts in Sri Lanka*

Intercrops	Yields (t per ha)
Cassava	11.0
Sweet potato	4.8
Taro	6.5
Cocoyam	19.6

Mulching is an important operation for these two crops and is supposed to be carried out immediately after planting with green leaves at a rate of 15 t per ha and repeated after 50 and 100 days from planting.

Table 53: *Planting methods and fertilizer doses for ginger and turmeric (Hedge et al. 1990)*

Intercrop	Planting materials	Spacing	Fertilizer dose (kg per ha) N	P_2O_5	K_2O
Ginger	Rhizome bits of 15 g weight	25 cm x 25 cm	50	45	45
Turmeric	Finger rhizome	25 cm x 25 cm	20	20	40

Varghese *et al.* (1978) obtained 8 t per ha of ginger and turmeric grown under coconuts. They also reported that there was a marginal increase (4.7 - 5.4 per cent) in the yield of coconuts after the introduction of these intercrops because of the overall improvement of soil fertility and effective control of weeds in the field. Under these two systems the generation of additional employment was 132 labour-days per ha. While in the case of coconut + ginger combination a total net return of Rs 10,520 (US$ 1,300) was achieved as against Rs. 5,500 (US$ 680) per ha at 1978 prices, the coconut + turmeric system ended up with a net loss of Rs. 1,800 (US$ 225) per ha as the market for turmeric was most unfavourable during 1978 due to an excess supply in the local market.

Ginger and turmeric are popular intercrops among smallholders of Sri Lanka, both in young coconut (1-5 years) plantations and mature coconut stands (over 35 years) because of their high cash returns, and their short pay-back period, which is an important decisive factor during the pre-bearing stage of coconuts. Studies at CRI and RARC have revealed that ginger yielded a good return, whereas turmeric did not come up to expectations as an intercrop under coconuts (Table 54).

According to Liyanage and Dassanayake (1993), annual rotation of the intercrops increased coconut yields by 27 per cent and these systems substantially increased the income of coconut holdings in the wet zone of Sri Lanka.

Table 54: *Yield performance of ginger and turmeric as intercrops under coconuts in wet zones of Sri Lanka. (Liyanage and Dassanayake 1993)*

Intercrops	Yield (t per ha)
Ginger	20.0
Turmeric	7.3

Grain Crops

The earliest report on the performance of upland rice (*Oryza sativa*) as an intercrop came from Nileshwar (Kerala) in 1931, which showed that cultivating rice under coconut, yielding only 140 kg per ha, was not remunerative. However, from a trial with upland rice (variety Kattamodan) conducted at Pilicode during the early 1940s, a grain yield of 750 kg per ha was obtained, compared to its yield level of 755 to 1200 kg per ha when grown in the open (Nambiar *et al.* 1988). However, when rice varieties PTB 29 and PTB 30 were tried as intercrops under coconuts their performance was poor due to inadequate summer rain. It is therefore recommended that upland rice should be grown immediately after the onset of monsoon at row spacing of 20 cm. A fertilizer dose of 80 kg N, 60 kg P_2O_5 and 80 kg K_2O per ha for this intercrop is considered optimal.

Gopalasundaram and Nelliat (1979) reported on a field trial conducted at Kasaragod between 1976 and 1978 to select suitable upland rice varieties for intercropping under coconuts. This feasibility study, involving three varieties of 110 days' duration, was carried out in the rainy season (June-September). The performance of the varieties are presented in Table 55.

Table 55: *Performance of three upland rice varieties as intercrops in a coconut garden at Kasaragod (kg per ha)*

Variety	1976	1977	1978	Mean yield
Rohini 1976	1586	1427	1646	1553
Chennulu	1840	1005		
Culture 12814	1703	1036	1170	1303

Factors such as the distribution and volume of rainfall and the number of sunshine hours during the growing season, and pests and disease incidents were responsible for their varied performance in different years. Nevertheless, the variety Rohini performed much better than the other two varieties in all three years and was therefore found to be the most suitable variety to be cultivated as an intercrop under coconuts. Initial studies on finger millet (*Eleusine coracana*) under coconuts, conducted during the early 1030s at Kasaragod were not very promising. It gave a yield of only 292 kg per ha grain in the first year and 236 kg per ha in the second year. Another trial was conducted between 1939 and 1942 at CRS, Pilicode, with five millet species, including finger millet. These millets were simultaneously raised as monocrops in the open and under coconuts as intercrops to study their comparative performances. Various sowing dates were tried to find out the optimal sowing time for these intercrops. Out of the five millets tried, *Panicum miliare* and *Echinochloa frumentacea* performed very badly. In the case of *Eleusine coracana,* the highest yield obtained under coconuts was 755 kg per ha, whereas it had yielded up to 1275 kg per ha in the open. *Pennisetum typhoides* yielded up to 2220 kg per ha under coconut, whereas in the open it yielded 2,250 kg/ha. Sowing at the beginning of the monsoon season resulted in higher yields than later sowing dates. Table 56 presents the average performance of three millets as monocrops and intercrops sown at different dates.

Table 56: *Yield of millets as intercrops in coconut garden and in the open at Pilicode (1939-40, 1940-41 and 1941-42)*

Scientific name	Variety	Sowing date	Under coconut	In the open
Eleusine coracana	EC 593	June '39	750	1,275
		June '40	352	738
		Aug. '40	207	154
		2 May '41	755	484
		28 May '41	483	815
Pennisetum typhoides	PT 17	June '39	1,148	1,102
		June '40	1,010	352
		Aug. '40	137	23
		2 May '41	831	161
		28 May '41	238	106
Paspalum scrobiculatum	PS 1	June '39	1,364	2,250
		June '40	874	1,352
		Aug. '41	716	465
		2 May '41	2,220	1,385
		28 May '41	906	2,094

Liyanage and Dassanayake (1993) reported that maize (*Zea mays*) and sugarcane (*Saccharum officinarum*) were tried as intercrops under coconuts at Lunuwila and Makandura. Maize yielded 800 kg per ha, and sugarcane produced an average of 45 t per ha.

Grain Legumes

In India, a number of trials were conducted at CRS, Pilicode; CPCRI, Kasaragod; the Regional Coconut Research Station (RCRS) Arisikere (Karnataka); and RCRS, Veppankulam (Tamil Nadu) with various grain legumes as intercrops under coconuts in different periods. The performance of the grain legumes showed mixed results. At Pilicode, among six crops tried, only pigeon pea (*Cajanus cajan*), which gave a yield ranging from 133 to 454 kg per ha was considered a remunerative intercrop, while the other five, among which black gram (*Vigna mungo*), green gram (*Vigna radiatus*), and horse gram (*Dolichos uniflorus*), failed completely. At Arsikere, cowpea (varieties C 152 and C 448) and pigeon pea (Hybrid 3C) gave satisfactory yields. At Veppankulam, cowpea (variety C 152) alone was promising, whereas other grain legumes gave very poor yields as they were not shade-tolerant. In 1974, at Kasaragod, horse gram gave a satisfactory yield of 355 kg per ha whereas another trial during 1976 revealed different performances for different varieties and different sowing times with respect to three grain legumes. Their yield figures are presented in Table 57.

It was evident from this trial that under Kasaragod conditions for these intercrops the middle of August was a more suitable sowing time than early September. Nevertheless, the performance of these intercrops mainly depends on a selection of suitable varieties and species for intercropping under coconuts (Das 1990).

Table 57: *Yield performance of three grain legumes grown under coconuts at Kasaragod. (Gopalasundaram and Nelliat 1979)*

Intercrop	Variety	Yields 16-08-1976	sowing dates 01-09-1976
Cowpea	779 New Era Kunnamkulan	75 300 500	83 267 467
Green gram	Philippines PS 7 PS 16	failed 177 300	failed 67 250
Black gram	S 1 T 9	58 317	58 233

In Sri Lanka, grain legumes are popular intercrops under coconuts in the low country semi-dry intermediate zone. Liyanage and Dassanayake (1993) reported the result of field trials carried out under mature coconut stands. While winged bean (*Psophocarpus palustris*) yielded 1117 kg per ha of grains, the average levels of productivity for pigeon pea and cowpea were 1000 and 815 Kg per ha, respectively. Although the economics of these trials was not worked out, these yields could be regarded as remunerative.

Annual Oilseeds

Although groundnut (*Arachis hypogea*) and soybean (*Glycine max*) are grain legumes, their prospects are discussed under oilseeds, as in this section they are more relevant from their end-use point point of view. Groundnuts (variety TMV 2) when grown under coconuts at CPCRI, Kayangulam yielded 600 kg per ha and gave an additional net return of Rs. 183 (US$ 30) per ha at 1964 prices. Kannan and Nambiar (1976) conducted a feasibility study at Pilicode and concluded that, among various oilseeds, only groundnut could be a remunerative intercrop under Kerala's agro-climatic conditions. When sown during the first fortnight of May and properly managed, a yield of 1364 kg per ha was obtained. The coconut + groundnut system provided an additional employment of 165 labour-days over the coconut monocrop. In another trial at CRS, Nileshwar, when groundnut (variety TMV 2) had been cultivated in the interspaces of a middle-aged coconut garden in red sandy loam soil during the May-September season for two years, it gave a yield of 1326 kg of pods and 1448 kg haulms per ha. This yield level was comparable with the productivity of groundnut when grown in the open under rain-fed conditions (Leela and Bhaskaran 1978).

At Kasaragod, sesame (*Sesamum indicum*) was tried during 1940-41, sunflower (*Helianthus annuus*) during 1973-74 and soybean (*Glycine max*) during 1990-92, in intercropping experiments under coconut during the rainy season. Although the early growth of these intercrops was satisfactory, rainwater dripping from coconut leaves affected their growth at a later stage, and ultimately they performed very poorly. But when soybean was raised during the post-rainy season under irrigated conditions it performed well. The highest yield of 980 kg per ha was obtained from the PR-472 variety (Hegde and Yusuf 1992). At Arsikere, growing groundnut in the rainy season and wheat in the following winter-season as two sequential intercropping activities under coconuts, resulted in a net additional income of Rs. 2450 (US$ 305) per ha over the coconut monocropping system. Similarly, growing soybean during the rainy season and wheat in the following winter season gave a net additional income of Rs. 2600 (US$ 325) per ha over coconut monocrop. In Sri Lanka, feasibility trials at CRI had indicated that groundnut and soybean could be grown successfully under coconuts. According to Liyanage and Dassanayake (1993) an average yield of 1900 kg per ha was obtained from groundnut, whereas 780 kg per ha was obtained from soybean.

Liyanage and Martin (1987) reported the result of a trial involving 15 soybean cultivars as intercrops under 60-year-old coconut gardens in Sri Lanka's intermediate rainfall (1875- 2500 mm) zone. When the yields obtained under coconuts were compared with those grown in the open, the cultivar Hampton was found to have suffered the lowest yield reduction (2.9 per cent) followed by

Forest (9.6 per cent). The study revealed that the cultivars Hampton, Hardee, Davis, Forest and Pb-1, which gave yields of 1.1 to 1.5 t per ha, were most suitable for intercropping under coconuts in Sri Lanka's intermediate zone.

Vegetables

Gopalasundaram *et al.* (1993) reported that only limited work had been undertaken on vegetable intercropping in coconut gardens in India. However, smallholders in South India, Sri Lanka and all other coconut growing areas in the world, grow varieties of vegetables under coconuts out of dire necessity. At Arsikere, a field study was conducted to investigate the feasibility of raising potato (*Solanum tuberosum*) during the monsoon season, followed by chillies (*Capsicum annuum*) during the winter season in the interspaces of coconut stands. Even though potato yields were lower than their normal yields in the open, the combination resulted in an additional net return of Rs. 5300 ($ 660) over the net return from coconut monocrops at 1978 prices. Rethinam (1989) reported that potato, French bean (*Phaseolus vulgaris*) and chilli cultivation was a profitable intercropping practice in Central Karnataka.

Hegde *et al.* (1993) initiated a field trial on intercropping of vegetables at CRCRI, Kasaragod in a 30-year-old coconut garden in June 1990. Vegetables were grown in all three seasons throughout the year, with irrigation in the winter and summer seasons. There were six intercropping models involving 12 vegetable species. These intercrops included *Amaranthus spp*, cowpea (*Vigna sinensis*), snakegourd, (*Trichosanthes anguina*), bottlegourd (*Lagenaria leucantha*), ridgegourd (*Luffa acutanguala*), water melon (*Citrullus lanatus*), eggplant/brinjal (*Solanum melongena*), chillies (*Capsicum spp.*), lady's finger/ okra (*Abelmoschus esculentus*), tomato (*lycopersicum esculentum*) and coccinia (*Coccinia indica*). This study revealed that the average yield of coconuts increased from 90.8 nuts per palm per year in 1985-89 to 98.7 nuts per palm per year during 1990-91, due to generally improved field conditions following the adoption of intercropping systems. Table 58 presents these systems' economic prospects through the gain in additional net returns, and employment creation compared to the coconut monocropping system. In Sri Lanka, a number of field trials were carried out at CRI/ RARC to investigate the suitability of cultivating various vegetables, including leafy vegetable species under coconuts. Table 59 shows the average yields of these intercrops as reported by Liyanage and Dassanayake (1993). Under today's market demand conditions these intercrops could give a fairly good return to coconut growers.

Table 58: *Economic aspects of intercropping systems at Kasaragod during the year 1990-91*

Vegetable crops	Additional net return		Additional employment labour days per ha
	Rs per ha	US$ per ha	
Amaranthus Bottlegourd + Eggplant	20.920	950	260
Bittergourd + Chillies + Chillies	1.829	83	290
Eggplant + Dolichos + Water melon	5.365	244	300
Snakegourd + Ridgegourd + Amaranthus	22.217	1.009	305
Cowpea + Lady's finger + Tomato	7.305	332	365
Coccinia + Coccinia + Coccinia	13.020	510	245

Table 59: *Yield performance of vegetable crops grown under coconuts in Sri Lanka*

Intercrop	Yield (t per ha)
Bushito (*Vigna spp.*)	2.0
Winged bean	3.2
Eggplant	9.5
Amaranthus spp.	28.3
Alternanthera spp.	79.3
Spinach	10.8

Fruits

A number of fruits are grown under coconuts by smallholders in India and Sri Lanka. Pineapple, (*Ananas sativa*), banana (*Musa paradisciaca*), mango (*Mangifera indica*), jack fruit (*Artocarpus heterophyllus*), bread fruit (*Artocarpus incisa*) and papaya (*Carica papaya*) are among the most popular fruits grown along with coconuts. In Sri Lanka, passion fruit (*Passiflora edulis*) is also found to be popular in coconut holdings.

Pineapple

Pineapples are planted at a distance of up to 1 m from the base of the palm in shallow trenches 2.7m apart and at 0.3 m spacing, totalling 8500-10000 plants per ha. The recommended fertilizer doses for this crop under Kerala conditions is 8g N, 4g P_2O_5 and 8g K_2O per plant. The application of fruiting-inducer hormones, when plants are about one year old, has become standard practice. About six months after this application, fruits may be ready for harvest. After harvest, the mother plant produces suckers which, in turn, will produce fruits (ratoon system). In India, generally two ratoons are taken, after which replanting is done to sustain higher yield levels. However, rodent pests, crows and mealy bugs (*Pseudococcus breripes*) are important pests and their control is imperative for obtaining good harvests from this intercrop.

At Kasaragod, an observational trial to study the comparative performance of pineapple (variety-Kew) was laid out under coconuts with and without summer irrigation. It was observed that both the number of fruits and the fruit weight were significantly higher in the irrigated plot compared to the unirrigated one. Gop lalasundaram *et al.* (1993) reported that this intercrop gave a mean yield of 1.54 kg per fruit per harvest. Hence, this system with irrigation was highly remunerative. Feasibility studies on pineapple intercropping conducted at RARC, Makandura in Sri Lanka showed an average yield of 14 t per ha. These studies also revealed that one double row and two single rows are the best planting systems for pineapple when raised under coconuts (Liyanage and Dassanayake 1993). Ohler (1984) has suggested that where intercropping under coconuts is to be encouraged, pineapple is one of the first crops to be considered, because not only does it give its first return within one-and-a-half years, but it also yields a fairly high return for the system.

Banana

According to the FAO (1966), banana gives one of the highest returns among the intercrops under coconuts. Its yield may increase from more than 1250 bunches in the first year to about 2200 bunches in the fourth year after planting. With the onset of the monsoon in June-July, three to four-month-old disease-free, vigorous sword suckers selected from high-yielding and healthy mother plants are planted in pits of 50 x 50 x 50 cm at a spacing of 2.7 m x 2.7 m in the interspaces of coconut rows. Wood ash, at a rate of 2 kg per pit is incorporated into the soil before planting the suckers. After one month from planting, compost or cattle manure at a rate of 10 kg per pit is applied. The recommended fertilizer dose is 160g N, 160g P_2O_5, and 320g K_2O per plant per annum, to be applied in two equal splits, one just two months after planting and the other two months thereafter. When rainfall distribution is not uniform throughout the year, irrigation once every three days is required in summer months.

The crop is harvested about 12-14 months after planting. After the first harvest, two to three ratoon crops are harvested and subsequent replanting is carried out in the fifth year. It has been observed that the first ratoon crop gives a higher yield than the first harvest.

In the Godavari delta of Andhra Pradesh, banana was the most profitable intercrop with a complementary effect on coconut yield. A study conducted by Ramanathan (1985) at Veppankulam (Tamil Nadu) showed that banana gave a mean yield of 6 kg per bunch and resulted in an additional net profit of Rs. 5748 ($ 475) at 1985 prices. Nambiar *et al.* (1988b) reported on a field trial carried out at RARS, Kumarakam (Kerala) in which they found that the Palayamkodan variety with an average yield of 32.93 t per ha in the first year and 57.84 t per ha in the second year was the most suitable for intercropping under coconuts in the Kuttanad area of Kerala. In Sri Lanka, banana and chillies are the only two crops that are extensively grown as intercrops under coconuts in all three climatic zones, viz, wet zone, intermediate wet zone and intermediate dry zone. Banana has yielded 49 t per ha per year under the intercropping system trial at CRI, Lunuwila.

Other Fruits

In Sri Lanka, papaya has yielded 10 tons per ha under coconuts, while passion fruit gave a yield of 600 kg per ha. Recent studies at RARC, Makandura, showed that lemon (*Citrus sp.*) with a plant density of 500 trees per ha under coconuts could yield 24 t per ha of fruits. Hence, it is one of the promising intercrops in the wet and intermediate zones of Sri Lanka (Liyanage and Dassanayake 1993). In India, although mango, jack fruits and bread fruits are grown in a mixed cropping system with coconuts in homegardens, the feasibility studies at Kasaragod could not establish their compatibility with palms, due to their size and spreading canopies. It is therefore suggested that such large canopy tree crops should not be raised in the interspaces of coconuts; rather, they can be grown in a corner of the garden, far away from the palms to avoid competition for light as well as for moisture.

Tree Spices

In the wet zone of Sri Lanka and some parts of South India, tree spices, such as cloves (*Eugenia caryophyllata*), cinnamon (*Cinnamomum verum*) and nutmeg (*Myristica fragrans*), are grown along with mature coconut stands. In long-time feasibility studies undertaken by the CRI it was observed that cloves with a density of 130 trees per ha of coconut garden yielded 136 kg per ha per year, while cinnamon at a density of 160 trees per ha under coconuts yielded 434 kg per ha per year.

Black Pepper

Among the perennials grown along with coconuts, the most common is black pepper (*Piper nigrum*). Being a climber, it needs some support to grow and coconut palm provides an ideal support. The rooted pepper vines are planted in 0.5 m^3 pits filled with a mixture of compost or farmyard manure and top soil, 1 m away from the coconut base, and they are allowed to grow up to 4-5 m height from ground level only, so that climbing on coconut palms for picking nuts may not be hampered. Recommended doses of fertilizers for this intercrop are 33g N, 13g P_2O_5 and 47g K_2O per vine for the first year; 66g N, 26g P_2O_5 and 94g K_2O per vine for the second year; and 100g N, 40g P_2O_5 and 140g K_2O per vine, from the third year onwards. Under growing conditions prevailing in Kerala, these fertilizers are to be applied in two equal splits in May and September. At the CPCRI, Kasaragod, in a trial, pepper (Pannyur 1) growing on 175 coconut palms gave a mean yield of 2 kg of dry berries per vine or 350 kg of dry berries per ha. This yield level was quite impressive (Nelliat *et. al* (1979).

Cocoa

Cocoa (*Theobroma cacao*) is an ideal intercrop for mature coconut gardens of the over 20 to 25 year age group. A number of studies in India and Sri Lanka revealed that when cocoa is grown under coconuts, the yield level of palms increases because of the improved microbial activity in the soil through the addition of leaf litter from cocoa, retention of soil moisture and suppression of weeds in the field. Coconut palms provide the optimum shade required by cocoa for good performance (Nair *et al.* 1991).

At Kasaragod, a field trial on cocoa (variety Forastero) was initiated in 1970 in a 16-year-old coconut plot. In this trial, 350 cocoa trees per ha were planted in a single hedge system and 600 cocoa trees per ha were planted in a double hedge system. The trial was irrigated at weekly intervals in summer, because cocoa cannot withstand the yearly 6-7 month long dry spell. Cocoa received fertilizers at a rate of 100g N, 40g P_2O_5 and 140g K_2O per plant in two equal splits per year, whereas coconut was provided with 1000g N, 670g P_2O_5 and 2400g K_2O per palm per year, applied in four equal splits. The single hedge cocoa plants showed better vigour and yield per plant than those under double hedge. But because of the higher plant density per ha the yield from 1 ha garden under double hedge was more than that from the single hedge system. The four-yearmean yield of wet beans of the single hedge system was 652 kg per ha per year, whereas it was 801 kg per ha per year for the double hedge system. Compared to pre-treatment level, the coconut yield had increased by 126 per cent under the single hedge cocoa treatment, whereas it rose by 167 per cent under double hedge cocoa treatment (Nelliat *et al.* 1979; and Das 1984).

A similar type of field trial was conducted at Pilicode in a 50-year-old coconut stand between 1970 and 1983. Nambiar *et al.* (1988b) reported that the double hedge system of raising cocoa was superior, yielding 378 kg dry beans per ha per year as against 165 kg per ha per year for the single hedge system. The net returns from cocoa were Rs. 2542 (US$ 250) and Rs. 5880 (US$ 580) at 1983 prices for the single and the double hedge systems, respectively. However, it was observed that there was no effect of cocoa intercropping on coconut productivity, unlike the situation in Kasaragod.

Coffee

Coffee (*Coffea spp.*) is a popular intercrop under mature coconut stands in the low country wet zone of Sri Lanka (Karunanayake 1989). The shade from coconut palms provides optimum conditions for coffee's growth and productivity. A study at CRI showed Robusta coffee under mature coconut stands yielded 650 kg per ha in the seventh year after planting. In addition, the coconut yield increased from 5171 nuts per ha per year to 7,183 nuts per ha per year under this system (Mathes 1986).

High-density Multi Species Cropping Systems

In an attempt to maximize the utilization of solar energy and soil nutrients, a few high-density multispecies cropping systems otherwise, known as 'Multistorey Cropping Systems', were experimented with at CPCRI, Kasaragod during the 1970s and 1980s. These models included 20-year- old 175 coconuts per ha, 175 pepper vines on coconuts, 350 cocoa or cinnamon trees in a single hedge system and 3500 pineapple plants. In double hedge models, 600 cocoa or cinnamon trees were planted. The entire block was given irrigation during the dry period with a perfo-spray. Recommended levels of fertilizers were applied to each crop species in these models.

Nelliat (1979) reported that the models having cinnamon as a component of intercrops were discontinued duc to thc poor performance of that species. In other models involving cocoa as one of the three intercrops, the yield of coconut increased from 10,500 nuts per ha per year in the pre-experiment stage to 14,000 nuts after a transition period of three years. Table 60 indicates the mean yield performance of each of the model's components with single hedge cocoa. The net annual income from this system was estimated as over Rs. 20,000 (US$ 2000) per ha per year at 1978 prices, while under coconut monocropping it was about Rs. 5000 (US$ 500) per ha per year. The total labour requirement for this system was 600 labour days per ha per year, whereas it was 150 labour days for coconut as a monocrop (Das 1989).

Table 60: *Crop mean yields in a multicropping system*

Crop	Mean yield
Coconut	14,000 nuts
Pepper	60 kg dry berries
Cocoa	300 kg dry beans
Pineapple	4000 kg fruits

Table 61: *Productivity of a high-density crop model in wet zone of Sri Lanka (9th year)*

Component crops	Plants per ha	Yield per ha
Coconut (talls)	156	13,628 nuts
Cocoa (trinitario)	305	307 kg
Coffee (robusta)	407	137 kg
Black pepper on gliricidia	508	1070 kg
Black pepper on coconut	156	136 kg

One such system was initiated in the wet zone of Sri Lanka by combining coffee, cocoa and black pepper under mature coconut stands. Liyanage and Dassanayake (1993) reported that cocoa was planted in single rows in the middle of coconut inter-rows, two rows of coffee were planted on either side of the cocoa row, two pepper vines were trained on to gliricidia support along the coconut rows, and another pepper vine was trained to each palm. This study revealed that the productivity of coconut holdings in the wet zone of Sri Lanka could be raised substantially without detriment to the coconut palms, whose yield had rather increased by 14.5 per cent in association with the high-density multispecies intercrops (Table 61).

Conclusion

Feasibility studies have shown that most of the annuals, biennials and perennials tried in coconut-based multiple cropping systems are compatible with coconuts. Not only were most of them found beneficial to coconut productivity because of consequential site enrichment, but also their productivity was often comparable to that when grown in the open. Hence, the coconut-based cropping systems are capable of improving the financial status of smallholders, while permitting them to use available resources more efficiently. But the success of these systems depends on the choice of component species and the availability of suitable shade-tolerant varieties of these species, time of planting, labour resources, access to capital and market outlets, price behaviour, availability of irrigation facilities in dry season and the theft problem. Nevertheless, the crop mix is better for smallholder coconut farming than monocropping.

14. Pastures and Livestock under Coconut

S.G. Reynolds

Introduction

The raising of livestock in association with tropical plantation crops, and especially coconut, is a well established practice with many advantages such as increased and diversified income, better use of scarce land resources and often, higher plantation crop yield through better weed control and nutrient recycling. As with all intercropping systems, concern has also been expressed about adverse effects on coconut yields, but with the increasing acceptance that for economic reasons coconut lands should be intercropped (Das 1990, 1991a; Opio 1990a), these concerns have lessened.

Over the years research and development activities in various countries have been undertaken which have been documented and reviewed. In particular, reference should be made to Plucknett (1979), Shelton *et al.* (1987a) and Reynolds (1988), as it is not the intention of this chapter to repeat all previous findings. In the last five or six years, interest in the integration of livestock and tree crops has increased and a number of workshops have been held and scientific papers published. Major findings having a practical bearing on the various coconut - livestock systems will be included here, but reference should also be made to de la Vina (1991), Halim (1989), Iniguez and Sanchez (1991), Mahindapala (1988), Shelton and Stur (1991), Silva (1990), Speedy and Pugliese (1993) and Tajuddin (1991).

Systems

Traditionally, cattle have been used as 'sweepers' or 'brushers' (CP 64), keeping the grass and weeds short, preventing excessive nutrient and moisture competition with the coconut palms, and ensuring easy locating and collection of fallen nuts. According to Arope *et al.* (1985), animals act as weeders or biological lawn mowers in the plantations, saving on part of the cost of weed-killer. Current emphasis in coconut areas is on planting high-yielding hybrids (mainly in large commercial plantations) and/or on coconut-based farming systems where complementary enterprises such as livestock are integrated with coconuts to increase productivity per unit area, increase employment opportunities and provide a buffer against low and fluctuating copra prices (Aguilar and Benard 1991). Increasingly, new management techniques have been adopted, improved grasses and legumes have been planted to increase the animal carrying capacity, and in smallholder systems increased use is being made of by-products, and forage production is being integrated with food crops. It is possible to identify a number of coconut systems involving livestock:

(i) - traditional system - where cattle are used to control the grasses and weeds, thus saving on labour costs and enabling fallen nuts to be located easily. Coconuts are the main concern, with the meat from grazing animals representing something of a bonus for feeding plantation workers. With the current low copra price the traditional plantation, often with ageing trees and declining productivity, is being forced to diversify, and unless intercropped with other cash crops such as cocoa or root crops, etc. is likely to be replaced by one of several more commercial systems;

(ii) - commercial (extensive) systems - where livestock are an important secondary (or possibly even primary) source of income. Size will range from large plantations where beef cattle (and sheep) are grazed, to smaller plantations running dairy herds near urban centres. Pastures will range from poor native species, through native species with a high proportion of naturalized legumes, to improved grasses and legumes;

(iii) - subsistence and semi-subsistence (intensive) smallholder systems - where individual smallholdings are very limited in size, cattle (and goats) are likely to be kept as draught animals and milk and meat sources, but of secondary importance to crops. Feed will consist of indigenous grasses collected on roadsides and crop residues and family scraps. Where the available land is sufficient, as well as raising food crops, forage crops such as Guinea grass (*Panicum maximum*), Napier (*Pennisetum purpureum*) and leucaena (*Leucaena leucocephala*) or gliricidia (*Gliricidia sepium*) are likely to be planted so that animals can be raised for milk or meat, produced for sale.

General Factors

Any attempt to grow two or more crops together, and particularly to grow one beneath the shading canopy of the other, requires some understanding of the environmental factors involved, and the degree of competition likely. Important factors affecting the growth of forage species under coconuts are the available soil moisture and nutrients, the amount of light, and the degree of competition between the forage species and coconuts. Humphreys (1991) stresses that the yield of plantation crops may be positively or negatively affected by the pasture system, depending on the nature of the interference which develops and the net effect on the crop environment. For more information on competition refer to Vandermeer, 1989; Gillespie, 1989; Snaydon and Harris, 1981; Huxley, 1985. The influence of the plantation tree canopy on the quantity and quality of light reaching the ground surface, on temperature and humidity and soil moisture levels, was recently reviewed by Wilson and Ludlow (1991).

Nutrients

In most situations, under-storey vegetation will compete with the coconut palms for nutrients (and water). Even though grazing animals recycle nutrients in a grazed pasture, a certain proportion of them will be immobilized in the standing pasture biomass, while others will be removed in animal products. Therefore the rule should be that sufficient fertilizer must be applied to meet the needs of a productive pasture and the palms, if coconut yield is to be maintained.

Moisture

The close relationship between rainfall and coconut yield and the effect of drought is well known. Nair (1989) reported on the water requirement of coconut palms, and Ramadasan *et al.* (1991) and Nambiar and Ayer (1991) emphasized the significance of effective rainfall and not just total rainfall. Pomier and de Taffin (1982) developed a drought index to assess the sensitivity of different coconut varieties to drought. Evans *et al.* (1992) noted that where rainfall falls below 2000 mm in the humid tropics, competition by understorey pastures can reduce copra yield. Liyanage (1985) suggests that below about 1750 mm there can be competition between the coconut and forage crops for soil moisture. However, Abeywardena (1979) suggested that even where rainfall is in excess of 2000 mm, dry periods may affect copra yields, therefore some competition between under-storey vegetation and the coconut is likely. The practical implication is that where rainfall is well distributed throughout the year and in excess of 2000 mm, there will be little competition; where rainfall is between 2000 and 1750 mm there could be a problem, and where annual rainfall is below 1750 mm and/or unevenly distributed there are likely to be problems.

Light

Shelton *et al.* (1987a) stressed that the level of shade is the most significant factor determining the output of pastures grown in plantations. Vandermeer (1989) concluded that the three variables which dictate the effective radiation intensity experienced by an under-storey plant are:

- the interception pattern of the tree crown;
- the time-course of the shadows;
- the diffuse radiation input.

The amount of light will especially be influenced by the age and spacing of the coconut trees, and for details of the changes in light transmission with age reference is made to Reynolds (1988). Moss

(1992) reported on seasonal changes in light transmission caused by drought effects on the palm fronds. Wilson and Ludlow (1991) suggested that the curve for light transmission in coconuts proposed by Nelliat *et al.* (1974) significantly overestimates the minimum light levels achieved, at least for the commonly used tall varieties of coconut (Figure 23). Except where dwarf or hybrid varieties are used, or palm density is high, in most of the larger plantations values rarely fall below about 40%. This was confirmed in the Solomon Islands where, according to Litscher and Whiteman (1982), light transmission ranged from 40-70% in coconuts ranging in age from 9-19. However, in many smallholder areas, where farmers are usually reluctant to remove any palms and where planting density is often higher, light transmission values may drop below 40%.

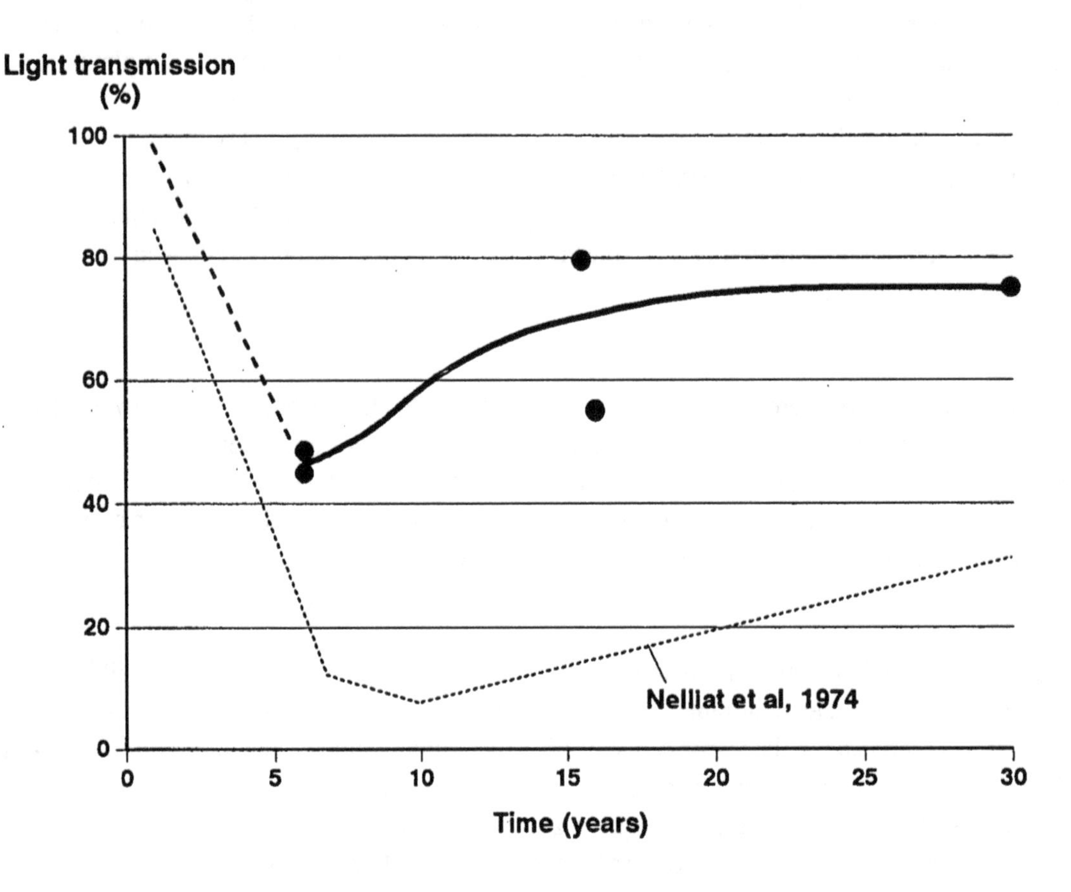

Figure 23: *Relative light transmission (%) profile of photon irradiance for coconut (Source: Wilson and Ludlow 1991)*

The light relations of pasture plants were discussed in some detail by various authors. Evans *et al.* (1992) noted that generally in legumes and grasses, high levels of shade will encourage plants to become more etiolated where they grow tall in an effort to gain better access to available light. Also, leaves become larger and thinner. This has the effect of decreasing the density of the pasture sward as well as the readily-digestible fraction of the leaf. This is one reason why cattle, when given a choice of open and heavily shaded pastures of the same grass or legume species, will usually graze the open sites first. In addition, heavily shaded pastures (*i.e.* less than 30-40% light transmission) have lower dry matter contents or higher water contents. This means that an animal has to take more bites to achieve a certain level of daily intake to achieve a certain level of performance. It has been shown that once dry matter percentages of intake drop below 18%, the potential for ruminant growth drops off markedly. According to Evans *et al.* (1992) the dry matter percentage of shaded T-grass in Vanuatu is often below 18%, especially under heavy shade! On the other hand, pasture quality for commonly available tropical species at moderate levels of shade (70% light transmission) appears to be similar to that grown in full sunlight and in some cases moderate shade will actually improve the quality (protein content and digestibility) of grasses.

The capacity of plants to accumulate soluble carbohydrate reserves is greatly diminished under shade (Wilson 1982), hence those species with a large reserve of biomass in roots and/or rhizomes and stolons which escape grazing may be more persistent under heavy shade than erect species which maximize leaf production (Wilson 1991). Although they may have a more conservative growth performance, stoloniferous species such as *Axonopus compressus, Brachiaria miliiformis (Brachiaria subquadripara), Paspalum conjugatum* and *Stenotaphrum secundatum* are reported to perform well in grazed pastures under strong shade (MacFarlane and Shelton 1986).

Table 62: *Shade tolerance of some tropical forages (after Wong, 1991 and Shelton et al., 1987a)*

Shade tolerance	Grasses	Legumes
High	*Axonopus compressus* *Brachiaria miliiformis* *Ischaemum aristatum* *Ottochloa nodosa* *Paspalum conjugatum* *Stenotaphrum secundatum*	*Calopogonium caeruleum* *Desmodium heterophyllum* *Desmodium ovalifolium* *Flemingia congesta* *Mimosa pudica*
Medium	*Brachiaria brizantha* *Brachiaria decumbens* *Brachiaria humidicola* *Digitaria setivalva* *Imperata cylindrica* *Panicum maximum* *Pennisetum purpureum* *Setaria sphacelata* *Urochloa mosambicensis*	*Arachis pintoi* *Calopogonium mucunoides* *Centrosema pubescens* *Desmodium triflorum* *Pueraria phaseoloides* *Desmodium intortum* *Leucaena leucocephala* *Desmodium canum* *Neonotonia wightii* *Vigna luteola*
Low	*Brachiaria mutica* *Cynodon plectostachyus* *Digitaria decumbens* *Digitaria pentzii*	*Stylosanthes hamata* *Stylosanthes guianensis* *Zornia diphylla* *Macroptilium atropurpureum*

Note: Based on animal performance on puero/T grass pastures in the Solomon Islands and data from Papua New Guinea. Data suggest the Animal Unit (A.U.) per ha capacity to decline linearly with light transmission down to 50%. It is suggested that below 50% the carrying capacity declines more rapidly.

Table 63: *The estimated animal unit carrying capacities of pastures and animal growth rates under various stand densities of 60-month-old E. deglupta (Anonymous 1981)*

Stem per ha	Light transm.%	A.U. per ha	Estim.daily gains kg per hd per day
>250	25	0.3	0.25
200	≤40	≤0.5	0.25 - 0.30
160 - 150	50	1.0	0.30 - 0.35
120 - 140	70	1.3	0.35 - 0.40
90 - 110	80	1.8	0.40
0 - 30	100	2.5 - 3.0	0.45 - 0.50

Some grasses and legumes are more shade-tolerant than others (Table 62). When light transmission values fall below 40 or 50%, both production values and the range of species are severely reduced. Najib (1989), Wong *et al.* (1989), Chong *et al.* (1991), Sanchez and Ibrahim (1991), Wong (1991) and Sophanodora and Tudsri (1991) all demonstrated the effect of decreasing photosynthetically active radiation on forage productivity, and Benjamin *et al.* (1991) investigated the shade tolerance of six tree legumes noting that *Leucaena leucocephala, Calliandra calothyrsus* and *Gliricidia sepium* were least affected by shade. In general, herbage production (and therefore carrying capacity) is inversely proportional to tree density (and light transmission values) as per the negative linear relationship (Figure 24; Table 63) demonstrated earlier by Smith and Whiteman (1983a) and Smith *et al.* (1983). Wong *et al.* (1985a) found that the critical level of shade for common guinea is about 52% light transmission.

Wong (1991), in his recent review of shade tolerance of tropical forages, defined shade tolerance (agronomically) as 'the relative growth performance of plants in shade compared to that in full sunlight as influenced by regular defoliation. It embodies the attributes of both dry matter productivity and persistence.' Grasses with high shade tolerance were found to have a higher specific leaf area and leaf area ratio than those with low shade tolerance (Table 64). Foliar nitrogen has been shown to increase in shaded grasses, but not in shaded legumes (Wong and Wilson 1980; Samarakoon, 1987). As nodulation in shaded legumes decreased with increasing shade intensity (Sophanodora 1989), this may explain the contrasting response of legumes.

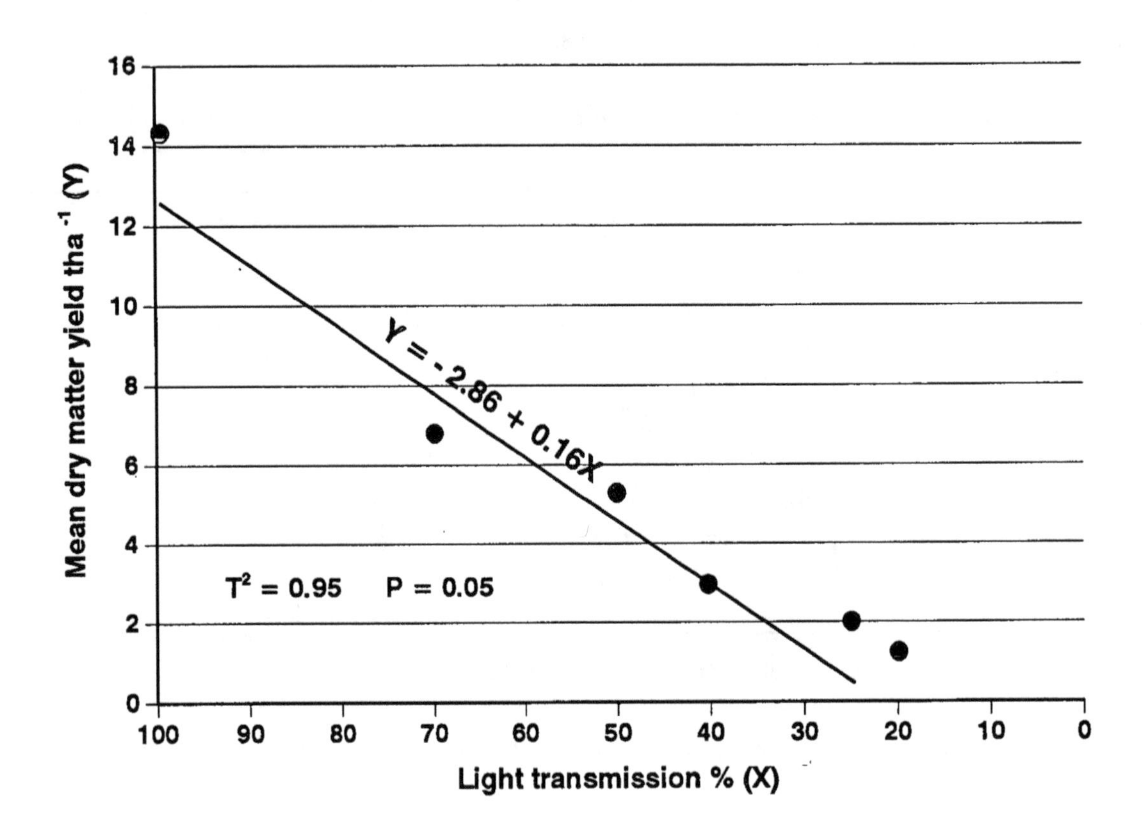

Figure 24: *Relationship between the mean dry matter yield of eight grass species and light transmission percentage under coconuts in the Solomon Islands (Based on data from Smith and Whiteman 1983a)*

Table 64: *Percentage composition of plant parts, specific leaf area and leaf area ratio of some tropical forages grown in shade (Wong 1991)*

Species	Composition (% of DM)			SLA cm^2/g	LAR cm^2/g
	Root	Stem	Leaf		
High shade tolerance					
Paspalum conjugatum	13	47	40	342	1.1
Axonopus compressus	24	30	46	296	1.2
Desmodium ovalifolium	11	37	52	437	2.3
Calopogonium caeruleum	12	35	53	407	2.2
Medium shade tolerance					
Panicum maximum	21	28	52	211	0.8
Digitaria setivalva	21	24	56	248	0.6
Centrosema pubescens	13	37	50	388	2.0
Low shade tolerance					
Digitaria decumbens cv. Transvala	13	57	30	176	0.4
Stylosanthes guianensis cv. Schofield	8	49	43	284	1.2
Stylosanthes hamata cv. Verano	12	41	48	222	1.7

SLA = specific leaf area; LAR = leaf area ratio.

Wong (1991) stresses that an important character in the selection of shade-tolerant species is their ability to persist and compete with shade-tolerant weeds under continual defoliation. The term 'persistence' includes both the survival of individual plants and seedling replacement. Indigenous shade species such as *A. compressus, S. secundatum, B. miliiformis* and *P. conjugatum* have been the most persistent and productive under low light levels. Any new shade-tolerant genotypes that are identified must be able to out-perform these species in dense shade. Persistence of forages is affected not only by their ability to tolerate shading but also by their ability to tolerate regular defoliation. A longer cutting interval has enhanced a number of grasses grown under the closed canopy of oil palms (Table 65). The shade-tolerant species *A. compressus* and *P. conjugatum* had a higher plant density at the end of the experiment than at the beginning, while less shade-tolerant grasses persisted poorly (Wong 1991). In other trials under shade the sown grasses and legumes *Brachiaria decumbens, B. mutica, B. humidicola, C. pubescens* and *Calopogonium mucunoides* did not persist well under regular grazing (Smith and Whiteman 1985). Ultimately they were replaced by naturalized species of lower productivity.

Table 65: *Persistence (expressed as % of initial plant density) of some tropical grasses as affected by defoliation frequency under the closed canopy of oil palm (Wong 1991)*

	Cutting interval (weeks)			Mean persistence
	8	12	16	
High shade-tolerance				
Paspalum conjugatum	172	156	510	279
Axonopus compressus	594	557	323	491
Medium shade-tolerance				
Panicum maximum	6	3	20	10
Digitaria setivalva	2	42	67	37
Brachiaria decumbens	4	44	56	35
Low shade-tolerance				
Setaria sphacelata cv Kazungula	5	7	1	4
Digitaria decumbens cv. Transvala	4	1	47	17

Nitrogen Economy of Shaded Pastures

The herbage yield of tropical grasses usually decreases with increasing shade, but according to Humphreys (1991) there are now well documented instances of pasture grasses, such as *Panicum maximum* var. trichoglume, exhibiting higher total biomass under moderate levels of shade than under full sunlight (Wong *et al.* 1985a; Wilson *et al.* 1986). Recently Samarakoon *et al.* (1990a) showed that the highest yields of *S. secundatum, A. compressus* and *Pennisetum clandestinum* occurred under shade rather than in full sun. This has occurred where N availability was limiting. Thus, Chen *et al.* (1991) reported that at low light intensity, signal grass production was lower at a high nitrogen rate than at a lower rate (Table 66).

Table 66: *Mean dry matter yield (kg per ha) of signal and guinea grass under various shade levels at 100 and 400 kg N per ha per yr (Chen et al. 1991)*

Light Intensity	Guinea Grass		Signal Grass	
	100 kg N	400 kg N	100 kg N	400 kg N
100% sunlight	32,770	34,428	40,633	44,407
40%	8,353	9,961	25,777	18,457
10%	652	1,029	7,173	1,473
$\bar{x}$	13,925	15,139	24,528	21,446

The source of this extra nitrogen does not appear to be from additional root nitrogen fixingability, or from redistribution of root nitrogen to plant tops; nor is it obviously related to improved nitrogen status of soils under a low-light environment (Wilson *et al.* 1986). It is hypothesized that the available soil nitrogen levels rather are increased by positive effects of shade on the rate of nitrification from soil organic matter sources (Shelton *et al.* 1987a). This theory appears to be supported by work that shows that the effect is most apparent in low nitrogen status soils and is negated when adequate fertilizer N is supplied (Samarakoon 1987; Samarakoon *et al.* 1990a).

Wilson *et al.* (1990) compared the growth of the grass *Paspalum notatum* under shade in a plantation of *Eucalyptus grandis* trees with that in full sun in an adjacent area. Dry matter yield in the summer period was 35% greater under the canopy (55% light transmission) than under full sunlight (even though there was a substantial reduction in radiation received by the grass under-storey). During winter, when shading by the trees was more intense, the herbage yield under the trees was similar to that in full sun. Throughout the year, grass under the trees had a higher proportion of green leaf, a higher concentration of nitrogen and potassium, and a lower dry matter content than the grass in full sun. Wilson *et al.* (1990) indicated that this should have a positive influence on its nutritive value for grazing animals. More recently, Wilson and Wild (1991) carried out a number of experiments to test the hypothesis (Wilson 1990) that shade (tree or artificial) increases the availability of soil nitrogen and that this leads to better growth of grass under shade than in full sunlight, when nitrogen is a limiting factor. The data (Table 67) clearly show that under some circumstances grass growth under shade can be significantly increased. The response occurs under conditions where growth in full sun is restricted by nitrogen deficiency. Shade increases the availability of soil nitrogen and this stimulates plant growth. However, no such effect was noted by Robinson (1991) and the relationship is clearly not a simple one. Differences in cutting interval may be partly responsible for the different responses. Torres (1983) mentions net radiation and soil temperature, as well as canopy precipitation interception and concentration under trees, as reasons why pasture growth may be improved under tree canopies compared to surrounding open areas.

Table 67: *Effect of shade on dry weight of tops (t per ha), leaf nitrogen concentration (%) and soil nitrate-N concentration (ppm of oven-dry soil). Wilson and Wild 1991*

Pasture Type	Brigalow clay soil				Spear grass sandy soil			
	Green Panic				Green Panic			
	I[a]	NI[a]	Buffel	Rhodes	I[a]	NI[a]	Buffel	Spear Grass
Top DW yield[b]								
Shade	23.41	15.65	15.76	15.09	13.41	9.73	13.58	7.51
Sun	16.31	12.39	16.59	11.40	9.07	6.59	10.81	6.59
Relative effect of shade	+44%	+26%	-5%	+32%	+48%	+48%	+26%	+14%[c]
Leaf nitrogen[d]								
Shade	2.81	2.89	2.08	1.53	2.64	2.64	1.79	1.38
Sun	2.21	2.23	1.64	1.34	2.39	2.58	1.49	1.49
Soil nitrate								
Shade	8.9	7.8	4.0	9.6	3.2	2.1	2.6	1.2
Sun	4.3	4.4	3.0	7.6	2.6	1.3	1.6	1.3

[a] I (irrigated); NI (non-irrigated)
[b] Based on cumulative yield totals over five harvests
[c] Response mainly due to increased weed growth
[d] Nitrogen in the youngest fully expanded leaf of each grass species at harvest on 22 May 1989

Nutritive Value of Shaded Pastures

There is evidence that low light intensities may adversely affect the nutritive value of forage species (Shelton *et al.* 1987a). Shade can result in a lowering of plant soluble carbohydrate level, higher silica content and lignification, lower cell wall digestibility, accentuated stem elongation and an increase in tissue percentage moisture content (which may reduce herbage intake by animals). However, although Hight *et al.* (1968) found that shaded ryegrass (*Lolium perenne*), produced under 22% light transmission, resulted in a reduction by 38% of liveweight gain compared to sheep fed on pasture grown in full sunlight, Norton *et al.* (1991) suggest that the short shading period used (of only a few days) was probably too short for the results to have much relevance in terms of the interpretation of the longer-term effects of shading on tropical pastures grown under plantation crops. A number of recent studies by Samarakoon *et al.* (1990a, 1990b) Wong *et al.* (1989), Wilson (1991) and Norton *et al.* (1991) failed to find any conclusive evidence of shade influencing forage quality to the extent that animal liveweight gains were seriously reduced when feeding on shaded pastures. Although some experiments were ongoing, as suggested by Wong *et al.* (1989), the lack of a consistent inverse relationship between shade and *in vitro* dry matter digestibility augers well for the integration of livestock with plantation crops.

Coconut Spacing and Pasture Performance

The coconut grower wishing to maximize pasture growth as well as yields of copra should use the least number of palms per hectare required to realize this goal (Whiteman 1980). For tall varieties, recommended spacings range from 7 x 7 m to 9.1 x 9.1 m and palm densities from 120 to 200 trees per ha (Table 68). Where pastures are established under palms at closer spacings and higher densities, light transmission values and pasture/animal production will be lower! Guzman and Allo (1975) suggested a minimum spacing of 8 x 8 m, while Steel *et al.* (1980) indicated that only coconut sites with light transmission values above 60% should be planted with pastures. In Western Samoa, Reynolds (1981) demonstrated continuing good liveweight gains where light transmission was around 50%.

Table 68: *Minimum recommended spacing and maximum planting density for coconut palms (tall variety) for pasture establishment*

Country	Spacing (m)	Density (trees per ha)	Source
Jamaica	7 x 7	200	Smith and Romney, 1969
Philippines	8 x 8		Guzman and Allo, 1975
Solomon Islands		160	Litscher and Whiteman, 1982
		150-200	Friend, 1990
Malaysia		148-197	Chen *et al.*, 1991
Sri Lanka		170-195	Anon, 1987
		180-193	de Silva and Tisdall, 1985
W. Samoa	9.1 x 9.1	120	Reynolds, 1988

However, a report by Manthriratna and Abeywardena (1979) has demonstrated that the same tree density can be achieved by various planting systems with little effect on the mean coconut yield per palm, and therefore yield per ha (Table 69). For intercropping (in this case with pasture) a rectangular system with a wide between-row spacing has many advantages over the square system. Thus, for group C (Table 69), 40 x 15 feet (12.19 x 4.57 m) and 25 x 24 feet (7.62 x 7.32 m) give the same density and the same yield per palm, but the more rectangular system would be superior for intercropping. For coconuts alone the highest net profit would appear to be a choice between 30 x 15 feet (9.14 x 4.57 m) and 25 x 18 feet (7.62 x 5.48 m), with the former being favoured for intercropping. However, Manthriratna and Abeywardena (1979) suggest that when the returns from the intercrop are considered, a lower plant density of 175-200 palms per ha and spacing of 35 or 40 x 15 feet might be the preferred spacing.

Child (1974), Liyanage (1955) and Plucknett (1979) had earlier described 'group', 'bouquet' and 'hedge' planting systems to allow space for intercrops. It is perhaps significant that trials are underway in a number of countries with several plantation crops to assess different planting systems. In Tonga, coconuts have been planted at wider spacings for intercropping for some time (CP 65). A spacing of 60 x 15 feet (18.27 x 4.57 m) was indicated by Havea (1989) and spacings of 13.6 x 6.06m (118 palms per ha) and 14.5 x 6.06 m (111 palms per ha) mentioned by Lavaka (1988). Opio (1990a) refers to the practice of 'hedge planting' which has been introduced in recent years to minimize the problem of poor light penetration so that continuous intercropping can be undertaken at any stage of development of the coconut trees. Hedge planting is based on wider spacing between the rows and closer spacing between the plants. He mentions that while there is no significant difference in plant population per unit area compared to the conventional spacing of 9 x 9 m, yields from hedge planting are generally higher (Table 70), sometimes by as much as 25%. This is mainly because inter-cultivation encourages the adoption of appropriate technology and improved husbandry practices. Hedge planting also allows for continuous inter-cultivation, which minimizes the problem of land shortages, encourages intensive land utilization and increases the productivity of land under coconut. In Sri Lanka (Anonymous 1989) the Coconut Research Institute has undertaken trials with a hedge-planting system of 30 x 18 feet (9.14 x 5.48 m) and in Indonesia on new plantings it is recommended that young coconuts be planted at 5 x 12 m, giving a tree population of 160 palms per ha and wide inter-row areas for cultivating other crops (Darwis 1988).

Table 69: *Yield per palm as influenced by density and rectangularity (Manthriratna and Abeywardena 1979)*

Group		Inter-row spacing (ft)	Intra-row spacing (ft)	Rectangularity Inter-row/ Intra-row	Density palms per ha	Mean yield nuts per palm
A	..	30	15	2.00	239	54.46
		25	18	1.39	239	53.89
B	..	35	15	2.33	205	61.88
		30	18	1.67	199	61.31
		25	21	1.19	205	60.75
C	..	40	15	2.67	179	69.30
		35	18	1.94	171	68.73
		30	21	1.43	171	68.17
		25	24	1.04	179	67.60
D	..	40	18	2.22	150	76.15
		35	21	1.67	146	75.59
		30	24	1.25	150	75.02
E	..	40	21	1.90	128	83.01
		35	24	1.04	128	82.44

Factors other than Shade that Reduce Pasture Area and Yield

Forage production is likely to be less under a coconut stand compared with open conditions, not only because of the obvious effects of shading and competition for nutrients and moisture, but also due to three other factors:

- the space occupied by the coconut palm basal stem and the surrounding root mass reduces the area available for pasture growth. In Western Samoa (Reynolds 1988) this was measured at 44.6m^2 per ha (for 16-20-year-old trees, but would be higher for older trees) for trees planted at 9.1 x 9.1m and would vary with planting density;
- coconut areas cultivated for sown pastures are restricted because of fear of damaging the root system. Reynolds (1988) measured the area occupied by coconut palm stems and the non-cultivated zone on improved pastures at 327.6 m^2 per ha or about 1/32rd of one hectare. If cultivation is not performed closer than 1.5 m from the base of the palms, as recommended by Guzman and Allo (1975) and Plucknett (1979), an area of 1290 m^2 per ha or about 1/8th of each hectare would remain uncultivated. If stoloniferous species are used then they will quickly cover this uncultivated zone, but where bunch grasses like Guinea grass are sown, this zone will remain in native species;
- Fallen fronds may affect pasture growth negatively unless removed and burnt. Although decay and fall of fronds are continuous processes, the numbers can be influenced by high wind, rain, nut collection (where collectors climb the trees and dislodge fronds) and whether or not fronds are collected frequently and removed from the pasture. Fronds appear to be more of a problem on local rather than on improved pastures because the latter quickly cover the fronds, and little grazing area is lost. Fronds not collected on native pastures could result in loss of grazing area. However, this has to be accepted as part of the nutrient circulation process, with a small portion of the pasture being tied up in frond decay, nutrient return and soil rebuilding. Frond burning results in nutrient loss, especially of sulphur and nitrogen, and slight damage to the pasture if

fronds are burnt in the paddock. Reynolds (1988) suggested that the area lost to grazing is about 120 m^2 per ha or 1/83rd of one hectar on native pastures and approximately 75 m^2 per ha or 1/133rd of one hectare on improved pastures.

Table 70: *Comparative yield levels for selected local talls under conventional spacing and hedge planting in Tonga (Opio 1990a)*

	Selected Local Tall Conventional Planting (Spacing 9 m x 9m)	Selected Local Tall Hedge Planting (Spacing 15 m x 5 m)	% Increase in Yields
Yr 1	0	0	0
Yr 2	0	0	0
Yr 3	0	0	0
Yr 4	0	0	0
Yr 5	0.37	0.41	20
Yr 6	0.49	0.60	21
Yr 7	0.68	0.84	27
Yr 8	0.81	0.96	19
Yr 9	0.98	1.15	17
Yr 10	1.10	1.30	18
Yr 11	1.37	1.63	19
Yrs 12-14	1.55	1.94	25
Yrs 15-19	1.63	NA	-
Yrs 20-24	1.40	NA	-
Yrs 25-29	1.59	NA	-
Yrs 30-34	1.44	NA	-
Yrs 35-40	1.19	NA	-
Yrs 40-50	0.93	NA	-
Over 50	0.73	NA	-
Av. Plant Pop per Ha	124	133	-

NB. Based on observations on 15 farms in Tonga; hedge planting was first introduced in Tonga in 1976 by the Department of Agriculture to encourage continuous intercropping.

Pasture Species

Grasses

Growth forms can vary from tufts, tussocks or bunches to prostrate, creeping or straggling types. The nature of the growth form usually determines the response of a grass to cutting and management and may determine the type of planting material available. Grasses with erect culms usually form tufts, tussocks or bunches, have the growing point near the level of defoliation and are propagated from seed or splits from the existing plant. Plants with stems that creep along the surface of the ground (stolons) or below the ground (rhizomes) tend to form a sward rather than tufts. Stoloniferous plants cover the ground surface with decumbent or prostrate culms, root at the nodes and can be vegetatively propagated by planting material taken from the existing plant. These are important characteristics if relatively inexpensive labour is available for pasture establishment and viable seed is either not available or is expensive to import and difficult to store.

There are 3 main types of grasses:

- bunch or tufted grasses which form clumps or tufts, and if allowed to set seed will spread when seed falls to the ground and germinates (*e.g.* Guinea grass - *Panicum maximum*);
- stoloniferous grasses which send out stems or stolons along the soil surface, root at the nodes and produce new shoots (*e.g.* Para grass - *Brachiaria mutica*);
- rhizomatous grasses which send out stems or rhizomes below the soil surface (*e.g.* Guatemala grass - *Tripsacum laxum*).

Legumes

Include creeping and erect types and also browse-plants and woody species. The creeping types such as Siratro (*Macroptilium atropurpureum*) root at the nodes of the stolons and can reach the sunlight by climbing in the associated grass. On the root systems of most legumes, nodules may develop in which nitrogen-fixing bacteria of the genus *Rhizobium* form a symbiotic relationship with the plants. Large amounts of nitrogen may be fixed through the action of these root nodules, which benefit the pasture. Peoples and Herridge (1990) and Evans *et al.* (1992) suggest that productive legumes can fix between 100-200 kg N per ha per yr. Reynolds (1988) reported that although not equal to the best exotic legumes, local legumes such as hetero (*D. heterophyllum*) and mimosa (*M. pudica*) achieved dry matter yields and N fixation rates in the same range as recognized exotic pasture legumes. Because of their high protein content, legumes improve the quality of tropical pastures and thus have a direct bearing on the level of animal production. Good grazing management is required to maintain a fair proportion of legumes in the pasture.

Pasture Species for the Coconut Environment

Indigenous species

Native vegetation under coconut varies according to the location and intensity of grazing. Unless there is control over the stocking pressure there may be changes in pasture composition over time, with undesirable weed species gradually dominating the sward (Table 71 and CP 66). Using cattle as 'sweepers' or 'weeders' without additional selective weed control measures, may control the weeds in the short term but allow tough unpalatable species to become dominant (Ohler 1984). The more promising of the native species include: Carpet or Mat grass (*Axonopus compressus*), Buffalo Couch grass (*Stenotaphrum secundatum*) (CP 67 and 68) , Pemba grass (*Stenotaphrum dimidiatum*), Cogon (*Imperata cylindrica*), T-grass (*Paspalum conjugatum*), as well as various legumes such as Alyce clover (*Alysicarpus vaginalis*), *Desmodium ovalifolium, Desmodium triflorum*, Hetero (*Desmodium heterophyllum*) - and sensitive plant (*Mimosa pudica*). Also refer to Stur and Shelton (1991a). Eng (1989) noted that some 60 different native species have been recorded under the plantation canopy, of which more than half have been found to be palatable.

Table 71: *Botanical composition (%) of native pasture before and after three years of grazing by buffalo under coconuts in Sorsogon (Moog and Faylon 1991)*

Species	Botanical composition (%)	
	Before grazing	After grazing
Imperata cylindrica	40	1
Paspalum and *Digitaria* spp.	6	22
Pueraria phaseoloides	33	0
Weeds	21	77

Productivity may vary from low to moderate depending on the relative percentage of productive grass, legume species and weeds, particularly bush weeds. For example, in Western Samoa local pastures dominated by *Mimosa pudica* and Hetero were considered to be particularly productive (Reynolds 1982) whereas in the Solomon Islands there was no significant difference in liveweight gains between improved pastures and naturalized pastures with a high legume content and consisting of *Axonopus compressus, Mimosa pudica, Centrosema pubescens* and *Calopogonium mucunoides* (Watson and Whiteman 1981a).

Exotic species

In situations where the aim is to do more than merely keeping weeds under control so that fallen nuts can be located, various exotic species are available. Each has certain characteristics and may be suited to particular physical, socio-economic and management conditions. Important characteristics include: yield; palatability; nutritive value; ease of propagation; rapidity of establishment; ability to compete with weeds; pest and disease resistance; ability to associate with other pasture species; adaptability to local soil, climatic conditions, management levels and, in terms of the coconut environment, shade tolerance and degree of competitiveness with coconut trees; ability to withstand high grazing pressure and persistence.

A wide range of different grass and legume species have been reported in early investigations carried out in different countries. According to Reynolds (1988) grass species most suited to the reduced light conditions under coconut palms are sod-forming stoloniferous grasses that form short to moderate height swards. They provide moderate carrying capacity, allow fallen nuts to be quickly located, are inexpensive and easy to establish from cuttings, compete well with aggressive weed species, maintain a reasonable balance with companion legumes under grazing, and do not compete excessively with coconut production. Such grasses include Angleton Grass or Alabang X (*D. aristatum*), Batiki (*I. aristatum*), Cori (*B. miliiformis*), Koronivia (*B. humidicola*), Palisade (*B. brizantha*), Signal (*B. decumbens*) and possibly Creeping Guinea (*P. maximum* cv. Embu). Although Para Grass (*B. mutica*), is popular in The Philippines, elsewhere it has been shown to be not very shade-tolerant and requires good management under the high light conditions (light transmission >75%) of old coconut plantations or where trees are widely spaced (9-10m square). Buffalo Couch (*S. secundatum*) and Pemba Grass (*S. dimidiatum*) are well adapted to heavy shade conditions in Vanuatu and Zanzibar, respectively.

In establishing pastures, the degree of shade will determine which of the recommended grass species is most suitable (Tables 62 and 72). Where light transmission is < 30%, dry matter yields of all species are low, so that grazing of existing species (such as *A. compressus*) may be most appropriate. In open plantations (light transmission >75%) the choice of species is wide but *B. brizantha*, *B. decumbens* and *B. humidicola* are particularly recommended. In more shady conditions (light transmission 50-75%) *I. aristatum* and *B. humidicola* should be used, whereas in heavier shade (light transmission 30-50%) *I. aristatum* may be suitable, but species such as *S. dimidiatum* and *S. secundatum* are probably most appropriate. For cut-and-carry *P. maximum* and *P. purpureum* are widely used.

The legumes most suited to coconut plantations include Centro (*C. pubescens*) and Siratro (*M. atropurpureum*, with Puero (*P. phaseoloides*) and sometimes Calopo (*C. mucunoides*) used as pioneers (and as cover crops). However, in some humid tropical environments Siratro is subject to *Rhizoctonia* leaf blight. Legumes that combine particularly well with *B. brizantha* and *B. decumbens* include Hetero (*D. heterophyllum*, *D. triflorum* and *A. vaginalis*. Sensitive plant (*M. pudica*) should be utilized where it is indigenous, but needs to be carefully controlled. In Zanzibar, *T. labialis* was found to combine well with Pemba Grass. Leucaena (*L. leucocephala*), or (on acid soils) Gliricidia (*G. sepium*), can be grown as a double-row hedge (rows 1 m apart) between every two rows of coconuts (see Reynolds, 1988).

Table 72: *Recommended grass species for different light conditions (Reynolds 1988)*

Light transmission (%)			
<30	30-50	50-75	>75
Establishment not generally recommended. Graze existing species.	*I. aristatum* *S. dimidiatum** *S. secundatum* *A. compressus*	*B. brizantha* *B. decumbens* *B. humidicola** *I. aristatum**	*B. brizantha*+ *B. decumbens*+ *B. humidicola*+ *B. milliformis* *P. maximum* *P. maximum* cv.Embu *D. aristatum* *I. aristatum* (*B. mutica*)**

* Especially recommended

** Only suitable in very open plantations with high light transmission.

How has the more recent work changed these recommendations? Recent relevant reports include those of: Benjamin *et al.* (1991), Chen (1991), Cheva-Isarakul (1991), Egara *et al.* (1989), Evans *et al.* (1992), Iniguez and Sanchez (1991), Kaligis and Sumolang (1991), Kaligis *et al.* (1991), Liyanage (1986, 1991), Liyanage *et al.* (1989), Moog and Faylon (1991), Oka Nurjaya *et al.* (1991), Parawan (1991), Parawan and Ovalo (1987), Rajaguru (1991), Sanchez and Ibrahim (1991), Sophanodora

(1989a), Sophanodora and Tudsri (1991), Stur (1991), Stur and Shelton (1991a), Vijchulata (1991), Wong (1989, 1991) and Wong *et al.* (1989).

Wong (1989) confirmed that for low light levels (< 50% sunlight) the shade-tolerant species *P. conjugatum, A. compressus, C. pubescens* and *D. ovalifolium* appear to be best. In moderate shade *P. maximum* and *B. decumbens* are suitable, and under old coconut plantations with high light transmission to these can be added *P. purpureum, S. sphacelata* cv. Kazungula, MARDI digit (*D. setivalva*), *D. pentzii, B. humidicola, Zornia diphylla* and *S. guianensis.*

Table 73: *Summary of adaptation of frequently occurring forages (Stur and Shelton 1991a)*

				Required		Resistance to						
	Tolerance	Forage-yield	Animal product.	Management level	Soil fertility	Grazing	Drought	Soil acidity	Water logging	Weed invasion	Fertilizer response	Potential competition with plantation crops
Naturally occurring												
Axonopus compressus	H	L	M	L	L	H	L	H	M	M	M	L
Paspalum conjugatum	H	L	L	L	L	M	L	H	H	L	L	L
Imperata cylindrica	M	L	L	-	M	L	H	H	H	-	M	M
Mimosa pudica	H	L	M	L	L	H	-	-	-	-	-	L
Desmodium heterophyllum	H	M	H	L	L	H	H	M	H	-	-	L
Cover crops												
Calopogonium mucunoides	M	H	L	M	L	L	L	H	M	L	L	L
Calopogonium caeruleum	H	M	L	L	L	-	-	H	-	L	-	L
Pueraria phaseoloides	M	H	H	H	L	L	M	H	M	L	M	L
Centrosema pubescens	H	M	H	M	L	M	M	H	L	M	M	L
Sown or planted												
Stenotaphrum secundatum	H	M	M	L	M	H	-	-	M	H	-	L
Ischaemum aristatum	H	M	M	M	L	M	-	M	M	H	M	L
Brachiaria decumbens	M	H	H	M	M	M	M	M	M	M	H	M
Brachiaria humidicola	M	H	H	L	M	H	H	M	H	H	H	M
Panicum maximum	M	H	H	H	H	L	M	M	L	L	H	H

L = low, M = moderate, H = high

In Vanuatu, Evans *et al.* (1992) noted that *B. decumbens*, Sabi Grass (*Urochloa mosambicensis) and B. humidicola* perform well under coconuts, provided at least 70% light reaches the ground - an average situation for a good stand of 60-year-old coconuts. 'Given very careful management involving undergrazing these grasses will probably persist down to 50% sunlight conditions. For replanted coconuts and particularly the hybrids at the recommended spacing of 9 m triangular, light conditions will be below 50% from 5-40 years of age. Under such conditions Buffalo Grass (*S. secundatum) (CP 67 and 68)* is the best available option at present, which combines high shade tolerance and a growth habit which is resistant to overgrazing'. Stur and Shelton (1991a) in reviewing available forage resources in plantations in Southeast Asia and the Pacific, summarized their main characteristics including their shade tolerance and potential competition with plantation crops (Table 73). Chen (1991) has summarized the dry matter productivity of the main tropical forage species under different light regimes. These are shown in Table 74; the dramatic decline in dry matter productivity with increased shade (decreased light transmission) is very apparent. At a workshop in Medan, North

Sumatra, Indonesia in September 1990 (Iniguez and Sanchez 1991) a Working Group, after reviewing past evaluations of germplasm, recommended the following as a starting point for future evaluations:

- productivity and compatibility with tree crops;
- ability to perform under defined light regimes;
- persistence under grazing;
- absence of adverse animal effects.

For cut-and-carry systems pure stands of *Pennisetum purpureum*, *Tripsacum laxum* and tree legumes were recommended.

Although there have been a number of studies on the shade tolerance of herbaceous legumes (*e.g.* Eriksen and Whitney 1982; Kaligis and Sumolang 1991; Ng 1991; Rika *et al.* 1991; Stur 1991; Wong 1991; Wong *et al.* 1985b) less information is available on tree legumes. Egara and Jones (1977) showed *Leucaena leucocephala* (CP 69) to have limited shade tolerance. More detailed studies were only undertaken recently under the ACIAR Forage Programme. Benjamin *et al.* (1991) reported on the response of six fodder tree legumes to a range of light intensities ranging from 100 to 20% of incident light inside a greenhouse in Australia. The relative order of shade tolerance was *Gliricidia sepium* > *Calliandra calothyrsus* > *Leucaena leucocephala* > *Sesbania grandiflora* > *Acacia villosa* > *Albizia chinensis*. With the psyllid insect causing serious damage to *Leucaena leucocephala* in Bali, Indonesia psyllid-resistant tree legumes are required. Oka Nurjaya *et al.* (1991) reported on a trial under coconuts (58% light transmission) to identify suitably adapted species. It was concluded that *Calliandra calothyrsus*, *Codariocalyx gyroides*, *Desmodium rensonii* and *Gliricidia sepium* warranted further study as forage species for use in the coconut plantations in Bali. A similar study was carried out in North Sulawesi where *Gliricidia sepium* and *Erythrina sp.* are commonly used as fences and live stakes under coconuts (Kaligis *et al.* 1991). Nine introduced and four local species were evaluated over the period December 1988 to June 1990 during which time 9 harvests were taken. *Calliandra* sp. CPI 108458 produced by far the highest leaf yields and other potentially useful species included *Flemingia macrophylla*, *Calliandra calothyrsus* (local), *Gliricidia sepium* (local), *Desmodium rensonii* and *Codariocalyx gyroides*. In the drier environment of South Sulawesi, Ella *et al.* (1989) had earlier found little difference between the leaf yield of *C. calothyrsus*, *L. leucocephala* and *G. sepium*. This was before the damage caused by the psyllid on Leucaena.

Key problems identified by Stur and Shelton (1991b) include:

- The lack of species available for low light situations (especially <30% light transmission).
- The need for a greater range of grasses and legumes which will persist and contribute to animal production in low management and input situations.

Wong (1991) suggests that species showing high productivity and forage quality in a wide range of light levels (90 to 20%) are desirable, but are presently not available. As part of the ACIAR Forage Programme, a total of 130 grass and legume accessions were screened for shade tolerance at the University of Queensland at five light levels (from 100 to 20%), with the 50% treatment meant to approximate the light in most coconut plantations, and the 20% light treatment similar to that in mature rubber plantations (Stur 1991).

Under 50% light transmission:

Grasses

Many of the commercially used tropical pasture grasses, such as the tall upright *Panicum maximum* cultivars, had the highest dry matter yields. Highest yielding of the lower growing grasses was

Paspalum wettsteinii. Also notable were *Paspalum malacophyllum, Urochloa stolonifera, Digitaria natalensis, Paspalum conjugatum* and *Axonopus compressus*. With the exception of the last two, these species have not been widely used under coconut.

Table 74: *Productivity (DM t per ha per yr) of some tropical forages in pure swards grown under the natural shade of coconuts (after Wong 1991)*

Species	Shade as % of sunlight				Reference
	0-25%	26-50%	51-75%	76-100%	
Coconut					
Brachiaria decumbens	0.7	4.4	9-11	28	Smith and Whiteman (1983a)
Brachiaria humidicola	0.7	4.1	9-12	22	
Brachiaria miliiformis	1.0	3.4	4-7	18	
Stenotaphrum secundatum	1.9	4.9	3-4	6	
Axonopus compressus	1.3	1.9	4-5	5	
Paspalum conjugatum	1.0	2.6	2	8	
Ischaemum aristatum	.03	5.5	7-8	14	
Stylosanthes guianensis	-	-	-	15.2	Steel and Humphreys (1974)
Centrosema pubescens	-	-	-	3.3	
Panicum maximum (tall)	-	-	15.0	-	Reynolds (1978)
Panicum maximum cv. Embu	-	-	10.6	-	
Brachiaria humidicola	-	-	10.5	-	
Brachiaria brizantha	-	-	8.9	-	
Panicum maximum	-	-	8.6	-	
Brachiaria miliiformis	-	-	8.2	-	
Ischaemum aristatum	-	-	8.0	-	
Paspalum conjugatum	-	-	8.5	-	
Brachiaria brizantha	-	-	5-11	-	Manidool (1984)*
Axonopus compressus	-	-	4-7	-	
Paspalum conjugatum	-	-	4-7	-	
Panicum maximum	-	-	1	-	

* Estimated.

Legumes

As with the grasses, the highest yielding tended to be the more upright species such as *Aeschynomene americana, Desmodium intortum* and *Mucuna* sp. Also performing well were *Vigna luteola, Stylosanthes humilis, Rynchosia minima, Centrosema macrocarpum* and *Desmodium heterophyllum*. The latter is already known to be shade tolerant.

Under 20% light transmission:

Grasses

The *Panicum maximum* cultivars remained the highest yielding grasses (although it is known that under low light conditions they are easily grazed out). Other species doing well were *B. humidicola (Koronivia grass)* (CP 70), *D. aristatum, A. compressus, D. pentzii, Acroceras macrum, Paspalum malacophyllum* and *D. milanjiana*.

Table 75: *Potential forage material for integrated tree cropping and small ruminant production systems (after Iniguez and Sanchez 1991)*

	Crop and Age		
	Young rubber/oil palm old coconut	3-6 yr rubber/oil palm young coconut	Mature rubber/ oil palm
Light Trans-mission (%)	100-70	60-30	30-10
	*Brachiaria decumbens** *Brachiaria humidicola* *Brachiaria mutica* *Digitaria setivalva (MARDI digit)* *Pueraria phaseoloides* *Centrosema pubescens* *Stylosanthes guinaensis*	*Arachis sp.* *Desmodium ovalifolium* *Paspalum notatum* *Paspalum wettsteinii* *Axonopus compressus*	*Arachis sp.* *Stenotaphrum secundatum*

* Can cause photosensitization in sheep

Legumes
Promising were *D. intortum, M. atropurpureum, C. mucunoides, Arachis pintoi* cv. Amarillo, *D. heterophyllum, D. gangeticum, C. pascuorum, D. heterocarpum, Neonotonia wightii* cv. Tinaroo and *Teramnus labialis*. While some of the species which performed well under shade were well known, a number of shade-tolerant species were identified which have not previously been used and need to be tested under plantation conditions taking into account particular environmental and socio-economic conditions and likely management levels.

At the same time, a series of trials have been undertaken under coconut in Bali (Rika *et al.* 1991) and North Sulawesi (Kaligis and Sumolang 1991) in Indonesia, and under rubber in Malaysia (Ng 1991) to evaluate the performance of 41 grass and 46 legume species selected for their assumed shade tolerance. Species which showed good growth and persistence (although initially low yielding) over the full trial period of up to eleven harvests in Bali and North Sulawesi included the legumes *Arachis pintoi, A. repens*, other *Arachis* sp., *D. ovalifolium, D. heterophyllum*, and the grasses *P. notatum, P. wettsteinii, A. compressus* and *D. milanjiana.*

In Malaysia species showing good regrowth and persistence under the declining light environment of maturing rubber were the grasses *P. maximum, B. brizantha, B. humidicola, B. dictyoneura* and *P. notatum* and the legumes *Stylosanthes scabra* cv. Seca and *S. guinanensis* CIAT 184. Other promising species included the grasses *S. secundatum* and *P. wettsteinii* and some *Arachis* sp. Further testing of some of the more promising species in grass - legume combinations and on farms is ongoing, with small grazing experiments underway in Indonesia and Malaysia to gather management and production data. Some of the *Arachis* sp. look particularly promising and *Paspalum notatum, Paspalum wettsteinii* and *S. secundatum* are being further evaluated (Shelton, personal communication).

The Effect of Management Levels on Species Selection
A factor which may have an important bearing on the selection of grass and legume species is the level of management and the simplicity/complexity of the proposed system (Table 76). A grass such as *B. miliformis* with Centro, regularly fertilized, may give excellent yields with a well managed dairy herd under good light transmission conditions of about 75-80%. Where beef steers are being grazed on the same area with low inputs and minimal levels of management, the more appropriate species may be *I. aristatum, S. dimidiatum* or secundatum, or possibly *B. humidicola* with *D. heterophyllum, A. vaginalis* or *T. labialis* or *Arachis pintoi* if it continues to show promise under grazing. Lack of persistence of grass species and invasion by unpalatable broadleaf weeds have been reported as key management problems in coconut plantations (Shelton *et al.* 1987a).

Table 76: *Level of management for selected grasses and legumes (Reynolds 1988)*

Level of Management		
Low	Moderate	High
Grasses		
A. compressus	*B. humidicola*	*B. miliiformis*
S. dimidiatum	*B. brizantha*	*B. mutica*
S. secundatum	*B. decumbens*	*P. maximum cv. Embu*
I. aristatum	*I. aristatum*	*P. maximum*
Legumes		
D. heterophyllum	*P. phaseoloides*	*S. guianensis*
A. vaginalis	*C. pubescens*	*M. atropurpureum*
T. labialis	*L. leucocephala*	*C. pubescens*

Unpalatable weed invasion is *the* major problem in overgrazed smallholder pastures, and Shelton (1991b) suggested that future research should aim at providing simple and robust systems suitable for smallholders with only recent experience in pasture and cattle management.

The Effect of Hedgerow Planting Systems on Species Selection
Although for the immediate future, in the large majority of coconut areas planted at traditional spacings, the need for shade-tolerant species remains, many coconut plantations comprise ageing stands of lower-yielding trees which have thinned over the years and which will need replacing (*e.g.* in The Philippines according to Ontolan (1988) it is estimated that 55% of the more than 3 million ha of coconut palms are older than 40 and 27% are beyond 50 years). With generally low productivity and profitability, decisions will have to be made about whether to diversify and intercrop or replant with pure coconut stands. In Vanuatu, according to Evans *et al.* (1992), increasingly under conditions of depressed copra prices, some commercial smallholders are showing interest in planting coconuts at lower densities so that the more productive improved pastures can persist under grazing.

If hedgerow planting systems are more widely adopted in future (CP 65) then with light transmission conditions in the inter-row areas being in excess of 80 or 90% throughout the life of the coconut trees, the need to identify shade-tolerant species will be less of a priority. The species which have already been identified, and are in widespread use in open areas, can be recommended and selected according to the particular environmental and socio-economic conditions and likely management levels.

Characteristics of the Main Pasture Species for Coconut Areas
These have been described in various publications, and were reviewed in detail by Reynolds (1988) and more recently by Reynolds (1994), therefore detailed descriptions are not given here. Reference should also be made to Evans and MacFarlane (1990), Pottier (1983) and Steel *et al.* (1980).

Pasture Establishment and Improvement
When the existing cover beneath coconuts has to be modified or replaced with exotic pasture species, the methods used will depend upon several factors: intensity and scale of development, nature of the existing vegetation, soil type and topography, and characteristics of species to be sown or planted. Where cropping has been practised with young coconuts, little land preparation will be necessary, whereas the tangled mass of undergrowth found in many older coconut areas may require considerable clearing.

Successful pasture establishment is largely a matter of common sense and practical experience. The principles and techniques have been described by a number of authors, and pasture establishment under coconuts was covered in some detail by Reynolds (1988) and recently revised and updated (Reynolds 1994). Reference should also be made to Evans *et al.* (1990, 1992), MacFarlane *et al.* (1991), Steel *et al.* (1980) and Walker (1992). Evans *et al.* (1992) suggest that decisions which have to be made include the following:

- which pasture species (grasses and legumes) should be used in different environments;
- how pastures can best be established;
- how these pastures should be managed;
- whether the economic returns will justify the investment in capital and labour.

Options may include the following:

- improving the productivity of existing shaded grass pastures through strategic weed management programmes and/or by increasing the amount of legume in the pasture;

- establishment or improvement of pastures under coconuts to improve income from an existing land-use system, overcome weed problems and enable a more efficient recovery of coconuts for copra production;
- establishment of improved grass/legume pastures into weed-infested or degraded native pasture under coconuts;
- establish pastures in a relatively recently planted coconut area.

Establishing Legumes in Grass-dominant Pastures

With the major objective of creating conditions that will favour the germination and establishment of legume seedlings, competition from the existing grass for light and moisture must be reduced and good soil/seed (or vegetative) contact ensured. Three steps are involved:

- heavily graze the pasture down to a height of 5-10 cm to reduce the leaf canopy and litter layer;
- establish legumes either by sowing into disced strips or by sod-seeding (zero-tilling) into rows using a disc seeder, preferably a triple disc seeder, and spraying along the rows when planting the seed with a non-residual, non-selective herbicide such as glyphosate;
- graze lightly about 6 weeks after oversowing, and graze fully after about 6 months. Evans *et al.* (1992) mention the case of zero-tilling or disc stripping legumes into 15-year-old signal grass where the legume content was increased from < 5% to at least 30% in one year.

Establishing Improved Grass/Legume Pastures into Weed-infested or Degraded Native Pastures under Coconuts

The major problem faced in most coconut plantations is that of replacing total weed infestations of Pistache (*Cassia tora*), Blue Rat's Tail (*Stachytarpheta urticifolia*), Fern (*Nephrolepis* sp.), Giant Mimosa (*Mimosa invisa*), Honolulu Rose (*Clerodendron fragans*), Lantana (*Lantana camara*), Mintweed (*Hyptis capitata*), Pico (*Solanum torvum*), Wild Tobacco Weed (*Pseudoelephantopus spicatus*) and other bush regrowth like Guava (*Psidium guajava*), with sown pastures. For plantations with access to machinery, Evans *et al.* (1992) indicated that the same principles which apply to rehabilitating open weed-infested pastures apply, except that costs are approximately 10% higher due to reduced efficiency and the need to weed around the coconuts. The most important aspect of sustainable weed control is to ensure that the pasture is competitive! Examples quoted by Evans *et al.* (1992) include slashing, disc harrowing, seeding, using twinning legumes to smother weed regrowth, zero-tilling where appropriate and ensuring that the pastures are not grazed too hard too soon. Where machinery is not available fire may be used or hand slashing followed by the use of twinning legumes such as Siratro, Glycine, Centro, or Puero to smother weed regrowth with grazing deferred long enough to ensure that smothering is successful.

Planting Time in Relation to Age of Coconut Trees and Cattle Damage to Young Trees

If pastures are established too soon under coconuts there may be a need for expensive ring weeding to control aggressive species from competing with the young coconuts and slowing their early growth. If cut-and-carry systems are planned at least initially, a grass such as Napier could be planted in the first year. Alternatives are catch crops such as groundnuts, pineapple, root crops, vegetables etc. or cover crops like *Calopogonium* or *Pueraria* to control weeds, enrich the soil and serve as pioneer legumes into which other legumes and grasses can be planted. Regular early fertilizer applications to coconuts will ensure rapid growth and allow early pasture establishment and grazing. Authorities differ as to how soon pastures under coconuts can be grazed, with some suggesting that cattle should not be introduced until the trees are 5 and even up to 8-years-old because of the likely damage in the form of chewed and damaged fronds and even damaged growing points. Much depends on the early coconut growth rate. Plucknett (1979) has suggested that weaner cattle can be grazed from the 3rd or 4th year onwards. Lane (1981) suggested that in Sri Lanka sheep could be grazed under coconuts after

2 years, with cattle being introduced at about 5 years. In Vanuatu, sheep have been successfully run on Carpet grass, T-grass and Buffalo grass plus *Mimosa pudica* pastures under coconuts at 10 ewes per ha in young 2-year-old coconuts and 6-7 ewes per ha in 15-year old coconuts (Simonnet 1990). No damage to young coconuts was reported, and the sheep controlled the *M. pudica*. Various tree guards, fencing and repellants have been suggested for protecting the young coconut trees, but for individual trees this is likely to be a costly option. If hedge planting systems are adopted it may be more economical to fence off the inter-row grazing areas from the coconut rows!

Types of Planting Material

Five main types of planting material can be used to establish a pasture:

- seeds;
- cuttings (stems and stolons);
- rhizomes;
- divided root stocks or pieces;
- stakes and bare stem seedlings.

Although this subject has been covered in detail elsewhere (Reynolds 1988, 1994) the cost aspect is stressed here. The choice of which type to use will depend on: species being used, resources available, type of establishment being undertaken and the availability of the planting material. Usually where sufficient funds are available, seed is likely to be purchased for both grass and legume establishment. However, as many smallholders are unlikely to have sufficient funds or access to seeds it is likely that vegetative propagation may be the most common method used. Fortunately, many of the shade-tolerant species mentioned earlier, like *B. brizantha, B. decumbens, B. humidicola* (CP 70), *I. aristatum, D. heterophyllum, G. sepium* and *L. leucocephala* can be vegetatively propagated. More labour will be required in the establishment process but this is likely to come from the family. Therefore, cash costs are kept to a minimum.

Pasture Management

Whether or not the pasture consists of native or exotic species, good management is required to obtain maximum production and maintain pasture productive capacity. Both types of pasture respond to good grazing control, but for a farmer to capitalize on his investment in improved pasture it is important that maximum skill is devoted to the management phase. Failure to do so may either result in the establishment of a poor quality improved pasture, or a weed-covered paddock where local species of lower feed-value will regrow. The two distinct periods in which management is particularly important are during establishment (*i.e.* following planting) and after establishment or in established pastures.

Early Grazing

Cattle can be used to assist in pasture establishment, the action of grazing and trampling, causing the grass to spread and to send out new stolons and reducing the competition of the grasses on the legume component. Evans *et al.* (1992) suggest that for fully sown weed-free pastures, the first grazing should occur 10-12 weeks after planting, or when the grasses are about 50-100 cm high in the case of tufted upright growing species (*e.g.* Guinea grass), or at 30-50 cm for stoloniferous species such as signal or Koronivia grass. Tufted grasses should be grazed down to about 30-50 cm and creeping grasses down to a height of 15-20 cm. Pastures should again be grazed 8-10 weeks later. As a general guideline, the stocking rate used in the first 6 to 12 months of pasture establishment should be only about 40% of the optimum rate for a fully established pasture. However, others have suggested that in early grazing, animals should be used at a high stocking rate for a short time to avoid any selective grazing! The purpose of early grazing is to conclude the cycle with a dense, permanent stand of the desired species after a year or so, and over the first three years the objective is to obtain a high legume content because the amount of nitrogen fixed by the legume is directly proportional to the quantity of legume in the pasture. Where there is a severe weed problem, grazing may have to be delayed for 6 months or so to allow climbing legumes to smother the weeds. However, in case there is also the risk of the sown grass being smothered, careful grazing management is required.

Weed Control

Methods have been described by Reynolds (1988, 1994), but the most recent comprehensive manual is the technical bulletin on '*Weed identification and management in Vanuatu pastures*' by Evans *et al.* (1990), which deals with the identification and management of pasture weeds, both in plantations and smallholder systems of production. Chee and Ahmad Faiz (1991) reviewed weed control methods in Malaysian rubber estates and showed that the complementary use of grazing sheep reduced the overall costs of weed control by between 16 and 36%. Sheep graze grass and some palatable broadleafed species selectively, resulting in a desirable purification of the legume *Calopogonium caeruleum* in cover crop mixtures under rubber.

Time Required for Establishment

When a good seed-bed is prepared and seed germinate well and/or cuttings are closely spaced, good establishment under wet season conditions may result in moderate grazing after 3 months, and full grazing after 6 months. However, as a general rule, full grazing is likely to be available only after about 12 months. The sooner full grazing can be undertaken, the quicker pasture establishment-costs begin to be repaid. A long establishment phase means that coconut production may be affected both through nuts lost in the long grass and possibly through reduced production due to competition from the ungrazed pasture.

Management of Established Pastures

The main objective is to strive to:

- obtain maximum herbage yield with the highest possible nutritive value throughout the year at the lowest possible cost, with no reduction in coconut yields;
- keep pastures productive and prevent any overall decline in quality. An optimum stocking rate is one that will maintain pasture stability and botanical composition, and produce consistent levels of animal production (Evans *et al.* 1992);
- maintain a good grass-legume balance with 20-30% legume content being the target;
- convert the feed to saleable products, such as meat and milk.

Management for sustained productivity has to be flexible because of changing circumstances arising from unexpected events such as climatic effects or economic (market) forces (Evans *et al.* 1992).

Stocking Rate

Usually defined as the number of grazing animals per unit (ha) of land at a particular time, although on extensive systems it becomes ha per animal. It is one of the most important variables influencing both the productivity per animal and per hectare. The long-term stocking rate is referred to as the carrying capacity, implying an optimum stocking rate that can be maintained without damage to the pasture. The stocking rate or carrying capacity of pastures under coconut may vary according to differences in:

- planting density, age of palms and therefore degree of shade;
- botanical composition of the pasture;
- climatic factors and time of year (stocking rate may be lower in very dry periods);
- soil fertility and amount of fertilizer used;
- type and age of animals;

- grazing system and pasture condition;
- availability of supplementary feeds;
- management level.

In the Solomon Islands, a relationship between age and density of coconut trees (and light transmission) and cattle-carrying capacity has been demonstrated. Humphreys (1991) suggested that recommended stocking rates in the Solomon Islands for differing levels of light transmission might be: 35% - 0.7 beasts per ha; 45% - 1.0 beast per ha; 50% - 1.3 beasts per ha; 60% - 1.6 beasts per ha; 80% - 2.5 beasts per ha. MacFarlane and Shelton (1986) found that smallholders in Vanuatu do not understand that shading reduces pasture growth and hence the number of animals that can be carried per ha, with the result that overgrazing was common, with stocking rates of 1-1.5 animals per ha under dense coconuts (25% light transmission) when rates of 0.5 per ha or less should be used. An indication of the carrying capacity of natural pastures under coconuts in different parts of the world is given in Table 77. Rates vary from 0.3 - 3.0 animals per ha. A summary of available data on stocking rates on different improved pastures is given in Table 78.

Liyanage (1983) estimated that under old coconuts in Sri Lanka, 4 milk cows could be maintained on 1 ha. where the coconuts were intercropped with improved pastures such as *B. miliiformis, B. brizantha*, and centro and fodder crops such as *P. maximum* and hybrid Napier NB 21.

Sheep have been run on native pastures under coconuts in Vanuatu at 10 ewes per ha under young 2-year-old trees and 6-7 ewes per ha under 15-year-old trees. In Sri Lanka, in the coconut triangle, sheep can be stocked at 5-15 ewes per ha per yr (Jayawardana 1988). Nugari (1984) reported 12 head of goats per ha being grazed on improved pastures under coconuts for 2 years in Indonesia. In The Philippines, the average carrying capacity per ha of good native pasture under coconuts is 7-15 head of goats or sheep under grazing and cut-and-carry systems. Improved pastures under coconuts, consisting of mixed swards of *Setaria splendida*-centro and *B. decumbens*-centro, had maximum carrying capacities ranging from 20-28 head per ha of goats and sheep (grazing). For *B. mutica*-Siratro pasture under coconuts, an optimum stocking rate of 20 goats per ha was noted for grazing and cut-and-carry systems (Parawan 1991; Parawan and Ovalo 1987). Buffalo (Carabao) have been grazed on native pastures under coconuts at 1-2 a.u. per ha.

Grazing System

There has been considerable debate over the years as to which system of grazing management - continuous or rotational - is the better system. While each has advantages and disadvantages, Reynolds (1988) and more recently Liyanage (1990), have suggested that under coconuts there are a number of reasons why rotational grazing systems may be more appropriate and, according to Payne (1985), why they should be preferred:

- Various practical tasks such as coconut collection, weed control and fertilizer application are more easily carried out with rotational grazing. Rotating the cattle in front of the nut pick-up labour, so that the forage crop is grazed as briefly as possible immediately before nut collection, obviously assists in ensuring a high pick-up percentage;
- Root weight data and observations on the rate of recovery of grasses after clipping indicated that shaded pastures require careful management to avoid excessive depletion of root reserves, either by lenient grazing (to maintain high levels of leaf area) or by allowing an extended recovery period in a rotational grazing system (Eriksen and Whitney 1981);
- Under the low light conditions of many coconut plantations, grasses lose their competitive growth advantage over legumes. Under continuous grazing, grasses may disappear and swards become legume dominant. In various plantations in Western Samoa, *M. pudica*-dominant pastures have resulted from continuous grazing, whereas good liveweight gains and grass persistence were noted under a 28-day rotational grazing regime;

- Some species, such as *L. leucocephala*, are more productive under rotational grazing than under continuous grazing, and others, like *P. maximum* are best rotationally grazed so that overgrazing does not quickly destroy this erect bunch grass (Payne 1985). Evans *et al.* (1992) suggest that similar management may be required for other shrub legumes presently being evaluated, such as *Gliricidia, Desmanthus, Acacia,* and *Aeschynomene,* or legumes with a similar growth habit.

Table 77: *Carrying capacity of natural pastures under coconuts (modified from Plucknett 1979)*

Country	Carrying capacity head (or a.u.) per ha	No. of palms per ha
Fiji	1.0	-
Indonesia	0.5 - 1.0	-
Jamaica	0.75	75 - 100
New Hebrides (Vanuatu)	1.5 - 3.0	(wide spacing, old palms)
Papua New Guinea	2.5	-
Philippines	0.5 - 2.3	-
Solomon Islands	1.0	260 (4-11 years old)
	1.25	260 (>11 years old)
	1.5 - 2.0 a.u.	175
Sri Lanka	0.3 - 1.25	-
Thailand	0.25 - 0.50	-
Trinidad & Tobago	0.75 - 2.5	125
Western Samoa	0.25 - 1.5	125 (20 years old)

The construction of fences in cattle/coconut systems should also be less costly than it is in many other systems, as growing (coconut) trees can be used as fence posts, or the erection of cheap live fencing is possible using species such as *Gliricidia* or *Erythrina*.

Table 78: *Stocking rates on improved pastures*

Country	Pasture	Stocking rates heads per ha	Notes
Fiji		up to 2.0	Pittaway (1990)
Ivory Coast	*C. pubescens*	0.75	Sandy soil, Ferdinandez (1968)
Indonesia	*B. decumbens* *C. pubescens*	2.7 - 6.3	Small Bali cattle (*Bos banteng* yearlings) Rika *et al.* 1981
Jamaica	*P. maximum* *D. decumbens*	1.75 0.75 - 1.0	Anon. (1971)
(New Hebrides (Solomon Islands (Western Samoa	Stoloniferous sp. *P. maximum*	up to 3.0) up to 3.5) or 4.0)	mainly Hereford steers and Brahman cross weaner steers, Frémond (1966) Reynolds (1980); Smith & Whiteman (1983b); Weightman (1977)
Vanuatu	*B. decumbens* *B. humidicola* *N. wightii*	3.0	Evans *et al.* (1992)
Sri Lanka	*B. brizantha*) *B. miliiformis*) *P. maximum*)	1.5 - 2.0) 1.5 - 6.0)	Small Sinhala cattle Ellewela (1956, 1957); Jayawardana, 1985
Philippines	improved grass-legume mixtures *B. mutica, P. maximum* *B. mutica C. pubescens* *B. mutica C. pubescens* (unfertilized)	>1.0 a.u. 3 a.u. 3.0 2.5 1.0 a.u.	Guzman and Allo (1975) Faylon (1982) Moog and Faylon (1991) Anon. (1978a) Sabutan *et al.* (1986)
Thailand	*B. decumbens C. pubescens*	1.5	Boonklinkajorn *et al.* (1982)

Health and Diseases of Grasses and Livestock

A subject which has so far not been investigated in any detail is that of the susceptibility of pasture species to various fungal diseases under the shady, humid conditions experienced under plantation crops such as coconut. Siratro (*M. atropurpureum*) is particularly susceptible to *Rhizoctonia* leaf rot, and in the Solomon Islands under low light conditions (light transmission of 31%) both *P. maximum* cv Embu and *B. miliiformis* suffered considerable insect damage and failed to recover (Steel and Whiteman 1980). With grazing livestock the major problem appears to be internal parasites. According to Parawan (1991) this problem is aggravated by the shading effect, which favours parasite egg survival and persistence. Respiratory diseases tend to become a problem when stocking rates are increased (especially in goats). In Malaysia, high mortality rates of up to 32% for sheep (Mohd. Nawi and Ahmad 1988) and the identification of high incidence (35-42%) of pneumonic pasteurella in the mortality list (Wan Mohamed *et al.* 1988) sustained by animals in the plantations, may relate closely to the very humid and hot environmental conditions under the tree canopies (Chen 1989). Sani and Rajamanicham (1991) noted that helminth infection in small ruminants is influenced by the micro-environment, being lower in open pastures and with high numbers of worms in sheep reared in oil palm estates.

Cut-and-Carry Systems

Cut-and-carry systems are widely utilized among smallholders because of:

- small size of holding with limited grazing area;
- fragmentation of land holdings;
- lack of fencing in mainly cropping areas (although tethering is also practised);
- low cost of labour.

Due to the limited land available in relation to the number of animals kept, a large proportion of the feed may be brought from outside the holding. The advantages and disadvantages of cut-and-carry (CP 71) or zero grazing or stall feeding systems are as follows:

Advantages

- efficient use of forage
- less wastage from trampling
- saving of grazing energy
- less soil damage
- less labour required to herd stock
- water reticulation not required
- less capital needed for fencing

Disadvantages

- higher labour input needed to cut-and carry fodder
- greater labour resources needed to dispose of excreta
- more capital required in structures, equipment and possibly fuel costs
- less opportunity for animals to select forage
- cutters may select low quality feed
- too little feed may be given to animals
- urine may be lost and dung may be returned to areas other than forage producing areas, resulting in a soil fertility decline
- animals may need supplementation with coconut cake, rice bran, etc.

Where land areas are limited, the smallholder has no alternative to the cut-and-carry system; where cut-and-carry and grazing systems have been compared, results have been mixed, with Devendra (1989) noting that a comparative study of both systems to examine potential milk production in Sahiwal x Friesian cows on a mixed pasture of *L. leucocephala* and *B. decumbens* indicated that rotational grazing was better than stall feeding (9180 v 8577 kg per ha for each lactation) (Table 79). Supplementation of rotationally grazed cows with concentrates at 4 and 6 per kg cow per day further increased milk production to 13,323 and 17,070 kg per ha for each lactation, respectively. The net profit per cow, with or without supplementation, was higher for rotational grazing due to the higher labour cost for the stall feeding system. In The Philippines, Posas (1981) noted that goats grazed under coconut compared to cut-and-carry fed goats have a much better liveweight gain per ha per year indicating that, at least for goats, grazing is a more efficient method of pasture utilization than cut-and-carry. However, a comparison of a cut-and-carry feedlot system, a semi-feedlot system and free grazing of beef cattle in Malaysia revealed higher daily liveweight gains for stall-fed animals (Sukri and Dahlan 1986). Trials were carried out with smallholders in West Jahore using native pastures and a feed ration consisting of coffee by-products (30%), copra cake (30%), palm kernel cake (37%), urea (2%) and mineral-vitamin premix (1%) with average daily gains per animal for the three systems of 0.48, 0.37 and 0.15 kg, respectively. An economic evaluation demonstrated that gross profit was highest for the feedlot animals and it was concluded that feedlot and semi-feedlot systems have great potential for increasing beef production among smallholder farmers and should overcome the major problem of low feed availability (and quality) in dry spells.

Table 79: *Milk yield in Sahiwal x Friesian cows and cost of milk production in two systems of production in Malaysia (Devendra 1989)*

Conc. supplementation (kg per day)	Rotational grazing*			Cut-and-Carry*
	0	4	6	0
Total milk yield (kg per cow per lactation)	1712	2733	3269	1418
Cost per kg of milk**	0.13	0.14	0.15	0.19
Gross income at 0.24 cts per kg milk**	412.14	657.96	787.00	341.29
Production cost per animal**	195.59	394.77	484.29	273.04
Net profit per cow**	216.55	263.18	302.70	68.26

* *L. leucocephala* and *B. decumbens* forages.
** US $.

Table 80: *A summary of cattle liveweight gain data from grazing experiments under coconuts (after Shelton 1991a)*

Country	Pasture	Light Transmission (%)	Liveweight gain (kg per ha per yr)	Stocking rate (b per ha)	Reference
Solomon Islands	natural	60	235-345	1.5-3.5	Watson and Whiteman 1981b
(2900 mm per year)	improved	60	227-348	1.5-3.5	
	natural	62	219-332	1.5-3.5	Smith and Whiteman 1985
	improved	62	206-309	1.5-3.5	
Western Samoa (2929 mm per year)	natural	50	148	1.8	Reynolds 1981
	improved	50	225-306	1.8-2.2	
	natural	70-84	127	2.5	Reynolds 1981
	improved	70-84	273-396	2.5	
	natural	70-84	401-466	4.0	Robinson 1981
	improved	70-84	421-744	4.0	
Indonesia (1709 mm per year)	improved	79	288-505	2.7-6.3	Rika *et al.* 1981
Thailand (1600 mm per year)	natural	n.a.	44	1.0	Manidool 1983
	improved	n.a.	94-142	1.0-2.5	
Vanuatu (1500 mm per year)	improved	n.a.	175	1.5	Macfarlane and Shelton 1986
	natural	n.a.	250-285	2.6-3.0	Evans *et al.* 1992
	improved	n.a.	550	3.0	

Animal Production from Pastures under Coconut

Reynolds (1988) presented details of liveweight gains for cattle grazing pastures under coconuts. These have recently been summarized by Shelton (1991a) and are presented in Table 80. Animal production ranged from a low of 44 kg per ha in Thailand to a high of 505 kg per ha in Indonesia, and a very high but probably unsustainable figure of 744 kg per ha for Western Samoa. Shelton (1991a) notes that the variation in gains was associated with a number of management and environmental differences such as light transmission, pasture species, soil type, fertilizer strategy, animal size and stocking rate whose relative influence is difficult to assess. However, it is clear that a number of factors can be identified:

- o plantation palm density and age: (and therefore light transmission) is an important factor as liveweight gains were highest in the more open (and usually older) plantations where forages received the highest amount of light. As suggested by Shelton *et al.* (1987a), liveweight gains under coconut plantations are generally considerably lower than those measured on unshaded pastures in comparable environments;
- o species: both Reynolds (1981) and Manidool (1983) demonstrated substantial improvement in liveweight gains from improved over native forages, indicating the desirability of replacing natural with improved species for maximum animal production. However, sown pastures have been shown not to persist in a number of long-term studies (Smith and Whiteman 1983b), especially where light transmission is less than 50%, and several studies have noted that differences are less obvious where there is a high proportion of naturalized legume in natural pastures, which clearly improves its quality for grazing animals;
- o legumes: the importance of legumes to pasture quality under coconuts was demonstrated in Vanuatu, where low liveweight gains in smallholder animals grazing pure *S. secundatum* pastures were replaced by gains of some 0.7 kg per head per day on buffalo grass containing the naturalized legumes *Desmodium canum* and *Vigna hosei* (Shelton 1991a). Moog and Faylon (1991) mention a study carried out in Sorsogon (Philippines) where buffalo grazing mature native pastures at 1 head per ha gained only 53 kg per ha per year, compared to 93 kg per ha per yr where the native pastures contained Centro and 151 kg per ha per yr where the stocking rate was raised to 2 head per ha;
- o stocking rate: a typical relationship between stocking rate and liveweight gain is shown in Figure 25 where Shelton (1991a) has analysed the data using the stocking rate model of Jones and Sandland (1974);
- o animal size: Zoby and Holmes (1983) indicated that small and medium sized animals achieved significantly larger daily gains than larger animals. Thus, Rika *et al.* (1981) used small Bali steers with initial liveweight of 97 kg, whereas Reynolds (1981) used steers of 250 kg liveweight.

Other Reported Liveweight Gain Data under Coconuts

In Vanuatu (Evans *et al.* 1992): Signal/Koronivia/Glycine pastures stocked at 3 an. per ha - 550 kg per ha per yr; Carpet/Mimosa/weed pastures stocked at 3 an. per ha - 250 kg per ha per yr; buf-Falo/Carpet/Mmimosa pastures stocked at 2.6 an. per ha - 285 kg per ha per yr. On Carpet grass, T-grass and Buffalo grass plus *Mimosa pudica* pastures under 2 and 15-year-old coconuts, sheep controlled the *M. pudica*, and with regular shearing (every 8 months) and avoidance of the less well drained soils so that foot infections were reduced, and regular drenching to reduce the worm burden, mortality was reduced to only 6% in 1985 and 1986. Prolificacy was 160%, fertility 90% and productivity 154%, with mean daily weight gains between 0 and 6 months of 178g per day. The sheep flock provided good weed control without damage to the palms and first results are described as encouraging in technical and economic terms. Longer-term studies are continuing to assess the effect of sheep on coconut production;

In Malaysia (Abdullah Sani Ramli and Basery 1982): yearling bulls grazed on native pastures by 12 coconut smallholders over a 368-day period made average daily liveweight gains of 0.39 kg. Yusof and Rejab (1988) noted that male Hereford x Kedan-Kelantan cattle on free grazing under coconuts made average daily weight gains of 242.3 g compared with gains made by animals receiving single supplements of fish meal (281.5 g), palm kernel cake (367.9 g) and copra cake (372.3 g). The performance of Malin sheep under coconuts over a 5-year period was assessed by Salleh *et al.* (1989), and H.K. Wong *et al.* (1988) reported high levels of copper in the liver of sheep grazing pastures where a high percentage of the fodder consisted of *Asystasia intrusa,* which contains high copper levels. Also it was noted that excessive uptake of *Mikania cordata* by sheep may be detrimental, and cause abortion in pregnant animals (Chen *et al.* 1991). C.C. Wong *et al.* (1988) noted that goats grazing on improved grasses under 18-year-old coconuts (light transmission 33%) gained on average 46.1 g per head per day (average initial body weight 10 kg; average final body weight 22.1 kg); however, the grasses failed to persist and weed percentage increased;

In Sri Lanka (Liyanage *et al.* 1989): liveweight gains of 0.31 kg per head per day were reported over a 1-year period on the model integrated unit stocked at 4 (Jersey x local crossbred) heifers per ha, with average heifer liveweight increasing from an initial weight of 70 kg at 6 months of age to 200 kg when grazed on *B. miliiformis, P. phaseoloides* and loppings of *Gliricidia* and *Leucaena* (but no concentrates except for some additional urea-treated rice straw in dry periods). In an earlier feeding trial *Gliricidia* leaves were fed with *B. miliiformis* at a ratio of 50:50 to Jersey x local heifers. Daily liveweight gains of 0.70 kg per head were produced during the wet season (Liyanage and Wijeratne 1987). In the coconut triangle Jayawardana (1988) reported Bannur and Red Madras sheep stocked at 5-15 ewes per ha per yr and carcass weights of 13.5 kg obtained at 12 months;

In The Philippines (Parawan 1991): it was noted that average daily gains for sheep and goats on native pastures (at stocking rates of 7-15 head per ha) were from 33-54 g per head. On improved pastures crossbred goats showed average daily gains of 61 g per head while native sheep produced gains of 44-56 g per head (both at stocking rates ranging from 20 to 28 head per ha). Moog and Faylon (1991) noted that in Albay cattle graze on fertilized T-grass (*Paspalum conjugatum*) for fattening and finishing. Grazing usually covers 60-80 days and cattle gain 0.7-1.0 kg per head per day at a stocking rate of 1 beast per ha. Although trial results demonstrate that good liveweight gains can be achieved from running cattle under coconuts, the major problem identified is the question of sustainability which appears to be related to the degree of persistence of species and control of unpalatable weed infestations (Shelton 1991b).

Data on the performance of dairy cows on pastures under coconuts are more limited. Details have been provided by Reynolds (1988) for Sri Lanka, Tanzania, the Caribbean and Western Samoa. Milk yields for Friesian crosses of 1000-1900 kg per lactation compared with milk yields of <500 kg for the small local Sinhala cattle on coconut lands in Sri Lanka. More recently, Jayawardana (1988) indicated that the indigenous x Jersey x Sindi cross cow that is recommended for the coconut triangle is expected to produce 4-5 l of milk per day at peak lactation on well-managed improved pastures with supplements. In the Tanga region of Tanzania a doubling of milk yields from about 340 l per lactation for local cows was reported when bushes and undergrowth were cleared from beneath the coconuts and better quality grasses replaced the unpalatable weeds. Other studies indicated that the average lactation milk yield ranged from about 1000 l for *Bos indicus* animals to 1900 l for half-bred *Bos taurus*. Mean lactation yields for Tanga East African Zebus and Boran/Jiddu cows in the same area were 1264 and 1050 kg milk respectively. In Zanzibar, Tanzania Friesians and Jerseys grazing under old coconuts spaced at 8 m^2 produced on average 4-5 l per cow per day and approximately 1200 l per lactation. In Western Samoa, Friesians grazing under coconuts (CP 72) averaged 3.36 kg milk per cow per day, while in Tonga, Friesian-Holsteins grazed under coconuts on *B. decumbens, B. mutica* and local species produce on average around 4 l per cow per day with a range from 2.6 l on the poorest farm to 8.3 l per cow on the best farm, where concentrates were also fed (Uotila 1992).

In The Philippines, Moog (1991) noted that smallholder dairy farmers raising Sahiwal-Holstein Friesians feed their animals with 5-19 kg of fresh Ipil-ipil (*Leucaena*) leaves in combination with fresh grass fodder and obtain 4-7 kg milk per cow per day. The *Leucaena* is planted in hedges around the home-lots and farm-lots, in evenly spaced rows (1-2m apart) under the coconuts. In Indonesia, Perdok *et al.* (1983) used urea-treated rice straw as feed for lactating Surti buffaloes. When a supplement of *Leucaena* leaves was fed, milk yield increased slightly from 2.41 to 2.73 kg per day, while when *Leucaena* plus coconut cake was fed the milk yield increased to 3.36 kg per day. Walker (1992) describes a farm in the Seychelles where under coconuts 'there was a good pasture of

Stenotaphrum secundatum and very good naturalized legumes *Desmodium canum, D. adscendens, D. auriculatum* and *D. triflorum.* The cattle were tethered and grazed most of the day and received no additional grasses. Milkers received about 2 kg per day of concentrates. Daily production was 25 litres from 3 cows.'

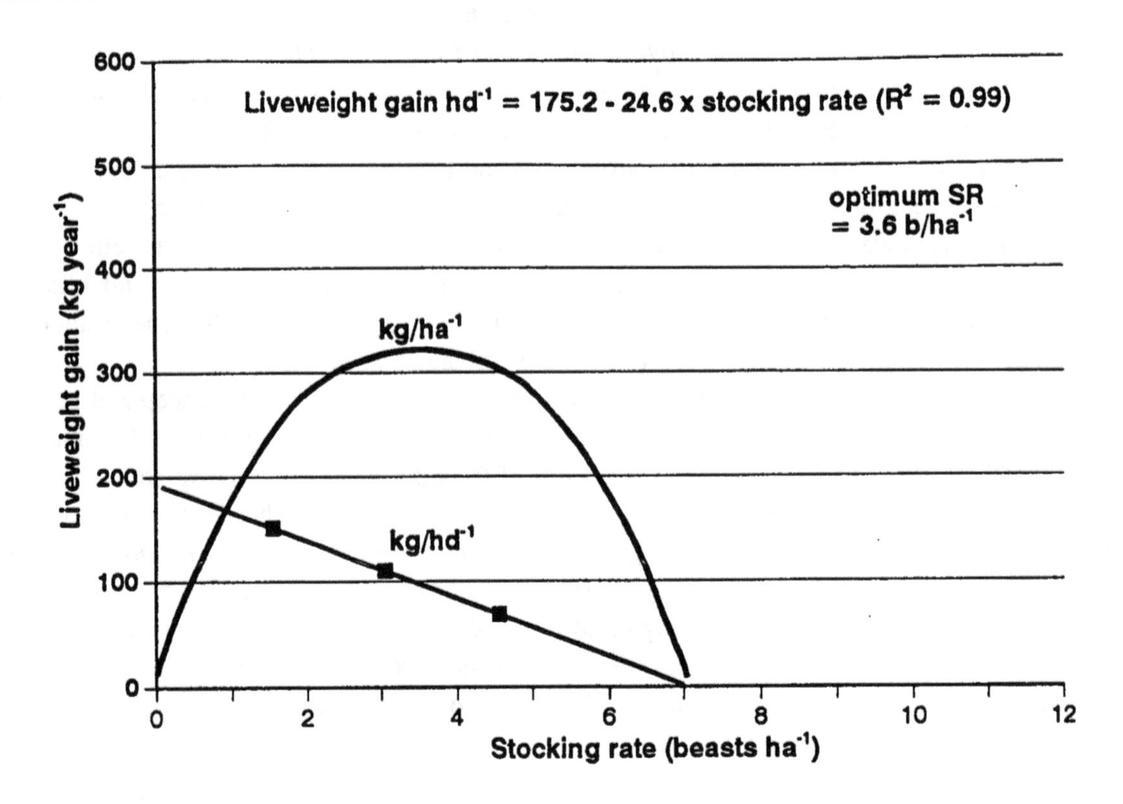

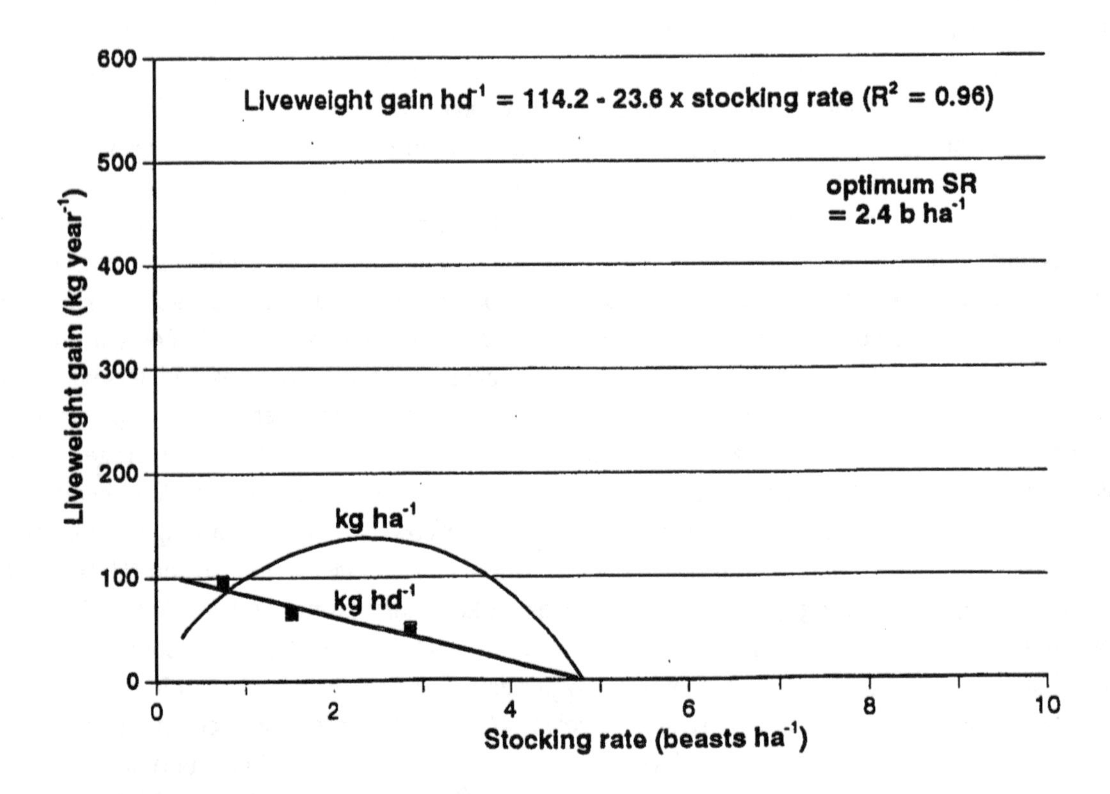

Figure 25: *Liveweight gain and optimum stocking rate under coconuts in Solomon Islands and Thailand (after Shelton 1991a)*

The need to provide shade for dairy cattle in the tropics has been stressed by many workers (*e.g.* Davison *et al,* 1988; Sorensen, 1989) and was briefly reviewed by Reynolds (1994) along with reference to the whole subject of milk production in the tropics. There have been several reports which suggest that the maximum level of milk production that can be obtained from unsupplemented tropical pasture is about 8-9 kg per cow per day (Stobbs and Thompson 1975; Stobbs 1978). Experiments to develop smallholder dairy units in Malaysia tended to support this (Wan Hassan *et al.* 1989), however, Archibald (1985) while agreeing that the 8-9 kg per cow per day applies to Jerseys, notes that the reviewers accepted that 12-14 kg milk per cow per day could be obtained from larger breeds such as Holstein-Friesians grazing tropical pastures as the sole diet. Stobbs and Thompson (1975) in fact recognized as more realistic an average annual yield per dairy cow of 2000 kg from improved tropical pastures!

Forage Production Seasonality and Supplementary Feed Sources

A major problem for livestock in many parts of the tropics is the irregular seasonal rainfall distribution and its effect on pasture production. In some areas there is a very pronounced dry season; in others there may be two dry periods separated by the 'long' and 'short' rains. Faylon (1982) mentioned the problem in his description of coconut-based livestock production systems in The Philippines: 'The situation becomes critical especially during summer months that do not favour the growth of grasses and legumes. During this dry, lean period, the ration should be supplemented with concentrates and crop residues to avoid costly losses in animal weight.' Arganosa (1991) has also referred to the problem, and Nagpala and Moog (1979) indicated that livestock raised under coconuts utilize forage from legume trees, leaves from Madre de Cacao (*Gliricidia sepium*), Acacia pods (*Samanea saman*), stalk of coconut fronds and banana leaves as feed. Also, *Leucaena* leaves and twigs are chopped and fed with rice bran water. In southern Luzon, dairy farmers usually tether their animals in the coconut plantation during the day and, before or after tethering, by-products (such as maize stover, pineapple leaves, rice straw or banana peelings) or a concentrate made from copra meal, salt and rice bran are provided (Arganosa *et al.* 1988). Sabutan *et al.* (1986) experimented with 98% molasses plus 2% urea for animals grazing under coconuts. In Vietnam, rice straw is commonly saved for use as a supplementary feed.

Table 81: *Liveweight and liveweight gain data for heifers grazing natural herbage under coconuts* and those fed rice straw and rice straw plus supplements (after Pathirana and Mangalika 1992)*

Parameter	G	GS	GSS[?]
LW (kg per head) Initial	62.4	63.0	61.9
Final	65.5	78.2	100.5
LW gain kg per head per yr	3.1[a]	15.2[b]	38.6[c]
kg per ha per yr	18.6[a]	91.2[b]	231.6[c]
g per head per day	9.0[a]	42.0[b]	106.0[c]

* 30 year old coconut plantation with palms spaced at 8.4 x 8.4 m (137 palms per ha)

G = heifers grazed (at 6 per ha) continuously on natural herbage

GS = as per G + unprocessed, unsupplemented rice straw ad. lib.

GSS = as per GS plus supplements of urea, molasses, rice polishings and a vitamin mineral mixture

abc. Values with each row bearing different letters are significantly different (P = 0.01) Annual rainfall was 1250-1750 mm.

In Sri Lanka, the University of Ruhuna examined the effect of feeding heifers grazing natural herbage under coconuts with rice straw and rice straw plus supplements (Pathirana and Mangalika 1992). Results after one year are shown in Table 81. Straw intake increased over time. Mean annual straw dry matter intake was 574 g per head per day on GS treatment compared to 1171 g per head per day on GSS treatment, with annual mean intakes of urea, molasses, rice polishings and mineral mixture at 29, 146, 125 and 15 g per head per day respectively. Supplementation not only increased the intake of straw but decreased the grazing pressure and resulted in heifers reaching a breedable stage much earlier.

Reynolds (1988) reviewed the various methods and feed sources for overcoming the problem of seasonal variations in forage production. Recent African experience with the utilization of agricultural by-products as livestock feed was reported in the proceedings of two workshops (ARNAB 1987; Said and Dzowela 1989). Sansoucy *et al.* (1988) provided up-to-date information on the use of sugarcane as feed, Speedy and Sansoucy (1991) reviewed the subject of feeding dairy cows in the tropics, and Speedy and Pugliese (1993) reported on the use of legume trees and other fodder trees as protein sources for livestock. One of the biggest problems is that while agricultural by-products and residues may be plentiful they may be very localized and may be found away from the centres of animal production. Also, they may be very unbalanced in composition (Chen 1989). Smallholders in particular may use them in complex combinations, as they use whatever resources available at the different times of year.

Effects of Pasture on Coconut Yields

The likely effects of any intercrops on coconut yields has been of considerable concern to plantation owners over the years and various studies were conducted. Reynolds (1988) has reviewed these in some detail. The various factors in the pasture-cattle-coconut system which influence coconut yields include: pasture species, fertilizer, moisture, grazing, grazing system, stocking rate, nut collection system and height of grass, legume introduction, weed control or up-keep methods and cultivation.

Much of the earlier work carried out in Sri Lanka demonstrated the effect of species on coconut yields; higher forage yields of *P. maximum* were achieved at the expense of coconuts, and *B. miliiformis* was shown to be less aggressive than *B. brizantha*. However, Humphreys (1991) suggested that the reduction in coconut yields by some grass species is not simply related to grass yields, as shown in the work of Waidyanatha *et al.* (1984) where higher-yielding grasses gave less reduction in the girth of rubber than lower-yielding grasses. In a recent study in Sri Lanka, Liyanage *et al.* (1989) demonstrated over a three-year period that integration of *B. miliiformis, P. phaseoloides, L. leucocephala* and *G. sepium* and grazing heifers with coconuts had no ill effects on nut and copra yields of palms in a 45-year-old plantation where palms were spaced at 8.4 x 8.4 m (137 palms per ha). Jayasundara and Marasinghe (1989) reporting on the same trial noted that the nut and copra yield of the integrated system was 11% higher than on the (coconut) monoculture system. Recently in mid-1992, on an ongoing University of Ruhuna research project at Tennahena estate near Hakmana in southern Sri Lanka, coconut yields on grazed paddocks (50 nuts per palm per yr) were reported to be much higher than those on ungrazed paddocks (32 nuts per palm per yr) (Dr.K.Pathirana, personal communication).

In The Philippines, Parawan and Ovalo (1987) noted that annual copra production was 2.10 t per ha where the coconuts were intercropped with native forage species, but increased to 2.36 t per ha under a *Setaria sphacelata*-centro cover, and decreased to 1.56 t per ha under *Brachiaria decumbens*-centro. Data reported by Moog and Faylon (1991) confirmed that the growing of forages and grazing livestock under coconuts may have a significant effect on nut yield (Table 82).

In India, Sahasranaman *et al.* (1983) reported a 28% increase in coconut yield with mixed farming, largely due to manuring and the better nutrient status of the soil with mixed farming practices.

In Malaysia, C.C. Wong *et al.* (1988) used Kambing-Katjang goats at 6 per ha to graze improved tropical grasses under mature coconuts on Bris soil. The yield of coconuts was increased by grazing (but it was not clear if this was due to the fertilizer applied to the pasture, reduced weed competition or the return of organic manures to the soil).

Table 82: *Nut yield in grazed and ungrazed pastures under coconut (Moog and Faylon 1991)*

Treatment	Nut Yield per Tree per Year
Ungrazed	10-30
Grazed + unimproved pasture	30-50
Grazed + improved pasture	80-100

In Papua New Guinea in West New Britain estates, copra yield started to rise in the late 1970s when there was a change of management technique and cattle were introduced (Table 83). It is clear that the intergrazing of cattle had a very beneficial effect (Ovasuru 1988).

When improved pastures are first established in existing coconut stands there is likely to be a slight initial depression in coconut yields due to soil/root disturbance and the nutrient demands of the sown pasture. However, as stressed by Rajaguru (1991), provided moisture and nutrient levels are adequate and if regular fertilization of both coconut and pasture is maintained, the pasture will have no adverse effects on the coconut. Long-term trials at the CRI in Sri Lanka have shown that there is usually an improvement in nut yields where coconuts are intercropped with well-managed pastures (L.V.K. Liyanage 1987a).

The Economics of Livestock under Coconuts

Adoption of any system of agricultural production will ultimately be determined by underlying social and economic factors. The incentive for intercropping is essentially economic, since this system not only provides higher gross returns per hectare but also plays an important role as an insurance against total crop loss. In recent years the marked fluctuation of copra prices, both monthly and yearly, has been a strong incentive for farmers to find a reliable secondary source of income. It is now generally accepted that for economic reasons coconut lands should be intercropped, although heavy establishment costs and long payback periods may favour short-duration crops rather than perennials (Ranatunga *et al.* 1988). In fact in Malaysia, Mahendranathan and Nor (1980) stressed that intercropping is compulsory under coconut replanting schemes, because coconut as a sole crop, even with an optimum planting density, provides very low incomes when compared to oil palm, rubber or padi. Also, a proposed coconut replanting project for Tonga (Anonymous 1987c) stressed that it would be integrated with a concurrent programme of intercropping. Where there is a choice of intercrop, it is undoubtedly true that in many situations intercrops of cocoa, coffee, banana, etc. and multi-storied cropping systems will produce much higher returns than grazing cattle. In other situations extensive grazing of cattle or smallholder cattle-raising systems, utilizing crop by-products and local feeds under coconuts, may be most appropriate. Sometimes complex crop-livestock mixtures will evolve. Thus in Indonesia, plantations which used to produce only coconuts now produce coconuts, vanilla and forage (from the *Erythrina, Gliricidia* and *Leucaena*, used to support the vanilla) for livestock (Nitis *et al.* 1991).

Table 83: *Copra production (t per ha) in West New Britain estates following introduction of cattle (after Ovasuru 1988)*

Year	Garua Estate	Numondo Estate	$\bar{x}$
78/79	0.84	0.81	0.83
79/80	1.06	0.82	0.94
80/81	1.30	0.96	1.13
81/82	1.55	1.18	1.37
82/83	1.68	1.39	1.54
83/84	1.57	1.47	1.52
84/85	1.71	1.54	1.63
85/86	1.73	1.56	1.65
86/87	1.81	1.51	1.66
87/88	1.60	1.45	1.53
$\bar{x}$	1.49	1.27	1.38

Table 84:

Total Area:	1,000 ha	
Coconut Area:	500 ha	
Pasture Area	500 ha (Under Coconuts)	
Livestock Nos:	180 Cows and supporting animals, based on 1 animal ha^{-1}.	
Production:		
Copra	600 kg ha^{-1}, i.e. 300 t $year^{-1}$	
Grading	80% I, 20% II	
Cattle	Calving 55%	
	Steers finished at 4 years old, 220 kg (Dressed)	
	cull cows sold at 400 kg Liveweight	
Machinery:		
2 Tractors (and Trailers)		
1 Utility Vehicle		
Labour:		
Copra	18 Staff (Cutters, Dryuing, etc.)	
Cattle	4 Staff	
General	1 Foreman	
	1 Clerk	
	Workshop/Driver (2)	
Financial Model		
Income		(Fijian dollars) $F
Copra*		
500 ha @ 500 kg ha^{-1} = 300 t		
240 t @ $ 300		72,000
60 t @ $ 270		16,200
Cattle		
Cows and Heifers 33 @ $280		9,240
Steers 40 @ $330		13,200
	Gross Income	$110,640
Expenditure		
Farm Working Expenses		
Labour		
Copra	4,500 man-day @ $6 $man\text{-}day^{-1}$	27,000
Livestock	4 Permanent @ $2,000 yr^{-1}	8,000
General	Foreman, Driver, etc.	9,000
Casual	Village Workers	3,000
Stock Purchases 2 bulls @ $400		800
Animal Health $10 Cow^{-1}		1,800
Freight		
Copra $40 t^{-1}		12,000
General		2,000
Weed Control (labour)		3,000
Pasture Maintenance (replanting		1,000
etc.)		10,000
Repairs and Maintenance		8,000
Vehicle Expenses (Fuel, R&M)		6,500
Administration		
	Total Farm Expenses	$92,100
	Farm Surplus	$18,540

* Based on Fiji Mill Prices, 1990

Country Studies

Fiji

A study reported by Pittaway (1990) provides a picture of the financial returns for a coconut estate on Taveuni, where cattle and copra are the main products, reasonable maintenance has been carried out and diversification and pasture improvement are planned/underway but have not yet come on stream (Table 84).

'The farm surplus is available for debt servicing, capital items, and living expenses. It is assumed that estate owners do not pay tax. The surplus of $ 18,540 is insufficient adequately to meet debt servicing commitments, reinvest in the business, and provide a reasonable standard of living for the owner. This highlights the difficult financial position of copra estates and the reason why those that remain as commercial operations are searching for suitable diversification enterprises and improvements, such as establishing better pastures.

The model shows that wages constitute a very large proportion of total costs. Any increase in these wages without a rise in income will further erode profitability. Attempts to cut costs on these systems only result in fairly quick falls in income, and the size of the surplus falls to where it is inadequate to meet servicing and living requirements.'

Returns were calculated to two levels of copra production (Table 85):

Table 85:

Estate Production	Returns		
	IRR	Gross Margin per ha	per man-day
700 kg per ha	-4%	$65	$4.25
400 kg per ha	-2%	$52	$4.75

'The negative IRR (internal rate of return) is caused by the initial establishment costs of the crop and the low copra yields (due to old coconut palms) plus low copra price. The low gross margins and returns per man-day highlight the low profitability of the crop'.

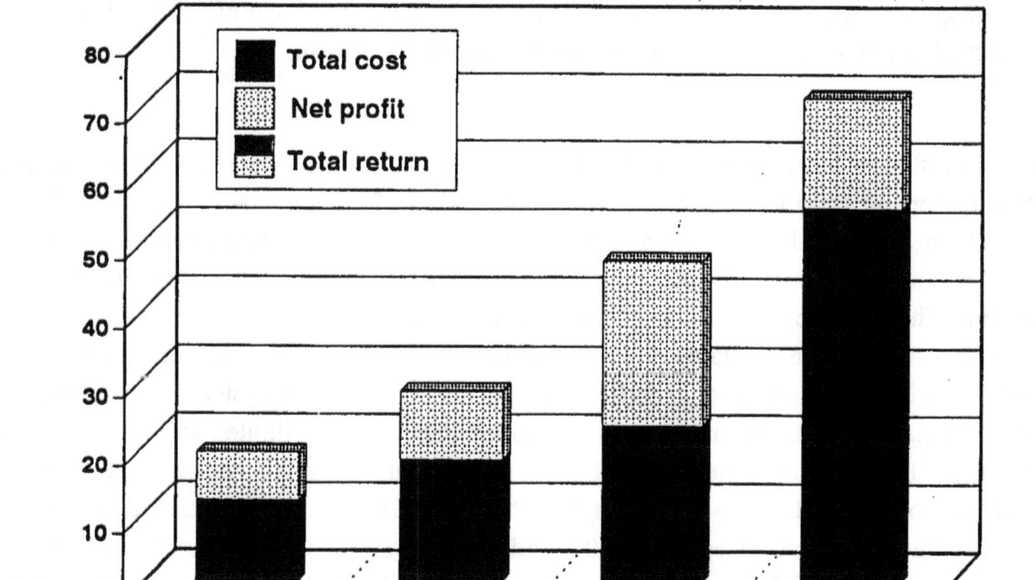

Figure 26: *Relative economics of coconut monocrop v coconut-based systems - 1 ha model (after Bavappa 1986)*

Clearly, monocropping of coconut is no longer profitable under these circumstances. The running of cattle under the coconuts provides important additional income, but further improvements are required to increase the livestock-carrying capacity as well as examining other diversification opportunities. It is perhaps significant that in Western Samoa, Opio (1989) indicates that the economic level of coconut production is 1.0 metric tons per yr. Analysis of yield streams indicate that coconut over 40 years of age give low economic returns due to declining yields, and the yields tend to drop below the figure of 1.0 metric tons per yr any time after year 40. Only with intensive intercropping can this uneconomic yield be supported, as intercropping will raise the productivity and returns from the land under the coconut.

India

Interplanting coconuts with fodder crops such as Napier grass for milk animals not only gave two crops instead of one, but the coconut palms also showed a 5% increase in yield (Kunhikrishnan 1972). At Kayanggulam the growing of hybrid Napier, stylo, puero and centro under coconuts for cutting and feeding to milk cows increased coconut yield by 23% within 3 years, largely because of the effects of manuring (Pillai 1974). The input requirements and economic analysis of the coconut/cattle mixed farming project at CPCRI reviewed by Nair (1979) indicates the economic viability of the system. This is also described by Sahasranaman *et al.* (1983), Nair (1983), and more recently Bavappa (1986) suggested that this system could be an ideal model for the self-reliant smallholder because of its high turnover and resource-use efficiency compared to monocropping (Figure 26). Das (1990) and Nair and Gopalasundaram (1990) also provided up-to-date information on the mixed farming (irrigated) system. Das (1990) provides an economic analysis of the system (Table 86), indicating that the gross cost was US$ 4144 per ha per yr, the gross return was US$ 5965 per ha per yr giving a net return of US$ 1821 per ha per yr at the 1988-89 factor-product costs. When the return to family labour and management was considered, the turnover was US$ 3102 per ha per yr. Although deficiencies were mentioned, such as low-yielding milk cows, the labour absorption potential of the model is clear (at 850 man-days per ha per yr this was a rise of 490% over irrigated monocrop coconuts and 608% over rain-fed monocrop coconuts). It is also clear that coconut-based farming systems are capable of distributing the labour inputs more or less uniformily over the year, unlike coconut monoculture. With its simple low-cost technology and farmers' interest in minimizing risk and maximising profit rather than maximizing yields, small farmers would not find it difficult to adopt (Das 1990).

Malaysia

A number of studies have shown that cattle-fattening using existing forages and by-products may provide a reasonable economic return to supplement income from the main crops and utilize excess labour. Data in Table 87 illustrate the returns from supplementing free grazing and cut-and-carry fed bulls with copra cake. An improvement in feed efficiency was noted when copra cake was used as a supplement. Where no supplement was given, yearling bulls grazed on native pastures under coconuts made average daily liveweight gains of 394.5 g which were comparable with the 0.38 kg per animal reported by Nitis and Rika (1978) in Bali. Although free grazing as a system of production slightly exceeded cut-and-carry fed animals it was concluded that with the limited areas available for free grazing on smallholder farms, the excellent gains under a cut-and-carry system meant that a system was available for smallholders to fatten cattle on their land. In Table 88 it is clear that increasing the content of coffee by-products from 40 to 60% had a negative effect on feed intake. The gross profit was highest with only 20% of coffee by-products in the ration. Hassan *et al.* (1989) carried out experiments over a 4-year period and concluded that smallholder dairy units could be economically viable and competitive with other agricultural enterprises, such as rubber and oil palm. Yusof and Rejab (1988) noted that the cost of weeding could be reduced by up to 50% in coconut stands grazed by ruminants. However, it was concluded that the rearing of livestock under coconut (intergrazing) is not as popular as intercropping (with cocoa, coffee, banana, pineapple and vegetables) mainly due to lower economic returns, higher capital and operational costs and due to social problems associated with livestock keeping. Devendra and Pun (1991) recently provided a specific example of the economic benefits of integrating ruminants with oil palm.

Papua New Guinea

In CPL plantations, the presence of cattle under coconuts reduces upkeep costs. With coconuts in monoculture it costs about US$ 96 (K 84) for upkeep costs, while with cattle it costs only US$ 27 (K 24) per ha, which is a significant reduction of about 70%. Ovasuru (1988) concludes that 'provided management is good and soil conditions are suitable, intergrazing is economically viable'.

Table 86: *Economics of coconut-based mixed farming experiment under irrigation at CPCRI, Kasaragod (after Das 1990)*

Particulars		US$ per ha per year
Cost	Labour wages $ 1.75 per day	1,488
	Fertilizers for the system including fodder grass	150
	Plant protection	34
	Cattle feed, dry fodder and concentrates	979
	Veterinary expenses	35
	Expenditure on rabbits	25
	Contingencies	152
Total variable cost		2,863
Annual Fixed cost including depreciation		1,281
Gross cost		4,144
Returns	Coconut @ $ 0.16 per nut and by-products	2,469
	Pepper @ 1.87 per kg for 235 kg	439
	Milk @ $ 0.31 per lit for 8760 litres	2,716
	Subsidiary crops and rabbits (30 nos.)	341
Gross returns		5,965
Net return		1,821
Imputed value of family labour, management and internal resources		3,102

US$ 1 = Indian Rupees 16.

The Philippines

The income of coconut farmers is among the lowest in The Philippines, with a monthly income of e average farmer being considerably below the poverty line (Ontolan 1988). The coconut-based farming system (CBFS) has been proposed as one solution to the low productivity and low income, with production alternatives ranging from single intercrops, a mixture of intercrops, to crop-livestock systems.

Various studies have compared the economics of coconuts grown in monoculture with pastures and cattle under coconuts. To illustrate the potential profit from adding cattle to a coconut operation, Barker and Nyberg (1968) presented an analysis of costs and returns for three case-studies:

A Coconut-beef cattle operation of 150 ha and approx. 1 a.u. per ha using traditional or typical management practices;

B Coconut-beef cattle operation of 150 ha, of which 100 ha is improved pasture. The assumed carrying capacity per ha of improved pasture is approx. 3 a.u.;

C Coconut-beef cattle operation of 300 ha, of which 200 ha is improved pasture. The assumed carrying capacity per ha of improved pasture is approx. 3 a.u..

Improved pastures of case-study B produced five times more beef than the natural pastures of case-study A for a net return of P 7830 to P 3690 for the unimproved farm. As farm and herd sizes are doubled, there are some economies of scale, and net returns in case-study C reached P 224,720 (Table 89).

Table 87: *Returns from three head of local Indian Dairy x Australian Milking Zebu bulls managed under coconuts (Abdullah Sani Ramli et al. 1982a)*

Parameters	Treatments*			
	I	II	III	IV
No. of participants	3	3	3	3
No. of bulls per participant	3	3	3	3
Initial cost of animals	1,015.49	1,043.19	1,020.60	993.63
Cost of supplementation	-	516.00	-	516.00
Sheds	-	42.85	57.14	57.14
Water trough	10.00	10.00	10.00	10.00
Salt licks	32.00	32.00	32.00	32.00
Miscellaneous	15.00	15.00	15.00	15.00
Labour input	828.50	931.50	1,105.20	1,138.50
Total Cost	1,930.50	2,621.05	2,269.94	2,792.77
Receipts from animal sale	2,480.28	3,094.13	2,314.20	2,795.54
Return to family labour	1,377.79	1,359.59	1,149.46	1,141.28
Net return	549.78	428.08	44.26	2.77
Net return workday^{-1}	5.98	4.14	0.36	0.02
Family return workday^{-1}	14.97	13.14	9.36	9.03
Net return day^{-1}	1.49	1.16	0.12	0.01
Family return day^{-1}	3.74	3.69	3.12	3.10

* Where I = free grazing
II = free grazing + copra cake supplement
III = cut-and-carry fed ad lib
IV = cut-and-carry fed ad lib and supplemented with copra cake
1 workday = 8 hours and daily labour input = 2.0-2.75 hours (for 368 days)
All costs and returns in Malaysian Ringgit.

Guzman and Allo (1975) surveyed 103 farms and observed that, through cattle sales, total income was increased by about 12%. One conclusion reached was that labour on coconuts appeared to be underutilized and could, with benefit, be devoted more to pasture establishment and management and animal husbandry, with no detrimental effects on nut production.

A comparison between estimated costs and returns of coconuts in monoculture and the integration of cattle (Anonymous 1982b) clearly indicated higher inputs but also increased returns with cattle (Table 90).

Table 88: *Growth performance of local Indian Dairy x Red Dane male calves fed with coffee by-products (Abdullah Sani Ramli et al. 1982)*

Parameter	Treatment				
	1	2	3	4	SE
Initial weight (kg)	101.0	105.5	108.3	102.3	4.36
Final weight (kg)	259.0	229.0	194.5	181.5	13.36
Av. daily gain (kg)	0.49	0.39	0.27	0.25	0.05
Feeding period (days)	322	322	322	322	
Daily DM intake (kg)	3.9	3.8	3.5	4.4	0.05
Feed per kg gain (kg)	7.9	9.8	12.9	17.7	1.87
Weight gain (M$)	758.40	592.80	413.76	38.16	
Cost of feed (M$)	288.06	247.49	117.47	50.09	
Gross profit (M$)	470.34	345.31	296.29	330.07	

Means in the same row with different subscripts are significantly different (P<0.05).
Treatments 1, 2 and 3 were fed with 20, 40 and 60 per cent coffee by-product in the concentrate ration. Treatment 4 was given 100 per cent Napier grass.

Table 89: *Costs and returns for cattle-coconut case-studies, Philippines (Barker and Nyberg 1968)*

	Case-Studies		
	A	B	C
Total Costs	2,560	24,060	46,940
Gross Returns	10,320	51,290	102,580
Net Return	7,760	27,230	55,640
Depreciation	610	6,090	7,290
Net return to Capital + Management	7,150	21,140	48,350
% return on Capital + Management	13	10.7	13.9
Interest on Capital at 6%	3,460	13,310	26,630
Net return to Management	3,690	7,830	24,720

* All costs in Philippine Pesos.

Studies of the economics of goat and sheep production with coconut tree smallholders indicated that 6 head of does or 6 head of ewes produced an annual income of P 2127 (US$ 97) and P 2295 (US$ 104) respectively. These incomes constituted 30 to 50% of the yearly gross income from copra per ha, depending on price fluctuations of copra (Parawan 1991).

Moog and Faylon (1991) noted that coconut farmers in Bicol, with an average land holding of 3.3 ha and an estimated annual net income of only P 4290, would receive an additional P 1800-4500 per yr from integrating buffalo-raising.

Parawan (1991) indicated that the main advantage of livestock integration, apart from the additional income from the livestock themselves, would be the savings on weeding costs (ranging from P 400 (US$ 19) per ha per yr for manual weeding and double if chemical herbicides are used), but he does not provide any indication of the percentage of these costs actually saved.

Table 90: *Estimated costs and returns of coconut plantations in monoculture and with cattle under traditional and improved management (P), 4 ha per farm*). (Anonymous 1982b)*

	Coconut Farm (trad. management)		Cattle-coconut Farm (trad. management)		Cattle-coconut Farm (improv. management)	
Income						
Cash income	4,870		10,400		35,170	
Copra sales		4,200		5,440		13,800
Cattle sales				4,120		19,440
Other cash receipts		670		840		1,930
Non-cash income						
Increase in inventory			720		2,310	
Total income	4,870		11,120		37,480	
Expenses						
Cash expenses	2,085		4,260		17,360	
Harvesting and hauling		560		700		1,540
Copra making		225		290		740
Weeding		1,200		600		300
Fertilization (coconut)						2,780
Purchase of cattle				1,000		2,250
Cattle care and up-keep						
Labour				1,365		2,675
Feed supplements						1,500
Veterinary supplies						500
Insemination cost				100		600
Pasture fertilization						2,795
Repair and maintenance						850
Miscellaneous cost		100		205		830
Non-cash expenses	315		1,760		7,730	
Depreciation						
Interest on cap. inv.				1,120		4,275
Interest on op. cap.		315		640		
Total expenses	2,400		6,020		25,090	
Net cash return	2,785		6,140		17,810	
Net return	2,470		5,100		12,390	
Net return per ha	620		1,275		3,100	

* Average of the first 10 years of operation.

Solomon Islands

As long ago as 1972, Walton indicated that the main economic advantages of introducing cattle into coconut plantations were: reduced weeding costs, increased copra production (from higher recovery of fallen nuts) and income from beef sales. Carrad (1977) concluded that the main benefits of using cattle to control undergrowth on Lever Brothers' copra estates in the Russell Islands were an increase in nut pick-up, a decrease in the number of man-days and hence reduced labour costs. Cattle-brushing was estimated to increase nut pick-up (thus reducing losses) from 75 to 90% and over the years there was a marked decline in the numbers of workers employed on most estates due to an increase in productivity and the use of cattle for most weeding (with some chemical control). For the effect of cattle introduction on both labour requirement and its main tasks, see Tables 91 and 92. Before introducing cattle, 69% of total man-days worked were applied on weeding (Table 92). From 1964 there was a rapid change, and after 2 years the division of labour was approximately equal; by 1974, 70% of labour time was spent on copra pick-up and only 30% on weeding. Substantial cost savings, the release of sufficient labour to replant 169 ha (which came into bearing in 1971) and efficiency improvement were the subsequent results (Table 93). In terms of profitability the maximum potential gain from cattle and coconuts, as opposed to coconuts only, was from 23% to more than 200%, depending on the yields of copra and beef, price of copra, discount rate, etc.

Sri Lanka

Early studies indicated that growing of *Brachiaria* species under coconuts would result in increased coconut yield, and pastures could be stocked at about 2.5 adult cows per ha, bringing in extra returns from meat and milk sales. Liyanage *et al.* (1989) observed that over a 3-year period the integration of pasture, fodder and cattle into a model coconut smallholding reduced the cost of inorganic fertilizer for the coconuts by about 69%, thus reducing the production cost of the coconuts.

Table 91: *The effect of cattle introduction on total labour in man-days per ha (Carrad 1977)*

Year	Estates	
	Pepesala	Banika
1961	14.59*	31.99
1962	10.00	37.07
1963	11.86	51.77
1964	11.29	33.34*
1965	11.75	30.79
1966	10.63	27.40
1967	11.46	25.49
1968	13.35	30.63
1969	11.62	26.43
1970	10.62	20.97
1971	10.70	16.06
1972	8.42	13.99
1973	8.92	12.98
1974	8.88	13.32

* cattle introduced

Table 92: *The influence of cattle on composition of labour's main tasks, Banika Estate (Carrad 1977)*

Year	Production (man-days)	Weeding (man-days)	Total (man-days)	Cattle numbers
1961	7,385	16,192	23,577	-
1962	8,457	18,865	27,322	-
1963	10,162	27,991	38,153	-
1964	9,840	14,728	24,568	209
1965	8,421	10,115	18,536	403
1966	8,365	8,132	16,497	234
1967	7,295	8,038	15,333	431
1968	7,409	11,031	18,440	337
1969	8,201	7,711	15,912	374
1970	5,753	6,873	12,626	1,037
1971	6,546	5,837	12,383	1,320
1972	7,027	3,761	10,788	1,030
1973	6,518	3,489	10,007	945
1974	7,233	3,040	10,273	1,079

Table 93: *Changes over time in type of weeding and weeding man per days per ha, Lever Brothers' Estate (Carrad 1977)*

Year	Weeding (man days per ha)	Type of Weeding
1961	5.4	Manual slashing only
1966	3.2	Decreased manual slashing, mechanical weeding, low cattle input
1974	2.7	Low manual weeding, some mechanical/spray weeding, high cattle input

Tanzania

Childs and Groom (1964) described the effects of undergrowth clearing under coconuts and stocking with local dairy cows in the coastal belt of Tanga region in Tanzania. With the establishment of better quality grasses, milk yields increased and coconut yields increased by 87%. A similar scheme with local farmers, resulted within 2 years in nut yield increases averaging 60% and a doubling of milk yields.

Vanuatu

The economic benefits of improving pastures under coconuts were examined in a series of studies at the Vanuatu Agricultural Research and Training Centre at Saraoutou. A comparison of liveweight gains on natural pastures (*Stenotaphrum secundatum*, without maintenance or fertilizers and with traditional management) and improved pastures (*Panicum maximum* plus fodder sorghum, with improved management and fertilizer) indicated that in spite of the higher investment involved in improving pastures, the profits (Table 94) made on the improved pasture were nine times higher than those on the natural pasture (Anonymous 1976). Simonnet (1990) indicated that the initial results from trials running sheep under coconuts were promising in economic terms.

Table 94: *Returns from cattle on natural and improved pastures under coconuts in Vanuatu (Anonymous 1976)*

Pasture species	Natural pastures *Stenotaphrum secundatum* (1 ha)	Improved pastures *Panicum maximum* (0.7 ha) Fodder sorghum (0.3 ha)
Stocking rate (head per ha)	1.7	7
Investments		
Land Preparation/sowing (0.7 ha)	-	2,740
Seed purchase (0.7 ha)	-	2,240
		4,980
Amortization over five years		996
Operation		
Land preparation/sowing (0.3 ha)	-	1,174
Seed purchase (0.3 ha)	-	440
Pasture maintenance	2,500	4,500
Fertilizer	-	7,380
Annual production cost	2,500 FNH	14,490 FNH
Meat production/steer sales		
Total weight of meat per ha (kg)	340	1,750
Revenue (at 50 FNH per kg)	17,000 (in 3 yrs.)	87,500 (in 2 yrs.)
Annual revenue	5,667 FNH	43,750 FNH
Profit	3,167 FNH	29,260 FNH

Western Samoa

Reynolds (1988) provided details of pasture establishment costs and of the value of additional beef produced on the improved pastures compared with local pastures. Hereford steers grazing Batiki grass produced an additional 73 kg per ha per yr, with steers on Cori and Palisade grass producing over 130 kg per ha of additional meat (Table 95). At US$ 1.05 per kg, this represented an additional income ranging from US$ 77 to US$ 141 per ha per yr (returns from additional copra would be in addition to these figures). From later trial results, Cori grass, grazed at a stocking rate of 4 steers per ha, produced beef valued at an extra US$ 180 per ha per yr (at the 1978 beef price).

Calving percentage was also increased on improved pastures to 75% (compared with values of around 55% on native pastures). While local pastures can be variable and in periods of drought may support very poor liveweight gains, where there is a high percentage of legume such as *Mimosa pudica* or *Desmodium heterophyllum* the production of beef in addition to coconuts is a very worthwhile enterprise as shown by the data in Table 95 where the value of beef produced on local pastures was US$ 67 per ha per yr compared with US$ 141 per ha per yr on the Cori pastures. In some later trials the value of beef on local pastures was US$ 211 per ha per yr compared with US$ 391 per ha per yr produced on Cori grass. With no establishment costs and only moderate maintenance costs, the income derived from cattle grazed on good local pastures is a considerable complement to that derived from copra sales.

In the light of fluctuating copra prices the importance of beef as a secondary source of income was demonstrated by Reynolds (1988). From a price equivalent to US$ 13.60 per 45.4 kg early in 1975, the local copra price dropped sharply to US$ 4.08 per 45.4 kg by the second half of the year. Based on figures for liveweight gains and copra production, gross income for coconut farmers, coconut farmers running cattle on local pastures, as well as for farmers running cattle on improved Guinea-Centro pastures, are presented in Table 96. The effect of the drop in copra price was to reduce gross income of the coconut farmer by 70%, whereas farmers with cattle on local pastures and improved pastures saw their income reduced by only 55.6 and 40.7%, respectively. The figures illustrate the importance of intercropping coconuts and the likely effects on farmer gross income. With the current low copra prices the relative returns from beef would be even more significant.

Table 95: *Value of extra beef produced on improved pastures, Western Samoa (Reynolds 1988)*

Species	Liveweight Gain (kg per ha per yr)	Beef Production* (kg per ha per yr)	Beef Production increment over local (kg per ha per yr)	Value** (US$ per ha per yr)
Local	127	63.5	--	67
Batiki	273	136.5	73.0	77
Palisade	389	194.5	131.0	138
Cori	396	198.0	134.5	141
Para	358	179.0	115.5	121
Tall Guinea	336	168.0	104.5	110

* Dressing percentage 50%.

** Meat value at US$ 1.05 per kg wholesale price.

Opio (1990b) described an intercropping/rotational grazing model (Figure 22) which on the basis of a cash-flow analysis gave much better returns in terms of gross revenues and returns to labour than monocrop coconuts. Although the return to labour and benefit/cost ratio figures for hybrid coconut are relatively high in Table 48, the fact that data for hybrid coconuts were not available beyond year 14 meant that Opio was uncertain of the calculated figures and he ignored this option in evaluating the systems. While local tall coconuts and cattle had a better return to labour figure than local tall alone, the returns to labour from rotational grazing and intercropping with taro and kava were higher than the minimum wage of WS$ 0.60, and this system would provide a reasonable alternative use of land during what traditionally would have been a fallow period, with a significant increase in income. Whether the light in years 8 to 16 actually falls as low as 20% is doubtful, but the approach adopted by Opio is worth following up. Data collected in the past 20 or 30 years have demonstrated that the introduction of livestock into coconut plantations can result in:

The effect of fluctuations in copra price on farmer gross income

Farmer Category	copra price		beef	Gross returns ha^{-1} $year^{-1}$ in US$ Total copra price		% reduction	% contribution of beef/copra high price		low price	
	high	low[1]		high	low		copra	beef	copra	beef
A. Coconuts only[2]	406.5	122.0	—	406.5	122.0	70.0	100	--	100	--
B. Cattle on local pastures under coconuts	406.5	122.0	104.9	511.4	226.9	55.6	79.5	20.5	53.8	46.2
C. Cattle on improved Guinea-Centro pastures under coconuts	345.6[3]	103.6	248.9	594.5	352.5	40.7	58.1	41.9	29.4	70.6

1) High price equivalent to US$ 13.60 per 45.4 kg; low price equivalent to US$ 4.08 per 45.4 kg.

2) Assuming copra production similar to farm where cattle raised on local pastures. In practice it is likely to be lower.

3) In absence of fertilizer, copra production was initially depressed where improved pastures of Guinea-Centro were established.

Table 96: *The Effect of Fluctuations in Copra Price on Farmer gross Income*

- reduced weeding costs;
- increased copra production from a better recovery of fallen nuts;
- new income from milk and beef sales;
- labour released for other tasks, such as replanting coconuts, pasture establishment and stock control;
- increased gross farm income;
- reduced dependence on one crop and better and more complete utilization of available feeds by cattle (crop by-products and crop residues utilized).

Perhaps the increasing importance in economic terms of integrating livestock, pastures and tree crops can be seen from the data in Table 97. In the past it has often been argued that livestock enterprises interfere with the productivity of the main crop, but with recent trends in costs and returns in coconuts (and oil palm and rubber), the situation appears to be changing. A Working Group at the Workshop on Research Methodologies Inherent to Integrated Tree Cropping and Small Ruminant Production Systems (Iniguez and Sanchez 1991) held in Medan, North Sumatra, Indonesia in 1990, estimated the relative net returns in various tree cropping and livestock production systems.

The relative returns from the livestock component under coconuts (compared to the returns from the copra) bears out the information reported from Western Samoa in the mid-1970s (Reynolds 1988). Investing in improved pastures usually results in increased returns from milk and beef (except where high quality local pastures already have a high legume content, or where the percentage light transmission is too low for the forage species used and the improved pasture fails to persist) and fertilizer use may increase copra production. However, an important factor is the level of management. Unless farmers understand and practise the fundamental principles of both coconut and stock management, the expected benefits may not result. A well run coconut enterprise may become a profitable livestock-coconut enterprise if the farmer practises both good coconut and stock management, but a poorly managed coconut enterprise is unlikely to show a profit simply because livestock are introduced, unless the introduction is accompanied by satisfactory stock management.

Table 97: *Estimates of the relative net return in various tree cropping and livestock production systems (Iniguez and Sanchez 1991)*

Systems	% contribution of animal to tree/animal system
Oil palm: sheep	5 - 10
Rubber : sheep	15 - 20
Coconut : cattle (Bali)	75
Coconut : sheep (Philippines)	50

NB. Weed control function has been taken into account, but animal impact on nutrient cycling has not been ascribed a value.

Mixed farming systems

Systems have been and are being developed in a number of countries and are presented here as examples of possible approaches to, and in some cases as starting points in, developing future systems. For descriptions of other systems refer to Reynolds (1988).

Colombia

Preston (1992) recently described the CIPAV Farming System (Figure 27) which could be modified for coconut areas provided good intercropping space is available. Multi-purpose trees play a crucial role in this intensive integrated farming system based on sugarcane developed by CIPAV in close consultation with local farmers in the Cauca valley in Colombia. The system is based on cane which provides the carbohydrate feed (juice and tops) and also fuel (bagasse). Multi-purpose trees and water plants supply the protein. Sugarcane and trees have well-developed systems of biological pest control, require minimum synthetic chemical inputs and are easily separated into high and low fibre fractions, as required for the different end uses of feed for monogastic and ruminant animals and fuel. The main trees used are *Gliricidia sepium, Trichantera gigantea, Erythrina glauca* and *edulis*. The foliage from *T. gigantea* is consumed by pigs, with that from other species being more appropriate for ruminants. Animals include pigs and ducks complemented by African hair sheep, with buffaloes and/or triple, purpose cattle to supply draught power as well as meat and milk. All livestock are managed in partial or total confinement to maximize nutrient recycling to the crops.

India

Various systems have been studied, such as the 5 ha dairy demonstration unit, 2 ha mixed farming unit, 1 ha buffalo unit and the 0.4 ha mini-dairy unit (Anonymous 1980). The system which has consistently given good results is the 1 ha mixed farming system.

CPCRI 1 ha mixed farming system - previously reviewed by Nair (1983). Bavappa (1986) provided a description of the system as well as further output and economic data. It was found that fodder grasses and legumes can be grown successfully in the interspaces of coconut palms, and with the fodder (mainly Hybrid Napier NB 21 and Guinea grass) produced from 1 ha of coconut plantation, 5 cross-bred milk cows could be maintained profitably. In the same plantation, subsidiary crops like pepper, tubers, vegetables and banana were also raised as well as 30 rabbits. The total output from the system has been found to be 15,900 coconuts, 218 kg of dry pepper, 695 kg of tubers, vegetables and banana, 7500 litres of milk and 550 m^3 of biogas per ha per yr, with a net annual return of Rs 14,500 per ha. Since this system is highly labour intensive, the total annual returns to the family including the family labour earnings was Rs 35,000 per ha (Table 98). The net returns of Rs 14,500 per ha (approx. US$ 1124) compares closely with the US$ 942 reported 10 years earlier (Reynolds 1988) and suggests that the 1 ha mixed farming unit is sustainable.

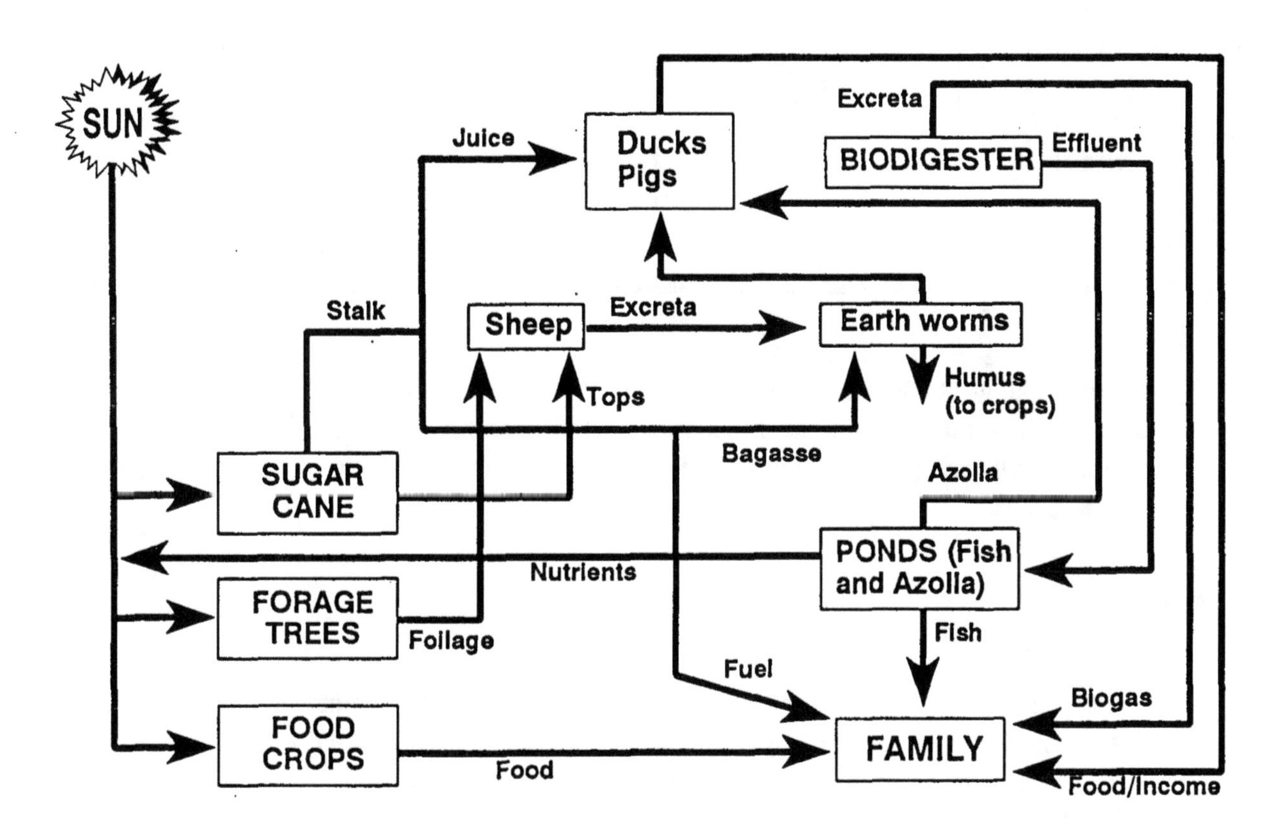

Figure 27: *The CIPAV Farming System (after Preston 1992)*

Table 98: *The economics of the CPCRI coconut-based farming system (Bavappa 1986)*

INPUTS	
Labour (man-days)	
Family labour	730
Hired labour	120
Total	850
Fertilizer cost for the system	Rs. 2,300
Plant protection cost	Rs. 500
Cattle feed cost	Rs. 12,600
YIELD	
Coconut (nuts)	15,900
Pepper (kg)	218
Milk (litres)	7,500
Subsidiary crops	Rs. 4,500
TOTAL COST	Rs. 56,500
TOTAL RETURN	Rs. 71,100
NET RETURN	Rs. 14,500
Earnings by family labour wages	Rs. 20,500
TOTAL RETURNS TO THE FAMILY	Rs. 35,000

Table 99: *Economics of mixed farming in homestead agriculture (0.28 ha with livestock)*

Item	Input					Output			Net profit
	Labour (in man days)	Cost (Rs)	Fertilizer Manure/Feed	Cost (Rs)	Total (Rs)	Produce	Rate	Amount (Rs)	
Coconut (60 nos) Maintenance	10	350	4 baskets cowdung @ Rs 5 per basket	1,200	1,550	8,400 nuts Byproducts leaf, fronds	2.25	2,100	16,250
Irrigation	90	3,150			3,150			350	
Harvesting	8**	400			400				
Guinea grass at intercrop	10	350			350	9,125 kg green fodder	0.40 per kg		
Miscellaneous crops***	50	1,750	50 baskets cowdung	250	2,000			2,000	Nil
Cow and Heifer (Maintenance)	90	3,150	Concentrate @ Rs 20/ day	7,300		Milk	5.00/l	15,000	3,160
			Fodder @ Rs 5kg/ day	3,650	14,300	Dung	5/basket	1,460	
Miscellanious				200		Calf/ heifer		1,000	

* Labour cost @ Rs 35 per man-day

** Hired labour @ Rs 50 per day for harvest

*** arecanut, jack, mango, tamarind, lime, breadfruit, pepper, banana, tapioca, colocasia, dioscoria, morphophallus, ginger, turmeric, bhindi, amaranthus, cucurbitaceous, solanaceous vegetables, anona, guava.

Kerala homestead system - a high level of productivity is achieved by combining a multi-tier cropping system, a livestock system and an irrigation system (Abdul Salam *et al.* 1990). The farmer and his family (total 7 persons) depend for their income on 0.28 ha of fertile land.

The system consists of 60 coconut palms with a high level of productivity (averaging 140 nuts per tree per yr), 1 milk cow (Jersey cross-bred) and a heifer to supply manure for the crops and milk for consumption and sale, and guinea grass for the cattle. With the coconuts comprising the top layer and tuber crops, vegetables and guinea grass the lower layer, the intermediate layers consist of arecanut, pepper, jackfruit, tamarind and mango and, below that, banana, tapioca and various fruit plants with a boundary fence of *Gliricidia*. The cow produces 10 litres of milk per day of which 7 litres are sold (at Rs 5 per litre). As well as the guinea grass, Rs 20 per day are spent on oil cakes for the animals.

The economics of the system are shown in Tables 99 and 100, with the main income derived from coconuts (the 60 coconut palms yield an average 8400 nuts per yr). The farm absorbs 258 man-days of labour per year and produces a net profit of Rs 22,710 which, if the family labour costs are included increases to Rs 31,460 per year or Rs 86.20 per day. After meeting the annual family expenditure there could even be a small saving of Rs 1460.

Indonesia

The three strata forage system (Devendra and Pun 1991; Nitis *et al.* 1991) has been developed in Bali as part of an intensification programme of plantation crop systems. Most of the smallholderfarmers practice integrated farming, with food crops for home consumption and plantation crops for inter-island trade and export. Cattle, goats and pigs are kept for draught purposes, as weeders, as suppliers of manure, and for food. Feed resources come mainly from native grasses, cereal straws and tree leaves, with livestock usually being permanently stalled or tethered for grazing during the day and stalled at night. Because of the small size of individual farms, smallholders usually do not have sufficient fodder for rapid animal growth. The roadside grasses and crop residues result in cattle liveweight gains of between 100 and 200 g per day and the fattening period could be reduced to < 2 years. These circumstances led to the development and successful demonstration of the three strata forage system (TSFS) which involves a first stratum of grasses and ground legumes; a second stratum of shrub legumes; and a third stratum of fodder trees (Devendra and Pun 1991). The TSFS is developed in the peripheral area, surrounding the core area and along the circumference (Figure 28). The 0.2 ha peripheral area comprises mixed pasture (first stratum), the 400 m circumference containing 4000 shrubs (second stratum) and 80 fodder trees (third stratum) have the potential to produce a forage yield of 15.5 t DM per year. Animals integrated with the plantation can be either stall-fed or tethered for grazing.

Table 100: *Details of labour requirement, input and output cost and returns from the homestead*

Items	Family labour (man-days)	Hired labour	Total	Total input cost Rs	Total returns Rs	Net profit Rs
Crops	160	8	168	7,450	27,000	19,550
Livestock	90	-	90	14,300	17,460	3,160
Total	250	8	258	21,750	44,460	22,710
Net profit per day			22,710 ÷ 365 = 62.23			
Amount available including family labour			Rs 22,710 + 8,750 + 31,460			
Actual income per day			Rs 86.20			
Annual expenditure of the family (7 members)			Rs 30,000			
Balance available			Rs 1,460			

Over a five and a half year period the project compared two systems, the TSFS and the non-TSFS (NTSFS) at two stocking rates (2 and 4 cattle per ha). Table 101 presents some of the results, and the main highlights were as follow (Nitis *et al.* 1991):

- the allocation of land as a forage boundary within the TSFS increased forage production by 98%. With the additional 4000 shrubs and 80 trees the total wet and dry season forage production in the TSFS was 91% more than the NTSFS;
- *Stylosanthes*, *Centrosema*, *Acacia*, *Gliricidia*, and *Leucaena* which were grown, provided increased dietary nutrients in the TSFS. Consequently, cattle raised in the TSFS gained 19% more liveweight and reached market weight 13% faster;
- the availability of increased forage also enabled higher stocking rates and liveweights to be achieved: 3.2 animal units (375 kg) per ha per yr in the TSFS compared to 2.1 animal units (122

kg) per ha per yr. Cattle in the TSFS were less infested by endoparasites. This was presumably due to less contact with the traditional cattle, since the TSFS cattle were always kept in confinement;

- the introduction of forage legumes into the TSFS reduced soil erosion by as much as 57% compared to the NTSFS. In addition, the fertility of the soil in the TSFS was considerably improved;
- with the presence of shrubs and trees which were lopped twice a year, firewood production was > 1.0 t per yr and TSFS supplied 64% of the farmer's requirement;
- farmers in the TSFS spent less time managing their cattle and also benefited from a 31% increase in income compared to NTSFS.

Kenya

A collaborative research programme has been underway near Mombasa since 1988 between the Kenya Agricultural Research Institute (KARI) and ILCA to develop smallholder dairy production in the coastal subhumid zone, where farming systems in the wetter areas (1,000 - 1,150 mm rainfall) are dominated by coconuts. In the study area farm size averages about 4 ha (to support a family averaging 12 persons). The farming system is being intensified through dairying with a 'package' of improved practices for smallholder dairying in a zero-grazing system. The package includes using cows 'improved' with Sahiwal and European genes together with disease control measures and improved feeding, in particular using planted forages such as Napier grass (*Pennisetum purpureum* cv. Bana) and *Leucaena*. The aim has been to develop feeding systems that will make the most of the milk production potential of the crossbred cows, and to develop feed production systems that will boost and stabilize feed and food-crop production on smallholdings. Work reported to date has included different rates and timings of applying slurry to maize and Napier, growing Napier with legumes such as *Clitoria ternatea*, the effects on milk yields of different cutting heights of Napier, immunizing cattle against East Coast fever, and various small ruminant studies (over half the households keep sheep or goats). Recently, Muinga (1992) reported on several trials using small quantities of maize bran as an energy supplement, *Leucaena* foliage as a protein supplement, and a basal diet of Napier grass. Cows used were Ayrshire/Brown Swiss x Sahiwals.

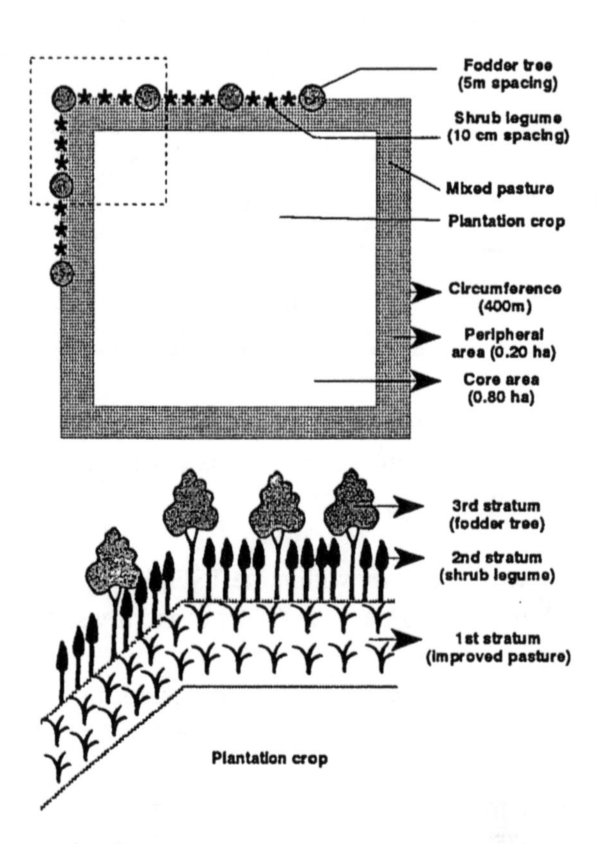

Figure 28: *Proposed integration of three-strata forage system with plantation crops (Nitis et al. 1991)*

Early lactation trial

Milk yield in unsupplemented cows during the dry season averaged 3.7 kg per day compared with 6.0 kg for cows supplemented with *Leucaena* alone (2 kg DM per day), 6.9 kg for those supplemented with maize bran alone (1 kg DM per day) and 8.6 kg for those receiving both *Leucaena* and maize bran. Supplementation also had an effect on the rate at which milk yield declined, and on weight loss. In a separate trial, cows fed with Napier grass harvested when 100 cm tall gave higher milk yields than cows fed Napier when 150 cm tall plus 2 kg of *Leucaena* foliage daily (Figure 29). It was suggested that during the rainy season farmers could feed their cows on young Napier alone, drying and storing the *Leucaena* foliage for use during the dry season.

Table 101: *Comparative productivity of TSFS and NTSFS plots (kg dry weight per plot per yr) (Nitis et al. 1990, after Devendra and Pun 1991)*

Parameter	TSFS*	NTSFS**
Food	853	1,268
Straw	750	1,218
First stratum	455	-
Second stratum	310	-
Third stratum	15	-
Shrubs	-	132
Trees	-	2
Improved grasses	-	10
Native grasses	-	242
Firewood	1,049	475
Cattle live weight gain (kg per 3 years)	186	166
Carrying capacity (cattle per ha)	4	2
Maximum live weight (kg per head)	300	200
Soil erosion (mm per 2 years)	11	20

* Three strata forage system

** Non-three strata forage system

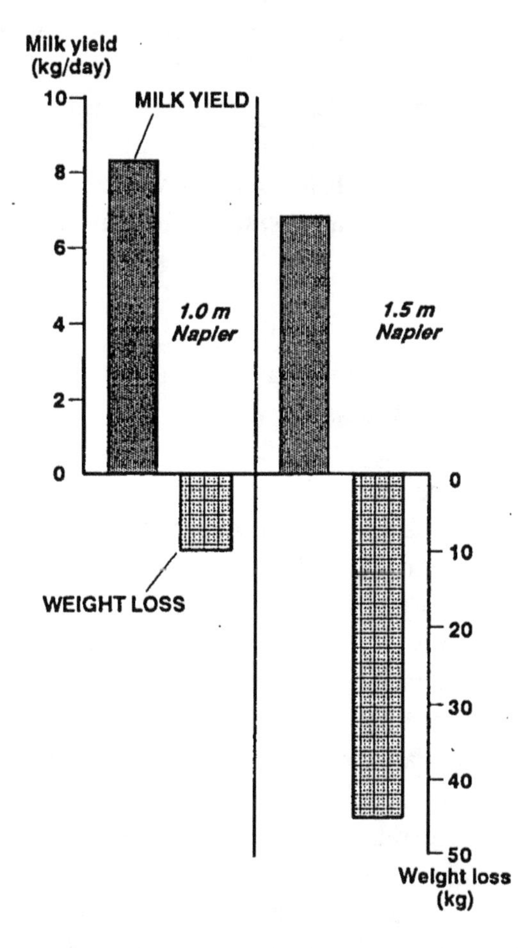

Figure 29: *Feeding dairy cows with young Napier grass gives a higher milk yield and lower weight losses than feeding older grass*

Mid to late lactation trial

In the rainy season, unsupplemented cows gave an average milk yield of 4.5 kg per day, compared with 6.7 kg per day for those receiving both maize bran and *Leucaena*. Supplementation had a large and significant effect on the rate at which milk yield declined over the 10-week trial. In unsupplemented cows, average milk yield during the last 4 weeks of the trial was only 32% of that in the first 3 weeks, whereas in cows receiving both maize bran and *Leucaena* there was no significant difference in milk yield between the two periods. Supplementation had a major effect on weight changes, with those supplemented with *Leucaena* maintaining their weight while animals in other treatments lost weight.

The immediate financial returns from supplementation were greatest in cows in early lactation. Supplementing the diet with 1 kg DM of maize bran and 2 kg of *Leucaena* daily increased milk yield by almost 5 kg per day. At 1992 prices on the Kenya coast this was worth about 80 Kenya shillings (Ksh). The supplements cost around Ksh 5 (maize bran sold for about Ksh 1 per kg and the estimated cost of *Leucaena* was Ksh per 2 kg DM). Thus, the return from supplementation was 16:1. The return on supplementation in mid to late lactation was more difficult to determine, but supplementation was likely to lead to earlier reconception, shorter calving intervals and therefore increased productivity. However, the *Leucaena* psyllid (*Heteropsylla cubana*) has now reached the African mainland (Anonymous 1992) and will influence future work with *Leucaena* and the relative importance of other multi-purpose trees such as *Gliricidia, Calliandra* and *Acacia* species. The likely effects can be judged from the report of Krisnawati Suryanata *et al.* (1988) from Timor, where the system of stall-fed cattle was hardest hit because *Leucaena* constituted 80% of the diet. Labour requirement for fodder collection of other species more than doubled. In parts of The Philippines, and especially in the intensive smallholder beef production sector, the effect of the psyllid was equally devastating. A survey of 31 farmers in Malimatoc, a village near the town of Mabini, Batangas Province, revealed that the number of animals being raised fell from 115 before the infestation occurred to 53 later! Moog (1992) suggests that there has been over-reliance on *Leucaena* and that the potential of other fodder species should be tapped. Perhaps lessons can be learned from experiments carried out at CATIE (Kass *et al.* 1992) using various fodder trees as part of the N-ration. Figure 30 shows the effect of supplementing milking cows with increasing levels of *Erythrina poeppigiana*, where the cows grazed African star grass (*Cynodon nlemfuensis*) and received a constant level of energy - 1 kg molasses per head per day).

Sri Lanka

A fodder production system for the coconut smallholder utilizing fodder grasses such as NB 21 *P. purpureum x P. americanum* and *P. maximum* cv. Hamil was described by Reynolds (1988). Stall feeding was recommended and it was suggested that this system could form the basis for integrating a dairy or beef fattening enterprise into a multiple cropping farming system, allowing crop by-products such as banana residues, cassava tops and peelings and sweet potato vines to be fully utilized while producing valuable manure. Since it was suggested by Lane in 1981, a great deal of work has been carried out with various fodder trees and the system needs revising along the lines of the three strata forage system from Bali (Nitis *et al.* 1991) or at least it should incorporate some fodder trees as protein sources. Another model integrated system for coconut smallholdings has been demonstrated by Liyanage *et al*, 1989). Established in 1985 in the wet intermediate rainfall zone (1595.5 mm per yr) under 45 year old tall coconut palms spaced at 8.4 x 8.4 m (137 palms per ha) this model was designed to increase the coconut production and productivity of land by introducing a pasture / fodder / cattle system with the major consideration of reducing expenditure to the bare minimum and making the system ecologically sound and economically viable. Covering approximately 1 ha, the model included 6 paddocks with 24 coconut palms in each. The recommended mixture was *Leucaena leucocephala* planted in double rows along the coconut avenue at a spacing of 2 x 1 m, with *Gliricidia sepium* and *Leucaena* planted alternately (1 m apart) along the boundary fence, and a pasture mixture of *Brachiaria miliiformis* and *Pueraria phaseoloides*. Coconuts were fertilized annually and fertilizer was applied to the grass-legume mixture at the time of planting. One year after establishment 4 Jersey x local crossbred heifers at six months of age were introduced and paddocks were grazed (animals being tethered and moved each day) on a 30-day rotation. In addition to grazing the grass-legume mixture (at the rate of around 35 kg of fresh matter per head per day, the cattle were fed loppings of *Gliricidia* and *Leucaena* at 2 kg of fresh matter per head per day rising to 5 kg as cattle grew). During drought periods, grazing depended on feed availability and urea-treated straw (4 kg urea dissolved in 100 l of water and mixed in 100 kg of straw) was fed ad lib supplemented with 750-1000 g (dry weight) of *Gliricidia/Leucaena* leaves. The amount of straw rose from an initial 4 kg per head per day to 10 kg. (For further information on the use of rice straw and supplements refer to the work of the University of Ruhuna team described by Pathirana and Mangalika (1992) and mentioned earlier). Drinking water and a commercial mineral mix were available.

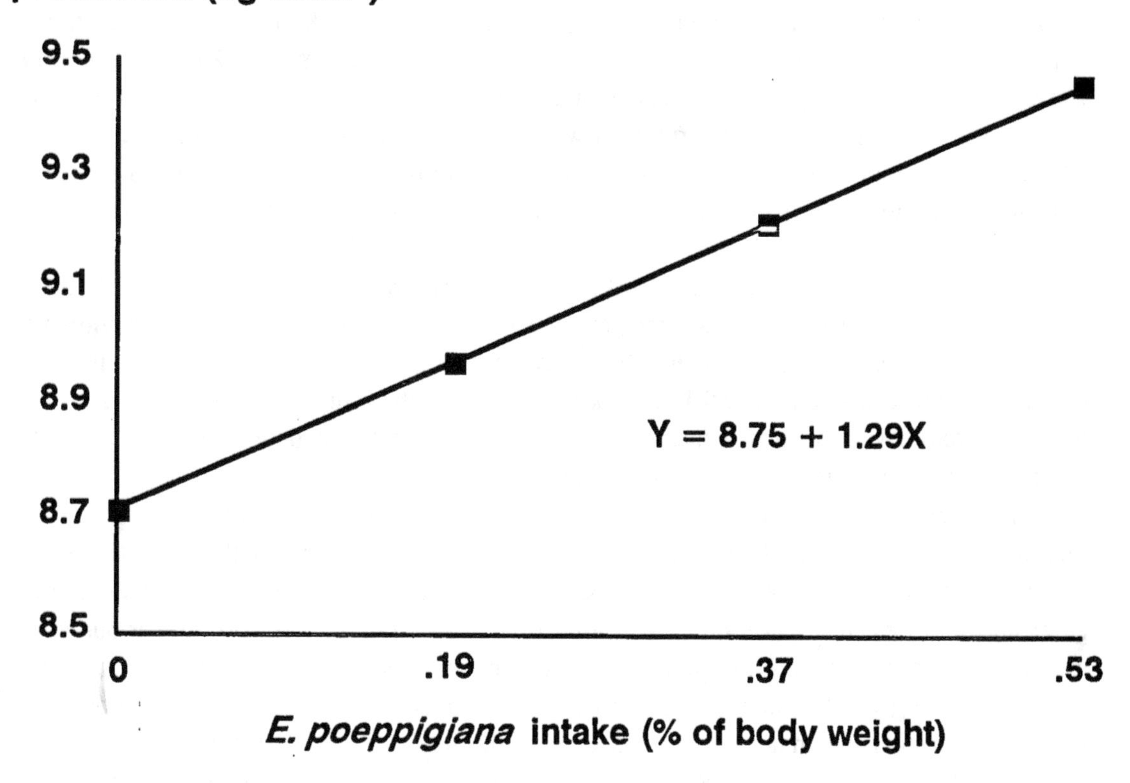

Figure 30: *Milk production from grazing cows in relation to levels of E. poeppigiana intake (after Kass et al.)*

Over a 3-year period, nut and copra yields were maintained (and according to Jayasundara and Marasinghe 1989) rose by 11% on the integrated system compared to the monoculture system), nutrient levels in coconut leaves were unchanged, nutrients were returned through cowdung and urine, the pasture mixture produced over 20,000 kg per ha dry matter each year with additional forage prunings from the *Gliricidia* and *Leucaena*, and heifer liveweights increased from an initial 70 kg to 200 kg at the end of the first year, giving a mean weight gain of 306 g per head per day. The cost of inorganic fertilizer per coconut palm was reduced from Rs 8.10 per palm per yr to Rs 2.49 per palm per yr (a saving of 69% in fertilizer cost) due to the nutrients returned in dung and urine. Although data collection was continuing, the model demonstrated that it is technically and economically viable and leads to considerable savings on inorganic fertilizers as well as increased copra yield.

Table 102: *The 'acre farm': gross margins per year (Rs)*

Income sources	1984	1985	1986	1987	1988
Dairy	5,000	7,000	7,000	7,000	7,000
Vegetables	4,000	4,000	3,000	3,000	2,000
Banana	4,000	7,000	5,000	5,000	4,000
Pepper, coffee and coconuts	-1,000	-1,000	4,000	6,000	7,000
Total	12,000	17,000	19,000	21,000	22,000

Earlier in 1983/4 a model 'acre farm' was established for farmer training at the mid-country livestock development centre Mahaberiyatenna, Digana (Westenbrink 1986). It had the following objectives: provision of gainful self-employment to the farmer and assurance of a satisfactory income as quickly as possible after commencement; provision of adequate food requirements for the farmer's family; a flexible cropping pattern and sustainability; optimizing of crop-stock integration and demonstration of improved land-use methods. Coconuts were planted at a wide spacing of 12 x 12m and the land area was divided into three segments for banana with underplanted grass, vegetables and tree legumes, and pepper, coffee and fruit trees underplanted with grass. The initial development cost included labour (Rs 30,000), house (Rs 10,000), cattle shed (Rs 4,000), bio-gas plant (Rs 3,000) and cattle (Rs 7,000). The unit consisted of 2 dairy cows and 2 offspring, 30 coconut trees, 187 banana clumps, tree legumes planted on 600 m of alley and fence, 800 m^2 of vegetables and 3000 m^2 of pasture. Records of all inputs and outputs were maintained and actual gross margins for 1984 and 1985 are shown in Table 102, with the projected gross margins for 1986, 1987 and 1988. As well as providing an annual income of Rs 12,000 in 1984 and Rs 17,000 in 1985, the farm also provided vegetables and starch foods for home consumption, 0.5 l of milk per day from the dairy and free energy for cooking and lighting. From 1988 onwards it had been expected that annual income would reach Rs 20-25,000; however it has not been possible to obtain any additional data.

Forage cultivation and livestock-raising under coconuts is now widely accepted as one of the methods by which the smallholder farmer can increase his income and food supply. Livestock feeding systems based on banana leaves and stems, sugarcane, fodder grasses, tree legumes like *Erythrina, Gliricidia* and *Leucaena*, rice straw, copra cake and rice bran, etc, as well as various conventional grasses and legumes, have given promising results in a number of countries and are routinely utilized by an increasing number of farmers. Improved integrated systems have been demonstrated which would appear to have many advantages over traditional systems. Whether they are adopted will depend on how they are perceived by the farmer and, in particular, his assessment of the risk factor involved (Beets 1990; Kirwan 1986; Reynolds 1989).

PART IV. POST-HARVEST TREATMENT, PROCESSING, RESEARCH AND DEVELOPMENT

15. Post-harvest Treatment

J.G. Ohler

Nut Storage

Storage before dehusking of the nuts is not actually required, although very green nuts may be dehusked easier after some storage. During periods of peak harvests, not all nuts harvested daily can be dehusked the same day, and storage will be necessary. To maintain a certain stock may also be advantageous to keep the dehusking team constantly supplied with material.

Sudaria and Pedro (1990) in The Philippines observed that dehusked nuts stored for 1, 2 and 3 months had 1, 9 and 27% germination, respectively. The moisture content of the meat decreased with increasing storage time. Compared to fresh nuts, the three groups of nuts had moisture contents of about 80%, 66% and 55%, respectively. The advantage of the reduction in time required for drying the copra was offset by copra weight losses due to germination. No significant differences were observed in colour, smell, and cleanliness of fresh meat and copra between stored and fresh nuts. Under very dry conditions nuts may dry out during storage without germinating. This may facilitate dehusking and scooping out of the copra from the shell.

According to Patil (1991), the advantages of storing or seasoning harvested nuts before they are further processed have been reported as:

- decrease in moisture content;
- increase in thickness of copra;
- increase in oil content;
- greater meat resistance to bacterial sliming during sundrying;
- easier husking;
- easier and cleaner shelling;
- uniform copra quality.

Trials indicated that whole, 12-month-old nuts dried sufficiently within two months to facilitate husking and copra extraction. Further drying reduced moisture content even further, and ball copra formation at the end of the 6th, 7th, 8th, and 9th month was 10, 33, 70, and 100%, respectively. The testa of the copra turned dark brown after 6 months' storage. There was no change in copra content during storage. This is in contradiction to what was mentioned above. Local experience will indicate the possibilities for storage during the wet and the dry seasons. Logically, storage should be undertaken in conditions as dry as possible, preferably under a roof, and close to the dehusking site. Nut storage under the trees for local dehusking, with the objective of keeping the husks in the field and reducing transportation costs, increases risks of nut theft and rodent attack.

Dehusking

The first problem in post-harvest treatment of coconuts is dehusking. Traditionally, this is done manually by skilled labour. The simplest method of removing the kernel from the nut is by splitting the nut into two halves or three parts with the use of a machete, without dehusking, after which the kernel can be scooped out to be sundried or taken to the kiln, or transported directly to a processing factory. This method is often applied where hot air dryers are used for copra manufacture. The wet endosperm sticks to the shell and cannot be removed in halves or large pieces. The kernel is scooped out with a flat metal implement which results in 'finger cut' kernels. The husk with the shell attached to it is used as fuel for the dryer. The increased surface of the cut endosperm exposed to the air increases deterioration. There is also an increased risk of contamination with dirt in the plantation. When the endosperm is transported in bags and pounded upon to reduce its volume, deterioration will be much increased, particularly if these bags have been used before for the same purpose. By this method, both husks and shells not used as fuel remain in the field. The shell will take a long time to decomposite and may become a nuisance.

The dehusking operation is often done in the neighbourhood of the copra kiln. Cost of nut transportation to the kiln, can be reduced by dehusking the nuts under the trees, so that only the unopened shells have to be transported to the kiln reducing weight by about 40 per cent and volume by about 60 per cent. The husks remain in the field for use as an organic mulch, rich in plant nutrients. When dehusking in the field, husked nuts must be shaded, otherwise they may burst when heated by sunshine. The most frequently used dehusking method is by the use of a pointed metal spike, secured in the ground in a slightly slanting position, with the point upwards. The nuts are brought down with force on the spike, followed by twisting the nut sidewards against the spike, loosening the husk. This movement is repeated one or twice for the total removal of the husk. Care is taken that the point of the spike enters the husk at the stalk end so as to avoid the damaging the shell.

Pomier (1984) described a portable husking spike as follows: 'The portable spike is tetrahedral in shape. One of the edges is made of 35-mm angle, with the lower end sharpened into a point, and the other extended by a 20-mm diameter round iron bar, forged into a spike for husking the nuts. The other five edges of the tetrahedron are made of 20-mm angular iron. Stability on the ground is further increased by two points made of 20-mm angular iron. The joints are electrically soldered, with each piece being positioned on a metal template. The apparatus is finished by a trimming machine, which is particularly necessary for the husking spike. It is important to note that the spike is slightly to right of centre compared to the symmetrical plan of the base triangle. The husker can go from one tree to the other. He puts down his spike, pushing the points into the ground and wedges it with his right leg and foot. He husks the nuts and throws the husks either into the windrow or in the middle of the inter-row.'

The number of nuts one man can dehusk per day depends very much on the type of the nuts, the thickness of the husk, and the skill and energy of the operator. Although numbers of up to 3000 nuts per day have been reported (5-6 nuts per minute), the average of 1500 nuts per day as observed by Risseeuw (1980) in Sri Lanka seems to be a more realistic figure. Pomier (1984) reported 1200 to 1300 per day. According to San and Pau (1986) in Malaysia, the skill and energy demand of the so-called 'chop' system is such that an average experienced worker is capable of efficient dehusking for about 4 hours. During this period he usually interrupts dehusking by splitting the husked nuts to catch his breath. An average worker generally manages to process 1000 Malayan Tall nuts, 1200 MAWA hybrids or 1500 Malayan Dwarf nuts per working day. In most countries, dehusking and splitting are performed by different labour.

From this description it is clear that dehusking is hard work, and this job is not very popular. Although it can provide jobs for the unemployed, it is often difficult to find labour for this operation. It may become more and more difficult in future to get skilled labour, especially for estates with large coconut production and also for areas where dehusking has never been practised. Labour costs also have a tendency to rise, especially for difficult jobs, increasing the coconut oil price and reducing its ability to compete with other oils on the local and world markets. Of course, it has been tried to facilitate the work with special tools or machines. So far however, with little success.

In Malaysia, a foot-operated dehusker that can be operated by female workers was developed in 1980. However, the number of nuts that could be dehusked per day by a woman was only 600, 1000, and 1200 for talls, hybrids and dwarfs, respectively. Nevertheless, after its introduction the existing backlog in dehusking was cleared and copra production quickly returned to normal (San and Pau 1986). Various mechanized systems have been developed during the past decades, but apparently no system really made an impact on this sector of the coconut industry. Major problems for mechanical dehusking include different sizes of nuts and shells and the different stages of maturity of the harvested nuts. The latter could be overcome by storing the nuts for a few weeks. Differences in nut shape and size will become less problematic when large areas are planted with uniform planting material, such as certain hybrids.

In 1983 it was announced that a British firm had developed a mechanized coconut dehusking machine. It had a capacity of 720 nuts per hour (Anonymous 1983b). However, not much was heard of it afterwards. This may have been the machine with a capacity of about 1,000 nuts per hour developed by the Natural Resources Institute. This machine was produced commercially by a private company after making modifications to the design that were not tested out in the field. Its performance did not come up to expectations.

In 1984, another dehusking machine was developed in Malaysia (San and Pau 1986). It was stated that this machine had a capacity of about 800 MAWA hybrid nuts per hour. It was expected that the machine, once perfected, could have a maximum capacity of 1000 nuts per hour. For use in the field, the machine could be adapted to utilize power from a tractor. The process is described as follows: Before dehusking, a sorting mechanism divides the nuts into different size groups. Subsequently, the coconut, with its stalk end facing down, is manually placed in the cavity of a rotating table. A plunger presses the coconut down to the preset level. As the rotating table advances, a circular saw slices off a piece of the husk at the stalk end of the coconut. The coconut then arrives at the dehusking unit where another plunger pushes it into the dehusking section where:

- vertical knives cut the husk into 9 vertical sections, before pushing it into a rotating cone;
- as the nut is pushed into the cone, it is firmly gripped by springloaded fingers and the rotation of the cone literally tears the husk from the nut.

In 1984, Fondacion de Coco y Palma (FONCOPAL) in Venezuela was reported to have received a quotation of a 'Simon Rosedowns' coconut opening machine from the U.K., that had a capacity of opening 360 nut fruits per hour (Anonymous 1984). In France, in 1990, a dehusking machine having a capacity of 1200 nuts per hour was developed by Biotropic in Montferrier-sur-Lez. The Central Plantation Crops Research Institute in India in 1991 tested a power-operated coconut dehusker with a designed capacity of 500-600 nuts per hour. The machine was working well except for the splitting part. (Anonymous 1991c).

Cummings (1993) described a coconut dehusker developed in Trinidad and Tobago. One was already operational in Malaysia, one in Trinidad and one in Dominica. It is a robust machine, existing in two versions - manual and electric. It has a capacity of 40 to 50 nuts per hour and it is easy to operate. Two to three persons suffice to feed, dehusk and stock the nuts. The breakage rate is 2 per cent, mainly green or old nuts, or nuts with thin or deteriorated shells. Hybrid nuts posed no problems. However, the output of this machine is rather low and the labour involved is relatively high.

Nut Transportation

The system of nut transportation depends on the volume of nuts to be handled and the distance over which this volume has to be transported. In smallholdings, transportation is often done by the farmer himself, as neither the number of nuts, nor the distance to the farmer's house is large. With the increase in either factor, other means of transport will be required. This can vary from large baskets on bicycles, horses or donkeys, to animal-drawn carts. In larger holdings, nuts are usually collected by animal or tractor-drawn carts and transported to the drying kiln where they are split and drained before being placed in the kiln. In some large plantations, nuts are broken immediately and placed in bags holding about 40 broken nuts each (Pomier 1984). However, this can be done only when there is no lack of transportation, and waiting hours at the kiln are kept low, because opened nuts start moulding within one day. If the nuts have to be transported to a central factory outside the plantation, this can be done by using lorries.

16. COCONUT PROCESSING

T.K.G. Ranasinghe

Introduction

This chapter deals with the utilization of coconut fruit, and products derived from the coconut palm. The newly emerging products from the coconut palm are also presented to encourage future commercial applications to increase income from coconut through trade in modern, value-added products. Products from the kernel (copra, oil, copra cake, copra meal, desiccated coconut, milk or cream) and husk (a series of white fibre and brown fibre products) and shell (charcoal, activated carbon, flour) enter international trade. Fresh coconuts and seednut are also exported from certain countries.

World Copra Production

Average world copra production during the period of 1988-92 was 8.634 million tonnes. The Philippines (40.1%) and Indonesia (25.4%) together with the other members of the APCC produced 88.4% of the world's total. Most of the copra is crushed in producing countries to obtain coconut oil and copra cake. The other major coconut industry is the manufacturing of desiccated coconut. The utilization of coconuts for production of copra and desiccated coconut is about 5 million tonnes copra equivalent. The remaining coconuts, equivalent to 3.634 million tonnes of copra, are consumed directly as foodnuts in the producing countries without industrial processing. This is an important and affordable source of food, nutrition and energy for people living in the producing areas.

World Coconut Oil Trade

In terms of trade, world coconut oil exports during the same period was 1.452 million tonnes, of which The Philippines and Indonesia contributed 61.8% and 15.7%, respectively. The APCC member countries are responsible for 86.6% of world exports of coconut oil. Export earnings for 1992 from all coconut products by The Philippines, Indonesia and Sri Lanka were US$ millions 700.4, 245.4 and 113.4, respectively out of the APCC total of US$ 1258 million. This corresponds to 55.7%, 19.5%, 9.0% of the APCC total export earnings. The livelihood of about 300 million people in coconut-producing countries depend on the coconut palm - be it cultivation, processing, trading or direct consumption of its products.

Coconut Preparation

Cracking

Dehusked coconuts are cracked or split into two halves along the 'equator' with a steel rod or heavy knife. This is done for making copra or domestic cooking to facilitate grating. Cracking is carried out manually, and there is no need for mechanization.

Shelling

Shelling is required for the manufacture of desiccated coconut, coconut cream or any wet process where large quantities of coconuts are to be milled into small particles. Husked coconuts are shelled (with a hatchet or a knife) to obtain a kernel in the shape of a ball without damage, to facilitate paring or peeling the brown skin (testa). This shelling operation is also labour intensive and somewhat dangerous to the operator, as he may injure his fingers. This work is usually undertaken by men. In Sri Lanka, a skilled man working 8-10 hours, shells 2000 nuts a day; in The Philippines the rate is 1500. This means that in a large desiccated coconut factory in The Philippines working up 300,000 nuts a day, a minimum of 200 men should be available for shelling alone. Coconut shelling is no problem in the Asian countries. In the Pacific, where this practice is not traditional, no skilled labour is available, resulting in low productivity and high labour costs.

Paring

Once the coconut is shelled, the brown skin (testa) is pared or peeled using a special knife. This process is also labour intensive and monotonous but not dangerous, and women are usually employed for this task. A skilled woman pares about 2000 nuts in a working day of 8-10 hours. Shelling and paring are undertaken by a pair of workers whose speeds match each other. After paring the nut to obtain a white ball of meat without brown specks, the nut is pierced to release its water and then transferred to the next stage in the process. Very hygienic conditions must be maintained during this operation, to prevent contamination of the kernel with micro-organisms such as salmonella.

Copra Manufacture and Storage

When making mill copra, the objective is to dry the kernel of the freshly opened nut from the 50% moisture level down to 6% as fast as is practically possible. The high moisture content and presence of protein and sugar makes the fresh kernel an ideal medium for the development of bacteria and fungi. It is therefore liable to deterioration and very susceptible to attack by micro-organisms, with the development of free fatty acids and rancidity, as well as the formation of aflatoxine. The care for the kernel therefore commences from the time the nut is opened. The lapse of time between opening the nut and commencement of drying should be as brief as is practically possible. Since deterioration depends upon the area of the kernel exposed, cracking or splitting the nuts into two halves is the best, though in some parts of the Pacific, kernels are 'finger cut' by twisting the kernel away from the shell with a metal implement (CP 78).

Sun Drying

Sun drying copra is the simplest and cheapest method available. It requires up to 5 consecutive days of sunshine and a moderately humid atmosphere (60 - 80% relative humidity) to facilitate evaporation of moisture. Excellent quality copra is made under such conditions which exist in certain parts of the world, such as Allepey in Kerala State in India, and Cebu in The Philippines. Unfortunately, most sun-dried copra is produced under unfavourable conditions, where intermittent rains or very humid air produce badly deteriorated copra.

Direct Heat Dryers using Husks and other Fuel

Where sun drying is not possible, artificial drying applying adequate techniques is the only way to produce good copra. There are various technologies for artificial drying, some producing poor quality and some producing excellent quality copra. The most common method involves the use of a dryer or kiln with direct heat and smoke generated from the combustion of fuel such as husks, shells and firewood which are available near the kiln at no cost. The kernels are placed on a platform 1.5 to 2 metres above the fire. Three to six fires are lit for about three days. The copra produced by this method is generally inferior due to discolouration from the smoke, and to inadequate drying (moisture levels of 8 - 15%). Such copra is liable to further deterioration during storage. With about 80% of the world's coconut land being smallholdings of less than 5 hectares, and most smallholders making their own copra, there is very limited scope for change in this system, particularly when farmers receive poor prices for their copra.

Direct Heat Dryers using only Shells for Fuel

In Sri Lanka, India, Malaysia and Papua New Guinea (large plantations), direct heat dryers using only coconut shells as a fuel, providing great heat and little smoke, produce excellent quality copra. These kilns are operated centrally, either by specialized copra manufacturers or by plantation owners with high levels of management. Drying takes about 4 or 5 days with about 6 or 7 fires, reducing the moisture content to the desired level of about 6%. In Sri Lanka, the opened nuts are sun dried on the first day for a few hours on a cemented barbeque (weather permitting) as this improves the copra colour. The Standard Sri Lanka copra kiln is illustrated in Figure 31.

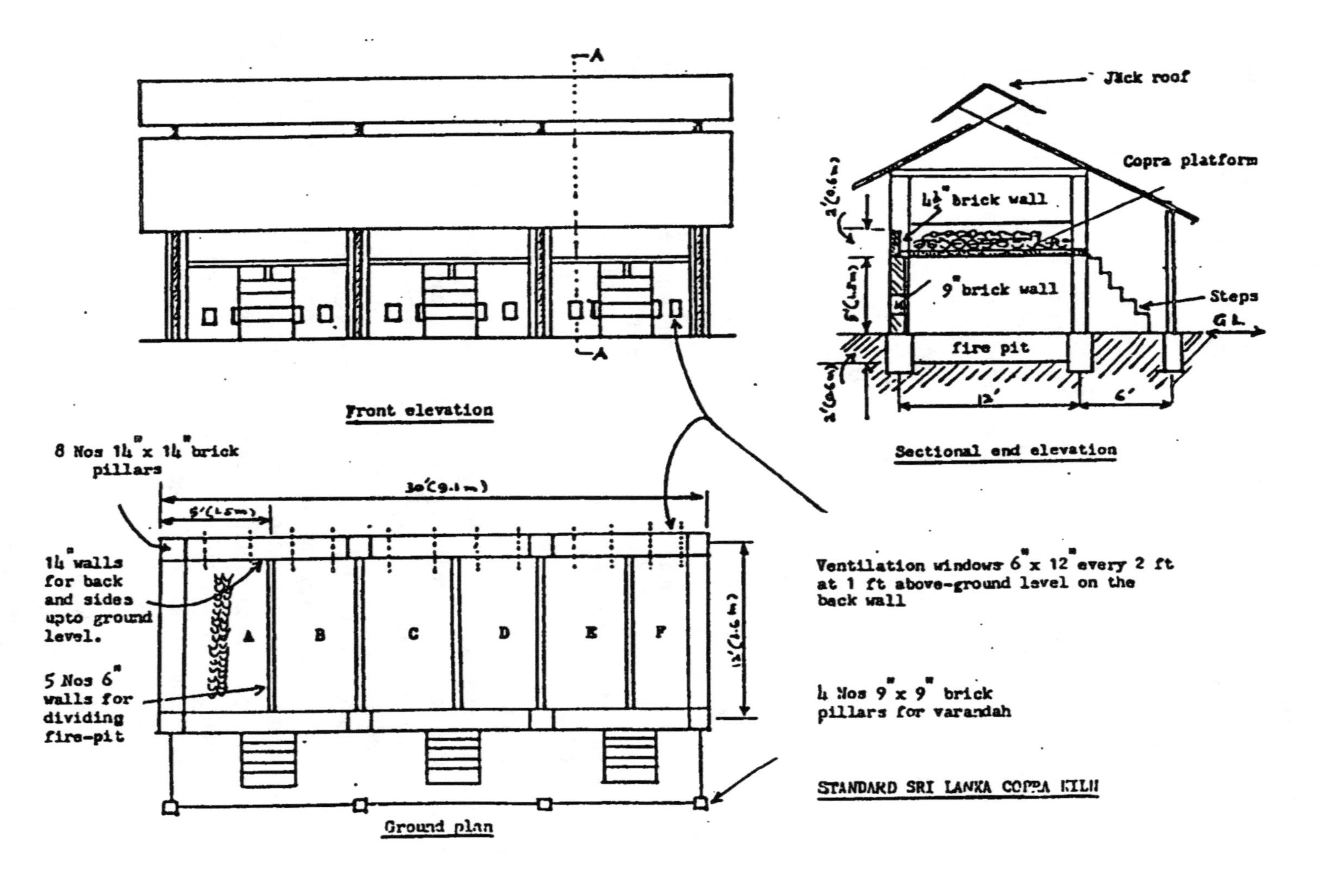

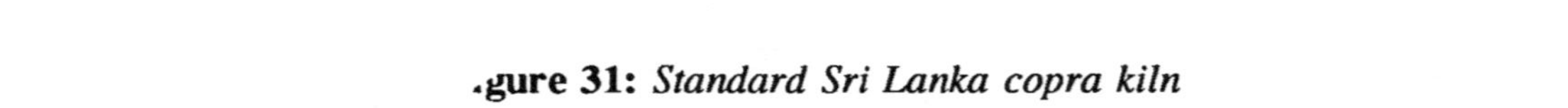
.gure 31: *Standard Sri Lanka copra kiln*

Indirect Heat or Hot Air Dryers

Copra is also manufactured using indirect heat, where the products of combustion exhaust without coming into contact with the copra. Therefore the quality of the combusted gases is not important; hence any fuel can be used. There are two basic types of indirect heat dryers, those where hot air moves up through the copra bed under natural draught, and those where hot air moves through the copra bed with a forced draught, using a motorized fan.

Indirect heat dryers, also known as hot-air dryers, produce good quality copra when properly operated. Whereas the ability to use any type of fuel is an advantage, there are two disadvantages:

- higher fuel cost compared to the direct type, as only about half the heat is transferred to the hot air;
- higher operating cost, as the steel heat exchanger has to be replaced from time to time.

There are various designs of smallholder hot air dryers working along the same principle. These can found in Western Samoa, Solomon Islands (Figure 31), The Philippines, Vanuatu, Marshall Islands and other countries. These dryers, including the 'Westec' estate copra dryer of Western Samoa, use natural draught. There are usually 2 or 3 firings over 2 days to obtain dry copra.

The forced-draught Chula dryer and the Pearson dryer of UK manufacture were used in large plantations in Sri Lanka during the British period, but there are none operating at present. They require electricity to operate the fan motor and the operating cost was high. The capacities were very large and the drying time was less than 24 hours on a continuous basis.

Ball Copra and Edible Copra

Whereas previous paragraphs dealt with mill copra which is milled for oil and cake, there are two other types of copra produced worth mentioning. In India, about 45,000 t of ball copra is manufactured annually by slow drying whole mature coconuts with occasional artificial drying. The kiln used is similar to the Sri Lanka copra kiln, but drying is carried out mainly by storage on the platform under complete shade for periods of 6 - 8 months. During the rainy season artificial drying is done by burning some paddy husk or other available fuel under the platform to accelerate drying. When the coconut is dried fully, the kernel shrinks and detaches from the shell and gives a rattling sound when shaken. At this stage, the nut is carefully husked and shelled to obtain copra in the form of a ball. This copra is consumed with sugar or made into chips for manufacturing sweets, etc. Ball copra is also used for religious and cultural ceremonies as well as for traditional medicines. In India, the demand for ball copra comes mainly from areas where coconut cannot be grown. In Sri Lanka, a limited quantity of 'edible white copra' is made for export at a very high premium. The manufacturing process is similar to the mill copra but the fuel used is coconut shell charcoal, which produces an even cleaner direct heat. Sometimes a small amount of sulphur is burnt to obtain an attractive white colour.

Copra Quality

Most copra-producing countries have quality specifications. General requirements (non-technical) for good quality mill copra stipulate white coloured cups, excluding wrinkled (immature), germinated, mouldy, charred (black) or broken cups. Technical specifications limit moisture content to 6% (sometimes up to 10%), minimum oil content of 68% on a dry basis, and a maximum free fatty acid content of 1% for the expelled oil. Only copra manufactured by the direct heat of coconut shells or indirect heat 'hot air' dryers under proper conditions could conform to these specifications. Generally, copra produced in Sri Lanka, India, Malaysia, Papua New Guinea and Pacific countries with hot air dryers conform to these specifications. In these countries grading practices exist, with a price premium for good quality. In The Philippines, Indonesia and other areas where copra is smoke dried, good white coloured copra cannot be produced, and the moisture content ranges between 8 and 15%, and the free fatty acid content of the expelled oil varies between 1 and 5%.

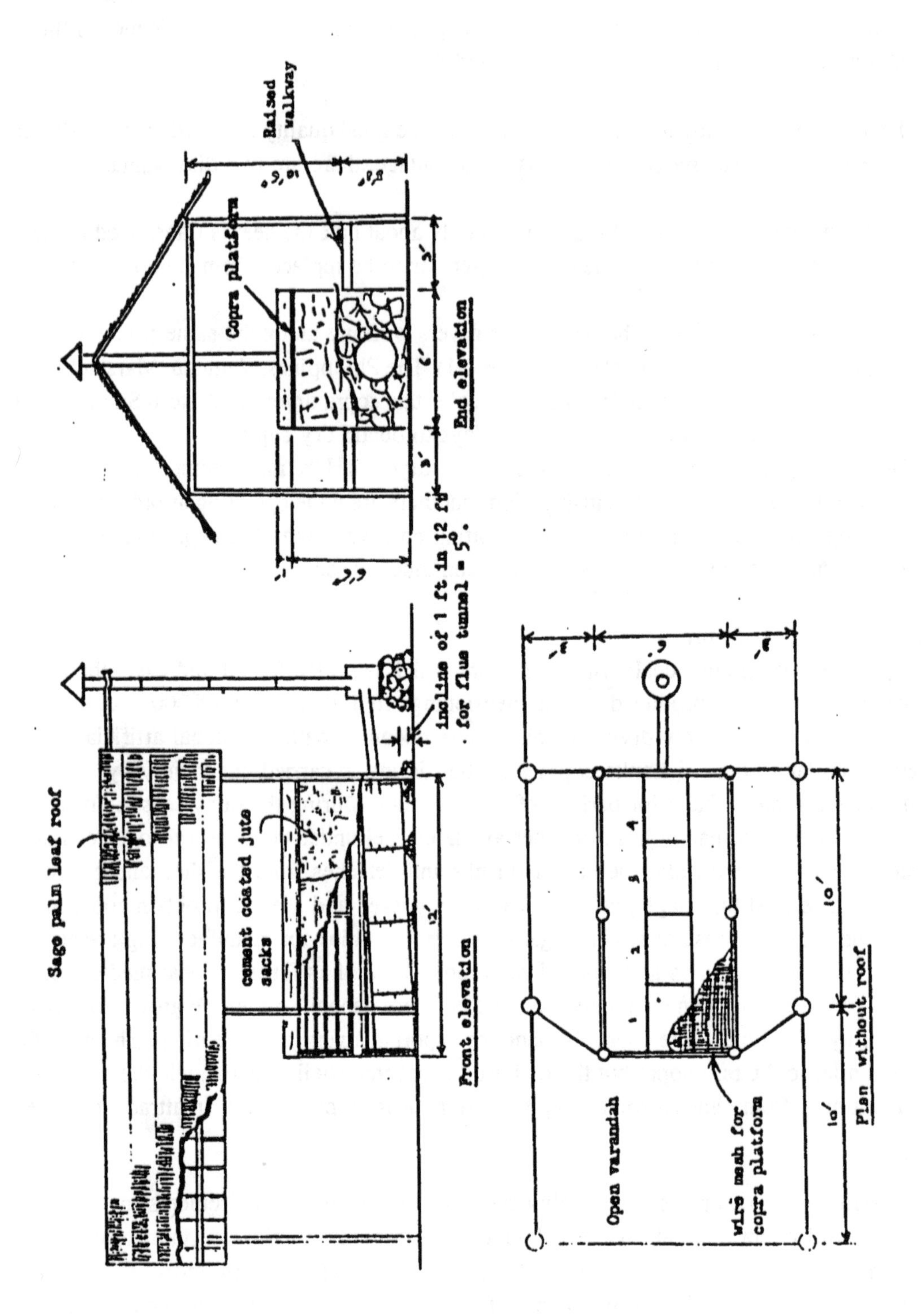

Figure 32: Kuleum dryer (Solomon Islands)

Copra Storage

The main objective of copra storage is to provide a buffer stock to make up for differences of receipts and issues to the mill. Technically, storage performs two important functions for oil extraction: to dry any excess moisture in the copra, and to equalize the moisture content in the entire copra stock prior to processing. During storage, copra must be protected from the elements and pests. Proper ventilation, and proper rotation of stocks when issues are made, should be practised.

Proper storage of copra after manufacture and during shipment is important, as even good copra will deteriorate if badly stored. Copra warehouses should have cemented floors elevated about 0.5 m above the outside ground level to minimize dampness during rainy weather; good ventilation at roof level to breathe out evaporated moisture; and an air gap between the floor and bagged copra with the use of pallets or wooden logs. The air gap provides ventilation for removing moisture 'sweating' on the floor during rainy weather. Good copra, when stored under poor conditions, and bad copra even under good storage conditions, deteriorate with consequent losses in both quality and quantity. Loss of quantity through drying of excess moisture is desirable, as the copra quality improves. Losses to be avoided are those due to decomposition and attack by fungi, bacteria, insect pests and rats. In case of insect pests, it is important to fumigate warehouses and returnable empty bags.

The Aflatoxin Problem

Poorly dried copra has occasionally been found to be contaminated with aflatoxins, which are a group of toxic chemicals produced by the *Aspergillus* mould, particularly yellow-green colour *A. flavus*, and *A. parasiticus*, and other penetrating mould. Although groundnuts and maize are most susceptible to aflatoxin contamination, copra, cottonseed and cassava are contaminated at lower levels. Aflatoxins found in these have been named B1, B2, and G1 and G2. Aflatoxin B1, which is the most abundant, is extremely poisonous and is the most powerful carcinogenic chemical known. It caused liver cancer in all test animals and is almost certainly one of the causes of cancer in humans. Animals vary in their susceptibility to the effects of aflatoxins, but the young and males are at greater risk. Aflatoxins cause death when present in high concentration. Investigations revealed that in 1880, 100,000 young turkeys died in the UK due to the presence of 10 ppm of aflatoxin in the feed. At lower levels, it causes stunted growth and poor feed efficiency.

Recent work undertaken in The Philippines by the Natural Resources Institute of the UK found that the safe moisture level for hot-air dried copra is below 8%, and for smoke dried copra it is below 11%. The higher level of moisture tolerable for smoke dried copra is due to smoke particles inhibiting mould growth, as in smoked meat, fish or rubber sheets. When contaminated copra is milled for oil, the aflatoxin passes into the oil. Edible coconut oil is usually chemically refined, which removes all contamination. However, the aflatoxin remaining in the copra cake has caused problems for its use as an ingredient in blending animal feeds. Copra cake is valued for its effect in enhancing butterfat content and increasing yields of milk in lactating cows. If contaminated cake is used for feeding lactating cows, the aflatoxin reappears in the milk as aflatoxin M1, which is also unacceptable.

The European Union introduced regulations on limits of aflatoxin levels in feed ingredients and feeding stuffs in 1976. These regulations were further tightened by the EC Commission directive of 13 February 1991, which for copra cake is 50 ppm. Prevention of aflatoxin contamination is best carried out by drying copra rapidly down to safe levels. Proper storage, handling, packing and transport of well-dried copra is equally important to prevent growth of mould spores during condensation of moisture, etc. Although experiments has been conducted in detoxification, none have been commercially acceptable to date.

Coconut Oil

Uses of Coconut Oil

Coconut oil is obtained by extracting the oil contained in copra. The residue is generally called copra cake, sometimes it is called copra meal. Coconut oil is used for food and for industrial purposes. Edible oil, by international standards should have less than 0.1% free fatty acid content. The oil obtained from good quality copra in Sri Lanka and India, with a free fatty acid content of < 0.1%, is used as cooking oil without any further processing. Even if the free fatty acid content is 3-5%, the international standard of < 0.1% can be obtained by refining and deodorizing. Edible oils are either used as cooking oil (frying oil) or processed further into filled milk, table margarine and baker's margarine.

In industry, coconut oil is used for manufacturing toilet and laundry soap. In its original or modified form, it is used as a vehicle in the paint and varnish industry. Coconut oil is also processed into methyl esters, fatty acids and fatty alcohols. These intermediate products are raw materials for manufacturing detergents, surfactants, emulsifiers, plasticizers and various other organic products which are bio-degradable. Copra cake is used in blending animal feed.

Oil Extraction Technology - Dry Processing

Oil extraction from copra is carried out by two methods: mechanically by pressing; or by use of solvents. Mechanical and solvent extraction uses standard technology that has been developed in the vegetable oil industry. For oil seeds such as copra which have a high oil content, mechanical extraction is efficient and economical. For oil seeds with low oil content or for further oil extraction from copra cake, the solvent extraction method is more suitable. The residual material from solvent extraction is called copra meal. Oil extraction involves five basic steps shown in the 'Simplified general flow diagram' (Figure 33):

- copra storage;
- preparation of copra;
- oil extraction - full press, prepress - solvent, and full solvent methods;
- processing of extracted oil;
- processing of cake or meal;
- storage of products.

Copra Storage

Copra preparation

The objective of copra preparation is to cut it down to the right size, dry it to the required moisture level, heat it to the right temperature, keep it at the right temperature for a sufficient period of time, and form it to the right shape before extraction. Grinding copra opens the oil cells to expose the oil for extraction. Moisture content affects the efficiency of extraction. Increasing the temperature reduces the viscosity of the oil for easier flow, while keeping it at a high temperature for a period of time coagulates the proteins to reduce resistance to oil flow through the material during extraction. For solvent extraction, the material should have maximum contact area with the solvent, the particles should be large enough to avoid erosion of the material or clogging the extractor basket perforations. The shape of the particles should facilitate good percolation of the solvent through the bed of material.

The equipment for size reduction and particle formation are: hammermills, peg mills, disc mills, rollers and flakers. The mills break the material into the desired sizes, rollers break and compress large particles left after crushing and grinding, whereas the flakers compress the material into thin and firm flakes suitable for solvent extraction. The size reduction equipment should not generate too much heat, which may cause an excessive temperature rise the material, resulting in darkening of the oil. The equipment used for drying and preheating include dryers, cookers and conditioners. The design of this equipment comprises two general types: the multi-decked kettle type with steam jacketed pans and revolving stirrer-scrapers; and the horizontal drum type with steam jacketed wall and rotating stirrer-conveyor screws. They are equipped with vapour ducts and exhaust fans.

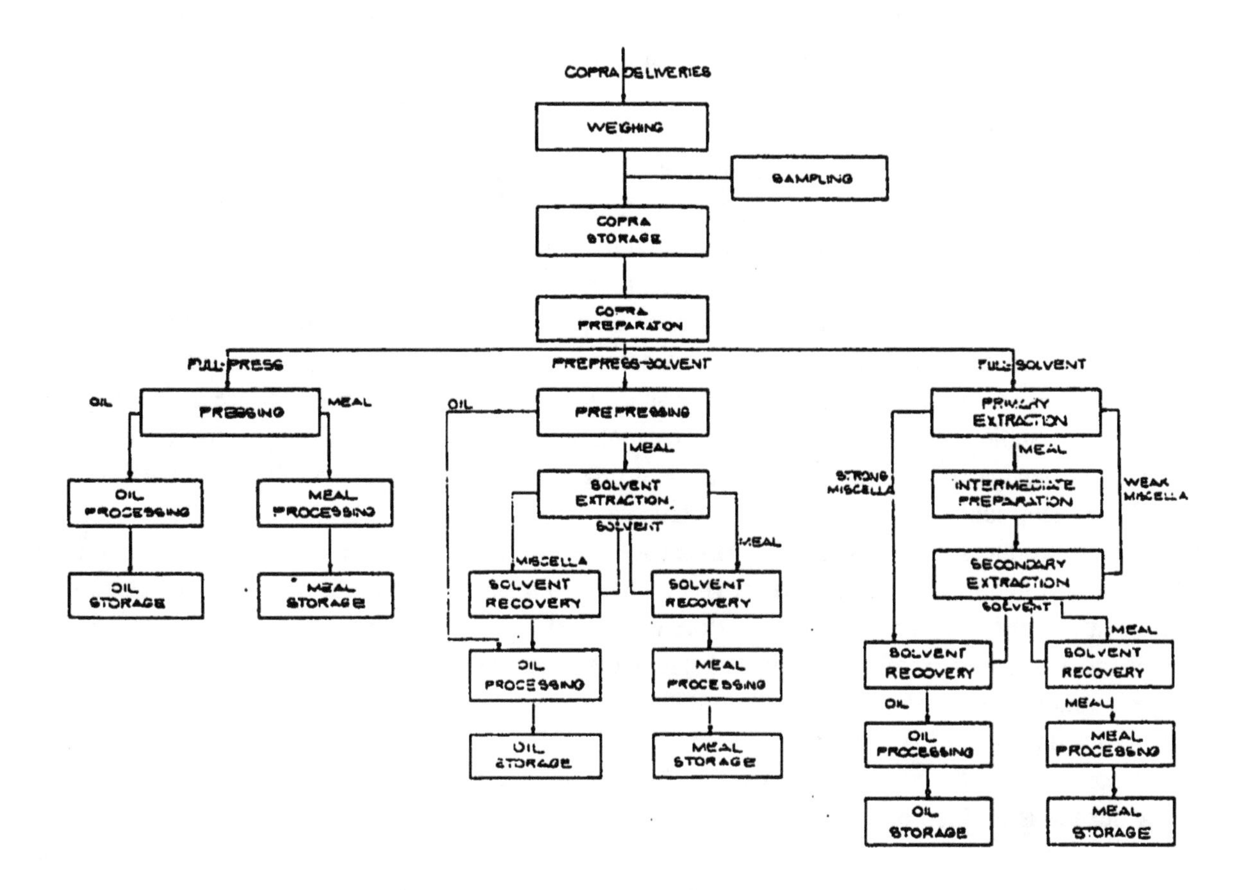

Figure 33: *Simplified general flow diagram of copra oil extraction*

Oil Extraction

Full press method

In the full press method of oil extraction, maximum pressure is applied to the material and sufficient time is allowed for the oil to escape. The temperature of the material should not be allowed to rise to a level where the oil darkens due to overheating. The pressure is applied to the material between the screws and the cage of slitted steel bars. Pressing may be done in a single or in double stages. By adjusting the clearance in the choking device, the thickness of the cake can be controlled. When the clearance is reduced to decrease the thickness of the cake, the pressure applied is increased. To prevent overheating the oil, the shaft is usually hollow for water cooling, and the oil is sprinkled over the cage bars.

Prepress-solvent method

In the prepress-solvent process, the oil is partially extracted by preliminary low pressure mechanical extraction, and then subjected to solvent extraction to remove most of the residual oil. The oil content of the initial copra material is around 70%; after the mechanical prepress extraction, the residual oil content in the prepressed cake is 16 - 20% for optimum operation. After full press extraction, the oil content in the residual copra cake is 6 - 10%, depending upon efficiency. The equipment used for prepressing is similar to expellers used for full pressing, but adjusted for less pressure and therefore it has a higher throughput.

Full solvent method

In solvent extraction, oil in the material is leached with a solvent (where the oil dissolves in the solvent) whereas the insoluble meal is retained unaffected. The extent of extraction depends upon the kind of solvent, the temperature of solvent, the ratio of solvent to meal, the number of extraction stages, the shape of particles, the porosity of the material, and the contact duration. The most common solvent used is hexane because of its price, low toxicity, suitable boiling point for recovery and handling, and its availability. Solubility of oil in hexane increases with temperature.

The resulting streams from solvent extraction are: the micella or oil, solvent solution, and the extracted meal which comprises the meal, some solvent and a little residual oil. The solvent in the meal is removed by heating to boil off the volatile solvent, and the solvent is recovered by condensation. The solvent from the micella is removed and recovered by evaporation and condensation. Traces of solvent left in the meal and the oil are removed by steam-stripping under reduced pressure.

The equipment for solvent extraction consists of two general types; the roto-cell type with cells revolving around a vertical axis, and the basket type with baskets travelling horizontally while the solvent is sprayed over the material in counter-current flow. In the full solvent process, the prepared copra is first subjected to a first extraction (percolation), using the weak micella from the second extractor as starting solvent, and producing the strong micella for oil recovery. The extracted meal is then flaked and subjected to a second extraction (immersion) which uses fresh solvent as starting solvent and produces the weak micella solvent for the first extractor. The solvent with the extracted oil is removed in a Desolventizer-Toaster (DT) which is either of the multidecked vertical design with steam heated pans and paddle scrapers, or the horizontal barrel type with rotating conveying paddles attached to a horizontal shaft, and steamjacketed walls. These are equipped with condensers for solvent recovery and scrubbers to remove dust entrained in the vapours. In some designs, the heat with the vapours leaving the DT is used to pre-concentrate the micella prior to evaporation of the micella.

The solvent with the micella is recovered, first by an evaporator, usually of the falling film type, where hexane is distilled off by indirect steam heating, and subsequently by a stripping column where traces of hexane in the oil are stripped by steam under vacuum. The water-hexane vapours from the stripper are condensed and collected in a water-hexane separator where hexane is separated by gravity from the water. It is decanted and reused in the extraction. Vent vapours of hexane from the extractors and condensers are recovered in a vent recovery system where hexane is absorbed by mineral oil. Hexane is recovered from the mineral oil by stripping, and recycled to the extractor.

Processing extracted oil

Processing the extracted oil is the next basic step in the oil extraction technology. Oil from the expellers contain substantial quantities of solids (foots) that should be removed before the oil is pumped into the storage tanks. The oil is cleaned first by settling and screening and subsequently by filtration. The screening equipment is a rectangular steel tank equipped with a continuous drag chain conveyor with scraper blades, which scoop the settled solids and lift them over a fine screen for drainage at one end of the screening tank, after which they are conveyed back to the expellers to be mixed with the copra. The filtering equipment generally consists of a plate and frame filter press with canvass filtering media. Some factories use leaf filters with perforated steel filtering leaves. The foots or filter cake from the filters are recycled to the expellers for oil extraction. The oil from solvent extraction is free of solids. It leaves the stripping column at 120°C and is cooled and pumped into the storage tanks through an oil meter.

Processing cake or meal

Copra cake leaving the expellers has a temperature of about 110 °C and is cooled in a cake cooler. The cake cascades down the cooler baffles and is cooled by a cross-flow of cooled air from blowers. After cooling, it is ground to fine particles by hammermills or disc mills. The ground cake is bagged for local use or pelletized for export. Pelletizing requires additional equipment which small plants cannot afford. However, cake or meal for export has to be pelletized for safer and easier handling and conveying. Prior to pelletizing, the cake has to be moistened to about 12% moisture and then fed to the pellet mill. Moistening in this manner improves pelleting property, and is required for export. The oil content of copra cake should be around 6 to 10% and moisture not higher than 12%.

Storage of products

Oil is stored in the open in vertical cylindrical steel tanks. These tanks have conical covers and manholes at the top and bottom for cleaning. The capacity of these tanks depends upon the capacity of the plant, frequency and volumes of withdrawals, and marketing factors. The unpelletized copra cake should be bagged in woven sacks and stacked about 10 high on wooden pallets. It is hazardous to store unpelletized cake in bulk for long periods. The piles of unbagged meal should be kept small and turned over as frequently as necessary to prevent spontaneous combustion.

Coconut oil quality

Typical quality specifications of coconut oil milled from good quality copra from Sri Lanka, Papua New Guinea and hot-air dried copra from the Pacific would have 1.0% maximum free fatty acid, 0.2% maximum moisture, and colour not deeper than 5. Typical free fatty acid value of oil extracted from copra from The Philippines and Indonesia will be about 3% and this is suitable for use as industrial-grade oil. The moisture level of stored coconut oil must be within reasonable limits as moisture acts as a catalyst to split oil into fatty acid and glycerine.

Typical operating data

Typical operating data for a small scale full press oil mill are as follows:

Moisture content of copra	6 to 10%
Oil content of copra	60 to 68%
Size of copra leaving crusher	3 to 6 mm
Moisture content of copra leaving cooker	3 to 4%
Oil content of cake after 1st expeller	18 to 22%
Oil content of cake after 2nd expeller	8 to 10%
Oil yield from copra	55 to 60%
Cake yield from copra	30 to 40%
Power consumption per t copra	80 to 100 kWh
Steam consumption per t copra	200 to 300 kg

Typical operating data for a large-scale full press oil mill are as follows:

Moisture in copra received	6 to 16%
Oil in copra received	55 to 65%
Moisture in copra after storage	6 to 8%
Oil in copra after storage	60 to 65%
Size of copra leaving crushers	6 to 12 mm
Size of copra leaving hammer mill	2 to 3 mm
Moisture in copra leaving dryer	4 to 5%
Cooker temperature	110 °C
Conditioner temperature	115 °C
Moisture in copra entering expeller	2 to 3%
Residual oil content in cake	6 to 8%
Moisture content in cake	1 to 2%
Oil yield from copra	60 to 63%
Cake yield from copra	37 to 40%
Power consumption per t copra	120 to 140 kWh
Steam consumption per t copra (6 atm)	100 to 150 kg
Water consumption per t copra	3 m^3

Typical operating data for a Prepress-solvent extraction plant (medium to large-scale) are as follows (Conditions until entering expeller same as for Full Press):

Residual oil in cake	16 to 20%
Moisture in cake	2 to 4%
Temp of cake entering cooler	90 to 100 °C
Temp of cake leaving cooler	60 to 65 °C
Size of cake entering cake breaker	5 to 25 mm
Size of cake leaving cake breaker	2 to 5 mm
Extraction temperature	50 to 55 °C
Number of stages	6 to 10
Oil in extracted cake	0.5 to 2%
Oil content in micella	10 to 15%
Oil recovery from copra	61 to 65%
Meal recovery from copra	32 to 36%
Power consumption per t copra	70 to 90 kWh
Steam consumption per t copra	450 to 550 kg
Water consumption per t copra	10 to 15 m^3
Solvent make up/prepress cake feed	0.5 to 1%

Typical operating data for a Full Solvent process (medium to large-scale) are as follows:

Residual oil in cake	0.5 to 2.0%
Moisture in oil (max)	0.2%
Oil recovery from copra	61 to 65%
Meal recovery from copra	32 to 36%
Power consumption per t copra	45 to 50 kWh
Steam consumption per t copra	600 to 650 kg
Water consumption per t copra	18 to 24 m^3
Solvent make up/copra fed	0.5 to 1.5%

Copra cake and meal

Copra cake or meal is the residue left after oil extraction from copra. With a global production of around 4,788,000 t of copra, annual copra cake production is around 1,700,000 t. Copra cake is used as an ingredient in blending animal feed, due to the presence of oil, protein and carbohydrates. Copra cake from poor quality copra has a limited market due to its degraded condition and the presence of aflatoxin.

Table 103: *Composition of copra expeller cake*

Components	percentage
Moisture	11.4
Oil	8.0
Protein	21.2
Carbohydrates	42.4
Fibre	11.5
Ash	5.5

Processing technology for copra cake has been dealt with in the paragraph on oil extraction, as copra cake is a by-product of oil extraction. Copra cake from expellers has a residual oil content of 6 - 10%, depending on the efficiency of these expellers. The moisture content is 1 - 2%. For copra meal, the oil content is 0.5 - 2% depending on the efficiency of solvent extraction. Due to absorption, the moisture level rises during storage. Expeller copra cake after moisture absorption is an excellent low-cost ingredient for blending animal feed, as is shown in the analysis presented in Table 103 (Crawford 1940):

Oil Extraction Technology - Wet Processing
Background
Processes which start with fresh coconuts (wet kernels) to obtain high quality oil and edible by-products are termed 'wet processes'. Traditionally, for centuries (and currently still in remote coconut producing areas), coconut farmers produced oil by gently heating coconut milk (or cream) obtained by squeezing grated fresh kernel. (The residue after extracting the oil is fed to animals). This is the traditional 'wet process', as the starting material is the wet kernel. Mention should be made of another traditional 'wet' process in Indonesia, where fresh kernel is grated and heated or fried in a pan for about an hour to recover the oil by straining. Further oil is recovered by pressing the residue left after straining. This oil is known as 'klentic oil'.

Since the 1920s, several technically feasible industrial processes have been developed (and some patented), for wet processing coconut to obtain high quality oil and edible quality coconut meal. Sometimes the meal is further separated into by-products such as proteins and flour for human consumption. With the adverse effects of malnutrition from protein deficiency, and the worsening food situation in developing countries, commercial exploitation of coconut for protein and other by-products for human consumption would be highly significant.

Early developments in wet processing
Early developments in wet processing were very innovative and paved the way for much of the subsequent work. However, these processes were either patented, or without detailed information on the composition or yields of the products. Much of the information on wet processing presented here was obtained from *Coconut Palm Products* (FAO 1975), and *Coconut Processing Technology Information Documents* (UNIDO/APCC 1980).

Alexander (1921), patented two innovations, viz.; centrifuging the milk to separate cream and water phases, and autoclaving the cream phase to destabilize the emulsion to separate the oil by centrifuging again. Lava (1937), patented the process of using acetic acid, ethyl alcohol and acid- producing micro-organisms to break down the milk emulsion. Gonzaga (1948), patented the process where coconut milk is heated to 100°C in a settling tank, to isolate the protein by heat coagulation. Cruz and Bernado (1949), patented a process for the extraction of milk, and separation of oil. Hiller (1952), (FAO/WHO-/UNICEF, 1963), patented a process by which husked coconuts are cooked to facilitate removal of kernels from the shell. These kernels are sliced and dried under vacuum at 65°C and the oil is extracted by using a screw expeller. Diakno (1960), patented a process in which comminuted coconut meat was centrifuged to obtain three fractions, viz.; oil, protein rich solids, and skim milk.

The azeotropic extraction process (FAO/WHO/UNICEF, 1963) uses a water-solvent mixture which boils at 120°C (hydrocarbons of hexane and benzene and a chlorinated solvent such as trichloroethylene). After extractive distillation, it is filtered and the filtrate distilled to recover oil. The meal is ground, solvent-washed and dried to obtain flour.

There are two other processes that should be included. Although they start with fresh coconuts, they are similar to the dry (copra) process. The end products are high quality oil and edible flour. The first refers to the Yenko Process (1954), which is patented. In this process, husked coconuts are cracked open, the kernels are scooped out, comminuted, and heated in a chamber at 80 to 100°C, until the moisture content is reduced to 1 to 8%. The granules are pressed to obtain oil and meal. The meal is ground into a flour. The second covers the 'Integrated Dry Process' by NIST (National Institute of Science and Technology, Philippines) which is also patented (UNIDO/APCC 1980). In this process, husked coconuts are cracked, shelled, and subsequently ground in a colloid mill. (In the integrated process, husks and shells for fibre and charcoal are processed separately). The milled kernel is dried in a fluidized bed dryer and the oil expelled. The meal is ground into a flour. Additional contributions were made by other workers - Cruz and West (1930), Leopold (1946), Floro (1952), Balce (1953), Birusel (1963), and Protein Compagne (1965). However, no technical information about these methods is available.

Major wet processes
The following can be termed major wet processes, and many of these processes have been operated at pilot plant level (where indicated).

The Chayen (impulse-rendering) process. (Pilot plant)
This was originally developed by British Glues and Chemicals Co Ltd (now Croda International Ltd). Impulse-rendering refers to mechanical rupture of membranes of the fat-containing cells by a series of high speed impulses transmitted through the medium of a liquid, whereby the fat in the cells is liberated and is removed by the liquid in violent movement. The impulse-renderer is in fact a flooded hammer mill. Fresh coconut kernels are fed into a hammer mill with ten times their weight of 0.15% sodium hydroxide solution. The fibre fraction is then removed on a vibrating screen, and the emulsion which passes through the screen is centrifuged to separate the free oil from the lipid-protein complex. The typical recoveries are 80% of the oil and 70% of the protein.

The Robledano-Luzuriage process (patented)
This process was developed in The Philippines following various previous attempts made over many years to obtain high quality oil and meal by processing wet kernels (Cruz and West 1930; Leopold, 1946; Floro, 1952; Balce, 1953). Fresh coconut kernel is comminuted and pressed to obtain approximately equal amounts of emulsion and residue (Nathanael 1964). The residue is pressed again to obtain more emulsion and residue. The emulsion is centrifuged to obtain cream, skim milk, and some solids (protein). The cream is subjected to enzymatic action under closely controlled temperature and pH conditions. After a freeze-thaw operation, the cream is centrifuged again to obtain oil. The protein in the skim milk is coagulated by heating, subsequently filtered and dried to produce a protein concentrate. Data on oil and protein recovery are not available.

The Sugarman process (patented)
This system was developed by Georgia Technological Research Institute, Atlanta, USA, and is applicable to all oil seeds (Sugarman 1956). The fresh kernel is reduced in size to the consistency of peanut butter in a grinding or flaking mill. Subsequently, it is transferred to a pebble mill, mixed with twice its weight of dilute alkali and milled for three hours. This is then transferred to a heated mixing tank, more water is added and subsequently agitated for one hour. The slurry is centrifuged to obtain cream, skim milk and meal. The emulsion of the concentrated cream is broken down by pH adjustment, followed by colloid milling and then centrifuged to recover the oil. The skim milk is treated with acid to precipitate the protein, which is filtered, washed and dried. Information on oil and protein recovery is not available.

Krauss-Maffei/CFTRI process (pilot plant)
This process was originally developed by Krauss-Maffei AG, of Germany. This plant was given to India and installed at the Central Food Technological Research Institute (CFTRI), Mysore. CFTRI modified the process to maximise the oil yield. In the original process (patent no 1,031,912 - 1958), husked coconuts are autoclaved at 3 kg per cm^3 for 10 minutes, shelled, and the meat first sent through a cutter, and subsequently through a roller mill. Then it is pressed in a hydraulic press, and the emulsion is centrifuged to obtain cream and skim milk. The cream is heated to 92°C, and filtered to obtain high quality oil. The skim milk is heated to 98°C in a flow heater to coagulate protein, which is separated by centrifuging, and then dried. The left-over whey is concentrated under vacuum

to obtain honey. The residue from the hydraulic press is dried and ground to obtain edible coconut flour. No data are available on the oil and protein yields.

Rajasekharan (1964) working at the Krauss-Maffei pilot plant, conducted an over-all study of wet processing. At CFTRI it was observed that autoclaving caused coagulation of protein within the kernel, making extraction more difficult (FAO 1968; FAO/WHO/UNICEF 1963). In the modified process, husked coconuts are shelled without autoclaving, and then milled. The milled meat is first squeezed in a dewaterer under reduced pressure (2:1) and subsequently fed into the K M press (12:1). The cream and skim milk are treated as in the original process to recover oil, protein and honey. The residue, after extraction of milk, is dried and fed into an expeller to extract the residual oil. In this modified process, the direct oil recovery was 90% and with the residual oil, 97%. The protein recovery was 85%.

Roxas process (patented)

In this process, shredded kernel is pasteurized before passing through the expeller. According to the patent, the purpose of heating the shredded kernel is twofold: to destroy bacteria, and to coagulate protein. The emulsion is subjected to a freeze-thaw operation before being centrifuged. The oil and protein are recovered as in the case of the Robledano-Luzuriage process. No data are available on oil and protein recovery.

Modified Solvol process (pilot plant)

In the early 1970s, the Chemical Construction Company Private Ltd, developed this process and installed a pilot plant at Faridabad, near New Delhi, India. At that time, a commercial plant was planned by PHILINDOIL in The Philippines in collaboration with the Indian company, but no information is available on the outcome of the project. In this process, husked nuts are cracked into two halves, and the water is collected after filtration for use in subsequent operations. The kernel is scooped out, the testa pared and kept in a container for separate oil extraction. The pared kernel is sliced and shredded into a pulpy mass and then centrifuged to separate milk from the meal. The meal is mixed with coconut water and pressed in a continuous screw press to recover more milk. This operation is repeated two more times to obtain all the milk from the kernel. After filtration, the milk is dried in a vacuum dryer. The resulting oil-protein-sugar mixture is centrifuged to separate the oil. Traces of oil in the protein-sugar mixture are removed by solvent extraction. The oil present in the meal and parings is also solvent-extracted. The extracted oil from the meal and parings is thoroughly desolventized, and added to the oil recovered directly from the milk. The finished products, viz.; coconut oil, coconut protein-sugar mixture and coconut flour are edible, and ready for packing and marketing. Oil recovery is 100%; protein yields are not stated.

Wet process by Tropical Products Institute (pilot plant)

In the early 1970s the Tropical Products Institute (TPI) of UK, currently known as the Natural Resources Institute, developed this wet process for coconuts. The coconuts are husked, shelled and pared, the kernel sliced and minced in a wedgemill with a 5 mm die plate. The minced mass is mixed with half its weight of water and passed through a second wedgemill having a 1 mm die plate. The resulting slurry is passed through a vibrating sieve to separate the cellulose material from the emulsion. The emulsion is acidified with acetic acid to pH 3.8 to cause rapid creaming, and this is allowed to stand for 6 hours. The aqueous layer is discarded by gravity flow or siphoning. The cream is centrifuged to obtain a high-grade oil and a slurry containing a high proportion of protein. The slurry is washed with isopropanol to give a buff-coloured powder containing 80% protein. The oil yield is 80-85% (FFA 0.1-0.2%, as lauric).

Coconut aqueous processing (pilot plant)

This system was developed by Hagenmaier (1977, 1980) in the 1970s at the University of San Carlos, Cebu, Philippines, where another Krauss-Maffei pilot plant was installed. The research accomplished was the most extensive work undertaken so far. In this process, fresh husked mature coconuts are shelled, the brown skin (testa) removed if desired, and the water discarded. The shells are separately converted into charcoal, except for about 20% used as boiler fuel. If the testa has been pared, these parings are dried separately and pressed to obtain parings oil and parings press cake. The kernel (pared or unpared) is milled to reduce particle size, mixed with hot water and the milk is extracted by passing this mass through two screw expellers in two stages. The second expeller is fed with the residue of the first after mixing it with hot water at 75°C, and the diluted hot milk expelled from the second expeller is mixed with the freshly milled kernel and fed into the first expeller. The residue from the second expeller is dried and passed through a separate expeller to obtain oil and residue press cake. The coconut milk from the first expeller is fed through a vibrating screen (150 mesh), from which solids collected are fed periodically into the second milk expeller. The screened milk is heated to 60°C while stirring and then centrifuged in a cream separator to obtain coconut cream, skim milk and a small amount of insoluble solids which accumulates as a sludge. The cream is diluted with coconut oil and the mixture heated, agitated and centrifuged to recover its oil. The oil is then extracted with water, and dried to produce a high quality oil called 'Natural oil'. The coconut skim milk and the insoluble solids are combined, the water removed in two steps (vacuum evaporation and spray-drying) to obtain a high-protein product called 'Cocopro 1' or 'Cocopro 3', depending on whether the kernel was pared or not at the beginning. As an alternative, if the skim milk and the insoluble solids are not combined, the skim milk is concentrated by vacuum drying to produce 'Cocopro syrup'; and the insoluble solids are spray-dried to obtain 'Cocotein'.

Natural oil is clear, almost colourless, has a pronounced coconut odour and a FFA content of 0.05%. The spray-dried Cocopro contains about 33% protein of good nutritive value. The Cocopro syrup has 28% moisture, and on a dry basis it has 28% protein. The Cocotein, which is a minor product, has about 50% protein. The residue press cake contains little protein and has a high crude-fibre content. Yields of end-products based on weight of kernel used (dry basis) are presented in Table 104.

Table 104: *End products of aqueous processing*

Component	Percentage
Natural oil	54.4
Expelled oil	8.5
Residue press cake	19.3
Cocopro 3	16.7
Total	98.9

Table 105: *Composition of spray-dried coconut milk powder*

Component	Percentage
Protein (N x 6.25)	14
Fat	62
Moisture	2.0
Ash	1.6

The total recovery of oil is 62.9% of the dry weight, and this is higher than what has been achieved by previous workers, and establishes that wet processing can be efficient. The above information has been based on *Coconut Aqueous Processing* by Hagenmaier (1977).

As an alternative, the extracted coconut milk can be processed further and marketed either as canned coconut milk or as spray-dried coconut milk powder without obtaining oil (Hagenmaier 1980). Spray-drying coconut milk is difficult due to its high oil content, viz. 79% dry basis. During preliminary trials, it became evident that inert additives have to be used not only to encapsulate oil globules, but also to make the dried product less deliquescent. The composition of spray-dried coconut milk with 8% commercial casein and 10% maize syrup solids is presented in Table 105.

High value coconut products by Alfa Laval

According to Leufstedt (1990), Alfa Laval successfully developed a plant and a technology for manufacturing and packaging high-value coconut products, which could lead to diversification of the coconut industry. With their dairy-standard hygienic plant and aseptic packaging equipment, it was possible to make full utilization of the edible parts of the coconut. In open air, the edible parts would normally degrade fast due to their low acid nature.

The process flow chart is given in Figure 34. Process parameters such as temperatures and storage time are controlled throughout the plant to prevent microbiological growth. All continuous process lines are designed to handle pasteurization, sterilization, production and cleaning-in-place (CIP). The plant consists of individual process lines according to the following descriptions:

- pre-treatment; fresh husked mature coconuts are shelled, pared, washed and cut to release the water, which is collected and pasteurized. The pared and cut kernel is prepared for milk extraction;
- extraction; after disintegrating the kernel, the milk is extracted by pressing. The collection of milk is an integral part of the pasteurization line. The fibre residue is then dried in a convection drier for low-fat desiccated coconut;
- pasteurization of coconut water the water is clarified and the small oil fraction removed. Pasteurization and chilling down to 5-8°C are included in the process line. The chilled coconut water has a shelf life of up to two days;
- pasteurization of coconut milk; the coconut milk is collected directly after extraction and continuously filtered, pasteurized and chilled to 5-8 °C. The chilled coconut milk has a shelf life of up to two days;
- virgin oil process line; coconut milk is converted into virgin oil (called natural oil in the aqueous process by Hagenmaier) by a centrifugal process involving purification, drying and polishing. Chilling of the protein-rich water phase and skim milk is included;
- sterilization; the low acid products, viz. coconut milk, water, and skim milk are formulated as desired, and subsequently sterilized at 140°C for 4 seconds, which is known as the Ultra High Temperature (UHT) process;
- aseptic packaging; the sterilized UHT products are subsequently packed aseptically at ambient temperature in either Bag-in-Box for bulk distribution (10-1300 litres) or in paper cartons (200 - 1000 ml) for the consumer market, like 'Tetra-Brik'. The aseptic packaging machines have to be designed for low acid products for long shelf-life at ambient temperature.

The single process lines can also be used for upgrading existing coconut plants. By integrating some of the above-mentioned process lines in a desiccated coconut plant for example, a wider range of products and better utilization of the raw material could be achieved. Instead of producing liquid milk, spray-dried coconut milk powder can be produced. Due to the heat-sensitive product and the high fat content, a fluidized type of spray drier is most suitable. Some information on the end products is presented below:

- virgin oil; has retained coconut flavour, low FFA value (less than 0.07% as lauric) without refining, maximum natural vitamin E (tocopherol) content, free from aflatoxin contamination, and no added chemicals;
- aseptic coconut cream; a white, smooth, liquid cream (easily poured), with excellent coconut flavour and 20-30% fat, aseptically packed;

- aseptic coconut skim milk; a white, protein-rich drink with good coconut flavour. For a drink, a protein content of 1-3% and a fat content of 0.5-3% is ideal. It is easy to flavour;
- aseptic coconut water; an almost clear drink from pure coconut water. To enhance its sweetness, some sugar can be added to reach a total Brix of 7-7.5. The pH of about 6 in the fresh coconut water is not altered. Flavouring is possible;
- aseptic ORS solution; a ready-to-drink product based on coconut water as an oral rehydration solution.

Fairchild wet process

This process was developed by E. Bradley Fairchild, and is covered by US patent no 3,587,696 issued on June 28, 1971. In this process, mature fresh coconuts are sorted, husked, and sorted again. The nut is shelled, pared, and opened to separate the water and the kernel. Up to this point, the same method is applied as for processing desiccated coconut or coconut cream. Nuts rejected before and after husking, and rejected kernels after opening the nuts, totalling about 25% of the intake of coconuts, are subjected to the traditional copra-making process. The husks are used for making fibre and the shells for making charcoal, or sold for this purpose.

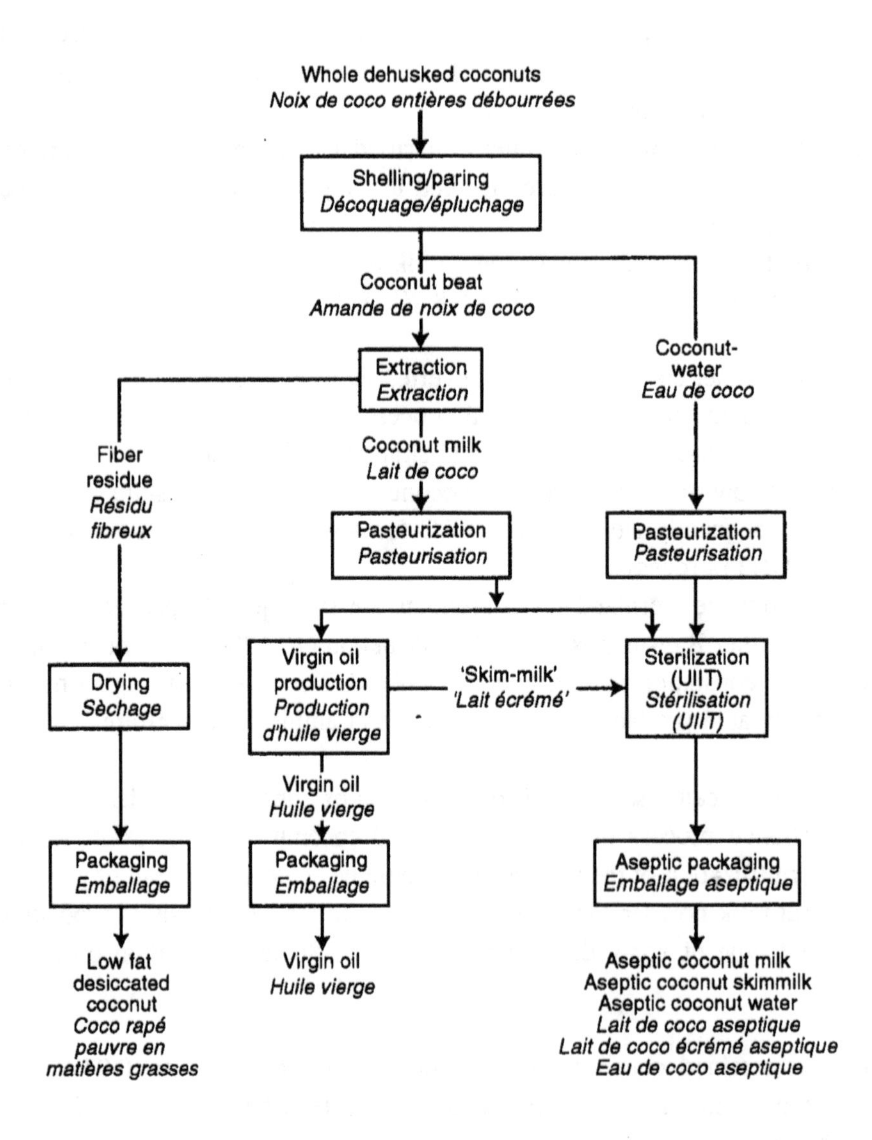

Figure 34: *High Value Coconut Products - Alfa Laval Process Flow Chart*

The kernel is converted into a slurry under aseptic conditions. To this purpose, distilled water from a subsequent operation is recycled by adding it to the kernels to which some water from an external source is also added. The products obtained are: pure bland coconut oil, aromatic essential oil, white bland coconut meal, testa fibrous meal, and coconut water. The coconut water is distilled to obtain a crystallized powder for soft drinks, and the distilled water is recycled as mentioned above. Information on the composition of the end products or yields is not available. The processing plants have two nominal capacity levels: 30 and 75 kg coconut kernels per minute. The 30 kg per minute capacity corresponds to an annual intake of 44 million coconuts, of which about 33 million (75%) would be processed into edible products and the balance would be culled and converted into copra.
During the late 1980s, complete processing plants with technology have been offered to interested parties. However, no information is available as to whether any of these plants were put to commercial operation.

Biotropic wet process
In the 1990s, Biotropic of France has been offering a complete plant for processing fresh coconut to obtain virgin coconut oil. Other products derived from the process are not indicated, but all are claimed to be of edible quality and ready for use. The husks are used for energy for the process. Details on yields or through-puts are not available. The above information is based on a leaflet prepared by Biotropic.

Commercialization of wet processing
Commercialization of technology for manufacture and marketing of high-value edible products (compared to the copra route) is very important for the diversification and development of the coconut industry. The market for crude coconut oil has suffered as a result of competition from other vegetable oils, whose production and productivity have increased steadily. On the other hand, losses suffered by the coconut industry in the copra route can be as high as 20% when very poor quality copra is stored under poor conditions for prolonged periods. However, in countries such as Sri Lanka, where optimum systems exist for producing good copra, these losses are marginal.

Serious attempts should be made to commercialize technology as developed above, following market research on the end products, and financial and economic feasibility studies. Further research and development of technology (and minor refinements) on a commercial plant, if required, need to be carried out in-house to improve commercial applicability and viability.

Desiccated Coconut

Uses of Desiccated Coconut
Desiccated coconut (DC) is the white kernel of fresh mature coconuts, shredded and dried down to about 2.5% moisture content under strict hygienic conditions. It is used for human consumption, and it retains the original oil and protein of a fresh nut. Four standard grades based on particle size are produced:

- extra fine;
- fine (macaroon);
- medium;
- coarse.

A limited amount of fancy cuts such as flakes, threads and chips are also produced for special markets. The main uses of DC are for:

- the confectionary industry, as a filling for chocolates and candies;
- the bakery industry for biscuits, cake and nut filling products;

- direct usage to decorate cakes, biscuits and ice cream;
- preparation of various snacks.

Major importing countries include the USA, Canada, Europe (UK, the Netherlands, Germany, France, Spain), South Africa, the Middle East and Australia. The Philippines, Sri Lanka and Malaysia together export 90% of the world demand of about 170,000 t (excluding re-exports). Other producers are Indonesia and Ivory Coast. Exports from Tonga and Fiji were recently discontinued.

Coconut Selection

The process of DC manufacture in Sri Lanka involves a series of steps: selecting, sorting and husking of coconuts, shelling, paring, washing, sterilizing, grinding, drying, sieving, packing, and storage. In The Philippines the process is similar, but there are some important differences which will be discussed later. The basic processes for the two countries are illustrated in Figure 35.

Selection and sorting of coconuts is very important for the process. Only fully mature nuts are suitable for DC, as immature nuts have a lower oil content, an off colour and poor keeping qualities. Germinated nuts have to be rejected as they will have begun to deteriorate due to biochemical changes. Whilst the foregoing selection is for technical reasons, selection of reasonably large nuts is for economic reasons to ensure a good yield of DC per nut, and to minimize the labour cost of the initial preparation of the nut, which is labour intensive. Nuts are husked outside the factory and some selection takes place during husking. In The Philippines, nuts are husked at the farm and transported to the factory to be used within a few days. Whilst in Sri Lanka, only seasoned nuts (stored for up to 2 months for drying out) are used for DC, in The Philippines, only green nuts are used. They are worked up soon after harvest. In Sri Lanka nuts are counted when received at the factory, whereas in The Philippines husked nuts are received on weight basis.

Wet Area of Process

Husked nuts are moved into the 'wet area' of the factory, where men remove the shells with small hatchet-like implements without damaging the balls of kernel. Subsequently, the shelled nut is pared by women, removing the outer brown skin known as testa. Here again, the kernel must remain undamaged in the form of a ball to facilitate rapid paring without leaving any brown specks. About 12 to 15% of the kernel weight is removed as parings. The parings are dried and used for oil extraction, together with the copra made from rejected nuts. The oil from this process is a by-product.

After shelling and paring, the kernel ball is cut open to release the water, after which the meat is washed thoroughly in clean water which has been treated with chlorine. At this point, kernels are inspected carefully for the removal of defective portions. After washing, the kernels are sterilized. In Sri Lanka the kernel pieces are immersed for 90 seconds in a sterilizing bath filled with boiling water. This is done to eliminate bacteria, including salmonellae which cause food poisoning. In The Philippines, sterilization using steam is carried out later in the process (after grinding). However, pasteurization is carried out after washing and subsequently the kernel pieces are immersed in a tank treated with sulphur dioxide (if required).

Dry Area of Process

The washed and sterilized coconut pieces are moved to the 'dry area' and fed into a grinder to reduce particle size to the desired level. The grinder has adjustable provisions, by which particles of varying sizes can be obtained.

In Sri Lanka, shredded coconut is dried using the old tray dryer or the more modern semi-automatic tipping dryer. In The Philippines, the shredded meat is sterilized by steam blanching, and subsequently dried on a continuous dryer with two stages, where the material is turned over from the first to the second stage. Major markets for DC maintaining sophisticated food standards have found The Philippine method more hygienic and acceptable. Sri Lanka has one such factory which recently started operating.

The dried material is then conveyed to a sifting or sieving machine, where different particle sizes are separated and then bagged. DC is packed into 5-ply kraft paper bags with an inner liner of polyethylene or polypropylene film. In Sri Lanka the net weight of a bag is 50 kg, whereas in The Philippines, bags containing 100 lb or 50 lb are shipped (mainly to the USA). The shelf-life of DC is about 6 months provided it is stored under cool dry conditions without stacking too high.

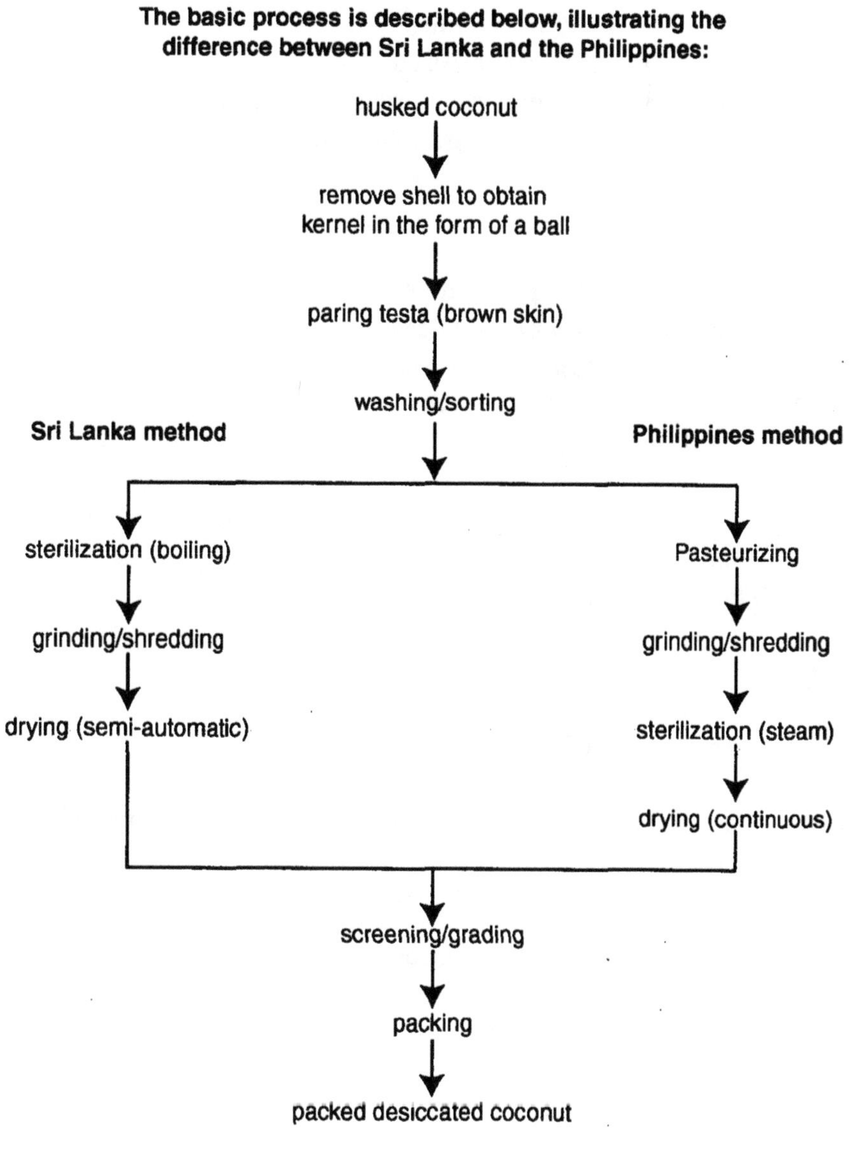

The basic process is described below, illustrating the difference between Sri Lanka and the Philippines:

husked coconut
↓
remove shell to obtain kernel in the form of a ball
↓
paring testa (brown skin)
↓
washing/sorting
↓

Sri Lanka method	**Philippines method**
sterilization (boiling)	Pasteurizing
↓	↓
grinding/shredding	grinding/shredding
↓	↓
drying (semi-automatic)	sterilization (steam)
	↓
	drying (continuous)

↓
screening/grading
↓
packing
↓
packed desiccated coconut

Figure 35: *Desiccated Coconut Manufacture*

Desiccated Coconut Quality

The composition of desiccated coconut is presented in Table 106.
Typical quality specifications for Sri Lanka DC are given below:

Physical Requirements;

Colour	-	Natural white;
Foreign matter	-	Shall be free from all foreign matter;
Taste and smell	-	Shall be sweet, pleasant and free from cheesy, smoky, soapy, sour or other undesirable flavours;
Size/grade	-	There are four grades. The sieve analysis for each grade based on the sieve aperture and BSS sieve is given below:

(a) Course grade

4.76 mm	-	100% pass through;
3.35 mm	5	15% max retained;
2.00 mm	8	15% max pass through;
1.40 mm	12	2.5% max pass through.

(b) Medium grade

2.80 mm	6	100% pass through;
2.00 mm	8	15% max retained;
1.40 mm	12	15% max pass through;
1.00 mm	16	2.5% max pass through;

(c) Fine grade

1.88 mm	10	100% pass through;
1.40 mm	12	15% max retained;

(d) Superfine grade

1.00 mm	16	100% pass through;

Special grades and cuts may be manufactured to meet customers' specifications, provided they comply with the other requirements.

Chemical Requirements;

Moisture	- 3% max for above 4 grades and 3.5% max for special grades.
Oil content	- 68% min.
Free fatty acids(FFA)	- 0.3% max (as lauric).

Note: In Sri Lanka no sulphite treatment is carried out and hence there is no residual SO_2.

Table 106: *Components of desiccated copra*

Components	Percentage
Moisture	2.5- 3.0
Fat/oil	58.0-59.0
Sugar	5.5- 6.5
Protein	6.0- 8.0
Carbohydrates, other than sugar	12.0-18.0
Crude fibre	2.0- 4.0
Ash	1.5- 2.0

Microbiological Requirements

Desiccated coconut (Sri Lanka) shall not contain bacteria of the Salmonellae group when 50 g. samples are tested.

In The Philippines, manufacturers maintain the quality stipulated by each buyer. Generally speaking, the physical requirements are similar to those stipulated in Sri Lanka, with the exception of particle sizes being specified according to US test sieves. Chemical analysis requirements are slightly different, as shown below for one particular USA buyer:

Moisture - 2.5 % max
Oil content - 65.0 % average
FFA - 0.1 % max
Residual SO_2 - 15 ppm (some buyers specify SO_2-free and others 100 ppm max).

Microbiological requirements in the USA and most advanced countries stipulate sophisticated levels, as presented in Table 107, to which The Philippines are able to conform, as a result of their superior technology and management levels.

Table 107: *Required microbiological levels in desiccated coconut*

Microbiological factor	Level
Total plate count	5,000 colonies per g max.
Mould and yeast count	50 colonies per g max.
Total coliform count	10 colonies per g max.
Salmonellae	negative
Staphylococcus	negative
Streptococcus	negative
E Coli	negative

Quality Control and Hygiene

To maintain the high quality standards required for DC, quality control and factory hygiene are of paramount importance. Quality control is carried out at every stage of the manufacturing process - inspection of nuts upon receipt to the factory stores, after preparation of the nut, and throughout the process. After processing, each bag is sampled, the samples bulked and then tested for physical, chemical and microbiological standards.

To conduct these tests, laboratory facilities are required. In Sri Lanka, the smaller factories do not have laboratories, hence these tests are conducted by the laboratory maintained by the Coconut Development Authority whereas in The Philippines, each factory is fully equipped for this task. The coconut kernel is an ideal medium for growth of micro-organisms including food poisoning organisms, hence, upon exposure it deteriorates very rapidly. The need for a high level of hygiene in such a factory cannot be over-stressed. Cleanliness of food factory building both inside and outside, washing and cleaning of the equipment after the day's production, and sterilization of equipment and surroundings before production are all very important. Washing and disinfecting of hands and feet (shoes), keeping nails cut short, clean clothes, body hygiene, screening for cuts and sores on hands and body, and for colds and other infectious diseases and routine medical examinations, apply to all personnel. Testing and controlling the quality of water used for washing coconut kernels, as well as equipment after production, is another important aspect of quality control. Chlorinators are used to ensure that the water is bacteria free.

Storage of Desiccated Coconut
Storage facilities for desiccated coconut prior to shipment is another important aspect of control. Storage should be on wooden pallets, not more than about 8 bags high (to avoid crushing). A clean cool dry store should be maintained for this purpose.

By-products
If coconuts are husked at the DC factory as in Sri Lanka, the husks will be collected as a by-product. These husks are normally sold to fibre mills for fibre extraction. Since coconuts are shelled at the factory, shells are collected in large quantities. In The Philippines, shells are used as fuel for the boiler, as steam is required for DC processing. In Sri Lanka, the shells are generally used to make charcoal. Rejected nuts and kernels are dried in a copra kiln to obtain copra, yielding oil and cake. Parings are dried to obtain poor quality copra, yielding industrial-grade oil and cake. Coconut water (mixed with wash water) is a waste product causing disposal problems. In Sri Lanka, the water is collected in a series of tanks, where traces of oil and small pieces of kernel are allowed to float and ferment. This material is collected periodically, and pressed to obtain acidulated oil (20-50% FFA), which is used for making laundry soap. In the very large DC factories in The Philippines, the waste-water is treated in a special tank from which the floating matter is skimmed by a rotary device, and oil is recovered. In a few factories, the waste-water is aerated as a secondary treatment to reduce the biological oxygen demand, before disposal.

Coconut Milk or Cream

Composition
Coconut milk or cream refers to the oil-protein-water emulsion obtained by squeezing grated fresh coconut kernel. Usually, in Asian and Pacific households, the undiluted and diluted forms are referred to as coconut milk and the concentrated form as coconut cream. With industrial processing and exports of this product during the past 10 years, all forms are currently referred to as coconut cream. The fat (oil) content in these three forms vary as shown in Table 108. The composition of the diluted cream, which is most common, is presented in Table 109.

Uses
In coconut-producing countries, coconut cream is processed in households (without industrial processing) for use in cooking and baking traditional food. In importing countries, it is mainly used by immigrants and expatriates from coconut-producing countries in preparing meat, fish and poultry dishes, sauces, salads, curries, island baking, seafood, desserts, cocktails, cakes, candies, cookies, coconut jam, and ice cream.

Table 108: *Coconut cream fat content*

Type of cream	Fat content (%)
Concentrated cream	40 - 70
Undiluted cream (or milk)	35 - 40
Diluted cream (or milk)	15 - 35

Table 109: *Composition of coconut cream*

Component	Percentage
Fat	15 - 35
Protein	2 - 4
Sugar	1 - 3
Water	60 - 80

Processing
Various techniques are used to manufacture coconut cream. A typical method is described below:

First, coconut is prepared as for desiccated coconut, as described above. Extraction and further processing are carried out in a restricted area under strict hygienic conditions. All equipment and floors are washed thoroughly and disinfected at the end of the previous day's production. Each day, before production, the equipment is flushed with hot water at 60°C. Sterilized kernels are fed into a mill to obtain fine particles to facilitate efficient cream extraction. The milled coconut is then pressed to release the cream. This is the first extraction. Additional equipment, if available, may be used to obtain a second extraction by wet milling the once-pressed residue, and pressing this again after sieving for the removal of excess liquid. This liquid, which is dilute cream, is treated again to obtain a more concentrated emulsion.

The first-pressed undiluted cream and second-pressed diluted cream (if available) are blended in the desired ratio to obtain a cream of desired fat content. An emulsifier is added and the cream is pasteurized and continuously homogenized. Subsequently, while still hot it is filled into cans seamed with coded lids and then retorted. After retorting, each retort load is left to cool overnight. The cooled cans are held in temporary storage for incubation and thorough testing to permit release for human consumption. Final packing is carried out prior to transfer to the finished product stores.

Packing
Coconut cream for export is usually packed in 400 g cans with internal lacquering, which guarantees at least 6 months' shelf-life. For local marketing, cheaper plastic pouches are used. However, their shelf-life is short. An innovation in coconut cream packing is the ultra high temperature (UHT) aseptic pack developed by Alfa Laval in Singapore. Plans are under-way for commercial introduction.

Trade
Coconut cream is imported mainly into the USA, Europe (UK, the Netherlands, Belgium, France), the Middle East, Australia and New Zealand. Producers and exporters of coconut cream include Indonesia, The Philippines, Sri Lanka, Malaysia, Thailand, Western Samoa, Fiji, the Cook Islands, Niue, Tahiti and Brazil. There are no international trade statistics available for coconut cream, as this is a relatively new product. Estimated annual quantities amount to about 20,000 tonnes.

Instant Coconut Milk Powder
A very recent development involves the manufacture of instant coconut milk powder, which is a spray-dried product similar to dairy milk powder. This is packed in triple laminate pouches and has a shelf-life of over 6 months. The major producers can be found in The Philippines, Sri Lanka and Malaysia. A by-product of this industry is de-fatted coconut (residue) which is a low-priced substitute or diluent for desiccated coconut.

Coconut Sap and Products
Fresh (unfermented) sap, obtained from the unopened spadix (toddy) is sweet due to the sugar content, and may be consumed as a beverage. For a discussion on its collection, the reader is referred to Chapter 9. The sap undergoes alcoholic fermentation due to the culture of yeast found in the environment. When fermenting, the alcohol content rises and finally the sap may turn sour due to acidification of the alcohol. If the sap is collected for its sugar, fermentation is inhibited by applying lime to the earthenware pots. The average composition of Sri Lanka toddy (Nathanael 1956) is presented in Table 110.

Table 110: *Average composition of Sri Lanka toddy (g per 100 ml)*

Component	Fresh	Fermented	Stale
Total solids	18.7	4.5	4.4
Sucrose	16.5	-	-
Ash	0.4	0.4	0.4
Acidity (acetic)	trace	0.6	1.8
Alcohol (v/v)	-	7.5	6.5

Coconut Syrup and Sugar

Unfermented sap, when gently boiled, yields a syrup or treacle or honey (65% total sugar min.) which is a traditional food item in coconut producing areas such as India, Sri Lanka, Indonesia, Thailand and Malaysia. By boiling, the volume of the liquid is reduced to about 1/6th of its original volume, at which stage a viscous syrup is obtained. It is golden yellow to reddish brown in colour due to caramelization of some sugar. It is consumed just like golden syrup from sugar cane. It is also used for preparing various traditional food items. In the Pacific Islands Tuvalu and Kiribati it is called 'Kamaimai' and when diluted with water it is taken as a beverage.

Coconut sugar is obtained by further boiling, reducing the liquid to about 1/7th of its original volume. This sugar is also called palm sugar, jaggery (India, Sri Lanka), Gula kelapa (Indonesia, Malaysia). It contains minimally 70% non-reducing sugars and maximally 13% reducing sugars. Palm sugar is also used in various traditional food preparations.

Fermented Coconut Sap, Vinegar and Spirits

The fermented sap (sometimes called coconut wine) is consumed as an alcoholic beverage in coconut producing areas. As shown above, it contains about 7.5% alcohol. If the fermented sap is allowed to stand for some time, acetic acid bacteria naturally present will cause acidification, forming acetic acid. Vinegar (meaning spoilt wine in French) of 6% acetic acid is commercially produced in Sri Lanka using the 'generator' process. In some coconut-producing countries, coconut vinegar of various strengths is used to prepare pickles, etc.

Alcohol can be made by distillation of fermented toddy. The alcohol may be diluted as desired for the production of alcoholic beverages. In Sri Lanka, about 30 million bottles of coconut 'arrack' are produced annually under the supervision of the Government Excise Department. This arrack contains about 32% alcohol. It has a typical brown colour, characteristic of other hard liquor such as whisky. In The Philippines, this alcoholic beverage is known as 'Lambanog' and it is illicitly brewed and consumed in the village areas. This also happens in Sri Lanka, to avoid paying duty.

Table 111: *Composition of tender coconut water*

Component	Percentage
Water	95.5
Protein	0.1
Fat	< 0.1
Carbohydrates	4.0
Mineral matter	0.4
Calcium	0.02
Phosphorus	< 0.01
Iron	5 ppm

Tender coconut water also contains vitamin B and vitamin C (ascorbic acid, 22-37 ppm).

Minor Food Products from Coconut

Coconut Water

The water contained inside tender coconuts (7 to 8 months old) is a refreshing drink. It is usually served to visitors visiting coconut plantations, particularly in remote areas. The custom in the Pacific region is that any stranger passing through a coconut grove, can help himself to as many coconuts as he likes for immediate consumption, but is not expected to carry away any without the blessings of the owner! In urban areas of coconut-producing regions, large numbers of tender coconuts are sold, competing with cooled bottles of aerated mineral water. A portion of the husk at the apex of the nut is previously sliced off with a knife.

The nut is cut open only when it is served. After the drink, the tender, partly-formed kernel is sometimes eaten as a snack, after splitting the nut into two halves. Unfortunately, when coconuts are consumed at this tender stage, no commercial or economic use can be made of the shell or husk. Tender coconut water is used against dehydration caused by vomiting or diarrhoea. It is also a useful substitute for saline glucose. After filtration, it was administered intravenously to the seriously wounded forces during World War II, when stocks of saline ran out in Sri Lanka. The water is sterile when the nut is opened. The water is also used as a culture medium in microbiological work. Concentration of total solids at its earliest stages of development is about 2.5 g per 100 ml, increasing gradually as the nut ripens, reaching a maximum of about 6 g per 100 ml at about the seventh month, after which it declines. The total solids will be about 2 g per 100 ml at full maturity (Child 1964). The composition of water of tender coconuts is presented in Table 111.

Coconut water from mature nuts is abundantly available at desiccated coconut factories, coconut milk/cream factories, and large copra processing centres. On the basis of about 140 ml of water per nut, a desiccated coconut factory utilizing 400,000 nuts per day, would have about 56,000 litres of coconut water daily. This is currently wasted, as the constituents of mature coconut water appear to be too diluted for any large-scale commercial use. In fact, coconut water from these factories has created disposal problems. Treatment of waste coconut water from desiccated coconut factories is dealt with under 'desiccated coconut' (above).

The approximate analysis of water from mature coconuts grown in Sri Lanka (Child and Nathanael 1947) is presented in Table 112.

Table 112: *Approximate analysis of mature coconut water in Sri Lanka*

Component	Percentage
Total solids	3.9 - 5.5
Reducing sugars (as invert sugar)	0.23 - 1.30
Additional reducing sugars after inversion (sucrose)	0.93 - 3.15
Ash	0.50 - 0.84

Carbohydrates are the most important constituents, and Child and Nathanael (1950) suggested that glucose, fructose, and sucrose were present in coconut water. This was confirmed by subsequent workers. Technically, but not economically it is possible to produce industrial alcohol and vinegar from mature coconut water. The acetic acid present in fermented coconut water was used in Sri Lanka during World War II for coagulating rubber in areas where coconuts and rubber were grown in adjacent lands (Coconut Research Scheme 1940). Coconut water from mature nuts is used to manufacture 'Nata de Coco' at home-industry and small-scale industry levels. This process is described (below).

In India (Cochin, Kerala State), bottled coconut water was being marketed in the late 1970s. This was actually mature coconut water with added sugars to increase the sweetness similar to tender coconut water. Alfa Laval has developed a technology for producing and packing coconut water from mature nuts after adding sugars to increase the Brix value, which is described in the section on wet processing (above). For further information on the use of coconut water as a growth medium, for production of protein foods, and medicinal uses, the reader is referred to *Coconut Palm Products*, published by FAO in 1975. The household and small-scale industry level utilization of coconut water indicates that large-scale utilization of this currently wasted by-product may be possible in the near future. What may be required at this stage, is development of suitable markets for selected products (such as sweetened mature coconut water), identification of locations of plants and suitable investors, after carrying out feasibility studies to determine the viability of the projects.

Nata de Coco

This is a gelatinous dessert produced by the action of bacteria on coconut water, developed in The Philippines. The 'nata', when formed, is cooked in thick sugar syrup and often served with fruit. It is believed to be composed mainly of polysaccharides, probably dextrose, and to be cellulous in nature. Bottled nata de coco is exported from The Philippines to Japan and the USA It is also produced for local consumption and export in Thailand, Indonesia and Sri Lanka. In Sri Lanka, it is called coconut cherry cubes.

To produce nata de coco, 'mother liquor' has to be obtained from a previous culture. Since this is not easy, stocks of freeze-dried mother liquor are usually held at government institutions engaged in industrial development or research. For example, in Sri Lanka, this is available at the Industrial Development Board, and at the Ceylon Institute for Scientific and Industrial Research. Ingredients required for culturing 1 kg of nata are; 12 cups (3 kg) of coconut water (or milk of 1 coconut in 12 cups of water), 2 cups of mother liquor, 1/4 cup of glacial acetic acid, and 3 cups of sugar. The coconut water or milk is first filtered, boiled and cooled. Subsequently, 1 cup of sugar is dissolved in it, and the mother liquor and acetic acid are added and mixed well. This mixture is poured into culture jars to a height of 60 mm, covered with clean paper and incubated at 28°C. The jars are left undisturbed, as otherwise the nata being formed at the surface will sink to the bottom. After 14 days, when it is about 25 mm thick, nata is picked out with a clean fork. Care should be taken not to contaminate the liquid below the nata formation. The liquid (mother liquor) will be used again as a culture for the next production of nata. The nata is cleaned by removing the creamy formation at the bottom and cut into squares of about 20 mm. Then it is washed and boiled for 1 minute in an open pan, drained, and soaked in water, whilst constantly changing the water. This is repeated until the acid taste is removed. The nata is finally drained for 2 hours. Sugar syrup is prepared, using 2 cups of sugar in 1 cup of water. Nata and colouring are added, and the product is kept overnight. Then the nata is cooked until the gummy texture is removed and the nata has become transparent. Flavouring is added and nata bottles are filled to 3/4 level and syrup is added to fill up the bottle. The bottles are sealed tightly with seal caps, and processed in boiling water for 30 minutes. They are then dried and cooled before storage. This information has been obtained from UNIDO/APCC Coconut Processing Technology Information Documents - Part 5 (1980).

Coconut Chips

Coconut chips or thin slices are a snack item. Originally coconut chips were made from special cuts of desiccated coconut which were roasted, salted and vacuum packed in cans. These chips have a ready market in certain tourist areas, such as Hawaii and the Caribbean.

Simple technology has been developed by the Food Research & Development Division of The Philippine Coconut Authority, to process coconut chips at household and small-scale industry levels, as reported in the UNIDO/APCC Coconut Processing Technology Information Documents - Part 5 (1980). About 6 freshly husked mature coconuts are required to make 1 kg of chips. The coconuts are cracked or split into two halves and water drained out. The coconut halves are baked in an oven at 176°C for 15 to 30 minutes until the kernel can be easily removed from the shell. After cooling, the kernel halves are pared with a sharp knife to remove the testa. The kernel is sliced with a potato peeler and then soaked in a solution of the desired flavour. The concentration and soaking time for the flavours are:

Salt - 2% solution for 2 hours
Barbecue - 4% solution for 2 hours
Sugar - 50% solution for 4 minutes.

The proportion, by weight, of the soaking solution to the sliced kernel is 2:1. After soaking, the solution is allowed to drain in a colander for 20 minutes. The slices are thinly spread on a baking sheet and placed inside an oven pre-heated to 300°C. The drying and roasting of the chips until they are golden-brown takes about 30 minutes. After cooling for 10 minutes, the chips are packed warm inside triple laminate packs, sealed and labelled. The moisture content should have been reduced to 3% for good keeping qualities.

Coconut Jam

Coconut jam is a traditional high-sugar coconut food product in The Philippines, commonly used as a dessert, bread spread, and rice cake topping (Gonzalez 1991). It is light to dark brown in colour, thick yet spreadable in consistency, with a rich, creamy coconut flavour. It is traditionally prepared by cooking sweetened coconut milk to a very thick consistency at low heat with constant stirring. Housewives and small-scale producers generally use brown sugar or 'panutsa' (brown sugar moulded in coconut shells).

The Food Research & Development Division of The Philippine Coconut Authority improved the process, as reported in the UNIDO/APCC Coconut Processing Technology Information Documents - Part 5 (1980). Twelve mature coconuts are required to prepare 6 standard jam bottles of 450 g. In this process, milk is extracted from grated coconut and water mixed in equal parts. A small portion (1/2 cup or 125 g) of the coconut milk is set aside and citric acid is added at a rate of 0.25% of the total weight of coconut milk. The bulk of the coconut milk is mixed with brown sugar and glucose, at 10.25% and 5.5% of the weight of coconut milk, respectively. The mixture is boiled over low heat with constant stirring for about 20 minutes. The mixture is strained to remove suspended matter, and then boiled over high heat with constant stirring. Just before the mixture thickens, the remaining coconut milk and citric acid mix are added, and cooking is continued over low heat until the mixture thickens. The hot product is poured into sterilized bottles and sealed hermetically. Coconut jam is a creamy, free-flowing, sweetened product, with a moisture content of about 25%, fat 4%, protein 6%, and total solids 75%.

White Soft Cheese from Coconut and Cow Milk

White soft cheese has been made from skim milk (SM) extended with coconut milk (Sanchez and Rasco 1983), at the Institute of Food Science and Technology, University of The Philippines at Los Baños, Laguna. The mixture of 40% SM and 60% coconut milk yielded white soft cheese comparable to that made from 100% cow milk in terms of flavour, aroma, texture, and general acceptability. (The mixture of 50% SM and 50% coconut milk, though less economical, also produced an equally good soft cheese). The coconut milk, therefore, has been able to extend cow milk, which is mostly imported into The Philippines (as non-fat dry milk powder), and thus reduced the cost of cheeese production. The importance of this development is manyfold. First of all, this is a new use for coconut and would benefit the coconut industry which is affected by poor export prices. Secondly, milk import for soft cheese production can be reduced, and finally, a low cost nutritious food can be made available in the coconut-producing rural areas of the world. This information has been based on the work of Sanchez and Rasco (1983).

In the preparation of coconut milk, husked coconuts were submerged in a 500 ppm solution of sodium hypochlorite for 1 hour and drained, split open, and grated with a motorized grater. The grated kernel was hand pressed through cheese cloth, without adding water. Reconstitution of SM with coconut milk was based on the beverage type formulation of Banzon (1978). The composition meets the average gross composition of milk and the recommended chemical composition for milk substitutes prescribed by the Protein Advisory Group of the United Nations (1972). In the beverage-type reconstituted milk formulation, the coconut milk furnishes 100% of the fat, and the protein source is SM. The coconut milk contributes 7.0% of the weight of the reconstituted milk, the SM 9.0%; the balance 84% is water.

The reconstituted milk was processed into white soft cheese according to the method of Dulay (1980). The starter consisted of *Streptococcus lactis* and *S. diacetilactis*. The rennet was prepared following the method of Dulay (1980). Previously boiled and cooled water (84%) at 72°C is added to coconut milk with constant stirring. At 60°C, skim milk is added in small quantities, and then cooled to 40°C. Then, 10% by weight of starter, and 0.1% by weight of aqueous solution of 25% calcium chloride are added. This is allowed to stand for 15 minutes, and 3% by weight of salt is added. This mixture is filtered through cheese cloth, pasteurized at 72°C for 5 minutes and cooled immediately to 40°C, using ice cold water. Subsequently, 3% by weight of rennet is added, stirred, and left undisturbed for 30 minutes for coagulation. The curd is cut, mixed for 5 minutes and poured into cheese moulds. The cheese is allowed to drain overnight at 5°C, and packed. The procedure in the preparation of white soft cheese is illustrated in Figure 36. For detailed information, readers are advised to refer the work of Sanchez and Rasco (1983).

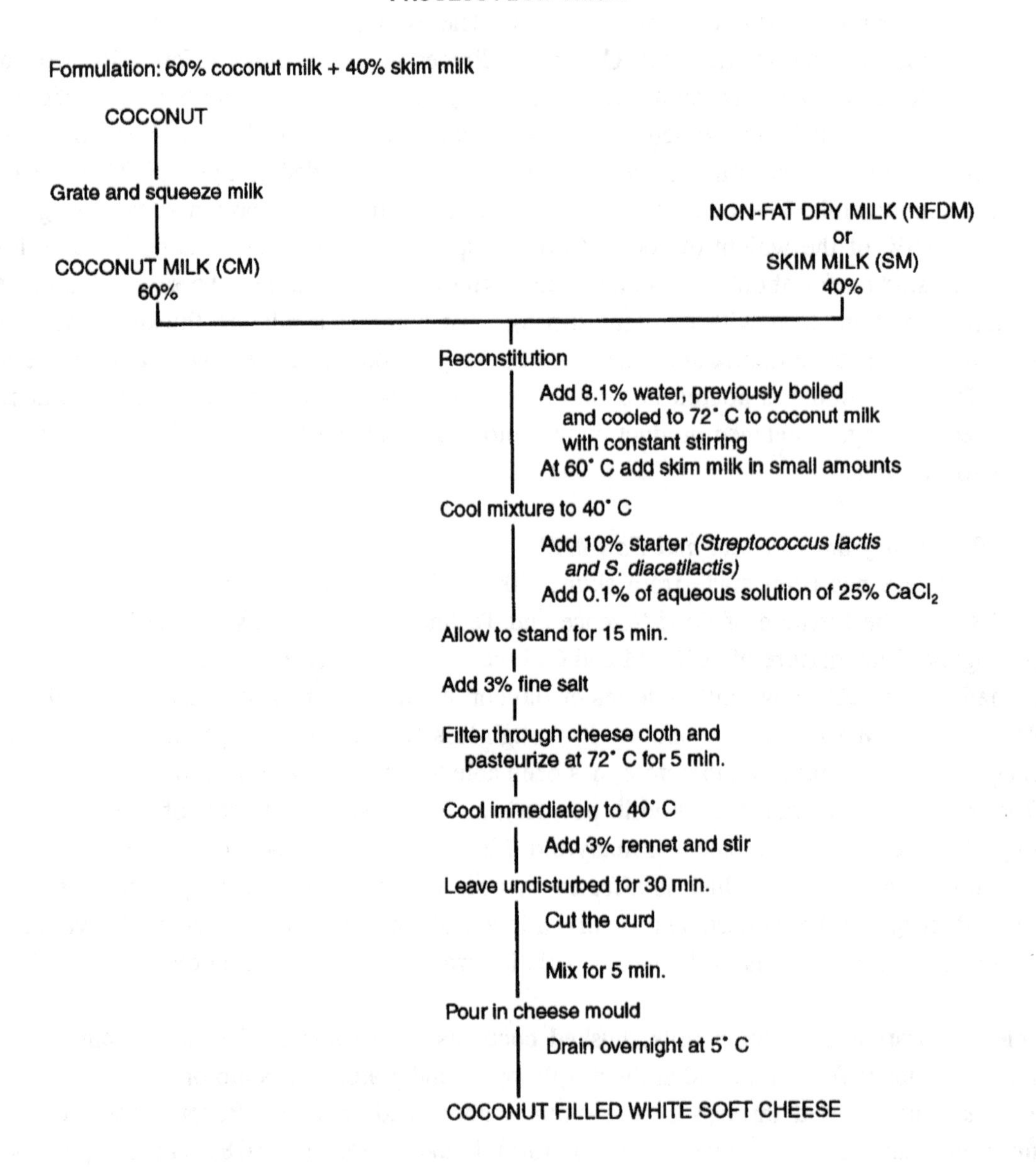

Figure 36: *White Soft Cheese from Coconut and Cow Milk (Sanchez and Rasco 1983)*

PROCESS FLOW CHART

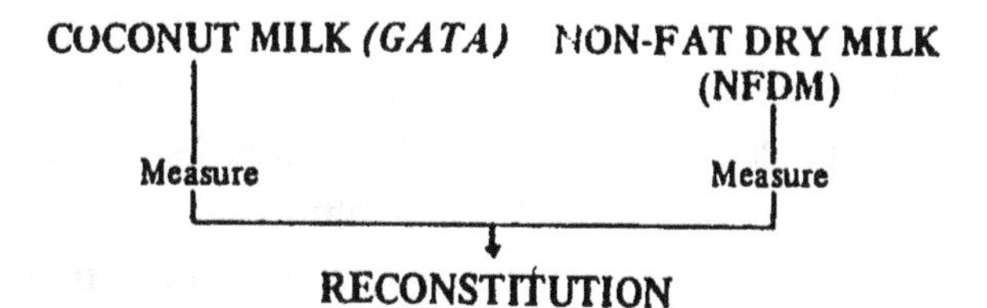

Add coconut milk to the previously measured water, boiled and cooled to 72 C. Stir to blend into a homogeneous mixture. Allow the temperature of the mixture to reach 60 C. Add NFDM, stirring continuously to prevent lumping.

↓

PASTEURIZATION

Heat mixture to 85 C in a water bath for 20 minutes and quickly cool in an ice-cold water bath.

↓

INOCULATION

Aseptically stir into the warm mixture 1.5% by weight *S. thermophilus* culture. Hold in a water bath for up to 30 minutes. Add *L. bulgaricus* and stir aseptically.

↓

PACKAGING

Transfer inoculated milk into yoghurt containers and seal.

↓

INCUBATION

Incubate the yoghurt containers to the desired pH (4.5–4.7) and acidity (0.8–0.9%), at 42 C in a water bath.

↓

CHILLING

Quickly cool containers in an ice-cold water bath to 5 C.

↓

STORAGE

Store the yoghurt containers at 5 C in a refrigerator.

↓

COCO-YOGHURT

Figure 37: *Yoghurt from Coconut and Cow Milk (Sanchez and Rasco 1984)*

Yoghurt from Coconut and Cow Milk

Yoghurt has been successfully processed at the Institute of Food Science and Technology, University of The Philippines, at Los Baños, Laguna, using reconstituted milk containing 50% non-fat dry milk (NFDM) and 50% coconut milk (Sanchez and Rasco 1984). Of the various combinations of NFDM and coconut milk that were tested, the above ratio approached the desired pH, acidity (% lactic acid), and viscosity necessary for high quality yoghurt. The coconut milk served the purpose of extending the cow milk, which is mostly imported into The Philippines (as powder). The significance of this development is that if it can be commercialized, most coconut-providing countries would be able to market yoghurt at a reduced price, thus provide the local rural population with an affordable, highly nutritious food item while reducing their dependence on imported cow milk powder. The information in this section was based on the work of Sanchez and Rasco (1984).

In the preparation of the starter culture, two lactic acid bacteria, *Streptococcus thermophilus* and *Lactobacillus bulgaricus*, were used because optimum temperature for their growth is prevalent in the tropics. *S. thermophilus* is a high-temperature-tolerant and a high acid-producing strain. In hot climates, lactic acid fermentation may often be initiated by *S. thermophilus* and continued by *L. bulgaricus* (Pederson 1971). The starter culture was prepared with 10% reconstituted NFDM, autoclaved at 10 psi for 10 minutes, then cooled to 42°C, and inoculated with 1% inoculum. The cultures were incubated at their optimum growth temperatures with *S. thermophilus* at 38°C, and *L. bulgaricus* at 43°C (Pederson 1971). After 14 hours' incubation, the cultures were transferred to 5°C storage until used. Reconstitution of NFDM with coconut milk was based on the beverage type reconstituted milk formula of Banzon (1978). In this formula, the coconut milk furnishes 100% of the fat, and the main protein source is NFDM. The coconut milk contributes 7.0% of the weight of the reconstituted milk, the NFDM 9.0%, and the balance 84% is water.

In the preparation of yoghurt, Kosikowski's method (1977) was modified by substituting coconut milk in the formula, as above. The milk base was pasteurized at 85°C for 20 minutes, and inoculated with 1.5% *S. thermophilus* culture 30 minutes prior to the addition of 1.5% *L. bulgaricus*. The procedure in the preparation of yoghurt is illustrated in Figure 37. For detailed information, readers are referred to the work of Sanchez and Rasco (1984).

Coconut Husk and Coir Products

The coconut husk (exocarp) is a by-product of the coconut industry and is wasted in many countries. However, in India and Sri Lanka these husks are used completely in three ways:

- for extraction of coir fibre;
- as domestic fuel;
- for soil moisture conservation by burying in coconut lands.

Coir fibre falls into two distinct categories: white coir and brown coir.

White Coir and Products

White coir fibre (also called yarn fibre or mat fibre or retted fibre) is the golden yellow-coloured fibre obtained (mainly in India) by retting fresh green husks in saline water for 6 to 12 months. Ideally, the husks should be obtained from nuts 1 month prior to their full maturity, as is done in India. After retting has been completed, the husks are beaten with wooden mallets to release the fibre, and then washed and dried. In India, this is done by women. The yield of white coir per 1000 husks is about 95 kg, whereas in Sri Lanka it is about 150 kg due to the larger husks. White coir is mostly spun into 2-ply yarn by hand spinning or by using wooden spinning wheels. This industry is labour-intensive and is basically a cottage industry. The yarn is used to manufacture products such as mats, mattings, carpets and rope. Mats are made in about 12 different types, such as creel mats, rod mats, rod inlaid mats, Dutch mats etc. Mattings are woven (as in textiles) in 2 treadle plain weave, 2 treadle basket weave, 3 treadle weave, 4 treadle weave, multi-treadle, and also on dobbies and jacquards. Yarn is dyed with attractive colours to obtain various designs in mats, mattings and carpets. These products are mainly made in India. Some products are made in Sri Lanka. A new development for coir is its use as a geotextile. White coir yarn netting of 50 mm x 50 mm and 2 metre width is now being used as a geofabric in the USA and Germany. These are gradually replacing jute and polypropylene fabrics which are used for soil erosion control in river embankments, road side slopes, landscaping, and plant protection, etc.

Brown Coir and Products

Brown coir is extracted mechanically (mainly in Sri Lanka and India) from brown husks by wet or dry milling. In Sri Lanka, wet milling involves soaking brown husks (from dried coconuts) for up to one month in fresh water, and combing on a pair of rotating spiked drums (called Sri Lanka drums), to obtain 'bristle', 'mattress' fibres and coir dust. The yield per 1000 husks is 50 kg bristle and 100 kg mattress fibre. The long, stiff bristle fibre is used mainly for brush manufacture. Mattress fibre consists of a mixture of medium and short fibres which are softer than bristle fibre and used mainly for filling mattresses and upholstery, for manufacture of needlefelt pads used in innerspring mattresses, and more recently in drainage filters with perforated pipes. Needlefelt pads and blankets made of mattress fibre recently found applications as geotextiles and are cheaper than white coir yarn netting.

Bristle fibre is exported in hanks (small bundles) after adding value through further processing such as hackling, bleaching, dyeing, cutting, drafting, and flagging. CP 79 shows a roof tile made of bristle fibre and cement in Zanzibar. Mattress fibre is further cleaned and exported in 125 kg high-pressure bales with a volume of 0.25 m^3 (density 500 kg per m3). Bristle and mattress fibres are blended in varying ratios, and spun and curled into a thick single-ply rope-like product called curled fibre (also called twisted fibre in Sri Lanka), which is exported. This is uncurled, formed into a sheet and sprayed with rubber latex to obtain rubberized coir mattresses, a layer for innerspring mattresses, furniture upholstery, car seats etc. Rubberized coir sheet as a geotextile, and as packaging material to protect electronic, photographic and scientific equipment during transportation was very recently developed. Bristle fibre is also spun mechanically into 2-ply yarn and exported to Europe and USA for stringing hop fields.

Wet milling is carried out using equipment comprising husk crushers, defibring machines, sifters and low-pressure baling machines. Here, mature dried brown husks are crushed, soaked for about 3 to 7 days, after which bristle and mattress fibre are extracted. The yield of coir per 1000 husks is 27 kg bristle and 63 kg mattress fibre in India (where husks are small), and 50 kg and 70 kg respectively in The Philippines. The end uses of the fibre in India are similar to those in Sri Lanka, except for the bristle fibre. Due to the reduced stiffness of the bristles, they are mainly used for curled fibre as they cannot compete with bristles from Sri Lanka for the brush industry. There are various reasons for the reduced stiffness. The husks are smaller and have shorter fibres than in Sri Lanka. The reduced soaking time, design of the equipment and mechanized operation (without hand feeding and judgement of an operator) make the bristles inferior. The baling equipment makes bales of about 50 kg with a density of 150 kg per m^3, which is adequate for internal transportation but inadequate for exports.

Dry milling of husks involves bursting or exploding husks by impact (in a decorticator) with little or no soaking to obtain an unseparated (mixed) brown fibre. Fibre yield per 1000 husks using this method is 110 kg in Sri Lanka and 90 kg in India. Dry milled fibre is stiffer than mattress fibre because of the presence of broken bristles. This mixed fibre is used mainly for spinning and curling to obtain curled fibre (used for rubberized coir). Sometimes it is used for spinning yarn, for drainage filters, for needlefelt pads or simply filling mattresses and upholstery. It can also be used to reinforce cement for the manufacture of such items as roof tiles (CP 79). Dry milling technology is in use in Thailand, The Philippines and Malaysia. Plants in Sri Lanka have closed down due to competition from the fibre produced by the traditional drum method, which is cleaner and does not have remaining small husk pieces.

Coconut Fibre Dust or Cocopeat

The coir fibre dust left unutilized when brown coir is extracted using 'Sri Lanka drums', has resulted in huge mountains, occupying vast areas of coconut land in Sri Lanka. The annual quantities of coir fibre dust produced is about 150,000 t. Traditionally, small quantities have been used in agriculture as a soil conditioner to improve the water-holding capacity of sandy soils. Since the 1980s, this material has been compressed into briquettes and blocks for export to Europe and Japan for use as an organic plant compound. The material is used in flower pots indoors after adding a stipulated amount of water which expands the volume of the briquettes seven to nine times. The exports in 1995 reached 26,700 t, earning US$ 6.6 million in foreign exchange.

The coir dust (also called coco peat), is now also being exported in the form of briquettes (200 mm x 100 mm x 50 mm - 600 g) and blocks (300 mm x 300 mm x 150 mm - 5 kg) for use as an organic plant compound in Europe. When water is added by the end user, the briquettes increase in volume by 8 times and the blocks by 5 times. Small quantities of coir dust were traditionally used as soil conditioners in Sri Lanka and, since the 1980s, have been exported to the Middle East for the same purpose in landscaping.

World Production and Trade

India is the largest producer of coir. According to information available from the Coir Board of India, for the 5 years 1987/88 through 1991/92, the average production of white and brown coir fibre have been 127,000 and 78,000 t per year, respectively.

Table 113: *Average coir exports 1988-92*

Country	Export (t)
Sri Lanka	76,700
India	28,200
Thailand	7,654
Malaysia	234
African countries	300
Others	1,000

Table 114: *Sri Lanka coir export destination*

Country	Import (t)
UK	13,700
Germany	12,500
Eastern Europe (mainly Yugoslavia)	9,600

The average exports of coir and coir products over the five-year period of 1988-92 (based on information from APCC member countries and FAO Coir Statistics) are presented in Table 113. Major importers of Sri Lanka brown coir fibre (based on information from the Coconut Development Authority) are; UK (13,700), Germany (12,500), Eastern Europe (9,600) - mostly Yugoslavia, and other countries such as Japan and China (Table 114).

Coconut Shells and Products

Coconut shells are commercially utilized mainly in Sri Lanka, The Philippines and Indonesia, where large amounts are available at desiccated coconut factories and copra processing units. In desiccated coconut factories all shells are available unless they are used as fuel for firing boilers, as in The Philippines. At copra processing centres where shells are used as fuel, only about 30 - 40% are left over for other purposes (Sri Lanka). If various fuels such as husks, shells, etc are used (Philippines), a higher percentage of shells can be saved, although in the case of direct-fired copra kilns, the copra quality is affected by the resulting heavy smoke. In the Pacific, however, shells are not saved as the whole coconut is axed into two halves and the shell attached to the husk is used as fuel for hot air copra dryers. All shells of nuts used for domestic consumption find their use as domestic fuel in the households, without effort or cost to the householder.

Shell Charcoal

The shells available in large quantities at desiccated coconut factories and copra processing units are carbonized with a limited supply of air (usually in a pit kiln) to obtain valuable charcoal. The yield of charcoal is 25 - 33% of the mass of raw shell. The number of shells per tonne of charcoal depends on the size of shell. In Sri Lanka and The Philippines, about 20,000 raw whole shells yield 1 t of charcoal, but in India about 50,000 shells are required as the coconuts (and the shells) are very small. Various factors are responsible for the quality of charcoal. First of all, the raw shells must come from fully mature coconuts, and be free of adhering fibre and kernel. Seasoned coconuts (stored for up to two months after harvesting) usually yield clean shells due to partial drying which facilitates easy release of the husk and testa with the kernel. Over-carbonizing not only results in over-burnt pieces, but reduces yield. Similarly, under-carbonizing leaves unburnt brown pieces which are undesirable. Experience and skill are required to obtain good quality and yield of charcoal. Good quality charcoal should be uniformly black in colour and be free from dust and dirt from husk and contaminants. It should be free of unburnt pieces and over-burnt pieces which are brittle. Broken edges should show a shiny black surface and a characteristic sharp fracture. When dropped on a cement floor, well carbonized charcoal lumps give a clear ring. The chemical requirements stipulated in 1982 in Sri Lanka Standard 571 are given below:

Moisture	10% max
Volatile matter on dry basis	20% max
Ash on dry basis	2% max
Fixed carbon on dry basis	79% min

Shell charcoal has been traditionally used for barbecues, blacksmith and goldsmith furnaces, smoothing irons in non-electric areas, bakeries, etc. (Sri Lanka). A major military use, and subsequent commercial use for shell charcoal was developed during World War II, when large quantities were exported from Sri Lanka to the UK for manufacturing activated carbon for use in gas masks. Other important uses of shell charcoal include reductant in smelting furnaces and for carbon electrodes in dry cells. Shell charcoal is one of the purest forms of charcoal, as hardwood charcoal and coal have much higher ash contents.

During the 1980s, in Sri Lanka, the Natural Resources Institute of U.K. developed a large steel vessel for carbonizing coconut shell into charcoal under controlled conditions. This eliminated the release of irritating and polluting gases emitted with the traditional method. The heat of carbonization was directed for drying desiccated coconut or copra. The Waste Heat Unit (WHU) had two sizes, namely for 1.5 t and for 3.4 t. of raw shells. Twelve units were installed for commercial operation in desiccated coconut factories. Owing to the failure of the steel heat exchangers, all these units have recently been put out of commission.

Shell charcoal activated carbon

Two types of activated carbon are distinguished. One refers to granular (made of shell charcoal, anthracite etc) activated carbon, mainly far gaseous phase adsorption applications, and the other is powdered activated charcoal (made of wood charcoal) for liquid phase applications. Examples of applications of shell charcoal activated carbon are military and industrial gas masks, solvent recovery plant in industry, recovery of petroleum gas, purifying recycled air in central air-conditioning, air pollution control, cigarette filters, tertiary treatment of water used in the brewery, liquor and pharmaceutical industry for the removal of flavours and odours, dechlorination of municipal water, and gold dust recovery. Examples of liquid phase applications of powdered grades include decolorizing edible oils and sugar refining.

Activated carbon from coconut shell charcoal is manufatured mainly in horizontal rotary kilns similar to, but much smaller than cement kilns. They are lined internally with refractory bricks. Granulated charcoal is fed continuously at a uniform rate into the kiln maintained at about 850°C and the bed of charcoal blanched with steam. This causes selective oxidation which results in the erosion of some carbon molecules to create a large surface area in the charcoal. After a residence time of several hours, activated carbon at about 400° C is discharged from the other end of the rotary kiln. This is allowed to cool without admitting air, as otherwise complete oxidation and burn off will take place. The different lots of activated carbon are tested for activity levels and blended as desired, and then sieved for size grading. Each lot is identified by its activity level and particle size range.

Shell flour

Coconut shell flour is obtained by initially breaking the shells into 5 or 6 mm granules with a hammer mill and subsequently reducing these granules to a flour (less than BSS 100 mesh size) in a pulverizer. The shell flour is then graded into particle sizes BSS-100+200, BSS-200+300, and BSS-300 (dust), using a cyclone separator.

Coconut shell flour is used for various purposes. It is used as a filler and extender for phenolic thermosetting plastics such as Bakelite electrical plugs and sockets to reduce 'ageing'. It is also used for phenolic glues in the manufacture of plywood and polyester laminated sheets, and as a flux coating for electric welding rods. Some importers require coconut shell in pieces or in the form of granules after the initial hammer milling operation. Shell flour is frequently in the coconut-producing countries used as a filler in the manufacture of mosquito coils (Indonesia, Malaysia, The Philippines, Sri Lanka, Thailand, Tanzania, etc). The particle size range for this application is BSS-200+300 mesh. A new development is its use for weather-resistant outdoor emulsion paints.

Miscellaneous uses of coconut shells

For centuries, coconut shells have traditionally been used in producing countries as spoons and cups for various purposes. In Sri Lanka, coconut spoons with handles made of arecanut (beetlenut) wood are available for sale even in cities. These are used in cooking and serving, due to their large size. Shell cups have been used for several decades to collect latex from tapped rubber trees.

Various handicraft items are made from coconut shells as they can be made smooth with sand paper and finished with lacquer to make them look attractive. At a handicraft training centre in Zanzibar, beautiful ornaments and souvenirs are made. Hair pins, ear rings, spice boxes with lids, coin collecting tills, are some of the items which attract tourists looking for souvenirs. In The Philippines too, very novel items such as buttons, ladies' handbags, etc. are made.

Trade in coconut shell products

Export of coconut shell charcoal and activated carbon by the three major producers, based on an average for the 5 year period 1988-92 (APCC Coconut Statistics Yearbook 1992) is presented in Table 115. Coconut shell flour exports from Sri Lanka have covered about 1000 to 1500 t per year in the past few years. This quantity includes coconut shell pieces and granules required by some importers.

Table 115: *Average shell charcoal export by three main producers during 1988-92 (t)*

Country	Shell Charcoal	Activated charcoal	Total charcoal equivalent
Philippines	35,196	12,505	76,879
Sri Lanka	13,054	11,013	49,764
Indonesia	11,084	3,050	21,251
Total	59,334	26,568	147,894

Coconut Wood

Traditional Processing and Utilization

Coconut stems have been used in all coconut-producing areas of the world as posts for traditional housing and various temporary buildings. The uppermost (immature) portion of the stem has been used as firewood for households, burning coral in the manufacture of lime, and for brick and tile kilns (mainly in Asia).

In countries where processing of coconut wood was traditional (Sri Lanka, India, Zanzibar), coconut wood has been used as rafters and beams for roofs of buildings, window and door frames, and boat building. In Sri Lanka, coconut wood rafters 4 inches x 2 inches section (100 mm x 50 mm) with lengths of 5 to 14 feet (1.5 to 4.3 m) can be obtained anywhere at Rs 14 per ft (US $ 0.92 per m). The most common lengths for roofing are 9 and 12 feet (2.7 and 3.7 m). From coconut wood, ridge plates (ridge rafters) 6 inchs x 2 inchs (150 x 50 mm) are also made when old, large-diameter coconut palms are cut down. These cost Rs 28 per ft (US $ 1.84 per m). The price of coconut wood rafters is about 30% of the price of conventional wood used for roof structures.

There are many houses and buildings in Sri Lanka with roof structures built of coconut wood that are still in use after 100 years. These are special cases where the bottom one-third of the stem of very old trees have been used without any treatment. In most cases, particularly when the middle third of the stem is used for roof structure, a preservative such as 'solignum' is applied. In some cases, used engine oil is used as a preservative to reduce cost. It is interesting to note that in the Pacific (Tuvalu, Kiribati etc), coconut stems cut to various lengths are tied together and immersed in seawater as a preservation treatment. During immersion, the salt water penetrates the coconut wood, substituting the sweet sap, which later prevents attack by termites or other insects, or fungal growth. The treatment is regarded as adequate when the wood sinks or is almost completely submerged. The lower-density stem portions take longer to submerging, as more salt water has to penetrate. This treatment can be applied where the sea is extremely calm. However, such treatment is also possible in saline back-waters of rivers and low-lying areas as in Kerala, South India. Thampan (1975) reports that the durability of the coconut wood stem increases if immersed in saline water before use.

Traditional processing of coconut timber in Sri Lanka involves the use of an axe and adze to obtain nearly rectangular cross-sections from the annular high density portions of the stem. Four or six pieces can be obtained from each annular section. Figure 38 illustrates how four pieces can be cut out from the annular section. The stem is cut to desired lengths within days of felling and debarked with the adze. Then it is marked out for 4 or 6 sections, depending on the stem diameter. 'V' grooves are cut out along the length using the axe, and the pieces are separated from the soft central portion by driving in wedgés. The rafters are smoothened on the 3 inner surfaces using the adze. The process is a manual one requiring skill, and has been carried out for centuries. It involves a manual process by which coconut stems cut down for housing and other purposes can be exploited. In the event of large-scale replanting to replace senile palms or to amend damage due to cyclones, etc, the manual process would be inadequate to exploit all of the stems.

Coconut Wood as a Renewable Resource

The tall coconut palm grown in tropical countries is considered to have reached the end of its economic life at 60 years due to declining productivity thereafter. At this age and older, these palms are most suitable for wood utilization. The extent of land under coconut increased during the first half of the 20th century due to increasing world demand for copra and coconut oil. These plantations currently contain large numbers of senile trees, which need to be substituted by high-yielding cultivars in order to improve productivity. The wood from felled palms is a renewable resource with a 60-year cycle.

In The Philippines, over 30% of the estimated 342 million coconut palms are older than 60. Therefore, about 100 million palms require felling for replanting. Subsequently, the remaining palms could be systematically substituted when they become senile. Without the backlog of senile palms, 1.6% of the palms should be cut down and substituted. This alone requires about 5.5 million palms to be felled annually, representing a huge volume of useful wood.

Proper utilization of palm wood is very important for three reasons:

- promotion of felling of old coconut trees, and be an incentive for replanting senile groves;
- removal of the palm stems from the plantation prevents the breeding of rhinoceros beetles in decaying stems;
- supply of conventional wood is rapidly decreasing everywhere and coconut wood could meet the deficit. In the case of coralline islands in the Pacific and elsewhere coconut wood is the only source of timber and conventional wood has to be imported. Commercial utilization of coconut wood will also provide income to farmers from the sale of stems, and generate employment.

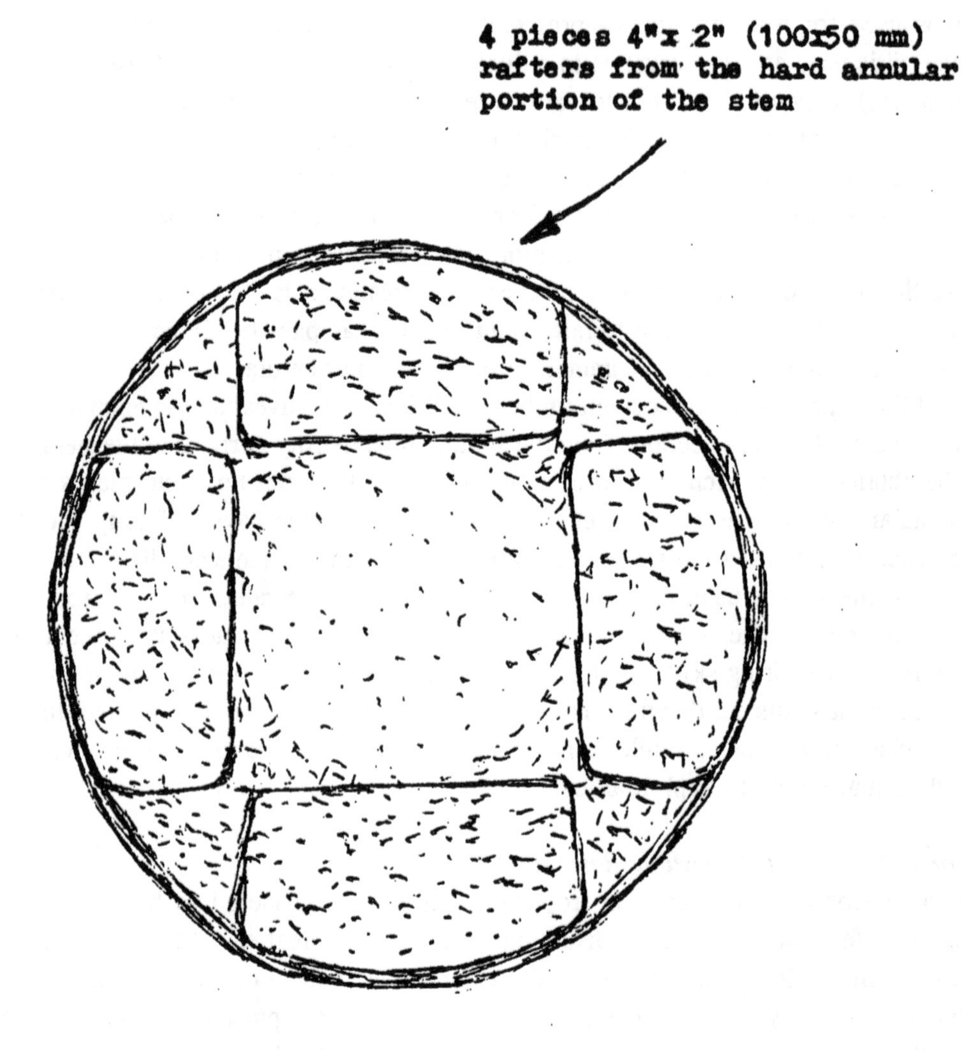

Figure 38: *Pattern of Extraction of Rafters from Coconut Stem in Sri Lanka (traditional manual process using axe and adze)*

Large-scale exploitation of coconut palm stems requires the adoption of mechanization to handle the large quantities that will be felled. In this respect, research activities were started in several countries

(Tonga, Fiji, The Philippines) in the late 1970s, funded by the Government of New Zealand and FAO. In Zamboanga, Philippines, a research station was established, and technology was developed to deal with all aspects of coconut wood utilization. The current state of the art is briefly described in this section for adoption by coconut-growing countries that may require mechanization when large-scale replanting projects are undertaken. It must be stated here that large-scale replanting schemes, though necessary, have not yet been implemented for various reasons, such as funding requirements, land ownership structure with about 80% of coconut land held by smallholders, and loss of farmers' income for a number of years until the new palms become productive.

Properties of Coconut Wood

Unlike conventional trees, the coconut stem has no regenerative tissue, hence, once formed it does not increase in diameter with age. The diameter varies from 200 to 300 mm, with a gradual reduction towards the top. Good, healthy palms have a larger average diameter compared with palms grown under poor conditions. The height of a coconut palm at 60 years of age will be about 25 metres.
The cross-section of the stem has three zones: dermal, subdermal and the central core. The dermal is the outer-most annular portion just below the bark. Physical properties of coconut wood depend on the density, moisture content and shrinkage. The basic density (oven-dry weight to green volume) decreases with increasing height of the stem, and increases from the core to the bark at any cross-section. Basic density varies from 110 kg per m^3 at the top core near the crown, to 850 kg per m^3 at the bottom-most annular portion of very old palms (Palomar 1990). The moisture content decreases with increasing basic density. The relationship between moisture and dry matter in the coconut stem varies from 1:1 at the bottom dermal portion to 4:1 at the top core. The dimensional stability of coconut wood depends on shrinkage or swelling as a result of decrease or increase in moisture content. Shrinkage and swelling cause defects such as cracks and splits. Tangential and radial shrinkage in coconut wood is of the same order, resulting in splits, whereas in conventional wood tangential shrinkage is about twice the radial shrinkage. Mechanical properties of coconut wood (and its end uses) depend upon basic density. There are three basic density groups, presented in Table 116.

Figure 39 illustrates the mechanical and related properties of coconut wood for the three density groups based on green and dry samples (Palomar 1990). All strength properties decrease with decreasing basic density. The strength properties of high-density coconut wood compare favourably with conventional hardwoods used as construction material.

Table 116: *Coconut wood density groups*

Group	Density (kg per m^3)
High density	> 599
Medium density	400 - 599
Low density	< 400

Chemical properties of coconut wood can be compared to conventional hardwoods and softwoods, as the holocellulose, lignin and pentosan contents are similar. However, the ash content of coconut wood is much higher.

Felling and Transport

It is important to cut palms as low as possible, so that the remaining stumps can easily be covered by cover crops for rapid deterioration. Small numbers of palms can be felled with an axe or with a two-man saw. Felling with an axe is a very slow process and requires regular sharpening, but an axe does not cost much. The two-man saw is faster and can fell about 10 palms a day. The mechanical method of felling involves a chain saw with a 600 mm guide bar. A two-man team (an operator and assistant) equipped with chain saw, axe, machete, wedges, repair kit and a medical kit, needs 2 to 3 working days to fell 100 palms and is thus about 3-5 times faster than the two-man saw. The mechanical method is essential for felling large coconut plantations and has proved to be most efficient and economical.

Felled coconut stems are transported by skidding them from the field to a suitable landing site on the roadside where they are aligned parallel to each other for easy bucking and loading. It is done by ground skidding with the butt portion of the stem raised. A towing bar mounted on the hydraulic lift of an agricultural tractor is useful in skidding after large-scale felling. When mechanized equipment is not available, draught animals with wooden skidding arches can be used, after cutting the stem into logs that can be hauled by the animals. Before the stem is cut into logs, the cutting points must be selected and marked, depending on the curvature. Cross-cutting logs is done with an axe, a two-man saw, or a chain saw. Wood residues must be disposed of by burning in the field to avoid perishing wood creating breeding grounds for rhinoceros beetles. The uppermost low-density portions of the palm stem should also be burnt, unless they are transported to the saw mill for conversion into firewood. From the landing site, whole stems or cut logs are transported to the sawmill by 2-wheel drive trucks, as coconut plantations usually have all-weather roads. If a portable saw mill is available at the landing site, transportation to a saw mill is not necessary.

Breaking down Logs and Saw Milling

The simple traditional method of breaking down coconut logs into rafters in Sri Lanka has already been described. This method involves the use of an axe, adze and wedges for splitting tangentially to separate the rafters from the core. The easiest way to break down coconut logs is to split them radially into boards using wedges. A line is drawn along the stem to mark where it is to be split. Along this line, wedges are driven into the stem approximately 0.5 m apart (Killman). After the log is split, the core portions are cut out with an axe. The split boards are rough and can be used for rafters or trusses in sheds and simple constructions. Although the boards are rough, the advantages are many: viz., low labour cost, low cost of equipment (axe, wedges and a mallet) and less time-consuming than hand sawing.

For large-scale coconut wood exploitation, sawing is essential. It gives the wood a smoother surface, required for furniture, etc. When sawing, the hard dermal and the medium subdermal portions have to be recovered. A sawing pattern is used to separate the three density groups. The round method of sawing provides for this separation.

The first cut is a thin slab to remove the roundness, and the second cut from the hard annular portion is 25 to 50 mm thick, depending on the diameter of the log. The log is then turned either 90 degrees or 180 degrees, and the same sequence of sawing is followed until the hard annular portion has been recovered. The optimum thickness and width of the hard portion thus recovered, are 50 mm and 125 mm, respectively. Figure 40 illustrates cutting patterns (with 180 degree turn) relative to selection and grading (FAO 1985). Saw milling can be performed using the two-man ripsaw, chain saw, circular saw or band saw.

The two-man ripsaw manual method (CP 76 and 77) has been applied in many coconut-growing countries for 'ripping' small quantities of short logs. This is carried out in the same manner as for conventional hardwoods, where the saw cuts in both directions. This method requires good skills, uses low-cost equipment, and is particularly adaptable in areas with difficult terrain not accessible by tractors, and it provides employment. The main disadvantage is the limited output.

The use of the chain saw for rip sawing coconut logs is faster than the manual rip saw, but not as fast as the circular saw or the band saw. The chain saw is also useful when the number of logs to be cut is not large. Two men working with one chain saw can handle 1 palm stem per day, producing 0.9 m^3 of boards of 50 mm thickness and widths of 50, 75, 100 and 125 mm. The chain saw is widely used in The Philippines for rip sawing coconut logs. Wood wastage is higher than with other methods, due to the wider cut. The advantages are relatively low initial costs, and portability.

BASIC DENSITY	MOISTURE CONTENT	STATIC BENDING			COMPRESSION PARALLEL TO GRAIN		COMPRESSION PERPENDICULAR TO GRAIN	IMPACT BENDING
		Modulus of Elasticity	Modulus of Rupture	Stress at Proportional Limit	Modulus of Elasticity	Maximum Crushing Strength	Stress at Proportional Limit	
(Kg/a3)	(%)	(MPa)	(MPa)	(MPa)	(MPa)	(MPa)	(MPa)	(N)
	57	10857	86	51.6	7988	49	8.3	20.2
600 and above								
	12	11414	104	61.7	9747	57	9.0	20.1
	107	6880	53	30.4	5151	31	2.8	18.3
400 to 599								
	12	7116	63	38.4	5282	38	3.4	10.1
	240	3100	26	13.1	2287	15	1.3	8.4
250 to 399								
	12	3633	33	15.4	2914	19	1.7	9.0

Figure 39: *Mechanical and Related Properties of Coconut Wood (Palomar 1990)*

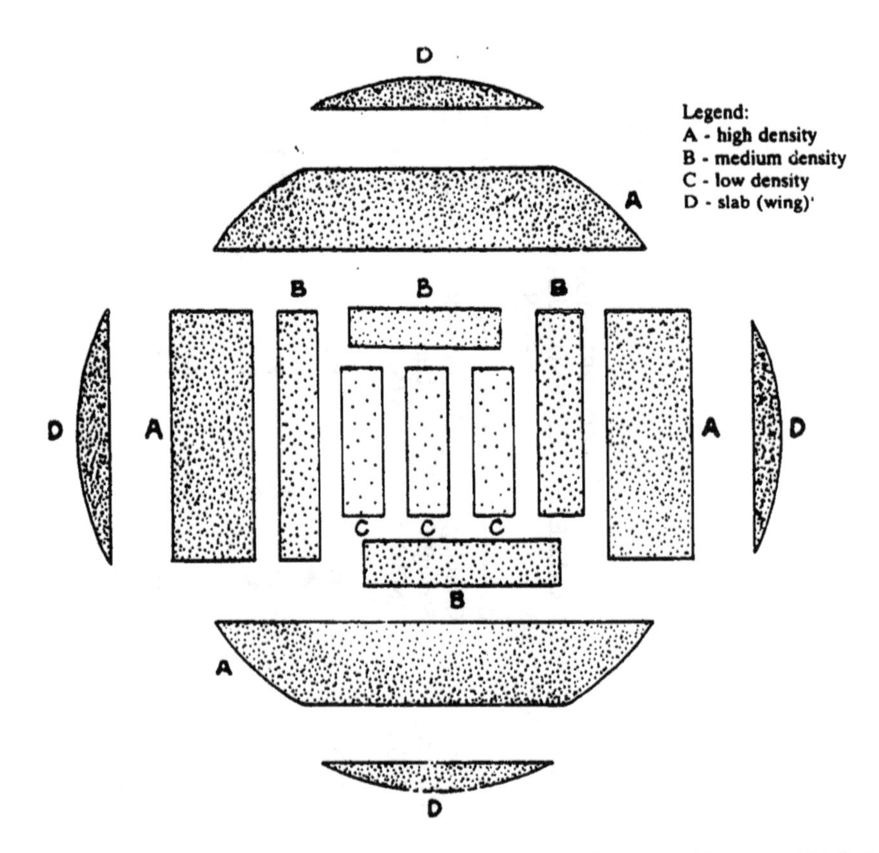

Figure 40: *Cutting Pattern Relating to Selection and Grading (Source: FAO 1985)*

The circular saw is being used successfully for sawing coconut logs in many countries. In Sri Lanka, stationary (circular) sawmills located in coconut-producing areas handle coconut logs in addition to conventional woods. Setting up a stationary sawmill merely for coconut logs would not be economic due to the limited supply of logs in the vicinity of the mill.

The use of mobile circular sawmills facilitates movement of the mill to sources where coconut stems are available after felling. It can be located within economic skidding distances from the felling areas and thus avoid transportation of stems by trucks. Two types of mobile circular saw mills have been used by The Philippine Coconut Authority - Zamboanga Research Centre. These are the Oregon mobile dimension sawmill and the Varteg combination sawmill. The Oregon mobile dimension sawmill can easily be towed by a 4-wheel-drive terrain vehicle. It has a 53 horsepower petrol engine which moves the saw along a 6.5 metre long metal beam containing a horizontal saw blade (320 mm) and a vertical saw blade (760 mm) which turn simultaneously. The log is firmly held stationary by two logholders. The frame with the engine and saw blades can be raised or lowered to adjust the thickness of the cut. Lateral movement is also possible for setting the width of the cut so that any desired section is sawn in one operation without having to re-saw. The mill can handle 4 m^3 round logs with a 49% recovery, producing 1.92 m^3 sawn wood per day (814 board feet) (Palomar 1990). The Varteg combination sawmill (mobile type) has a 1120 mm diameter circular breakdown saw and a breast bench type 920 mm re-saw. The mill is much heavier than the Oregon dimension sawmill but it can be towed by a tractor. The mill is available with a 100 horsepower diesel engine or can be powered by a tractor. Both saws rotate simultaneously. The coconut log is fixed on the saw table which moves the log through the breakdown saw. Subsequently, the boards and slabs are re-sawn in the breast bench to produce the desired sections. The mill has a capacity of 10 m^3 round logs per day with 48% recovery or 4.8 m^3 (2035 board feet) sawn timber (Palomar 1990). Band sawmills existing in coconut-producing areas could be used for sawing coconut logs. A 1300 to 1500 mm band saw powered by a 75-100 kW motor is adequate for this purpose. The saw teeth require hard facing material like stellite to saw high-density coconut wood effectively.

Saw blade maintenance is an important aspect in sawing hard coconut wood, which is similar to conventional tropical hardwoods. The butt end of the stem is the hardest portion and conventional saws easily become blunt. The bark contains sand granules which also cause blunting of saws. For this reason, the teeth have to be hard faced. Hard facing is done by either welding stellite tips on to the saw teeth or by brazing tungsten carbide tips into the face of the teeth. Brazing tungsten carbide tips is not possible for band saws.

Sharpening stellite teeth can be done using grinding wheels. For grinding at site, the manual jockey grinder is used. Tungsten carbide tips require expensive impregnated diamond wheels for sharpening. The skill involved is more demanding than for stellite teeth. However, tungsten carbide tips in good condition have a much longer service life compared to stellite tips.

Drying Coconut Wood

Differences in density and moisture content across the stem create difficulties in drying sawn coconut timber. The high-density outer annular portions develop cracks and twists, whereas the soft core portions (used for non-load-bearing purposes) tend to collapse. Drying methods include air drying, forced air drying and kiln drying.

Air drying is the simplest and most economical method. The sawn timber is fillet-stacked in an open shed. The drying rate depends on the humidity, temperature and natural movement of the air. Generally, 25 mm and 50 mm boards take 4-11 weeks and 16-21 weeks' air drying, respectively, to attain equilibrium moisture content of 17-19% (Palomar 1990). Due to the slow movement of moisture in air drying, mould and stain fungi appear, making it unsuitable for furniture, etc. This is overcome by dipping the newly-sawn timber in anti-sapstain chemicals for 2 or 3 minutes. The chemicals successfully used at the Zamboanga Research Centre include:

- Basilit PN and Pentabrite with standard sodium pentachlorophenate;
- Difolatin (fungicide) with tetrachloroethylthio;
- Daconil (fungicide) with tetrachloroisopthalnitrile.

Their concentrations range from 0.5 to 2.0%, depending on the weather conditions. It is important to note that some chemicals are banned in certain countries because they are toxic. Forced-air drying with a blower requires the sawn timber to be stacked in an enclosed shed so that the air is discharged at the other end. Drying in this manner reduces the drying time by half. The drying time can be further reduced if the air is slightly heated with the use of a heat exchanger where waste matter can be burnt. In this case, the efficiency can be improved if air is recirculated while discharging part of it and making up with freshly heated air. This system is also referred to as kiln drying. In the 1980s a commercially operating sawmill installed in Dunkannawa, Sri Lanka dried sawn coconut timber in a solar-heated (kiln) forced air drying shed with a recirculating system. This portable sawmill was originally installed in Batticoloa on the east coast to exploit the large number of coconut palms brought down by a cyclone in 1978.

Kiln drying operations at Zamboanga Research Centre showed that 25 mm boards can be dried to 13% moisture content in 14 days, employing initial dry bulb temperature (DBT) of 45°C and wet bulb temperature (WBT) of 41°C to final DBT of 50°C and WBT of 37°C. The 50 mm boards can be dried to 16% moisture content in 19 days, with initial DBT of 48°C and WBT of 39°C to final DBT of 50°C and WBT of 45°C. Combinated drying methods have been found to be effective, particularly with 50 mm boards because of reduced drying costs and defects. Air and kiln drying is one method, the other is the forced-air and kiln drying method. The use of air at atmospheric temperature initially, and application of heat in the second stage, facilitates obtaining good quality timber.

Secondary processing of Coconut Wood

Secondary processing is necessary for manufacturing furniture or other wood products becoming increasingly popular in The Philippines. High density coconut wood is not easy to work on as the tools and equipment become dull and blunt, similar to conventional hardwoods. To produce various shapes and sizes, the woodworking equipment required are sawing machines, planer, jointer, boring and mortising machine, lathe and sander. All these processes are standard operations in the woodworking industry. Palomar (1990) gives the technical details for the equipment used at the Zamboanga Research Centre. All furniture and wooden products require finishing off, to enhance the appearance and service life through preservative action. Available applications include lacquer, shellac, wax and polyurethane, similar to finishing off conventional wood. Since these are transparent, dyes or stains can be applied if the wood needs to look darker.

Preservation of Coconut Wood

Coconut wood is classified as non-durable. However, when used in the interior, treatment is not required except for low density material. All coconut timber exposed to the weather requires suitable treatment due to attack by moulds or insects. Timber in ground contact requires treatment to avoid rapid deterioration. Preservation requires proper preparation of selected good timber after secondary processing, and must be free from applications of any finishing coats. The timber must also be properly dried. However, in the case of treatment by diffusion process, the timber must be in the green or freshly sawn condition to enable free movement of preservative solution into the wood. Wood preservatives cover two types; oil-borne and water-borne. Creosote, pentachlorophenol, cuprinol and solignum are oil borne, and chromated copper arsenate (Greensalt, Tanalith, Boliden K-33 and Wolman CCA are trade names) and disodium octaborate tetrahydrate (trade name Timbor) are water-borne. The preservative can be applied in one of two ways: the pressure method, in which the wood is impregnated in a closed cylinder, with high investment cost, requiring skilled personnel; the non-pressure method, which is simple and low cost, and quite suitable for rural areas.

There are various techniques for the non-pressure method. The simplest is to brush or spray up to three coats over dry wood, whether the preservative is oil-borne or water-borne. The other method involves dipping the dry wood in a hot or cold preservative solution for 3 to 5 minutes. This method is more effective than brushing or spraying. The soaking method requires dry wood to be immersed in an oil-borne preservative solution for a few days or weeks. When the solution is water-borne, this method is known as steeping.

The dip diffusion method involves immersion of green coconut wood for 2 or 3 minutes in a vessel containing 20 to 30% Timbor. This water-borne preservative diffuses from the higher to lower concentration in the green wood. After treatment, the coconut wood is stacked, covered with a polyethylene sheet and stored for 4 to 6 weeks. The covering avoids evaporation of water during diffusion. Double diffusion involves two water-borne chemical solutions, where the green wood is first immersed in one and then in the other. The chemicals react with each other to precipitate the compound which is the desired preservative. An example is copper sulphate solution and a solution of sodium dichromate mixed with arsenic pentoxide. The treatment time is 2 or 3 days for the first chemical and 3 days or more for the second compound.

The hot and cold bath methods involve heating an oil-borne preservative, with the coconut wood immersed for several hours (80 to 100°C) and then transferring it to a cold bath of the same preservative. While heating, air inside the wood expands and some air escapes. During the heating process a small amount of preservative is absorbed. When immersed in the cold bath, the air in the wood contracts, causing the sucking in of much more preservative. The treatment time ranges from 1 to 12 hours and depends on the condition of the dry wood, and the penetration required. Recommendations for treatment based on the work carried out at the Zamboanga Research Centre are listed in Figure 41 (Palomar 1990).

Utilization of Coconut Wood

Coconut wood has many applications as building material, particularly for housing in rural coconut areas. High density wood can be used for structural purposes such as posts, trusses, rafters, purlins, secondary beams, window and door frames as well as floors and floor joists. Medium density coconut wood can be used for walls, horizontal studs and ceiling joists. Low density material can be used only for non-load-bearing applications, such as wall panelling and for ceilings. CP 73 and CP 75 show a very effective low-cost rice huller made of coconut stem portions, made in Zanzibar. Furniture (CP 80), handicraft and novelty items can be made from coconut wood as it has an unusual but attractive grain, hence its name 'porcupine' timber. For furniture which has to bear load, high density wood is required. Medium density can be used for handicraft and novelty items. Finishing of coconut wood with lacquer etc. is important in such applications. The use of coconut wood (in the round or sawn form) as posts and poles is possible if suitably protected against weather conditions and ground contact. For temporary purposes, untreated wood can be used.

For further detailed information on coconut timber technology developed during the past decade, the reader is referred to the work of Palomar, and Sulc, in *Coconut Wood Utilization*, published by APCC in 1990, and the work of Killman in *How to Process Coconut Palm Wood*, published by GATE of GTZ GmbH, Eschborn, Germany (undated).

SERVICE CONDITION	PRESERVATIVE AND CONCENTRATION	PROCESSES & TREATING SCHEDULE	TIMBER CONDITION	RETENTION (Kg/cu.m.)
Ground contact: (poles/posts)	CCA: 4–6%	**Pressure:** Ist vacuum, 45min; Pressure, 120 min; 2nd vacuum, 10 min.	Dry	14–20
	Creosote – bunker oil: 70:30 mixture	**Pressure:** Ist vacuum I to 1 hrs.; Pressure, 2-3 hrs.; Temp., 160–180 F; 2nd vacuum, I hr.	Dry	160–192
	Creosote - bunker oil: 70:30 mixture	**Hot and Cold Bath:** 8–10 hrs. heating and overnight cooling.	Dry	128–192
Outdoor, not in contact with ground: (sign board, benches, roof shingles, etc.)	CCA: 2–3%	**Pressure:** Ist vacuum. 30 min.; Pressure. 60 min., 2nd vacuum, 10 min.	Dry	7–12
	Copper sulphate, 3%; arsenic pentoxide plus sodium dichromate, 3%	**Double Diffusion:** Ist soaking in 3% copper sulphate for 2–3 days and 2nd soaking in 3% arsenic pentoxide plus sodium dichromate for 3 days.	Green	7–12
Indoor, not in contact with ground: (beam, rafters, jambs, etc.)	PCP: 5% in oil: cuprinol or solignum: ready mixed.	**Dipping/Brushing:** Dip for 3–5 min., or brush for 3 coatings.	Dry	1.8–2.0
	CCA: 2%	**Steeping/Dipping /Brushing:** Dip for 10-20 min. or brush for 3 coatings.	Dry	
	Timbor: 20–30%	**Dip Diffusion:** Dip for 2-3 min., block-stack and cover with polethelene sheet for 4–6 weeks.	Green	8–10

Figure 41: *Recommended Treatment for Coconut Timber (Palomar 1990)*

Energy and Minor Uses of Coconut Palm

Energy from Coconut Palm

Biomass energy from the coconut palm is available on a regular and renewable basis, as it produces 12 to 16 times per year a leaf (petiole/leafstalk, leaflets, mid-ribs) and a bunch of nuts (husks, shells, bunch stalks, spathes). Other sources of renewable energy are coconut oil from the dried kernel of the nut, and alcohol from distillation of fermented sap of the inflorescence. When a coconut palm is cut down, the timber is available and this may also be considered as a source of energy. According to Banzon (1984), the harvestable energy-rich parts of the coconut palm on a sustained yield basis are the nut's husk and shell, petiole of the leaf, oil from the nut kernel and sap from the inflorescence. Representative weights from Philippine coconuts and corresponding energy contents are summarized in Table 117.

Table 117: *Energy contents of coconut parts*

Part	Component	Weight	Energy
Fruit	oil	0.148 kg per nut	5.56 MJ per nut
	husk	0.242 kg per nut	4.04 MJ per nut
	shell	0.193 kg per nut	4.44 MJ per nut
Leaf	petiole	2.00 kg per petiole	33.50 MJ per petiole
Inflorescence	sap	1.38 l per day	3.64 MJ per day

Coconut Leaf (petiole, leaflets and mid-ribs)

The coconut leaf when tender (yellow), is used for decorative purposes at religious and cultural festivals in most coconut areas. The mature leaf when green is used for weaving baskets and bags for transportation of fruits, vegetables etc. Mature leaves (green or dry) are woven (plaited) into thatch used for roofing and walls in rural areas. It is frequently used as roofing in temporary structures on the seaside at beach resorts, as it does not get heated like conventional roofing during the day. In South Asia, leaves which are partially dry when they fall off the tree, are soaked (retted) in saline or fresh water pools for a month or two to remove perishable matter before weaving for thatch. In this manner, the keeping quality of thatch increases so that the roof lasts for one year. Leaves are used as domestic fuel in rural coconut areas. Their smaller ends are used as a torch after tying them into a bundle, because it burns very slowly. In large plantations, leaves are placed around the base of the palms as a mulch and for fertilizer, gradually releasing the nutrients it contains when it deteriorates. A leaf comprising petiole, about 200 leaflets and mid-ribs, weighs 10-15 kg when green.

Petioles (leafstalks) are used as fuel in the households in coconut areas. They are used extensively for walls and fencing in remote Pacific Islands. Mid-ribs of leaflets (called ekels in Asia) are used for making stiff garden brooms (with wooden handle), brooms for sweeping dry and wet floors (when made into small bundles), fish and lobster traps, bird cages, and a series of handicraft items such as baskets and souvenirs. Sri Lanka has developed an export market for mid-ribs, the export of which in 1993 reached 12,000 t, earning US$ 2.4 million.

Bunch Stalks and Spathes

Bunch stalks and spathes (one each) are produced with each bunch of nuts. These are used as domestic fuel in the vicinity where they are produced. They have no commercial value. The weight of each of these is about 5% of that of a leaf.

Coconut Oil as Diesel Fuel

After petroleum crude oil prices were raised by the petroleum exporting countries in 1975 there was much enthusiasm for developing coconut oil as a substitute for diesel oil, particularly for vehicles. Various experiments were conducted in The Philippines, and 5 to 10% substitution was found to be workable. Higher levels of coconut oil mixed with diesel oil had problems of ignition and clogging of filters. Other problems in using high levels of crude coconut oil include corrosion caused by the high free fatty acid content, and high wastage of moving parts as coconut oil behaves as a 'cutting oil'. The National Institute of Science and Technology of The Philippines experimented with certain derivatives (methyl esters) of coconut oil and found that these performed better than coconut oil mixed with diesel. No published information on these experiments is available and it may be that the results were not entirely satisfactory. In any case, the price of crude coconut oil is still too high to make it worthwhile as a substitute.

Coconut Oil as a Lighting Fuel

Coconut oil has been traditionally used as a lighting fuel in rural areas without electricity in Sri Lanka, India and elsewhere, particularly before kerosene oil became available. It is currently still used for religious and cultural ceremonies. It is burnt with a wick as a naked flame. Its light is bright and white, unlike kerosene oil which produces a reddish colour and soot. When coconut oil is burnt, it also acts as an insect repellant. If a lamp could be developed similar to the kerosene pressure lamp (petromax) to burn coconut oil it could be an asset in remote non-electric coconut areas. It would be particularly useful if the price for kerosene rises in future.

17. RESEARCH AND DEVELOPMENT

J.G. Ohler

Research

History

The earliest systematic coconut research started in India in 1916 with the establishment of four research stations on four different soil types. The research period of 50 years may not seem long when it is realized that the economic life of a coconut palm under very favourable conditions may be 80-100 years (Swaminathan 1991).

In 1970, the Central Plantations Crop Institute (CPCRI) in India was established, working in a nation-wide research network under the Indian Council of Agricultural Research on various crops such as coconuts, arecanuts, oil palm, cashewnuts and spices. In addition to research, the various symposia on coconut organized by this Institute also have contributed much to the international co-operation between coconut research workers and institutes. A great deal of our knowledge of coconut has come from India. The research on spices and cashew were delinked from CPCRI. CPCRI has at present two Regional Stations (Vital and Kayangulam), five research centres, a seed farm and the World Coconut Germplasm Centre (Sipighat) (Nair and Menon 1991).

In Sri Lanka, the Coconut Research Institute was established in 1929 and this Institute has expanded considerably since 1945. It has also contributed in an important manner to our general knowledge of coconut. In Indonesia, studies of coconut started at the very beginning of the 20th century. The first germplasm survey was conducted in 1926-27 in North Sulawesi and in Maluku provinces (Tarigans 1989). However, the first coconut research station in Manado, Sulawesi, was not established until 1930. This station suffered considerably during the war but was re-activated with UNDP/FAO. The Agency for Agricultural Research and Development (AARD), with international funding, has strengthened coconut research since 1973. In The Philippines, coconut scattered research has been undertaken by various organizations and agencies since the beginning of the century. In 1973, The Philippine Coconut Authority was established, co-ordinating all coconut research and development activities.

Other centres of research include:

- Jamaica, where important work was undertaken on the Lethal Yellowing disease;
- Trinidad and Tobago, where important work was carried out on the Red Ring disease and Hartrot;
- the Solomon Islands, where the Protectorate and Levers' Pacific Plantations have been operating a joint Coconut Research Scheme;
- in Tanzania, where an extensive research programme was set up in 1980. Its two major objectives include control of Lethal Disease (LD) and production of high-yielding tolerant genotypes. Various CBFS systems are also being studied. This programme is presently funded by the World Bank and the Government of Germany;
- last but not least, the French Institute IRHO (Institut de Recherches pour les Huiles et Oléagineux), currently CIRAD-CP, which has also contributed substantially to our knowledge of coconut, especially in the selection and breeding sector. It has been active at two research stations, one in Ivory Coast and one in Vanuatu, where in association with local goverments extensive coconut research has been conducted over many years, emphasizing the breeding of high-yielding planting material, agronomy, and pest and disease control.

Furthermore, the organization's staff is involved with national programmes in various countries. At the Marc Delorme Coconut Research Station in Ivory Coast, established in 1951, the world's most important breeding programme is still in progress. Laboratory-based research is mainly conducted in industrialized countries, such as in the UK and France. Work related to processing and marketing is sometimes handled by government industrial, commercial or trade departments. Research on the processing of coconut products is undertaken by various industrial organizations or enterprises, universities and other institutions such as the Natural Resources Institute in the UK, and the Royal Tropical Institute in the Netherlands. Most private-sector research activities with respect to coconut processing aim at improved quality of the products. Results may not always be available to others. International agencies such as the Food and Agriculture Organization of the United Nations (FAO), the United Nations Industrial Development Organization (UNIDO), ITC and UNCTAD, all contribute in various ways to coconut industry research. In addition, bilateral aid donors are also largely involved in specific problems or projects.

FAO has played an important role in increasing world-wide interest in coconut research. The five technical working parties that were organized between 1961 and 1979 provided important incentives to coconut research in various countries. The meeting of scientists from many countries and the exchange of opinions and ideas have led to the solution of serious problems in many countries. They have also contributed to increased exchange of information and intensified international co-operation in research programmes, safe and accelerated exchange of germplasm, training of research workers, etc. FAO's direct contribution to coconut research has also been substantial, including research on the control of rhinoceros beetle, studies on Cadang-cadang and Lethal Yellowing diseases, and technical programmes implemented in various countries, sponsored and sometimes staffed by FAO. FAO and UNDP have also promoted and supported the establishment of regional technical cooperation networks. However, regional networks of coconut-producing countries mean regional networks of low budget countries and lack of funds may be major constraints on their functioning. Exchange of knowledge and specialists between these countries may be the most beneficial activity of these networks.

Constraints and Progress

As coconut is cultivated mainly by smallholders, it has always been considered a poor man's crop, and as such has not attracted large capital investments. As a result, research on coconut has not been regarded as a top priority for a long time. The fact that coconut is mainly grown in developing countries with low national budgets, lacking the funds for adequate research, is another reason for relatively low research activity. For the same reason, international contact between research workers has not always been easy.

In addition to the general lack of funds for research, coconut also faces technical constraints inherent to the tree itself, hampering research programmes, especially breeding programmes. According to Bourdeix (see Chapter 8) these are:

- compared to other crops, the coconut palm's annual seed production is very low;
- seeds are very large, have a relatively short viability which is a constraint on their transportation, storage and seed exchange programmes. This situation has been much improved by the development of embryo conservation and cultivation. Although vegetative propagation of coconut has been undertaken successfully, it has not yet been possible to do so on such a large-scale that trials could be conducted with identical trees;
- studies on pests and diseases or physiological studies must be undertaken in the canopy at considerable height above the ground;
- in addition, due to the size of the tree and low planting density, large areas are required for scientific trials. A genetic experiment frequently covers an area of eight hectares for a minimum period of twelve years. It took many years before it was recognized that the widely used breeding technique based on mass selection could not contribute to substantial improvements, due to the small genetic variability within local coconut populations. This has considerably retarded results in breeding programmes;
- serious diseases for which no control methods are known hamper the exchange of planting material with the affected areas. Only with the use of very sophisticated scientific methods, such as *in vitro* culture, genetic engineering and cryopreservation may some of these problems be solved.

In spite of these problems, it may be concluded that coconut research has gradually intensified and has made considerable progress, referring to varietal improvement, drought resistance, disease identification and control, light interception by coconut and undercrops, better land-use by intercropping and low input production systems, with agroforestry types of intercropping making proper use of the beneficial actions of rhizobacteria, mycorrhizas and other soilmicro-organisms, and the various fields of processing. The *in vitro* production of planting material may be the most important field of coconut research, but has been only sporadically successful. The final technique suitable for large-scale application has still to be developed, but expectations are high. Clonal propagation of outstanding material such as very high-yielding or disease-resistant palms could have a great impact on coconut production.

At the experimental stations, yields of almost 7 t of copra per ha have been realized. Plantation yields of more than 4 tons copra per ha have already been obtained. Nucé de Lamothe (1990) expects that under the most favourable soil and climatic conditions, it should be possible to exceed yields of 6 t of copra per ha. Assuming an oil percentage of 65%, this would mean an oil yield of about 4 t per ha.

Programmes and Prospects

Serious understaffing and underfunding are major constraints on most national research programmes. Growers are unable to support sufficiently extensive research programmes, so reliance is placed on uncertain and short-term government subvention, which is often a serious disincentive to research initiatives. Even in some major producing countries of Asia, the national programmes are not supported in a manner commensurate with the economic importance of the crop (Anonymous 1990b). National research programmes are mostly related to local matters such as soil and climate, and social and economic conditions.

National research programmes often lack the necessary expertise to solve all problems, and are not even adequately addressing major problems facing crops. They are often not producing sufficient substantive results directly relevant to smallholders. Duplication of research may occur in various countries, whereas other fields of research remain untouched. Nowhere are the needs of the crop worldwide being addressed and there are insufficient means by which small producing countries, unable to mount their own research effort, can have access to new technologies, including higher-yielding varieties. These countries, particularly would benefit most from research results. Or, citing Persley (1992): 'On distributive or equity grounds, virtually all of the benefits of coconut research accrue to developing country producers and consumers. Over half of the benefits accrue directly to producers in developing countries, of which approximately 95% are smallholders, and the remaining benefits accrue to consumers. Thus, the major beneficiaries of coconut research are the millions of smallholder coconut growers. Another consideration is that of income security for smallholders. The present price fluctuations are partially a result of erratic supply. Research should aim to develop technologies to stabilize as well as to increase production. This includes identifying varieties able to remain productive under poor environmental conditions such as drought. This research would help to stabilize the income level for coconut producers, and reduce poverty.'

International research programmes

Persley (1989; 1992) strongly advocated international initiatives in coconut research. These international initiatives should be additional and complementary to existing national, regional and international efforts, and be perceived neither as a replacement, nor as competitive with them. An international initiative would help to bring together a critical mass of expertise, resources and research capacity, focused on the needs of the commodity and it would provide training opportunities for national programme staff. Persley proposed the establishment of an International Coconut Research Council (ICRC) or any other international organization that can fund, promote, direct and supervise coconut research, avoiding overlapping studies and waste of money, and supporting local research institutions wherever needed. The ideal scenario for the proposed International Research Council was described as follows:

'The purpose of the proposed ICRC would be to identify, support, promote and undertake research on priority problems of international significance. The new body would:

- conduct research itself on a limited scale, especially in relation to germplasm conservation and utilization;
- enable additional research to be undertaken on a commissioned basis by national programmes, regional or international organizations, or other advanced laboratories, within an agreed global programme of priority problems;
- organize subject-specific networks among active research workers, on problems of international significance;
- establish regional networks, to identify priority problems requiring additional research efforts, and to facilitate the distribution of research results to all coconut-producing countries.'

This idea deserves full support. It needs no explanation that such a programme would require substantial international funding, which should come mainly from international development institutions, but also from industrial and commercial organizations dealing with coconut products. Support from international funds and organizations is imperative for the future of the crop. Basic research should preferably be conducted in international, well equipped laboratories and research stations for the benefit of all coconut countries and organizations.

Very important developments in this aspect are the projects of the Science and Technology for Development Programme (STD), an initiative of the European Community. These projects include: 'Improvement of Coconut by Biotechnology: Application to Breeding and Crop Protection' started in 1994. This project is being co-ordinated by institutions from four different countries: Max Planck Institut für Züchtungsforschung (Germany); Centro de Investigacion y Mejoria Agraria (CIMA) (Spain), National Coconut Development Programme (NCDP) (Tanzania), and The Philippine Coconut Authority Albay Research Centre (ARC). 'Coconut: Development of Methods for Clonal Propagation of Elite, Disease Resistant Palms by Somatic Embryogenesis'. This project is co-ordinated by institutions from five different countries: ORSTOM-CIRAD (France); Philippine Coconut Authority (PCA); Centro de Investigacion Cientifica de Yucatan (CICY) (Mexico); University of Hannover (Germany), and Wye College, University of London (UK).

Co-operation is especially important in such programmes as tissue culture, marker-assisted selection in breeding programmes, sensitive and non-radioactive pathogen diagnosis, the use of bio-insecticides and crop improvement by genetic engineering (Rohde *et al.* 1995). Development of new techniques based on molecular and cell biology may save much time and money. It can also substantially reduce the extensive field trials needed to obtain similar results. According to Swaminathan (1991), conventional breeding methods, using naturally occurring sources of disease resistance/tolerance will have to be continued; however, quick results can be achieved only when sensitive biochemical and molecular genetic methods can be evolved for identifying such resistance at earlier stages in terms of DNA fingerprinting and use of RFLP (Restriction Fragment Length Polymorphism) and RAPD (Randomly Amplified Polymorphic DNA) mapping to identify the resistant genotypes precisely.

The areas of biotechnology relevant to co-operation with developing countries, according to Rohde *et al.*, 1995) are:

- tissue culture;
- marker-assisted selection in breeding programmes;
- sensitive and non-radioactive pathogen diagnosis;
- the use of bio-insecticides;
- crop improvement by genetic engineering.

Coconut tissue culture has been established at ARC and NCDP. The project now focuses on the development of DNA marker technology for the characterization of coconut germplasm, and on the provision of highly sensitive techniques (and the appropriate equipment) for the detection of the most important coconut pathogens (MLOs, Cadang-cadang viroid and related sequences, and coconut Foliar decay virus). The molecular techniques for pathogen diagnosis established within the project will facilitate the safe movement of germplasm by providing sensitive diagnostic procedures. Genetic engineering of coconut for drought resistance or tolerance to MLO may represent an alternative approach to the evaluation of coconut accessions for these traits, once reproducible tissue culture protocols for coconut regeneration have been established.

Results obtained so far, and expected to accumulate within the course of the project, are expected to have a general impact on coconut breeding by accelerating the selection process through the provision of DNA markers co-segregating with important agronomic traits. The controlled crosses at NCDP and ARC not only serve to construct a coconut linkage map, but the unselected offspring will be analysed to find molecular markers for *e.g.* growth habit and early flowering. Genome analyses will also be applied to coconut palms resisting abiotic (drought) and biotic stresses (MLO).

Research areas have also been identified by collaboration with the tissue culture laboratories at ARC and NCDP. These will include investigations into DNA marker identification for the 'filled' type of Makapuno nut which - if successful - will be of direct benefit to farmers by ensuring the highest yield of jelly endosperm (Rohde *et al.* 1995). Research in Makapuno coconut production can open up important new markets for coconut growers. Once this special coconut taste has become known, countries such as the USA, Europe, Japan and China may become important consumers of this product (Green 1991). Other research activities in which an international research organization could play an important co-ordinating role include:

1 Multi-local field trials

Trial results from different countries are often not comparable due to the use of different planting material, different fertilizers used, etc. Preferably, local trials should be combined into multi-local trials, using the same planting material on comparable soils and in comparable climates. Such multilocal trials could involve, among others: variety comparison trials, fertilizer trials, and root development trials. For fertilizer trials, the most homogeneous planting material available should be used to avoid differences due to planting material variability. Once clonal material becomes available, this problem will be solved. Multi-local trials can give additional information of interactions between genotypes and environment. Standardization of experimental methods is recommended to permit a comparison of trial results obtained in different locations.

2 Root system studies

Studies on physiological characteristics determining drought tolerance are very important, as a large part of the coconut population grows in areas subject to dry periods. The development of drought-resistant cultivars could improve the situation of the coconut farmers substantially in such regions. Special attention should be given to root system development under different conditions of fertilizer application in different soils and climates. Total root volume, root length, rooting depth and activity can be important factors in determining drought tolerance and the results of low-input farming. Knowledge of root system characteristics of coconut cultivars can also be of importance in the choice of an intercropping system. In this aspect, possible differences in coconut root development between broadcast application and basin application of fertilizers are also important. Further research is also needed into the influence of certain fertilizer elements on the drought tolerance of coconut palms.

3 Low-input farming systems

An important breeding programme should be the production of varieties specially adapted to low-input farming, producing reasonably under conditions of low fertilizer application and unfavourable climatic conditions such as drought. In addition, low-input farming systems, combining the use of leguminous cover crops for nitrogen supply, with favourable conditions for mycorrhiza development for additional natural phosphorus supply should be developed for different ecologic conditions. The role of mycorrhizas and other soil organisms that have a favourable influence on coconut nutrition should be further studied. Outstanding mycorrhiza species should be isolated and propagated for introduction in coconut plantations. The possible positive influence of mycorrhizas on coconut embryo cultivation merits attention. Practical propagation methods of mycorrhizas should be developed.

4 Toddy production and processing

Toddy and its derivatives such as palm sugar, alcoholic beverages and vinegar, may provide important additional income for coconut farmers, especially in situations where copra marketing is difficult and when coconut oil prices are low. Toddy is produced by traditional methods. Research into the improvement of the tapping methods and selection of palms with a high sap and sugar production may increase the income generated by these activities substantially.

5 Pest control

Pest control is often hampered by smallholders failing to apply the necessary measures, thereby endangering the surrounding holdings. Chemical control is often practically impossible due to the size of the palms. Spraying or dusting machines that can reach the canopies of tall palms are costly and treatments involve the danger of drifting insecticides causing harm elsewhere. Most chemicals are expensive and some may be handled by trained personnel only. In a few cases only, chemotherapy can be replaced by certain cultural practices. For all these reasons, biological control is preferable wherever possible. However, biological control is a complex matter and the controlling agent cannot always be found locally. This requires international co-operation. An international institution that could co-ordinate research and provide expertise, laboratories and other equipment could be of great help in this matter, especially to the smaller developing countries with insufficient means for this type of research. Research on control of widely spread insects such as *Oryctes* spp., *Eriophyes guerreronis* and *Rhynchophorus* spp. should also be co-ordinated internationally.

6 Disease control

Regional research programmes on control of diseases may have global importance and deserve international encouragement and support. In addition to basic research on MLOs, thorough studies should be undertaken into vectors and their control in regions where such diseases occur. Vector control should be based not only on the application of insecticides, but especially on the elimination of host plants, the change of environmental conditions favourable for their development, and introduction of vector predators. A potential threat to coconut is the Cadang-cadang disease of viroid nature. Although its spread is slower than that of most yellowing diseases, it is continuous. According to Randles (1991), the areas of investigation include: the epidemiology and spread of Cadang-cadang; identification and selection of resistant varieties; the possibility of using mild strain protection; early and rapid diagnosis of infection in coconuts and other hosts; the improvement of inoculation procedures. Rapid detection methods based on the use of a nucleic acid probe developed for Cadang-cadang and related viroids should be further investigated, as not all palms in which these methods indicated the presence of such viroids showed disease symptoms.

7 Harvesting and post-harvest treatment

Harvesting, dehusking and traditional copra production are labour intensive, unpopular and increasingly expensive. In view of the reducing availability of labour for harvesting and dehusking and the increase in labour costs, it is imperative that mechanized devices be developed to substitute manpower for these laborious tasks.

8 Processing
Notwithstanding the existence of a wide variety of copra drying methods, varying from sun drying to smoke drying and kiln drying, the copra produced is often of low quality, affected by dirt, moulds and insects, and having a high FFA content. Some of the moulds produce toxic substances such as aflatoxin, affecting the attractiveness of copra and coconut oil in international trade, resulting in lower prices. The low quality is often not so much the result of a failure of the method used but more a failure of the operators. Wherever possible, the copra manufacturing stage should be bypassed. Domestic manufacture of coconut cream and oil from fresh coconut meat is often very inefficient, and as much as half of the oil and proteins may remain in the residue, which is used as animal feed. It could better be employed as valuable human food. The development of efficient simple devices for coconut cream production at village level could substantially reduce the considerable losses in this industry. Research on the quality of coconut oil as a human food could best be conducted by independent institutions supervised by an international organization. Great harm has been done to the trade in coconut oil by insufficiently-founded accusations claiming the oil to be dangerous for human health, causing coronary and heart problems.

National Research Programmes
Apart from the above mentioned research items that deserve international co-operation, there is also wide scope for national research programmes dealing with local problems, the solving of which would directly benefit local farmers. It is generally recognized that coconut as a monoculture is no longer economically viable. Therefore, the most important field of investigation would be the development of sustainable coconut-based farming systems. The technical and economic results of such systems depend on many local variables such as soils and climate, consumers' preferences for food crops, prices, marketing possibilities, etc. These research programmes can be divided into programmes for existing plantations with an already established spacing between the palms, and programmes for new plantings for which different spacings may be adopted.

Interactions between intercrops and coconut palms are important, especially those related to light interception of palms and intercrops. Different palm varieties have different light interception patterns, depending on factors such as leaf length, petiole length and leaf area. Different intercrops have different light requirements. Some crops grow better than others in the shade of palms. The influence of light interception by palms on intercrop development increases with increasing palm planting density. Research in these fields can be done in farmers' fields, trials also serving as demonstration and training plots.

Socio-demographic studies may contribute much to the success of the adoption of new cultivation techniques by the farmers. Whether an individual will adopt or reject an innovation will eventually depend on how he perceives the innovation. An individual may perceive only the characteristics or qualities of the objects that satisfy his needs and values. Hence, the action towards the object is determined by the perceived use of the object in attaining the individual's goal. The more valuable and the more favourable or adoptable the meaning of an innovation, the more likely it is that the innovation will be acceptable to many. There remains a research gap concerning the social aspects of the adoption of the farming systems technology. Present studies on farming systems have not given due attention to the views and perception of farmers towards CBFS technology options as a resourse for increasing coconut productivity (Argañosa and Gomez 1991).

Of all research disciplines, agronomy comes closest to the actual world of the farmer, and although much basic and applied research remains to be done, the greatest short-term advance in coconut production is likely to come from adaptive trials, demonstrations, and adequate extension services to get existing knowledge incorporated into farming practice. There will have to be close collaboration between agronomists, agro-economists, and extension officers if this objective is to be realized (Green 1991).

Overview of Actual International Support in Coconut Research

Although not yet sufficient, international co-operation in coconut research and development has increased considerably during the past forty years. For many years FAO and UNIDO have been involved in the support and sponsoring of research and development of the coconut industry. Such sponsoring is also provided through the support of inter-regional country-to-country assistance programmes on a TCDC basis, as part of national field-oriented action programmes.

IPGRI (formerly IBPGR), CGIAR, TAC and ITC have also been involved in the establishment and financing of research on coconut. TAC recommended that CGIAR should support a small initiative in coconut research within the mandate of the proposed institute for forestry and agroforestry. IPGRI has been separated from FAO. It has funded collecting expeditions to expand the genetic base of ongoing national breeding programmes. IPGRI also supports the Coconut Genetic Resources Network (COGENT).

The World Bank's Special Programme for African Agricultural Research (SPAAR) envisages additional support to research on perennial crops, including coconut. In 1986, the European Economic Community expressed its concern when the DG XII - the division responsible for promoting science and technology in support of development - commissioned a study to evaluate proposals for co-ordinating and strengthening the ongoing coconut and oil palm research funded by the European Committee under bilateral aid programmes. As a result, the Bureau for the Development of Research on Tropical Perennial Oil Crops (BUROTROP) was established.

The Asian and Pacific Coconut Community (APCC) has been active at regional level, mostly in the co-ordination of research and data collection. COCOTECH is a permanent panel on Coconut Techno-Economic Studies of APCC. The main tasks of these promoting and financing organizations include the solution of various production problems with which all coconut growers are confronted, such as:

- need for an adequate supply of disease-free, pest-free, improved planting material;
- low productivity of old plantations;
- control of a number of diseases of unknown etiology;
- need for adequate quarantine measures;
- inadequacy of available information on nutrition, physiology and adaptability of coconut;
- need for research on the possibility of intercropping and of developing integrated farming systems under different agro-ecological and socio-economic conditions;
- need for improved inputs, such as fertilizer;
- a shortage of suitable scientific and extension manpower, particularly breeders;
- need for developing small-scale, on-farm processing techniques for crop production;
- general poverty of the population associated with coconut farming, and the desire of governments to improve the standard of living of coconut-growing farmers.

Compared to the situation before World War II, currently an additional number of coconut research centres have been established, mostly concentrating on local problems. Rather than again summing up the achievements of coconut research, already discussed in the various chapters, the various organizations involved in coconut research will be briefly discussed in Appendix I. These organizations can be divided into two main groups, the international and the national organizations, the latter including non-government institutions involved in coconut research. Although these lists may be incomplete they may contribute to intensified contacts between staff members working at the various stations.

Development

The increasing importance of coconut to meet the growing demand for vegetable oils in many tropical developing countries requires increased yields and reduced production costs and the improvement of the coconut industry. A great disadvantage of tree-crops is that once planted, the planting pattern can not be rearranged. Substitution of low-productive trees by improved planting material is costly and time consuming and will usually be done only after the palms have reached a stage at which almost no nuts are produced anymore. Therefore, rapid changes in coconut production may not be expected.

Most smallholders apply inadequate production techniques, resulting in low yields and low income. Due to senility of palms, soil depletion, and inadequate management, yields often decline steadily. However, Fowler and Teskey (1985) stated that the objective function of the smallholder is very different from that of the plantation manager. The former is, among other things, seeking to maximize returns (in the widest sense) from his and his family's labour; the latter, on the other hand, attempts to maximize returns from land and/or capital employed. This crucial distinction should be borne in mind in discussing proposals for the development of smallholder farming; smallholders have very different priorities and value systems. For example, because of the commonly low levels of production and the limited physical and financial infrastructures in most rural areas in the Pacific (and elsewhere) food supply and risk play a large part in their decision-making. Smallholders tend to be risk-averse and hence change their methods and crops only in small steps, thus minimizing uncertainties and risk of capital loss. It was concluded that for this reason much of the agricultural research carried out, supposedly for the benefit of smallholders, has in fact been highly irrelevant to their needs, has not been adopted and has thus involved a considerable waste of both funds and time. In order to be effective, agricultural research, and the development programmes that flow from it, need to better satisfy the smallholders' needs and priorities and be designed in such a manner that meets their land, labour and capital resources, and is within the acceptable bounds of risk.

According to Amrizal *et al.* (1989), farmer's receptivity to new technology of coconut cultivation in Indonesia is the aspect of diffusion that cannot be explained only by the nature of individual behaviour, but also through the social system. The system has a direct effect on diffusion, and also an indirect influence through its individual members. There are also differences in the rate of adoption of the same innovation in different social systems. An adoption decision is made by a system rather than by an individual. The same innovation may be desirable for one adopter in one situation, but undesirable for another potential adopter in a different situation.

Negative attitude of smallholders to improved cultivation practices in coconut is often caused by their failure to see the causes of decline and the effect of improvement measures. One of the reasons for this situation is the long time required for the improvement measures to show their effect on coconut. This long time also involves higher risks, such as that of bad weather conditions or pest attacks, and problems of payment of interest and repayment of loans during the transition period. Combining coconuts with other crops that respond quicker to improvement measures such as fertilizing may be more comprehensible to smallholders. The main motivation for farmers to undertake additional activities on their coconut lands is increasing their income or providing essential products for their households, such as food, animal traction and stable manure. The sooner they get their reward for additional investment of labour and/or capital, the sooner they will recognize the benefits of a multicropping system and the effect of fertilizer application. The only way by which existing plantations can be adapted to other farming systems is by gradual thinning out of unproductive coconut palms combined with gradual interplanting of seedlings of improved varieties, and the growing of intercrops to provide farmers with an additional income during the replanting process and to establish a sustainable farming system based on a multicropping system, including animal production.

With the application of actual knowledge of coconut, it would be theoretically possible to increase world coconut production by a factor of four or more. Further research may even increase this factor. However, the emphasis in the near future should be as much on the application of research results as on the development of new techniques. Much more international assistance is needed in the field of improving existing coconut plantations or replanting senile plantations. By no means should research be stopped, but all its investments will be in vain if results are not applied. The only beneficiaries might be large estates.

Replanting and new planting programmes can be successful only if all conditions for development are available, including the availability of properly trained extension staff. The supply of necessary materials, such as improved planting material, fertilizers and other chemicals at the right time and in the right place at reasonable prices, is imperative. Credit facilities should be available to farmers to overcome the first year(s) after planting when income from coconut and intercrops might be insufficient for their needs. Adequate transportation facilities are also important.

Public opinion is not always on a level with technical development. An example may be a citation from a former director of the Central Plantation Crops Institute in Kasaragod, India: 'After I took charge of the post in January 1970, I started reorganizing the work, and one decision that I had to take related to the field experiments laid out in the main block. The seedlings in these experiments had been underplanted, and to ensure precision, the old trees had to be cut and removed. In the seventies, felling of coconut trees was still being considered as a crime and the reaction of public was beyond imagination. Reports appeared in many papers that the new director is doing 'a mass murder' as a part of his cultural revolution at the Institute!' (Bavappa 1991). Many smallholders still adopt this view about felling senile coconut trees.

Development programmes should be integral programmes in which all production factors are available and supported by properly trained extension staff. Planning and implementation of coconut development programmes require many years of financial and technical support and therefore should be of a long-term nature. Participating farmers should not be abandoned after a few years. Short-term projects will fail. Bilateral donors can play an important role in these programmes. According to Amrizal *et al.* (1989) the problems encountered by coconut growers in India, Indonesia, The Philippines, Sri Lanka, Thailand, Kiribati, Papua New Guinea and Vanuatu with application of new technologies would seem to originate from the following:

- shortage or insufficient expertise of extensions' staff;
- absence of social and/or economic incentives;
- illiteracy;
- low farm income and inadequate credit facilities preventing the purchase of necessary inputs;
- inadequate distribution systems for planting material, fertilizers and pesticides;
- inadequate transport facilities;
- tenure forms restricting the farmer's freedom in his choice of land-use;
- reluctance to remove spontaneous seedlings and other 'useful' plants from mature groves;
- reluctance to remove over-age palms before the young palms come into bearing;
- reluctance to try unknown crops and technologies;
- insufficient manpower to ensure regular harvesting and dehusking.

Governments often fail to provide the required factors necessary for development. Often there is a lack of incentives for smallholders to improve their cultivation techniques or to replant their plantations with high-yielding varieties. Governments do not always have the right perception of coconut cultivation problems and their solutions. For instance, in Indonesia, intercropping in the early stage of coconut plantation in government-sponsored hybrid schemes was prohibited, although elsewhere the introduction of intercrops in coconut revived a dying industry. The coconut hybrids MYD x WAT, MRD x WAT and NYD x WAT were introduced to the farmers through the Smallholder Coconut Development Project (SCDP) under monoculture farming systems. The rate of adoption of hybrids compared with the total coconut area in the country accounted for 6%. However, in areas outside the government-sponsored projects, the cultivation of hybrids was insignificant (Tarigans 1989). Within the SCDP, during the early growth stages the palms were maintained by the project. Maintenance was handed over to farmers when the palms reached the productive stage, but the project controlled and guided the maintenance by the farmers. An assessment of the results indicated that 78% of the farmers was satisfied with the hybrids. Early bearing and high-yields were the reasons for the the preference. The rest (22%) of the farmers, who were not satisfied, complained about too small fruits, low prices for the nuts, lower yields than expected, high expenditure required for fertilizers and chemicals, and immature nutfall. The low level of hybrid planting outside the project areas implies the difficulty of such an investment by smallholders (Tarigans 1989). According to Amrizal *et al.* (1989), all farmers were concerned about the removal of senile coconut palms as a precondition to becoming a participant in the SDCP. Moreover, the removal of the old palms made the smallholders more dependent on credit during the non-bearing stage of the palms. The discouraging of intercropping under hybrid coconuts was another reason for the farmers to reject the new technology. Normally, most farmers received an additional income of 45% or more from intercrops. The farmers surveyed felt that coconut hybrids are high-cost new cultivars and that the technology package was unsuitable. This technology package was oriented to high production by using high-cost inputs. They believed that hybrid coconut would not increase their income significantly compared to what they already had. The farmers in general complained about the insufficiency of extension services.

Most of the non-project farmers received new technology from their friends or salesman of agro-inputs. The survey also indicated that middlemen supplied credit to the farmers to solve their domestic problems, harvesting costs, etc. Farmers prefered middlemen to banks, as money could be obtained quickly without any collateral. Lack of knowledge is not always the cause of lack of farmers' co-operation in coconut improvement schemes. When the owners of coconut groves have other sources of income, such as fishery, they may lack motivation to take better care of their trees. In some South American countries the author observed the deterioration of the smallholder coconut industry as a result of severe *Rynchophorus palmarum* attack. A government campaign in one of these countries, lasting several years, assisting in the control of this pest, providing extension and establishing demonstration plots had no effect whatsoever. As soon as the campaign was over, the old situation was reestablished. Driving along the coast, one could see that in most coconut areas the majority of the coconut palms had been killed. When one observed a healthy-looking plantation this invariably was not the property of a farmer, but of an educated person of an other profession. These remaining wealthy 'coconut-farmers' profited substantially from steeply rising coconut prices resulting from scarcity, whereas smallholders could not benefit from the high prices and became even poorer than before. Apparently, a socio-demographic study preceding the implementation of the control campaign might have resulted in more positive and lasting results.

A study on the co-operators and non-co-operators of the National Coconut-based Farmers Systems programme in The Philippines showed that farmers' organizations were potent conduits for development. Similarly, attendance at seminars led to co-operators' positive perception of the advantages of intercropping and simplicity of livestock integration and CBFS in general (Argañosa and Gomez 1991). The study suggests that the programme should strengthen its promotional/informational campaigns to bring about positive perceptions of CBFS. Although a multi-media approach could be an effective means to create awareness, and consequently adoption of the technology, interpersonal communication through seminars/training and farm visits is more effective in bringing about adoption of modern management techniques. Thus, the programme should consider producing more informative materials about CBFS through magazines, comics, or by allotting a prime slot to CBFS on the radio to reinforce the decision of farmers to adopt a certain technology or a combination of options.

A constraint on improving or rejuvenating a plantation is the risk involved, and the uncertainty farmers have to face before their income rises again (Silva 1988). Usually, credit facilities are not abundant in coconut areas. Smallholders are also afraid to lose their land in case loans cannot be repaid, which in agriculture can be caused by uncontrollable factors such as storms, floods, serious droughts or heavy pest or disease attack. In addition, farmers may be afraid to lose their social status if their neighbours knew that they were in debt. Smallholders who are tenants and not land owners may find it harder to get credit for their enterprises. On the other hand, they will be much less interested in replanting or fertilizing coconut palms that will yield favourable results only after a period of some years. All such factors should be given ample attention before setting up a 'subsidized replanting scheme'.

As is indicated in the above paragraphs, development depends much on the attitude of farmers, which in turn is related to their level of education. Therefore, education and training in developing countries are fundamental for agro-economic progress, and should be supported by adequate extension services and credit facilities. Adequate extension involves the availability of a sufficient number of extension officers with a certain degree of knowledge of modern management of coconut and coconut-based farming systems. Without these factors, development projects will fail and coconut production will stagnate or deteriorate, notwithstanding great research efforts and achievements.

The establishment of internationally funded, regional training centres for senior extension officers could contribute substantially to the improvement of this situation. Although extension has to be adapted to the local cultural, social and economic situation as well as to local ecological conditions, this should not hamper the establishment of regional training centres. Each extension officer himself is aware of conditions in his home country, and can adapt measures accordingly. Regional training of extension officers can greatly increase the effect of regional coconut development programmes. Training centres should be an integral part of development programmes. They should always be combined with the establishment of local training schools. Instruction books and brochures could be printed in different languages at an international centre.

In countries where research on coconut is conducted, participation of the extension staff in the research activities, in co-operation with research staff can be recommended. Extension officers could pass a practical stage at a research station at regular intervals. Extension staff involved with research will be much more interested in its results and at the same time will be trained in modern cultivation techniques. On the other hand, the extension staff can inform the research staff about farmers' problems and wishes.

Extension service and demonstration plots are very important tools for successful intercropping schemes. Demonstrations should be based on proven practices. Combined research and demonstration plots with different intercropping programmes, with and without fertilizer use, should be simple, so the farmer can understand cause and effects. The set-up of the trials should be adapted to local crop priorities, farmers' investments capacities, local food requirements, and markets for commercial products. Introduction of new, profitable crops can be done on demonstration plots, but farmers should also be able to observe the performance of traditional crops under conditions of improved management. On-farmtrials involve risks of failure that can have serious consequences for a smallholder with very little resources, unless all investment costs are carried by the institution monitoring the trials. Demonstration plots should be sufficient in number and be situated at localities that will give the majority of the farmers an opportunity to visit them. Demonstration plots are for farmers, not for visiting experts only. Establishment of secondary processing industries in developing countries might shorten the line between coconut producer and the industries, permitting higher prices to be paid to farmers. This is of great importance in remote areas such as the Pacific islands. However, the production volume in such areas is often too small for the establishment of processing industries. In the initial stages of such industries, international support with expertise and funding will often be required. Coconut development programmes should preferably include the provision of industrial plants for the processing of coconut wood so as not to waste this valuable product and to compensate for the cost of felling by the sale of the wood. It may be concluded that much can still be done to improve the situation of coconut products on the international market and, consequently, to the improvement of the situation of the coconut farmers. May this book be of use to promote such a development.

Appendix I

Institutions and experimental stations involved with coconut research and development (with addresses where available).

International Organizations

Asian and Pacific Coconut Community (APCC).

3rd Floor, Lina Building, Jl. Rasuna Said Kav. 7, Kuningan, Jakarta, Indonesia. 12920; P.O. Box 1343 Jakarta 10013, Indonesia. Tel: (62-21) 522-1711 to 13;

Fax: (62-21) 522-1714; Telex: 62863 APCC IA; Cable: COCOMUN.

This community is the only intergovernmental organization which deals exclusively with coconut. Its objectives are to promote, co-ordinate and harmonize all activities of the coconut industry. All developing countries within the geographical limits of Asia and the Pacific may apply for membership of the community. Together, the APCC countries account for about 85% of the world's coconut production.

In research, a stated aim of APCC is to assist member states in intensifying their own research activities by mobilizing both internal and external resources and technical assistance and by promoting the co-ordination of research activities of member states and undertaking intensive research into all aspects of the coconut industry, seeking the help of national and international organizations (Anonymous, 1991). It has 14 member countries, each of which has identified an institution or agency as its national focal point or national centre. Non-member countries can and do participate in certain activities. It has a widespread programme of activities, mainly concerning information on the various aspects of the coconut situation in its member countries to improve the flow of information to farmers and researchers. Its permanent panel on coconut techno-economic studies, the so-called COCOTECH (Permanent Panel on Coconut Technology) meetings, financially supported by UNDP, have permitted member countries to exchange information and assess regional priorities for research and development. APCC, with the assistance of the International Development Research Centre (IDRC), has established an integrated Coconut Information Programme. It is meant to further strengthen information activities within APCC by establishing a coherent sustainable information system which would directly assist in production improvement, and in processing and marketing of coconut. The Coconut Information Centre in Sri Lanka, for which IDRC funding has tapered off, is also one of the beneficiaries of the current on-going APCC/IDRC integrated coconut information programme. It also publishes a statistical yearbook, and a coconut newsletter-COCOINFO.

Bureau for the Development of Research on Tropical Perennial Oil Crops (BUROTROP).

17, Rue de la Tour, 75116, Paris, France. Tel.: 33-1-40507129; Fax. 33-1-40507130.

BUROTROP is an international non-profit association, whose activities are established by its Board of Directors where both donor and producer countries from Africa, Asia, Latin America and the Pacific are represented. Representatives from international aid agencies, research institutions and private companies also attend in an observer capacity.

BUROTROP was created on the initiative of a group of European scientists and administrators to help strengthen and co-ordinate coconut and oil palm research and development activities. Financial assistance was requested from the EC, thus enabling various European institutes to co-operate in achieving a common goal. A study of the oil market was carried out under the aegis of the EC which produced a broad outline of the situation, identified research and development requirements and proposed institutional options to help solve these problems. Following this study, six EC member countries (Belgium, France, Germany, Netherlands, Portugal and the United Kingdom) decided to collaborate within the framework of the EC in supporting coconut and oil palm research and development (Anonymous 1995).

Initially, it operated under the wing of CIRAD, and formerly of IRHO. It was inaugurated in 1990 and on January 1, 1995 it became an independent non-profit international association.

Its Executive Committee consists of 15 members representing donor countries and producer countries from Asia, Africa, Latin America and the Pacific. Its objectives are to:

1 promote the constitution of research and development on priority themes specified by producer countries. It does not conduct research, but aims to strengthen these activities in producer countries
2 improve training needs, analyze and strengthen existing means

3 co-ordinate donor country programmes
4 support requests for financing of projects of mutual interest in bi- and multilateral frameworks.

Its aims are to encourage the development of tropical perennial oilseed crops and to contribute to the increase of the income of producers, in particular smallholders. It considers the subsistence aspect of the crops primordial. Although it is a European initiative, it has an international vocation, and wishes to collaborate with all producers and donors. It is open to any organization willing to participate in the reinforcement of coconut and oil palm research and development activities. It has also started a database project with respect to research in progress on coconut and oil palm, which would answer the question: 'who is doing what and where?'.

Coconut Genetic Resources Network (COGENT).
c/o IPGRI APO. 30, Orange Grove Road, 1025 Singapore. Fax.: 65 7389636

When the Technical Advisory Committee of CGIAR recommended that coconut should be included in the CGIAR portfolio of activities it also recommended that first priority should be given to germplasm aspects, and that IBPGR be invited to strengthen its work on coconut genetic resources. In 1991, participants in the International Workshop on Coconut Genetic Resources in Cipanas, Indonesia, called for the implementation of an international coconut genetic resources network. For the first five years, the International Board of Plant Genetic Resources (IBPGR) was invited to act as the executing agency for the first five years. TAC supported the establishment of such a network. The network was set up in 1992.

Its objective is to improve coconut production on a sustainable basis and to increase income in developing countries through improved cultivation of coconut and efficient exploitation of its products. It aims to further develop an International Coconut Genetic Resources Network to co-ordinate research activities of national, regional and global significance, particularly on germplasm exploration, conservation and enhancement with the objective of constituting the basis for more collaborative initiatives on the broader aspects of coconut research and development. It is set to function at three levels: national, regional and global, to support and strengthen the capacity of national programmes to conserve and utilize coconut genetic resources. Its ongoing activities include a common database for passport data, characterization and evaluation of coconut germplasm around the world and the development of guidelines for the safe movement of coconut germplasm. It is developing network strategies and specific activities in priority areas for collecting, a conservation strategy, assessment of the various new methods for measuring coconut genetic diversity, a plan for multi-locational trials, support to research on diseases that threaten the safe movement of coconut germplasm, and liaison with ongoing coconut physiology investigations. (Riley, 1993).

Coconut Information Centre (CIC).
c/o Coconut Research Institute, Lunuwila, Sri Lanka.

This centre was established in 1979 with the objective of ensuring a better flow of information of interest to scientists and other groups engaged in work on coconuts around the world. It collects all available published and unpublished material on coconut production and post-harvest technology. This information is regularly updated.

Consultative Group on International Agricultural Research (CGIAR).
Mailing address: 1818 H Street, N.W., Washington D.C. 20433 U.S.A.

In 1985 the TAC (the Technical Advisory Committee of CGIAR) review of CGIAR Priorities and Strategies identified coconut as a priority commodity for international support. Subsequently, CGIAR requested TAC to explore the desirability of establishing an international research initiative on coconut, and the form such an initiative might take. Through TAC, a proposal was prepared, the 'Role of the CGIAR in Coconut Research', which summarizes the salient points. Results suggested that a well-organized and adequately funded international research effort could yield high returns on investments. The long-term nature of coconut research, the history of discontinuity and lack of support in its funding, the prospects of high returns from research investments, and the likely benefits to smallholder producers, make coconut a particularly suitable target for an international research initiative. Priority research areas that warrant an international effort were: germplasm improvement; disease and pest control; sustainability of coconut-based farming systems; post-harvest handling and utilization and socio-economics (Anonymous, 1990)

Food and Agriculture Organization of the United Nations (FAO).
Via delle Terme di Caracalla, 00100 Rome. Tel.: 39-6-52251; Fax.: 39-6-5225-5312.

FAO activities are widely known and need little elaboration. FAO and UNDP have promoted and financed national and international crop research and development programmes in developing countries. FAO organizes meetings and symposia, and is helpful in co-ordinating working groups and technical co-operation programmes; it sponsors the establishment of national and regional research and development organizations.

Institute for Research, Extension and Training in Agriculture (IRETA). (South Pacific). USP-Alafua Campus, Apia, Western Samoa. Tel: 21674; Telex: USP SX.

Inter-American Institute for Co-operation on Agriculture (IICA).
Apartado Postal 55, 2200 Coronado, San José, Costa Rica. Tel: 506 290222; Fax: 506 292652/294741.

A specialist agency of the Inter-American system for agriculture. It is renowned for its experience in agricultural research and development and has permanent offices in its member states, through which it implements its technical co-operation programmes and activities.

Intergovernmental Group on Oilseeds, Oils and Fats (IGG/OOF).
Address: IGO/OOF; FAO (see above).

This is an organization under the umbrella of FAO. It was established in 1965 and provides the only international forum for countries to discuss global oilseeds, oils and the oil meals market situation and outlook, to exchange information and take up policy issues that can have implications for production, trade, consumption and food aid. One of the objectives is to identify specific problems calling for short-term action and consider measures which could contribute to the solution of medium and long-term problems. It is open to all FAO member nations and associate members substantially interested in oilseeds, oils and the oil meals market.

International Plant Genetic Resources Institute (IPGRI), formerly International Board of Plant Genetic Resources (IBPGR).
Via delle Sette Chiese 142, 00145 Rome, Italy. Tel: (39.6) 518921; Fax: (39.6) 5750309.

The basic function of IBPGR is to promote and co-ordinate an international network of genetic resource centres to further the collection, conservation, documentation, evaluation and use of plant germplasm, and thereby contribute to raising the standard of living and welfare of people throughout the world. It has also commissioned research on coconut embryo culture and *in vitro* crop preservation. IBPGR is financed by the governments of various countries, as well as by the United Nations Environment Programme and the World Bank. (Anonymous, 1992). It is the co-ordinator of the Coconut Genetic Resources Network (COGENT), established in 1992.

International Coconut Cultivar Registration Authority (ICCRA).
c/o Dr. H.C. Harries, P.O. Box 6226, Dar-es-Salaam, Tanzania.

ICCRA has been established under the auspices of the International Society of Horticultural Science (ISHS). It has been set up by the appointment of an International Registrar. Its work programme involves three stages:

1 to compile and publish a checklist of coconut cultivars;
2 to regularly add to, and amend, this checklist;
3 to provide registration facilities for new cultivars and hybrids.

It has no intention of merely duplicating other national and international efforts, but it would encourage responsible individuals from these national organizations to appoint one of their members to act as a national registrar, affiliated to the International Registration Authority (Harries, 1991). A checklist of coconut cultivars and information on locations, origins, descriptions and so on, will greatly assist further development of the IBPGR coconut database. ICCRA also concerns itself with aspects such as the sources used by, and stocks held in, private and commercial seed gardens and nurseries; the interests of palm society amateur gardeners as well as professional botanical garden superintendents; the information needs of high-school teachers as well as those of university taxonomists. By using commonly available computer hard and software, listings can be readily exchanged. Information will appear in scientific, technical or popular publications.

International Council for Research on Agroforestry (ICRAF).
P.O. Box 30677, Nairobi, Kenya. Tel: 254 2 521450
This organization is particularly interested in coconut-based agroforestry systems.

International Trade Centre: UNCTAD/GATT (ITC).

Science and Technology for Development Programme (STD).
The STD programme is an initiative of the European Community (EC) which primarily supports co-operative research activities between institutions in the EC and those in developing countries. The programme covers two main areas: agriculture (including forestry and fisheries) and medicine, health and nutrition in tropical and sub-tropical areas. Its main objectives have included strengthening research capacities in both developing countries and in the EC by means of joint research activities, the improvement of co-ordination within the EC, and the development of close co-operative links between developing countries and the EC. (Hall, 1991).

STD supports various coconut research projects, such as: the physiological mechanisms for adaptation to drought and the creation of adapted planting material in coconut and oil palm; the study of trypanosome diseases *(Phytomonas)* of coconuts and other crops; the study of *Phytophtora* diseases of coconut, characterization of the species involved, epidemiology and control strategies; and coconut micro-propagation using immature coconut inflorescences.

South Pacific Commission (SPC).

This is an international organization for the Pacific, with 27 member countries and territories. It is a technical assistance agency serving 22 countries and territories of Melanesia, Micronesia and Polynesia. It is not a scientific organization and does not carry out research programmes. It offers small awards and grants in selected areas; it is not a funding organization *per se.*

Technical Advisory Committee (TAC).

In 1985, the Technical Advisory Committee (TAC), reviewing the priorities and future strategies of the Consultative Group on International Agricultural Research (CGIAR), recommended that coconut be included in an expanding overall programme. In 1990 the CGIAR accepted TAC's recommendations on the desirability of having an international initiative on coconut research.

Technical Co-operation among Developing Countries (TCDC).

TCDCs are regional organizations, sponsored by FAO, for the co-ordination of research and development efforts with inter-regional exchange of information, expert service, etc. through a regional network. At national levels, interdepartmental and inter-agency groups on important socio-economic and technical aspects of coconut research and development should be established by the respective governments. These groups are chaired by national co-ordinators. Each participating country is represented in the organization by its national co-ordinator. Together with representatives from UNDP and FAO, they form the Consultative Board. The organization formulates workplans for co-operative efforts by the proposed regional network by sharing expertise and skills to improve coconut production. It prepares programmes and detailed workplans and the network's budget. It identifies priority areas of research, needs for training and additional research facilities. By definition, a TCDC network must be self-sustaining and provide on-going services for participating countries, gradually reducing the burden of external assistance. Where research and development often collapse following the departure of foreign experts, TCDC programmes would have a better life expectancy. In practice, funding of TCDCs is the weak point of the organizations.

National Institutes and Research Stations

Australia

Australian Centre for International Agricultural Research (ACIAR).
ACIAR supports coconut research in the South Pacific, and is involved in various coconut research programmes in this region. It sponsors collaborative research between Australia, Oceania and The Philippines.

Commonwealth Scientific and Industrial Research Organization (CSIRO).
306, Camody Road, St. Lucia, 4067, Brisbane, Australia.

Benin

Station de recherche sur le cocotier. Sémé Podji, Via Porto Novo.

Brazil

Empresa Brasileira de Pesquisa Agropecuária (EMBRAPA) - Centro de Pesquisa Agropecuária dos Tabuleiros Costeiros (CPATC). Avenida Beira Mar, 3250, Caixa Postal 44, CEP 44001-970, Aracaju-SE.
Centro Naciocal de Pesquisa do Coco. Praia 13 de Julho, 49.000 Aracaju SE, Brazil. Tel.: (99) 2319116; telex: 2318.

China

Wencháng Coconut Research Centre, Hainan Island.

Colombia

Instituto Colombiano Agropecuario (ICA). Colombia.
Calle 37 n° 8-43, Piso 5, Apartado Aérco 7984, Bogotá, Colombia.
Tel: (57) 1 285 8948; Fax: (57) 1 285 4351
Federation of Colombian Oil Palm Growers (FEDEPALMA).
Cra. 9 No. 71-42, Santa Fé de Bogotá, Colombia.

Cook Islands

Totokoitu Research Station.

Cuba

Estación Nacional de Frutales - IICF.
Apdo. 37, Guira de Melena; La Habana 33600.
Tel: 218908
Fax: (19537) 335086

Ecuador

Instituto Nacional de Investigaciones Agropecuarias (INIAP), Ecuador.
Eloy Alfaro y Avenida Amazonas, Idif. MAG, Piso 4, Apartado 2600, Quito, Ecuador.
Tel: (593) 252 8650 ; Fax (593) 250 4240
Estación Experimental 'Santo Domingo'. Santo Domingo de los Colorados; Casilla 101, Ecuador. Tel: (593) 250 4520; Fax: 593 250 4240.

El Salvador

Programma de Frutales.
Centro de Technologla Agricola; Tarial, Ahuachapán. Tel: 503 282066.

Fiji

Koronivia Research Station.
P.O. Box 77, Nausori, Fiji Island.
Tel: 679 477 044; Fax: 679 400 262.
Wainigata Research Station.

France

Centre de Coopération Internationale en Recherche Agronomique pour le Développement (CIRAD). Avenue du Val de Montferrand, BP 5035, 34032 Montpellier Cedex 1. Tel. of Coconut Programme: 67 61 71 31/32; Fax: 67 61 7120.

A department of CIRAD working especially with perennial crops, is CIRAD-CP. Some sections of the former IRHO (Institut de Recherches pour les Huiles et Oléagineux)) have been incorporated into CIRAD-CP. This organization has associated experimental coconut stations in Ivory Coast and in Vanuatu and is also involved in many collaborative programmes in various countries.

Institut Français de Recherche Scientifique pour le Développement et Coopération (ORSTOM).
B.P. 5045, 34032 Montpelier, Cedex, France.

Ghana

Crop Research Institute.
P.O. Box 3785, Kumasi. Telex: 2536 GOPDEC GH.
Oil Palm Research Institute (OPRI).
P.O. Box 74, Kadé.

India

Indian Council of Agricultural Research (ICAR).
Krishi Bhawan, New Delhi - 110001, India.
Coconut Development Board, Ministry of Agriculture, Department of Agriculture and Co-operation, Cochin-682 011, P.B. 1027, Kerala, India.
Indian Society for Plantation Crops. Secretary: c/o Central Plantation Crops Research Institute, Kasaragod 671 124, Kerala, India. Fax: 499 523 300
Central Plantation Crops Research Institute (CPCRI)
Kasaragod 671 124 Kerala, India. Fax: 499 523 300.
CPCRI Regional Station Kayamkulam, Krishnapuram P.O. 690 533, Kerala.
CPCRI Regional Station Vittal - 574 243, Karnataka.
Regional Agricultural Research Station (KAU), Pilicode - 671 353, Kasaragod (Dt), Kerala.
ICAR Research Complex for Goa (CPCRI), Margao 403 602, Maharashtra.
World Coconut Germplasm Centre, Port Blair - 744 101, Andaman, India.

Indonesia

Agency for Agricultural Research and Development (AARD), Jalan Ragunan 29, Jakarta 12540, Indonesia. Tel: (62) 21 780 6202; Fax: (62) 21 780 0644.
Research and Development Centre for Industrial Crops (RDCIC); Jl. Tentara Pelajar 1, Bogor. Tel: 0251 326194
Coconut Research Institute. P.O. Box 4, Manado, Sulawesi Utara.
Coconut Research Institute Sub Balitka Pakuwon, Parangkuda, Sukabumi 43157, Java.

Ivory Coast

Station de recherche sur le Cocotier Marc Delorme, Port Bouet, 07 BP 13, Abidjan 07.
Tel.: (225) 248 873/ 248 067; Fax: (225) 248 572.

Jamaica

Coconut Industry Board
P.O. Box 204; 18, Waterloo Road, Kingston 10, Jamaica.
Tel: 809 92 61770; Fax: 968-1360.

Kenya

Coast Agricultural Research Station. P.O. Box 6, Kikambala.
Mtwapa Research Centre, Mtwapa.

Malaysia

Malaysian Agricultural Research and Development Institute (MARDI).

Jalan Sungai Dulang, Sungai Sumun, Hutan Melintang, Perak. P.O. Box 25, 36307 Sungai Suman Post Office, Perak. Tel: 05-669242, 669249, Tlx. MA 37115.
Highlands Research Unit.
P.O. Box 2009, Kelang, Selangor.

Mexico
Instituto Nacional de Investigaciones Agropecuarias y Forestales (INIFAP). Calle 62, # 462 X 55 Depto. 209, Mèrida, Yucatan, Mexico.
Centro Investigacion Cientifica de Yucatan A.C. (CICY).
Apartado Postal 87 - Cordemex 97310 Merida, Yucatan. Tel.: (99) 44035/440309; Fax: (99) 44097.

Nicaragua
Centro Experimental El Recreo del Instituto Nicaraguense de Technologia Agropecuaria (INTA).
Apartado 5735, Managua, Nicaragua. Tel: 505 2-27526; Fax: 505 2-27853.

Nigeria
Nigerian Institute for Oil Palm Research. P.M.B. 1030, Benin City.

Papua New Guinea
Cocoa and Coconut Research Institute (CCRI).
P.O. Box 1846, Rabaul. Tel: 675 92 3031; Fax: 675 92 3057.
Jim Grose Coconut Research Station, CCRI. P.O. Box 642 Madang.
Tel: 82 33 60; Fax: 82 33 60.
Omuru Hybrid Coconut Centre.
P.O. Box 296, Madang.
Lowland Agricultural Experiment Station, Keravat. P.O. Keravat, via Rabaul, East New Britain Province. Tel. 926251, Fax. 926237.

The Philippines
Philippines Coconut Authority (PCA).
Don Mariano Marcos Avenue, Diliman, Quezon City, Philippines.
Tel: 99450106; Fax: 29200415.
Albay Research Centre.
Banao, Guinobatan, Albay - 4503. This station is focused on Cadang-cadang disease and its control.
Davao Research Centre.
P.O. Box 295, Bago-Oshiro, 8000 Davao City, Mindanao. This station deals primarily with agronomy and entomology.
Zamboanga Research Centre.
P.O. Box 356, San Ramon, Zamboanga City, Mindanao. Dealing mainly with breeding and selection, and coconut wood processing and uses.

Seychelles
Grand Anse Experimental and Food Production Center.
P.O. Box 166, Mahe.

Solomon Islands
Dodo Creek Research Station.
P.O. Box G13, Honiara. Telex: Honiara 31111.
Yandina Research Station, Lever Solomons Plantation. (OOi, 1987)

Sri Lanka
Coconut Development Authority (CDA).
Navala Road, Colombo - 5, Sri Lanka.
Coconut Research Board (of CDA).

Coconut Research Institute.
Bandirippuwa Estate, Lunuwila. Tel.: 031 5300; Fax: 94 31 7195

Surinam
Palm Research Centre. Agricultural Experiment Station.
P.O. Box 160, Paramaribo. Tel: 597 474177; Fax: 597 470301.

Tanzania
National Coconut Development Programme (NCDP).
P.O. Box 6226, Dar-es-Salaam.
Tel: 255 51 74 834; Fax: 255 51 75 549; Telex: 41456NCDP

Thailand
Chumpon Horticultural Research Centre. Amphoe Sawi, Chumpon.
Sawi Agricultural Station, Amphoe Sawi.

Tonga
Vaini Research Station.

Trinidad & Tobago
Central Experiment Station.
Centeno Via Arima. P.O. Box 4763, Arima. Tel: 646 4335 - 7/ 642 1872.

United Kingdom
Natural Resources Institute (NRI)
Central Avenue, Chatham, Kent ME 4 TB.
Tel: 0 1634 880088; Fax: 0 1634 880066.
Tropical Products Institute.
Culham, Abingdon, Oxfordshire OX14 3DA.
Wye College (University of London).
Wye, Ashford, Kent TNs5 5 AH, UK.

United States of America
Agricultural Research & Education Center
3205 College Avenue, Fort Lauerdale, Florida 33314.
Agricultural Experiment Station.
Agana 96910, P.O. Box EK, Guam.

Vanuatu
Vanuatu Agricultural and Training Centre (VARTIG).
P.O. Box 231, Santo Tel: (678) 36 320/130; Fax: (678) 36 355.

Venezuela
FONAIAP.
Avenida Universidad, Via El Limon, Ed. Aragua, Maracay, Venezuela.
Tel: 043 453075/452491; Fax: 043 454320
Fondo para el Desarollo del Coco y de la Palma Aceitera (FONCOPAL).
Ed. Nuevo Centro, Avenida Libertador 405-407, Oficina3-Piso 9, Chacao, Caracas, Venezuela.

Vietnam
Centre for Research on Oil and Oil plants (CRHO). Ho Chi Minh City.

Western Samoa
School of Agriculture and Institute for Research. University of South Pacific, Alafua Campus, Apia.16.
Olamanu Crop Development Centre.

Bibliography

Abad, R.G. (1983) Generation of pest management schemes in coconut-based intercropping systems. *Philippine Journal of Coconut Studies* (Philippines) 8, 1-2: 31-36.

Abad, R.G. (1985) Management of important coconut diseases. *Philippine Journal of Coconut Studies* (Philippines) 10, 1: 11-16.

Abdul Salam, M., Sreekumar, D. and M.K. Mammen (1990) Mixed farming on homestead agriculture - an economical approach. *Indian Farming* 40, 5: 14-16.

Abdullah Sani Ramli, and Basery, M. (1982) The integration of cattle with coconut cultivation. I. Growth performance and production systems. *MARDI Research Bulletin* (Malaysia) 10, 3: 384-392.

Abdullah Sani Ramli, Mohd Shukri Haji Idris, and Basery M. (1982) *Beef cattle production in coconut smallholdings.* Paper presented at the National Coconut Conference, 25-26 May 1982, Kuala Lumpur, MARDI (Malaysia).

Abeywardena, V. (1979) Influence of watering on the yield of coconut. *Ceylon Coconut Quarterly* (Sri Lanka) 30: 91-100.

Abilay (1983) *Breeding selected upland crops for partial shade tolerance.* Paper presented at the Symposium on Coconut Based Farming Systems. Visca, Leyte (Philippines), June 1-4 1983.

Abrahams, A., and Thomas, K.J. (1962) A note of the *in vitro* culture of exised coconut embryos. *Indian Coconut Journal* (India) 12, 2: 84-87.

Adriano, F.T., and Manahan, M. (1931) The nutritive value of green, ripe and sport coconut (buka, niyog, and macapungo). *Philippine Agriculturist* (Philippines) 20, 3.

Aguilar, E.A., and Benard, G.H. (1991) *CBFS experiments in The Philippines: A status report.* Proceedings International Symposium on Coconut Research and Development, 26-29 Nov. 1991, CPCRI, Kerala, India.

Aldaba, F.R., (1995) Philippines et cocotiers: Problèmes et perspectives. *OCL* (France) 2: 203-206.

Alexander, W. (1921) A process for separation of food products from fresh coconuts. *US Patent* no 1: 366, 339.

Akpan, E.E.J., and Obisesan, I.Q. (1984) The effect of the climate factors upon nut characteristics of the coconut palm (*Cocos nucifera* L.). In: Pushparajah, E. and Chew Poh Soon (eds), *Cocoa and Coconuts; Progress and Outlook.* Incorporated Society of Planters, Kuala Lumpur (Malaysia) 733-744.

Alforja, L.M., *et al.* (1985) *Some productivity factors affecting the yield performance of MYD x WAT hybrid in The Philippines.* Paper presented during the 6th PCA PHF Conference. Lapu Lapu City (Philippines) September 16-18.

Amalu, U.C., Omoti, U., and Ataga, D.O. (1987) A note on the growth characteristics of two cultivars of Nigerian tall coconuts. *Nigerian Journal of Palm and Oil Seeds* (Nigeria) 8, 1: 115-121.

Amrizal, M.D., Androecia, D., Damanik, S., and Karmawati, E. (1989) Indonesian farmers' receptivity to new technologies in coconut. *Industrial Crops Research Journal* (Indonesia) 1, 2: 27-36.

Andrew M.H. (1972) *A Century of Coconuts.* Calvert, Vavasseur & Co Inc, New York, USA.

Androecia, D., and Damanik, S. (1989) Indonesia among the coconut producing countries. *Industrial Crops Research Journal* (Indonesia) 1, 2: 78-84.

Anilkumar, K.S., and Wahid, P.A. (1988) Root activity pattern of coconut palm. *Oléagineux* (France) 43, 8-9: 337-342.

— (1989) Impact of long-term inorganic fertilization on soil nutrient availability and nutrition of coconut palm. *Oléagineux* (France) 44, 6: 281-286.

Anonymous *Biotropic leaflets on coconut dehusking machine, and coconut wet process.* Biotropic, Parc Agropolis Bat 1 & 2, 2214 bd de la Lironde, 34980 Montferrier-sur-Lez, France.

— (1940) *Coconut Research Scheme,* Ceylon 1940. Report for 1939. Government Sessional Paper, no 8: 6.

— (1963) *Use of coconut preparation as a protein supplement in child feeding; prospects.* Protein Advisory Group, FAO/WHO/UNICEF, Rome (Italy) Report N4-R9/add 5.

— (1971) *Eleventh report of Research Dept. of Coconut Industry Board (C.I.B),* Jamaica, West Indies, July 1970-June 1971.

— (1972) *Guideline no. 13 for the preparation of milk substitutes of vegetable origin.* Protein Advisory Group of United Nations.

— (1976) *Progress in the improvement of pastures under coconuts in the New Hebrides at Saraoutou Experimental Station,* Saraoutou, Santo, New Hebrides, mimeo report.

— (1978a) Coconut cattle integrated farming in The Philippines. *Asian Livestock* 3, 5, 13.

— (1978b) *Coconut smallholders in Peninsular Malaysia.* International Conference on Cocoa and Coconuts, 14-17 June 1978, Kuala Lumpur, Malaysia, 721-730.

— (1979) *Tanzania Pilot Project 20/79.* FAO/World Bank Cooperative Programme. FAO, Rome, Italy.

— (1980a) Dairy demonstration units at National Dairy Research Institute, Karnal. *NDRI/ICAR* (India) Publication No. 177, 38 p.

— (1980b) *Coconut Processing Technology Information Documents* Parts 1 to 7. UNIDO/APCC, UNIDO/IOD, 377 with addenda 1 to 6.
— (1981) *Annual report.* Solomon Islands Pasture Research Project. University of Queensland, Australia, mimeo report.
— (1982a) *Coconut Statistics 1981 Annual (for Philippines).* United Coconut Association of Philippines Inc., Manila.
— (1982b) *The Philippines recommends for integrated cattle-coconut farming.* Philippines Council for Agriculture and Resources Research and Development, Technical Bulletin 51, Los Baños, Laguna, Philippines.
— (1983a) *Sri Lanka Coconut Statistics.* Ministry of Coconut Industry, Colombo.
— (1983b) *Coconut dehusker. Coconis* (Sri Lanka) 12, 6: 2.
— (1984a) Maquina cortadora de coco ofrece Inglaterra a Venezuela. *Coco y Palma* (Venezuela) 12: 4.
— (1985a) Coconut tree climber. *The Cocomunity* (Indonesia) 15, 2: 8.
— (1985b) Coconut Wood - Processing and use. *FAO Forestry Paper* no. 57, Rome, Italy.
— (1986a) Nitrogen, potassium and placement trial with coconuts. *Technical Report Tree Crops Research* (Tuvalu) 36-38.
— (1986b) *The effects of land tenure and fragmentation of farm holdings on agricultural development.* FAO COAG Dec. 1986. Provisional Agenda item 8, for the 9th Session of COAG, 23 March-April 1987, FAO, Rome, Italy, 28p.
— (1987a) *Terminal Statement of Project TCP/TON/6654 - Formulation Mission - Coconut Replanting.* FAO, Rome, Italy, 8p.
— (1987b) *Summary of the Fifth South Pacific Islands Regional Meeting on Agricultural Research,* Development, Extension and Training in Coconut. (IRETA, Western Samoa).
— (1987c) Planting Coconut Seedlings. *Advisory Circular* No. A1, Coconut Research Institute of Sri Lanka.
— (1988) World supplies of copra, coconut oil and meal will remain tight until early 1989. *Oil World* 29, 31: 241-244.
— (1989a) Production recovery in copra complex slower than expected. *Oil World* 32, 22: 173-176.
— (1989b) Cocotier. In: *Institut de Recherches pour les Huiles et Oléagineux.* Rapport d'Activité. Oléagineux (France) 44, 4: 72-95.
— (1989c) *The role of Phytophtora palmivora Butler, Phytophtora katsurae Ko & Chang and Thielaviopsis paradoxa Dade in premature nutfall of coconuts in Jamaica.* Twenty-fifth report of the Research Department (for calender year 1989) Coconut Industry Board of Jamaica W.I. (Jamaica) 34-53.
— (1989d) *Report on the meeting of the working group on Coconut-Based Farming Systems, held at Chumpon, Thailand,* September 1989. UNDP/FAO Project RAS/80/032. FAO (Italy).
— (1989e) Diversifying coconut plantations. *Article in Daily News* (Sri Lanka), 12 April 1989.
— (1990a) *Agronomy/Crop physiology. Report of the Research Department.* Coconut Industry Board (Jamaica) 25: 16-23.
— (1990b) *Improved coconut production. Terminal Report AGDP/RAS/80/032,* FAO (Italy) 20 pp.
— (1991a) *Project proposal for the detection and diagnosis of Lethal Yellowing and related diseases of coconut palm in Africa and the Caribbean Region.* Paper presented at the BUROTROP First African Coconut Seminar, Arusha, Tanzania.
— (1991b) Coconut industry in Tanzania. *Bulletin Burotrop* (France) 1, 7-11.
— (1991c) *Natural Resources Institute - Draft Outline Project Proposal Coconut Dehusking Machine.* Paper presented at BUROTROP First African Coconut Seminar, Arusha, Tanzania.
— (1992a) IRHO-CIRAD Activity Report 1989-1991 of the Institut de Recherches pour les Huiles et Oléagineux. *Oléagineux* (France) 47, 6: 327-374.
— (1992b) *Descriptors for Coconut (Cocos nucifera L.).* International Board for Plant Genetic Resources, Rome, Italy, 60 pp.
— (1992c) Psyllid strikes in Africa. *ILCA Newsletter* 11, 4: 5. (October).
— (1993a) Central Plantation Crops Research Institute. *Bulletin Burotrop* (France) 5: 21.
— (1993b) Now what? US study says margarine may be harmful. *Bulletin Burotrop* (France) 5: 27.
— (1983c) Production physiology in coconut. *Coconis* (Sri Lanka) 11: 26.
— (1993c) Canola could replace coconut and palm kernels. *Bulletin Burotrop* (France) 5: 27.
— (1993d) *APCC Coconut Statistics Year Book 1992.* Asian & Pacific Coconut Community, Jakarta, Indonesia.
— (1995) *Burotrop 1995-2005.* Achievements & Strategy.
Anupap Thiracul, Chulaphan Petchipiroon, and Maliwan Rattanapruk (1992) Existing Status of Coconut Genetic Resources of Thailand. In: *Report of a working session for implementation of the international Coconut Database held at CIRAD,* Montpellier, France 19-22, May 1992 (mimeographed).
Apacible, A.R. (1968) Selection of coconut. *Sugar news* 44: 93-98.
APCC (1987) *Statistical Year Book,* APCC Publication, Jakarta, Indonesia.
— (1990) *Coconut based farming system.* Proceedings of the APCC COCOTECH conference

Appert, J. (1974) Sur deux coleoptères Hispines du genre *Gestronella* nuisibles au cocotier à Madagascar. *Oléagineux* (France) 29, 12: 559-564.

Aravindakshan, M. (1991) Thrust areas in coconut management. In: Silas, E.G., Aravindakshan, M., and Jose, A.I., (eds), *Coconut breeding and Management*, Kerala Agricultural University (India) 175-181.

Arboleda, N.P., Floria, R.N., and Candy, C.S. (1986) *Coconut problems and approaches. The Philippine experience.* Paper for Workshop proceedings of the International Consultation on Plantation Crop Diversification. FAO (Italy) November 1986: 51-62.

Archibald, K.A.E. (1985) *Dairy cattle feeding in the humid or high rainfall tropics. In: Milk Production in Developing Countries* (ed. Smith, A.J.). Proceedings of the Conference held in Edinburgh from 2-6 April 1984. Univ. of Edinburgh, CTVM, 110-132.

Arganosa, A.S. (1991) *Research and Development status of livestock integration under coconut in the Philippines.* Proceedings of the FAO/MARDI International Livestock - Tree Cropping Workshop, 5-9 Dec. 1988, Serdang, Malaysia, 90-100.

Argañosa, E.L.O., and Gomez, E.D. (1991) Perception of the attributes of coconut - based farming systema technology options by coconut farmers in selected barangays of Sorsogon and Camarines Norte, Philippines. *The Philippine Journal of Coconut Studies* (Philippines) 16, 1: 16-21.

Arganosa, A.S., Trung, L.T., and Rigor, E.M. (1988) Reproductive performance of imported Sahiwal-Holstein Crosses in The Philippines under three management systems. *Philippine Journal of Veterinary and Animal Science* (Philippines) 14.

ARNAB (1987) *Utilization of Agricultural By-Products as Livestock Feeds in Africa.* Proceedings of a Workshop held in Blantyre, Malawi, Sept. 1986. ILCA. 179p.

Arope, A., Ismail, T., and Chong, D.T. (1985) Sheep rearing under rubber. *Planter*, Kuala Lumpur (Malaysia) 61: 70-77.

Asgarali, J., and Ramkalup, P. (1985) Study of *Lincus* sp. (Pentatomidae) as the possible vector of hartrot in coconut. *Surinaamse Landbouw* (Surinam) 33, 2: 55-61.

Ashburner, G.R., Thompson, W.K., Maheswaran, G., and Burch J.M. (1991) The effect of solid and liquid phase in the basal medium of coconut (*Cocos nucifera* L.) embryo culture. *Oléagineux* (France) 46, 4: 149-152.

Asmono D., Hartana A., Guhardja E. and Yahya S. (1993) Genetic diversity and similarity of 35 coconut populations based on isoenzyme banding pattern analysis. *Buletin Pusat Penelitian Kepala Sawit* 1, 1: 39-54.

Assy Bah, B. (1992) *Utilisation de la culture in vitro d'embryons zygotiques pour la collecte et la conservation des ressources génétiques du cocotier (Cocos nucifera L.).* Thèse de doctorat en sciences, Université Pierre et Marie Curie Paris VI (France), 147 p.

Assy Bah, B., and Engelmann, F. (1992a) Cryopreservation of immature embryos of coconut (*Cocos nucifera* L.). *Cryo-Letters* 13: 67-74.

— (1992b) Cryopreservation of mature embryos of coconut (*Cocos nucifera* L.) and subsequent regeneration of plantlets. *Cryo-Letters* 13: 117-126.

— (1993) *Medium-term conservation of mature embryos of coconut (Cocos nucifera L.).* Plant Cell, submitted.

Assy Bah, B., Durand-Gasselin, F., Engelmann, F., and Pannetier, C. (1989) Culture *in vitro* d'embryons zygotiques de cocotier (*Cocos nucifera* L.) Méthode, révisée et simplifiée, d'obtention de plants de cocotier transférables au champ. *Oléagineux* (France) 44, 11: 515-523.

Avilan,L., Rivas, N., and Sucre, R. (1984) Estudio del systema radical del cocotero (*Cocos nucifera* L.). (Study of the root system of coconut, *Cocos nucifera* L.). *Oléagineux* (France) 39, 1: 13-23.

Balce, S. (1953) *A survey of potential industrial products of the coconut.* Paper read at the Symposium: Problems of the coconut industry, 8th Pacific Science Congress, Quezon City, Philippines. 27 pp.

Ballad, M.E. (1991) Response of coconuts grown on three soils of Leyte, Philippines to organic and inorganic sources of nitrogen. *The Philippine Journal of Coconut Studies* (Philippines), 16, 2, 1-5.

Banzon, J.A. (1978) Reconstitution of milk using coconut milk and non fat dry milk. *Philippine Journal of Coconut Studies* 3, 2: 1-8.

— (1984) Harvestable energy from the coconut palm. *Energy in Agriculture*, 3, 337- 344.

Baranwal, V.K., Manikandan, P., and Ray, A.K. (1989) Crown choking disorder of coconut. A case of boron deficiency. *Journal of Plantation Crops* (India) 17, 2: 114-120.

Barile, E.R., and Sangalang, J.B. (1990) Variation in sunlight reduction under the canopy of different coconut cultivars and hybrids. *The Philippine Agriculturist* (Philippines) 73, 3-4: 287-295.

Barker, R., and Nyberg, A.J. (1968) Coconut-cattle enterprise in The Philippines. *Philippine Agriculturist* (Philippines) 52, 1: 49-60.

Barr, J.J.F. (1993) *Technical Report on Coconut Research in Kiribati 1990-1992.* Ministry of the Environment & Natural Resources Development. Division of Agriculture (Republic of Kiribati). 195 pp.

Bastine, C.L., Narayanan Nair, E.R., and Abdurazak, M.P. (1991) Adoption rate and constraints of adoption of newer technologies by coconut farmers in Northern Kerala. In: Silas, E.G., Aravindakshan, M., and Jose, A.I.

(eds), *Proceedings National Symposium on Coconut Breeding and Management*, Kerala University (India), pp. 251-254.

Baudouin, L., Asmady, Noiret, J.M. (1987) Importance des facteurs de l'environnement dans le choix de têtes de clones chez le palmier à huile. *Oléagineux* (France) 42, 7: 263-269.

Bavappa, K.V.A. (1986) *Research at Central Plantation Crops Research Institute* (CPCRI), Kasaragod, Kerala, India.

— (1991) Reminiscences. In: *Souvenir, Platinum Jubilee of Coconut Research and Development in India.* Central Plantation Crops Research Institute, Kasaragod, India.

Bavappa, K.V.A and Jacob, V.J. (1982) High intensity multispecies cropping: A new approach to small scale farming in the tropics. *World crops* 34: 47-50.

Bedford, G.O. (1975) Observations on the biology of *Xylotrupes gideon* in Melanesia. *Journal of the Australian Entomological Society* (Australia) 14: 216-16.

— (1976) Observations on the biology and ecology of *Oryctes rhinoceros* and *Scapanes australis* (Col. Scarabeidae Dynastidae), pests of coconut palms in Melanesia. *Journal of the Australian Entomological Society* (Australia) 15: 241-251.

— (1980) Biology, ecology and control of palm Rhinoceros beetles. *Ann. Rev. Entomol.* 25: 309-339.

Been, B.O. (1981) Observations on field resistance to lethal yellowing in coconut varieties and hybrids in Jamaica. Cyclostyled. Research Dept. Coconut Industry Board, Kingston (Jamaica) *Oléagineux* (France), 36, 1: 9-12.

— (1992) Status of existing coconut collection in Jamaica. In: *Report of a working session for implementation of the international Coconut Database held at CIRAD*, Montpellier, France 19-22 may 1992 (France) (mimeographed).

Beets, W.C. (1990) *Raising and Sustaining Productivity of Smallholder Farming Systems in the Tropics.* AgBe Publishing, Alkmaar, The Netherlands.

Benjamin, A., Shelton, H.M., and Gutteridge, R.C. (1991) *Shade tolerance of some tree legumes.* In: Proceedings of the Workshop on Forages for Plantation Crops, Shelton, H.M. and Stur, W.W. (ed.). Sanur Beach, Bali, Indonesia, 27-29 June, 1990. ACIAR Proceedings (Australia) 32, 75-76.

Bennet, C.P.A., Roboth, O, Sitepu, G. and Lolong, A. (1986) Pathogenicity of Phytophthora palmivora (Butler) causing premature nutfall disease of coconut (*Cocos nucifera* L.). *Indonesian Journal of Crop Science* (Indonesia) 2, 2, 59-70.

Benoit, H., and Ghesquiere, M. (1984) Electrophorèse, compte rendu cocotier. IV. Déterminisme génétique (Electrophoresis, report on coconut. IV. Genetic determinism). *Rapport interne IRHO-CIRAD*, (France). 11 pp.

Bhaskara Rao, E.V.V., and Koshi, P.K. (1981) *Coconut germplasm collection in Pacific Ocean Islands.* IBPGR/-FAO/CPCRI Project. CPCRI, Kasaragod (India). 47 pp. (mimeographed).

Bhaskara Rao, E.V.V., Pillai, R.V., and Ratnambal, M.J. (1993) Current Status of genetic resources research in India. In: *Advances in Coconut Research and Development.* Platinium Jubilee of Coconut Research in India. M.K. NAIR *et al.* (eds), Oxford & IBH Publishing CO. PVT. LTD. (India), pp. 15-22.

Bhaskaran, R., Ramadoss, N., and Ramachandran, T.K. (1988) Biological control of Thanjavur Wilt disease of coconut. *Indian Coconut Journal* (India) 19, 6: 3-8.

Bhaskaran, R., Rethinam, P., and Nambiar, K.K.K. (1989) Thanjavur Wilt of coconut. *Journal of Plantation Crops* (India) 17, 2: 69-79.

Bhaskaran, R., Suriachandraselvan, M., and Ramachandran, T.K. (1990) *Ganoderma* Wilt disease of coconut. A threat to coconut cultivation in India. *Planter* (Malaysia) 66, 774: 467-471.

Bhaskaran, U.P., and Leela, K. (1978) Response of coconut to irrigation in relation to production status of palms and soil type. In: *PLACROSYM - I, Proceedings of the First Annual Symposium on Plantation Crops* (India) pp. 201-206.

Bhat, S.K., and Sujathna, A. (1989) Relative acceptance of different baits by Rattus rattus wroughtoni Hinton. *Journal of Plantation Crops* (India) 17, 2: 121-127.

— (1993) Rodent and other vertebrate pest management in coconut and cocoa. *Technical Bulletin no. 26, Central Plantation Crops Research Institute* (India), 13 pp.

Biberson, O., and Duhamel, G. (1987) Poisoning of coconuts with MSMA (Monosodic methylarsonate). *Oléagineux* (France) 42, 10: 385-386.

Biddappa, C.C. and Bopaiah, M.G. (1989) Effect of heavy metals on the distribution of P, K, Ca, Mg, and micronutrients in the cellular constituents of coconut leaf. *Journal of Plantation Crops* (India) 17, 1: 1-9.

Biddappa, C.C., Khan, H.H., Joshi, O.P. and Manikandan, P. (1988) Effect of heavy metals on micronutrient nutrition of coconut. *Current Science* 57, 20: 1111-1113.

Bin Mohd Kamil, N.F., and Bin Ahmed, M.S. (1978) *Socio-economic status of coconut smallholders in Lower Perak.* Proceedings International Conference on Cocoa and Coconuts. Kuala Lumpur (Malaysia) pp. 731-741.

Birusel, D.M. (1963) Physiomechanical process of obtaining high quality coconut oil. *US Patent* no 3, 106: 571 (8 October 1963).

Blaak, G., (1983) *Report on an advisory mission on the agronomy of the coconut palm on Hainan Island, Peoples Republic China.* Consultant report to Food and Agricultural Organization of the United Nations, Rome, Italy. Koninklijk Instituut voor de Tropen, Department of Agricultural Research. Amsterdam, The Netherlands.

— (1986) *Coconut diversification problems and approaches. Considerations on management of light and mycorrhizas to reduce N and P fertilizer costs.* In: Proceedings of the Workshop of the International Consultation on Plantation Crop Diversification. FAO (Italy), November 1986.

Blackburn et al. (1987) *Testimony before the U.S. Senate Committee on labor and human resources on S.1109, a bill to amend the Federal Food, Drug and Cosmetic act to require new labels on foods containing Coconut, Palm and Palm kernel oil.* Harvard Medical School (U.S.A.) Nutrition Coordinating Center New England Deaconess Hospital, Boston MA, December 1, 1987.

Blackburn, G.L., Kater, G., Maxioli, E.A., Kowalchuk, M., Babayan, V.K., and Bistrian, B.R. (1992) A revelation of coconut oil's effect on serum cholesterol and atherogenesis. *Philippine Journal of Coconut Studies* (Philippines) 17, 2: 21-28.

Blair, G.P. (1970) Studies on Red Ring disease of the coconut palm. *Oléagineux* (France) 5, 1: 19-22 and 5. 2: 79-83.

Blake, J. (1989) Coconut (*Cocos nucifera* L.) Micropropagation. In: Bajaj, Y.P.S. (ed.) *Biotechnology in Agriculture and Forestry* 10, *Legumes and Oil Seed Crops* pp. 538-554.

Bock, K.R., Ivory, M.H., Adams, B.R. (1970) Lethal Bole Rot disease of coconut in East Africa. *Ann. appl. Biol.* 6: 453-464.

Bondar, G. (1940a) Notas entomologicas da Bahia. VI. *Revista Entomologica de Rio de Janeiro* (Brazil) 11, 3: 842-861.

— (1940b) *Insetos nocivos e molestias do coqueiro (Cocos nucifera) no Brasil.* Tipografia naval, Bahia (Brazil) 156 pp.

Bonneau, X., Ochs, R., Kitu, W.T., and Yuswohadi (1993a) Chlorine, an essential element in the mineral nutrition of hybrid coconuts in Lampung (Indonesia). *Oléagineux* (France) 48, 4: 179-190.

Bonneau, X., Ochs., R., Qusari, L., and Nurlaini Lubis, L. (1993b) Hybrid coconut mineral nutrition on peat, from the nursery to the start of production. *Oléagineux* (France) 48, 1: 9-26.

Boonklinkajorn, P., Duriyaprapan, S., and Pattanavibul, S. (1982) *Grazing trial on improved pasture under coconuts.* Report No. 12, Institute of Scientific Research and Technology of Thailand.

Bopaiah, B.M., (1988) *Microbiological studies in relation to high density multispecies cropping systems in coconut.* PH.D. thesis. University of Myore, India.

— (1990) Microbiological and enzyme activities profile in the root zone and interspace soils of coconut and arecanut palms. *Journal of Plantation Crops* (India) 18, 1, 50-54.

— (1991) Recycling the coconut wastes to improve the soil fertility in coconut gardens. *Indian Coconut Journal* (India) 22, 3: 2-3.

Bopaiah, B.M., Shetty, H.S., and Nagaraja, K.V. (1987) Biochemical characterization of the root exudates of coconut palm. *Current Science* (India) 56, 16: 832-833.

Borah, S.C. (1991) Standardization of nursery techniques in coconut. In: Silas, E.G., Aravindakshan, M., and Jose, A.I. (eds), *Coconut breeding and management.* Kerala Agricultural University (India) 276-277.

Barker, R., and Nyberg, A.J. (1968) Coconut-cattle enterprise in The Philippines. *Philippine Agriculturist* 52, 1: 49-60.

Bourdeix, R. (1988a) Efficacité de la sélection massale sur les composantes du rendement chez le cocotier. *Oléagineux* (France) 43, 7: 283-295.

— (1988b) Déterminisme génétique de la couleur du germe chez les cocotiers Nains. *Oléagineux* (France), 43, 10: 371-374.

— (1989) *La sélection du cocotier Cocos nucifera L. Etude théorique et pratique, optimisation des stratégies d'amélioration génétique.* Thèse de doctorat en sciences, Université de Paris-Sud Centre d'Orsay (France).

Bourdeix, R. (1995) *Personal communication.*

Bourdeix, R., Meunier, J., and N'Cho, Y.P. (1991a) Une stratégie de sélection du cocotier *Cocos nucifera* L. II. Amélioration des hybrides Grand x Grand. *Oléagineux* (France) 46, 7: 267-282.

Bourdeix, R., Meunier, J., and N'Cho, Y.P. (1991b) Une stratégie de sélection du cocotier *Cocos nucifera* L. III. Amélioration des hybrides Nain x Grand. *Oléagineux* (France) 46, 10: 361-374.

Bourdeix, R., and N'Cho, Y.P. (1992) Influence du sens du croisement sur la production de huit hybrides entre écotypes Nains et Grands. *Oléagineux* (France) 47, 3: 113-118.

Bourdeix, R., N'Cho, Y.P., and Le Saint, J.P. (1990) Une stratégie de sélection du cocotier. Synthèse des acquis. *Oléagineux* (France) 45, 8-9: 359-371.

Bourdeix, R., N'Cho, Y.P., and Sangare, A. (1995) *Amélioration de l'hybride de cocotier Grand Ouest Africain x Grand Rennell. Comparaison avec un témoin PB121 (Nain Jaune Malaisie x Grand Ouest Africain).*

Bourdeix, R., N'Cho, Y.P., Sangare, A., and Baudouin, L. (1994) Résultats et perspectives de l'amélioration génétique du cocotier. In: *Actes du colloque sur la recherche européenne au service du cocotier*. Montpellier, Septembre 1993. CIRAD-CP ed. France).

Bourdeix, R., N'Cho, Y.P., Sangare, A., Baudouin, L., and Nucé de Lamothe, M. de (1992) L'hybride de cocotier PB121 amélioré, croisement de Nain Jaune Malaisie et de géniteurs Grand Ouest Africain améliorés. *Oléagineux* (France) 47, 11: 619-633.

Bourdeix, R., N'Cho, Y.P., Sangare, A., Le Saint, J.P., and Nucé de Lamothe, M. de (1994) Rythmes de production chez le cocotier Nain (*Cocos nucifera* L.); étude de l'alternance castration-production comme mode de gestion des champs semenciers. *Agronomie Africaine* (France).

Bourdeix, R., Sangare, A., and Le Saint, J.P. (1989) Efficacité des tests hybrides d'aptitude individuelle à la combinaison chez le cocotier: premiers résultats. *Oléagineux* (France) 44, 5: 209-214.

Bourdeix, R., Sangare, A., Le Saint, J.P., and Nucé de Lamothe, M. de (1991) La sélection du cocotier à l'IRHO et son application pour la production de semences. In: *Platinium Jubilee of Coconut Research in India Kerala Agricultural University*, Kasaragod (India).

Bourgoing, R. (1989) Setting up hybrid coconut smallholder plantations on peat soils in tidal zones. Example of Indonesia. *Oléagineux* (France) 44, 6: 287-293.

— (1990) Choice of cover crop and planting method for hybrid coconut growing on smallholdings. *Oléagineux* (France) 45, 1: 23-30.

Bourgoing, R., and Boutin, D. (1987) Méthode de contrôle de l'imperata par l'utilisation d'un rouleau en bois leger et implantation de la plante de couverture (Pueraria) sur jeunes plantations hybrides de cocotiers en milieu villageois. *Oléagineux* (France) 42, 1: 19-23.

Braconnier, S., and d'Auzac, J. (1985) Anatomical study and cytological demonstration of potassium and chlorine flux associated with oil palm and coconut stomatal opening. *Oléagineux* (France) 40, 11: 547-551.

— (1989) Effect of soil chlorine deficiency on the coconut hybrid PB 121. *Oléagineux* (France) 44, 10: 467-474.

— (1990) Chloride and stomatal conductance in coconut. *Plant Physiology and Biochemistry* 28, 1: 105-111.

Braconnier, S., Zakra, N., Weaver, R., and Ouvrier, M. (1992) Fertilizer nitrogen distribution in coconut hybrid PB 121, using ^{15}N. *Oléagineux* (France) 47, 2: 63-69.

Bradley, J.D., (1965) A comparative study of the coconut flat moth (*Agonoxena argaula* Meyr.) and its allies, including a new species (Lepidoptera Agonoxenidae). *Bulletin of Entomological Research* 56: 453-472.

Branton, R.L., and Blake, J. (1983a) Development of organized structures in callus derived from explants of Cocos nucifera *L. Ann. Bot.* 52: 673-678.

— (1983b) A lovely clone of coconuts. *New Scientist* 26: 554-557.

Briton-Jones, H.R. (1940) *The diseases of the coconut palm.* Baillière, Tindal and Cox, London (UK).

Brookfield, H.C. (1985) *Land, cane and coconuts.* The Australian National University, Research School of Pacific Studies, Canberra, Australia, 251 pp.

Brown, E.S. (1954) The biology of the coconut pest *Melittomma insulare* (Col. Lymexylonidae) and its control in the Seychelles. *Bulletin of Entomological Research* 45: 1-66.

— (1959) Immature nutfall of coconuts in the Solomon Islands. *Bulletin of Entomological Research* 50, 96-133: 523-566.

Brunin, C., and Coomans, P. (1973) Boron deficiency in young coconuts in Ivory Coast. *Oléagineux* (France) 28, 5: 229-234.

Brunin, C., and Ouvrier, M. (1973) Fluor toxicity from certain phosphates in coconut in the nursery. *Oléagineux* (France) 28, 11: 509-512.

Buckley, R., and Harries, H.C. (1984) Self-sown wild-type coconuts from Australia. *Biotropica* 16, 148-151.

Buffard-Morel, J., Verdeil, J.L., and Pannetier, C. (1988) *Vegetative propagation of coconut palm (Cocos nucifera L.) through somatic embryogenesis.* Proceedings 8th International Biotechnical Symposium, Paris, (France) p. 117.

Burgess, R.J. (1981) *The intercropping of smallholder coconuts in Western Samoa: An analysis using multi-stage linear programming.* Australian National University Press, Canberra, Austalia.

Cachan, P. (1959) Etude épidémiologique de la zygène en Côte d'Ivoire. *J. Agric. Trop. Bot. Appl.* 6, 12: 653-674.

Calub, D.J. (1989) *Coconut replanting in The Philippines.* Report of the Workshop on Coconut Replanting, 23-27 November, 1988, MARDI, Serdang, Selangor, Malaysia (ed. Sivapragasam, A. and Jamil, M.M. b.M.). UNDP FAO Project RAS/80/032 Improved Coconut Production in Asia and Pacific, Annex 1,5, Country Report, 102-126.

Calvez, C., Julia, J.F., and Nucé de Lamothe, M. de (1985) L'amélioration du cocotier au Vanuatu et son intérêt pour la région Pacifique. *Oléagineux* (France), 40, 10: 477-490.

Calvez, C., Renard, J.L., and Marty, G. (1980) Tolerance of the coconut hybrid Local x Rennell Tall to New Hebrides Disease. *Oléagineux* (France) 35, 10: 443-451.

Carandang, D.A. (1975). *Yield response of arrowroot (Maranta arundinacea) grown under coconut trees to N, P and K fertilization.* In: Research Reports 1975. University of The Philippines at Los Baños pp 270.

— (1977) *Multiple cropping: a means of increasing income in coconut and upland farms in The Philippines.* Unpublished PhD Thesis. University of The Philippines at Los Baños.

Carcallas, C.D and Aparra, N.O. (1983) *Coconut based cropping involving annual crops in Eastern Visayas.* Paper presented at the National Symposium on Coconut Based Farming Systems, Visca, Leyte, June 1-4, 1983.

CARFV (1993) *Rapport d'activité 1991-1992 du Centre Agronomique de Recherche et de Formation du Vanuatu.* (mimeotyped).

Carpio, C.B. (1983) Biochemical Studies of *Cocos nucifera* L. *Philippine Journal of Biology* (Philippines) 11, 2-3: 319-338.

Carrad, B. (1977) *Cattle and coconuts: a study of copra estates in the Solomon Islands.* Draft report for South Pacific Commission, ANU, Canberra, Australia.

Cassidy, N.G. (1968) A note on the salt tolerance of the coconut palm (*Cocos nucifera* l.). *Tropical Agriculture* (Trinidad) 45: 217-250.

Cecil, S.R. (1991) Calcium and magnesium nutrition of coconut palm. In: Silas, E.G., Aravindakshan, M., and Jose, A.I. (eds), *Coconut breeding and management,* Kerala Agricultural University (India) pp. 219-224.

Cecil, S.R., and Khan, H.H. (1993) Nutritional requirement of coconut and coconut based farming systems in India. In: Nair M.K., Khan, H.H., Gopalasundaram, P., and Bhaskara Rao, E.V.V. (eds), *Advances in Coconut Research and Development.* Indian Society for Plantation Crops (India) 257-275.

Chaillard, H., Daniel, C., Houeto, V., and Ochs, R. (1983) Oil palm and coconut irrigation. A 900 ha 'experiment' in the Benin's Republic. *Oléagineux* (France) 38, 10: 519-533.

Chakra Borthy, B.K., Nath, B.K., Dhar, P.B., and Gosnami, R.N. (1970) Note on Crown Rot disease in coconut. *Indian Journal of Agricultural Research* (India) 40, 6: 502-504.

Chan, E. (1981) Coconut research in United Plantation Berhad, 1981 report. In: *1981 Yearly Progress report on Coconut research and development.* FAO, Rome (Italy) 1982: 6 pp.

Chandar Rao, S., Patil, K.D., and Dhandar, D.G. (1993) Stem Bleeding Disease of coconut (*Cocos nucifera*) in Goa - Present status and strategy for its management. *Indian Coconut Journal* (India), 23, 8: 2-4.

Charles, A.E., (1961) Selection and breeding of the coconut palm. *Tropical Agriculture* (Trinidad) 38: 283-296.

Chee, Y.K., and Ahmad Faiz (1991) Sheep grazing reduces chemical weed control in rubber. In: *Forages for Plantation Crops. Proceedings of a Workshop* (ed. Shelton, H.M. and Stur, W.W.), Sanur Beach, Bali, Indonesia, 27-29 June, 1990, pp. 120-123.

Chen, C.P. (1989) *Problems and prospects of integration of forage into permanent crops. In: Grasslands and Forage Production in South-East Asia* (ed. R.A. Halim). Proceedings First Meeting of Regional Working Group on Grazing and Feed Resources of South-East Asia, 27 Feb.- 3 March 1989, Serdang, Malaysia, pp. 128-139.

— (1991) *Management of forages for animal production under tree crops.* In: Proceedings International Work shop on Integrated Tree Cropping and Small Ruminant Production Systems (ed. Iniguez, L.C. and Sanchez, M.D.), Medan, North Sumatra, Indonesia, 9-14 Sept., 1990, pp. 10-23.

Chen, C.P., Tajuddin Ahmed, Z., Wan Mohamed, W.E., Tajuddin I., Ibrahim, C.E., and Salleh, Mod. R. (1991) *Research and Development on Integrated Systems in Livestock, Forage and Tree Crops Production in Malysia.* In: Proceedings of the International Livestock-Tree Cropping Workshop, 5-9 December 1988, FAO/MARDI, Serdang, Malaysia, pp. 55-72.

Cheva-Isarakul, B.S. (1991) *Livestock production under tree crops in Thailand.* In: Proceedings of the FAO/-MARDI International Livestock-Tree Cropping Workshop, 5-9 December 1988, FAO/MARDI, Serdang, Malaysia, pp. 112-119.

Chew, P.S., and Ooi, S.C. (1982) *Studies on the productivity of coconut palm (Cocos nucifera L.) in Peninsular Malaysia.* Paper presented at the National Coconut Conference. Kuala Lumpur (Malaysia), IRRI, September 1980.

Chhabra, N. (1991) Performance must continue in coconut. *Indian Horticulture* (India) 1.

Chikkasubbanna, V., Jayaprasad, K.V., Subbaiah, T., and Poonasha, N.M. (1990) Effect of maturity on the chemical composition of tender coconut (*Cocos nucifera* L. var. Arsikere Tall) water. *Indian Coconut Journal* (India) 20, 12: 10-12.

Child, R. (1964) *Coconuts.* Longman, London, UK, 216 pp.

— (1974) *Coconuts.* 2nd. ed. Longman, London (UK), 335 pp.

— (1975) Coconut breeding in Fiji. In: *Proceedings Fourth session FAO Technical Working Party on Coconut Production,* Production and Processing, Kingston. Rome (Italy).

Childs, A.H.B., and Groom, C.G. (1964) Balanced farming with cattle and coconuts. *East African Agricultural and Forestry Journal* 29, 3: 206-207.

Child, R., and Nathanael, W R N. (1947) Utilization of coconut water. *Tropical Agriculture Magazine, Ceylon Agricultural Society* 103: 85-89.

— (1950) Changes in the sugar composition of coconut water during maturation and germination. *Journal of Science of Food and Agriculture* 1: 326-329.

Chong, Y.H. (1989) Dietary fats, classification and cholesterolic effects. *The Planter* (Malaysia) 65: 563-571.

Chong, D.T., Tajuddin, I., and Abd. Samat, M.S. (1991) *Productivity of cover crops and natural vegetation under rubber in Malaysia.* In: Proceedings of the Workshop on Forages for Plantation Crops (ed. Shelton, H.M. and Stur, W.W.), Sanur Beach, Bali, Indonesia, 27-29 June 1990. ACIAR Proceedings 32, 36-37.

CIB (1973) *Coconut Industry Board of Jamaica.* Thirteenth report of the Research Department for the period July 1972-June 1973.

CIRAD-CP (1993) *Rapport de mission au Brésil.* Document interne (France) 78/93.

Cintra, F.L.D., Leal, M. de L. da S., (1992) Distribuição do sistema radicular de coqueiros anões. *Oléagineux* (France) 47, 5: 225-233.

Cintra, F.L.D., Passos, E.E. de M., and Leal, M. de L. da S. (1993) Avaliação do sistema radicular de cultivares do coqueiro gigante. *Oléagineux* (France) 48, 11: 453-461.

Cochereau, P. (1965) Contrôle biologique d'*Aspidiotus destructor* Sign. dans l'Île Vaté, Nouvelles-Hebrides au moyen de *Lindorus lophantae* Blaisd. C.R. *Acad. Agric. France* (France) 51, 5: 318-321.

Cock, M.J.W., and Perera, P.A.C.R. (1987) Biological control of *Opisina arenosella* Walter (Lepidoptera Oecophoridae). *Biocontrol News and Information* 8: 283-316.

Comstock, R.E., Robinson, H.F., and Harvey, P.H. (1949) A breeding procedure to make maximum use of both General and Specific Combining Ability. *Agronomic Journal* 41: 360-367.

Concibido, E. (1985) Wilt disease of coconut in Socorro, Oriental Mindoro, Philippines. *Philippine Journal of Coconut Studies* (Philippines) 10, 1: 28-30.

Coomans, P. (1975) Influence of climate factors on seasonal fluctuations of coconut production. *Oléagineux* (France) 30, 4: 153-157.

Coomans, P., and Ochs, R. (1976) Rentabilité des fumures minérales sur cocotier dans les conditions du Sud-Est Ivoirien. *Oléagineux* (France) 31, 8-9: 375-382.

Corbett, G.H. (1923) The two coloured coconut leaf-beetle. Department of Agriculture. *Straits Settlements and Federated Malay States Bulletin* 74.

— (1932) Insects of coconut in Malaya. Department of Agriculture. *Straits Settlement and Federated Malay States Bulletin* 10.

Corley, R.H.V. (1983) Potential productivity of tropical perennial crops. *Experimental Agriculture* (UK) 19, 3: 217-237.

Cosico, W.C. (1983) *Land and soil evaluation of areas planted to coconut (Cocos nucifera L.) in The Philippines.* Unpublished PhD thesis. U.P. at Los Banos.

Cosico, W.C., and Fernandez, N.C. (1983) Effect of some land qualities and soil properties on productivity of coconut in The Philippines. *Philippine Journal of Coconut Studies* (Philippines) 8, 1-2: 25-30.

Costa, R.G., Passos, E.E.M., and Gheyi, H.R. (1986) *Aplicação de água salina na irrigação de plantas jovens de coqueiro (Cocos nucifera L.). (Cocos nucifera L.). Pesquisa em Andamento.* Ministério de Agricultura (Brazil) 1-5.

Crawford, M. (1940) Coconut poonac as a food for livestock. *Tropical Agriculture Magazine,* Ceylon Agricul tural Society 44: 168.

Creencia, R. (1978) *Cultural management of coconuts.* Paper presented at the 5th. National Coconut Consulta tion, Bago-Oshiro, Davao, Philippines.

Cruz, A.O., and West, A. P. (1930) Water-white coconut milk and coconut flour. *Philippine Journal of Science* 41: 51-58.

Cruz, M.T., and Bernardo, M. V. (1949) The extraction of liquid contents from raw coconuts and the separation of the oil therefrom. *Philippines patent* 39. (1.12.49).

Cuavas, S.E. (1975) *Annual crop species under coconuts.* National Coconut Research Symposium. Tacloban City Philippines). November 17-19 1975.

Cummings, J. (1993) *Coconut processing material available from AJAX.* Proceedings of the Seminar on Small Scale and Medium Scale Technology for oil palm and coconut. Burotrop (France) 29.

Cundall, E.P. (1987) *Coconut breeding programme of the PNG.* Cocoa and Coconut Research Institute. Proceedings of the Fifth South Pacific Islands Regional Meeting on Agricultural Research, Development, Extension and Training in Coconut (IRETA, Western Samoa) 82-93.

Cutter, V.M., and Wilson, K.S. (1954) Effects of coconut endosperm and other growth stimulants upon the development *in vitro* of embryos of *Cocos nucifera. Bot. Gazette* 115: 234-240.

Dabek, A.J. (1993) Lethal diseases of coconut, oil and rafia palm in Madagascar. *FAO Plant Protection Bulletin* (Italy) 41, 1: 15-21.

Dabek, A.J., Johnson, C.G., and Harries, H.C. (1976) Mycoplasma-like organisms associated with Kaincopé and Cape St. Paul Wilt diseases of coconut palm in West Africa. *PANS* 22, 3: 354-358.

Daniel, C. (1991) Behaviour of coconut in drought conditions. *Bulletin Burotrop* (France) 2: 15-21.

Daniel, C., Adje, I., and Vihoundje, F. (1991) Dwarf x Tall coconut hybrid performance in a dry climate with supplemental irrigation. *Oléagineux* (France) 46, 1: 13-22.

Daniel, C., and Manciot, R. (1973) The chlorine nutrition of young coconuts in the New-Hebrides. *Oléagineux* (France) 28, 2: 71-72.

Darbin, T., Pomier,M., and Taffin, G. de (1983) The coconut palm and the improvement of valley bottoms in the Middle Ivory Coast. *Oléagineux* (France) 38, 4: 231-242.

Darwis, S.N. (1988) *Status of intercropping on coconut lands in Indonesia.* In: Proceedings Workshop on Intercropping and Intergrazing in Coconut Areas, 7-11 September, 1988, Colombo, Sri Lanka.

— (1991) Indonesian coconut research highlights. *Indian Coconut Journal* (India) 22, 6-7: 22-32.

Darwis, S.N., and Luntungan, H.T. (1992) Status of existing coconut collection in South East Asia. *Industrial Crops Research Journal* (Indonesia) 4, 2: 27-39.

Das, P.K. (1984) *Economics of cocoa mixed cropping with coconuts in India.* In proceedings of the Sixth Symposium on Plantation Crops. PLACROSYM VI Steering Committee, Kasaragod, India. pp. 397-408.

— (1985) Trends in oil production and trade in the world. *Oléagineux* (France) 40, 2: 85-90.

— (1990) Economics of coconut based farming systems. XXVII COCOTECH Meeting, 25-29 June, 1990, Manila, Philippines, 539-554.

— (1991a) *Coconut intercropping with cassava: An economic analysis.* CORD (Indonesia) 7, 2: 58-65.

— (1991b) Economic viability of coconut based farming systems in India. *Journal of Plantation Crops* (India) 19, 2: 191-201.

Das, P.K., Yusuf, M., and Hedge, M.R. (1991) Reducing the risk in rainfed coconut cultivation. *Indian Farming* (India) 1: 29-34.

Davis, T.A., and Altevogt, R (1988) The myths and facts of *Birgus*. *Coconut Today* (Philippines): June: 9-33.

Davis, T.A., Kaat, H., and Upara, A.R. (1985) A fertilizer trial to correct foliar yellowing in Ternate, Indonesia. *Philippine Journal of Coconut Studies* (Philippines) 10, 2: 14-23.

Davison, T.M., Silver, B.A., Lisle, A.T., and Orr, W.N. (1988) The influence of shade on milk production of Holstein-Friesian cows in a tropical upland environment. *Australian Journal of Experimental Agriculture* (Australia) 28: 149-154.

Dayrit, C.S., Florentino, R., Blackburn, G., Maxioli, E., and Babayan, V.K. (1992) Coconut oil revisited. *Philippine Journal of Coconut Studies* (Philippines) 17, 2: 11-14.

Demarly, Y. (1977) *Génétique et amélioration des plantes.* Masson (France) 287 pp.

Desmier de Chenon, R. (1975) Présence en Indonésie et Malaisie d'un lépidoptère mineur des racines du palmier à huile, *Sufetula sunidesalis* Walker et relations avec les attaques de *Ganoderma*. *Oléagineux* (France) 30: 449-456.

— (1982) *Latoia (Parasa) lepida* (Cramer), Lepidoptera limacodidae, ravageur du cocotier en Indonésie. *Oléagineux* (France) 37: 177-183.

Devendra, C. (1989) *Ruminant production systems in developing countries: resource utilization.* In: Feeding Strategies for Improving Productivity of Ruminant Livestock in Developing Countries. Proceedings of a combined advisory group meeting and research co-ordination meeting, Vienna 13-17 March 1989. IAEA, Vienna (Austria) pp. 5-30.

— (1993) Sustainable animal production from small farm systems in South-East Asia. *FAO Animal Productivity and Health Paper* No. 106, 143 pp.

Devendra, C., and Pun, H.L. (1991) *Practical technologies for mixed small farm systems in developing countries.* In: Proceedings Expert consultation on strategies for sustainable animal agriculture in developing countries. Held 10-14 Dec. 1990, FAO, Rome, Italy, pp. 135-156.

Dhanapal, R., Yusuf, M., and Bopaiah, M.G. (1995) Moisture movement studies under drip irrigation in coconut basins. *Journal of Plantation Crops* (India) 23, 1: 28-34.

Dharmarajn, E. (1962) Parasites, the hyperparasites, predators and pathogens on the coconut leaf caterpillar *Nephantis serinopa* Meyrick recorded in Ceylon and in India and their distribution in these countries. *Ceylon Coconut Quarterly* (Sri Lanka) 13: 102-111.

Diakno Process (1960) - *Philippines patent* 7,674.

Diaz, M.L.R., and Villareal, D.Z. (eds), (1990) *La problematica del Amarillamento Letal del cocotero en Mexico.* Centro de Investigacion Cientifica de Yucata, A.C. (Mexico).

Ditablan, E.G., and Astete, L.M. (1986) The coconut based multi-story cropping system. Asian Pacific Coconut Community, *Quarterly Supplement* (Indonesia) 5: 45-55.

Dollet, M., et al. (1977) Etude d'un jaunissement léthal des cocotiers au Cameroun: la maladie de Kribi. Observations d'organismes de type mycoplasmes. *Oléagineux* (France) 32, 7: 317-322.

Dollet, M., Gianotti, J., and Czarnecky (1976) Maladie de Kaincopé: présence de mycoplasmes dans le phloème des cocotiers malades. *Oléagineux* (France) 31, 4: 169-171.
Dollet, M., and Wallace, F.G. (1987) Compte rendu du premier *Phytomonas* Workshop - Cayenne, Mars 1987. *Oléagineux* (France) 42, 12: 461-468.
Dootson, J., Ratanapruk, M., and Suwannawuth, W. (1987) Coconut development in Thailand and its stimulation by cocoa. *Planter* (Malaysia) 62, 729: 519-530.
Dufour, F.O., Quyillec, G., Olivin, J., and Benard, J.L. (1984) Revelation of a calcium deficiency in coconut. *Oléagineux* (France) 39, 3: 133-142.
Duhamel, G. (1987) Piquetage des cocoteraies. *Oléagineux* (France) 8-9: 325-326.
— (1993) Crop improvement programmes in South Pacific region. In: NAIR M.K., *et al.* (eds), *Advances in Coconut Research and development. Platinium Jubilee of Coconut Research in India.* Oxford & IBH Publishing CO. PVT. LTD. (India) 95-100.
Dulay, T.A. (1980) *Laboratory manual in dairy science*, 135. Dairy Training and Research Institute, U.P., Los Baños College, Laguna, Philippines. 99 pp.
Dupuy, B., and N'Guessan, K. (1991) Utilisation des acasias pour régénerer les anciennes cocoteraies. *Bois et Forêts des Tropiques* (France) 230: 15-29.

Ebert, A.W., Rillo, E.P., Orense, O.D., Areza, M.B.B., and Cueto, C.A. (1991) Philippine-German Project on coconut tissue culture - First results. *The Philippine Journal of Coconut Studies* (Philippines) 16, 1: 12-15.
Eden-Green, S.J., and Schuiling, M. (1978) Bole rots in pre-bearing coconut palms apparently affected by the Lethal Yellowing disease. *FAO Plant Protection Bulletin* (Italy) 26, 1: 13-15.
Eeuwens, C.J. (1976) Mineral requirements for growth and callus initiation of tissue explants from mature coconut palms (*Cocos nucifera* L.) and date (*Phoenix dactylifera*) palm cultured *in vitro. Physiol. Plant* 36: 23-28.
Egara, K., and Jones, R.J. (1977) Effect of shading on the seedling growth of the leguminous shrub *Leucaena leucocephala. Australian Journal of Experimental Agriculture and Animal Husbandry* (Australia) 17: 976-981.
Egara, K., Kodpat, W., Manidool, C., Intaramanee, S., Srichoo, C., Krongyuti, P., and Sukkasame, P. (1989) *Development of Technology for Pasture Establishment in Thailand.* Department of Livestock Development, Ministry of Agriculture and Cooperatives, Thailand, 139 pp.
Ella, A., Jacobsen, C., Stur, W.W., and Blair, G. (1989) Effect of plant density and cutting frequency on the productivity of four tree legumes. *Tropical Grasslands* 23, 1: 28-34.
Ellewela, D.C. (1956) Report of the Animal Husbandry Officer - 1955. *Ceylon Coc. Quart.* 7(1+2): 53-55.
— (1957) *Report of the Animal Husbandry Officer.* Annual Report Coconut Research Board, Ceylon Coconut Research Institute (Sri Lanka) 1955, pp. 46-48.
Ekpo, E.N., and Ojomo, E.E. (1990) The spread of lethal coconut diseases in West Africa: Incidence of Awka Disease (or Bronze Leaf Wilt) in the Ishan area of Bendel State of Nigeria. *Principes* 34, 3: 143-146.
Eng, P.K. (1989) *Forage development and research in Malaysia.* In: Grasslands and Forage Production in South-East Asia. Proceedings of Meeting held 27 Feb.-3 March 1989, Serdang, Malaysia, 20-29.
Enig, M.G. (1991) Fat 'facts' aren't always facts. *The Philippine Journal of Coconut Studies* (Philippines) 16, 1: 26-28.
Enonuya, D.O.M. (1988) High performance liquid chromatographic analysis of nutwater syrup fractions from two varieties of Nigerian coconuts (*Cocos nucifera* L.) *Nigerian Journal of Palm and Oil Seeds* (Nigeria) 9, 48-58.
Eriksen, F.I., and Whitney, A.S. (1981) Effects of light intensity on growth of some tropical forage species. I. Interaction of light intensity and nitrogen fertilization on six forage grasses. *Agron. Journal* 73: 427-433.
— (1982) Growth and N. fixation of some tropical forage legumes as influenced by solar radiation regimes. *Agron. Journal* 74: 703-709.
Eschbach, J.M., and Manicot, R. (1981) Micronutrients in coconut nutrition. *Oléagineux* (France) 36, 6: 291-304.
Eschbach, J.M., Massimino, D., and Mendoza, A.M.R. (1982) Effect of chlorine deficiency on the germination, growth and photosynthesis of coconut. *Oléagineux* (France) 37, 3: 116-123.
Escamilla, J.A., Harrison, N.A., Alpizar, L., and Oropeza, C. (1991) Detection of Lethal Yellowing mycoplasma-like organisms by DNA probes in palms of Yucatan, Mexico. In: Nair, M.K., Khan, H.h., Gopalasundaram, P., and Bhaskara Rao, E.V.V. (eds), (1991); *Advances in coconut research and development. Indian Society for Plantation Crops* (india) 597-604.
Eskafi, F.M., Basham, H.G., and McCoy, R.E. (1986) Decreased water transport in Lethal Yellowing-diseased coconut palms. *Tropical Agriculture* (Trinidad) 63, 2: 225-228.
Evans, T.R., and MacFarlane, D. (1990) Pasture species identification and Potential Adaptation. Vanuatu Pasture Improvement Project, *Technical Bulletin* No. 1. AIDAB/Vanuatu Government/GRM Int. Pty Ltd./CSIRO. 44 pp.
Evans, T.R., Macfarlane, D.C. and B. Mullen (1990) Weed identification and management in Vanuatu pastures. Vanuatu Pasture Improvement Project, *Technical Bulletin* 2, 40 pp.

— (1992) Sustainable Commercial Beef Production in Vanuatu. Vanuatu Pasture Improvement Project, *Technical Bulletin* 4. Department of Agriculture, Livestock and Horticulture, Port Vila, Vanuatu. 68 pp.

Fagan, H.J. (1985) Effect of nitrogen and potassium on severity of *Drechslera* leaf spot and growth of coconut seedlings in sand culture. *Oléagineux* (France) 40, 5: 245-260.

Fairchild wet process (1971) *US patent no 3,587,696* (28 June 1971). Patent holder- E Bradley Fairchild Ltd, P.O. Box 21252, Seattle, WA 98111-3252, USA. Fax (206) 625 9782.

FAO (1966) *Coconut as part of a mixed farming system.* FAO commodity Report. New series 3, Food and Agriculture Organization of the United Nations, Rome, Italy, 29 pp.

Faure, M. (1994) *Papua New Guinea: National Coconut Breeding Programme.* In: Proceedings of the Workshop on Standardization of Coconut Breeding Research Techniques held at Abidjan, Ivory Coast, June 20-25, 1994.

Faylon, P.S. (1982) Coconut based livestock production systems in The Philippines. *Asian Livestock* 7, 7: 51-53.

Felizardo, B.C. (1982). *Intercropping coconut with sweet corn and sweet potato.* PCRDF Professorial Chair Lecture. Department of Soil Science, University of The Philippines at Los Baños.

Fenwick, D.W., and Mohammed, S. (1964) Artificial infections of seednuts and young seedlings of the coconut palm with Red Ring nematode *Rhadinaphelengus cocophilus* (Cobb.). *Nematologica* 10: 459-463.

Ferdinandez, D.E.F. (1968) *Report of the agrostologist - 1967.* Ceylon Coconut Quarterly 19, 1+2: 54-67.

Fernandez, W.L. (1988) Microbial examination of mature coconut fruit. *Philippine Agriculturist* (Philippines) 71: 13-20.

Fernando, A.M.A. (1989) A mulch roller for the management of cover-crops. *Coconut Bulletin* (Sri Lanka) 6, 1: 11-13.

Fernando, L.H., Asghar, M. and Opio, F.A. (1984). 'Review of small scale production and marketing in coconut in Western Samoa'. *Journal of South Pacific Agriculture* (Western Samoa) 9, 1: 1-29.

Ferreira, J.M.S., and Morin, J.P. (1986) A barata do coqueiro *Coraliomela brunnea* Thunb. (1981) (Coleoptera Chrysomelidae). *Embrapa CPNPCO, Aracaju, Circula Técnica* (Brazil) 1: 5-10.

Ferreira, J.M.S., Warwick, D.R.N., and Siqueira, L.A. (eds), (1994) *Cultura do coqueiro no Brasil.* EMBRAPA Centro de Pesquisa Agropecuária dos Tabuleiros Costeiros - CPATC, Aracaju, Sergipe, Brazil.

Fisher, B.J., and Tsai, J.H. (1978) *In vitro* growth of embryos and callus of coconut palm. *In vitro* 14, 3: 307-311.

Floro, M. (1952) *The future of the coconut industry: complete utilization of fresh coconuts.* Meeting of Coconut Planters Association, Manila, 1952.

Foale, M.A. (1968a) Early results of coconut density and variety trial. *Oléagineux* (France) 23, 12: 721-722.

— (1968b) The growth of the young coconut palm (*Cocos nucifera* L.) II. The influence of nut size on seedling growth in three cultivars. *Australian Journal of Agricultural Research* (Australia) 19, 6: 927-937.

— (1986) Tabular descriptions of crops grown in the tropics, 10. Coconut (Cocos nucifera L.). *Technical Memorandum CSIRO Institute of Biological Resources; Division of water and land resources, Canberra.* (Australia) 86, 4, 53 pp.

— (1991a) *Coconut genetic diversity - present knowledge and future needs.* Paper presented at the Workshop on International Network for Coconut Genetic Resources. Cipanas, Indonesia, 10 pp.

— (1991b) *Physiological basis for yield in coconut.* Paper presented at the Indian Symposium on Coconut Research and Development II, Kasaragod (India).

— (1991c) *There is a glimmer of hope for coconut oil.* Paper presented at the International Workshop on International Network for Coconut Genetic Resources. Cipanas, Indonesia. 2 pp.

— (1991d) *The effect of exposing the germspore on germination of coconut.* Poster Paper, presented at the Indian Symposium on Coconut Research and Development II. Kasaragod (India).

— (1992) *The coconut ecosystem of Zambesia Province, Mozambique. Personal communication.*

Forde, S.C.M., and Leyritz, M.J.P. (1968) A study of confluent orange spotting of the oil palm in Nigeria. *Journal of the Nigerian Institute for Oil Palm Research* (Nigeria) 4: 371-380.

Fowler, M.H., and Teskey, G. (1985) *The role of smallholder farming in rural development.* Asian Pacific Coconut Community, Quarterly Supplement (Indonesia) 30 March 1985, pp. 29-35.

Franqueville, H., and Renard, J.L. (1989) Effectiveness of Fosetyl-Al in coconut *Phytophthora* control. Appli cation methods (I). *Oléagineux* (France) 44, 7: 351-358.

Franqueville, H. de, Taffin, G. de, Sangare, A., Saint, J.P. Le, Pomier, M., and Renard, J.L. (1989) Detection of *Phytophthora heveae* tolerance characters in coconut in Côte d'Ivoire (1). *Oléagineux* (France) 44, 2: 93-103.

Franssen, C.J.H. (1954) Biological control of *Sexava nubila* in Talaud Is. *Ent. Ber.* 15, 4: 99-102.

Fremond, Y. (1966) Leguminous cover crops in coconut plantations. *Oleagineux* (France) 21: 437-440.

Frémond, Y., and Brunin, Ch. (1966) Production de feuilles et precocité chez le jeune cocotier. *Oléagineux* (France) 21, 4: 213-216.

Frémond, Y., Ziller, R., and Nucé de Lamothe, M. de (1966) *Le cocotier.* G.P. Maisonneuve & Larose, Paris (France) 265 pp.

Friend, D. (1990) The coconut industry, growing, harvesting and extraction. *The Cocomunity Quarterly Supplement* (Sri Lanka) Dec.: 1-33.

— (1991) The coconut industry. Growing, harvesting and extraction. *The Planter* (Malaysia) 67, 780: 107-129.

Friend, D., and Corley, R.H.V. (1994) Measuring coconut palm dry matter production. *Experimental Agriculture* 30: 223-235.

Frison, M. (1992) *Appendix XII Safe movement of Coconut Germplasm: presentation of guidelines. In: Coconut Genetic Resources.* Papers of an IBPGR Workshop, Cipanas, Indonesia, 8-11 October, 1991. International Crop Network Series n°8. International Board for Plant Genetic Resources, Rome, (Italy) 35-40.

Gascon, J.P., Noiret, J.M., and Benard, G. (1966) Contribution à l'étude de l'hérédité de la production de régimes d'*Elaeis guineensis* Jacq. *Oléagineux* (France) 21, 11: 657-661.

Gascon, J.P., and Nucé de Lamothe, M. de (1976) Amélioration du cocotier. Méthode et suggestions pour une coopération internationale. *Oléagineux* (France) 31, 11: 479-482.

Gascon, J.P., and Meunier, J. (1979); Anomalies of genetic origin in oil palm, *Elaeis*. Description and results. *Oléagineux* (France) 34: 437-447.

George, M.V., Kumar, K.V., and Mathew, J. (1991) Trend in area, production and productivity of coconut in India. *Indian Coconut Journal* (India) 22, 6-7: 45-48.

Geus, J. de (1973) *Fertilizer guide for the tropics and subtropics*, 2nd ed. Centre d'Etude de l'Azote, Zürich (Switzerland) 328 pp.

Ghai, S.K., and Thomas, G.V. (1989) Occurrence of *Azospirillum* spp in coconut-based farming systems. *Plant and Soil* (Netherlands) 114: 235-241.

Gianotti, J., Arnaud, F., Dollet, M., Delatte, R., and Taffin, G. de (1975) Mise en culture de mycoplasmes à partir de racines et d'inflorescences de cocotiers atteints par la maladie de Kaincopé. *Oléagineux* (France) 30, 1: 13-18.

Giblin-Davis, R.M. (1991) The potential for introduction and establishment of the Red Ring nematode in Florida. *Principes* 35, 3: 147-153.

Gillespie, A.R. (1989) Modelling nutrient flux and interspecies root competition in agroforestry interplantings. *Agroforestry Systems* 8: 257-265.

Gomez, E.D., and Orozco, R.C. (1979) *Communication patterns of crops and livestock technology in selected provinces in The Philippines.* Vol. I, II, III.

Gomez, A.A., and K.A Gomez (1980). *Varietal screening for intensive cropping*, University of The Philippines at Los Baños.

Gonzaga, L.G. (1948) *Process of recovery of oil from ripe coconuts.* Philippines patent no 3.

Gonzalez, O.N. (1991) Coconut as food, *Philippine Journal of Coconut Studies* (Philippines) 16,.2.

Gopalasundram, P., and Nelliat E.V. (1979) Intercropping in coconut. In: *Multiple Cropping in coconut and Arecanut Gardens. Technical Bulletin* 3, Central Plantation Crops Research Institute, Kasaragod, India. 6-23.

Gopalasundaram, P., Varghese, P.T., Hegde, M.R., Nair, M.G.K., and Das, P.K., (1993) Experiences in coconut based farming systems in India. In: *Advances in Coconut Research and Development.* Indian Society for Plantation Crops, Kasaragod, India. pp. 383-393.

Green, A.H. (1991) Coconut production. Present status and priorities for Research. *World Bank Technical Paper* (USA) 136.

Gressit, J.L. (1953) The coconut rhinoceros beetle (*Oryctes rhinoceros*) with particular reference to the Palau islands. *Bernice P. Bishop Museum Bulletin*, Honolulu, Hawaii (USA) 212, 157 pp.

— (1959) The coconut leaf-mining beetle Promecotheca papuana. *Papua New Guinea Agricultural Journal* (Papua New Guinea) 12, 2-2: 119-148.

Griffith, R. (1987) Red Ring Disease of coconut palm. *Pl. Dis.* 71: 193-196.

— (1993) Red Ring Disease of the coconut palm. In: Nambiar M.K., Khan, H.H., Gopalasundaram, P., and Bhaskara Rao, E.V.V. (eds), *Advances in coconut research and development.* Indian Society for Plantation Crops, New Delhi (India) pp. 477-484.

Gunasekera, T.G.L.G. (1989) Rehabilitation of low yielding coconut palms. *Coconut Bulletin* (Sri Lanka) 6, 1: 14-15.

Guzman, E.V. de (1970) The growth and development of coconut 'Makapuno' embryo *in vitro*. The induction of rooting. *Philippine Agriculturist* (Philippines) 53, 2: 65-78.

Guzman, E.V. de, Rosarion, A.G. del, and Eusebio, E.C. (1971) The growth and development of coconut 'Makapuno' embryo *in vitro*. Resumption of root growth in high sugar media. *Philipine Agriculturist* (Philippines) 53, 10: 566-579.

Guzman, M.R., and Allo, A.V. (1975) *Pasture production under coconut palms.* ASPAC/FFTC, Taipei, Taiwan.

Habana, H.A., Magat, S.S., and Padrones, G.D. (1987) Magnesium fertilization of bearing palm grown on inland-upland area of Davao. *Annual Report Coconut Authority* (Philippines): 8-15.

Hagenmaier, R.D. (1977) *Coconut Aqueous Processing*. University of San Carlos Publications.
— (1980) *Coconut Aqueous Processing*. University of San Carlos Publications, second revised edition, 1980.
Halim, R.A. (1989) *Grassland and Forage production in South-East Asia*. Proceedings of First Meeting of the Regional Working Group on Grazing and Feed Resources of South-East Asia, held at Serdang, Malaysia 27 Feb. - 3 March 1989. Organized by University Pertanian Malaysia, MARDI, Dept. of Veterinary Services, Malaysia and sponsored by FAO (TCP/RAS/8853), 219 p.
Hall, T.J. (1991) Science and Technology for Development. A research programme of the E.E.C.. *Bulletin BuroTrop* (France) 2: 19.
Hanold, D. and Randles, J.W. (1991-a) Detection of coconut Cadang-cadang viroid-like sequences in oil and coconut palm and other monocotyledons in the south-west Pacific. *Ann. of applied Biology* (UK) 118: 139-151.
— (1991-b) Coconut Cadang-cadang disease and its viroid agent. *Plant Disease* (USA) 75, 4: 330-335.
Harikumar, V.S., and Thomas, G.V. (1991) Effect of fertilizer and irrigation on vesicular-arbuscular mycorrhizal association in coconut. *Philippine Journal of Coconut Studies* (Philippines) 16, 2: 20-24.
Harlan, J.R. (1970) The evolution of cultivated plants. In: *Genetic resources in plants, their exploration and conservation*. Frankel O.H., and Bennett, E. Blackwell, Oxford (UK) pp. 19-23.
— (1973) *Problèmes pratiques de prospection: plantes à graines*. Conférence technique de la F.A.O. sur les ressources végétales. Rome (Italy).
Harland, S.C. (1957) The improvement of the coconut palm by breeding and selection. *Bulletin Coconut Research Institute* 15: 1-14.
Harries, H.C. (1967) Coconut hybridization by the polycaps and mascopol systems. *Principes* 20, 4: 136-147.
— (1970) The Malayan Dwarf supersedes the Jamaica Tall Coconut. I. Reputation and performance. *Oléagineux* (France) 25, 10: 527-531.
— (1971) The Malayan Dwarf supersedes the Jamaica Tall Coconut I. Reputation and performance. *Oléagineux* (France) 25: 527-531.
— (1978) The evolution, dissemination and classification of *Cocos nucifera* L. *The Botanical Review* 44, 3: 265-319.
— (1981) Germination and taxonomy of the coconut. *Annals of Botany* 48: 873-883.
— (1990) Malesian origin for a domestic *Cocos Nucifera*. In: Baas, P. *et al.* (eds), *The plant diversity of Malesia*, Kluwer academic publishers (Netherlands).
— (1991) The promise, performance and problems of F1 hybrid Coconut. In: *Coconut Breeding and Manage ment*. Proceedings of the National Symposium held from 23rd to 26th November, 1988. 380 pp. Kerala Agricultural University, Vellanikkara 680 654, Trichur, India.
— (1992) Biogeography of the coconut *Cocos nucifera* L. Principes 36, 3, 155-162.
Harries, H.C., and Romney D.H. (1974) Maypan: an F1 hybrid coconut variety for commercial production. *World Crops* 36, 3: 110-111.
Harris, W.V. (1958) Termites of the Solomon Islands. *Bulletin of Entomological Research* 49: 737-750.
Hassan, W.E.W., Phipps, R.H., and Owen, E. (1989) Development of smallholder dairy units in Malaysia. *Tropical Animal Health and Production* 21, 3: 175-182.
Havea, L. (1989) *Coconut replanting in the Kingdom of Tonga. Country report. In: Report of the Workshop on Coconut Replanting* (ed. Sivapragasam, A. and Jamil, M.M.), part of UNDP/FAO Project RAS/80/032 Improved Coconut Production in Asia and Pacific. Held 23-27 Nov., 1988 at MARDI, Serdang, Selangor, Malaysia, pp. 25-54.
— (1991) *The economics of coconut intercropping practices for smallholders in Tonga. Research Project Report submitted in partial fulfilment for the degree of Bachelor of Agriculture*, University of the South Pacific, School of Agriculture, October 1991, pp. 82.
Hegde, M.R., Gopalasundram. P., and Yusuf, M., (1990) Intercropping in coconut gardens. *Technical Bulletin* 23, Central Plantation Crops Research Institute, Kasaragod, India 7 pp.
Hegde, M.R., and Yusuf, M. (1992) Performance of soybean varieties as intercrops in coconut. In: *Abstract of papers PLACROSYM X*. Indian Society for Plantation Crops, Kasaragod, India. 75.
Hegde, M.R., Yusuf M., and Gopalasundaram, P. (1993) Intercropping of vegetables in coconut gardens. In: *Advances in Coconut Research and Development*. Indian Society for Plantation Crops, Kasaragod, India pp. 407-412.
Hight, G.K., Sinclair, D.P., and Lancaster, R.J. (1968) Some effects of shading and of nitrogen fertilizer on the chemical composition of freeze-dried and oven-dried herbage, and on the nutritive value of oven-dried herbage fed to sheep. *New Zealand Journal of Agricultural Research* 11: 286-302.
Hiller, (1952, 1954) *US Patent 2583-1952 and Philippines patent 186-1954*.
Hinckley, A.D. (1963) *Parasitation of Agonoxena argaula Mey. (Lep. Agonoxenidae)*. Proceedings Hawaiian Entomological Society 1962 (USA) 18, 2, 267-272.

Hoffmann, W. (1929) Die Kultur der Kokospalme. In: *Neues Handbuch der tropischen Agrikultur.* Hamburg (Germany).
Holliday, P. (1980) *Fungus diseases of tropical crops.* Cambridge University Press, Cambridge (UK).
Hoof, H.A., van, and Seinhorst, J.W. (1962) *Rhadinaphelenchus cocophilus* associated with Little Leaf of coconut and oil palm. *Netherlands Journal of Plant Pathology* (Netherlands) 68, 252-256.
Hopkins, G.H.E. (1927) Pests of economic plants in Samoa and other island groups. *Bulletin of Entomological Research* 18: 18-32.
Hory, J.P. (1989) *Chimiotaxonomie et organisation génétique dans le genre Musa.* Thèse de doctorat, Université de Paris-Sud, Centre d'Orsay (France) 105 pp.
Howard, F.W. (1989) Lethal Yellowing: 'A very great problem'. *The Palm Enthusiast* (USA) 6, 2: 10-24.
Howard, F.W., Atilano, R., Barrant, C.I., Harrison, N.A., Theobold, W.F., and Williams, D.S. (1987) Unusually high Lethal Yellowing disease incidence in Malayan Dwarf coconut palms on localized sites in Jamaica and Florida. *Journal of Plantation Crops* (India) 15, 2: 86-100.
Howard, F.W., Norris, R.C., and Thomas, D.C. (1983) Evidence of transmission of palm lethal yellowing agent by a plant-hopper *Myndus crudus* (Homoptera Cixiidae). *Tropical Agriculture* (Trinidad) 60: 168-171.
Hüger, R.M. (1966) A virus disease of the Indian rhinoceros beetle *Oryctes rhinoceros* caused by a new type of insect virus *Rhabdionvirus oryctes* gen. n., sp. n. *Journal of Invertebrate Pathology* 8: 38-51.
Huguenot, R., and Vera, J. (1981) Description et lutte contre *Castnia daedalus* Cr (Lep. Castniidae) ravageur du palmier à huile en Amerique du Sud. *Oléagineux* (France) 36: 543-598.
Humphreys, L.R. (1991) *Tropical Pasture Utilization.* Cambridge University Press. Cambridge, England.
Hurpin, B., and Mariau, D. (1966) *Contribution à la lutte contre les Oryctes nuisibles aux palmiers. Mise au point d'un élevage permanent au laboratoire.* C.R. Séance Acad. Agric. Fr. 1966 (France) 178-186.
Huxley, P.A. (1985) The tree/crop interface - or simplifying the biological/environmental study of mixed cropping agroforestry systems. *Agroforestry Systems* 3, 3: 251-266.

IBPGR (1992) *Descriptors for coconut.* International Board for Plant Genetic Resources, Rome (Italy) 61 p.
IDEFOR/DPO (1992) Rapport d'activité 1991-1992: recherches de la station de recherches 'Marc Delorme'. 361 pp.
Igbinosa, I.B. (1985) Life table studies of the nettle caterpillar *Latoia viridissima* Holland on the oil palm *Elaeis guineensis* Jacq. and the coconut palm, *Cocos nucifera* L. *Agriculture, Ecosystems and Environment* 14: 77-93.
Ikram, A. (1990) Rhizosphere microorganisms and crop growth. *The Planter* (Malaysia) 66, 630-638.
Illingworth, R. (1991) Hybrid Coconut Seed Garden in Costa Rica. *Principes* 35: 104-106.
Imperial, J.S. (1985) The viroid etiology of the Cadang-cadang disease. *Philippine Journal of Coconut Studies* (Philippines) 9, 1-2: 10-17.
Iniguez, L.C., and Sanchez, M.D. (1991) *Integrated Tree Cropping and Small Ruminant Production Systems.* Proceedings Workshop on Research Methodologies, Medan, North Sumatra, Indonesia, Sept. 9-14, 1990, 329p, USAID.
Intengan, C. Ll. (1987) Fats and serum cholesterol. *Philippine Journal of Coconut Studies* (Philippines) 12, 2: 30-36.
Intengan, C. Ll., Dayrit, C.S., Pesigan, J.S., Canvaling, T., and Zakamea, I. (1992) Structural lipid of coconut and corn oils vs. soybean oil in the rehabilitation of malnourished children - a field study. *Philippine Journal of Coconut Studies* (Philippines) 17, 2: 15-17.
Iremiren, G.O. (1986) Response of coconut nursery seedlings to some cultural treatments in Nigeria. In: Pushparajah, E. and Chew Poh Soon (eds), *Cocoa and Coconuts*, pp. 861-868. Incorporated Society of Planters, Kuala Lumpur (Malaysia).
IRETA (1988) *Institute for Research, Extension and Training in Agriculture (W. Samoa) Annual Report.*
IRHO-CIRAD (1980) *Coconut seeds from the Institut de Recherches pour les Huiles et Oléagineux.* (France) unpublished.
IRHO-CIRAD (1989) *Rapport d'activités 1988: recherches de la station de recherches 'Marc Delorme'.* 361 pp.
IRHO-CIRAD (1992) Rapport d'activité de l'Institut de Recherches pour les Huiles et Oléagineux. *Oléagineux* (France) 47, 6: 265-456.
Iyer, R.D. (1981) Embryo and tissue culture for crop improvement, especially of perennials, germplasm conservation and exchange. In: *Tissue Culture of Economically Important Plants.* Proc. Int. Symp. Singapore, RAO A.N. ed., Costed, ANBS, pp. 229-230.
Iyer, R., Moosa, H., and Kalpana Sastry (1993) VA mycorrhizal status of a coconut based high density multi-species cropping system. In: Nair, M.K., Khan, H.H., Gopalasundaram, P., and Bhaskara Rao, E.V.V.(eds), *Advances in coconut research and development.* Indian Society for Plantation Crops, New Delhi (India) 429-431.

Jackson, G.V.H., and McKenzie, E.H.C. (1988) *Marasmiellus cocophilus* on coconuts in Solomon Islands. *FAO Plant Protection Bulletin* (Italy) 36, 2: 91-97.

Jacob, P.M., Nair, R.V., Rawther, T.S.S., Muralidharan, A., Govindankutty, M.P., and Sasikala, M. (1991) Screening for coconut germplasm for resistance/tolerance to Root (Wilt) disease. In: *Annual Report of Central Plantation Crops Research Institute*, Codeword Process & Printer, Mangalore (India) 154 p.

Jacob, P.M., and Rawther, T.S.S. (1991) Varietal resistance. In: Nair, M.K., (eds) (1991) *Coconut Root(wilt) disease. Central Plantation Crops Research Institute* (India), Monograph Series - no 3, 67-72.

Jay, M., Bourdeix, R., Potier, F., and Sanlaville, C. (1989) First results from the study on the polymorphism of coconut leaf polyphenols. *Oléagineux* (France) 44, 3: 151-161.

Jayasankar, N.P. (1991) Combating important diseases of coconut. *Indian Horticulturist* (India) 36, 3: 25-29.

Jayasekara, K.S., and Jayasekara, C. (1993) Efficiency of water use in coconut under different soil/plant management systems. In: Nair, M.K., Khan, H.H., Gopalasundaram, P., and Bhaskara Rao, E.V.V. (eds), *Advances in coconut research and development.* Indian Society for Plantation Crops (India) 427.

Jayasekara, K.S., and Loganathan, P. (1988) Boron deficiency in young coconut (*Cocos nucifera* L.) in Sri Lanka. Symptoms and corrective measures. *Cocos* (Sri Lanka) 6: 31-37.

Jayasekara, K.S., and Mahindapala, R. (1988) An irrigation system for a five-acre coconut plantation. *Coconut Bulletin* (Sri Lanka) 5, 1: 14-17.

Jayasekara, C., Ranasinghe, C.S., and Mathes, D.T. (1993) Screening for high yield and drought tolerance in coconut. In: Nair, M.K., Kahn, H.H., Gopalasundram, P., and Bhaskara Rao, E.V.V. (eds), *Advances in coconut research and development. Indian Society for Plantation Crops.* New Delhi, (India).

Jayasundara, H.P.S., and Marasinghe, R. (1989) A model for integration of pasture, tree fodder and cattle in coconut smallholdings. *Coconut Bulletin* (Sri Lanka), 6, 2: 15-18.

Jayasuriya, V.U. de S., and Perera, R.K.I.S. (1985) Growth, development and dry matter accumulation in the fruit of *Cocos nucifera* L. var. *nana* form *pumila. Cocos* (Sri Lanka) 3: 16-21.

Jayalekshmy, A., Arumughan, C., Narayanan, C.S., and Mathew, A.G. (1988) Changes in the chemical composition of coconut water during maturation. *Oléagineux* (France) 43, 11: 409-414.

Jayawardana, A.B.P. (1985) *Pastures in Sri Lanka.* In: Proceedings of International Symposium on Pastures in the Tropics and Subtropics, 2-6 October 1984, Tsukuba, Japan. TARC Tropical Agriculture Research Series No. 18, 71-85.

— (1988) *Integration of coconut with animal husbandry.* In: Proceedings Workshop on Intercropping and Intergrazing in Coconut Areas, 7-11 September, 1988, Colombo, Sri Lanka.

Jeganathan, M. (1974) Toddy yields from hybrid coconut palms. *Ceylon Coconut Quarterly* (Sri Lanka) 25, 3-4: 139-148.

Jodha, N. S. (1981) Intercropping in traditional farming systems. *Journal of Development Studies* 16, 4: 427-447.

Johnston, C.F., Fielding, W.J., and Been, B.(1994) Hurricane damage to different coconut varieties. *Tropical Agriculture* (Trinidad) 71, 3: 239-242.

Jones, R.J., and Sandland, R.L. (1974) The relation between animal gain and stocking rate. Derivation of the relation from the results of grazing trials. *Journal Agricultural Sciense Camb.* 83: 335-342.

Jose, A.I., Krishnakumar, N., and Gopi, C.S. (1991) Yield prediction in coconut based on foliar nutrient levels. In: Silas, E.G., Aravindakshan, M., and Jose, A.I., *Coconut breeding and management*, Kerala Agricultural University (India): 212-218.

Joseph, T., and Radha, K. (1975) Role of *Phytophthora palmivora* in bud rot of coconut. *Plant Disease Reporter* 59, 12: 1014-1017.

Joshi, O.P., Khan, H.H., Bindappa, C.C., and Manikandan, P. (1985) Effect of coir dust on mineralisation of urea nitrogen in coconut growing red sandy loam soils (Arenuc Paleustults). *Journal of Plantation Crops* (India) 13, 2: 88-95.

Julia, J.F. (1979) Determination and identification of insects responsible for juvenile diseases of coconut and oil palm in Ivory Coast. *Oléagineux* (France) 33, 3: 113-118.

— (1982) *Myndus taffini* (Homoptera Cixiidae) vecteur du deprissement foliaire des cocotiers au Vanuatu. *Oléagineux* (France) 37: 109-114.

Julia, J.F., Dollet, M., Randles, J., and Calvez, C. (1985) Foliar Decay of coconut by *Myndus taffini* (FDMT) New Results. *Oléagineux* (France) 40, 1: 19-27.

Julia, J.F., and Mariau, D. (1976a) Recherches sur l'*Oryctes monoceros* Ol. en Côte d'Ivoire. I Lutte biologique. Le rôle de la plante de couverture. Oléagineux (France) 31, 2, 63-68.

— (1976b) id. III Piégeage olfactif à l'aide de chrysantémate d'ethyl. *Oléagineux* (France) 31, 6: 263-272.

— (1978) La punaise du cocotier *Pseudotheraptussp.* en Côte d'Ivoire. I. Etudes prélables à la mise au point d'une méthode de lutte intégrée. *Oléagineux* (France) 33, 65-75.

— (1982) Deux espèces de *Sogatella* (Homoptère Delphacidae) vectrices de la maladie de la pourriture sèche du coeur des jeunes cocotiers en Côte d'Ivoire. *Oléagineux* (France) 37, 11: 517-520.

Kaligis, D.A., and Sumolang, C. (1991) *Forage species for coconut plantations in North Sulawesi.* In: Proceedings of Workshop on Forages for Plantation Crops. Shelton, H.M. and Stur, W.W. (ed.). Sanur Beach, Bali, Indonesia 27-29 June 1990. ACIAR Proceedings 32, pp. 45-48.

Kaligis, D.A., Stur, W.W. and Mamonto, S. (1991) *Growth of tree legumes under coconuts in North Sulawesi.* In: Proceedings of Workshop on Forages for Plantation Crops. Shelton, H.M. and Stur, W.W. (ed.). Sanur Beach, Bali, Indonesia 27-29 June 1990. ACIAR Proceedings 32, pp. 72-74.

Kalshoven, L.G.E. (1957) Analysis of ethological, ecological and taxonomic data on oriental Hispinae. *Tijdschrift voor Entomologie* (Netherlands) 100: 5-24.

Kannan, K., and Nambiar, K.P.P. (1976) Studies on intercropping coconut gardens with annual crops. *Coconut Bulletin* 5, 9: 1-3.

Karunanayake K. (1989) *Coconut based farming systems in Sri Lanka.* In: Coconut Based Farming Systems. Proceedings of 27th COCOTECH Meeting. Asian and Pacific Coconut Community, Jakarta, Indonesia, pp. 183-197.

Kass, M., Benavides, J. Romero, F., and Pezo, D. (1992) *Lessons from main feeding experiments conducted at CATIE using fodder trees as part of the N-ration.* In: Legume trees and other fodder trees as protein sources for livestock (ed. Speedy, A. and Pugliese, P.). Proceedings of FAO Expert Consultation, MARDI, Kuala Lumpur, Malaysia 14-18 October 1991. FAO Animal Production and Health Paper No. 102, 161-175.

Kastelein, P. (1986) Observations on Red Ring disease of coconut palms in Surinam. *Surinaamse Landbouw* (Surinam) 34, 1-3: 40-45.

Kastelein, P., Sanchit-Bekker, M.L., and Dipotaroeno, M.S. (1985) Confusion in the recognition of Hartrot and Red Ring disease in coconut. *Surinaamse Landbouw* (Surinam) 33, 2: 56-61.

Kasturi Bai, K.V., Voleti, S.R., and Rajagopal, V. (1988) Water relations of coconut palms as influenced by environmental variables. *Agricultural and Forest Meteorology* 43, 3-4: 193-199.

Kasturi Bai, K.K., and Ramadasan, A. (1990) Growth studies in coconut seedlings. *Journal of Plantation Crops* (India) 18, 2: 130-133.

Kaunitz, H., and Dayrit, C.C. (1992) Coconut oil consumption and coronary heart disease. *Philippine Journal of Coconut Studies* (Philippines) 17, 2: 18-20.

Khalfaoui, J.L.B. (1985) Approach to genetic improvement of adaptation to drought depending on physiological mechanisms. *Oléagineux* (France) 40, 6: 329-334.

Khan, H.H., Biddappa, C.C., and Cecil, S.R. (1990) Improving the coconut production: future needs related to nutritional aspects. *Indian Coconut Journal* (India) 20, 12: 2-6.

Khan, H.H., Bidappa, C.C., and Joshi, O.P. (1985a) A review of Indian work on phosphorus nutrition of coconut. *Journal of Plantation Crops* (India) 13, 1: 11-21.

Khan, H.H., Gopalasundaram, P., Joshi, O.P., and Nelliat, E.V. (1986) *Effect of NPK fertilization on the mineral nutrition and yield of three coconut genotypes. Fertilizer Research* (Netherlands) 10: 185-190.

Khan, H.H., Sankaranarayanan, M.P., George, M.V., and Narayana, K.B. (1983) Effect of phosphorus skipping on the yield and nutrition of coconut palm (*Cocos nucifera* L.). *Journal of Plantation Crops* (India) 11, 2: 129-134.

Khan, H.H., Sankaranarayanan M.P., Joshi, O.P., George, M.K., and Narayana, K.B. (1985b) Comparative efficiency of selected phosphates as P-carriers for coconut (*Cocos nucifera* L.). *Tropical Agriculture* (Trinidad) 62, 1: 57-61.

Killman, W. (undated) *How to Process Coconut Palm Wood - A Handbook.* GATE, GTZ GmbH, Eschborn, Germany.

Kintanar, Q.L. (1987) Absorption of fats and oils and lipid metabolism. *Philippine J. of Coconut Studies* (Philippines) 12, 2: 18-29.

Kirwan, A.K. (1986) Farmer acceptance/rejection of credit for multiple cropping technology in Zamboanga del Sur, The Philippines, *Journal of Australian Institute of Agricultural Science* (Australia) 52, 3: 144-148.

Koshy, P.K., Sosamma, V.K., and Sundaraju, P. (1991) *Radopholus similis*, the burrowing nematode of coconut. *Journal of Plantation Crops* (India) 19, 2: 139-152.

— (1993) Nematode management in coconut and coconut-based cropping system. In: Nair, M.K., Khan, H.H., Gopalasundaram, P., and Bhaskara Rao, E.V.V. (eds), *Advances in coconut research and develpment.* Indian Society for Plantation Crops, New Delhi (India), pp. 465-467.

Kosikowski, F. (1977) *Cheese and Fermented Cheese Products.* 2nd edition. Edwards Brothers Inc, Ann Arbor, Michigan, USA.

Kovoor, A. (1981) *Palm tissue culture.* FAO Plant Production and Protection Paper (Italy) 30, 69pp.

Kowalski, J. (1917) Un ennemi du cocotier aux Nouvelles-Hébrides. *Le Promecotheca opacicollis Gestro. Ann. Epiphyties* 4: 286-327.

Krisnawati Suryanara, W.I.I., Mella, A.P.Y., Djogo, and Andi Renggana (1988) An analysis of a village agroecosystem in Timor. In: *Agroecosystem Research for Rural Development* (ed. Kanok Perkasem and A.T. Rambo) Chiang Mai University and SUAN.

Kullaya, A. (1991) *The state of coconut production in Tanzania.* Country Report, presented at the First African Coconut Seminar, Arusha (Tanzania).

Kumar, A.S.A. and Pillai, S.J. (1990) Germination, seedling vigour and recovery of quality seedlings as influenced by pre-sowing preparations of seed coconuts. *Indian Coconut Journal* (India) 20, 10: 6-7.

Kunhikrishnan, K. (1972) Filling in around the coconut palm. *World Farming* 14, 10: 12.

Kurup, V.V.G.K., Voleti, S.R., and Rajagopal, V. (1993) Influence of weather variables on the coconut and composition of leaf surface wax in coconut. *Journal of Plantation Crops* (India) 21, 2: 71-80.

Lane, I.R. (1981) The use of cultivated pastures for intensive animal production in developing countries. In: *Intensive Animal Production in Developing Countries* (ed. Smith, A.J. and Gunn, R.G.). Occas. Pub. No. 4, BSAP, pp. 105-143.

Lava, V.G. (1937) *Oil recovery.* US Patent no 2, 101: 571.

Lavaka, S. (1988) *Status of intercropping and intergrazing in coconut lands in the Kingdom of Tonga.* In: Proceedings Workshop on Intercropping and Intergrazing in Coconut Areas, 7-11 September, 1988, Colombo, Sri Lanka.

Leal, E.C., Santos, Z.G. dos, Ram, C., Warwick, D.R.N., Leal, M.L. da S., and Renard, J.L. (1994) Efeito da adubação mineral sobre a incidência das lixas *Spaerodothis torriendella* e *Spaerodothis acrocomiae* no coqueiro *Cocos nucifera* L. *Oléagineux* (France) 49, 5: 213-220.

Lecoustre, R., and Reffye, P. de (1986) Theory of regional variables, its possible applications to agronomic research, in particular for oil palm and coconut. *Oléagineux* (France) 41, 12: 541-548.

Lee Heng Lye, and Jerry (1985) *Susceptibility of coconut varieties to Scapanes australis and Oryctes Rhinoceros attack in Papua New Guinea* (cited by Ovasuru, 1992).

Leela K., and Bhaskaran, U.P. (1978) *Effect of intercropping coconut stands with groundnut on soil fertility and plantation management.* In: Proceedings of the First Symposium on Plantation Corps. Indian Society for Plantation Crops, Kasaragod, India. 393-397.

Leo, C. C. (1968) *A survey of home consumption of fresh coconuts in some Far Eastern Countries.* FAO/WHO/UNICEF PAG Document no. 9/7.

León, R., Sánchez, G., Alpizar, L., Escamilla, A., Santamaria, J., and Oropeza, C. (1993) Studies on the physiology of *Cocos nucifera* palm affected by Lethal Yellowing in Mexico. In: Nair, M.K., Khan, H.H., Gopalasundaram, P., and Bhaskara Rao, E.V.V. (eds), *Advances in coconut research and development.* Indian Society for Plantation Crops (India) 621-628.

Leopold, H. (1946) Coconut palms - 120,000,000 of them. *Food Industry* 18: 1535.

Lepesme, P. (1947) *Les insectes des palmiers.* Paul Lechevalier. Paris (France) 903 pp.

Le Saint, J.P., Nucé de Lamothe, M. de, and Sangare, A. (1983) Les cocotiers Nains à Port Bouët (Côte d'Ivoire). II. Nain Vert Sri Lanka et complément d'information sur les Nains Jaune et Rouge Malaisie, Nain Vert Guinée Equatoriale, Nain Rouge Cameroun. *Oléagineux* (France) 38, 11: 595-606.

Le Saint, J.P., and Nucé de Lamothe, M. de (1987) Les hybrides de cocotiers Nains: performance et intérêt. *Oléagineux* (France) 42, 10: 353- 362.

Le Saint, J.P., Taffin, G. de, and Benard, G. (1989) Conservation des semences de cocotier en emballage étanche. *Oléagineux* (France) 44, 1: 15-25.

Leufstedt, G. (1990) Opportunities of future diversification of the coconut industry. *Oléagineux*, vol 45, 11, 90.

Levang, P. (1988) Le cocotier est aussi une plante sucrière. *Olaéagineux* (France) 43, 4: 159-164.

Lever, R.J.A.W. (1951) A new coconut pest in Singapore. *Malayan Agricultural Journal* (Malaysia) 34, 2: 78-82.

— (1953) Notes on outbreaks, parasites and habits of the coconut moth (*Artona catoxantha. Malayan Agricultural Journal* (Malaysia) 36, 1: 20-27.

— (1969) Les ravageurs du cocotier. *FAO Agricultural Studies* (Italy) 77, 190 pp.

Lim-Sylianco, C.Y. (1987) Anticarcinogenic effect of coconut oil. *Philippine Journal of Coconut Studies* (Philippines) 12, 2: 89-102.

Lim-Sylianco, C.Y., Mallorca, R., Serrame, E., and Wu, L.S. (1992a) A comparison of germ cell antigenotoxic activity on non-dietary and dietary oil and soybean oil. *Philippine Journal of Coconut Studies* (Philippines) 17, 2: 1-5.

Lim-Sylianco, C.Y., Balboa, J., Casareno, R., Mallorca, R., Serrame, E., and Wu, L.S. (1992b) Antigenotoxic effect of bone marrow cells of coconut oil versus soybean oil. *Philippine Journal of Coconut Studies* (Philippines) 17, 2: 6-10.

Litscher, T., and Whiteman, P.C. (1982) Light transmission and pasture composition under smallholder coconut plantations in Malaita, Solomon Islands. *Experimental Agriculture* 18: 383-391.

Liyanage, D.V. (1955) Hedge planting for coconuts? *Ceylon Coconut Quarterly* 6, 1/2: 24-48.

— (1967) Identification of genotype of coconut suitable for breeding. *Experimental Agriculture* 3: 205-210.

— (1969) Effect of inbreeding on some characters of the coconut palm. *Ceylon Coconut Quarterly* (Sri Lanka) 20: 161-167.

— (1972) Production of improved coconut seed by hybridisation. *Oléagineux* (France) 27, 12: 597-599.

Liyanage, D.V. and Azis, H. (1983) A new technique for establishing coconut seed gardens. *Cocos* (Sri Lanka) 1, 1-6: 1983.

Liyanage, D.V., and Luntungan, H.T. (1978) Choice of coconut seed for planting in smallholdings in Indonesia. *Pemberitaan - Lembaga Penelitian Tanaman Industri* (Indonesia) 29: 1-11. Cited by Ohler, 1984.

Liyanage, D.V., Mankey, T., Luntungan, H., Djisbar, A., and Sufiani, S. (1986) Coconut Breeding in Indonesia. *Cocos* (Indonesia) 4: 1-10.

Liyanage, D.V., and Sakai, I. (1961) Heritabilities of certain yield characters of the coconut palm. *Journal of genetics* 57: 245-252.

Liyanage, D.V., Wickramaratne, M.R.T., and Jayasekara, C. (1988) Coconut Breeding in Sri Lanka: A Review. *Cocos* (Indonesia) 6: 1-26.

Liyanage, L.V.K. (1985) Rationale for intercropping. *Coconut Bulletin* 2, 2: 31-35.

— (1986) Pasture Management and Animal Husbandry in Coconut Lands. *Coconut Bulletin* 3, 1: 19-22.

— (1987a) *Pasture/cattle integrated system in coconut lands (an extension paper).* Agronomy Division, CRI, Sri Lanka.

— (1987b) Moisture conservation in coconut lands. *Coconut Bulletin* (Sri Lanka) 4: 1-4.

— (1990) *Techno-economic feasibility of pasture/fodder/tree/animal integration in coconut lands.* XXVII COCOTECH Meeting, 25-29 June, 1990, Manila, Philippines, pp. 103-113.

— (1991) *Forages for plantation crops in Sri Lanka.* In: Proceedings of Workshop on Forages for Plantation Crops, Shelton, H.M. and Stur, W.W. (ed.). Sanur Beach, Bali, Indonesia 27-29 June 1990. ACIAR Proceedings 32, pp. 157-161.

Liyanage, L.V.K., and Wijeratne, A.M.U. (1987) *Uses and management of Gliricidia sepium in coconut plantations in Sri Lanka.* In NFTA 1987. Gliricidia sepium (Jacq.) Walp.: Management and Improvement. Proceedings of a workshop (ed. Withington, D. *et al.*) held at CATIE, Turrialba, Costa Rica, June 1987. Nitrogen Fixing Tree Association Species. Pub. 87-01, 95-101.

Liyanage, L.V.K., Jayasundera, H.P.S., Mathes, D.T., and Fernando, D.N.S. (1989) Integration of pasture, fodder and cattle in coconut small holdings. *CORD* (Indonesia) 5, 2: 53-66.

Liyanage, M. de S. (1983) Agroforestry systems associated with coconuts. *The Sri Lanka Forester* 16, 1+2: 25-27.

— (1988) Use of coir dust for moisture conservation. *Coconut Bulletin* (Sri Lanka) 5, 1: 18-19.

— (1993) The role of MPTS in coconut-based farming systems in Sri Lanka. *Agroforestry Today* (Kenya), 5, 3: 7-9.

Liyanage, M. de S., and Dassanayake, K.B. (1988) *Coconut intercropping research and development in Sri Lanka.* In: Proceedings Workshop on Intercropping and Intergrazing in Coconut Areas, 7-11 September, 1988, Colombo, Sri Lanka.

Liyanage, M. de S., and Dassanayake, K.B. (1993) *Experiences in coconut based farming systems in Sri Lanka.* In: Advances in Coconut Research and Development. Indian Society for Plantation Crops, Kasaragod, India, pp. 357-367.

Liyanaga, L.V.K., Jayasundaran, H.P.S., and Gunasekara, T.G.L.G. (1987) Potential use of nitrogen-fixing trees on small coconut plantations in Sri Lanka. In: Withington, D., *et al.*: *Multipurpose tree species for small-farm use.* Proceedings of an International Workshop. Pattaya (Thailand) pp. 251-253.

Liyanage, M. De. S., and Martin, M.P.L.D (1987) *Soybean-coconut intercropping.* In: Proceedings of Tropical and Subtropical Cropping Systems, pp. 57-60.

Liyanaga, L.V.K., and Mathes, D.T. (1989) Effect of irrigation on establishment and early growth of coconut (var, CRIC 60) in the dry zone of Sri Lanka. *Cocos* (Sri Lanka) 7: 1-13.

Loganathan, P., Dayaratne, P.M.N., and Shanmuganathan (1984) Evaluation of the phosphorus status of some coconut growing soils of Sri Lanka. *Cocos* (Sri Lanka) pp. 29-34.

Long, V.V. (1993) Coconut Selection and Breeding Programme in Vietnam. In: Nair, M.K., *et al.* (eds), *Advances in Coconut Research and Development. Platinium Jubilee of Coconut Research in India.* Oxford & IBH Publishing CO. PVT. LTD. (India) pp. 107-113.

Louis, H. (1989) Studies on the root system in coconut palm. *Indian Coconut Journal* (India) 19, 12: 8-15.

Louise, C., Dollet, M., and Mariau D. (1986) Research into Hartrot of the coconut, a disease caused by *Phytomonas (Trypanosomatidae)*, and into its vector *Lincus* sp. *(Pentatomidae)* in Guiana. *Oléagineux* (France) 41, 10: 438-449.

Lourduraj, A.C., Geethalakshmi, V., Rajamanickam, K., and Kennedy, F.J.S. (1992) Soybean - A suitable intercrop in Coconut gardens. *Indian Coconut Journal* (India) 22, 11: 8-9.

Maas, P.W.Th. *et al.* (1970) Rode Ring ziekte van kokos en oliepalm. *Mededelingen Landbouwproefstation Suriname* (Surinam) 43: 114-115.

MacFarlane, D., and Shelton, M. (1986) *Pastures in Vanuatu. ACIAR Tech. Reports* No. 2, 32 pp.

MacFarlane, D.C., Mullen, B.F., Kamphorst, J., Banga, T., William, M., and Evans, T.R. (1991) Lukaotem Gud Pasja Mo Buluk Long Vanuatu. Vanuatu Pasture Improvement Project/Smallholder Cattle Improvement Project/Animal Health and Production Division/Department of Agriculture, Livestock and Horticulture. *Technical Bulletin 3*, 111 pp.

Mac Key, J. (1974) *Genetic and evolutionary principles of heterosis. Heterosis in Plant Breeding.* Proceedings of 7th Congress of Eucarpia. Elsevier (Netherlands) 364 pp.

Magat, S.S. (1993a) Coconut nutrition: experiences in The Philippines. In: Nair M.K., Khan, H.H., Gopalasun daram, P., and Bhaskara Rao E.V.V. (eds), *Advances in coconut research and development. Indian Society for Plantation Crops, New Delhi* (India) pp. 277-298.

— (1993b) Coconut production and productivity in The Philippines: realities and opportunities early 90's. *Bulletin Burotrop* (France) 6: 11-12.

Magat, S.S., Margate, R.Z., and Habana, J.A. (1986) Sodium chloride (common salt) fertilization of bearing coconuts. I: Early yield response. *Philippine Journal of Coconut Studies* (Philippines) 11, 1: 37-43.

— (1988a) Effects of increasing rates of sodium chloride (common salt) fertilization on coconut palms grown under an inland soil (Tropudalfs), of Mindanao, Philippines. *Oléagineux* (France) 43, 1: 13-19.

Magat, S.S., Alforja, L.M., and Margate, R.Z. (1988b) Influence of soil, climatic factors, and nutritional status on the productivity of YMD x WAT 'MAWA' coconut hybrids in The Philippines. *The Philippine Journal of Coconut Studies* (Philippines) 13, 2: 1-5.

Magat, S.S., Alforja, L.M., and Oguis, L.G. (1988c) An estimation of the critical and optimal levels of leaf-chlorine in bearing coconuts: a guide for foliar diagnosis. *The Philippine Journal of Coconut Studies* (Philippines) 13, 2: 6-10.

Magat, S.S., Alforja, L.M., and Margate, R.Z. (1988d) An estimate of the optimum and critical levels of leaf-sulfur concentration in bearing coconuts (Local Tall). *The Philippine Journal of Coconut Studies* (Philippines) 13: 6-9.

Magat, S.S., and Habana, J.A. (1991) Effect of leaf pruning on the yield of coconut (A research note). *The Philippine Journal of Coconut Studies* (Philippines) 16, 1: 9-11.

Magat, S.S., and Oguis, L.G. (1979) *Early results of a study on the chlorine nutritional needs of coconut in the Philippines.* Paper presented at the Fifth Session of the FAO Technical Working Party on Coconut Production, Protection and Processing. Manila (Philippines).

Magat, S.S., and Padrones, G.D. (1984) Initial response of young bearing coconut palms to various fertilizer sources of chlorine. *The Philippines Journal of Coconut Studies* 12: 21-29.

Magat, S.S., Padrones, G.D., Habana, J.A., and Alforja, L.M. (1991) Residual effects of chloride fertilizers on yield of coconuts grown on an inland soil of Davao. *The Philippine Journal of Coconut Studies* (Philippines) 16, 1: 1-8.

Mahendranathan, T. (1976) *The role of smallholder in beef and milk production.* In: Proceedings Symposium on Smallholder Livestock Production and Development (eds), Devendra, C. and Thamutaram, S.). Min. of Agriculture Malaysia, Bull. 144: 16-28.

Mahendranathan, T., and Nor Mohd. Nordin bin Mohd (1980) *Country paper: Malaysia.* In: Report of the International Workshop on Integrated Livestock/Fish/Crop/Forestry Production Systems for Small Farmers and Fishermen. Chiang Mai, Thailand 9-14 April. Vol. II: 35-49.

— (1980) *The role of smallholder in beef and milk production.* In: Proceedings Symposium on Smallholder Livestock Production and Development (eds), Devendra, C., and Thamutaram, S.). Ministry of Agriculture, Malaysia, Bull. No. 144, 16-28.

Mahindapala, R. (1987) Irrigation in Coconut. *Coconut Bulletin* (Sri Lanka) 4: 6-12.

— (1988) *Report of the Workshop on Intercropping and Intergrazing in Coconut Areas,* held 7-11 September 1988 in Colombo, Sri Lanka (under UNDP/FAO Project RAS/80/032 Improved Coconut Production in Asia and the Pacific). 152 pp.

Mahmud, Z., Akuba, R.H., and Amrizal (1992) Industrial crops for oilseed in Indonesia. *Industrial Crops Research Journal* (Indonesia) 5, 1: 22-30.

Maliwan Rattanapruk, Howl, J.C. Anupap Thirakul, Chulaphan Petchipiroon, and Dootson, J. (1985) Comparison of precocity and yield of hybrid coconut varieties in Thailand. *Oléagineux* (France) 40, 3: 125-131.

Malosu, D. (1987) *Progress on coconut development and research in Vanuatu.* Proceedings of the Fifth South Pacific Islands Regional Meeting on Agricultural Research, Development, Extension and Training in Coconut, IRETA (Western Samoa) pp. 107-117.

Manciot, R., Ollagnier, M., and Ochs, R. (1980) Nutrition minérale et fertilisation du cocotier dans le monde. *Oléagineux* (France) 35, hors série, pp. 3-55.

Manciot, R., and Sivan P. (1988) *Coconut hybrids for the South Pacific Islands. In: Coconut Breeding and Management.* Proceedings of the National Symposium held from 23rd to 26th November, 1988. 380 pp. Kerala Agricultural University, Vellanikkara 680 654, Trichur (India).

Manidool, C. (1983) *Pastures under coconut in Thailand*. Misc. paper in Seminar on Recent Advances in Pasture Research and Development in Asian Countries. Aug. 1983, Khonkaen University, Thailand.

— (1984) Pastures under coconut in Thailand. In: *Asian Pastures: Recent Advances in Pasture Research in Southeast Asia*. FFTC Book Series No. 25, ASPAC, Taipei, Taiwan.

Mannil, T.M. (1989) Coconut palms in sandy soils need adequate crop manangement. *Indian Coconut Journal* (India) 20, 2: 13-14.

Manthriratne, M.A.P. (1971). In: *Annual Report of Coconut Research Institute for 1971*.

Manthriratna, M.A., and Abeywardena, V. (1979) Planting densities and planting systems for coconut, Cocos nucifera L. 2. Study of yield characters and the economics of planting at different densities. *Ceylon Coconut Quarterly* (Sri Lanka) 30: 107-115.

Maramorosh, K. (1964) *A survey of coconut diseases of unknown ethiology*. FAO, Rome (Italy), 38 pp.

— (1993) The threat of Cadang-cadang disease. *Principes* 37, 4: 187-196.

Marconi, F.A.M. (1952) As gargatas das palmeiras. *Biologico* (Brazil), 18, 6: 103-107.

Marechal, H. (1926) Observation and preliminary experiments on the coconut palm with a view to developing improved seed for Fiji. *Fiji Agricultural Journal* (Fiji) 1: 16-45.

Margate, R.Z., and Magat, S.S. (1988) Growth response of seedlings from seednuts collected from palms fertilized with sodium chloride (common salt). *Philippine Journal of Coconut Studies* (Philippines) 13, 1: 6-9.

Margate, R.Z, Magat, S.S., and Prudente, R.L. (1979a) Coconut intercropping system: Multicropping trials underbearing palms. *Philippine Coconut Authority Annual Report 1979*.

Margate, R.Z., Magat, S.S., and Abad, R.G. (1979b) Boron requirement of hybrid coconut seedlings grown in an inland coconut soil of Davao. *Philippine Journal of Coconut Studies* (Philippines) 4, 4: 6-14.

Mariano, R.L.R., Lira, R.V.F. de, Padovan, I.P., and Allana, E. (1990) Ocorrência da murcha de Phytomonas em coqueiro no estado de Pernambuco, Brasil. *Fitopathologia Brasileira* (Brazil) 15, 1: 80-82.

Mariau, D. (1967) Les fluctuations des populations d'*Oryctes* en Côte d'Ivoire. *Oléagineux* (France) 22, 7: 451-454.

— (1969) Pseudotheraptus, un nouveau ravageur du cocotier en Afrique occidentale. *Oléagineux* 24: 21-25.

— (1971) Les ravageurs et maladies du palmier à huile et du cocotier. Méthodes de lutte contre les termites attaquant les jeunes plants de cocotier. *Oléagineux* (France) 26: 233-234.

— (1974) Hispine du genre *Coelaenomenodera* ravageurs du cocotiers à Madagascar. *Oléagineux* (France) 30, 7: 303-309.

— (1977) *Aceria (Eriophyes) guerreronis* un important ravageur des cocoteraies africaines et américaines. *Oléagineux* (France) 32, 3: 101-111.

— (1986) Comportement de *Eriophyes guerreronis* Keifer à légard de différentes variétés de cocotiers. *Oléagineux* (France) 41, 11: 499-505.

Mariau, D., and Calvez, C. (1973) Méthode de lutte contre l'*Oryctes* en replantation de palmier àa huile. *Oléagineux* (France) 28, 5: 215-218.

Mariau, D., Desmier de Chenon, R., Julia, J.F., and Philippe, R. (1981) Les ravageurs du palmier à huile et du cocotier en Afrique occidentale. *Oléagineux* (France) 36: 170-228.

Mariau, D., Desmier de Chenon R. and Sudharto, P.S. (1991) Les insectes ravageurs du palmier à huile et leurs ennemies naturels en Asie du Sud-Est. *Oléagineux* (France) 46: 400-476.

Mariau, D., and Julia, J.F. (1977) Nouvelles recherches sur la cochenille du cocotier *Aspidiotus destructor* Signoret. *Oléagineux* (France) 32: 65-75.

Mariau, D., Renoux, J., and Desmier de Chenon R. (1992) Coptotermes curvignathus Holangren Rhinotermitidae, principal ravageur du cocotier planté sur tourbe à Sumatra. *Oléagineux* (France) 47: 561-568.

Marshall, D.R., and Brown, A.H.D. (1973) *Définition d'une stratégie d'échantillonnage optimum pour la conservation génétique*. Conférence technique de la FAO sur les ressources génétiques végétales, Rome (Italy). (Cited by Nucé de Lamothe M. de, and Wuidart, W. (1982).

Marty, G., Guen, V. de Le, and Fournial, T. (1986) Cyclone effects on coconut plantations in Vanuatu. *Oléagineux* (France) 41, 2: 63-69.

Mataora, T. (1987) *Coconut research and development in the Cook Islands*. Proceedings of the Fifth South Pacific Islands Regional Meeting on Agricultural Research, Development, Extension and Training in Coconut, IRETA (Western Samoa) pp. 118-122.

Mathai, G., Mathew, J., and Balakrishnan, B. (1991) Reaction of exotic cultivars of coconut (*Cocos nucifera* L.) to root(wilt) disease of Kerala. In: Silas, E.G., Aravindakshan, M., and Jose, A.I.; *Coconut breeding and management*. Kerala Agricultural University (India) pp. 161-162.

Mathen, K. (1960) Observations on *Stephanitis typicus*, a pest of coconut palms. *Indian Coconut Journal* (India) 14, 1: 8-27.

— (1985) Lace bug abundance in Root(wilt) disease affected coconut palms - cause or effect? *Journal of Plantation Crops* (India) 13, 1: 56-59.

Mathen, K., Rajan, P., Nair, C.P.R., Sasikala, M., Givndankutty, M.P., and Solomon, J.J. (1990) Transmission of Root(wilt) disease to coconut seedlings through *Stephanitis typica* (Distant) (Heteroptera: Tingidae). *Tropical Agriculture* (Trinidad) 67, 1: 69-73.

Mathes, D.T. (1984) Effect of extraction of inflorescence of immature fruit bunches on the production of female flowers in coconut. *Cocos* (Sri Lanka) 2: 44-47.

— (1986) Coffee, cocoa and pepper under coconut, *Coconut Bulletin* (Sri Lanka) 3: 9-11.

— (1988) Influence of weather and climate on coconut yield. *Coconut Bulletin* (Sri Lanka) 5, 1: 8-10.

Mathew, J., Nambiar, K.K.N., Jose, C.T., and Anil Kumar (1989) Stem bleeding disease of coconut - a method for indexing the disease severity. *Journal of Plantation Crops* (India) 17, 2: 80-84.

Mathew, J., Pillay, G.R., Santhakumari, G., and Varghese, K. (1993) Influence of supplemental irrigation on the productivity and water use of adult coconut palms. *Journal of Plantation Crops* (India) 21, 2: 81-87.

Mathew, C., *et al.* (1991) Physiology and biochemistry. In: Nair, M.K. (eds); *Coconut Root(Wilt) disease.* Central Plantation Crops Research Institute (India), Monograph Series - no 3: 61-66.

McDowell, R.E., and Hildebrand, P.E. (1980) *Integrated crop and animal production: Making the most of resources available to small farms in developing countries.* A Bellagio Conference, October 18-23, 1978. Rockefeller Foundation, New York.

Menon, K.P.V., and Pandalai, K.M. (1958) *The coconut palm, a monograph.* Indian Central Coconut Committee, Ernakulam (India) 384 pp.

Mercado, B.T., and Velasco, J.R. (1961) Progress report: Effect of aluminium on the growth of coconut. *The Philippine Agriculturist* (Philippines) 45, 5: 268-274.

Merilyn, V.J., and Thomas, G.V. (1992) Distribution of nitrogen fixing *Beijerinckia* in the rhizosphere of coconut. *Journal of Plantation Crops* (India) 20, 2: 146-149.

Merino, G. (1938) Report on coconut *Zygaena* in province of Palawan. *Philippine Journal of Agriculture* (Philippines) 9, 1: 31-35.

Meunier J. (1976) Les prospections de palmacées. Une nécessité pour l'amélioration des palmiers oléagineux. *Oléagineux* (France) 31, 4: 153-157.

— (1986) Report of mission for the study of coconut germplasm in Indonesia. *IRHO Document* (France) no. 2026 December 1986.

— (1992) *Genetic diversity in Coconut.* A brief survey of IRHO's Work I. Isozyme Electrophoresis. In: Coconut Genetic Resources. Papers of an IBPGR Workshop, Cipanas, Indonesia, 8-11 October, 1991. International Crop Network Series no. 8. International Board for Plant Genetic Resources, Rome (Italy) pp. 35-40.

Meunier, J., Rognon, F., and Nucé de Lamothe, M. de (1977) L'analyse des composantes de la noix du cocotier. Etude de l'échantillonage. *Oléagineux* (France) 32, 1: 9-14.

Meunier, J., Baudouin, L., Nouy, B., and Noiret, M. (1988) Estimation de la valeur des clones de palmier à huile. *Oléagineux* (France) 43, 5: 195-200.

Meunier, J., Le Saint, P., Nucé de Lamothe, M. de, Rognon, F., and Sangare, A. (1991) Genetic improvement and planting material production. In: Green, A.H. (ed.) (1991) *Coconut production. Present status and priorities for research.*

Meunier, J., Sangare, A., Le Saint, J.P., and Bonnot, F. (1984) Analyse génétique des caractères du rendement chez quelques hybrides de cocotier *Cocos nucifera* L. *Oléagineux* (France) 39, 12: 581-586.

Mkumbo, K.E., and Kullaya, A. (1994) *Coconut Breeding in Tanzania, the current and future prospects.* In: Proceedings of the Workshop on Standardization of Coconut Breeding Research Techniques held at Abidjan (Ivory Coast) June 20-25, 1994. In press.

Mohd. Nawi, A.L., and Ahmad, H. (1988) *Projek ternakan bebiri di rancangan Felcra Pauh Manis.* In: Proceedings Workshop on 'Pengurusan dan penggunaan foraj secara efisien di kalangan penternak' at Pusat Latihan Ternakan, Sg. Siput, Perak Darul Ridzuan on 28-30 June, 1988, pp. 75-83.

Moll, R.H., and Stuberg, W. (1971) Comparison of response of alternative selection procedure initiated with two populations of maize. *Crop Science* 11: 706-711.

Montfort, S. (1984) *Recherche d'une méthode d'obtention d'haploïdes in vitro de Cocos nucifera L. in vitro.* Thèse de doctorat en sciences, Université de Paris-Sud Centre d'Orsay (France).

Moog, F.A. (1991) Forage and legumes as protein supplements for pasture based systems. In: *Feeding Dairy Cows in the Tropics* (ed. Speedy, A. and Sansoucy, R.). Proc. of the FAO Expert Consultation held in Bangkok, Thailand 7-11 July 1989. FAO Animal Production and Health Paper 86, pp. 142-148.

— (1992) *Heteropsylla cubana: Impact on feeding systems in Southwest Asia and the Pacific.* In: Proceedings FAO Expert Consultation on Legume Trees and Other Fooder Trees as Protein Sources for Livestock. 14-18 October 1991, MARDI, Kuala Lumpur, Malaysia, pp. 233-243.

Moog, F.A., and Faylon, P.S. (1991) Integrated forage-livestock systems under coconuts in The Philippines. In: *Proceedings of Workshop Forages for Plantation Crops*, Shelton, H.M. and Stur, W.W. (ed.). Sanur Beach, Bali, Indonesia, 27-29 June. ACIAR Proc. No. 32: 144-146.

Morin, J.P., Lucchini, F., Araujo, J.C., Ferrera, J.M., and Fraga, L.S. (1986) Estudo de comp ortamento olfactivo de *Rhynchophorus palmarum* (L.) (Coleoptera, Curculionidae) no campo. *Annuais da Sociedade Entomologica do Brasil* (Brazil) 18: 267- 273.

Moses, T. (1962) Palms of Brasil. *Principes* 6: 26-37.

Moss, J.R.J. (1992) Measuring light interception and the efficiency of light utilization by the coconut palm (*Cocos nucifera*). *Experimental Agriculture* (UK) 28: 273-285.

Moura, J.I.L., Mariau, D., and Delabie, J.H.L. (1993) Efficiência de *Paratheresia menezesi* Townsend (Diptera, Tachinidae) no controle biologico natural de *Rhynchophorus palmarum* (L.) *Oléagineux* (France) 48, 5: 219-223.

Moura, J.I.L., Mariau, D., and Delabie, J.H.L. (1994) Taticas para o controle de *Amerrhinus ynca* Sahlb. 1823 (Coleoptera, Curculionidae). Broca da raque foliar do coqueiro (*Cocos nucifera* L.) *Oléagineux* (France) 49, 5: 221-226.

Moutia, L.A. (1958) Contribution to the study of some phytophagous acarina and their predators in Mauritius. *Bulletin of Entomological Research* 49: 59-75.

Muinga, R.W. (1992) Feeding systems for smallholder milk production. *ILCA Newsletter* 11, 4: 4-5.

Muliyar, M.K., and Rethinam, P. (1991) Production of coconut hybrids in India - present and future. In: *Coconut Breeding and Management*. Proceedings of the National Symposium held from 23rd to 26th November, 1988. 380 pp. Kerala Agricultural University, Vellanikkara 680 654, *Trichur* (India) 208-211.

Murashige, T., and Skoog, F. (1962) A revised medium for rapid growth and bioessays with tobacco tissue culture. *Physiol. Plant.* 15: 473-479.

Murry, D.V. (1977) Coconut palm. In: Alvin P. de T., and Kozlowski, T.T. (eds), *Ecophysiological tropical crops*. Academy Press, New York (USA) pp. 24-27.

Nagarajan, R., Manickam, T.S., Kothandaraman, G.V., Ramaswamy, K., and Palaniswamy, G. (1985) Manurial value of coir pith. *Madras Agricultural Journal* (India) 72, 9: 533-555.

Nagarajan, R., Manickam, T.S., and Kothandaraman, G.V. (1988) Composting coir dust for better utilization as manure in agricultural farms. *Journal of Plantation Crops* (India) 16, 1: 59-62.

Nagpala, R.G,. and Moog F.A. (1979) *Feeds and feeding practices under different cropping patterns in Batangas and Quezon*. PCARRD Research Highlights. Los Baños, Laguna, Philippines.

Naim, S.H., and Husin, A. (1984) Coconut palm sugar. In: Pushparajah, E., and Chew Poh Soon (eds), *Cocoa and coconuts, progress and outlook*. Incorporated Society of Planters, Kuala Lumpur (Malaysia) 943-946.

Nair, M.G.K., Hegde, M.R., Yusuf M., and Das, P.K. (1991b) Mixed cropping in coconut gardens. *Technical Bulletin* 24, Central Plantation Crops Research Institute, Kasaragod, India. 8 pp.

Nair, M.K., and Gopalasundaram, P. (1990) *Coconut based farm family models in India*. XXVII COCOTECH Meeting, 25-29 June, 1990, Manila, Philippines, pp. 360-369.

Nair, M.K., and Menon, A.R.S. (1991) Central Plantation Research Institute Kasagarod. *Indian Coconut Journal* (India) 22, 6 and 7: 3-9.

Nair, M.K., Nampoothiri, K.U.K., and Dhamodaran, S. (1991a) Coconut Breeding - Past achievements and future strategies. In: *Coconut Breeding and Management*. Proceedings of the National Symposium held from 23rd to 26th November, 1988. 380 pp. Kerala Agricultural University, Vellanikkara 680 654, Trichur (India) pp. 17-25.

Nair, P.K.R. (1977). Multi-species crop combination with tree crops for increased productivity in the tropics. *Gartenbauwissenchaft* (Germany) 42, 4: 145-150.

— (1979) *Intensive multiple cropping with coconut in India: principles, programmes, prospects*. Advancing Agronomy and Crop Science, Verlag Paul Parey, Berlin (Germany) 6, 145 pp.

— (1983) Agroforestry with coconuts and other tropical plantation crops. In: *Plant Research and Agroforestry* (ed. P.A. Huxley). ICRAF, Nairobi, Kenya, 617 pp.

— (1986) Coconut diversification: problems and approaches. The Indian experience. In: *Diverse Peoples, Diverse Farms*. Proceedings Workshop of the International Consultation on Plantation Crops Diversification. FAO (Italy) pp. 55-90.

Nair, P.K.R, and Varghise, T.P. (1976) Crop diversification in coconut plantations. *Indian Farming* (India) 25 11: 17-19.

Nair, R.R. (1989) Summer irrigation requirement of the coconut palm. *Indian Coconut Journal* 19, 12: 3-7.

Najib, M.M.A. (1989) Growth performance of two erect grasses under the canopy of rubber for livestock integration system. In: *Modernization in Livestock and Poultry Production*, Proceedings 12th MSAP Annual Conference, March 29-31, 1989, pp. 284-289.

Nambiar, C.K.B., Bidappa, C.C., and Khan, H.H. (1988a) Effect of coir dust blended fertilizers on carbon and nitrogen changes in a coastal sandy soil. *Journal of Plantation Crops* (India) 16, 2: 100-104.

Nambiar, K.K.N., and Ayer, R. (1991) Current status of research on the stem bleeding disease of coconut in India. *Cord* (Indonesia) 7, 2: 31.

Nambiar, I.P.S., Nambiar, P.K.R., and Rajan, K.C. (1988b) Coconut based farming Systems. In: *Six Decades of Coconut Research,* Kerala Agriculture University, Trichur, India, pp. 137-141.

Nambiar, K.K.N., Joshi, Y., Venugopal, M.N., and Mohan, R.C. (1986) Stem Bleeding disease of coconut: Reproduction of symptoms by inoculation with *Thielaviopsis paradoxa. Journal of Plantation Crops* (India) 14, 2: 130-133.

Nambiar, M.G., Thamkamma Pillai, P.K., and Vijaya Kumar, G. (1970) Cytological behaviour of first inbred generation of coconut. *Indian Journal of Genet.* (India) 30: 744-752.

Nambiar, P.K.N., and Rao, G.S.L.H.V. P. (1991) Varietal and seasonal variations in oil content of coconut. In: Silas, E.G., Aravindakshan, M., and Jose, A.I. (eds), *Coconut breeding and management.* Kerala Agricultural University (India), pp. 283-286.

Nambiar, P.K.N., and Rawther, T.S.S. (1993) Fungal diseases of coconut in the world. In: Nair, M.K., Khan, H.H., Gopalasundaram, P., and Bhaskara Rao, E.V.V.; *Advances in coconut research and development.* Indian Society for Plantation Crops (India) pp. 545-561.

Nanden-Amattaram, T.L., and Parsadi-Sewkaransing, M. (1989) Some preliminary observations on the occur rence of *Phytomonas* flagellates in coconuts from coconut palm (*Cocos nucifera)* infected by 'Hartrot' disease in Suriname. *De Surinaamse Landbouw* (Surinam) 37, 1-4: 14-20.

Napiere, C.M. (1985) Diseases of coconut in The Philippines and their control. *Philippine Journal of Coconut Studies* (Philippines) 10, 1: 1-10.

Naranjo, B.R. (1991) Principales enfermedades del cocotero. ICA-Informa (Colombia) 25, 5-13.

Narayanan Kutty, M.G., and Gopalakrishnan, P.K. (1991) Yield components in coconut palms. In: Silas, E.G., Aravindakshan, M., and Jose, A. (eds), *Coconut breeding and management.* Kerala Agricultural University (India) pp. 94-98.

Nasayao, L.Z., and Malasaga, E.M. (1989) Male flowers and pollen characteristics of some coconut populations in The Philippines. *Philippine Journal of Coconut Studies* 9, 2: 2-6.

Nathanael, W.R.N. (1960a) Coconut nutrition and fertilizer requirements - the plant approach. *Ceylon Coconut Quarterly* (Sri Lanka) 1, 3-4: 101-120.

Nathanael, W.R.N. (1960b) Chemical and technological investigations on coconut products. *Ceylon Coconut Q.* II, 1/2: 31-54.

Nathanael, W.R.N. (1961) Coconut nutrition and fertilizer requirements - the plant approach. *Ceylon Coconut Quarterly* (Sri Lanka) 1, 3-4: 101-120.

— (1964) *The processing and utilization of coconut products and byproducts.* FAO Technical Working Party on Coconut Production, Protection, and Processing, 2nd Session, Colombo, 1964.

N'Cho, Y.P., Le Saint, J.P., and Sangare, A. (1988) Les cocotiers Nains à Port Bouët (Côte d'Ivoire). III. Nain Brun Nouvelle-Guinée, Nain Vert Thaïlande, Nain Rouge Polynésie. *Oléagineux* (France) 43, 3: 55-66.

N'Cho, Y.P., Sangare, A., Bourdeix, R., Bonnot, F., and Baudouin, L. (1993) Evaluation de quelques écotypes de cocotier par une approche biométrique. I. Etude des populations de Grands populations). *Oléagineux* (France) 48, 3: 121-132.

Nelliat, E.V. (1972) NPK nutrition of coconut palm - A review. *Journal of Plantation Crops* (India) 1 (Suppl.) pp. 70-80.

— (1979) Multistoreyed Cropping, In: *Multiple cropping in coconut and arecanut Gardens. Technical Bulletin* 3, Central Plantation Crops Research Institute, Kasaragod, India.

Nelliat, E.V, Bavappa, K.V., and Nair, P. (1974) Multi-storeyed cropping a new dimension in multiple cropping for coconuts. *World Crops* 26, 6: 262-266.

Nelliat, E.V., Gopalasundram.P., Varghese. P.T., and Sivaraman K. (1979) Mixed cropping in coconut. In: *Multiple cropping in coconut and arecanut gardens. Technical Bulletin* 3, Central Plantation Crops Research Institute, Kasaragod, India. 28-34.

Nelliat, E.V., and Krishna, J.L.N. (1976) Intensive cropping in coconut gardens. *Indian Farming* (India) 27, 9: 9-12.

Ng, K.F. (1991) *Forage species for rubber plantations in Malaysia.* In: Proceedings of Workshop on Forages for Plantation Crops. Shelton, H.M. and Stur, W.W. (ed.). Sanur Beach, Bali Indonesia, 27-29 June 1990. ACIAR Proc. No. 32, 49-53.

Nguyen, T., Thanh-Tuyen, and Aourillo, D.I. (1992) Plant regeneration through somatic embryogenesis from cultured zygotic embryos of coconut. *Philippine Journal of Coconut Studies* (Philippines) 17, 1: 1-7.

Nienhaus, F., and Steiner, K.G. (1976) Mycoplasmalike organisms associated with Kainkopé disease of coconut palms in Togo. *Plant Disease Reporter* 60, 12: 1000-1002.

NIFOR (1989) *NIFOR: History, activities and achievements.* Sadoh Press Ltd 47 Eregie Street, off Ighomo Aihie Str., by Uwasota St. Ugbowo Benin City (Nigeria) 32 pp.

Nimal, P.A.H. (1989) Place of coconut in home gardens. *Coconut Bulletin* (Sri Lanka) 6, 1: 16-19.

Ninan, C.A., and Pandalai, K.M. (1961) *Recent trend in coconut breeding in India*. In: Proceedings First Meeting of the FAO Technical Working Party on Coconut Production, Protection and Processing, Trivandrum, India. FAO, Rome (Italy).

Nipah, J.O., and Dery, S.K. (1994) *Coconut Breeding in Ghana*. In: Proceedings of the Workshop on Standardization of Coconut Breeding Research Techniques held at Abidjan, Ivory Coast, June 20-25, 1994.

Nirula, K.K., Antony, J., and Menon, K.P.V. (1956) Investigation on the pests of coconut palm: Nephantis serinopa. *Indian Coconut Journal* (India) 9, 2: 101-131.

— (1953) Some investigations on the control of termites. *Indian Coconut Journal* (India) 7, 1: 26-34.

Nitis, I.M., Lana, K., Sukanten, W., Suarna, M., and Putra, S. (1990) The concept and development of the three strata-forage system. In: *Shrubs and Tree Fodders for Farm Animals* (ed. Devendra, C.). IDRC, IDRC-276e, Ottawa, Canada, pp. 92-102.

Nitis, I.M., Putra, S., Sukanten, W., Suarna M., and Lana, K. (1991) Prospects for increasing forage supply in intensive plantation crop systems in Bali. In: *Forages for Plantation Crops*, Proceedings of a workshop, Sanur Beach, Bali, Indonesia 27-29 June, 1990 (ed. Shelton, H.M. and Stur, W.W.). ACIAR Proceedings 32, 134-139.

Nitis, I.M., and Rika, K. (1978) *Bali cattle grazing improved pasture under coconuts. I. Effect of stocking rate on steer performance and coconut yield.* In: Seminar on Integration of Animals with Plantation Crops, Pulau Pinang, 13-15 April, Malaysian Society Animal Production and Rubber Research Institute of Malaysia.

Noiret, J.M. (1981) Application de la culture in vitro à l'amélioration et à la production de matériel clonal chez le palmier à huile (Application of *in vitro* cultivation for the improvement of clonal material of oil palm. *Oléagineux* (France) 36: 123-126.

Norton, B.W., Wilson, J.R., Shelton, H.M., and Hill, K.D. (1991) The effect of shade on forage quality. In: Proceedings of Workshop on Forages for Plantation Crops, Shelton, H.M. and Stur, W.W. (ed). Sanur Beach, Bali, Indonesia, 27-29 June, 1990. *ACIAR Proceedings* 32: 83-88.

Novarianto, H., *et al.* (1994) *Coconut Breeding Programme in Indonesia*. In: Proceedings of the Workshop on Standardization of Coconut Breeding Research Techniques held at Abidjan (Ivory Coast) June 20-25, 1994.

Nowell, W. (1923) *Diseases of crop plants in the Lesser Antilles*. The West India Committee, London (England).

Nucé de Lamothe, M. de (1970) Application du principe des croisements interorigines au cocotier. Premiers résultats obtenus en Côte d'Ivoire. *Oléagineux* (France) 25, 4: 207-210.

— (1990) La recherche sur le cocotier: progrès réalisés et perspectives. *Oléagineux* (France) 45, 3: 119-129.

— (1993) La recherche sur le cocotier en amélioration génétique: vers quel type d'organisation...? *Bulletin BUROTROP* (France) 5, 2 pp.

Nucé de Lamothe, M. de, and Rognon, F. (1977) Dwarf coconuts at Port-Bouet. 1. Ghana Yellow Dwarf, Malayan Red Dwarf, Equatorial Guinea Green Dwarf, Cameroon Red Dwarf. *Oléagineux* (France) 32, 11: 462-466.

Nucé de Lamothe, M. de, and Benard, G. (1985a) L'hybride de cocotier PB121 (ou Mawa) (NJM x GOA). *Oléagineux* (France) 40, 5: 261-266.

Nucé de Lamothe, M. de, and Benard, G. (1985b) L'hybride de cocotier PB213 (GOA x GRL). *Oléagineux* (France) 40, 10: 491-496.

Nucé de Lamothe, M. de, and Rognon, F. (1972a) La production de semences hybrides chez le cocotier par fécondation naturelle dirigée (The production of hybrid seednuts in coconut by controlled natural pollination). *Oléagineux* (France) 27, 10: 483-488.

Nucé de Lamothe, M. de, and Rognon, F. (1972b) La production de semences hybrides chez le cocotier par pollinisation assistée. *Oléagineux* (France) 27, 11: 539-544.

Nucé de Lamothe, M. de, and Rognon, F. (1975) Pollinisation assistée et contamination par des pollens indésirables. *Oléagineux* (France) 30, 8-9: 359-364.

Nucé de Lamothe, M. de, and Rognon. F. (1977) Les cocotiers Nains à Port Bouët (Côte d'Ivoire). I. Nain Jaune Ghana, Nain Rouge Malaisie, Nain Vert Guinée Equatoriale et Nain Rouge Cameroun. *Oléagineux* (France) 32, 8-9: 367-375.

Nucé de Lamothe, M. de, Sangare, A., Meunier, J., and Le Saint, J.P. (1991) Coconut hybrids, Interest and prospects; IRHO contribution to research and development. In: *Coconut Breeding and Management.* Proceedings of the National Symposium held from 23rd to 26th November, 1988. 380 pp. Kerala Agricultural University, Vellanikkara 680 654, Trichur (India).

Nucé de Lamothe, M. de, and Wuidart, W. (1979) Les cocotiers Grands à Port Bouët (Côte d'Ivoire). I. Grand Ouest Africain, Grand de Mozambique, Grand de Polynésie, Grand de Malaisie. *Oléagineux* (France) 34, 7: 339-349.

Nucé de Lamothe, M., and Wuidart, W. (1992) La production de semences hybrides de cocotier: cas des semences hybrides Nain x Grand. *Oléagineux* (France) 47, 2: 93-102.

Nucé de Lamothe, M. de, Wuidart, W., Rognon, F, and Sangare, A. (1980) La fécondation artificielle du cocotier. *Oléagineux* (France) 35, 4: 193-205.

Nucé de Lamothe, M. de, and Wuidart, W. (1981) Les cocotiers Grands à Port Bouët (Côte d'Ivoire). II. Grand Rennell, Grand Salomon, Grand Thaïlande, Grand Nouvelles Hébrides. *Oléagineux* (France) 36, 7: 353-365.

Nucé de Lamothe, M. de, and Wuidart, W. (1982) L'observation des caractéristiques de développement végétatif, de floraison et de production chez le cocotier. *Oléagineux* (France) 37, 6: 291-300.

Nucé de Lamothe, M. de, and Wuidart, W. (1994) La production de semences hybrides de cocotier: cas des semences hybrides Nain x Grand. *Oléagineux* (France) 47, 2: 93-102.

Nugari (1984) *Pengaruh pengembalaan ternak kambing terhadap kandungan N, P dan K dan bahan organik tanah.* Fapet UNUD Denpasar.

Nunez, T.C., Edosema, D.B., Mesorado, R.B., and Kabristante, M.P. (1991) Trichodermal action on inoculated coconut husk (a research note). *Philipine Journal of Coconut Studies* (Philippines) 16, 2: 31-32.

Ochs, R., (1977) Ecologic constraints of perennial oilcrops (oil palm and coconuts) in West Africa. Choice of crop in function of climate and soil. *Oléagineux* (France) 32, 11: 461-466.

Ochs, R., Bengy, A. de, and Bonneau, X. (1992) Setting up coconut plantations on deep peat soils. *Oléagineux* (France) 47, 1: 9-22.

Ochs, R., and Bonneau, X. (1988) Copper and iron deficiency symptoms in coconut on peat soils in Indonesia. *Oléagineux* (France) 43, 12: 455-457.

Ochs, R., Bonneau, X., Olivin, J., Ouvrier, M., Pomier, M., Taffin, G. de, and Zakra, N. (1991) Mineral nutrition and manuring with particular reference to the new hybrid varieties. In: Green, A.H. (ed.) (1991) *Coconut production. Present status and priorities for research.* The World Bank (USA) pp. 61-70.

Ochs, R., Bonneau, X., and Qusairi, L. (1993a) Copper mineral nutrition in hybrid coconuts on peat soils. *Oléagineux* (France) 48, 2: 65-76.

Ochs, R., Olivin, J., Daniel, C., Pomier, M., Ouvrier, M., Bonneau, X., and Zakra, N. (1993) Coconut nutrition: IRHO experience in different countries. In: Nair M.K., Khan, H.H., Gopalasundaram,. P., and Bhaskara Rao, E,V.V. (eds) *Advances in Coconut Research and Development.* Indian Society for Plantation Crops (India) pp. 313-327.

O'Connor, B.A. (1954) Notes on the coconut stick insect, *Graeffea crouani. Agricultural Journal of Fiji* (Fiji) 25, 3-4: 89-92.

Odi Link Hoak (1981) Evaluation of some hybrid coconuts. In: *1981 Yearly Progress report on Coconut Research and Development.* FAO, Rome (Italy) 1982, pp. 6-7.

Ohler, J.G. (1984) *Coconut: Tree of Life. FAO Plant Production and Protection Paper* 57, Food and Agriculture Organization, Rome, Italy, 446 pp.

Oka Nurjaya, M.G., Mendra, I.K., Gusti Oka, M., Kaca, I.N., and Sukarji, W. (1991) *Growth of tree legumes under coconuts in Bali.* In: Proceedings of Workshop on Forages for Plantation Crops, Shelton, H.M. and Stur, W.W. (ed). Sanur Beach, Bali, Indonesia, 27-29 June, 1990. ACIAR Proceedings, pp. 70-71.

Ollivier, J. (1993a) Magnesium deficiency symptoms in coconut. *Oléagineux* (France) 48, 7: 339-342.

— (1993b) Potassium deficiency symptoms in coconut. *Oléagineux* (France) 48, 11: 483-486.

Ollagnier, M. (1985) Ionic reactions and fertilizer management in relation to drought resistance or perennial oil crops (oil palm and coconut). *Oléagineux* (France) 40, 1: 1-10.

Ollagier, M., and Ochs, R. (1971) The chlorine nutrition of oil palm and coconut. *Oléagineux* (France) 26, 6: 367-372.

— (1972) Sulphur deficiency of oil palm and coconut. *Oléagineux* (France) 4, 193-198.

Ollagnier, M., Ochs, R., Pomier, M., and Taffin, G. de (1983) Effect of chlorine on the hybrid coconut PB-121 in the Ivory Coast and Indonesia. Growth, tolerance to drought, yield. *Oléagineux* (France) 38, 5: 309-321.

Ollivier, J. (1993a) Magnesium deficiency symptoms in coconut. *Oléagineux* (France) 48, 7: 339-342.

— (1993b) Potassium deficiency symptoms in coconut. *Oléagineux* (France) 48, 11: 483-486.

Omoti, U., Amalu, U.C., and Ataga, D.O. (1986b) Distribution and morphology of roots of the Nigerian Tall coconut in relation to nutrient absorption and fertilizer placement. In: Pushparah, E., and Soon, C.P. (eds), *Cocoa and Coconuts: progress and outlook* pp. 773-770. Incorporated Society of Planters (Malaysia).

Ontolan, D.G. (1988) *Status report on intercropping and intergrazing in The Philippines.* In: Proceedings Workshop on Intercropping and Intergrazing in Coconut Areas, 7-11 September, 1988, Colombo, Sri Lanka.

Ooi, S.C. (1987) *The coconut sector in some South Pacific countries and a consideration of some priority areas of research.* Proceedings of the Fifth South Pacific Islands Regional Meeting on Agricultural Research, Development, Extension and Training in Coconut, IRETA (Western Samoa) pp. 44-72.

Opio, F.A. (1986) *Coconut intercropping/rotational grazing potential in the South Pacific.* A paper presented at the Coconut Industry Interest Meeting, October 2-5, 1986, Suva, Fiji.

— (1988) Farmers receptivity to new technology of coconut cultivation in Western Samoa. Asian Pacific Coconut Community, *Technical Bulletin* (Indonesia) 7, 40 pp.

— (1989) *Coconut Industry in Western Samoa.* APCC Occasional Publication Series 7, 25 p.

— (1990a) *The need for coconut based farming systems.* In: Proceeding of APCC XXVII COCOTECH Confer ence on Coconut Based Farming Systems, 25-29, June 1990, Manila, Philippines, 1-16.

— (1990b) *Simple benefit-cost analysis method for determining optimum intercrop combination in coconut based system.* In: Proceeding of APCC XXVII COCOTECH Conference on Coconut Based Farming Systems, 25-29, June 1990, Manila, Philippines, 292-299.

— (1990c) *Coconut intercropping/rotational grazing model in the South Pacific.* XXVII COCOTECH Meeting, 25-29 June, 1990, Manila, Philippines, pp. 342-359.

— (1992) An economic assessment of coconut based farming systems in the South Pacific. *Journal of Agro forestry Systems.*

Oswald, S., and Rashid, M.K. (1992) *Integrated control of the coconut bug Pseudotherapthus wayi (Brown) in Zanzibar.* National Coconut Development Programme (Tanzania) 1/92.

Otanes, F.Q. (1956) *Noteworthy campaigns against locusts during a quarter of a century, 1930-1954.* Depart ment of Agriculture, Manila (Philippines).

Ouvrier, M. (1984a) Study of the growth and development of young PB-121 (MYD xWAT) hybrid coconuts. *Oléagineux* (France) 39, 2: 73-82.

— (1984b) Nutrient removal in the harvest of the hybrid coconut, PB-121 depending on potassium and magnesium fertilizing. *Oléagineux* (France) 39, 5: 263-271.

— (1984c) Exportation par la récolte du cocotier hybride PB121 en fonction de la fumure potassique et magné sienne. *Oléagineux* (France) 39, 5: 263-271.

— (1987) Nutrient removal in the harvest of the PB 111 coconut depending on potassium and magnesium fertilizing. *Oléagineux* (France) 42, 7: 271-280.

— (1990) Evolution of mineral composition in the young PB 121 coconut hybrid. *Oléagineux* (France) 45, 2: 69-80.

Ouvrier, M., and Brunin, C. (1974) Densités racinaires dans une cocoteraie industrielle et technique d'epamdage des engrais. *Oléagineux* (France) 29, 1: 15-17.

Ouvrier, M., and Ochs, R. (1978) Mineral exportations by the hybrid coconut PB 121. *Oléagineux* (France) 33, 8-9: 437-443.

Ouvrier, M., and Taffin, G. de (1985a) How to fertilize coconut groves. I. Smallholdings. *Oléagineux* (France) 40, 2: 74-76.

— (1985b) Evolution of mineral elements of coconuts left in the field. *Oléagineux* (France) 40, 8-9, 431-436.

Ovasuru, T. (1988) *Intercropping and intergrazing in Papua New Guinea.* In: Proceedings Workshop on Intercropping and Intergrazing in Coconut Areas, 7-11 September, 1988, Colombo, Sri Lanka.

— (1992) Coconut germplasm collection in Papua New Guinea. In: *Report of a working session for implementation of the international Coconut Database held at CIRAD,* Montpellier (France) 19-22 May 1992 (mimeographed).

Pablo, S.J. (1983) *Multi-storey cropping project under coconut.* Paper presented at the Symposium on Coconut Based Farming Systems. Visca, Leyte, (Philippines) June 1-4, 1983.

Pacumbaba, E.P. (1985) The search for a vector of Cadang-cadang disease in coconut. *Philippine Journal of Coconut Studies* (Philippines) 10, 1: 17-20.

Pacumbaba, E.P., and Alfiler, A.R.R. (1987) The effect of removing Cadang-cadang infected palms on the spread of the disease. *Philippine Coconut Authority, Annual Report* (Philippines) 1987, 100-101.

Padmasiri, M.H.L. (1986) Nursery practices. *Coconut Bulletin* (Sri Lanka) 3, 2: 39-42.

Padua-Resurreccion, A.B., and Banzon, J.A. (1979) Fatty acid composition of the oil from progressively maturing bunches of coconut. *Philippine Journal of Coconut Studies* (Philippines) 4, 3, 1-17.

Pai, L.H., Pau, T.Y., Arikiah, A., and Chan, E. (1984) Replanting coconut in the cocoa coconut intercropping system. In: Pushpajah, E. and Chew Poh Soons (eds), *Cocoa and Coconuts: Progress and Outlook* pp. 891-897. Incorporated Society of Planters, Kuala Lumpur (Malaysia).

Palomar, R.N. (1990) *State of the art: Cocowood Utilization.* Proceedings of the workshop on 'Coconut Wood Utilization' at Zamboanga, April 1990. Asian Pacific Coconut Community, Jakarta, Indonesia.

Paner, V.E. (1975) *Multiple cropping research in The Philippines.* Proceeding of the Workshop on Cropping Systems. International Rice Research Institute, Los Banos, Philippines, pp 188-202.

Pannetier, C., and Buffard-Morel, J. (1982) Premiers résultats concernant la production d'embryons somatiques à partir de tissus foliaires de cocotier *Cocos nucifera* L. *Oléagineux* (France) 37, 7: 349-354.

Parawan, O.O. (1991) *Integration of livestock under tree crops in The Philippines.* In: Proceedings of the Internat. Livestock-Tree Cropping Workshop, 5-9 December, 1988, FAO/MARDI, Serdang, Malaysia, pp. 78-89.

Parawan, O.O., and Ovalo, H.B. (1987) Integration of small ruminants with coconuts in The Philippines. In: *Small Ruminant Production Systems in South and Southeast Asia* (ed. Devendra, C.). Proceedings of a Workshop 6-10 October, 1986, Bogor, Indonesia, pp. 221-227.

Passos, E.E.M. (1994) Ecofisiologia do coqueiro. In: Ferreira, J.M.S., Warwick, D.R.N., and Siqueira, L.A. (eds), *A cultura do coqueiro no Brasil.* EMBRAPA - Centro de Pesquisa Agropecuária dos Tabuleiros - CPATC, Aracaju, Sergipe (Brazil) pp. 74-84.

Patel, J.S. (1938) *The Coconut. A monograph.* Government Press, Madras, India.

Pathirana, K.K., and Mangalika, U.L.P. (1992) *Rice straw for cattle grazing natural herbage under coconut.* Unpub. Paper. University of Ruhuna, Sri Lanka.

Patil, R.T. (1991) Post-harvest technology of coconut. In: Silas, E.G., Aravindakshan, M., and Jose, A.I.; *Coconut Breeding and Management.* Proceedings of the National Symposium on Coconut Breeding and Management, Kerala University (India) November 1988, pp. 298-312.

Pau, T.Y., and Chan, E. (1985) Optimum density for Mawa hybrid coconuts (PB 121). *Oléagineux* (France) 40, 4: 189-193.

Payne, W.J.A. (1985) A review of the possibilities for integrating cattle and tree crop production systems in the tropics. *Forest Ecology and Management* 12: 1-36.

Pederson, C.S. (1971) *Microbiology of Food Fermentation.* AVI Publishing Co, West Port, Connecticut, USA.

Peiris, T.S.G. (1993) The degree of influence of rainfall on coconut. In: Nair, M.K., Khan, H.H., Gopalasun daram, P., and Bhaskara Rao, E.V.V. (eds), *Advances in coconut research and development.* Indian Society for Plantation Crops (India) pp. 413-420.

Peoples, M., and D. Herridge (1990) How much nitrogen is fixed by legumes? *Journal of Australian Institute of Agricultural Science* 3, 3: 24-29.

Perdock, H.B., Muttettewgama, G.S., Kaasschieter, G.A., Boon, H.M., Van Vageningen, N.M., Arumugum, V., Linders, M.G.F.A., and Jayasuriya, M.C.N. (1983) *Production responses of lactating or growing ruminants fed urea-ammonia treated paddy straw with or without supplements.* In: Proceedings of the Third Australian-Fibrous Agricultural Residues Workshop, April 12-17, 1983. Sri Lanka, pp. 213-230.

Perera, U.V.H. (1989) Fertilizer use in coconut small holdings: is it remunerative? *Coconut Bulletin* (Sri Lanka) 36, 3: 1-5.

Peries, R.R.A. (1984) Some observations on the pre-nursery system for raising coconut seedlings. *Cocos* (Sri Lanka) 2, 10: 10-17.

— (1993) Seedling Selection in coconut based on juvenile and adult palm correlations. In: Nair, M.K. *et al.* (eds), *Advances in Coconut Research and Development. Platinium Jubilee of Coconut Research in India.* Oxford & IBH Publishing CO. PVT. LTD. (India) pp. 71-72.

— (1994) Coconut Breeding in Sri Lanka. In: *Proceedings of the Workshop on Standardization of Coconut Breeding Research Techniques* held at Abidjan (Ivory Coast) June 20-25, 1994.

Peries, O.S., *et al.* (1975) The incidence of ganoderma root and bole rot of coconut in Sri Lanka. *Ceylon Coconut Journal* (Sri Lanka) 26, 3-4: 99-103.

Pernes, J. (1984) Gestion des ressources génétiques des plantes. Tome 2: manuel. Agence de coopération culturelle et technique, Paris (France) ISBN 92-9028-043-3, éditions Lavoisier. Perry, M. (1992) Appendix X Documentation : General Introduction Paper. In: *Coconut Genetic Resources. Papers of an IBPGR Workshop,* Cipanas, Indonesia, 8-11 October, 1991. International Crop Network Series n°8. International Board for Plant Genetic Resources, Rome (Italy) pp. 35-40.

Perry, M. (1992) Appendix X Documentation: General Introduction Paper. In: *Coconut Genetic Resources.* Papers of an IBPGR Workshop, Cipanas, Indonesia, 8-11 October 1991. International Crop Network Series no. 8. International Board for Plant Genetic Resources, Rome, Italy.

Persley, G.J. (1989) *Coconut international research priorities.* Paper presented for consideration by the Technical Advisory Committee of the Consultative Group on International Agricultural Research. TAC Secretariat FAO, Rome, 73 pp.

Persley, G.J. (1992) *Replanting the tree of life. Towards an international agenda for coconut palm research.* C.A.B. International, Wallingford (UK), 156 pp.

Perthuis, B. Desmier de Chenon, R., and Merland, E. (1985) Revelation of the Marchitez sorpresiva vector of the oil palm - the bug *Lincus lethifer* Dolling (*Hemiptera Pentatomidae Discocephalinae*). *Oléagineux* (France) 40, 10: 473-476.

Phillips, J.S. (1956) Immature nutfall of coconuts in the British Solomon Islands Protectorate. *Bulletin of Entomological Research* 47: 875-895.

Philippines Coconut Authority (1984) *Coconut Based Farming Systems: status and prospects,* Diliman, Quezon, City, (Philippines).

Pillai, N.G. (1974) Mixed farming in coconut gardens. In: *Summer Inst. on Improvement and Management of Plantation Crops.* CPCRI, Kasaragod, India.

Pillai, N.G., and Rawther, T.S.S. (1991) Symptomatology. In: Nair, M.K. *et al.* (eds), *Coconut Root(wilt) disease. Monograph Series* - no 3. Central Plantation Crops Research Institute (India).

Pillai, N.G., Chowdappa, P., Solomon, J.J. and Methew, J. (1991) Remission of symptoms of Root(wilt) disease of coconut injected with oxytetracycline HCl. *Journal of Plantation Crops* (India) 19, 1: 14-20.

Pillai, R.V. (1991) Economic life-span of coconut hybrids. *Indian Coconut Journal* (India) 21, 10: 2-6.
Pittaway, S. (1990) *Macadamia Nut Project: Report of the Farming Systems Specialist.* FAO/TCP Project TCP/FIJ/8954, 24 pp.
Plavsic-Banjac, B., Hunt, B.P., and Maramorosh, K. (1972) Mycoplasmalike bodies associated with Lethal Yellowing disease of coconut palms. *Phytopathology* 62: 298-299.
Plucknett, D.L. (1979) *Managing pastures and cattle under coconuts. Westview Tropical Agriculture Series* 2, Westview Press, Colorado, USA.
Pomier, M. (1979) Plantation des cocotiers élevés en sacs de plastique. Conseils de l'IRHO no: 189. *Oléagineux* (France) 34, 1: 17-20.
— (1984) The portable husking spike. *Oléagineux* (France) 39, 6: 322-323.
Pomier, M., Beligne, V., Bonneau, X., and Taffin, G. de (1986) Restauration de la fertilité des sols lors de la replantation d'une cocoteraie. *Oléagineux* (France) 41, 5: 223-230.
Pomier, M., and Benard, G. (1988) Nitrogen deficiency symptoms in coconut. *Oléagineux* (France) 43, 5: 375-378.
Pomier, M., and Bonneau, X. (1987) Développement du système racinaire du cocotier en fonction du milieu en Côte d'Ivoire. *Oléagineux* (France) 42, 11: 409-421.
Pomier, M., and Taffin, G. de (1982a) The tolerance to drought of some coconut hybrids. *Oléagineux* (France) 37, 2: 55-62.
— (1982b) Study of the fertilization of soils in the replanting of coconut palms. *Oléagineux* (France), 37, 10: 455-461.
Ponto, S.A.S., and Mo, T.T. (1950) De bestrijding van een slakrups *Darna (Orthocraspeda) catenatus* (Sn.), op klapper in het Palu-dal. *Landbouw* (Netherlands) 22: 69-81.
Pordesimo, L.O., and Noble, D.H. (1990) Evaluation of alternative replanting strategies for small coconut farms in The Philippines using a simulation model. *Agricultural Systems* (U.K.) 32, 1: 27-39.
Posas, O.B. (1981) *Grazing versus cut and carry trials on goats under coconuts.* PhD thesis. University of the Philippines, College, Laguna, Philippines.
Pottier, D. (1983) *A pasture handbook for Samoa.* UNDP/FAO project SAM/76/003 IRETA, USP, School of Agriculture, Alafua Campus, Apia, Western Samoa.
Potty, N.N., and Radhakrishnan, T.C. (1978) Stem Bleeding disease of coconut - nutritional relationship. In: *Proceedings Placrosym I. Indian Society for Plantation Crops,* Kerala (India) 347-350.
Prasada Rao, G.S.L.H.V. (1986) Effect of drought on coconut production. *Indian Coconut Journal* (India) 17, 8: 11-12.
— (1991) Agrometeorological aspects in relation to coconut production. *Journal of Plantation Crops* (India) 19, 2: 120-126.
Preston, T.R. (1992) *The role of multipurpose trees in integrated farming systems for the wet tropics.* FAO Expert Consultation on Legume Trees and Other Fodder Trees as Protein Sources for Livestock. 14-18 October 1991, MARDI, Kuala Lumpur, Malaysia, pp. 193-209.
Protein Compagne GMBH (1965) *Improvements in or relating to the extraction of edible proteins from oil seed residues.* British patent 996, 984. (30 June 1965).
Prudente, R.L, Margate, R.Z. and Maravilla, J.N. (1979) *Coconut Intercropping Systems: Screening for shade-tolerant intercrops.* Philippines Coconut Authority, Annual Report 1979: 128-129.
Punchihewa, P.G. (1991) Coconut industry - current situation and prospects. *The Planter* (Malaysia) 779: 63-73.

Quillec, G., and Renard, J.L. (1984) *Phytophthora* rot of coconut. *Oléagineux* (France) 39, 3: 143-145.
Quillec, G., Renard, J,L., and Ghesqueire, H. (1984) *Phytophtora hevea* of coconut: Role in Bud Rot and nutfall. *Oléagineux* (France) 39, 10: 477-484.

Radha, K., Kochu Babu, M., Rethinam, P., Antony, K.J., and Sukumaran, C.K. (1985) 'Contain' Coconut Root (Wilt)disease by eradication of diseased palms. *Indian Coconut Journal* (India) 16: 3-10.
Radhakrishnan, T.C. (1990) Control of stem bleeding disease of coconut. *Indian Coconut Journal* (India) 20, 4: 13-14.
Rajagopal, V. (1991) Is water stress associated with Root(Wilt) disease of coconut? *Journal of Plantation Crops* (India) 19, 2: 91-101.
Rajagopal, V., Kasturi Bai, K.V., and Voleti, S.R. (1990) Screening of coconut genotypes for drought tolerance. *Oléagineux* (France) 45, 5: 215-223.
Rajagopal, V., Mathes, C., Patil, K.D., and Abraham, J. (1986) Studies on water uptake by Root(wilt) diseased coconut palms. *Journal of Plantation Crops* (India) 14, 1: 19-24.
Rajagopalan, A., Nair, R.R., Prasada Rao, G.S.L.H.V., and Nambiar, I.P.S. (1991) Influence of irrigation and fertilizer levels on flowering behaviour of T x GB coconut hybrids. In: Silas, E.G., Aravindakshan, M., and

Jose, A.I. (eds), (1991) *Coconut breeding and management.* Kerala Agricultural University (India) pp. 278-279.

Rajagopal, V., Ramadasan, A., Kasturi Bai, K.V., and Balasimha, D. (1989) Influence of irrigation on leaf water relations and dry matter production in coconut palms. *Irrigation Science* (Germany) 10, 1: 73-81.

Rajagopal, V., Sasikala, M., Sumathykuttyamma, B., Chempakan, B., and Rawther, T.S.S. (1988) Early diagnostic techniques on the Root(wilt) disease of coconut in India. *Philippine Journal of Coconut Studies* (Philippines) 13, 2: 31-35.

Rajagopal, V., Shivashankar, S., and Kasturi Bai, K.V. (1993) Characterization of drought tolerance in coconut. In: Nair, M.K., Khan, H.H., Gopalasundaram, P., and Bhaskara Rao, E.V.V. (eds), *Advances in coconut research and development.* Indian Society for Plantation Crops, New Delhi (India).

Rajagopal, V., Voleti, S.R., Kasturibai, K.V., and Shivashankar, S. (1991) Physiological and biochemical criteria for breeding for drought tolerance in coconut. In: Silas, E.G., Aravindakshan, M., and Jose, A.I. (eds), *Coconut breeding and management.* Kerala Agricultural University (India) pp. 136-143.

Rajaguru, A.S.B. (1991) *Integration of livestock under tree crops in Sri Lanka.* In: Proceedings of the FAO/MARDI International Livestock-Tree Cropping Workshop, 5-9 December, 1988, Serdang, Malaysia, pp. 101-111.

Rajamannar, M., Prasadji, J.K., and Rethinam, P. (1993) Tatipaka disease of coconut - Current status. In Nair: M.K., Khan, H.H., Gopalasundaram, P., and Bhaskara Rao, E.V.V. *Advances in coconut research and development.* Indian Society for Plantation Crops, New Delhi (India) pp. 591-595.

Rajasekharan, N. (1964) *Chemical and Technological Investigations on Coconut Products.* Thesis, Banaras Hindu University, India.

Raju, C.R., Prakash Kumar, P., Mini Chandra Mohan, and Iyer R.D. (1984) Coconut plantlet from leaf tissue cultures. *Journal of Plantation Crops* (India) 12, 1: 75-81.

Ram, C. (1989a) Microflora associada à Queima-das-folhas do coqueiro. *Fitopatologia Brasileira* (Brazil) 14: 136-138.

— (1989b) Epidemiologia e controle quimíco da 'Queima-das-folhas' (*Botryodiplodia theobromae*) do coqueiro (*Cocos nucifera*). *Fitopatologia Brasileira* (Brazil) 14, 3-4: 215-220.

Ramadasan, A., Balakrishnan, T.K., and Rajagopal, V. (1991) Response of coconut genotypes to drought. *Indian Coconut Journal* (India) 21, 12: 2-5.

Ramadasan, A., Kasturi Bai, K.V., Shivashankar, S., and Vijayakumar, K. (1985) Heritability of seedling vigour in coconut palm. *Journal of Plantation Crops* (India) 13, 2: 136-138.

Ramadasan, A., and Mathew, J. (1987) Leaf area and dry matter production in adult coconut palms. *Journal of Plantation Crops* (India), 15, 1: 59-63.

Ramadasan, A., Satheesan, K.V., and Balakrishnan, R. (1980) Leaf area and shoot dry weight in coconut seedling selection. *Indian Journal of Agricultural Science* (India) 50, 7: 553-554.

Ramaiah, K.S. (1990) Lightning damages on coconut palms. *Indian Coconut Journal* (India) 20, 9: 7-9.

Ramanathan, T. (1985) Banana is a profitable intercrop in coconut. *Indian Coconut Journal* (India) 16, 5: 6-11.

— (1987) Technology for increasing coconut production in Tamil Nadu. *Indian Coconut Journal* (India) 17, 10: 3-8.

Ranatunga, A.S., Liyanage, L.V.K., and Perera, R.A.J.R. (1988) Coconut-based cropping systems in the wet and intermediate zones: present constraints and prospects. *Coconut Bulletin* (Sri Lanka) 5, 2: 8-12.

Randles, J.W. (1991) Cadang-cadang disease. In: Green, A.H. (ed.) (1991) *Coconut production. Present status and priorities for research.* The World Bank, Washington D.C. (USA) pp. 97-101.

Randles J.W., and Hanold, D. (1992) Indexing of coconut germplasm for viroïd and virus. In: *Coconut Genetic Resources.* Papers of an IBPGR Workshop, Cipanas, Indonesia, 8-11 October, 1991. International Crop Network Series no. 8. International Board for Plant Genetic Resources Rome (Italy) 44-45.

Randles, J.W., Julia, J.F., Calvez, C., and Dollet, M. (1987) Association of single-stranded DNA with the Foliar Decay disease of coconut palm in Vanuatu. *Oléagineux* (France) 42, 1: 11-18.

Rao, A.S. (1989) Water requirements of young coconut palms in a humid tropical climate. *Irrigation Science* (Germany) 10: 245-249.

Rao, M.B.S., and Koyamu, K. (1955) The dwarf coconut. *Indian Coconut Journal* (India) 8: 106-112.

Rapp, G. (1989) *Coconut stick insect Graeffea crouani (Orthoptera: Phasmida), coconut flatmoth Agonoxena argaula (Lepidoptera: Agonoxenidae). Studies, outpest status, biology, ecology and control carried out in the Kingdom of Tonga, South Pacific.* - Stutgart-Hohenheim (Germany) 1989, 114 p.

Raveendran, T.S., Vijayaraghaven, H., and Ramashandran, T.K. (1989) Some physiological aspects and production trends of certain coconut hybrids and their parents. *Cocos* (Sri Lanka) 7: 36-41.

Rawther, T.S.S., and Pillai, N.G. (1991) Origin, distribution and production loss. In: Nair, M.K. *et al.* (eds), *Coconut Root(wilt) disease.* Central Plantation Crops Research Institute (India), Monograph Series - no 3: 1-6.

Remison, S.U. (1988) Survey of the effect of sea spray on coconuts. *Nigerian Journal of Palms and Oil Seeds* (Nigeria) 9: 116-120.

Remison, S. U., Iremiren, G.O., and Thomas, G.O. (1988) Effect of salinity on nutrient content of the leaves of coconut seedlings. *Plant and Soil* (Netherlands) 109: 135-138.

Remison, S.U., and Iremiren, G.O. (1990) Effect of salinity on the performance of coconut seedlings in two contrasting soils. *Cocos* (Sri Lanka) 8: 33-39.

Remison, S.U., and Mgbeze, C.C. (1988) Effects of storage and planting methods on the germination of coconut. *Nigerian Journal of Palms and Oil Seeds* (Nigeria) 9: 59-70.

Renard, J.L. (1989) Coconut Hartrot characterization and control methods. *Oléagineux* (France) 44, 10: 476-481, 475-482.

Renard, J.L., and Darwis, S.N. (1993) Report on the coconut *Phytophtora* disease seminar. Manado, Indonesia, October 1992. *Oléagineux* (France) 48, 6: 301-305.

Renard, J.L., and Dollet, M. (1991) Phytopathology. In: Green, A.H. (ed.) (1991) *Coconut production. Present status and priorities for research.* The World Bank, Washington D.C. (USA) pp. 86-97.

Renard, J.L., Gianotti, M., and Ghosh, S.K. (1977) Etude d'un jaunissement léthal. *Oléagineux* (France) 32, 7: 317-322.

Renard, J.L., and Quillec, G. (1984) *Phytophthora heveae* of coconut. II Control methods. *Oléagineux* (France) 39, 11: 529-534.

Renard, J.L., Quillec, G., and Arnaud., F. (1975) Une nouvelle maladie du cocotier en pepinière. Symptômes, moyens de lutte. *Oléagineux* (France) 30, 3: 109-112.

Repellin, A., Daniel, C., and Zuily-Fodil, Y. (1994) Merits of physiological tests for characterizing the perfor mance of different coconut varieties subjected to drought. *Oléagineux* (France) 49, 4: 155-167.

Resende, M.L.V. de, Bezerra, J.L., Jose, C., Santana, L. do, and Sobral, L.F. (1991) Efeito de fertilizantes e de Benomil no controle da Lixa-pequena do coqueiro. *Agrotrópica* (Brazil) 3, 1: 45-51.

Rethinam P. (1989) *Research highlights of AICRP on plants. Technical Bulletin 21*, Central Plantation Crops Research Institute, Kasaragod, India. 17 pp.

— (1991) The coconut...now to the non traditional regions. *Indian Horticulture* (India) 36, 3: 4-9, 46.

Rethinam, P., Antony, K.J., and Muralidharan, A. (1991) Management of coconut Root(wilt) disease. In: Nair, M.K., Khan, H.H., Gopalasundaram, P., and Bhaskara Rao, E.V.V. (eds), *Advances in coconut research and development.* Indian Society for Plantation Crops New Delhi (India) pp. 575-584.

Rethinam, P., Roy, A.K., Baranwal, V.K., Chakra Borthy, B.K., and Nambiar, K.K.N. (1990) Crown Choke disease of coconut. *Indian Coconut Journal* (India) 20, 9: 3-6.

Reyne, A. (1948) De cocos palm. In: Hall, C.J.J. van, and Koppel, C. van de; *De Landbouw in de Indische Archipel*, II A: 427-525. Van Hoeve, The Hague (Netherlands).

Reynolds, S.G. (1978a) *A pasture handbook for Samoa.* Unpublished report, South Pacific College of Tropical Agriculture (SPRACTA), Apia (W. Samoa).

— (1978b) Evaluation of pasture grasses under coconuts in Western Samoa. *Tropical Grasslands* 12, 3: 146-151.

— (1980) Grazing cattle under coconuts. *World Animal Review* 35: 40-45.

— (1981) Grazing trials under coconuts in Western Samoa. *Tropical Grasslands* 15, 1: 3-10.

— (1982) Contributions to yield, nitrogen fixation and transfer by local and exotic legumes in tropical grass-legume mixtures in Western Samoa. *Tropical Grasslands* 16, 2: 76-80.

— (1988) *Pastures and cattle under coconuts.* FAO Plant Production and Protection Paper (Italy) 91. First Edition, 321 pp.

— (1989) Possible reasons for non-adoption of pasture and forage crop research findings at farmer level. In: *Grasslands and Forage Production in South-East Asia* (ed. Halim R.A.). Proceedings First Meeting Regional Working Group on Grazing and Feed Resources of South-East Asia 27 Feb.-3 March 1989, Serdang, Malaysia, pp. 181-191.

— (1994) *Pastures and Cattle under Coconuts.* FAO Plant Production and Protection Paper (Italy) 91. Second Edition.

Rhanasinghe, T.K.G. (1995) *Personal communication.*

Rika, I.K., Nitis, I.M., and Humphreys, L.R. (1981) Effects of stocking rate on cattle growth, pasture production and coconut yield in Bali. *Tropical Grasslands* 15, 3: 149-157.

Rika, I.K., Mendra, I.K., Gusti Oka, M., and Oka Nurjaya, M.G. (1991) New forage species for coconut plantations in Bali. In: *Proceedings of Workshop on Forages for Plantation Crops,* Shelton, H.M. and Stur, W.W. (ed.). Sanur Beach, Bali, Indonesia, 27-29 June, 1990. ACIAR Proceedings 32: 41-44.

Rillo, E.P. (1991) *Current status of coconut embryo and tissue culture in The Philippines.* Paper presented at the International Symposium of Coconut Research and Development II (India).

Rillo, E.P., and Paloma, M.B.F. (1989) *Reaction of some coconut populations and hybrids to Phytophthora disease in the Bicol region.*

Risbec, J. (1937) Observations sur les parasites des plantes cultivées des Nouvelles-Hebrides. *Faune Colon.* Franc. Paris (France) 6, 1: 214.

Risseeuw, C. (1980) *The wrong end of the rope. Women coir workers in Sri Lanka.* Research Project Women and Development. University of Leiden (Netherlands) 253 pp.

Roberts, E.H., *et al.* (1984) Recalcitrant seeds: their recognition and storage. In: Holden J.H.W. and Williams J.T. (eds), *Crop Genetic Resources, conservation and evaluation.* George Allen and Unwin, London, (UK) pp. 38-52.

Robinson, A.C. (1981) *End of Assignment Report.* UNDP/FAO Animal Health and Production Project SA M/76/003.

Robinson, J.B. (1991) The growth of *Chloris gayana* within and adjacent to a plantation of *Eucalyptus grandis. Tropical Grasslands* 25: 287-290.

Rochat, D., Descoins, Malosse, C., Nagnan, P., Zagatti, P., Akamouf, F.C., and Mariau, D. (1991) Ecologie chimique des charançons de palmiers, *Rhynchophorus* spp. (Coleoptera). *Oléagineux* (France) 48, 5: 225-236.

Rochat, D., Gonzalez, A.V., Mariau, D., Villanueva, A., and Zagatti, P., (1993) Evidence for male produced aggregation pheromone in American palm weevil *Rynchophorus palmarum* Coleoptera, Curculionidae. *Journal of Chemical Ecology* 17, 6: 1221-1230.

Rodriguez, M.J.B., Ignacio-Namia, M.T.R., and Estokio, L.P. (1989) A mobile laboratory for Cadang-cadang disease of coconut. *Philippine Journal of Coconut Studies* (Philippines) 14, 2: 21-23.

Rodriguez, M.J.B. and Estioko, L.P. (1987) *Studies on the molecular forms of coconut Cadang-cadang viroid (CCCV).* Philippine Coconut Authority, Annual Report (Philippines) pp. 94-95.

Rognon, F., and Boutin, D. (1988) Girth size in the PB-121 coconut hybrid, a practical way to measure growth. *Oléagineux* (France) 43, 4: 165-172.

Rognon, F., and Bourgoing, R. (1992) La production de semences hybrides de cocotier: cas des semences hybrides Nain x Grand. II. Exploitation du champ semencier. *Oléagineux* (France) 47, 7: 481-489.

Rognon, F., Boutin, D., and Bourgoing, R. (1990) Intensification and rehabilitation in tall coconut smallholdings. *Oléagineux* (France) 45, 1: 13-21.

Rognon, Amblard, P., and Boutin, D. (1984a) Chemical land preparation for oil palm and coconut in Indonesia. Imperata eradication. I - Principles and organization. *Oléagineux* (France) 39, 11: 519-527.

— (1984b) Chemical land preparation for oil palm and coconut in Indonesia. Imperata eradication. II - Practising the treatments. *Oléagineux* (France) 39, 12: 575-580.

Rohde, W., *et al.* (1993) Rapid and sensitive diagnosis of mycoplasmalike organisms associated with Lethal Disease of coconut palm by a specifically primed polymerase chain reaction for the amplification of 16S, DNA. *Oléagineux* (France) 48, 7: 319-322.

Rohde, W., Ritter, E., Kullaya, A., and Rodriguez, J. (1995) Improvement of coconut by biotechnology: application to plant breeding and crop protection. *Bulletin Burotrop* (France) 8, 13-15.

Romney, D.H. (1987) Competition from mature coconut trees on underplanted coconut varieties and hybrids. *Oléagineux* (France) 42, 5: 191-194.

Romney, D.D. (1991) Rehabilitation and replanting schemes for smallholders. In: Green, A.H. (ed.) (1991) *Coconut production. Present status and priorities for research.* The World Bank, Washington D.C.(USA) 34-42.

Rouziere, A. (1994) What technologies can be used to revitalize the coconut sector? A few proposals from CIRAD. *Oléagineux* (France) 49, 3: 115-124.

— (1995) Coprah, huile, tourteau, oléochimie: analyse de la filiere philippine du cocotier. *OCL* (France) 2, 3: 206-209.

Rungrueng, S. (1988) *Increased coconut production on small-holder plantations in Southern Thailand.* In: Proceedings Workshop on Intercropping and Intergrazing in Coconut Areas, 7-11 September, 1988, Colombo, Sri Lanka.

Sabutan, M.G., Organas, Q., and Rabanal, J.T. (1986) *Coconut-based forage crops and beef cattle production farming systems under Central Mindanao conditions.* Terminal Report, Univ. of Southern Mindanao, North Cotabato.

Sadakathullas, S., and Abdul Kareem, A. (1994) Management of rodents in coconut plantations. *The Planter* (Malaysia) 70: 395-399.

Sahasranaman, K.N., Pillai, N.G., Jayasankar, N.P., Potti, V.P., Varkey, T., Kamalakshy Amma, P.G., and Radha, K. (1983) Mixed farming in coconut gardens: economics and its effect on Root(wilt) disease. In: *Coconut Research and Development* (ed. Nayar, N.M.) pp. 160-165, Wiley Eastern Ltd. New Delhi, India.

Said, A.N., and Dzowela, B.H. (1989) *Overcoming Constraints to the Efficient Utilization of Agricultural By-Products as Animal Feed.* Proceedings of the Fourth Annual Workshop held at the Institute of Animal Research, Mankon Station, Bamenda, Cameroun 20-27 October 1987. ARNAB. ILCA. 474p.

Saint, J.P. Le, Taffin, G. de, and Bénard, G. (1989) Coconut seed preservation in sealed packages. *Oléagineux* (France) 44, 1: 15-25.

Sakai, K.I. (1960) Method of breeding of coconut palm. A comment on 'The improvement of coconut palm by breeding and selection of Dr S. C. Harland'. *Tropical Agriculturist* 116: 185-189.

Salam, M.R., and Sreekumar, D. (1990) Coconut based mixed farming system to sustain productivity. *Indian Coconut Journal* 20, 10: 3-5.

Salleh, Moh. R., Kamal Hizat, A., and Khusahry, M.Y.M. (1989) *Performance of Malin sheep integrated with coconut plantation at Hilir Perak - a preliminary report.* In: Modernization in Tropical Livestock and Poultry Production, Proceedings 12th MSAP Annual Conference, March 29-31, 1989, pp. 270-276.

Samarakoon, S.P. (1987) *The effects of shade on quality, dry matter and nitrogen economy of Stenotaphrum secundatum compared with Axonopus compressus and Pennisetum clandestinum.* M. Agricultural Science, thesis, University of Queensland, Australia.

Samarakoon, S.P., Wilson, J.R., and Shelton, H.M. (1990a) Growth, morphology and nutritive quality of shaded *Stenotaphrum secundatum, Axonopus compressus* and *Pennisetum clandestinum. Journal of Agricultural Science*, Cambridge, 114: 161-169.

Samarakoon S.P., Shelton, H.M., and Wilson, J.R. (1990b) Voluntary feed intake by sheep and digestibility of shaded *Stenotaphrum secundatum* and *Pennisetum clandestinum* herbage. *Journal of Agricultural Science*, Cambridge, 114: 143-150.

San, T.T., and Pau, T.Y. (1986) UPB's coconut dehusking machine. In: Pushparajah, E., and Chew Poh Soon, (eds), *Cocoa and Coconuts, progress and outlook.* Incorporated Society of Planters, Kuala Lumpur (Malaysia) pp. 911-916.

Sanchez and Rasco (1983) Coconut in white soft cheese. *Philippine Journal of Coconut Studies* (Philippines) 8: 1-2.

— (1984) Coconut milk in yoghurt manufacture. *Philippine Journal of Coconut Studies* (Philippines) 9: 1-2.

Sanchez, M.D., and Ibrahim, T.H. (1991) Forage species for rubber plantations in Indonesia. In: *Proceedings of Workshop on Forages for Plantation Crops.* Shelton, H.M. and Stur, W.W. (ed.). Sanur Beach, Bali, Indonesia. ACIAR Proceedings 32: 54-57.

Sangare, A. (1981) Compétition pollinique et légitimité des semences produites dans des champs semenciers de cocotiers. *Oléagineux* (France) 36, 8-9: 423-427.

Sangare, A., and N'Cho, (1990) *Rapport de Mission Tanzanie* (Report of a mission to Tanzania). IRHO/ CIRAD-CP, station Marc Delorme, document interne n°2295 (Ivory Coast).

Sangare, A., Le Saint, J.P., and Nucé de Lamothe, M. de (1984) Les cocotiers Grands à Port Bouët (Côte d'Ivoire). 3. Grand Cambodge, Grand Tonga, Grand Rotuma. *Oléagineux* (France) 39, 4: 205-215.

Sangare, A., Taffin, G. de, Franqueville, H.de, Arkhurst, E.D., and Pomier, M. (1992) Le jaunissement mortel du cocotier au Ghana. Premiers résultats sur le comportement au champ du matériel végétal. *Oléagineux* (France) 47, 12: 699-704.

Sangare, A., Wuidart, W., and Nucé de Lamothe, M. de (1978) Les phases mâles et femelles de l'inflorescence du cocotier. Influence sur le mode de reproduction. *Oléagineux* (France) 33, 12: 609-617.

Sani, R.A., and Rajamanicham, C. (1991) Gastrointestinal parisitism in small ruminants. In: *Integrated Tree Cropping and Small Ruminant Production Systems* (ed. Iniguez, L.C. and Sanchez, M.D.). Proceedings of a workshop on research methodologies, Medan, North Sumatra, Indonesia, Sept. 9-14, 1990, pp. 197-201.

San Juan, N.C., *et al.* (1986-a) Epidemiology of Wilt Disease of coconut in Socorro, Oriental Mindoro, I. Disease increase with space and time. *Philippine Journal of Coconut Studies* (Philippines) 11, 1: 8-22.

— (1986b) Epidemiology of the Wilt Disease of coconut in Socorro, Oriental Mindore, II. Ecological factors affecting disease development. *Philippine Journal of Coconut Studies* (Philippines) 11, 2: 23-28.

Sansoucy, R., Aarts, G., and Preston, T.R. (1988) Sugarcane as feed. *FAO Animal Production and Health Paper* (Italy) 72, 319 pp.

Santiago, R.M. (1989) Coconut toddy yield in relation to leaf water status and some climate factors. Philippine *Journal of Coconut Studies* (Philippines) 9, 1: 32-37.

Santos, G.A. (1990) Activities in coconut genetic resources collection, conservation and genetic improvement in The Philippines. *Philippine Journal of Coconut Studies* (Philippines) 15, 1: 16-20.

Santos, G.A., Carpio, C.B., Ilagan, M.C. (1980) Evaluation of various yield groups and their progenies. *PCA Agricultural Research* 1980, Annual Report (Philippines) pp. 46-49.

Santos, G.A., Carpio, C.B., Ilagan, M.C., Cano, S.B., and de la Cruz, B.V. (1986a) Yield and agronomic traits of four variety hybrids and some local tall coconut populations in The Philippines. *Oléagineux* (France) 41, 6: 269-280.

Santos, G.A., Bahala, R.T., Cano, S.B., and Cruz, B.V. de la (1986b) Yield and agronomic traits of four variety hybrids and some local tall coconut populations in The Philippines. *Philippine Journal of Coconut Studies* (Philippines) 11, 2: 13-23.

Saseendran, S. A., and Jayakumar, M. (1988) Consumptive use and irrigation requirement of coconut plantations in Kerala. *Journal of Plantation Crops* (India) 16, 2: 119-125.

Sasidharan, N.K., Varghese, A., and Sivaprasad, P. (1991) Impact of weed control methods on the occurrence and intensity of weed flora and mycorrhizal association in coconut. In: Silas, E.G., Aravindakshan, M., and Jose, A.I (eds), *Coconut breeding and management*. Kerala Agricultural University, New Delhi (India) 262-267.

Satyabalan, K. (1990) Early germination and seedling selection in coconut *Indian Coconut Journal* (Indfia) 20, 11: 2-6.

Satyabalan, K., and Lakshmanachar, M.S. (1960) Coconut breeding: effects of some breeding procedures. *Indian Coconut Journal* (India) 8: 113-115.

Satyabalan, K., and Mathew, J. (1983) Identification of prepotent palms in West Coast Tall coconut based on the early stages of growth of the progeny nursery. In: Nambiar, N.M. (ed.), *Coconut Research and Development*. Wiley Eastern Ltd., New Delhi (India) pp. 15-22.

Savithri, P., and Khan, H. (1994) Characteristics of coconut coir pith and its utilization in Agriculture. *Journal of Plantation Crops* (India) 22, 1: 1-18.

Schut, B. (1975) A possible phytotoxic effect of copper on dwarf coconut palms. *Surinaamse Landbouw* (Suriname) 23, 1: 26-30.

Schuiling, M., Mpunami, A., Kaiza, D.A., and Harries, H.C. (1992) Lethal disease of coconut palm in Tanzania - III low resistance of imported germplasm. *Oléagineux* (France) 47, 12: 693-698.

Schuiling, M., and Mpunami, A. (1992) Lethal Disease of coconut palm in Tanzania. Comparison with other coconut diseases in East Africa. *Oléagineux* (France) 47, 8-9: 511-515.

Schuiling, M., and Kaiza, D.A., and Mpunami, A. (1992a) Lethal Disease of coconut palm in Tanzania. II - History, distribution and epidemiology. *Oléagineux* (France) 47, 8-9: 516-521.

Schuiling, M., Kaiza, D.A., and Harries, H.C. (1992b) Lethal Disease of coconut palm in Tanzania. III - Low resistance of imported germplasm. *Oléagineux* (France) 47, 12: 693-697.

Sebastian, L.C., Mujer, C.V., and Mendoza, E.M.T. (1987) A comparative cytochemical study on mature Makapuno and normal coconut endosperm. *Philippine Journal of Coconut Studies* (Philippines) 12, 1: 14-22.

— (1991a) *Productivity of cattle under coconuts*. In: Proceedings of Workshop on Forages for Plantation Crops, Shelton, H.M. and Stur, W.W. (ed.), *Sanur Beach*, Bali, Indonesia, 27-29 June 1990. ACIAR Proc. No. 32: 92-96.

Shelton, H.M. (1991a) *Prospects for improving forage supply in coconut plantations of the South Pacific*. In: Forages for Plantation Crops (ed. Shelton, H.M. and Stur, W.W.), Proceedings of a workshop, Sanur Beach, Bali, Indonesia, 27-29 June 1990. ACIAR Proceedings no. 32: 92-96.

— (1991b) *Prospects for improving forage supply in coconut plantations of the South Pacific*. In: Forages for Plantation Crops (ed. Shelton, H.M. and Stur, W.W.), Proceedings of a workshop, Sanur Beach, Bali Indonesia, 27-29 June, 1990. ACIAR Proceedings 32, 151-156.

Shelton, H.M., Humphreys, L.R., and Batello, C. (1987a) Pastures in the plantations of Asia and the Pacific: performance and prospect. *Tropical Grasslands* 21, 4: 159-168.

Shelton, H.M., Schottler, J.H., and Chaplin, G. (1987b) *Cattle under trees in the Solomon Islands*. Report prepared at Department of Agriculture, University of Queensland, for Mininstry of Agriculture and Lands, Solomon Islands. Nov. 1987, 22p.

Shelton, H.M., and Stur, W.W. (1991) *Forages for Plantation Crops*. Proceedings of a workshop, Sanur Beach, Bali, Indonesia, 27-29 June, 1990. ACIAR Proceedings 32, 168 pp.

Shenoi, P.V. (1991) *Coconut, past, present, and future. Souvenir, Platinum Jubilee of Coconut Research and Development (1916-1991)*, Coconut Plantation Crops Research Institute, Kerala, India.

Shivashankar, S. (1988) Polyphenol oxydase isozymes in coconut genotypes under water stress. *Plant Physiology and Biochemistry* 15, 1: 87-91.

— (1991) Biochemical changes during fruit maturation in coconut. *Journal of Plantation Crops* (India) 19, 2: 102-109.

Shivashankar, S., and Kasturi Bai, K.V. (1988) A comparative study of the growth and nitrogen accumulation capacity of hybrid coconut (*Cocos nucifera*) seedlings. *Philippine Journal of Coconut Studies* (Philippines) 13, 1: 24-29.

Shivashankar, S., Nagaraja, K.V., Voleti, S.R., and Kasturi Bai, K.V. (1993) Biochemical changes and leafwater status of coconut genotypes differing in drought tolerance. In: Nair, M.K., Khan, H.H., Gopalasundaram, P., and Bhaskara Rao, E.V.V. (eds), *Advances in coconut research and development*. Indian Society for Plantation Crops, New Delhi (ndia).

Shivashankar, S., and Ramadasan,, A. (1985) Chlorophylls and nitrate reduction activity in relation to heterosis in coconut seedlings. *Annals of Botany* 55: 755-758.

Shuhaimi Shamsudin, Paiman Ahmad, M.D., Palaniappan, S. (1990) Early performance of several coconut varieties in PPPTR. *Kemajuan Penyelidikan* (Indonesia) 15: 24-27.

Silva, S. de (1988) *Replacement of senile coconut palms in small holdings, field experience of a replacement model*. Asian Pacific Coconut Community (Indonesia) 75 pp.

— (1990) *Coconut based farming systems: a bibliography.* APCC (Indonesia) 95 pp.

Silva, M.A.T. de, *et al.* (1973) Nutritional studies on initial flowering of coconut (var. typica). I. Effect of magnesium deficiency and Mg-P relationship. *Ceylon Coconut Quarterly* (Sri Lanka) 24, 3-4: 107-113.

Silva, M.A.T. de, Greta, M., Anthonypillai and Mathes, D.T. (1985) The sulphur nutrition of coconut. *Cocos* (Sri Lanka) 3: 22-28.

Silva, N.T.H.H., de and Tisdell, C.A. (1985) Spacing of a perennial crop for economic gain: towards an operational approach for coconut palms (*Cocos nucifera* L.). *Indian Journal of Agricultural Economics* XL, 1: 42-52.

— (1986) An assessment of alternative technologies for increasing coconut production in Sri Lanka. Asian Pacific Coconut Community, *Quarterly Supplement* (Indonesia) 8: 1-16.

Simonnet, P. (1990) Sheep flock management in a tropical environment under coconut. *Oléagineux* (France) 45, 10: 451-455.

Sitepu, D. (1983) Coconut Wilt in Natuna Islands of Indonesia. In: D. Sitepu, *Exotic plant quarantine pests and procedures for introduction of plant materials*, 81-85.

Sitepu, D., and Darwis, S.N. (1989) Management of major coconut diseases in Indonesia. *Industrial Crops Research Journal* (Indonesia) 1, 2: 66-77.

Slobbe, W.G., van (1977) *Hart rot disease in Surinam.* Paper Second Meeting of the International Council on Lethal Yellowing. Florida (USA).

Smith, R.W. (1970) The Malayan Dwarf supersedes the Jamaica Tall Coconut. 2. Change farming in practice. *Oléagineux* (France), 25, 11: 593-598.

— (1972) The optimum spacing for coconuts In: *Cocoa and Coconuts in Malaysia,* pp. 429-443. The Incorpor ated Society of Planters, Kuala Lumpur Malaysia).

Smith, M.A., and Whiteman, P.C. (1983a) *Evaluation of tropical grasses in increasing shade under coconut canopies.* Experimental Agriculture, Cambridge, 19, 2, 153-161.

— (1983b) *Rotational grazing experiment under coconuts at Lingatu Estate, Russell Islands.* Technical Report, Solomon Islands Pasture Research Project, University of Queensland (Australia).

— (1985) Animal production from rotationally-grazed natural and sown pastures under coconuts at three stocking rates in the Solomon Islands. *Journal of Agricultural Science*, Cambridge 104: 173-180.

Smith, M.A., MacFarlane, D.C., and Whiteman, P.C. (1983) *Species evaluation, soil fertility and weed control.* Technical report, Solomon Islands Pasture Research Project, University of Queensland. (Australia), mimeo report.

Smith, R.W. and Romney, D.H. (1969) The spacing of coconuts. *The Farmer* 74, 12: 411-414.

Snaydon, R.W., and Harris, P.M. (1981) Interactions below ground - the use of nutrients and water. In: *ICRISAT.* Proceedings of the International Workshop on Intercropping, 10-13 Jan. 1979, Hyderabad, India, 188-201.

Soekarjoto, S., Sudasrip, H., and Davis, T.A. (1980) *Setora nitens* a serious sporadic pest of coconut in Indo nesia. *The Planter* (Malaysia) 56: 167-182.

Solomon, J.J., and Govindankutty, M.P. (1991) Etiology - E. Mycoplasma-like organisms. In: Nair, M.K. *et al.* (eds), *Coconut Root(wilt) disease.* Central Plantation Crops Research Institute (India), Monograph Series - no 3: 31-40.

Solomon, J.J., and Pillai, N.G. (1991) Root(wilt) disease of coconut - current status. In Nair, M.K., Khan, H.H., Gopalasundaram, P., and Bhaskara Rao, E.V.V. (eds), *Advances in coconut research and development.* Indian Society for Plantation Crops (India) pp. 563-573.

Sophanodora, P. (1989a) Pasture under plantation crops in southern Thailand. *Forage Newsletter, ACIAR* 12: 2-3.

— (1989b) *Productivity and nitrogen nutrition of some tropical pasture species under low radiation environments.* Ph.D. thesis, Univ. of Queensland, Australia.

Sophanodora, P., and Tudsri, S. (1991) Integration of forages for cattle and goats into plantation systems in Thailand. In: *Proceedings of Workshop on Forages for Plantation Crops*, Shelton, H.M and Stur, W.W. (ed.). Sanur Beach, Bali, Indonesia 27-29 June 1990. ACIAR Proc. No. 32: 147-150.

Sorensen, P.H. (1989) Strategies for the feeding of dairy cattle in hot environments. In: *Modernization in Tropical Livestock and Poultry Production*, 12th MSAP Annual Conference 1989. Held at Genting Highlands, Pahang, Malaysia, March 29-31. MSAP and MARDI. 310-317.

Sosamma, V.K., and Koshy, P.K. (1991a) Etiology-C. Nematodes. In: Nair, M.K. *et al.* (eds), *Coconut Root(wilt) disease.* Central Plantation Crops Research Institute (India), Monograph Series - no. 3, 19-26.

— (1991b) *Interactions between vesicular-arbuscular mycorrhizae (VAM) and burrowing nematode on coconut.* Central Plantation Crops Research Institute (India), Annual Report 1990-1991, 102.

Southern, P.J. (1967a) Sulphur deficiency in coconuts, a widespread field condition in Papua and New Guinea. Part I: The field and chemical diagnosis of sulphur deficiency in coconuts. *Papua and New Guinea Agricultural Journal* (Papua and New Guinea) 19, 1: 18-37.

– (1967b) Sulphur deficiency in coconuts, a widespread field condition in Papua and New Guinea, Part II: The effect of sulphur deficiency on copra quality. *Papua and New Guinea Agricultural Journal* (Papua and New Guinea) 19, 1: 38-44.

Southern, P.J., and Dick, K. (1967) The distribution of trace elements in the leaves of the coconut palm, and the effect of trace element injections. *Papua and New Guinea Agricultural Journal* (Papua and New Guinea) 19, 3: 125-137.

Speedy, A., and Pugliese, P. (1993) *Legume trees and other fodder trees as protein sources for livestock.* Proceedings of FAO Expert Consultation. MARDI, Kuala Lumpur, Malaysia 14-18 October 1991. FAO Animal Production and Health Paper (Italy) 102, 339 pp.

Speedy A., and Sansoucy, R. (1991) *Feeding Dairy Cows in the Tropics. FAO Animal Production and Health Paper* (Italy) 86, Proceedings of the FAO Expert Consultation held in Bangkok, Thailand 7-11 July, 244 pp.

Steel, R.J.H., and Humphreys, L.R. (1974) Growth and phosphorus response of some pasture legumes grown under coconuts in Bali. *Tropical Grasslands* 8, 3: 171-178.

Steel, R.J.H., and Whiteman, P.C. (1980) *Pasture species evaluation, pasture fertilizer requirements and weed control in the Solomon Islands.* Technical Report, Solomon Islands Pasture Research Project, University of Queensland, mimeo rept.

Steel, R.J.H., Smith, M.A., and Whiteman, P.C. (1980) *A pasture handbook for the Solomon Islands.* University of Queensland/ADAB, Ministry of Agriculture and Lands, Honiara, Solomon Islands.

Steer, J.D. (1989) *Lethal Yellowing. Twenty-fifth report of the Research Department for calendar year 1989.* Coconut Industry Board Jamaica W.I. (Jamaica) 54.

Steer, J., and Coates-Beckford, P.L. (1990) Role of *Phytophthora katsurae, P. palmivora, Thielaviopsis paradoxa* and *Enterobacter* sp. in Budrot disease of coconuts in Jamaica. *Oléagineux* (France) 45, 12: 539-545.

Stobbs, T.H. (1978) Milk production, milk composition, rate of milking and grazing behaviour of dairy cows grazing two tropical pastures under a leader and follower system. *Australian Journal of Experimental Agriculture and Animal Husbandry* 18: 5-11.

Stobbs, T.H., and Thompson, P.A.C. (1975) Milk Production from tropical pastures. *World Animal Review* 13: 27-31.

Stur, W.W. (1991) Screening forage species for shade tolerance - a preliminary report. In: *Proceedings of Workshop on Forages for Plantation Crops*, Shelton, H.M and Stur, W.W. (ed.). Sanur Beach, Bali, Indonesia, 27-29 June, 1990. ACIAR Proceedings 32: 58-63.

Stur, W.W., and Shelton, H.M (1991a) Review of Forage Resources in Plantation Crops of Southeast Asia and the Pacific. In: *Forages for Plantation Crops* (ed. Shelton, H.M and Stur, W.W.), Sanur Beach, Bali, Indonesia, 27-29 June, 1990. ACIAR Proceedings 32: 25-31.

– (1991b) *Compatibility of forages and livestock with plantation crops. In: Forages for Plantation Crops* (ed. Shelton, H.M and Stur, W.W.), Proceedings of a workshop, Sanur Beach, Bali, Indonesia, 27-29 June, 1990. ACIAR Proceedings 32: 112-116.

Sudaria, E.E., and Pedro Jr., R.C. de (1990) The effect of storing dehusked mature coconuts on the quality of copra (a research note). *Philippine Journal of Coconut Studies* (Philippines) 15, 1: 24-26.

Sudhakara, K. (1990) Button shedding and premature nutfall in coconut. *Journal of Plantation Crops* (India) 18, 2: 66-77.

Sugarman, N. (1956) *Process for simultaneously extracting oil and protein oleaginous materials.* US patent 2, 762, 820 (11 September 1956).

Sukri, M.I., and Dahlan, I. (1986) *Feedlot and semi-feedlot systems for beef cattle fattening among smallholders.* In: Proceedings 8th Annual Conference MSAP (ed. Hutagalung, R.I. *et al.*) held at Genting Highlands (Malaysia) 13-14 March 1984: 74-78.

Sukumaran Nair, and Balkrishnan (1991) *Inbreeding depression in coconut (Cocos nucifera L.).* In: Coconut Breeding and Management. Proceedings of the National Symposium held from 23rd to 26th November, 1988. 380 pp. Kerala Agricultural University, Trichur, India: 51-54.

Swaminathan, M.S. (1991) Coconut research - challenges ahead. In: *Souvenir, Platinum Jubilee of Coconut Research and Development in India.* Central Plantation Crops Research Institute, Kasaragod, India.

Swan, D.I. (1974) *A review of the work on predators, parasites and pathogens for the control of Oryctes rhinoceros (L.) in the Pacific area.* Commonwealth Institute of Biological Control, Miscellaneous Publications 7, 64 pp.

Szent-Ivany, J.J.H. (1956) New insect pest and host plant records in Papua and New Guinea. *Papua and New Guinea Agricultural Journal* (Papua New Guinea) 11, 3: 82-87.

Taffin, G., and Wuidart, W. (1981) Précautions à prendre avec des semences de cocotier ayant effectué un long voyage. *Oléagineux* (France), 8-9: 429-432.

Taffin, G. de, and Ouvrier, M. (1985) Harvesting coconuts by cutting bunches. *Oléagineux* (France) 40, 4: 197-201.

Taffin, G. de, and Sangare, A. (1989) Intérêt du cocotier en Afrique de l'Ouest. *Oléagineux* (France) 44, 12: 579-585.

Taffin, G. de, and Rognon F. (1991) Le diagnostic foliaire du cocotier. Conseils de l'IRHO no: 318. *Oléagineux* (France), 46, 4: 169-173.

Taffin, G. de, Zakra, N., Pomier, M., Braconnier, S., and Weaver, R.W. (1981) Search for a stable cropping system combining coconut and nitrogen fixing trees. *Oléagineux* (France) 46, 12: 489-499.

Taffin, G. de, Zakra, N., and Bonny, C.P. (1991) Dwarf x Tall coconut hybrid performance under commercial conditions in Côte-d'Ivoire. *Oléagineux* (France) 46, 5: 194-195.

Tajuddin, Z.A. (1991) *Proceedings of the International Livestock-Tree Cropping Workshop*, held at Serdang, Malaysia 5-9 December 1988. Organized by MARDI and FAO, 134 pp.

Tammes, P.L.M. (1955) Review of coconut selection in Indonesia. *Euphytica* (Netherlands) 4: 17-24.

Tarigans, D.D. (1989) Assessment of experience with high yielding coconut varieties in Indonesia. *Industrial Crops Research Journal* (Indonesia) 1, 2: 46-59.

— (1991) Effect of foliar sprays of urea on the growth of coconut. *Industrial Crops Research Journal* (Indonesia) 3, 2: 8-14.

Tarigans, D.D., and Darwis, S.A. (1989) Country Report Indonesia. In: *Coconut Based Farming Systems*. UNDP/FAO Project RAS/80/032. Report of the Meeting of the Working Group held at Chumphon, Thailand, September 1989, pp. 28-35.

Taulu, D.B., *et al*., (1980) Coconut trichromes: Their significance in classification and insect resistance. *The Philippine Journal of Coconut Studies* (Philippines) 5, 2: 39-43.

Taylor, T.H.C. (1937) *The biological control of an insect in Fiji*. Department of Agriculture (Fiji).

Taysum, D.H. (1981) *A report on a tour of coconut plantations in Southern Nampula and Zambezia Provinces (Mozambique)*. Internal Report, UNDP/FAO Project: Moz/75/009 Rome (Italy).

Teoh, K.C., Chan, K.S., and Chew, P.S. (1986) Dry matter and nutrient composition in hybrid coconuts (MAWA) and cocoa on coastal clay soils. In: Pushparajah, E., and Soon, C.P.: *Cocoa and Coconuts; Progress and Outlook*, pp. 819-836. Incorporated Society of Planters (Malaysia).

Thampan, P.K. (1975) *The Coconut Palm and its Products*. Green Villa Publishing House, Vytilla, Cochin-19, Kerala, India, 242 pp.

— (1981) *Handbook on coconut palm*. Mohani Primlani, Oxford & IBH Publishing, New Delhi (India), 311 pp.

Thankamma Pillai, P.K., Vijayakumar, G., and Nambiar, M.C. (1976) Cytogenetic and genetic studies in coconut: A review. In: *International Symposium on Coconut Research and Development*, December 1976. Abstract of paper. Indian Society of Plantation Crops, Kasaragod, Kerala (India).

Thirumalaiswamy, K., Vijayaraghavan, H., and Joseph-Savery (1992) Early diagnosis of Thanjavur Wilt of Coconut. *The Planter* (Malaysia) 68: 599-601.

Thomas, G.V. (1988) Vesicular-arbuscular mycorrhizal symbiosis in coconut in relation to Root(Wilt) disease and intercropping or mixed cropping. *Indian Journal of Agricultural Sciences* (India) 57: 145-147.

— (1991) *Increasing nutrient availability and disease alleviation by micro-organisms in plantation crops*. Central Plantation Crops Research Institute (India) Annual Report 1990-1991, 49-51.

Thomas, G.V., and Ghai, S., K. (1987) Genotype dependent variation in vesicular-arbuscular mycorrhizal colonization of coconut seedlings. *Proceedings Indian Academy of Science (Plant Sci)*, 97: 289-294.

Thomas, G.V., and Shantaram, M.V. (1984) *In situ* cultivation and incorporation of green manure legumes in coconut basins. *Plant and Soil* (Netherlands) 80: 373-380.

— (1986) Solubilisation of inorganic phosphates by bacteria from coconut soils. *Journal of Plantation Crops* (India) 14, 1: 42-48.

Thomas, G.V., Shantaram, M.V., and Saraswathy, N. (1985) Occurrence and activity of phosphate solubilising fungi in coconut soils. *Plant and Soil* (Netherlands) 87, 3: 357-369.

Thomas, G.V., and Ghai, S.K. (1991) Influence of *Rhizobium* inoculation and seed pelleting on nodulation of green manure legumes in an acidic coconut soil. *Journal of Plantation Crops* (India) 18 (Supplement), pp. 72-77.

Thomas, G.V., Iyer, R., and Bopaiah, B.M. (1991) Beneficial microbes in the nutrition of coconut. *Journal of Plantation Crops* (India) 19, 2: 127-138.

Torres, J.P. (1937) Makapuno: a distinct type of coconut. *Agricultural and Industrial Monthly* (Indonesia) 4, 10: 18-19.

Torres, F. (1983) Role of woody perennials in animal agroforestry. *Agroforestry Systems* 1: 131-163.

Tothill, J.D., Taylor, T.H.C., and Paine, R.W. (1930) *The coconut moth in Fiji*. Imperial Institute of Entomology (U.K.).

Trewren, K. (1987) *Tuvalu technical report on tree crops research*, 1986. Ministry of Commerce and Natural Resources, Department of Agriculture (Tuvalu).

Trewren, K. (1991) Coconut production on coral soils. In: Green, A.H. (ed.), *Coconut production*. Present status and priorities for research, 43-51. The World Bank, Washington D.C. (USA).

Tsai, J.H. (1988) Lethal Yellowing of coconut palms. In: Hiruki, C., *Tree mycoplasmas and mycoplasma diseases*, pp. 99-107. The University of Alberta Press (Canada).

Tulner, H.H. (1933) Ongepubliceerde onderzoekingen van Ir. H.H. Tulner aan het Klapperproefstation te Menado (1931-1932).*(not published)*

Turner, P.D., *et al.* (1979) Coconut stem necrosis, a disease of hybrid and Malayan Dwarf coconuts in North Sumatra and Penninsular Malaysia. *The Planter* (Malaysia) 55, 644: 768-784.

— (1980) Coconut stem necrosis and dry bud rot. *Cocomunity* (Indonesia) Quarterly Supplement, APCC/QS/33/-80, 3-4.

Ubaitoi, I. (1987) *Coconut research and development in Kiribati.* Proceedings of the Fifth South Pacific Islands Regional Meeting on Agricultural Research, Development, Extension and Training in Coconut (IRETA, Western Samoa, pp. 132-134.

Uchida, J.Y., Ooka, J.J., Nagata, M., and Kadooka, C.Y. (1992) *A new Phytophthora fruit and heart rot of coconut.* University of Hawaii (USA), College of Tropical Agriculture and Human Resources, Research and Extension Services 138, 8pp.

Uexkull, H.R., von (1972) Response of coconuts to (potassium) chloride in The Philippines. *Oléagineux* (France) 27, 1: 13-19.

— (1985) Chlorine in the nutrition of palm trees. *Oléagineux* (France) 40, 2: 67-74.

Uotila, M.E. (1992) *Tonga: Back-to-Office Report.* Report of the Regional Animal Production Officer (Dairy Development) following a visit to Tonga 17-26 November 1992. FAO/RAPA, Bangkok (Thailand). 9p.

Uthaiah, B.C., Lingaiah, B., and Balakrishna Rao, K. (1989) Studies on the effect of mulches on coconut seedling establishment. *Indian Coconut Journal* (India) 20, 4: 7-8.

Uyenco, F.R., and Ochoa, J.A.K.(1984) Microbial degradation of coconut coir dust for biomass production. *Philippine Journal of Coconut Studies* (Philippines) 9, 1-2: 51-54.

Van, N.H., Olivin, J., and Ochs, J. (1984) Oil palm and coconut soils in West Africa. *Oléagineux* (France) 39, 3: 117-129.

Vandermeer, J. (1989) *The Ecology of Intercropping.* CUP, UK. 237 pp.

Vanderplank, P.L. (1960) Bionomics and ecology of the red tree ant (*Oecophylla smaragdina*). *Journal of Animal Ecology* 19, 1: 15-53.

Van der Vecht, J. (1950) *The coconut leaf moth (Artona catoxantha Hamps.) Part I. Life history and habits of Artona catoxantha, its parasites and hyperparasites.* Bogor Agricultural Research Station, General Contribution 110 (Indonesia).

Varadan, K.M., and Chandran, K.M. (1991) Irrigation requirement of coconut under drip method. In: Silas, E.G., Aravindakshan, M., and Jose, A.I. (eds), *Coconut breeding and management.* Kerala Agricultural University (India) pp. 281-282.

Varadan, K.M., Madhavachandran, K., and Lakshmanan, K. (1990) Effect of irrigation and mulching on soil moisture and soil temperature under coconut. *Journal of Plantation Crops* (India) 18, 1: 55-65.

Varghese, P.T., Nair, P.K.R., Nelliat, E.V., Ram Varma and Gopalasundram, P. (1978) *Intercropping with tubercrops in coconut gardens.* In: Proceedings of the First Symposium on Plantation Crops. Indian Society for Plantation Crops, Kasaragod, India, pp. 399-415.

Vavilov, N.I. (1951) The origin, variation, immunity and breeding of cultivated plants. In: Selected writings of N.I. Vavilov (transl. by K. Starr Chester) *Chronica botanica* 13, 364 pp.

Velasco, J.R. (1960) Growth of young coconut plant in sand culture. *Philippine Agriculturist* (Philippines) 43, 9: 548-567.

Velasco, J., R., and Sierra, Z.N. (1993) The effects of the rare earths and chrromium on the growth of coconuts. *The Philippine Agriculturist* (Philippines) 76, 1: 49-71.

Venard-Combes P., and Mariau, D. (1983) *Augosoma centaurus* Fabricius (Coleoptera, Scarabeidae) important ravageur du cocotier en Afrique. Descriptions, biologie, méthode de lutte. *Oléagineux* (France) 38, 12: 651-657.

Verdeil, J.L., Huet, C., Grosddemanges, F., Rival, A., and Buffard-Morel, J. (1992) Embryogenèse somatique du cocotier (*Cocos nucifera* L.) obtention de plusieurs clones de vitroplants. *Oléagineux* (France), 47, 7: 465-469.

Vernon A.J., Emose, P.N., and Mudallar, T. (1975) Coconut varietal selection and Breeding, part 2, recent work in Fiji. *Fiji Agricultural Journal* (Fiji) 37: 47-52.

Vesey-Fitzgerald D. (1941) *Melittomma insulare*, a coconut pest in the Seychelles. Bulletin of Entomological Research 32, 383-402.

Vijayaraghavan, H., Ramadoss, N., Ramanathan, T., and Rethinam, P. (1987) Effect of toddy tapping on Thanjavur Wilt Disease of coconut. *Indian Coconut Journal* (India), 17, 12: 1-3.

Vijayaraghavan, H., Manmohanlal, S., and Ramanathan, T. (1988) Effect of Zinc on yield. *Indian Coconut Journal* (India) 18, 11: 3-7.

Vijayaraghavan, H., Raveendran, T.S., and Ramanathan, T. (1989) Effect of certain growth regulators in preventing button shedding and increasing yield in coconut. *Indian Coconut Journal* (India) 20, 2: 3-7.

Vijayaraghavan, H., and Ramachandran, T.K. (1989) Effect of *in situ* cultivation and incorporation of green manure crops on yield of coconut. *Cocos* (Sri Lanka) 7: 26-29.

Vijchulata, P. (1991) Integrated small ruminant and tree cropping production systems in Thailand. In: *Proceedings Workshop on Integrated Tree Cropping and Small Ruminant Production Systems* (ed. Iniguez, L.C. and Sanchez, M.D.), Medan, North Sumatra, Indonesia, Sept. 9-14, 1990, pp. 280-288. USAID.

Villegas, L.G. (1991) Coconut intercropping. In: *Coco-based Farming Systems: State of Knowledge and Practice,* BARDA. Quezon City, Philippines.

Vina de la, A.C. (1991) (ed.), *Utilization of Native Forages for Animal Production,* Proceedings of Second Meeting of the Regional Working Group on Grazing and Feed Resources in Southeast Asia. 26 Feb.-5 March, 1991, UP Los Baños, organized by Bureau of Animal Industry, Dept. of Agric., Philippines (sponsored by FAO), 229p.

Voleti, S.R., Kasturi Bai, K.V., Rajagopal, V., and Shivashankar, S. (1990) Relative water content and proline accumulation in coconut genotypes under moisture stress. *Journal of Plantation Crops* (India) 18, 2: 88-95.

Voleti, S.R., Kasturi Bai, K.V., Rajagopal, V., and Nambiar, C.K.B. (1993a) Influence of soil type on the development of moisture stress in coconut (*Cocos nucifera* L.) genotypes. *Oléagineux* (France) 48, 12: 505-509.

— (1993b) Water potential in the leaves of coconut (*Cocos nucifera* l.) under rainfed and irrigated conditions. In: Nair, M.K., Khan, H.H., Gopalasundram, P., and Bhaskara Rao, E.V.V. (eds), *Advances in coconut research and development.* Indian Society for Plantation Crops, New Delhi (India).

Vo Van Long (1994) *Coconut Breeding Programme in Vietnam.* In: Proceedings of the Workshop on Standardization of Coconut Breeding Research Techniques held at Abidjan, Ivory Coast, June 20-25, 1994.

Wahid, P.A. (1984) Diagnosis and correction of nutrient deficiencies in coconut palm. *Journal of Plantation Crops* (India) 12, 2: 98-111.

Wahid, P.A., and Kamalam, N.V. (1988) Nutrient distribution in the crown of healthy and Root(wilt) affected coconut palms. *Indian Coconut Journal* (India) 18, 11: 8-12.

Waidyanatha, U.P. de S., Wijesinghe, D.S., and Stauss, R. (1984) Zero grazed pasture under immature *Hevea* rubber: productivity of some grasses and grass-legume mixtures and their competition with *Hevea.* Tropical. *Grasslands* 18, 1: 21-26.

Walker, B. (1992) *Pastures for beef and milk production for the Seychelles. Pasture Manual prepared under TCP/SEY/0053.* FAO, Rome (Italy), 47 pp.

Wallace, T. *et al.*, (1961) *The diagnosis of mineral deficiencies in plants by visual symptoms,* 2nd. ed. Chemical Publishing Co. Inc., New York (USA).

Walton, J. (1972) Cattle under coconuts. In: *Cocoa and Coconuts in Malaysia* (ed. R.L. Wastie and D.A. Earp). Incorp. Soc. of Planters, Kuala Lumpur, Malaysia, pp. 422-425.

Wan Hassan, W.E., Phipps, R.H., and Owen, E. (1989) Development of smallholder dairy units in Malaysia. *Tropical Animal Health and Production* 21: 175-182.

Wan Mohamed, W.E., Mohamed, N., and Abdul Rahman, A. (1988) *Perspectives and progress of sheep industry at Kumpulan Guthrie.* In: Proceedings Symposium on Sheep Production in Malaysia, 15-16 Nov. 1988, Univ. Pert. Malaysia, pp. 98-108.

Warwick, D.R.N., Bezerra, A.P.O., and Renard, J.L. (1991) Reaction of coconut hybrids to Leaf Blight (*Bot ryodiplodia theobromae* Pat.). Field observations. *Oléagineux* (France) 46, 3: 100-108.

— (1991) Root transmission of Red Ring nematode (*Rhadinaphelengus cocophilus*) in coconut plants. In: *Abstracts of papers International Symposium on Coconut Research and Development-II.* Central Plantation Crops Institute (India) 7.

Warwick, D.R.N., Bezerra, A.P.O., and Renard, J.L. (1991) Reaction of coconut hybrids to leaf blight (1) *Botryodiplodia theobromae* Pat.). Field observations. *Oléagineux* (France) 46, 3: 100-108.

Warwick, D.R.N., Passos, E.E.M., Leal, M.L.S., and Bezerra, A.P.O. (1993a) Influence of water stress on the severity of coconut Leaf Blight caused by *Lasiodiplodia theobromae. Oléagineux* (France) 48, 6: 279-282.

Warwick, D.R.N., Leal, E.C., Bezerra, A.P.T., and Tupinamba, E.A. (1993b) Field resistance of coconuts to 'lixa' and Leaf Blight in Brazil. In: Nair *et al.* (1993) *Advances in Coconut Research and Development* (India), pp. 585-589.

Warwick, D.R.N., Leal, E.C., and Ram, C. (1994) Doenças do coqueiro. In: Ferreira, J.M.S., Warwick, D.R.N., and Siqueira, A.S., *Cultura do coqueiro no Brasil.* EMBRAPA-CPATC, Aracaju, Sergipe, (Brazil) 282-306.

Watson, S.E., and Whiteman, P.C. (1981a) Animal production from naturalized and sown pastures at three stocking rates under coconuts in the Solomon Islands. *Journal of Agricultural Science*, Cambridge 97: 669-676.

Watson, S.E., and Whiteman, P.C. (1981b) Grazing studies on the Guadalcanal Plains, Solomon Islands. 2. Effects of pasture mixtures and stocking rate on animal production and pasture components. *Journal of Agricultural Science*, Cambridge 97: 353-364.

Way, M.J. (1953) Studies of *Theraptus sp.* (Coreidae), the cause of the gumming disease of coconuts in East Africa. *Bulletin of Entomological Research* 44: 657-667.

— (1954) Studies of the life history and ecology of the ant *Oecophylla longinoda* (Latreille). *Bulletin of Entomological Research* 45: 93-112.

Weightman, B.L. (1977) *The background to cattle and pasture development in the New Hebrides.* In: Proceedings Regional Seminar on Pasture Research and Development in Solomon Islands and Pacific Region, Ministry of Agriculture and Lands, Honiara, Solomon Islands, pp. 252-257.

Westenbrink, G. (1986) Livestock and cash crops integration. In: *Integrated Farming Systems* (ed. Gunasena and Herath), National Agricultural Society of Sri Lanka, University of Peradeniya, pp. 1-20.

Whitehead, R.A. (1963) The processing of coconut pollen. *Euphytica* (The Netherlands) 12, 2: 167-177.

— (1964) The processing of coconut pollen. *Oléagineux* 19 (France) 7: 477-483.

— (1968) Selecting and breeding coconuts palms (*Cocos nucifera* L.) resistant to Lethal Yellowing disease. A review of recent work in Jamaica. *Euphytica* (Netherlands) 17: 81-101.

Whiteman, P.C. (1980) Tropical pasture science. Oxford University Press, Oxford.

Wickramaratne, M.R.T., and Padmasari, M.H.L. (1986) Some observations on the position of the soft eye in Cocos nucifera. *Cocos* (Sri Lanka) 4: 35-37.

Wickramaratne, M.R.T., Coe, R., and Fernando, S. (1987) Evaluation of criteria for selection of seed coconuts (Cocos nucifera L.). *Cocos* (Sri Lanka) 5: 1-7.

— (1987a) More palms, more nuts, more copra. *Coconut Bulletin* (Sri Lanka) 4, 1: 1-5.

— (1987b) Breeding coconuts for adaptation to drought. *Coconut Bulletin* (Sri Lanka) 4, 1: 16-23.

Williams, J.B. (1976) The social importance of cattle in Asia and the influence of social attitudes on beef production. In: *Beef Cattle Production in Developing Countries* (ed. Smith, A.J.) University of Edinburgh CTVM, pp. 375-387.

Wilson, J.R. (1982) Environmental and nutritional factors affecting herbage quality. In: *Nutritional limits to animal production from pastures* (ed. J.B. Hacker), CAB, pp. 111-131.

— (1990) Agroforestry and soil fertility - the eleventh hypothesis: shade. *Agroforestry Today* 2: 14-15.

— (1991) *Ecophysiological constraints to production and nutritive quality of pastures under tree crops.* In: Proceedings of the International Livestock-Tree Cropping workshop (ed. Tajuddin, Z.A.), 5-9 December, 1988, Serdang, Malaysia, pp. 39-54.

Wilson, J.R., and Ludlow, M.M (1991) The environment and potential growth of herbage under plantations. In: *Forages for Plantation Crops* (ed. Shelton, H.M. and Stur, W.W.). Proceedings of a workshop, Sanur Beach, Bali, Indonesia, 27-29 June 1990. ACIAR Proceedings 32: 10-24.

Wilson, J.R., and Wild, D.W.M. (1991) Improvement of nitrogen nutrition and grass growth under shading. In: *Forages for Plantation crops* (ed. Shelton, H.M. and Stur, W.W.). Proceedings of a workshop, Sanur Beach, Bali, Indonesia, 27-29 June 1990. ACIAR Proceedings 32: 77-82.

Wilson, J.R. Catchpoole, V.R., and Weier, K.L. (1986) Stimulation of growth and nitrogen uptake by shading a rundown green panic pasture on brigalow clay soil. *Tropical Grasslands* 20: 134-143.

Wilson, J.R., Hill, K., Cameron, D.M., and Shelton, H.M. (1990) The growth of *Paspalum notatum* under the shade of a *Eucalyptus grandis* plantation canopy or in full sun. *Tropical Grasslands* 24: 24-28.

Wong, C.C. (1989) Review of forage screening and evaluation in Malaysia. In: *Grasslands and Forage Production in South-East Asia.* Proceedings of First Meeting of the Regional Working Group on Grazing and Feed Resources of South East Asia, 27 Feb.-3 March 1989, Serdang, Malaysia, pp. 51-68.

— (1991) *Shade tolerance of tropical forages: a review.* In: Proceedings of Workshop on Forages for Plantation Crops, Shelton, H.M. and Stur, W.W. (ed.). Sanur Beach, Bali, Indonesia, 27-29 June 1990. ACIAR Proceedings 32: 64-69.

Wong, C.C., and Wilson, J.R. (1980) Effects of shading on the growth and nitrogen content of green panic and Siratro in pure and mixed swards defoliated at two frequencies. *Australian Journal of Agricultural Research* 31: 269-285.

Wong, C.C., Rahim, H., and Mohd. Sharudin, M.A. (1985a) Shade tolerance potential of some tropical forages for integration with plantations. 1. Grasses. *MARDI Research Bulletin* (Malaysia) 13, 3: 225-247.

Wong, C.C., Mohd. Sharudin, M.A., and Rahim, H. (1985b) Shade tolerance potential of some tropical forages for integration with plantations. 2. Legumes. *MARDI Research Bulletin* (Malaysia) 13, 3: 249-269.

— (1979b) Production de matériel végétal cocotier. Sélection en pépinière. Conseils de l'IRHO no. 197. *Oléagineux* (France), 34, 10: 453-456.
— (1981a) Production de matériel végétal cocotier. Tenue d'un germoir. Conseils de l'IRHO no. 215. *Oléagineux* (France), 36, 6: 305-309.
— (1981b) - Production de matériel végétal cocotier. Pépinière en sac de plastique Conseils de l'IRHO no. 216. *Oléagineux* (France) 36, 7: 367-376.
— (1981c) Production de matériel végétal cocotier. Sélection des hybrides en germoir. Conseils de l'IRHO no. 218. *Oléagineux* (France) 36, 10: 495-500.
— (1981d) Production of coconut planting material. Conduct of a seedbed. *Oléagineux* (France) 36, 6: 305-309.
— (1981e) Production of coconut planting materioal. The polybag nursery. *Oléagineux* (France) 36, 6: 367-376.
— (1994a) Sulphur deficiency symptoms in coconut. *Oléagineux* (France) 49, 1: 31-34.
— (1994b) Iron deficiency symptoms in coconut on coral soils. *Oléagineux* (France) 49, 2: 67-71.
Wuidart, W., and Nucé de Lamothe, M. de (1981) Seed germination and development of coconut plants in function of nut position. *Oléagineux* (France) 36, 12: 599-604.
Wuidart, W. and Rognon, F. (1978) L'analyse des composantes de la noix du cocotier. Méthode de détermination du coprah. *Oléagineux* (France) 33, 5: 225- 233.
— (1981) La production de semences hybrides de cocotier. *Oléagineux* (France) 36, 3: 131-137.
— (1993) La production de semences hybrides de cocotier: cas des semences hybrides Nain x Grand. III. Les semences. *Oléagineux* (France) 48, 1: 41-49.

Yenko (1954) *Dry process by Yenko.* Philippines patent no 187 of 1954.
Yusof, O., and Rejab, M.S. (1988) *Status of intercropping and intergrazing in coconut lands in Malaysia.* In: Proceedings Workshop on Intercropping and Intergrazing in Coconut Areas, 7-11 September, 1988, Colombo, Sri Lanka.

Zakra, A.N., Pomier, M., and Taffin, G. de (1986) Initial results of an intercropping experiment with food crops in the Middle Ivory Coast. *Oléagineux* (France) 41, 8-9: 381-389.
Ziller, R. (1960) Study on some factors influencing the copra content of coconuts. *Oléagineux* (France) 15, 2: 73-81.
— (1962) La sélection du cocotier dans le monde. *Oléagineux* (France) 17, 11: 837-846.
Zimmerman, E.C. (1958) *Insects of Hawaii. 8. Lepidoptera: Pyraloidae.* University of Hawaii (USA).
Zizumbo Villarreal, D., Hemandez Roque, F. and Harries, H.C. (1993) Coconut varieties in Mexico. *Economic Botanic* 47, 1: 65-78.
Zoby, J.L.F., and Holmes, W. (1983) The influence of size of animal and stocking rate on the herbage intake and grazing behaviour of cattle. *Journal of Agricultural Science*, Cambridge 100: 139-148.
Zuniga L.C. (1953) The probable inheritance of the Makapuno character of the coconut. *Philippine Agriculturist*, 36: 402-413.
Zuniga, L.C., Armedilla, A.L., and Gala, D. de (1969) Maternal and paternal selection on coconut. Philippine *Journal Plantation Industry* (Philippines) 34: 9-16.
Zushun, M. (1986) An investigation on meteorological indices for coconut cultivation in China. *Oléagineux* (France) 41, 3: 119-126.
— (1994) A new coconut hybrid, WY78F1. *Oléagineux* (France) 49, 2: 49-54.

Index

www.ingramcontent.com/pod-product-compliance
Lightning Source LLC
Jackson TN
JSHW050345140125
77033JS00020B/611
* 9 7 8 1 8 5 3 3 9 4 6 7 6 *